PRINCIPLES OF CROP PRODUCTION
THEORY, TECHNIQUES, AND TECHNOLOGY

George Acquaah
LANGSTON UNIVERSITY

PEARSON
Prentice Hall

Upper Saddle River, New Jersey 07458

Library of Congress Cataloging-in-Publication Data

Acquaah, George.
  Principles of crop production : theory, techniques, and technology/George Acquaah.
        p. cm.
  Includes bibliographical references.
    ISBN 0-13-114556-8
      1. Field crops. 2. Food crops. 3. Crop yields. I.[nbsp]Title.
  SB185 .A27 2001
  631—dc22
2001021271

To all my children—
Parry, Kwasi, Bozuma, Tina

Director of Production and Manufacturing: Bruce Johnson
Executive Editor: Debbie Yarnell
Development Editor: Kate Linsner
Marketing Manager: Jimmy Stephens
Managing Editor: Mary Carnis
Manufacturing Manager: Ilene Sanford
Manufacturing Buyer: Cathleen Petersen
Production Editor: Melissa Scott, Carlisle Publishers Services
Production Liaison: Janice Stangel
Creative Director: Cheryl Asherman
Senior Design Coordinator: Christopher Weigand
Cover Design: Christopher Weigand
Cover Art: Courtesy of Getty Images, Inc.-Stone Allstock
Chapter Opener Photo: Courtesy of Getty Images, Inc.-Photodisc.
Composition: Carlisle Communications, Ltd.
Printer/Binder: Courier Westford

Pearson Prentice Hall™ is a trademark of Pearson Education, Inc.
Pearson® is a registered trademark of Pearson plc
Prentice Hall® is a registered trademark of Pearson Education, Inc.

Pearson Education LTD.
Pearson Education Singapore, Pte. Ltd.
Pearson Education Canada, Ltd.
Pearson Education—Japan

Pearson Education Australia PTY, Limited
Pearson Education North Asia Ltd.
Pearson Educación de Mexico, S.A. de C.V.
Pearson Education Malaysia, Pte. Ltd.

10 9 8 7 6 5 4 3 2 1
ISBN 0-13-114556-8

Contents

Chapter 2:   Plant Morphology   51

Chapter 3: Fundamental Plant Growth Processes   99

Chapter 4: Plant Growth and Development   123

Chapter 8: Plant Nutrients and Fertilizers   246

Chapter 9:   Plant and Soil Water   278

Chapter 10:   Pests in Crop Production   310

Chapter 11: Agricultural Production Systems   361

Chapter 12:   Organic Crop Production   393

Chapter 15:   Tillage Systems and Farm Energy   444

Chapter 16:   Seed, Seedling, and Seeding   467

Chapter 17: Harvesting and Storage of Crops   495

Chapter 18: Marketing and Handling Grain Crops 521

PART TWO   Commercial Production of Selected Field Crops   549

Chapter 19:   Wheat (Common)   551

Chapter 22:   Sorghum   603

Chapter 23:   Barley   615

Chapter 24:   Soybean   626

Chapter 25:   Peanut   639

Chapter 26:   Cotton   651

Chapter 27:   Potato   666

Preface

In many agricultural instructional programs, students are required to take at least an introductory course in each of the major areas of the field of agriculture, irrespective of their areas of concentration or major. The major areas are crop science, animal science, soil science, and economics or business. Introductory courses in crop science may have titles such as "Elements of Crops," "Introduction to Plant Science," and "Introduction to Agronomy." This text, *Principles of Crop Production: Theory, Techniques, and Technology,* was designed for use at the introductory and intermediate levels of study. Industry professionals will find this a comprehensive text for preparing for the Certified Crop Advisor (CCA) examination.

The specific objectives of the author in designing this text were:

1. To develop a comprehensive text on crop production.
2. To emphasize the general principles of crop production.
3. To present crop production as a science, an art, and a business.
4. To introduce and emphasize some of the new concepts and trends in crop production.
5. To engage the student and facilitate the learning of the principles of crop production.
6. To provide a good reference on the subject of crop production for industry professionals.
7. To provide a detailed overview of the production practices pertaining to the major field crops.

In terms of the audience, the author's experience is that the backgrounds of students who enroll in this course vary from one institution to another, and even within the same institution. Some programs attract students with farm backgrounds, while others attract urban dwellers to whom a good example of cereal is cornflakes and soil is dirt! In some programs, students normally enroll in this course after completing their general education requirements in basic sciences. Further, instructors vary greatly in their opinions concerning depth of coverage needed for certain topics. The author is of the opinion that it is better to have a little more than a little less. The instructor can choose to deemphasize a topic or even leave it out altogether, rather than want it and not have it. The book is divided into two parts, the first dealing with the Underlying Principles of Crop Production and the second discussing the Commercial Production of the Major Field Crops.

The scientific principles are applicable anywhere in the world, and for that matter the textbook can be used worldwide. However, special reference is made to the North American experience, and as such most examples are from that region. Crop production is also a business. From this perspective, the text devotes some time to the role of the crop producer as a manager of resources. Agriculture is risky business and production is tied to marketing. As such, risk management and produce marketing are discussed in detail.

Modern agriculture is technology-driven. The text discusses the evolution of crop production, the key technological advances, and their impact on crop production. New and emerging concepts are also mentioned to some extent. Current trends in general agriculture and society that directly impact crop production are discussed. For example, the strong environmental safety consciousness of society cannot be overlooked. There is a strong move to deemphasize agrochemicals and to use them more responsibly so as not to endanger the environment. Thus, sustainable agricultural concepts are emphasized in the text. Crop production is presented as a managed ecosystem or an agroecosytem. Production practices such as organic farming that exclude or deemphasize the use of inorganic inputs into crop production are presented.

The material is presented in a deliberately straight-to-the-point fashion. The student is directly engaged in a personal way, starting with a direct introduction of the subject of the chapter to the student, followed by expected outcomes. The sections are introduced with statements that reflect some of the key messages in the chapters. Questions are also posed throughout the text to get the reader more involved in the material being presented. At the end of the chapter, the student is invited to participate in an outcomes assessment, which includes multi-structured questions to test the understanding of the material just studied. Whereas most questions can be answered by using materials directly in the text, Part D usually contains general questions that may need additional material to answer completely. There are boxed readings that present additional information to enhance the topics presented in the chapter.

In this information age, students have access to a tremendous volume and variety of information via the Internet. At the end of each chapter, the author has provided web addresses for a selected number of relevant sites with additional information that would enhance the understanding of the materials presented in the chapter. Certain websites update their information regularly, thus enabling the reader who needs to have the most current data, beyond what is practical in a textbook situation, to continually update the textbook information. The author is also of the opinion that a text should be a good source of reference for the student after completion of the course. To this end, additional information such as a glossary, conversion tables, and other general information has been included.

ACKNOWLEDGMENTS

The author would like to acknowledge with gratitude the guidance and assistance provided by Dr. Greg Ruark, director of the USDA National Agroforestry Center, in the preparation of the section on agroforestry. The author thanks Otumfuo Nana Nyame for technical and financial support, as well as encouragement and guidance at all stages of the project.

The author also wishes to thank the following reviewers for their valuable assistance at various stages of development: Louis Harper, California Polytechnic University; Dean Kopsell, The University of New Hampshire; Bradley Lange, Central Community College; Robert Peregoy, Spokane Community College; Randy Rosiere, Tarleton State University; Jonathan Shaver, Oklahoma State University; Craig Sheaffer, University of Minnesota; and Leon Slaughter, University of Maryland.

George Acquaah

INTERNET RESOURCES

AGRICULTURE SUPERSITE

This site is a free online resource center for students and instructors in the field of Agriculture. Located at http://www.prenhall.com/agsite, this site contains numerous resources for students including additional study questions, job search links, photo galleries, PowerPoint™ slides, *The New York Times* eThemes archive, and other agriculture-related links.

On this supersite, instructors will find a complete listing of Prentice Hall's agriculture texts, as well as instructor supplements that are available for immediate download. Please contact your Prentice Hall sales representative for password information.

THE NEW YORK TIMES eTHEMES OF THE TIMES FOR AGRICULTURE AND *THE NEW YORK TIMES* eTHEMES OF THE TIMES FOR AGRIBUSINESS

Taken directly from the pages of *The New York Times,* these carefully edited collections of articles offer students insight into the hottest issues facing the industry today. These free supplements can be accessed by logging onto the Agriculture Supersite at: http://www.prenhall.com/agsite.

AGRIBOOKS: A CUSTOM PUBLISHING PROGRAM FOR AGRICULTURE

Just can't find the textbook that fits your class? Here is your chance to create your own ideal book by mixing and matching chapters from Prentice Hall's agriculture textbooks. Up to 20% of your custom book can be your own writing or come from outside sources. Visit us at: http://www.prenhall.com/agribooks.

PART ONE
UNDERLYING PRINCIPLES

Crop Production and Society

PURPOSE

This chapter explores the evolution of crop production, from its primitive beginnings as an unorganized activity to its present status as a planned and managed activity. The role of crops in society and the factors that affect their production are briefly discussed. Modern crop production is presented as a science, an art, and a business. Also, the structure of U.S agriculture, with emphasis on crop production, and focusing on the producer in terms of background, opportunities, and the environment of operation, is presented.

EXPECTED OUTCOMES

After studying this chapter, the student should be able to:

1. Discuss the different views on the evolution of crop production.
2. Discuss the importance of crops to society as sources of food, feed, and fiber.
3. Discuss crop production as a science, an art, and a business.
4. List important cereal, feed grains, oil, and fiber crops.
5. Discuss factors determining the dynamic structure of agriculture.
6. Trace a brief history of agriculture in the United States.
7. Discuss the structure of U.S. crop farms, including size, crop types, and sales.
8. Discuss the characteristics of the U.S. farmer, including age, education, income, land tenure adopted, and cropland use.
9. Describe the farming regions of the United States.
10. Describe specific government programs available to producers.

KEY TERMS

Agriculture	Expert systems	Morrill Act of 1862
Agronomy	Farm	Small family farm
Bast fiber	Forage	Vegetable fiber

TO THE STUDENT

Crop production has not always been an organized and planned activity, as it is known today. In the pre-civilization era, people ate what they could find that was safe to eat. At some period in time, crop production assumed some structure and became an organized and purposeful activity. Its evolution has been guided by art and science, the practice becoming more science-based with time. Crop production today is big business. Notwithstanding the era, the fundamental principles of crop production have essentially stayed unchanged. The methods are more effective and efficient with better results. But when did crop production as a planned activity begin? What crops were the first formally cultivated by humans? How have methods of crop production changed over the years? These and other related issues are addressed in this chapter. As you study this chapter, pay attention to the evolution of crop production and the factors that guide the transformation processes. Note the fairly distinct eras modern agriculture has gone through. Since modern agriculture is a planned activity, note the role decision making by the producer plays in the activity. Note also the importance of crop production to society.

The farm is the heart of agriculture. Farming is practiced in a very complex environment in which natural and human-made factors interact in a dynamic fashion to impact its structure. In addition to the role of the macroeconomic landscape and the political environment, the characteristics of the farmer impact the structure of farming by influencing how the farmer responds to the opportunities and challenges presented by the factors of change. In this chapter, you will learn about the U.S. farm structure, including aspects such as farming regions, size of farms, and the kinds and acreages of crops grown. You will also learn about the U.S. farm sector, including aspects such as credit for agriculture, and specific characteristics of the U.S. farmer. Agriculture is a dynamic activity, and its evolution is driven by scientific and technological advancement. Farmers differ in how they embrace technological innovation.

1.1: BEFORE AGRICULTURE, PEOPLE SURVIVED BY HUNTING AND GATHERING FOOD

Agriculture The science and art of producing crops or animals under supervision of humans in a specific location.

What is **agriculture**? Before agriculture, the practice by which people began to grow plants on purpose and domesticate animals, ancient humans survived by hunting animals, fishing, and gathering plants for food. Modern humans usually view this lifestyle in unfavorable light. *Hunter-gatherers* are stereotyped as primitive, unskilled, and ignorant, among other perceptions. Much of this prejudicial view of these ancient people sometimes betrays the lack of adequate understanding of the lifestyle of these people by modern people. Recent work by anthropologists on hunter-gatherer communities, such as the !Kung bush people and Australian aborigines by E. N. Wilmsen, is credited with the favorable image these ancient cultures are now beginning to enjoy. A critical look at the lifestyle of hunter-gatherers indicates that they had to be resourceful in order to survive and that they, in their own mind at least, lived in a Golden Age.

The art of hunting and gathering did not appear to have completely vanished into oblivion with the advent of modern crop culture. Studies show that pockets of surviving hunter-gatherers still exist in areas where the land and climate do not favor deliberate crop culture as a profitable enterprise. These areas include parts of the Kalahari Desert, the Arctic region, and pockets of the tropical rain forest.

It is estimated that the bulk (60 to 80%) of the food of hunter-gatherers consisted of plants and plant products. The wild plant resources of these people consisted of grass seeds such as wild rice *(Zizania aquatica)* and wild progenitors of common Asian rice *(Oryza sativa)* (e.g., *O. barthii, O. longistaminats,* and *O. sublata)*. Other grass species used included *Panicum* spp. (e.g., *P. capillare, P. obtusum, P. miliaceum,* and *P. laetum)*, mannagrass *(Glyceria fluitans)* and wild oats *(Avena barbata* and *A. fatua)*. Legumes, both tuberous and non-tuberous, were also sources of food for ancient people. They used the seeds, pods, and leaves of these plants for food. Examples of wild legumes are *Acacia, Vigna, Dolichos,* and *Canavalia.* Leguminous materials have been known to contain highly toxic substances. In order for these technologically challenged people to have used them for food, they had to discover methods of detoxifying and processing these plants and their products to make them safe for consumption. Ancient peoples had to be ingenious to accomplish this. These discoveries had to be based on trial and error. Poisonous extracts were used as poisons for arrowheads and darts that were utilized for hunting. Should we then say that technology is relative and that, to the primitive cultures, what they discovered and used then was high-tech to them?

Roots and tuber plants, especially of the genus *Dioscorea* (yam), were widely used in the tropical regions for food. Bulbs from the family Liliaceae, wild onions *(Allium)*, and sweet potato *(Ipomea batatas)* were important food sources. Just like legumes, roots and tuber plants frequently contained toxic substances that had to be detoxified to make them safe for use.

Hunter-gatherers obtained oil from animal sources. Plant oils were also obtained from species including coconut *(Cocos nucifera)*, oil palm *(Elaeis guineensis)*, shea butter *(Butyrospermum)*, and olive *(Olea)*. Wild fruits and nuts are some of the readily visible and attractive sources of food under any circumstance. Examples of these sources of food are bramble fruits *(Rubus)*, grape *(Vitis), Citrus, Musa,* chestnut *(Castanea)*, and *Carica.*

Vegetables and spices from the families Solanaceae and Cucurbitaceae were particularly important to hunter-gatherers. Species of *Solanum, Capsicum,* and *Lycopersicon* are widely distributed. Wild watermelon *(Colocynthus citrulus)* occurs in southern Africa.

What did it take to be a hunter-gatherer? Hunter-gatherers were very knowledgeable. To survive, they had to understand their environment very well. To be nomadic, they had to know when and where to migrate. As already indicated, they acquired botanical knowledge, which helped them use toxic plants for food through detoxification by heating or leaching. They learned to use plants for medicinal purposes. Non-agricultural communities have been known to employ various practices, such as burning, to alter vegetation. Burning of the brush favored certain species. The hunter-gatherers understood the lifecycles of plants, their adaptation, and their distribution well enough to be able to anticipate and locate abundant food sites. It is clear from the foregoing that the hunter-gatherer was the "real survivor," not the modern-day made-for-TV imitators.

1.2: AGRICULTURE IS AN INVENTION

Is it really possible to determine precisely where, how, and when agriculture started? Views of agricultural origins are diverse, ranging from mythological to ecological. Whatever prompted humans to purposely raise their plants in a specific area and confine their animals did not likely happen overnight. It was an evolutionary process that eventually transformed plants from being independent, wild progenitors to fully dependent and domesticated cultivars, with the concomitant evolution of agricultural economies.

Various mythologies suggest that agriculture is a divine gift to humans. Others explain that the shift from nomadic lifestyle to sedentary plant and animal culture was necessitated by religious obligations, which required animals and certain plants to be kept for ritual sacrifices to gods. The *propinquity theory* of Gordon Childe suggests that wild animals, and for that matter plants, were forced into closer cohabitation when advancing dry climatic conditions pushed humans and animals into more habitable areas such as the banks of perennial rivers. This is also referred to as the *theory of domestication by crowding.*

Of all the views of agricultural origins, suggesting agriculture as an invention and discovery appears to be the most enduring. There are various views as to how this discovery was made. Charles Darwin suggested that first humans had to become sedentary for this discovery to occur. Byproducts of their lifestyle, specifically garbage, had to be disposed of in a refuse dump near their habitation. Out of this discarded heap of waste, seeds from a useful plant (that had been previously gathered for food or medicine) are likely to have germinated and grown in the fertile soils of the refuse dump to produce bountifully. Some wise person was bound to have made the observation of these plants and the connection between planting seeds and producing desired plants in a certain location in the vicinity of his or her abode. This is the *plant domestication by happy accident model.*

Carl Sauer further developed this theory. He suggested that, not only did the inventors of agriculture have to be sedentary, but also primitive agriculturists were most likely to have hailed from fishing communities. Such communities had to be located in wooded lands where it was relatively easier to till the soil, not having to contend with the soil-binding effect of grass roots in grasslands. Further, these discoverers would have previously acquired certain skills in other aspects of living in the wild that would predispose them to agricultural experimentation.

From this and other presuppositions, the most likely candidate for an area of origin of agriculture is Southeast Asia, from where it spread to China, India, the Near East, Africa, the Mediterranean, and last North and West Europe. Edgar Anderson later strengthened the Sauer model with a genetic component. He suggested that hybridization would have produced new genetic recombinants from which useful selections could be found.

What do you think were factors that influenced the use of certain plants for food by hunter-gatherers? The hunter-gatherers in pre-agricultural times were influenced by certain characteristics of the plants they selected, which certainly impacted the domestication process. The plants that were frequently gathered for food were likely to have been easier to harvest, available in large quantities, and easy to transport back to their habitation. It should have been important that plants were available with great predictability and seasonal distribution. The plant parts harvested would have been more preferred if they were larger in size, as in the case of yams. For grains with smaller seed, they were likely to have been attractive if they could be harvested in large numbers from dense stands. As for grains, a plant with large naked grains (hull-less) such as wheat would have been favored over smaller and coarse-grained plants (hulled) such as barley and oats. Hulled or covered grains have to be processed to remove the cover to obtain the grains. However, the smaller grains would have been easier to harvest and hence a possible reason that oat was domesticated before wheat and barley.

In view of the fact that these theories are not proofs of the origin of agriculture, you may find one more attractive than the others. Suppose you were not privy to these theories; what would you suggest as a theory to explain the origin of agriculture?

1.3: Agriculture Provides Food, Feed, and Fiber for Society

Can modern societies survive without crop plants? Crop plants are totally depended upon by humans for survival in certain societies. They provide food and fiber for humans and feed for livestock.

1.3.1 FOOD VALUE OF CROPS

Humans and other animals require energy, amino acids, vitamins, and minerals for proper growth and development. Energy is obtained from carbohydrates, fats, and proteins. The food type used as source of energy varies from one part of the world to another. In many developing countries, up to about 75% of energy is obtained from cereal carbohydrates. In the Arctic regions, the other extreme occurs whereby most of the energy used by humans is derived from animal fat and protein. As indicated by the U.S. Department of Agriculture (USDA) food guide pyramid, cereals represent the base of the triangle. Hence, cereals are the food types that should be consumed in the biggest servings each day (Figure 1–1).

Proteins are made up of amino acids. There are 20 amino acids, of which 9 are deemed essential for human survival and must be obtained from external sources. These are leucine, valine, phenylalanine, threonine, isoluecine, lysine, methionine, cysteine, and tryptophan. The others are synthesized within the body. The major food crops of the world and their nutritional value regarding essential amino acids is presented in Table 1–1. Legumes tend to be low in cysteine and methionine, both sulfur-containing amino acids. Cereal grains, on the other hand, are low in lysine and threonine. Eating a combination of cereal and legume improves the overall nutritional quality of the mixture through protein complementation. Such mixtures are culture-dependent. One of the key breeding objectives for cereal crops is improving their nutritional value.

Utilization of crop plants is influenced by many social factors. In the United States, for example, corn and sorghum are grown primarily for animal feed. On the other hand, these crops are the primary staples of many countries, especially those in the developing economies (one person's food is another person's feed?). However, because of the limited nutritional value of these crops, an overdependence on them leads to nutritional deficiency problems, unless dietary supplementation is adopted. Vitamin A–induced blindness is common in regions of the world where rice, a crop that is deficient in vitamin A, is a staple food.

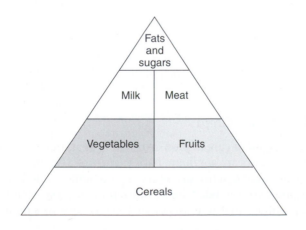

FIGURE 1–1 The USDA food guide pyramid. USDA recommends the most (6 to 11) servings from this group each day because foods in this group are an excellent source of complex carbohydrates (important for energy, especially in low-fat diets), vitamins, minerals and fiber. (Source: Figure modified from the original USDA diagram)

	Essential Amino Acids		**Net Protein Utilization (%)**
Crop	**Low**	**Adequate**	
Corn	Tryptophan Lysine	-	72
Wheat (grain)	Lysine	-	63
Rye	Tryptophan Lysine	-	60
Rice (polished)	Lysine Threonine	Tryptophan	60
Millet	Lysine	Tryptophan Methionine Cystine	58
Soybean	Methionine Cystine Valine	Lysine Tryptophan	63
Lima bean	Methionine Cystine	Tryptophan Lysine	50
Peanut	Lysine Methionine Cystine Threonine	-	38
Potato	Methionine Cystine	Tryptophan	63

Table 1–1 Nutritional Value of Major World Food Crops

Source: Extracted from N. S. Scrimshaw and V. R. Young. 1976. The requirements of human nutrition. In *Food and Agriculture, a Scientific American Book,* W.H. Freeman and Company, San Francisco. Percentages are close approximations.

1.3.2 THE CROPS THAT FEED THE WORLD

Are there certain crops that can be described as "lifesavers" in the world today? The 30 most important food crops in the world include cereals, roots, fruits, vegetables, and legumes (Table 1–2). Some of these—for example, corn and wheat—are very widely distributed and are grown to some degree in North America, South America, Europe, Africa, Asia, and Australia. Others, such as yams and cassava, are grown only in certain tropical regions. From the list in Table 1–2, it is clear that cereal and root crops are the most important crops that feed the world. Also, from this list, indicate the crops that are agriculturally important in your geographic area. What crops are important in your region but are not included in the list? Is there a reason that cereals and root crops are important to feeding the world?

Overview of Major Grain Crops

The top three most important cereal grain crops are wheat, rice, and corn, in that order of world production. Others are sorghum, barley, and oats. Of these, corn, oats, barley, and grain sorghum are used extensively as feed grain in the United States. In 2000, the world food grain production totaled 1,871 million metric tons, the United States accounting for 332.2 million metric tons (Table 1–3). In terms of human consumption,

Table 1–2 Thirty Major Food Crops of the World

1. Wheat	11. Sorghum	21. Apples
2. Rice	12. Sugarcane	22. Yam
3. Corn	13. Millet	23. Peanut
4. Potato	14. Banana	24. Watermelon
5. Barley	15. Tomato	25. Cabbage
6. Sweet potato	16. Sugar beet	26. Onion
7. Cassava	17. Rye	27. Beans
8. Grapes	18. Oranges	28. Peas
9. Soybean	19. Coconut	29. Sunflower
10. Oats	20. Cottonseed oil	30. Mango

Source: Extracted from J. R. Harlan. 1976. Plants and animals that nourish man. In *Food and Agriculture, a Scientific American Book,* W. H. Freeman and Company, San Francisco. The ranking is according to total tonnage produced annually. The tonnage of the top 7 is more than two times the tonnage of the remaining 23.

Table 1–3 World Total Food Grain Production Trends: 1994–1999

Region	1994/95	1995/96	1996/97	1997/98	1998/99
			(Million metric tons)		
All foreign countries					
Production	1,408.5	1,436.8	1,535.8	1,541.5	1,499.6
Consumption	1,528.6	1,548.4	1,591.1	1,596.2	1,608.6
USA					
Production	354.7	274.5	335.2	340.0	349.2
Imports	5.7	4.4	6.1	5.8	5.4
Exports	101.3	95.0	82.4	75.3	82.0
World total, trade	219.6	205.9	213.2	212.2	206.4

Source: USDA.

about 75% of the world's sorghum (*Sorghum bicolor* L.) is used for human consumption, making it third in overall importance of food crops in the world. Corn and wheat are widely distributed; they are grown in high concentration in certain regions in North America, Central America, and South America (Figure 1–2). Rice, on the other hand, is mostly concentrated in Asia. Leading producers of these principal cereal grains are shown in Table 1–4.

Cereal crops are relatively cheaper and easier to produce and are suitable for combating hunger and population explosion. Cereal and potato together supply over 75% of the world's food calories. This proportion is even greater in certain parts of Asia, reaching 90% in some regions. The great dependence on cereals for food has led to an improvement in the fortification of cereal products with minerals and vitamins (e.g., iron, thiamine, riboflavin, and niacin). Plant breeders have genetically improved the lysine and tryptophan content of corn with the development of opaque-2 high-lysine mutant corn. On average, cereal grains have a high caloric value of above 300 cal/100 g. The average protein content is between 7.5 and 14.2%, while the average fat content is between 1.0 and 7.4%, oats having the highest average in each case as compared with the major cereal grains.

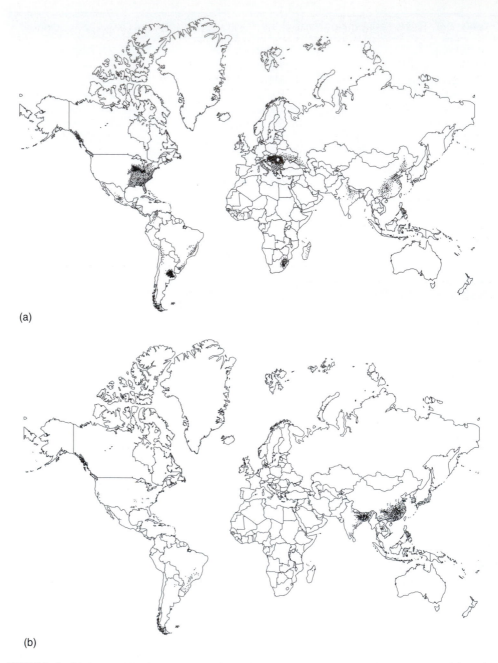

(a)

(b)

FIGURE 1–2 Major production regions of selected major cereal crops in the world: (a) major corn production regions, (b) major rice production regions. Whereas corn is widely distributed, rice production is concentrated in Asia. Field crops do not lend themselves to indoor cultivation and hence are restricted to zones of natural adaptation. However, plant breeders are able to expand this zone through genetic manipulation to develop cultivars that can be produced in unconventional areas. (Source: USDA)

Table 1–4 Leading Producers of the Three Most Important Cereal Crops of the World

Wheat

Country	Proportion (%)
Former Soviet Union	30.0
United States	12.7
China	7.6
India	6.7
France	4.9
Canada	4.5
Turkey	2.2
Pakistan	2.1

Rice

Country	Proportion (%)
China	33.1
India	21.1
Indonesia	7.3
Bangladesh	5.9
Japan	4.9
Sri Lanka	4.6
Thailand	3.9
Italy	3.3
Brazil	2.0
South Korea	1.9
United States	1.3

Corn

Country	Proportion (%)
United States	46.0
China	8.0
Brazil	4.8
Former Soviet Union	4.2
Mexico	3.9
South Africa	3.5
France	3.4
Argentina	3.2
Yugoslavia	2.6
Rumania	2.2

Source: USDA.

Monocots store food primarily in their endosperm. This tissue contains starch and small quantities of protein and other nutrients. On the other hand, dicots (legumes and oil seeds) that are utilized as grain crops store food in their cotyledons. The cotyledons lack starch and instead have high amounts of protein, oil, and some carbohydrates. The most important food and feed legume or oil seed grains are soybean, peanut, bean, and pea (Table 1–5). Dicot protein, however, is low in certain essential amino acids, especially methionine. Sesame seed protein is high in methionine.

Table 1–5 Major Vegetable Oils and Their Characteristics

Crop		Oil Content (%)	Iodine Number
Drying Oil			
Flax	*Linum usitatissimum* L.	35–45	170–195
Tung	*Aleurites fordii* L.	40–58	160–170
Safflower	*Carthamus tinctorius* L.	24–36	140–150
Semidrying Oil			
Soybean	*Glycine max* L. Merr.	17–18	115–140
Sunflower	*Helianthus annuus* L.	29–35	120–135
Corn (germ)	*Zea mays* L.	50–57	115–130
Cottonseed	*Gossypium hirsutum* L.	15–25	100–116
Rapeseed	*Brassica napus* L.	33–45	96–106
Nondrying Oil			
Sesame	*Sesamum indicum* L.	52–57	104–118
Peanut	*Arachis hypogeae* L.	47–50	92–100
Castor bean	*Ricinus communis* L.	35–55	82–90
Coconut	*Cocos nucifera* L.	67–70	8–12
Olive	*Olea europaea* L.	—	86–90
Palm	*Elaeis guineensis* L.	—	49–59
Palm kernel	*Elaeis guineensis* L.	—	204–207

Source: Extracted from J. H. Martin and W. H. Leonard (1976).
Note: the larger the iodine number, the more unsaturated the fatty acid. Principles of field crop production. 2nd ed. Macmillan Co. London.

Overview of Major Feed Grains

Cereal grains are widely used to feed livestock and poultry as an energy source. As already mentioned, most of the cereal grain in developed nations is used for feed, while in the developing countries, most of it is used for food. Corn, sorghum, barley, and oats are usually fed to livestock as whole grain or after a little processing. To improve the nutritional quality of cereal grains, they are fed to animals as mixtures with protein supplements. Sometimes roughage such as hay or silage is added to slow the otherwise rapid passage of cereal meal through the digestive system of animals. Apart from whole grain, the byproducts of grain processing (e.g., bran) and spent grain (from breweries) are used as feed.

Overview of Major Oil Crops

Major oil crops include olive *(Olea europaea)*, linseed *(Linum usitatissimum)*, sesame *(Sesamum indicum* L.), sunflower *(Helianthus annus* L.), soybean *(Glycine max* L. Merr.), coconut *(Cocos nucifera* L.), palm *(Elaeis guineensis)*, corn *(Zea mays* L.), and peanut *(Arachis hypogeae* L.).

Plant-derived oils are used for food as well as industrial purposes. Oils supply three essential fatty acids (arachidonic, linoleic, and linolenic acids), and vitamins A, K, D, and E. The most important vegetable oils are presented in Table 1–5. The quality of oil, based on a determination of the iodine value, is also presented. Oils and fats are chemically made up of units called *triglycerides*. A triglyceride consists of glycerine and three fatty acids. A fatty acid also consists of a long chain of carbon, hydrogen, and oxygen atoms. A fatty acid is described as *saturated* if the carbon atoms are bonded to hydrogen or other carbon atoms. In some cases, not enough hydrogen atoms are incorporated, resulting in double bonds. The fatty acid is then described as *unsaturated*. The degree of saturation is measured by the iodine value: the lower the value, the more saturated the fatty acid. Unsaturated oils are also called *drying oils,* while those with low iodine number (saturated oils) are called *nondrying oils.* Saturated fats and oils are known to increase blood cholesterol. Drying oils (e.g., linseed) have industrial value and are widely used in paints and varnishes. Give examples of commercially marketed cooking oils that are high in saturated oils and those high in unsaturated oils.

Oils are extracted from seeds by the method of extrusion (uses mechanical press to squeeze oil out) or solvent extraction. The byproduct of these processes is called the *oil seed meal.* It is high in protein and widely used as protein supplement in feeds for poultry and livestock.

Overview of Major Fiber Crops

Plant fibers are used for clothing and textiles. Cotton lint is the most important fiber for clothing. Based on the plant part from which the fiber is obtained, there are three types of plant fiber: cotton, **bast fiber**, and **vegetable fiber**.

Bast fiber. Fiber obtained from the bark of certain herbaceous plants (also called **soft fiber**).

Cotton Cotton (*Gossypium* spp) fiber is obtained from the seed of the plant. There are three types of cotton fibers. *Asiatic cotton (G. arboreum* and *G. herbaceum)* produces short fibers of less than 1 inch (25 mm) long. They are used mainly in manufacturing surgical supplies. *Egyptian cotton (G. barbadense)* has long, fine, and strong fibers and is used in manufacturing sewing threads. The third type, *upland cotton (G. hirsutum)*, can produce fibers of varying lengths and fineness. Most (99%) of U.S. cotton is upland cotton.

Vegetable fiber. Fiber obtained from the leaf and sheath of plants (also called **hard fiber**).

Bast Fiber *Bast* or *soft fibers* are obtained from the bark of certain herbaceous plants. The four most common sources of bast fibers are flax *(Linum usitatissimum* L.), hemp *(Cannabis sativa* L.), jute *(Corchorus* spp), and kenaf *(Hibiscus cannabinus* L.). The first two types of fiber are used for textiles, whereas the last two are used mainly for manufacturing packing materials.

Vegetable Fiber Vegetable fiber from leaves is also called *hard fiber.* It is obtained from the leaves and leaf sheaths of monocots such as Abaca or manila hemp *(Musa textilis* Nee), sisal *(Agave sisalina* Perr.), and henquen *(Agave fourcroydes* Lem.). The primary use of hard fibers is in the manufacture of rope and cordage.

Overview of Major Forage Crops

Important **forage** crops include alfalfa, clovers, lespedeza, timothy, sudangrass, and johnsongrass. Cereal grain crops may be grown and used for livestock feed during its early vegetative stages. Forage crops are grown primarily for their vegetative parts that are used for feeding livestock. The quality of forage is determined by the voluntary intake of digestible energy. Feeds are generally classified by their crude protein and crude fiber content. The common forage crops and their crude protein, crude fiber, and digestible energy are presented in Table 1–6. Forage crops are sources of vitamins A and E and numerous essential nutrients, including sodium, potassium, calcium, phosphorus, magnesium, sulfur, copper, zinc, and iron. Forages are very important in ruminant production, accounting for about 75% digestible energy.

Forage. Vegetative matter, fresh or preserved, that is used as feed for livestock.

Table 1-6 Feed Value of Selected Forage Crops

Crop	Crude Protein	Crude Fiber	Digestible Energy
Alfalfa hay (vegetative)	25	20	2.6
Barley hay	9	26	2.5
Bermudagrass hay	9	30	2.2
Red clover hay	15	30	2.6
Corn fodder	9	26	2.8
Fescue hay	10	31	2.5
Oats hay	9	31	2.4
Timothy hay (vegetative)	12	33	2.6

Source: Extracted from National Research Council data.

1.4: CROP PRODUCTION: AN ART, A SCIENCE, AND A BUSINESS

Crop production is a complex operation. Its success depends on both the crop and environmental factors, coupled with socioeconomic and political factors. What kind of person (in terms of training, experience, knowledge, etc.) do you think would be very successful as a crop producer in these modern times?

1.4.1 THE ART OF CROP PRODUCTION

Is formal education required to be successful as a farmer? The art of crop production has evolved over the ages, taking on the sophistication of the day. Crop producers in primitive cultures selected specific crops, chose specific varieties, and prepared the land prior to planting. They planted in the right season, protected the crop from pests (weeds, insects, and diseases), and adopted techniques to increase productivity.

These artistic values are perpetrated in modern agriculture but at improved levels. Farmers still exchange ideas and experiences, only this time they do it through magazines, rural papers, and the Internet. They have access to improved cultivars or varieties and production practices (e.g., use of manure, fertilizers, and effective pest control) and better harvesting and storage facilities. Instead of removing insect pests manually, they have learned to use domestic products such as ash to control them. Modern growers also rotate crops, fallow the land, and use conservative measures to protect the soil.

Just how is modern crop production technology-dependent? A significant difference between the primitive art and modern art of crop production is technology that enables modern crop producers to perform tasks more efficiently and effectively. Technology used in farming is constantly evolving. Countries and individuals with economic resources tend to adopt advanced technology, while those with limited resources utilize ordinary technology. Primitive agriculturists or farmers used primitive tools that required great human power to operate. This limited production to small acreages, because it was a tedious operation. As technology advanced, some of the tedium in production was transferred to draft animals, which were used in various ways, including transportation and tillage. The development and use of machines has reduced the need for labor. An individual using a variety of machines and implements can single-handedly operate a large farm.

Technology is not only in terms of mechanical innovations but also chemical (agrochemicals), biological (e.g., improved cultivars), cultural (improved practices), and general knowledge and expertise for management. In terms of technology, U.S. agriculture, and for that matter that of all industrialized nations, may be classified into three eras:

1. The mechanical era (1930–1950)
2. The chemical era (1950–1970)
3. The biotechnology and information technology era (1970–present)

The introduction of machines into crop production provided power for production. They replaced human labor and the use of draft power (animal power). The impact of machines includes the ability of producers to cultivate larger acreages. Work is done more rapidly and on time. Timeliness of farm operations has resulted in planting on time to avoid environmental stresses, thereby reducing field losses from delayed harvesting. The use of machines has increased crop productivity several times over. Land preparation is much easier with machines. Land can be restructured through grading, leveling, terracing, and other modifications that conserve soil or facilitate the use of machines and other technologies. Automation has improved timeliness and efficiency of operations such as irrigation and reduced the need for human labor. Production of certain major world crops such as corn and wheat has been so mechanized that all operations, from planting to harvesting, can be done with machines.

In 1986, the Office of Technology Assessment of the U.S. Congress conducted a study of technology, public policy, and the changing structure of American agriculture. It published a list of emerging agricultural production technology areas. The study considered the development, commercial adoption, and impact of the 28 areas of technology that would likely emerge by the year 2000 (Table 1–7). One of the predictions made was that the impact of biotechnology on plant agriculture would be most significant after the year 2000. Agrochemicals provide additional nutrients (by way of fertilizers) to boost crop productivity through yield enhancement and protection of plants from diseases and pests that cause reduced harvest and storage losses. Improved cultivars have higher yields and have more desirable agronomic qualities. Access to production information is facilitated by innovations in information technology. The information, coupled with computer technology, has improved the effectiveness and efficiency of management practices.

Much of the sophisticated technology in use in crop production is possible because of computers (Figure 1–3). These electronic machines enabled the design and production of technologies or facilitated their operation. Computers are central to equipment automation and automated guided vehicles (robots) that are being developed for various crop production operations (especially in the horticultural industry, such as for fruit harvesting).

Computers enable crop producers to access databases for information pertaining to crop production. By accessing the Internet, producers have access to weather information, government legislation, government programs, markets, and numerous other pieces of information. They can communicate among themselves and with experts to share and obtain information for enhancing crop production.

Record keeping and management are critical to a successful operation. Crop producers can use computers to keep inventory and manage the financial aspects of their operations. They can keep records of all their production activities (fertilizer application, pesticide application, irrigation scheduling). Computers are also at the heart of **expert systems**, the knowledge-based decision aids that use artificial intelligence technologies.

Expert systems.
Knowledge-based decision aids that make use of artificial intelligence technologies.

Animal Technologies
Animal genetic engineering
Animal reproduction
Regulation of growth and development
Animal nutrition
Disease control
Pest control
Environment of animal behavior
Crop residue and animal wastes use
Monitoring and control in animals

Plant Technologies
Plant genetic engineering
Enhancement of photosynthetic efficiency
Plant growth regulators
Plant disease and nematode control
Management of insects and mites
Weed control
Biological nitrogen fixation
Chemical fertilizers
Water and soil-water-plant relations
Soil erosion, productivity, and tillage
Multiple cropping
Organic farming
Monitoring and control of plants
Engine and fuels
Land management
Crop separation, cleaning, and processing

Common Technologies
Communication and information management
Telecommunications
Labor saving

Source: Office of Technology Assessment.

Note: These technologies were given a 50:50 chance of emerging before 2000. Some of them are already in the marketplace; others are still in the experimental stages. Further, specific new technologies will continue to be developed and implemented in these areas as knowledge in science advances.

Using commercial software, producers, for example, can perform disease diagnosis on their farms. Computers make the variable rate technology possible. The training of agricultural scientists in schools and colleges involves the use of computers.

Agricultural *information technologies* are grouped into three general categories:

1. Communication and information management
2. Monitoring and control technologies
3. Telecommunications

Agricultural information technology is facilitated by computer technology. Communication and information management is comprised of on-farm digital communication systems (i.e., local area networks, or LANs), and computer-based information processing

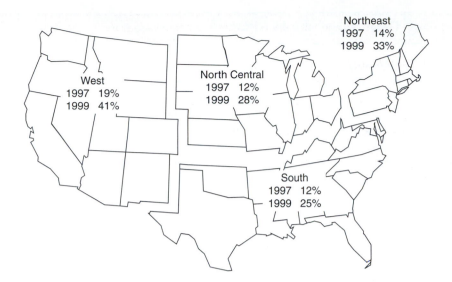

FIGURE 1–3 The role of computers in modern crop production is increasing. Producers use computers in planning production, in actual production, and in marketing of produce. Access to computers in the continental United States more than doubled in each region in 1999 from the previous survey results in 1997. For the entire nation, 47% of farms had access to a computer in 1999, up from 38% in 1997. (Source: USDA)

technology. Producers use this setup to monitor and control production operations (e.g., irrigation schedule, temperature control of storage houses). Telecommunication occurs either in voice mode or via data. Producers communicate via telephone and two-way radio communication. In precision farming, producers utilize satellites for various production operations. Satellite communication is also utilized in cellular phone communication. Do you think crop production in industrialized nations is approaching a time when robots will feature prominently in field operations?

1.4.2 CROP PRODUCTION AS A SCIENCE

What is the advantage of formal training in agricultural science to a modern crop producer? As a science, crop production involves applying theories from various scientific disciplines, chiefly agronomy. The term *agronomy* is derived from the Greek words *agros* ("field") and *nomos* ("to manage"). Agronomy can be defined as the branch of agriculture concerned with the principles and practices of crop production and field management. It became a distinct and recognized branch of agricultural science in about 1900. As a branch of agriculture, it combines soil and crop science. Agronomists have a strong foundation in basic science disciplines, such as chemistry, botany, and physics, applying the principles of these disciplines to the production of crops. The ultimate goal of crop production is the yield of an economic plant product. This is accomplished through the management of plant morphological and physiological responses within a given production environment. Plant breeders utilize genetic principles to develop new and improved cultivars that are high-yielding, environmentally responsive, disease-resistant, and adapted to environmental stresses. They manipulate plant morphology to make them more physiologically responsive to the growing environment. The yield (quantity and quality) of economic products is enhanced.

Modern producers use science and technology to reduce the adverse impact of the vagaries of the weather in crop production. For example, farmers have access to weather forecast information, providing guidance for the best time to plant crops in the field. Soils scientists conduct research to determine plants' nutritional needs for optimal yield. Fertilizer research provides guidance for optimal rates of application of plant soil nutrients. Soil analysis is used to determine the precise needs of crops to meet yield goals. The use

Two scientists, Alphonse de Candolle and Vavilov, were in the forefront in describing the geography of plant domestication and crop origins. De Candolle published a landmark book on the subject called the *Origin of Cultivated Plants.* Vavilov was interested in plant genetic diversity. He also published an essay on the origin of cultivated plants in 1926. His thesis was that one could reliably determine the center of origin of cultivated plants by analyzing patterns of variation in plants. Essentially, he proposed that the region of origin of a plant would coincide with the geographic region of greatest genetic diversity of this plant. This region would have wild progenitors of the plant.

Vavilov pursued his study that eventually culminated in his proposing eight *centers* of origin of most cultivated plants. His delineation was based on areas where indigenous cultures arose and where agriculture had been practiced for a long time. Centers of diversity, however, are not the same as centers of plant origin. It turns out that many crop plants do not originate in any of Vavilov's centers of plant origins. Eventually, Vavilov was compelled to modify his theory, following intense disagreement from the intellectual community. He proposed an addendum to his original theory by describing what he called *secondary centers of plant origin*. This model, however, was not adequate, either.

Harlan proposed an alternative to the existing Vavilov model that he called the *no model* model. The rationale was that human nature is very complex. As such, no one model will have universal application, since the motives of humans for doing one thing are varied and complex. Domestication of plants possibly began in different regions of the world for different reasons. People domesticated plants as the need arose.

The process of domestication of crop plants is an evolutionary one in which crop plants are changed from wild progenitors to the fully domesticated cultivated species that form the basis of modern agricultural economies. Wild species are very adapted to their natural environments and are able to survive independent of human intervention. On the other hand, fully domesticated species are able to survive only with the aid of humans.

of new crop varieties that are adapted to new production areas has expanded crop production worldwide.

Scientists develop chemical products such as growth hormones, pesticides, and fertilizers that enhance crop productivity. They also develop cropping systems, which are the production packages comprised of crop or pasture communities and sets of management practices of farming systems (pertaining to particular farms). Agricultural engineers develop machinery and equipment to facilitate production operations. These include equipment for land preparation, seeding, cultivating, fertilizing, irrigating, spraying to protect against pests, and harvesting.

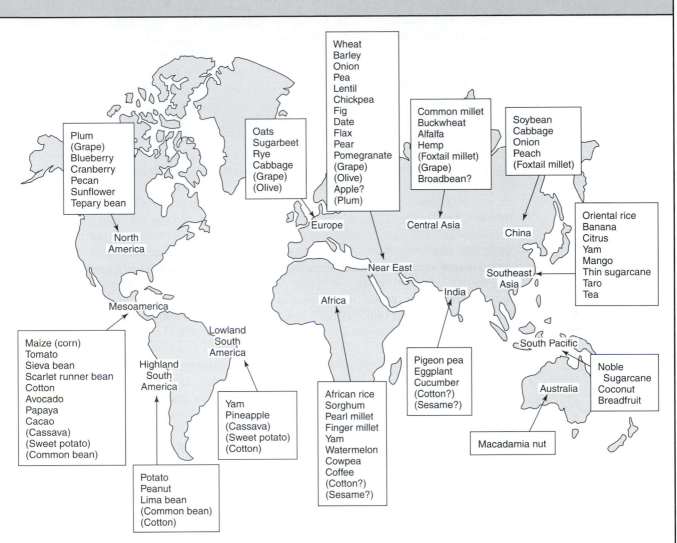

Plum
(Grape)
Blueberry
Cranberry
Pecan
Sunflower
Tepary bean

North America

Oats
Sugarbeet
Rye
Cabbage
(Grape)
(Olive)

Europe

Wheat
Barley
Onion
Pea
Lentil
Chickpea
Fig
Date
Flax
Pear
Pomegranate
(Grape)
(Olive)
Apple?
(Plum)

Central Asia

Common millet
Buckwheat
Alfalfa
Hemp
(Foxtail millet)
(Grape)
Broadbean?

Soybean
Cabbage
Onion
Peach
(Foxtail millet)

China

Oriental rice
Banana
Citrus
Yam
Mango
Thin sugarcane
Taro
Tea

Southeast Asia

Near East

Africa

India

Mesoamerica

Maize (corn)
Tomato
Sieva bean
Scarlet runner bean
Cotton
Avocado
Papaya
Cacao
(Cassava)
(Sweet potato)
(Common bean)

Highland South America

Lowland South America

Yam
Pineapple
(Cassava)
(Sweet potato)
(Cotton)

Potato
Peanut
Lima bean
(Common bean)
(Cotton)

African rice
Sorghum
Pearl millet
Finger millet
Yam
Watermelon
Cowpea
Coffee
(Cotton?)
(Sesame?)

Pigeon pea
Eggplant
Cucumber
(Cotton?)
(Sesame?)

South Pacific

Australia

Noble Sugarcane
Coconut
Breadfruit

Macadamia nut

The cost of marketing is largely responsible for the high cost of agricultural produce. In 1999, it accounted for 80% of the cost of the produce. (Source: USDA)

1.4.3 CROP PRODUCTION AS A BUSINESS

In what way is the crop producer a manager? Crop production and productivity (production efficiency) entails managing production inputs to produce outputs. The crop producer is a decision maker.

The Crop Producer as a Decision Maker

A crop production operation requires careful planning. The producer should be able to make the right choices in terms of selecting and managing the appropriate production inputs. The goal of crop production is high productivity for profitability. Even in farming systems where production is primarily for domestic use (selling only when there is surplus), the producer still aims for high productivity. The crop producer is faced with critical decisions throughout the enterprise.

There are three categories of factors that the producer has to be concerned with in planning and producing a crop. Some factors are within the producer's total control, others can be manipulated for better results, while yet others are totally outside his or her control (Figure 1–4).

Factors Within Total Control   What aspects of the crop production operation are within the control of the operator? The crop producer is responsible for selecting the site for the enterprise. This requires knowledge of the requirements of the crop plant of interest, in terms of adaptation, suitability of the soil type, and other growth requirements. The soil type needed for tuber and root crop production is different from that for cereal grains. The producer has to select the best crop that can grow on the soil and find the best cultivar suited to the region. The cultivar should be high-yielding, regarding the plant product of interest to the producer, and should be adapted to the cultural method to be used. The producer has control over the acreage to produce. This will be based in part on the anticipated market demand. The producer also decides on the best cultural practice to adopt for high productivity. This includes time of planting, planting density, type and amounts of production inputs, pest control, time of harvesting, and others.

Factors That Can Be Manipulated   What natural production resources can be modified in a crop production undertaking? Some factors in the production operation can be modified for best results. Sometimes, the best site cannot be found for a crop plant. Certain soil amendments may be required to improve soil physical and chemical conditions. This includes drainage, leveling, terracing, and liming. Even if the soil conditions are adequate to start with, the producer may want to increase crop productivity by providing supplemental nutrition and other factors that promote growth and development (e.g., fertilizers and irrigation). Sometimes, temperature can be manipulated within certain limits for early planting. This could be done by simply using raised beds. Producers also use various devices to protect plants against frost (e.g., row covers, heat caps, and wind machines). Production under a controlled environment (greenhouse) allows temperature, light, and humidity to be more effectively controlled for optimal production environment for high productivity.

FIGURE 1–4 Factors affecting modern crop production. Certain factors are within the control of the operator, whereas others are not. In addition, some factors can be modified through the use of technology and other modern resources. For example, crops can be cultivated in arid regions with irrigation.

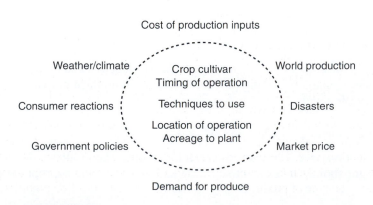

Cost of production inputs

Weather/climate

World production

Crop cultivar
Timing of operation

Consumer reactions

Techniques to use

Disasters

Location of operation
Acreage to plant

Government policies

Market price

Demand for produce

Factors Outside the Producer's Control Are there any crop production factors that the producer can do absolutely nothing about? The weather factor is considered as the chance element in crop production. Shall we then blame it on the weather when we have a crop failure? The producer may select the best site and cultivar and adopt the best cultural practices and still be unsuccessful if the weather does not cooperate. Mild fluctuations in the weather are usually not difficult to overcome or plan against (e.g., a brief drought period or a mild unexpected cold spell). Supplemental irrigation or using drought-resistant cultivars can rectify a mild drought. However, a severe and protracted drought may make irrigation impractical, leading to heavy or total crop failure. Acts of nature such as strong winds, hail, and floods are usually devastating to a crop production enterprise. Too much rain during the crop harvesting time may lead to significant field and even storage losses. The producer does not have control over the incidence of pests. Certain pests are endemic in some regions, whereas others are associated with certain weather systems. For example, a grower does not have control over a locust invasion.

The acreage devoted to production is related to anticipated demand for the crop plant product for which the operation is initiated. The producer is not in control of the total acreage of the crop produced in other regions or nations of the world. As such, there could be overproduction at the end of the production season and consequently a loss of potential revenue to falling prices. Agricultural production is a high-risk activity. Failure to make the right decision in a timely fashion can lead to significant losses. You might think about ways by which decision making in crop production can be improved.

Crop production is not immune to politics. The government may implement legislation that discourages crop producers from allocating adequate resources to a production activity. Farmers allocate resources based on anticipated profitability of the crop to be produced. Political influences on crop production may come in the form of trade agreements, application of tariffs, taxes, subsidies, and financial assistance. Administrations implement policies according to a certain political agenda. Since political administrations usually have a limited term in office, there is always the potential for a drastic policy change in the short term. Agricultural production is not as resilient as many manufacturing industries. Agricultural production adapts slowly to significant changes in demand and price, and it often suffers adverse consequences as a result. This is largely because agricultural production is slow and tedious, often requiring a full year to complete the lifecycle of most field crops. Agriculture is not able to compete for land with other high-return manufacturing enterprises.

Since a government often cannot directly create the supply of a particular commodity, it has to stimulate production by implementing incentives to producers. Field production is subject to the vagaries of the weather. Returns to field crop production are unstable. Farmers prefer to grow crops for which the price is predictable and stable. To encourage the production of specific field crops, the government may introduce price stabilization measures. The government may also implement policies to assist selected agricultural industries by direct subsidy or price support. The government may also enter into trade agreements such as the General Agreement on Tariffs and Trade (GATT) to stabilize world trade with respect to certain agricultural commodities.

What is the driving force behind decision making in crop production? Farmers make production decisions based on expected profitability. If the outlook and world price are poor for a crop, producers will divert resources into alternative and more viable enterprises. Agriculture is generally a risky enterprise. Certain crops are high-risk and have large variations in yield and market price. These are less attractive to producers, especially those with limited resources. Crops such as cotton are capital-intensive and require high-technology input. This limits the type of producers who are able to undertake cotton production.

Producers must first identify markets for their produce. Access to markets may be local or even international. When the crop production center is located far from the processing center, a good and efficient transportation network is required to encourage distant farmers to produce the crop.

1.5: EVOLUTION OF MODERN AGRICULTURE

Crop production has not always been what it is today. How has it changed over the years? Modern agriculture benefits from research and technological advancement. As economics permits, producers are able to implement new techniques and technologies to facilitate agricultural production. World agriculture has experienced changes that can be categorized into four fairly distinct eras, based on how production resources were utilized and managed.

1.5.1 ERA OF RESOURCE EXPLOITATION (BEFORE 1900)

Modern agriculture started with the identification and exploitation of production resources. First, prime lands were identified and cleared for farming. The land was tilled and prepared for planting and crop establishment. The basic rationale of this era was that soil productivity was a resource to be exploited for crop productivity. The soil nutrients were repeatedly removed without replacement. This is the mining approach to crop production. This led to the progressive depletion of organic and inorganic soil factors and consequently reduced crop productivity. When the soil was deemed unproductive (just like an old mine), it was abandoned for a new fertile site. This was a non-sustainable production approach, requiring new lands to be exploited on a regular basis.

Some exploitative activities continued even to the twenty-first century. A major production activity of this kind is the continued mining of groundwater of the Great Plains for crop irrigation. The consequence of this exploitation is the decline in water levels of the Ogallala aquifer, as discussed later in this book. Furthermore, the implementation of the fiber subsidies encouraged overgrazing of the land in sheep and goat production regions.

1.5.2 ERA OF RESOURCE CONSERVATION AND REGENERATION (EARLY 1900s)

In the early 1900s, producers began adopting production techniques that were less exploitative of the soil. Abandoned farmlands were eventually restored to health, though only after many years in fallow. Crop rotation provided a strategy for a more efficient use of soil moisture that the previous continuous cropping system did not offer. Legumes, through symbiosis, contributed to regeneration of the soil by nitrogen fixation. Organic manure from green manuring (growing a crop for the purpose of plowing under while still green) and deliberate use of barnyard or livestock manure replenished soil fertility.

1.5.3 ERA OF RESOURCE SUBSTITUTION (MID-1900s)

The activities of this era made the industrialization of agriculture possible. The period started roughly in the mid-1900s (a little before 1950). This period saw the exploitation of mechanization that has continued to this day. Machines replaced farm draft animals. Supplemental moisture was supplied through crop irrigation. Chemical fertilizers were used more commonly than the bulky organic fertilizers. Pesticides were used more frequently to control weeds and other crop pests. Plant breeders developed more improved cultivars for farmers. Technological advances in this era enabled producers to cultivate large tracts of land. The late 1900s saw the introduction of biotechnology and other revolutionary technologies, such as precision farming, into crop production.

1.5.4 ERA OF INFORMATION

Technology will continue to advance. New resources will be discovered. The future of agricultural production will probably be shaped by the ability of the producer, as a manager, to translate information and knowledge into value. The use of the Internet and other regional and national information networks will continue to make crop production information more readily accessible to producers.

1.6: THE RICH HISTORY OF U.S. AGRICULTURE

What are some of the major landmarks along the evolutionary path of U.S. agriculture? Like other countries, U.S. agriculture began modestly as a low input enterprise. Land in the early days was abundant and labor cheap. Use of technology was very limited. Agricultural chemicals were not commonly used. Farms expanded rapidly because of the abundance of productive land. However, as the United States expanded to the Pacific, lands became limited.

The agricultural policy of the nineteenth century revolutionized U.S. agriculture by refocusing it from the trend of expansionism to one of concentration and enhanced productivity. The strategy adopted under this policy was to develop and provide farmers new capital resources in the form of scientific knowledge and new technologies.

This began with the establishment of the land-grant university system through the **Morrill Act of 1862**. Agricultural scientists were trained, while researchers at these institutions developed new technologies and techniques to boost agricultural production. Producers were able to produce more on a smaller acreage. Between 1910 and 1970, the United States doubled its agricultural output, using smaller total acreage. Technology, however, soon displaced workers, leading to significant migration of farmworkers between 1950 and 1955 into other sectors of the U.S. economy. Producers were also aided by government programs such as the Federal Farm Loan of 1916, which enabled them to acquire farm machinery at attractive interest rates.

Morrill Act of 1862. The act of Congress that established the land-grant university system in the United States.

U.S. agriculture suffered severe setbacks between 1920 and 1930 during the Great Depression, but it rebounded by 1950. The government introduced programs to compensate farmers for income losses during periods of inelastic demand for agricultural products. Farmers received compensation for idling their land.

American crop yields were at record highs in 1974 and 1975. Exports also increased when the Russian wheat crop failed in 1972. Farm incomes soared. Farms became larger and more specialized in crops produced. Most acreage was devoted to wheat, corn, and soybeans. Such specialization also called for increases in nutrients that were continually removed with the intensive cultivation of the land. The adverse effect of intensive crop production was increased environmental pollution from the use of pesticides and fertilizers entering groundwater and other bodies of water. Intensive tillage also increased surface runoff of silt into bodies of water.

The growth of U.S. agriculture also spawned an extensive system of support agribusinesses. These industries focusing on the input side of agriculture include those producing machinery, chemicals, and seed, and those at the output side, such as transportation, processing, packaging, and marketing.

The reader can consult the USDA website listed at the end of the chapter for historical accounts of various aspects of U.S. agriculture.

1.6.1 THE SUCCESS OF U.S. AGRICULTURE IS BASED ON SOUND POLICY

What has made U.S. agriculture so successful? First, America is blessed with a diversity of natural resources that suit a wide variety of agricultural production enterprises. Second, the tremendous success of American agriculture depends on the nineteenth-century implementation of an effective agricultural development policy that combines research, education, farm credits, and other governmental policies. These factors together created capital technologies that producers were then encouraged to exploit for increased productivity. The result is currently high agricultural productivity and surpluses. U.S. agriculture contributes to solving world food problems through export and food aid programs.

The U.S. agricultural policy is successful because it incorporates certain elements:

1. It enlarges the producer's (farmer's) supply of key production resources and keeps prices low.
2. Prices of agricultural products, on the other hand, are kept stable and high to encourage their continued production.
3. Innovation is promoted through the land tenure system.
4. Research and technology are encouraged. It should be pointed out that from time to time the U.S. government changes the specific methods by which it implements its agricultural policy. These changes notwithstanding, agricultural development does not appear to be hampered.

1.7: STRUCTURE OF U.S. AGRICULTURE★

1.7.1 MOST U.S. FARMS ARE CLASSIFIED AS "SMALL," GROSSING LESS THAN $50,000 ANNUALLY

Farm. Any location from which $1,000 or more in agricultural products were sold, or normally would have been sold, during the census year.

What is a **farm**? For statistical purposes, the U.S. Department of Agriculture (USDA), Office of Management and Budget, and Bureau of Census define a farm as any place from which $1,000 or more of agricultural products were sold, or normally would have been sold, during the census year. The number of farms is on the decline, while farm size is on the rise (Figure 1–5). There appears to have been some stabilization in these trends in the 1970s. In 1993, the United States had about 2 million farms (2,063,300), down from the peak of 6.8 million in 1935. The number of farms in 1997 was 2,057,910. It should be pointed out that, in spite of the general decline in number of farms, certain states actually experienced an increase in the number of farms (Figure 1–6). Farms with the largest sizes are found in the mountain states and the northern plains.

Farms may be classified according to several criteria, including size, sales, and specialization. In this book, the classification used before 1999 and the 2000 classification are discussed. The size of farms can be determined by gross sales. In this regard, farms

*NOTE: The data presented in this section (and other parts) are drawn primarily from the U.S. federal government system. Various federal agencies collect and maintain comprehensive data on various aspects of U.S. agriculture. Because of the time it takes to collect, analyze, and publish results from these data, final reports frequently include data that are several years old. The author endeavored to use the latest possible data at the time of printing to show trends. The reader is encouraged to visit the website of the referenced agencies (provided at the end of the chapter) for the most current data.

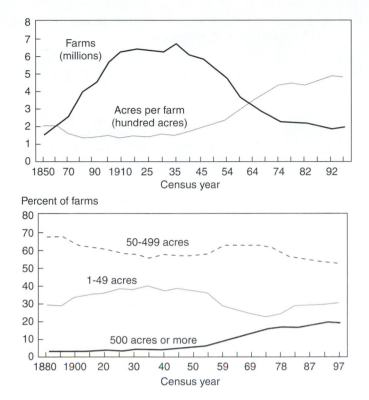

FIGURE 1–5 Number of farms and average farm size in the United States: 1975 to 1999. The definition of a farm was amended in 1993 to include equine, maple syrup, and short-rotation woody crop production, hence the change in the trends since that time, shown as a break in the figure. Prior to this time, the trend was a steady decline in the number of farms and an increase in farm size. (Source: USDA)

may be classified into two broad groups: *commercial* and *noncommercial.* A commercial farm will have gross sales of more than $50,000, while a noncommercial farm will have sales of less than $50,000. Most farms in the United States were classified in 1997 as noncommercial farms. However, they accounted for only about 10% of production as measured by gross sales. In 1997, 61% of farms reported gross farm sales of less than $20,000 and accounted for 16.9% of the total acreage operated. Only 2.8% of farms in 1997 reported gross farm sales of $500,000 or more. These large farms accounted for 16.5% of the acreage operated. Most commercial farmers (90%) practiced farming as their primary occupation, as compared with only 29% of noncommercial farmers. Most noncommercial farmers supplement their farm incomes from other side activities such as the rental of part of their land to other producers, grazing, and custom work. In 2000, the U.S. Department of Agriculture released a new classification of farms (Table 1–8).

Physical size (acreage) and economic size (gross sales) are not directly related. **Small family farms** may pose higher economic sales than large farms by producing high-value products in a small area. A case in point is the greenhouse industry, where operators utilize relatively small areas but experience high economic returns on investment. For example, in 1994, cash grain farms averaged $93,093 on an average acreage of 584, while greenhouse and nursery operators produced products on an average of 56 acres and averaged $176,396 gross value.

Small family farm.
A farm with sales of less than $250,000 during the year.

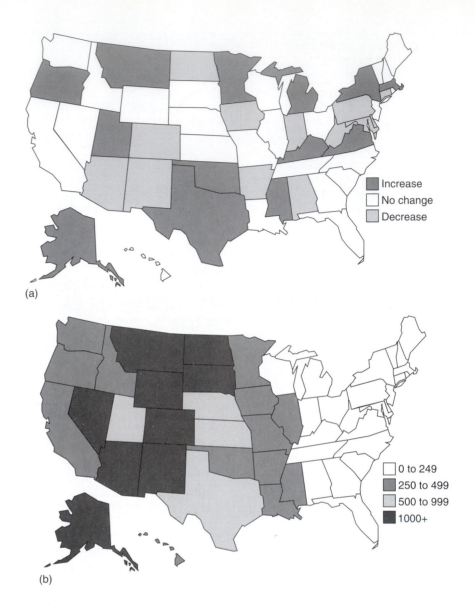

FIGURE 1–6 Changes in farm number and size in 1999: (a) areas with a net change in number of farms and (b) distribution of farm sizes. Most states did not lose or gain in number of farms. Farms with the largest acreage were mostly in the mountain states and the northern plains. (Source: USDA)

Increase
No change
Decrease

(a)

0 to 249
250 to 499
500 to 999
1000+

(b)

To increase gross sales, field crop producers may utilize the land intensely by adopting production practices such as double-cropping of wheat, as occurs in the southeast farming region.

1.7.2 LESS THAN 25% OF U.S. LAND IS CROPLAND

How is land used in the United States? Are there any trends in land use? The mainland (the 48 contiguous states) constitutes about 1.9 billion acres of land area. The major uses of cropland include cropland harvested, summer fallow, land idled in federal programs, and crop failure. The United States lost 25.8 million acres of cropland devoted to crops between 1945 and 1992 (Table 1–9). Cropland use does not have definite trends. However, certain significant events have impacted the acreage devoted to crop production. Foreign demand for American grain in the mid-1940s was responsible for the increase in acreage that peaked in 1949. After the war, the importing countries embarked on economic recovery activities, including the agricultural sector. This was responsible for the

Table 1-8 USDA-Defined Farm Typology Groups

SMALL FARMS

Small family farms (sales of less than $250,000)
Limited-resource: Any small farm with gross sales less than $100,000, total farm assets less than $150,000, and total operator household income less than $20,000. Limited-resource farmers may report farming, a nonfarm occupation, or retirement as their major occupation.
Retirement: Small farms whose operators report they are retired (excludes limited-resource farms operated by retired farmers)
Residential/lifestyle: Small farms whose operators report a major occupation other than farming (excludes limited-resource farms with operators reporting a non-farm major operation)
Farming occupation/lower sales: Small farms with sales less than $100,000 whose operators report farming as their major occupation (excludes limited-resource farms whose operators report farming as their major occupation)
Farming occupation/higher sales: Small farms with sales between $100,000 and $249,999 whose operators report farming as their major occupation

OTHER FARMS

Large family farms: Farms with sales between $250,000 and $499,999
Very large farms: Farms with sales of $500,000 or more
Nonfamily farms: Farms organized as non-family corporations or cooperatives, as well as farms operated by hired managers

Note: The $250,000 cutoff for small farms was suggested by the National Commission on Small Farms.

Table 1-9 Net Change in Major Use of Land in the United States Between 1945 and 1992

Land Use	Net Change (Million Acres)
Cropland	+9.0
Cropland used for crops	−25.8
Cropland idled	+15.4
Cropland used for pasture	+19.3
Grassland pasture and range	−70.5
Forest-use land	−43.0
Forestland grazed	−200.0
Forestland not grazed	+156.9
Special uses	+94.2
Urban land	+42.8
Transportation	+2.1
Recreation and wildlife areas	+64.3
National defense areas	−6.2
Miscellaneous farmland uses	−8.9
Miscellaneous other land	−0.9
Total change	−11.3

Source: Extracted from USDA, ERS data (includes 48 contiguous states only).

The structure of agriculture (as determined by number and size of farms, contractual arrangements, control of management decisions, off-farm employment, extent of tenancy, ownership of farmland, etc.) is determined by three principal factors:

1. Technical factors
2. Institutional factors
3. Economic environment

These three factors interact in a dynamic fashion to influence the structure of agriculture.

Technical factors

The chief technical factors impacting the structure of agriculture are technological forces, economies of size, specialization, and capital requirements. New technology influences the size of farm operation and capital requirements. Farmers differ in terms of innovation. Also, incentives for innovation vary from one situation to another. Early adopters of technology are often able to reduce per-unit production costs and increase the profitability of their enterprise, at least in the short run, over their counterparts. Adoption of more powerful machinery enables innovators to expand their operations for increased profitability. Studies have shown that most economies of size are captured by moderate-size farms. Further, while the lowest-average cost of production may be attainable on a moderate-size farm, average cost of production tends to remain constant over a wide range of farm sizes. This indicates that there is a strong incentive for farmers to increase the size of their operation for increased profitability.

Technology also tends to foster specialization and influence regional patterns of crop production. For example, as farm power increased, cotton production in the United States shifted westward to areas where the land was flat and most conducive to the use of large machinery for optimal advantage. Technology has also impacted diversification of enterprise. In low input agriculture, farmers use technologies such as crop rotation to manage pests and conserve soil fertility. The introduction of agrochemicals (fertilizers and pesticides) into farming caused a shift from diversified cropping to monocropping. Even though technology-induced specialization is applicable to all sizes of farm operation, improvements in farm machinery have tended to encourage large-scale specialized operations. It follows that, if a producer invests in a specialized piece of equipment, he or she would focus more on the crop for which the equipment was designed. Further, there will be a tendency to expand the selected operation, thus making specialization and farm growth occur simultaneously.

New technology is expensive. In fact, agriculture is among the most capital-intensive industries in the

decline that ensued and lasted through the early 1970s, when there was resurgence in the demand for U.S. grains. The economic recession in 1982 led to agricultural surplus in the grain industry as demand for grain declined. Government programs and regulations, such as the *Conservation Reserve Program (CRP)* implemented in 1986 and the *Acreage Reduction Program (ARP)*, idled croplands. In 1980, no land was idled. However, in 1988, soon after the implementation of the CRP, about 78 million acres of land were idled. CRP is a voluntary program whereby participants receive monetary incentives for idling environmentally sensitive cropland (highly erodible, or HEL) for a period of 10 to 15 years. Producers who enroll in the program may receive up to 50% cost share assistance for establishing vegetative cover on the land to reduce soil erosion. In ARP, producers may voluntarily idle cropland devoted to a program crop (i.e., not all crops are eligible) by a specific proportion of that crop's acreage base (the average of the acreage planted and considered planted to each program crop in a certain period such as 5 years for wheat).

There are four principal harvested crops in U.S. crop production: corn for grain, wheat, soybeans, and hay (Figure 1–7). Cropland harvested attained a peak of 351 million acres in 1981 but declined to 287 million acres in 1988. The projected acreage in 1997 was 321 million acres. Acreages devoted to summer fallows range between 22 mil-

United States. As such, while technology promotes farm growth, the cost of technology often poses a problem for beginning farmers. Further, the cost of farmland has increased because of increased competition for land as high-technology producers seek to expand their operations.

Institutional factors

Institutional forces (from both private and public sectors) influence the structure of agriculture mainly by influencing the cost of production inputs, the prices of products, and the generation of new technology for agriculture. The cost of production inputs depends on the level of competition in the agribusiness sector. Monopoly tends to drive up prices of farming inputs. Research is the way new technologies are developed. Extension is the way technological innovations are evaluated and transferred in a timely fashion to producers. Basic agricultural research in the United States is conducted largely by the USDA and land-grant universities. It appears that mechanical innovations have favored and thus encouraged the growth of large farms. Larger operators tend to avail themselves of research and extension information more than smaller farmers.

Public policies affect the structure of agriculture directly or indirectly. They affect such aspects of the industry as resource use, capital requirements, techno-logical development and adoption, risk, cost, and profits. Specific areas of policy affecting agriculture are commodity programs, tax policy, and agricultural credit policy. The goal of commodity programs is mainly to stabilize and increase farm prices and incomes of producers. Crops covered by crop programs include wheat, feed grains, cotton, rice, peanuts, and tobacco. These programs utilize a variety of mechanisms to accomplish their goal, including price support, direct payments, acreage allotments, set-asides, conservation reserves, surplus disposal, and stock accumulation.

The adoption of capital-intensive innovations is affected by policy provisions such as investment tax credit and accelerated depreciation. Agricultural credit policies such as those administered through the Farm Services Agency of the USDA ensure appropriate capital availability for agriculture.

Economic environment

Agriculture is impacted by the macroeconomic climate of the day. The political leadership influences the economic policies of the day as well. The economic and political environment impacts agriculture by determining economic factors such as interest rate and inflation, both of which impact costs and prices of agricultural products.

lion acres and 34 million acres annually. Crop failure in U.S. crop production varies between 5 million and 11 million acres annually. Much of this loss is attributable to drought or flood and wet weather at planting.

Production of leading crops in the United States is regionalized. The Great Plains in the Northwest is the major wheat area, while corn production is concentrated in the upper Midwest. Soybean is produced mainly in the Corn Belt (Missouri, Iowa, Illinois, Indiana, and Ohio) and Minnesota, while the Southeast dominates in peanut production. Cotton is grown predominantly in the South and Southwest, while rice production is concentrated in six states in the South and on the West Coast.

1.7.3 MOST U.S. FARMS SPECIALIZE IN ANIMAL PRODUCTION

Farms may be classified according to the production specialization that accounts for most of the gross sales from the production activity. In 1999, cattle production was the principal activity on farms in the lower sales bracket, whereas those in the higher sales bracket focused on cash grains (Figure 1–8). Cash receipts from crops continued to increase in

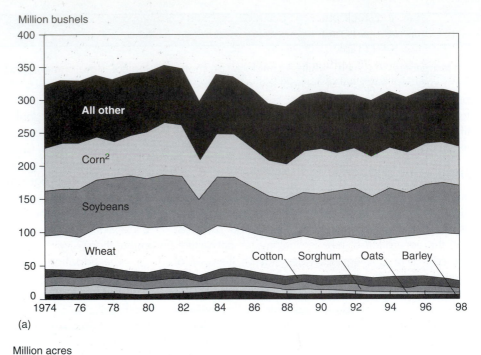

(a)

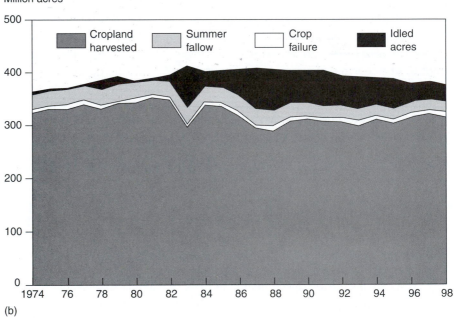

(b)

FIGURE 1–7 Production of major crops in the United States. (a) Average acreage harvested between 1974 and 1998. Except for oat production that is shrinking, the output for other major crops held fairly steady over the period, fluctuating in concert with response to factors that affect modern crop production. (b) More than 300 million acres of land annually are cultivated and harvested for various crops. Idled acreage is on the decline. (Source: USDA)

the 1990s while livestock proceeds declined in the mid-1990s. The value of crop production of major crops is on the decline. They fell from $88.4 billion in 1996 to about $65.5 billion in 1999 (Table 1–10).

In 1997, about 70 million acres of soybean were harvested. This represented the second highest acreage harvested for the crop on record. This increase is attributable to

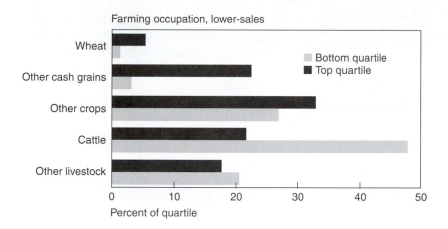

FIGURE 1–8 Cattle production was the most important activity for farms in the lower sales bracket in 1999, whereas producers in the higher sales bracket tend to focus more on cash grains. (Source: USDA)

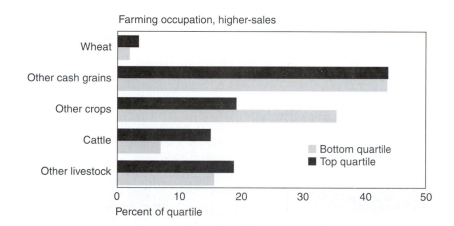

Table 1–10 The Value of Crop Production (Billion Dollars) of Major Crops in the United States Between 1994 and 1999

Year	Field and Miscellaneous Crops	Fruits and Nuts	Commercial Vegetables	Total Value
1994	78,334	10,121	8,347	96,803
1995	82,176	10,859	9,167	102,203
1996	88,452	10,446	8,353	108,253
1997	83,886	12,835	9,321	106,041
1998	70,572	11,212	9,426	91,211
1999	65,572	12,293	9,208	87,073

Source: National Agricultural Statistics Service.

One of the earliest dramatic impacts of biological technology on crop production was dubbed the *Green Revolution*. Between 1950 and 1970, the Mexican government and the Rockefeller Foundation cosponsored a project to increase food production in Mexico. The first target crop was wheat, and the goal was to increase wheat production by a large margin. Using an interdisciplinary approach, the scientific team, headed by Norman Borlaug of the Rockefeller Foundation, started to assemble genetic resources (germplasm) of wheat from all over the world (East Africa, Middle East, South Asia, Western Hemisphere). These introductions were crossed with indigenous (Mexican) wheat that had adaptability (to temperature and photoperiod) to the region and was disease-resistant but was low-yielding and prone to lodging. The team was able to develop lodging-resistant cultivars through introgression of dwarf genes from semi-dwarf cultivars from North America. This breakthrough occurred in 1953. Norman Borlaug eventually received the Nobel Peace Prize for his efforts at curbing hunger.

During the period, wheat production increased from 300,000 tons per year to 2.6 million tons per year. Yields per unit area increased from 750 kg per hectare to 3,200 kg per hectare. The most dramatic increase (250%) was observed in corn, in which average yield per hectare increased from 700 to 1,300 kg. Similar significant increases were observed in other crops, including bean, sorghum, and soybean.

Technology and the Green Revolution

Several factors were responsible for this outstanding success, the most significant being the technological component. Two specific technological strategies were employed in the Green Revolution: plant improvement and agronomic package:

1. Plant improvement: the Green Revolution centered around the breeding of high-yielding, disease-resistant, and environmentally responsive (adapted, responsive to fertilizer, irrigation, etc.) cultivars
2. Agronomic package: improved cultivars are as good as their environment. To realize the full potential of the newly created genotype, certain production packages were developed to complement the improved genotype. This agronomic package included tillage, fertilization, irrigation, and pest control.

Other factors that influenced this transformation were political and socioeconomic. Coupled with the two technological inputs was the profitability of agricultural production.

Transformation of tropical agriculture

The Green Revolution gained international reputation through its impact on tropical agriculture in the developing countries. The successful wheat cultivars were introduced into Pakistan, India, and Turkey in 1966, with similar results of outstanding performance. The Mexican model (interdisciplinary approach, international team) for agricultural transformation was duplicated in rice in the Philippines in 1960. This occurred at the International Rice Research Institute (IRRI). The goal of the IRRI team was to increase productivity of rice in the field. Rice germplasm

increased flexibility provided by the federal *Agricultural Improvement and Reform Act of 1996*. This act provided producers planting flexibility that, coupled with relatively favorable prices, enticed many farmers to increase the acreage devoted to soybean. Unlike soybean, acreage has shifted away from crops such as barley and oats in favor of more profitable crops. Wheat and corn acreages also fluctuate. For example, wheat acreage in 1981 was 80.6 million acres but was only 53.2 million acres in 1988.

1.7.4   MOST U.S. FARMS ARE OPERATED BY INDIVIDUALS

Legal structure of the farm refers to the farm's form of business organization. Farms may be owned and operated by *individuals* (or *proprietorships*), *partnerships*, or *corporations*. The average acreages of farms in these three categories were 1,165 acres for corporate farms, 856 acres for partnerships, and 373 acres for proprietorships. Most

was assembled. Scientists determined that, like wheat, a dwarf cultivar that was resistant to lodging, amenable to high-density crop stand, responsive to fertilization, and highly efficient in partitioning of photosynthates or dry matter to the grain was the cultivar to breed.

In 1966, IRRI released a number of dwarf rice cultivars to farmers in the Philippines. The most successful was IR8, a cultivar that was early maturing (120 days), thus allowing double-cropping in certain regions. The key to the high yield of the IR series was its responsiveness to heavy fertilization. The short, stiff stalk of the improved dwarf cultivar resisted lodging under heavy fertilization. Unimproved indigenous genotypes experienced severe lodging with heavy fertilization, resulting in drastic reduction in grain yield. This effective technology was exported to Africa and Asia and other tropical low-income, low-productivity regions.

The creation of International Agricultural Research Centers (IARC)

The Green Revolution initially involved only a select number of crops (especially rice, corn, and wheat), which produced yield increases at a rate that equaled or exceeded population growth in the tropical regions of the world. The Rockefeller Foundation continued its effort at sustaining the momentum in food production that was sparked by the Green Revolution. A research report on the state of food technology in low-income countries by one of its scientists, Ralph W. Cummings, revealed a lack of local research that was capable of transforming the productivity of other major world crops such as sorghum, dry bean, sweet potato, millet,

peanut, pigeon pea, and chickpea. The nations in dire need of high yield could ill afford the resources needed to support food crop research. Through a collaborative effort between international agencies, the United Nations Development Program (UNDP), World Bank, Food and Agricultural Organization (FAO), private philanthropic foundations, and several nations, a network of regional agricultural research centers (called the International Agricultural Research Centers) have been established in strategic places throughout the tropics to develop technical packages for production of important world food crops.

Shortcomings of the Green Revolution

As astounding a success as the Green Revolution was, it was not without criticism. It produced adverse social effects. An improved cultivar is as good as its environment. High-yielding cultivars required high input (capital) to realize their potential. Producing such crops required more investment in terms of capital, work, and time. The technology was not readily affordable to the poor farmer, thus widening the gap between the rich and the poor. Farmers who accumulated wealth from the adoption of the improved technology soon had resources to acquire machinery. This resulted in rampant eviction of tenant farmers, resulting in landlessness. Many farmers changed from mixed culture to monoculture. This change altered the lives of farming communities. Another negative impact of the Green Revolution was that women were marginalized in many developing countries. As machines displaced laborers, women were the first to be fired from their jobs.

farms (9 out of 10) in 1996 were classified as operated by individuals. Of those operated by corporations, 89% were family corporations. The share of farmland operated by individuals was 71%, with gross farm sales of 74%. However, corporate farms had the highest gross sales, averaging $246,826 as compared with $201,205 for partnership operations, in 1996. Gross sales for proprietorship farms averaged about $63,159.

1.7.5 MOST U.S. FARMLAND IS UNDER SOME FORM OF PRIVATE OWNERSHIP, FULL, OR PART OWNERSHIP

Who owns cropland in the United States? Land is the main asset in agricultural production. As such, its ownership is central to the production activity, influencing how it is used, improved, conserved, held, and transferred. Land tenure is the system of rights and institutions that defines access to land. The system involves land ownership, leasing, zoning ordinances, subsurface mineral rights, conservation easements, and other factors. Land tenure is important in crop production because it affects how the producer allocates land and other resources in a production enterprise for high crop productivity. Further, the actions of the producer in this regard impact the environment and eventually the society at large. There are three main categories of land tenure or ownership interests in the farmland: *full owners* (own all the land in their operation), *part-owners* (own part of the land and rent the remainder), and *tenants* (rent all the land they operate or work on shares for others).

Land tenure can be a complicated system, sometimes embroiled in great controversy when private, corporate, and government interests clash. Landowners have property rights and may use their property as they deem fit and profitable. However, in the exercise of such liberties, they may infringe upon the rights of neighbors and the society in general. Therefore, the federal government imposes certain restrictions on the use of both public and private lands in order to balance the rights of owners and those of the general public or other parties that may be affected by how the property is handled. This is the reason for the existence of zoning ordinances, conservation easements, contracts, and other such governmental impositions.

In 1996, 53% of cropland was under full ownership, while 38% was part-owned, with only 9% operated on rental basis. However, full owners tended to operate small acreages, averaging 227 acres per farm, while part-owners operated the largest farms, averaging 732 acres per farm. Tenants also operated large farms, averaging 636 acres per farm in 1996. In all, part-owners, who also represented 51% of the gross farm sales, operated 61% of cropped land in 1996. In terms of gross sales, tenants reported the highest average of $146,335, followed by part-owners with $114,443. Full owners reported average gross sales of $47,708.

1.7.6 ABOUT HALF OF THE PEOPLE WHO OWN U.S. FARMLANDS DO NOT OPERATE A FARM

Land ownership in the United States has been changing over the years. Once the federal government owned most of the land. In 1992, 60% (1.4 billion acres) of United States land (a total of 2.3 billion acres) was privately owned; the federal government owned only 29%. Nearly all cropland is privately owned. Similarly, most grassland pasture and

Table 1-11 Federal Landholdings by Agency

Department/Agency	Area (Million Acres)	Percent of Total
Department of Agriculture	184.9	28.4
Forest service	184.5	28.4
Other agencies	0.4	0.1
Department of Defense	20.8	3.2
Department of the Interior	443.4	68.2
Bureau of Land Management	271.2	41.7
Fish and Wildlife Service	90.4	13.9
National Parks Service	73.2	11.3
Other agencies	8.6	1.3
Other departments	1.2	0.2
Total	650.3	100.0

Source: USDA, ERS (1993).

rangeland are privately owned. In spite of the vastness of agricultural land, only 1% of the U.S. population, or 3% of the households, own two-thirds of all privately owned land. Of about 3 million farmland owners, nearly half (44%) do not operate a farm.

Federal landholdings are managed by four agencies: USDA's Forest Service, Department of Interior's Bureau of Land Management (BLM), Fish and Wildlife Service, and National Parks Service (Table 1–11). The objectives of these agencies vary. Forest Service and BLM manage lands for grazing, wildlife preservation, and timber harvest. The National Parks Service and Fish and Wildlife Service manage federal land for recreational or preservation purposes. Most federal lands are concentrated in the western United States.

Unlike residential property that experiences rapid ownership turnovers, farmland ownership transfer occurs at the rate of 3.5% per year (one-quarter the rate of residential). Only about 1% of the 1.3 billion acres of U.S. agricultural land in 1993 was foreign-owned.

1.8: The USDA Recognizes 10 Farming Sections

The USDA has divided the United States into 10 sections comprised of adjoining states in geographic locations. These are *Northeast, Lake States, Corn Belt, Northern Plains, Appalachia, Southeast, Delta States, Southern Plains, Mountain,* and *Pacific* (Figure 1–9). The major uses of land data in these sections show that the Northern Plains, Corn Belt, and Southern Plains have most of the cropland acreage, while cropland in the Mountain region and Southern Plains has the highest concentration of grassland pasture and range. The major use in cropland has changed over the years for various reasons that differ from one farming section to another (Table 1–12). The Corn Belt, Northern Plains, and Mountain region have experienced gains in land devoted to crop production, while the Northeast, Appalachia,

FIGURE 1–9 The 10 USDA farm production sections of the United States. The Mountain states have most of the land area but account for a relatively small amount of total agricultural cash receipts for the United States.

Table 1-12 Net Change in Major Uses of Land in the Contiguous 48 States of the United States Between 1945 and 1992

Farming Region	Net Change
Northeast	−10.7
Lake States	−3.7
Corn Belt	+7.4
Northern Plains	+11.1
Appalachia	−5.9
Southeast	−8.9
Delta	+1.5
Southern Plains	+3.3
Mountain	+14.3
Pacific	+0.5
United States	+9.0

Source: USDA, ERS.

Southeast, and Lake States have lost cropland. The gain in croplands in the Western states is attributable in part to government programs, specifically subsidy for irrigation water. Shifts in land use resulting from losses in cropland in the eastern half of the United States have several major causes, including the constraints imposed by geography and climate and the predominantly small farms.

The needs of society associated with urbanization often outcompetes agricultural enterprises for land when this happens. Land prices soar and agricultural profitability declines, resulting in cropland lost to urban development.

1.9: AGRICULTURAL REGIONS
OF THE UNITED STATES

There are three principal geographic crop areas in the United States—from the Atlantic Coast to the Great Plains, the Great Plains, and the Pacific Coast (Figure 1–10):

1. *Atlantic Coast to the Great Plains.* This region, which constitutes about 50% of the country to the east, has most of the cropland areas in the United States (Figure 1–11). The vegetation types include deciduous forest and prairie grassland. Rainfall in this region is moderate to high. The major concentrations of crops are corn (in the Corn Belt), wheat (in the Wheat-Corn Belt), cotton (in the Cotton Belt), and the humid-subtropical crop belt. The soils are richer in the northern parts, declining in productivity toward the south. Even though these areas have high concentrations of their respective crops, the crops are usually grown in rotation with other minor crops. The northern evergreen forest of the north and east is devoted to pasture and hay crops.

2. *The Great Plains.* The Great Plains region occupies the middle strip of the United States, where evaporation usually exceeds precipitation two- or threefold. Crop producers in this region have to cope with moisture deficits by adopting cultural practices that conserve moisture. The crops suited to this region include wheat (the major crop), cotton, and sorghum in the Southern Plains. Crop rotation is also practiced in this region.

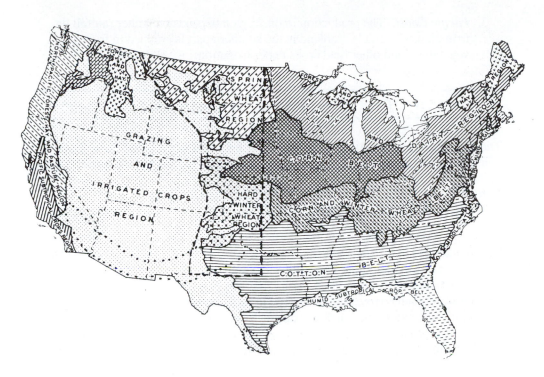

FIGURE 1–10 The three geographic crop areas of the U.S. are the Atlantic Coast, the Great Plains on the west, the Great Plains in the center, and the Pacific coast on the east coast. (Source: USDA)

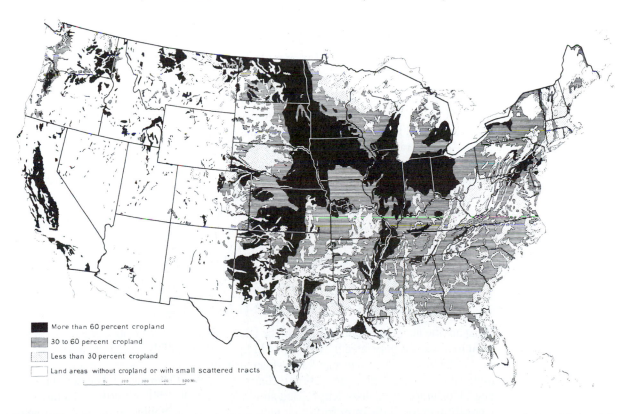

■ More than 60 percent cropland
30 to 60 percent cropland
Less than 30 percent cropland
Land areas without cropland or with small scattered tracts

FIGURE 1–11 Distribution of cropland in the U.S. indicates that the western part has the least amount of arable land, because of the deserts and mountains. The Great Plains and the Midwest Regions have the most arable lands. (Source: USDA)

3. *Pacific Coast.* The production in this region depends on winter rainfall or irrigation. The arable cropland in the southern part is cropped to fruits, vegetables, and other field crops under irrigation.

1.10: THE U.S. FARM SECTOR IS DYNAMIC

The U.S. farm sector activities discussed in this section are labor, agricultural credit, household farm income, and cash receipts from agricultural activities by the states.

1.10.1 U.S. FARM LABOR CONSISTS LARGELY OF FAMILY WORKERS

A large portion (about 12%) of total farm production expense is from labor. U.S. agriculture utilizes labor from a variety of sources. In 1997, 69% of farm labor consisted of *family workers* (including farm operators and unpaid workers). Service workers (custom crews) constituted 9% of the U.S. farm labor force. The trend toward larger and fewer farms is accountable largely to the decline in farm employment from 9.9 million in 1950 to 2.9 million in 1997. The decline appears to have been arrested by the leveling off of mechanization and labor-saving technology. The role of labor in crop production varies according to the type of farm and the size of the operation. Labor expenses in cash grain production is about 5% of the total operational cost, whereas in horticultural specialty production and fruit farming the labor cost may be 40 to 45% of total production cost. Larger farms use most (70%) of the farm labor. How important do you think migrant workers are to U.S. crop production?

1.10.2 U.S. PRODUCERS HAVE ACCESS TO AGRICULTURAL CREDIT FROM COMMERCIAL AND GOVERNMENT SOURCES

The sustainability of profitability of farm enterprises depends to a large extent on the availability of agricultural credit. Loans made to agricultural producers are classified as *real estate loans* (used to purchase farmland or major capital improvement; they have terms of 10 to 40 years) or *non-real estate loan* (for variable purposes; they have terms of less than 10 years). Producers may also apply for *seasonal operating loans* (have terms of less than 1 year) or *machinery and equipment loans* (have terms of 7 or more years). Farm business debt increased from $151.1 billion in 1995 to $156.2 billion in 1996. Farm real estate debt similarly rose from $79.5 billion in 1995 to $81.9 billion by the end of 1996, while the non-real estate farm debt rose by 4% to $74.2 billion in the same period. These loans are obtained through the Farm Credit System or commercial lenders. The Farm Services Agency also provides direct loans to producers (Table 1–13).

1.10.3 ABOUT A DOZEN STATES GENERATE MORE THAN HALF THE NET U.S. FARM INCOME

What are the major agricultural states in the United States? In 1996, 60% of the nation's net farm income was generated by 12 states, led by California (Table 1–14). California's continued dominance in cash receipts stems from its large land mass and its commodity mix of high value crops and other products (e.g., dairy products, greenhouse and nursery products, eggs, hay, grapes, tomatoes, lettuce, and almonds). In 1996, 75% of California's farm sales were from crops, 27% from fruits and nuts, 23% from vegetables, and 10% from green-

Table 1-13 Lending Agencies Involved in U.S. Agricultural Production and Farm Debt in Selected Years

						Farm Debt Outstanding, December 31						
	1950	1960	1970	1980	1985	1990	1991	1992	1993	1994	1995	1996
Real Estate Debt						($ Billion)						
Farm Credit System	0.8	2.2	6.4	33.2	42.2	26.2	25.4	25.5	25.0	24.7	24.9	25.8
Life Insurance Co.	1.1	2.7	5.1	12.0	11.3	9.7	9.6	8.8	9.0	9.0	9.1	9.4
Banks	0.8	1.4	3.3	7.8	10.7	16.3	17.5	18.8	19.7	21.2	22.4	23.4
Farm Service Agency	0.2	0.8	2.2	7.4	9.0	7.7	7.1	6.4	5.9	5.5	5.1	4.7
Individuals/Others	2.1	4.5	10.5	29.8	28.1	15.2	15.7	16.1	16.8	17.6	18.1	18.5
Total	**5.2**	**11.3**	**27.5**	**89.7**	**100.1**	**74.9**	**75.1**	**75.6**	**76.3**	**78.0**	**79.6**	**81.9**
Non-Real Estate Debt												
Farm Credit System	0.5	1.5	5.3	19.8	14.0	9.8	10.2	10.3	10.5	11.2	12.5	14.0
Banks	2.4	4.5	10.5	30.0	33.7	31.3	32.9	32.9	34.9	36.7	37.7	38.3
Farm Service Agency	0.3	0.4	0.7	10.0	14.7	9.4	8.2	7.1	6.2	6.0	5.1	4.4
Individuals/Others	2.5	4.5	4.8	17.4	15.1	12.7	13.0	13.2	14.2	15.2	16.2	17.4
Total	**5.7**	**11.1**	**21.3**	**77.1**	**77.6**	**63.2**	**64.3**	**63.6**	**65.9**	**69.1**	**71.5**	**74.2**
Grand total	10.9	22.4	48.8	168.8	177.6	138.1	139.4	139.3	142.2	147.1	151.0	156.2

Source: USDA, ERS.

house and nursery. In 1996, California generated cash receipts totaling $5.6 billion, followed by Iowa with $4.0 billion. This top-12 list is not static but varies from year to year. The ranking of the states by cash receipts for 1999 is shown in Figure 1–12.

1.10.4 THE BALANCE SHEET FOR U.S. AGRICULTURE SHOWS A LOWER DEBT-TO-ASSET RATIO IN THE 1990s THAN THE 1980s

Is U.S. agriculture debt-laden? Farm business asset in 1996 totaled $1,034.9 billion, with a debt total of $156.2 billion. The average equity per farm was $426,000. In 1985, the debt-to-asset ratio peaked at 24%. However, in 1996, this ratio was down to 15.1%. Most (78%) of the farm business asset in 1996 was attributed to real estate assets, the average real estate value per farm at year's end being $390,000.

1.10.5 MANY U.S. FARMERS DERIVE CONSIDERABLE INCOME FROM OFF-FARM SOURCES

Is agriculture a profitable undertaking in the United States? In 1996, the average farm household (excludes non-family corporations, cooperatives, and institutional farms) income was $50,360, of which about 84% came from off-farm sources. This means that many operators spent most of their work efforts in non-farming activities. Most U.S. farms are classified as small farms (i.e., they gross less than $50,000 per year). In 1996, about 39% of small operators indicated non-farm activity as a major occupation, while 27% were retired. On the other hand, most large farm operators reported farming as their major occupation. However, large operators reported significant off-farm income, averaging $34,950 per household. The 1999 income distribution showed only 11.4% for direct farm income, down from 16% in

Table 1-14 Net Income from Crop Production of Top 20 Crop Production States in the United States

Rank	State	Cash Receipts from Crops (Million Dollars)	Cash Receipts from Agriculture (Million Dollars)	Major Crops 1	Major Crops 2	Top Agricultural Product
1.	California	17,096	23,310 (1)	Greenhouse/ nursery	Grapes	Dairy products
2.	Iowa	7,396	12,853 (3)	Corn	Soybean	Corn
3.	Illinois	6,989	9,050 (5)	Corn	Soybean	Corn
4.	Texas	5,295	13,053 (2)	Cotton	Greenhouse/ nursery	Cattle/calves
5.	Florida	4,942	6,131 (9)	Oranges	Greenhouse/ nursery	Oranges
6.	Minnesota	4,641	6,809 (6)	Corn	Soybean	Corn
7.	Nebraska	4,177	9,454 (4)	Corn	Soybean	Cattle/calves
8.	Washington	4,017	5,681 (13)	Apples	Wheat	Apples
9.	Indiana	3,663	5,558 (14)	Corn	Soybean	Corn
10.	North Carolina	3,404	7,831 (8)	Tobacco	Greenhouse/ nursery	Hogs
11.	Kansas	3,299	7,869 (7)	Wheat	Corn	Cattle/calves
12.	Ohio	3,177	5,122 (15)	Soybean	Corn	Soybean
13.	North Dakota	2,996	3,532 (23)	Wheat	Barley	Wheat
14.	Arkansas	2,530	5,887 (11)	Soybean	Rice	Broilers
15.	Missouri	2,500	4,950 (16)	Soybean	Corn	Soybean
16.	Georgia	2,408	5,687 (12)	Cotton	Peanuts	Broilers
17.	Oregon	2,320	2,977 (28)	Greenhouse/ nursery	Wheat	Greenhouse/ nursery
18.	Michigan	2,195	3,643 (20)	Corn	Greenhouse/ nursery	Dairy products
19.	Idaho	2,061	3,410 (25)	Potatoes	Wheat	Dairy products
20.	South Dakota	2,051	3,684 (19)	Corn	Soybean	Cattle/calves

Source: Compiled from USDA/ERS 1996 data. Ranking for total receipts from agriculture is in parentheses.

1996 (Figure 1–13). According to the 2000 classification of farms, some small family farms posted losses, whereas the very large farms showed significant profits (Figure 1–14).

Average household income and dependence on off-farm incomes depend on characteristics of the operator (Table 1–15). In 1996, 6% of operators reported a loss in household income averaging $36,060 per household. Of the 28% of households that reported income of $50,000 or more, an average of $41,509 was derived from farming activities. Households whose operations involving retirees and those with other occupations (largely dependent on off-farm income) had the highest average household income. Further, household incomes were lowest for the youngest and oldest operators, while income tended to increase with higher educational levels of the operators.

1.10.6 PEOPLE WITH ALL KINDS OF BACKGROUNDS UNDERTAKE FARMING

Can anyone be a farmer? People from all walks of life and age groups undertake farming in the United States for various reasons.

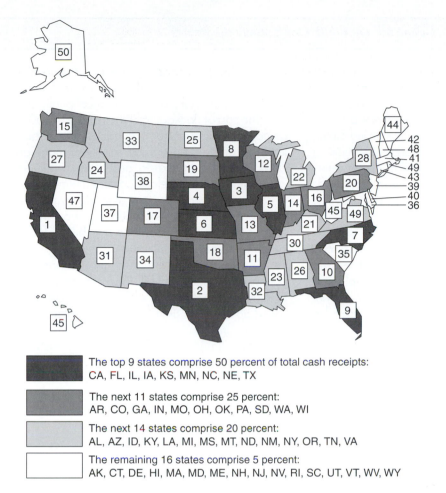

The top 9 states comprise 50 percent of total cash receipts:
CA, FL, IL, IA, KS, MN, NC, NE, TX

The next 11 states comprise 25 percent:
AR, CO, GA, IN, MO, OH, OK, PA, SD, WA, WI

The next 14 states comprise 20 percent:
AL, AZ, ID, KY, LA, MI, MS, MT, ND, NM, NY, OR, TN, VA

The remaining 16 states comprise 5 percent:
AK, CT, DE, HI, MA, MD, ME, NH, NJ, NV, RI, SC, UT, VT, WV, WY

FIGURE 1–12 Nine states in 1999 accounted for 50% of U.S. total cash receipts from agricultural production. The Northeast, Mountain, Appalachia, and Delta states generally have low cash receipts from agricultural production, while California, the Corn Belt states, Texas, and Florida are major agricultural states. The boxes show National Rank in total cash receipts from agricultural production. (Source: USDA)

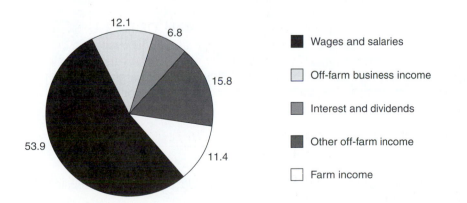

- Wages and salaries
- Off-farm business income
- Interest and dividends
- Other off-farm income
- Farm income

FIGURE 1–13 Income distribution for U.S. farmers in 1999. On the average, more than 50% of farm proceeds are devoted to labor costs. About 11% of total farm proceeds constituted the average farm income to the U.S. producer. (Source: USDA)

Have you ever wondered how much of your grocery bill goes to the producer—the farmer? Would it surprise you to know that less than 25% of the cost of food is attributable to farm costs? Well, the current estimates are that, for every dollar you spend on food in the United States, only 20¢ can be charged to the producer, so who or what is responsible for the balance of 80¢? Marketing factors responsible for the cost of marketing include packaging, transportation, energy, advertising, business taxes, net interest, depreciation, rent, and repairs.

The cost of marketing food has soared due primarily to the rising cost of labor, transportation, food packaging materials, and other inputs used in marketing agricultural products. Other factors implicated in the marketing cost of food include the growing volume of food and the increase in services provided with the food.

In just one decade, the cost of marketing farm food rose by about 54% from its value of $302 billion in 1988. Between 1988 and 1998, consumer expenditure for farm food rose $186 billion. It is estimated that the marketing bill alone accounted for about 88% of

this rise. The largest contributor to the marketing bill was cost of labor. Labor is used by manufacturers, wholesalers, retailers, and operators of public eateries.

The cost of packaging stands at about $50 billion a year, and accounts for about 8.5% of the food dollar bill. Packaging utilizes a variety of materials, including paper, can, plastic, and glass containers.

Natural gas and electricity are the major sources of energy involved in the cost of food services and distribution. Public eateries account for about 40% of the energy budget. In 1998, the energy bill for food marketing costs was $21 billion.

Transportation is the fourth most important component of the marketing bill. The cost of transportation is impacted by the costs of labor and fuel. Retailers spend about 4% of their budget on transportation.

Consumer advertising is a critical aspect of marketing. In 1998, food advertising costs totaled $22 billion, with the manufacturing sector accounting for about 50% of this bill. The balance of the marketing cost was spread over items such as depreciation, rent, repairs, and loan servicing.

The cost of marketing is largely responsible for the high cost of agricultural produce. In 1999, it accounted for 80% of the cost of the produce. (Source: USDA)

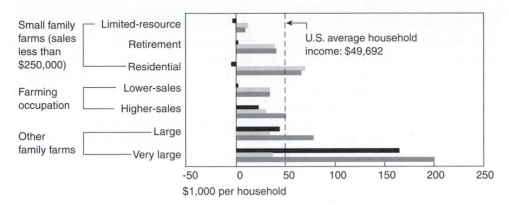

FIGURE 1–14 In 1996, very large and large producers operated very profitable enterprises, while some small farmers posted negative farm incomes, having to depend more on off-farm earnings. (Source: USDA)

Table 1–15 Farm Operator Households and Household Income Based on Selected Characteristics

Item	Number of Households	Average Household Income ($)	Share of Off-Farm Sources (%)
All operator's households	2,036,810	44,392	89.4
Household Income Class			
Negative	170,331	(28,968)	(40.4)
0–$9,999	210,182	5,470	183.0
$10,000–$24,999	443,779	17,643	112.7
$25,000–$49,999	668,579	36,507	96.2
$50,000 and over	543,938	113,918	71.7
Operator's Major Occupation			
Farmwork or ranch work	903,820	40,342	64.8
Other	797,718	53,425	108.9
Retired	335,272	33,815	94.9
Operator's Age Class			
Less than 35	168,825	32,506	93.4
35–44	407,345	47,266	89.3
45–54	476,807	51,953	91.6
55–64	469,052	50,421	87.7
65 or older	514,780	33,518	87.2
Operator's Educational Level			
Less than high school	425,612	30,173	94.4
High school	819,087	41,479	87.3
Some college	443,374	48,726	85.8
College	348,736	63,075	93.1

Source: USDA, ERS 1996 data. Income from off-farm sources is more than 100% of total household income if farm income is negative.

Primary Occupation

Farm operators may be in the business full-time or part-time. In 1994, 44% of all farm operators declared farming as their primary occupation, while 2% were full-time farm managers. These two categories together accounted for most of the gross value sales. Full-time operators tended to operate higher mean acreages than other categories of farm operators. However, retirees or people who had other principal types of occupations, farming only as a part-time activity, operated more than 50% of U.S. farms. The latter groups of farm operators generally had smaller farms (165 acres) as compared with the national average of 448 acres. The 2000 classification paints a similar picture, with small farms outnumbering large ones (Figure 1–15). Large and very large farms operated less than 25% of the total cropland acreage.

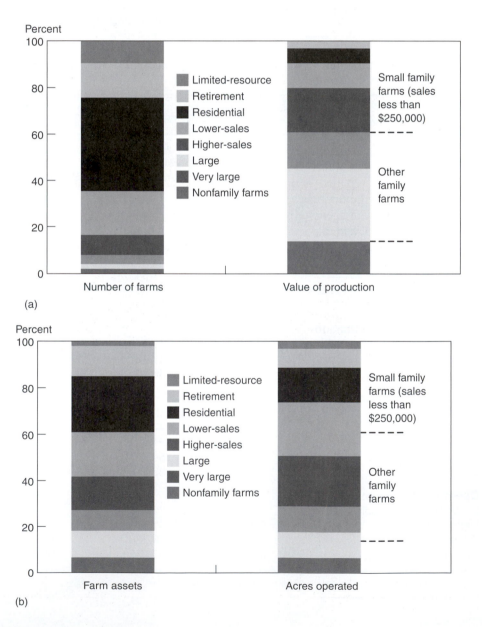

FIGURE 1–15 Classification of U.S. farms: the criteria for classification of farms is often revisited by the USDA. The classification in the figure was released in 2000. (a) Most farms are classified as residential. Only a small number of U.S. farms are classified as very large. However, the large and very large farms account for most of the production. (b) Large and very large farms operate less than 25% of the total crop acreage. (Source: USDA)

Age Group

Less than 10% of U.S. farm operators in 1994 were 35 years old or younger. About 20% of farm operators were in the 35- to 44-year age group. More than 55% of all farm operators were less than 55 years old. Older farmers (65 years and older) accounted for 24% of farm producers but 11% of all farm sales. You may want to think about what can be done to increase the participation of youth in agriculture.

Education

Farm operators with less than a high school education generally operate smaller farms and experience less income than their higher-educated counterparts. About 50% of the older farmers (65 years or older) have less than a high school education. Most farm operators (four out of five) have at least a high school education. College graduates tend to operate larger mean acreages and have the highest income from farming.

Household Income

Is bigger better in agricultural production? Generally, the larger the farming operation, the higher the earnings from farming activities. Small operators, thus, tend to depend heavily on off-farm income to support the household. Most U.S. farmers are small operators.

Use of Information

U.S. agriculture has tremendous information support to assist producers in decision making. Much of the information is free of charge and available in a variety of formats— television, radio, newspaper, and electronic. For example, weather information is available to assist in the timing of crop production activities. The sources of information include government agencies and private businesses.

About 50% of U.S. operators in 1994 availed themselves of information from three of the major sources of information. In this activity, the more formally educated (college

Table 1–16 Use of Computers for Farm Business in Selected States

State	Number of Farms	Computer (%)	Internet Access (%)
Alabama	49,000	24	28
Arizona	7,900	39	44
California	89,000	40	46
Florida	45,000	27	34
Georgia	50,000	18	20
Illinois	79,000	32	30
Indiana	66,000	22	25
Iowa	97,000	32	30
Michigan	52,000	25	28
Missouri	42,000	11	18
Montana	27,000	36	38
Oklahoma	83,000	17	28
Texas	226,000	23	31
Washington	40,000	39	50
West Virginia	21,000	11	19
United States	2,185,450	24	29

Source: USDA 1999.

level) the operator, the more the variety of sources of information consulted. Farmers in general consult electronic information less often. About one out of every eight farm operators utilizes information from electronic sources. In 1994, the use of this medium was highest among those for whom farming was the primary occupation and especially among operators in the 35 to 44 age group and those with higher education. Older farmers (65 and older) were less inclined to use information from electronic sources. In 1999, a survey showed that an average of 24% of U.S. farms utilized computers in their farm business (Table 1–16).

SUMMARY

1. Crop production (and thus agriculture in general) is an evolving activity.
2. Agriculture emerged when ancient cultures switched from life as hunter-gatherers to a sedentary lifestyle, growing crops of choice in a fixed location.
3. Domestication of crops is likely to have occurred simultaneously in different regions of the world and for different reasons.
4. Crop production is an art, a science, and a business. Producers depend on experiences and share information among themselves. Crop productivity has dramatically improved through the application of knowledge from scientific research and technological advancements. The producer, to be successful, should be able to manage production resources effectively.
5. Certain production factors are outside the control of the producer. These include the vagaries of the weather, political factors, and the world production scene.
6. Crops are a major source of food, feed, and fiber for humans and their livestock. Cereal crops (especially wheat, corn, and rice) are the most important food crops in the world. Plant fibers for clothing and textiles include cotton, bast fibers (e.g., flax), and vegetable fibers (sisal). Forage crops include alfalfa and sudangrass. Crops are an important source of nutrients, especially carbohydrates.
7. Gains in agricultural productivity are affected by factors including research and development, extension, education, and government programs.
8. Modern agriculture has gone through certain distinct eras: (a) era of resource exploitation during which production resources were mined without replacement; (b) era of resource conservation and regeneration in which the techniques employed replenished soil fertility; and (c) era of resource substitution in which technological advances (biotechnology, mechanical) revolutionized production through use of improved inputs, techniques, and technologies.
9. The structure of agriculture is dynamic. It is influenced by three key factors— technical, institutional, and economic.
10. The success of American agriculture was sparked by the implementation of an effective agricultural policy in the nineteenth century. This policy ensures that producers have adequate resources and enjoy high and stable prices for their products. Further, through research and extension, producers receive information about new and improved technologies and techniques.
11. U.S. agriculture continues to evolve. Starting as a low-input, labor-intensive activity, U.S. agriculture was revolutionized in the nineteenth century to become a science- and technology-based production. The Morrill Act of 1862 established educational and research programs to promote agriculture.
12. The Great Depression (1929 to early 1940s) dealt a setback to U.S. agriculture. By 1950, there had been a successful rebounding of production, with record yields occurring in the mid-1970s.
13. The success of U.S. agriculture spawned numerous industries on both the input and output sides of production.

14. The USDA recognizes farming sections in the United States: Northeast, Lake States, Corn Belt, Northern Plains, Appalachia, Southeast, Delta States, Southern Plains, Mountain, and Pacific.
15. A farm is defined as any place from which $1,000 or more of agricultural products were sold, or normally would have been sold, during the census year.
16. Farms may be classified by sales (commercial or noncommercial), specialization of production, and acreage. Most (75%) U.S. farms are smaller than 500 acres.
17. People from a variety of backgrounds undertake farming in the United States. Farming is a primary occupation for about 44% of practitioners. Only about 10% of farmers are 35 years old or younger. About four out of five farmers have at least a high school education.
18. Nearly all U.S. cropland is privately owned.
19. The USDA administers a number of conservation and environmental programs aimed at sustaining productivity while protecting the soil and general environment from destruction. These include the Conservation Reserve Program, Conservation Compliance, and the Wetland Programs. Environmentally sensitive land is protected from destruction.

REFERENCES AND SUGGESTED READING

Chrispeels, M. J., and D. E. Sadava. 1994. *Plants, genes, and agriculture.* Boston: Jones and Bartlett.

Economic Research Service/USDA. 1993. *Structural and financial characteristics of U.S. farms.* Washington, DC: U.S. Government Printing Office.

Economic Research Service/USDA. 1994. *Structural and financial characteristics of U.S. farms.* Washington, DC: U.S. Government Printing Office.

Economic Research Service/USDA. 1997. Agricultural resources and environmental indicators, 1996–97. Agricultural handbook No. 712. Washington, DC: U.S. Government Printing Office.

GREAN (Global Research on the Environmental and Agricultural Nexus) for the 21st Century: A Proposal for Collaborative Research Among U.S. Universities, CGIAR Centers and Developing Country Institutions. University of Florida, Gainesville.

Harlan, J. R. 1976. The plants and animals that nourish man. In *Food and agriculture.* San Francisco: W.H. Freeman and Company.

Jennings, P. R. 1976. The amplification of agricultural production. In *Food and agriculture.* San Francisco: W.H. Freeman and Company.

Martin, J. H., and W. H. Leonard. 1971. *Principles of field crop production.* New York: Macmillan.

Metcalfe, D. S., and D. M. Elkins. 1980. *Crop production: Principles and practices.* 4th ed. New York: Macmillan.

Pimentel, D., and C. W. Hall, eds. 1989. *Food and natural resources.* San Diego: Academic Press.

U.S. Congress, Office of Technology Assessment. 1986. *Technology, public policy, and the changing structure of American agriculture.* Washington, DC: U.S. Government Printing Office.

U.S. Department of Agriculture. 1998. *Agriculture facts book 98.* Washington, DC: U.S. Government Printing Office.

U.S. Department of Agriculture/Economic Research Service. 1999. Food security assessment. GFA-11, Washington, DC: US Government Printing Office.

Wortman, S. 1976. Food and agriculture. In *Food and agriculture.* San Francisco: W.H. Freeman and Company.

SELECTED INTERNET SITES FOR FURTHER REVIEW

http://www.usda.gov/factbook

Agricultural Facts Book 1998 edition, with summaries on U.S. agricultural structure and sector, numerous figures and tables, and other agricultural information. Check for the most current issue of *Agricultural Facts Book.*

http://www.ers.usda.gov/data/AgResources

Homepage of ERS (Economic Research Services); publishes annual Agricultural Resources and Environmental Indicators—land, water, production inputs, production management, technology, conservation and environmental programs; and numerous tables and figures on these topics. Check for the current edition of the report.

http://www.ers.usda.gov

Up-to-date situation and outlook coverage of major commodities, the U.S. farm economy, agricultural trade, and monthly reports on major field crops and livestock.

http://www.nhq.nrcs.usda.gov/land/index/croplnd.html

Cropland figures.

http://www.usda.gov/history2/text1.htm

Historical accounts on various aspects of U.S. agriculture.

OUTCOMES ASSESSMENT

PART A

Answer the following questions true or false.

1. T F   There are 10 farming regions in the United States.
2. T F   The majority of U.S. production comes from large farms.
3. T F   Most U.S. farms are larger than 500 acres.
4. T F   Most U.S. farmers have at least a high school education.
5. T F   Most U.S. land is currently privately owned.
6. T F   A commercial farm has sales of less than $50,000.
7. T F   There are about 2 million farms in the United States.
8. T F   Agriculture is an invention.
9. T F   Vavilov proposed eight centers of crop origin.

10. T F   Cereal grains tend to be high in lysine.
11. T F   Legumes tend to be low in cysteine.
12. T F   Drying oils have a high iodine number.
13. T F   Vegetable fibers are called hard fibers.

PART B

Answer the following questions.

1. List the top 10 crops that feed the world.

   _____   _____

   _____   _____

   _____   _____

   _____   _____

   _____   _____

2. Define productivity.

   _____

3. Give two major feed grains in the United States.

   _____   _____

4. Give the two amino acids that are frequently low in legumes.

   _____

   _____

5. The acronym CRP stands for _____.

6. According to sales, there are two broad groups of farms:
   a _____; b _____.

7. About what percentage of U.S. farmers are 35 years or younger?
   _____.

8. What is a highly erodible land (HEL)? _____.

PART C

Write a brief essay on each of the following topics.

1. Briefly discuss the factors that make the United States agricultural policy successful.

2. Give a brief historical account of U.S. agriculture.

3. Define a farm.

4. Classify U.S. farms according to (a) size and (b) specialization.

5. Describe the U.S. producer (farmer) according to (a) occupation, (b) education.

6. Discuss the Conservation Reserve Program (CRP).

7. Describe the strategy (approach) of implementation of USDA conservation programs.

8. What is a wetland?

9. Briefly discuss the life of hunter-gatherers.

10. Discuss the theories of the emergence of agriculture.

11. Discuss crop production as a science.

12. Discuss crop production as a business.

13. Discuss the nutritional value of food crops.

14. What are bast fibers?

PART D

Discuss or explain the following topics in detail.

1. In view of current trends, predict the U.S. agricultural scene by 2020 (the crops produced, the characteristics of the producers, the way crops will be produced, etc.).

2. How is the U.S. farmer perceived in society today?

3. How important is agriculture to the U.S. economy today?

4. How important is research to U.S. agriculture?

5. What should be done to interest the youth in farming?

6. How important is world trade to the success of U.S. agriculture?

7. In agricultural production, is bigger always better?

8. Discuss the impact of the USDA on U.S. agriculture.

9. Without the involvement of the government, could U.S. agriculture have been so successful?

2

Plant Morphology

PURPOSE

The purpose of this chapter is to discuss the science of plant taxonomy and how plants are identified. The discussion focuses on plant morphological attributes used in plant identification. Anatomical structure and function of plant parts are also discussed.

EXPECTED OUTCOMES

After studying this chapter, the student should be able to:

1. Define the term *plant taxonomy*.
2. Discuss the rules pertaining to taxonomy.
3. Classify plants on an operational basis, including agronomic use, adaptation, growth form, stem type, and growth cycle.
4. List and give examples of important crop plant families.
5. Describe how plants are identified using morphological features of the leaf, stem, flower, and seed in dicots and monocots.
6. List the major cellular organelles and describe their functions.
7. List the different types of plant tissues.
8. Discuss the structure and functions of simple and complex tissues.
9. Describe the structure and function of the leaf.
10. Describe the structure and function of the stem.
11. Describe the structure and function of the root.
12. Describe the structure and function of the flower.

KEY TERMS

Adventitious roots	Annual plant	Caryopsis
Androecium	Biennial plant	Cell

Chlorophyll	Mitochondria	Rhizome
Cool season plant	Monocarp	Stolon
Dicot	Monocot	Taxon
Dioecious plant	Monoecious plant	Testa
Epidermis	Perennial plant	Tissue
Gynoecium	Phloem	Warm season plant
Meristem	Plant taxonomy	Xylem

TO THE STUDENT

There is enormous biological diversity in nature. Some crop plants are distributed over a wide range of environments in the world. A particular crop has different culture-based nomenclature. There is a need for consensus in classifying and naming plants. Classifying and naming plants facilitates their use by researchers and consumers. A unifying system of classifying and naming plants is especially critical to international collaboration in research and plant use. Whereas a particular plant is called corn in the United States and maize in the United Kingdom, the scientific name *Zea mays* identifies the same plant in all parts of the world. Classification is a work in progress. As new information becomes available, scientists review and reassign plants to more appropriate groups. You will therefore find that older textbooks sometimes have different classes and names assigned to certain organisms. The crop producer may not appreciate the scientific names of plants, which sometimes are rather difficult to pronounce, much less memorize. In this chapter, you will also learn some of the common operational ways of classifying plants that make more sense to the ordinary producer. Plant anatomy is the science of cataloging, describing, and understanding the function of plant structures. It involves the study of the structure of cells, tissues, and tissue systems. An understanding of the anatomy of crop plants helps producers properly allocate and distribute plants in the field. The plant breeder needs to understand plant anatomy to more effectively manipulate plant structure for higher productivity. Your goal in studying this chapter should not be to memorize botanical terminologies and jargon of plant structures. Try to link structure to function. Crop production involves applying basic scientific information to producing crops.

2.1: WHAT IS PLANT TAXONOMY?

Plant taxonomy. The science of identifying, naming, and classifying plants.

Plant taxonomy is the science of identifying, classifying, and naming plants. This task is accomplished by using data from a variety of sources, including morphological, anatomical, ultrastructural, physiological, phytochemical, cytological, and evolutionary. Plants are grouped according to relationships based on characteristics from these sources. The flower plays a significant role in plant taxonomy because it is a very stable organ across different environments.

2.2: TAXONOMY IS A WORK IN PROGRESS

Taxon. A taxonomic group of organisms (e.g., family, genus, or species).

What is the basis for assigning plants to certain categories? There are seven general taxonomic groups in botanical or scientific classification of plants (Figure 2–1). *Kingdom* is the most inclusive group; *species* is the least inclusive. Each group is called a **taxon**

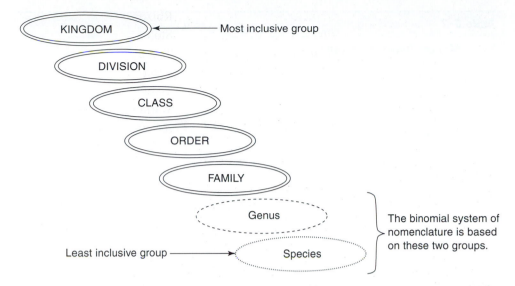

KINGDOM ← Most inclusive group
DIVISION
CLASS
ORDER
FAMILY
Genus
Least inclusive group → Species

The binomial system of nomenclature is based on these two groups.

FIGURE 2–1   The seven basic taxonomic groups of organisms. Between kingdom and division, there are *subkingdom* and *superdivision.* Similarly, *subclass* may follow class. Conventional plant breeding is limited to sexual gene transfer that is possible mainly within species. Recombinant DNA technology enables gene transfer to occur across the whole spectrum of categories.

(plural is *taxa*). Carolus Linnaeus developed a two-part name, called the *binomial nomenclature,* for plants. This consists of the *genus* and the *species* names. Taxonomy is a work in progress. As new information becomes available, scientists revise the existing classification, reassigning certain plants to new groups. Similarly, scientists assign new names to old names (e.g., the grass family used to be called Gramineae but is now called Poaceae). An example of scientific or binomial nomenclature is presented in Table 2–1.

There are five major groups (kingdoms) of all organisms: *Plantae, Animalia, Protoctista, Fungi,* and *Monera* (Table 2–2). Plants belong to the kingdom Plantae (plant kingdom). Should crop science students be concerned about the other four kingdoms apart from kingdom Plantae? Yes. Crop production is also concerned about the other four kingdoms because organisms in these groups pose significant problems for crops by causing diseases and being pests. In this role, these organisms decrease crop productivity. Organisms that are beneficial to plants occur in these groups as well.

The kingdom Plantae consists of two major groups—**bryophytes** *(non-vascular plants)* and **trachaeophytes** *(vascular plants). Non-vascular plants are also called lower*

Table 2–1   Examples of Scientific Classification of Plants

Plantae	**Kingdom**	Plantae
Magnoliophyta	**Division**	Magnoliophyta
Liliopsida	**Class**	Magnoliopsida
Cyperales	**Order**	Rosales
Poaceae	**Family**	Fabaceae
Zea	**Genus**	*Glycine*
mays	**Species**	*max*
"PJ457"	**Cultivar**	"Kent"

Table 2–2 The Five Kingdoms of Organisms as Described by Whitaker

1. **Monera** (have prokaryotic cells)
    Bacteria
2. **Protoctista** (have eukaryotic cells)
    Algae
    Slime molds
    Flagellate fungi
    Protozoa
    Sponges
3. **Fungi** (absorb food in solution)
    True fungi
4. **Plantae** (produce own food by the process of photosynthesis)
    Bryophytes
    Vascular plants
5. **Animalia** (ingest their food)
    Multicellular animals

Divisions in the Kingdom Plantae

	Division	*Common Name*
Bryophytes	Hepaticophyta	Liverworts
(non-vascular;	Anthocerotophyta	Hornworts
no seed)	Bryophyta	Mosses
Vascular plants	Trachaeophyta	
(a) Seedless	Psilotophyta	Whisk ferns
(spore-bearing)	Lycophyta	Club mosses
	Sphenophyta	Horsetails
	Pterophyta	Ferns
(b) Seeded	Pinophyta	Gymnosperms
(cone-bearing		(non-flowering)
naked seed)	Subdivision: Cycadicae	Cycads
	Subdivision: Pinicae	
	Class: Ginkgoatae	*Ginko*
	Class: Pinatae	Conifers
	Subdivision: Gneticae	*Gnetum*
(c) Seeded	**Magnoliophyta**	**Angiosperms (flowering)**
(seeds borne	Class: Liliopsida	Monocots
in fruits)	Class: Magnoliopsida	Dicots

---

plants. Similarly, vascular plants are also called *higher plants* and may bear seeds or be seedless. They are large-bodied and comprise three major vegetative organs—stem, leaf, and root. Higher plants have *conducting tissues* (or *vascular tissue*). More than 80% of all species in the plant kingdom are flowering plants. There is enormous variation in plants. Some are naturally occurring and maintained and are called *botanical varieties,* while others are human-made (created through plant breeding) and are called *cultivars,* a contraction of two terms—*culti*vated *var*iety. This term is often used synonymously with *variety.*

2.2.1 FLOWERING PLANTS BELONG TO THE DIVISION MAGNOLIOPHYTA

Division Magnoliophyta consists of plants that bear true seeds that are contained in fruits. This is the most important division in terms of crop production, because practically all the economically important plants used for food, feed, and fiber belong to this group. The division is further separated into two classes: Liliopsida (have one cotyledon—monocot) and Magnoliopsida (have two cotyledons—dicot). A common classification of crop plants is on the basis of the number of cotyledons (i.e., either monocots or dicots). The distinguishing external and internal features of these two groups of subclasses of plants are described later in this chapter. Selected field crop families in the division Magnoliophyta are presented in Table 2–3.

Table 2–3 Selected Field Crop Families in the Division Magnoliophyta (Flowering Plants)

Monocots

Poaceae (Gramineae) *(grass family)*
In terms of numbers, the grass family is the largest of flowering plants. It is also the most widely distributed.
Examples of species: wheat, barley, oats, rice, corn, fescues, bluegrass
Aracaceae *(palm family)*
The palm family is tropical and subtropical in adaptation.
Examples of species: oil palm *(Elaeis guineensis),* coconut palm *(Cocos nucifera)*
Amaryllidaceae *(amaryllis family)*
Plants with tunicate bulbs characterize this family.
Examples of species: onion, garlic, chives

Dicots

Brassicaceae (Cruciferae) *(mustard family)*
The mustard family is noted for its pungent herbs.
Examples of species: cabbage, radish, cauliflower, turnip, broccoli
Fabaceae (Leguminosae) *(legume family)*
The legume family is characterized by flowers that may be regular or irregular.
The species in this family are an important source of protein for humans and livestock.
Examples of species: dry bean, mung bean, cowpea, pea, peanut, soybean, clover
Solanaceae *(nightshade family)*
This family is noted for the poisonous alkaloids many of them produce (e.g., belladonna, nicotine, atropine, solanine).
Examples of species: tobacco, potato, tomato, pepper, eggplant
Euphobiaceae *(spurge family)*
Members of the spurge family produce milky latex and include a number of poisonous species.
Examples of species: cassava *(Manihot esculenta),* castor bean
Asteraceae (Compositae) *(sunflower family)*
The sunflower family has the second-largest number of flowering plant species.
Example of species: sunflower, lettuce
Apiaceae (Umberliferae) *(carrot family)*
Plants in this family usually produce flowers that are arranged in umbels.
Examples of species: carrot, parsley, celery
Cucurbitaceae *(pumpkin family)*
The pumpkin or gourd family is characterized by prostrate or climbing herbaceous vines with tendrils and large, fleshy fruits containing numerous seeds.
Examples of species: pumpkin, melon, watermelon, cucumber

2.2.2 CLASSIFYING AND NAMING PLANTS IS A SCIENCE GOVERNED BY INTERNATIONAL RULES

Who decides what name to give a particular plant so it is universally recognizable? The science of plant taxonomy is coordinated by the *International Board of Plant Nomenclature,* which makes the rules. The Latin or Greek language is used in naming plants. Sometimes, the names given reflect specific plant attributes or use of the plant. For example, some specific epithets indicate color, such as, *alba* (white), *variegata* (variegated), *rubrum* (red), and *aureum* (golden); others are *vulgaris* (common), *esculentus* (edible), *sativus* (cultivated), *tuberosum* (tuber bearing), and *officinalis* (medicinal). The ending of a name is often characteristic of the taxon. Class names often end in *opsida* (e.g., Magnoliopsida), orders in *ales* (e.g., Rosales), and families in *aceae* (e.g., Rosaceae). Certain specific ways of writing the binomial name are strictly adhered to in scientific communication. These rules are as follows:

1. It must be underlined or written in italics (because the words are non-English).
2. The genus name must start with an uppercase letter, and the species name (specific epithet) always starts with a lowercase letter. The term *species* is both singular and plural and may be shortened to sp. or spp.
3. Frequently, the scientist who first named the plant adds his or her initial to the binary name. The letter *L* indicates that Linnaeus first named the plant. If revised later, the person responsible is identified after the *L,* for example, *Glycine max* L. Merr (for Merrill).
4. The generic name may be abbreviated and can also stand alone. However, the specific epithet cannot stand alone. Valid examples are *Zea mays,* and *Zea, Z. mays,* but not *mays.*
5. The cultivar or variety name may be included in the binomial name—for example, *Lycopersicon esculentum* Mill cv. "Big Red," or L. *esculentum* "Big Red." The cultivar name (cv), however, is not written in italics.

2.2.3 OPERATIONAL CLASSIFICATION SYSTEMS

What if one does not have an interest or a need to communicate in scientific terms, using names that are often difficult to pronounce, let alone memorize? Crop plants may be classified for specific purposes—for example, according to seasonal growth, kinds of stem, growth form, and economic part or agronomic use.

Seasonal Growth Cycle

Plants may be classified according to the duration of their lifecycle (i.e., from seed, to seedling, to flowering, to fruiting, to death, and back to seed). On this basis, crop plants may be classified as **annual**, **biennial**, **perennial**, or **monocarp** (Figure 2–2).

Annual   Annual plants (annuals) complete their lifecycle in one growing season. Examples of such plants include corn, wheat, and sorghum. Annuals may be further categorized into *winter annuals* or *summer annuals*. Winter annuals (e.g., wheat) utilize parts of two seasons. They are planted in fall and undergo a critical physiological inductive change called *vernalization* that is required for flowering and fruiting in spring. In cultivation, certain non-annuals (e.g., cotton) are produced as though they were annuals.

Biennial   A biennial is an herb that completes its lifecycle in two growing seasons. In the first season, it produces only basal leaves; then it grows a stem, produces

Annual plant.   A plant that completes its lifecycle in one growing season.

Biennial plant.   A plant that completes its lifecycle in two growing seasons.

Perennial plant.   A plant that grows year after year without replanting and usually produces seed each year.

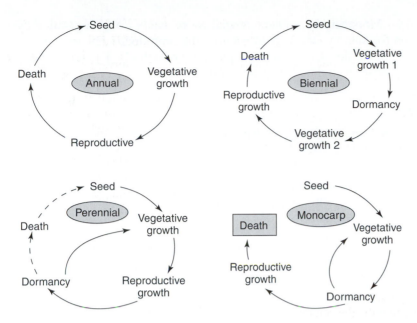

FIGURE 2–2 Classification of flowering plants according to the duration of their growth cycle from seed to seed. Variations occur within each category, even for the same species, due in part to the activities of plant breeders.

flowers and fruits, and dies in the second season. The plant usually requires a special environmental condition or treatment, such as exposure to a cold temperature (vernalization), to be induced to enter the reproductive phase. Examples are sugar beet and onion. Even though annuals and biennials rarely become woody in temperature regions, these plants may sometimes produce secondary growth in their stems and roots.

Perennial Perennials may be herbaceous or woody. They persist all year round through the adverse weather of the non-growing seasons (winter, drought) and then flower and fruit after a variable number of years of vegetative growth beyond the second year. Herbaceous perennials survive the unfavorable season as dormant underground structures (e.g., roots, rhizomes, bulbs, and tubers) that are modified primary vegetative parts of the plant. Examples of herbaceous perennials are grasses with rhizomes (e.g., indiangrass) or stolons (e.g., buffalograss).

Woody perennials may be vines (e.g., grape), shrubs, or trees. These plants do not die back in adverse seasons but usually suspend active growth. Although some perennials may flower in the first year of planting, woody perennials flower only when they become adult plants. Woody perennials may be categorized into two types:

1. *Evergreen.* These are perennials that have green leaves all year round. Some leaves may be lost, but not all at one time. Examples are citrus and pines.
2. *Deciduous.* These plants shed their leaves during one of the seasons of the year (dry, cold). New leaves are developed from dormant buds upon the return of favorable growing conditions. Examples are oak and elm. Intermediate conditions occur in these groups of plants, whereby only partial loss of leaves occurs *(semideciduous).*

Monocarp Monocarps are characterized by repeated, long vegetative cycles that may go on for many years without entering the reproductive phase. Once flowering occurs, the plant dies. Common examples are bromeliads. The top part dies, so that new plants arise from the root system of the old plant.

Stem Type

There are three general classes of plants based on stem type. However, intermediates do occur between these classes.

Herbs These are plants with soft, non-woody stems. They have primary vegetative parts and are not perennials. Examples are corn, many potted plants, many annual bedding plants, and many vegetables.

Shrubs A shrub has no main trunk. It is woody and has secondary tissues. Branches arise from the ground level on shrubs. Shrubs are perennials and are usually smaller than trees. Examples are dogwood, kalmia, and azalea.

Trees Trees are large plants that are characterized by one main trunk. They branch on the upper part of the plant, are woody, and have secondary tissues. Examples are pine and orange.

Common Stem Growth Form

Certain plants can stand upright without artificial support; others cannot. Based on this characteristic, plants may be classified into groups. The common groups are as follows:

1. *Erect.* Erect plants can stand upright without physical support, growing at about a 90-degree angle to the ground. This feature is needed for mechanization of certain crops. Plant breeders develop erect (bush) forms of non-erect (pole) cultivars for this purpose. There are both pole and bush cultivars of crops such as bean (*Phaseolus vulgaris* L.) in cultivation.
2. *Decumbent.* Plants with decumbent stem growth form, such as peanuts *(Arachis hypogea),* are extremely inclined with raised tips.
3. *Creeping (repent).* Plants in this category, such as strawberry (*Fragaria* spp. white clover *(Trifolium repens)*), have stems that grow horizontally on the ground.
4. *Climbing.* Climbers are plants with modified vegetative parts (stems or leaves) that enable them to wrap around a nearby physical support, so they do not have to creep on the ground. An example is yam (*Dioscorea* spp.). Climbers are vines that, without additional support, will creep on the ground. There are three general modes of climbing. Twiners are climbing plants that simply wrap their stringy stems around their support, as occurs in sweet potato. Another group of climbers develops cylindrical structures, called tendrils, that are used to coil around the support on physical contact (e.g., garden pea). The third mode of climbing is by adventitious roots formed on aerial parts of the plant (e.g., English ivy).
5. *Despitose (bunch or tufted).* Grass species, such as buffalograss, have a creeping form, whereas others, such as tall fescue, have a bunch form and hence do not spread by horizontal growing stems.

Agronomic Use

Crop plants may be classified according to agronomic use. Examples are as follows:

1. *Cereals.* These are grasses such as wheat, barley, and oats that are grown for their edible seed.
2. *Pulses (grain legumes).* These are legumes grown for their edible seed. (e.g., peas, beans).
3. *Grains.* Crop plants grown for their edible dry seed or caryopsis (e.g., corn, soybean, cereals).
4. *Small grains.* Grain crops with small seed (e.g., wheat, oats, barley).
5. *Forage.* Plants grown for their vegetable matter that is harvested and used fresh or preserved as animal feed (e.g., alfalfa, red clover).
6. *Roots.* Crops grown for their edible (swollen) roots (e.g., sweet potato, cassava).
7. *Tubers.* Crops grown for their edible modified (swollen) stem (e.g., Irish potato, yam).
8. *Oil crops.* Plant grown for their oil content (e.g., soybean, peanut, sunflower).
9. *Fiber crops.* Crop plant grown for use in fiber production (e.g., jute, flax, cotton).
10. *Sugar crops.* Crops grown for use in making sugar (e.g., sugar cane, sugar beet).
11. *Green manure crops.* Crop plants grown and plowed under the soil while still young and green, for the purpose of improving soil fertility (e.g., many leguminous species).
12. *Cover crops.* Crops grown between regular cropping cycles, for the purpose of protecting the soil from erosion and other adverse weather factors (e.g., many annuals).
13. *Hay.* Grasses or legume plants that are grown, harvested, and cured for feeding animals (e.g., alfalfa, buffalograss).
14. *Silage crops.* Crops preserved in succulent condition by the process of fermentation.
15. *Green chop (soilage crop).* Forage that is harvested daily and brought fresh to livestock.
16. *Drug crops.* Crops grown for their medicinal value.
17. *Trap crops.* Crops planted to protect the main cash crop from a pest.
18. *Companion crops.* Crops along with another crops for mutually beneficial impact, and harvested separately.
19. *Rubber crops.* Crops grown for their latex.

2.3: VISUAL IDENTIFICATION OF PLANTS REQUIRES FAMILIARITY WITH PLANT MORPHOLOGY

Student agronomy clubs may participate in crop judging contests. To be successful, participants need to study morphological attributes that distinguish among plant species. Crop inspectors and crop extension agents also need to be able to identify plant species. Certain morphological features of plant parts, such as the leaf, flower, and stem, may be used to identify crop plants. The descriptive features are categorized into those for dicots and those for monocots. One of the most visible distinguishing features of these two subclasses of flowering plants is the leaf.

The leaf is the primary photosynthetic organ of plants. A typical leaf has three components: thickened *leaf base (pulvinus),* a slender *petiole* (or *leaf stalk*), and a flat *lamina*

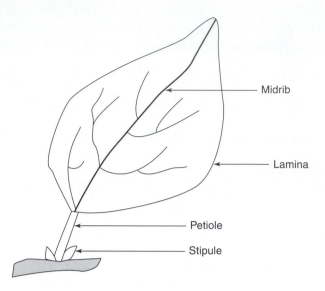

Midrib

Lamina

Petiole

Stipule

FIGURE 2–3 A simple leaf. The prominent part of the leaf is the lamina. It varies in size, shape, thickness, and other characteristics among species. Some leaves may not have a petiole and are attached directly to the branch.

(or *leaf blade*) (Figure 2–3). A leaf is sessile if it lacks a petiole. The leaf blade consists of veins, the middle one, or midrib, normally being larger than the rest. The pattern of veins *(venation)* in dicots is branched and rebranched to form a weblike or netted venation (called *reticulate venation*). Monocot leaves normally have *parallel venation,* in which the veins are not webbed but run parallel to the dominant midrib. In dicots, a pair of scalelike structures called *stipules* occurs at the leafbase. Further, an axillary bud occurs in the leaf axil. Monocots are generally called *narrow leaf* plants because of the lamina shape, while dicots are generally called *broadleaf* plants for a similar reason.

2.3.1   *IDENTIFYING BROADLEAF (FOLIAGE) PLANTS*

Broadleaf plants may be described by foliage leaf form, shape, margin, arrangement, attachment, tip, and base.

1. *Leaf form.* The *form* of the leaf refers to the shape of a single lamina. Leaf forms range from needle shape to circular (Figure 2–4).
2. *Leaf shape.* Leaf *shape* refers to the complexity of the leaf. The leaf may have an undivided lamina and is said to be a *simple leaf.* On the other hand, the leaf blade may be divided into several smaller leaflets arranged on either side of the midrib, or *rachis,* or secondary veins *(rachilla).* Such a leaf is described as a *compound leaf.* There are several arrangements of leaflets on a compound leaf (Figure 2–5). There are two basic types of compound leaves: *palmate* or *digitate* (leaflets arising from one point on the tip of the petiole) and *pinnate* (leaflets arranged like a feather). Second-degree pinnate arrangement is called *bipinnate,* while third-degree pinnate is called *tripinnate.*
3. *Leaf margin.* Lamina margin types range from *unindented* and smooth to *toothed* (slight indentation) to *lobed* (deeply incised) (Figure 2–6). Intermediate margin types occur.

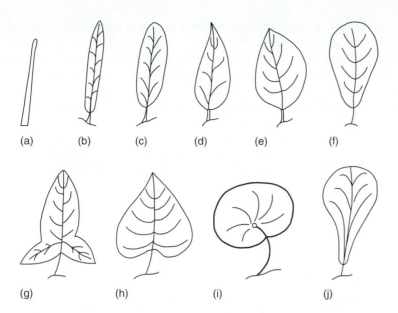

FIGURE 2–4 Selected leaf forms. The shape of the lamina varies from narrow and needlelike to round: (a) filiform, (b) linear, (c) oblong, (d) lanceolate, (e) ovate, (f) abovate (g) sagittate, (h) cordate, (i) peltate (j) spatulate.

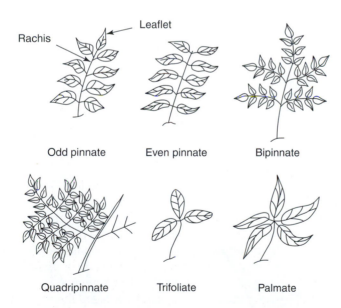

FIGURE 2–5 Compound leaf shapes. Instead of one solid lamina, some leaves consist of several to many small leaflets. Variation in the palmate and other shapes occur.

4. *Leaf arrangement.* Leaf arrangement may be one of three basic types: *alternate, opposite,* or *whorl* (Figure 2–7).
5. *Leaf attachment.* Bud leaves attach to stems several ways. It may be petioled or sessile (i.e., with or without petiole). The leaf may also clasp round the stem (Figure 2–8).
6. *Tips and bases.* Plant leaves differ in the leaf tip and base shape (Figure 2–9).

FIGURE 2–6 Variation in leaf margins. The leaf margin may be intact and smooth or be rough and incised to varying extents.

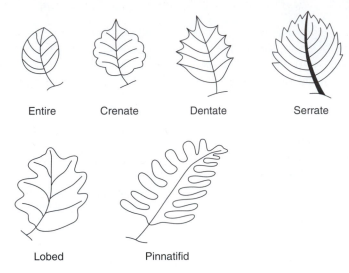

Entire   Crenate   Dentate   Serrate

Lobed   Pinnatifid

FIGURE 2–7 Common types of leaf arrangement.

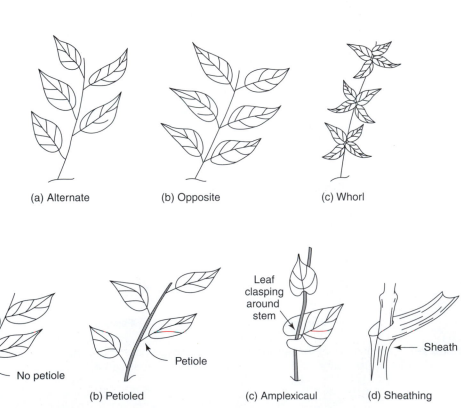

(a) Alternate   (b) Opposite   (c) Whorl

(a) Sessile   (b) Petioled   (c) Amplexicaul   (d) Sheathing

No petiole

Petiole

Leaf clasping around stem

Sheath

FIGURE 2–8 Common types of leaf attachment.

2.3.2 IDENTIFYING GRASS PLANTS

Just like broadleaf plants, grasses are identified on the basis of the characteristics of plant parts.

Leaf

Grass leaves have one basic form—linear. However, they differ in other features, including venation, blade (or lamina), sheath, collar, ligule, and auricles (Figure 2–10).

1. *Venation.* This is the arrangement of the young leaf in the bud shoot. The leaf may be folded (e.g., Kentucky bluegrass) or rolled (e.g., tall fescue).
2. *Blade (or lamina).* This is the often broad and expanded main body of the leaf. This is the primary site of *photosynthesis,* the process by which plants manufacture food.

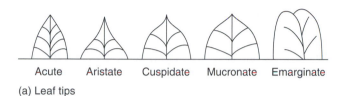

Acute   Aristate   Cuspidate   Mucronate   Emarginate

(a) Leaf tips

FIGURE 2–9 Selected common leaf tips and bases.

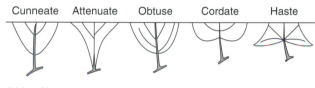

Cunneate   Attenuate   Obtuse   Cordate   Haste

(b) Leaf bases

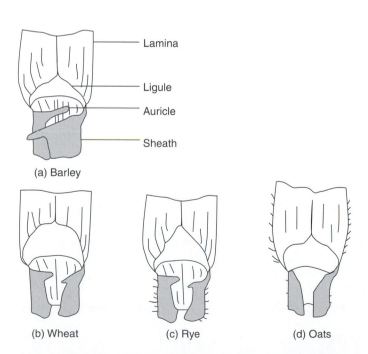

Lamina

Ligule

Auricle

Sheath

(a) Barley

(b) Wheat   (c) Rye   (d) Oats

FIGURE 2–10 Grasses have certain unique features at the junction between the lamina and the stem: sheath, collar, and ligule. The sheaths in these small grains are split. The sheaths of rye and oats and the leaf margins of oats have pubescence. They may overlap or be closed in other species. The ligules vary in shape and size. Wheat has a large and rounded ligule. The auricles in barley are long and clasping, while rye has very short auricles. Oats have no auricles.

3. *Sheath.* The lower portion of the grass blade that usually encloses the stem is called the *sheath.* Grass blades attach to the stem by a sheathing. The sheath may be split and just touching, split and overlapping, or closed and entirely fused.
4. *Collar.* The leaf blade and sheath join in a region called the *collar.* The collar may be narrow or broad. It may also be divided into two sections by the midvein.
5. *Ligule.* This is a translucent membrane or a ring of hairs that occurs at the leaf junction on the inside of the leaf blade and the sheath. The ligule may be membranous or ciliate (hairy).
6. *Auricles. Auricles* are clawlike appendages that project from either side of the collar to the inside of the leaf blade and extend partially around the stem in a clasping fashion.

2.4: FLOWERS ARE VERY RELIABLE MEANS OF PLANT IDENTIFICATION

Flowers are very stable plant organs under varying environmental conditions. They are thus very important in taxonomic studies. Flowers may occur individually *(solitary)* and are called *simple flowers.* They may also occur in groups and are then called *inflorescence.*

2.4.1 SIMPLE FLOWER

Gynoecium. The female part of a flower or pistil formed by one or more carpels and composed of the stigma, style, and ovary.

Androecium. The male part of a flower composed of the anther and filament.

The typical simple flower has four main parts: a *petal* (the showy and colorful part), *sepal* (the protective cover for the flower in bud stage), *pistil* (the female reproductive organ), and *stamen* (the male reproductive organ) (Figure 2–11). A pistil consists of a basal *ovary* (the part that usually forms the fruit), a median *style* (passage connecting the ovary to the stigma), and the terminal *stigma* (the pollen receptive area). The ovule-bearing unit that is part of the pistil is called the *carpel.* Some flowers have more than one carpel. The carpels together form the **gynoecium.** The ovules of the carpels develop into seed. The petals collectively form the *corolla,* while the collectivity of sepals is called the *calyx.* The stamens consist of a stalk, or *filament,* and pollen-bearing structure called *anther.* The anther consists of pollen sacs. The stamens together form the **androecium.**

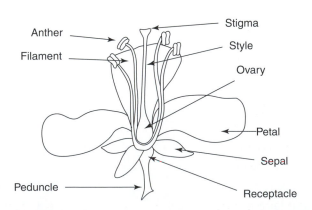

FIGURE 2–11 Parts of a typical flower. A complete flower has four parts: sepal, petal, stamen, and pistil.

A flower that has the full complement of appendages (sepals, petals, stamens, carpels) is said to be complete. A flower in which one or more of these appendages are missing is called an incomplete flower. Similarly, a flower in which both the stamen and carpels are present is called a perfect flower, whereas a flower in which either the stamens or the carpels are missing is called an imperfect flower.

2.4.2 INFLORESCENCE

There are two groups of inflorescence: *indeterminate* and *determinate*. In indeterminate inflorescence, the apical bud continues to grow for an indefinite period. The different types of indeterminate inflorescence are *raceme* (elongated axis with pedicelled flowers), *spike* (like a raceme but with sessile flowers), *corymb* (pedicels gradate in length), and *panicle* (a compound inflorescence consisting of clusters of other inflorescence types) (Figure 2–12). In determinate inflorescence, the terminal bud becomes a flower. This type of inflorescence is called a *cyme*. There are different types of cymes. A third group of inflorescence produces a flat-top but without a definite central axis. The types in this group are *umbel* and *head*. The flowers in a head are sessile.

2.4.3 A TYPICAL LEGUME FLOWER

Certain flowers have unique characteristics that readily and exclusively identify the particular family of plants. The subfamily of the legume family, Papillionoideae, is the most important of legume subfamilies in terms of agronomic importance. Examples of papillionacious crops are peanut, soybean, and dry bean. Members of this subfamily are capable of nodulation. Also, they are characterized by flowers that have a five-petal corolla—a standard, two wings, and one keel petal (Figure 2–13). The flower has five petals: one standard, two wing, and two fused together to form a *keel*. The female structure has one carpel that may contain many ovules. The mature fruit of members of the family Fabaceae is called a legume (has one carpel and splits on both sides).

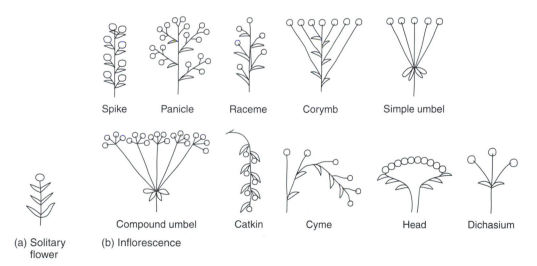

Spike   Panicle   Raceme   Corymb   Simple umbel

Compound umbel   Catkin   Cyme   Head   Dichasium

(a) Solitary flower

(b) Inflorescence

FIGURE 2–12 Types of inflorescence. A simple flower is also solitary, occurring alone on a pedicel. An inflorescence consists of numerous florets arranged in a variety of ways. There are three basic types of inflorescence: head, spike, and umbel.

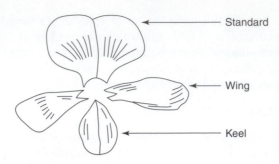

FIGURE 2–13 The petals of a papillionacious legume flower are unique, having a standard, two wings, and one folded keel petal.

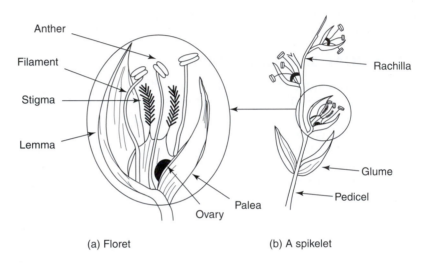

(a) Floret

(b) A spikelet

FIGURE 2–14 The grass family has three basic kinds of inflorescence: panicle, raceme, and spike. The flower consists of subunits called spikelets that bear florets.

2.4.4   TYPICAL GRASS FLOWER

The grass family (Poaceae) is characterized by a flower with a spike inflorescence (Figure 2–14). The basic unit of a grass inflorescence is a spikelet, which is comprised of flowers (florets) that are surrounded by sterile bracts. The grass floret consists of a *lemma* and *palea* (both structures are equivalent to petals) that enclose the reproductive parts of the plant. The glumes constitute the sepals. Grass flowers are *incomplete*. In plants such as corn, the sexes are located on different parts of the plant. Thus, flowers of corn are *imperfect*. Corn is also called a **monoecious plant** (both types of flowers occur on the same plant but at different locations). A **dioecious plant** has imperfect flowers but the sexes occur on separate plants.

2.4.5   TYPICAL COMPOSITE FLOWER

The sunflower plant (family Asteraceae) has two kinds of florets that are arranged on a receptacle (Figure 2–15).

Monoecious plant. A plant with separate male and female flowers on the same plant.

Dioecious plant. Plant species in which individual plants may have either staminate (male) or pistillate (female) flowers, but not both.

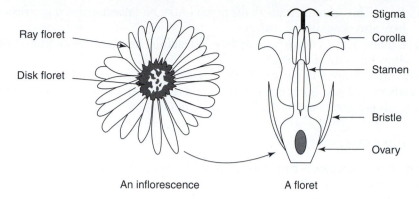

FIGURE 2–15 The inflorescence of the family Asteraceae (sunflower family) is a head with numerous florets.

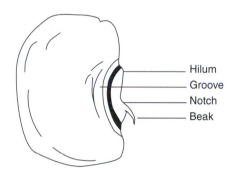

FIGURE 2–16 Features of a dicot seed. The hilum may have distinct color in a ring form, as occurs in black eye pea. A notch is an indentation in the region of the hilum. The beak is a pointed structure protruding from the notch, while a groove is a troughlike depression near the hilum.

2.5: SEED IDENTIFICATION

Dicot seed and monocot seed have certain fundamental differences.

2.5.1 LEGUME SEED

Legume seed can be identified on the basis of morphology by the **testa** (seed coat) color, texture, shape, and size:

1. *Color.* This characteristic is reliable in fresh seed. As seed ages, its color deteriorates.
2. *Texture. Seed texture* refers to the seed coat appearance. It may be smooth, rough, dull, or shiny.
3. *Shape.* This is the most stable and reliable identification characteristic. It is defined by three elements (Figure 2–16):
   a. *Notch.* An indentation in the region of the *hilum* (a scar representing the point of attachment of the seed to the ovary wall). It may be shallow or deep.

Testa. The seed coat formed from the integument(s) of the ovule.

b. *Beak.* This feature occurs in the notch of the hilum as a pointed protrusion. It may also occur on the surface.

c. *Groove.* The groove is a troughlike indentation that leads away from the hilum down the side of the seed.

4. *Size.* Size is highly influenced by the environment. Seeds shrivel under adverse weather. Seed size is determined by weight (e.g., 100 seed weight, or the number of seeds per gram), rather than by visual examination.

2.5.2   GRASS GRAIN

Caryopsis. A small, one-seeded, dry fruit with a thin pericarp surrounding and adhering to the seed occuring in grasses (commonly called grain or kernel)

The fruit of grass species is called the *grain*. Even though the term *seed* is used on occasion (e.g., seed corn), it is an imprecise reference. Grass grain (kernel or **caryopsis**) is identified on the basis of color, endosperm (food storage tissue of grain) type, shape, and size.

1. *Color.* The color of the grain is found in the pericarp (outer covering), aleurone layer (protein-rich area at the outer edge of the endosperm), and the endosperm. For example, the corn may have red or colorless pericarp, blue or colorless aleurone layer, while the endosperm may be yellow or white.

2. *Type of endosperm.* The endosperm characteristics differ in chemistry and physical structure (Figure 2–17). In corn, the endosperm may be described as sugary, starchy, or flinty. In wheat, the endosperm may be of soft or hard starch (vitreous or glassy).

3. *Texture.* The kernel may be dull, shiny, or glossy in texture.

4. *Shape.* In rice, for example, classification for the market is also based on shape and length of grain. There are slender, long, medium, and short grain rice types. The kernel may be pointed at both ends, as in barley, while that is not the case in wheat.

5. *Brush.* Tiny bristles or brush may cap one end of the kernel. Barley has no brush, but wheat does. However, the Durum wheat has no brush.

2.5.3   FORAGE GRASS GRAIN

Identification of forage grass grain is based on the characteristics of the spikelet (e.g., glumes, lemma, and palea). The lemma varies in size, shape, texture, color, nerves (or veins), and other factors. These structures are usually attached to the grain.

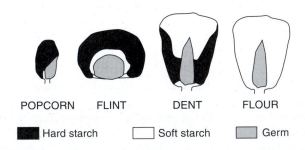

FIGURE 2–17 Types of endosperm in corn. Popcorn has very little soft starch, while flint corn is mostly hard starch. On the other hand, dent corn is about 50% soft starch, while flour corn is all soft starch. (Source: USDA)

2.6: PLANT INTERNAL STRUCTURE

2.6.1 FUNDAMENTAL UNIT OF ORGANIZATION

The **cell** is the fundamental unit of organization of organisms (Figure 2–18). Some organisms consist entirely of one cell *(unicellular)*. Other organisms consist of many cells working together *(multicellular)*. In crop production, the producer cultivates multicellular organisms only, even though unicellular ones (e.g., bacteria) are of importance to production. Except bacterial cells, which lack compartmentalization into organelles with specific functions (called *prokaryotes*), all other cells consist of a membrane-bound nucleus and other membrane-enclosed organelles. These organisms are called *eukaryotes*. Animal and plant cells differ in significant ways (Table 2–4).

Cell. The basic structure and physiological unit of plants and animals.

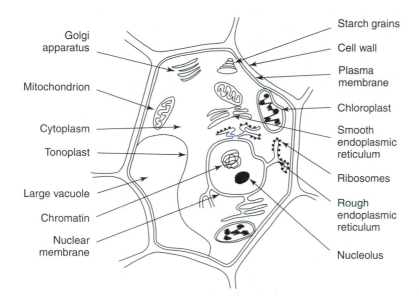

FIGURE 2–18 The parts of a plant cell.

Table 2–4 A Comparison of Animal and Plant Cells

Cell Part	Animal Cell	Plant Cell
Cell wall	−	+
Plasma membrane	+	+
Nucleus	+	+
Peroxisome	+	+
Mitochondrion	+	+
Chloroplast	−	−
Central vacuole	−	−
Ribosomes	+	+
Golgi apparatus	+	+
Cytoskeleton	+	+
Rough endoplasmic reticulum	+	+
Lysosome	+	Not in most
Centriole	+	Not in most
Flagellum	+	Not in most

Cell Structure

The eukaryotic cell consists of many organelles and structures with distinct as well as interrelated functions.

Cell Wall   How is the plant cell wall different from the animal cell wall? The outer membrane of the cell is called the *plasma membrane* (or *plasmalemma*). In plants, an additional structure called the *cell wall* forms around the plasma membrane. The cell wall consists of cellulose, hemicellulose, protein, and other pectic substances. Cellulose, the most abundant of the cell wall constituents, is not digestible by animals. As the cell matures and ages, the cell wall becomes rigid and inelastic through the process of *lignification,* or lignin deposition. All cells have a standard, or *primary, cell wall.* In addition, certain cells develop another wall *(secondary cell wall)* inside the primary one that consists of cellulose and lignin. Adjacent cells are held together by a pectin-rich material called the *middle lamella.* This layer breaks down during fruit rot to release the characteristic slimy fluid.

Nucleus   The *nucleus* is a densely staining body that is usually spheroidal and is the primary location of hereditary material, deoxyribose nucleic acid (DNA). It is composed of DNA, ribose nucleic acid (RNA), protein, and water. The DNA occurs in structures called *chromosomes.* Each species has a specific number of chromosomes per cell (Table 2–5). The number of chromosomes in the sex cell (gametic cell—e.g., pollen grain) is half (haploid, *n* number) that of the somatic or body cell (diploid, 2*n* number).

Vacuoles   Vacuoles are cavities in the cell that contain a liquid called vacuolar sap, or *cell sap.* They also store water-soluble pigments called *anthocyanins* that are responsible for the red and blue colors of many flowers and fruits and the fall colors of some leaves. Vacuoles also absorb water to create a turgor pressure that is neces-

Table 2–5   Number of Chromosomes per Cell Possessed by a Variety of Plant Species

Species	Scientific Name	Chromosome Number (n)
Broad bean	Vicia faba	6
Potato	Solanum tuberosum	24
Corn	Zea mays	10
Bean	Phaseolus vulgaris	11
Cucumber	Cucumis sativus	7
Sorghum	Sorghum bicolor	10
Sugarcane	Saccharum officinarum	40
Wheat	Triticum aestivum	7
Oat	Avena sativa	7
Field pea	Pisum arvense	7
Flax	Linum usitatissimum	15
Sweet potato	Ipomea batatas	15
Peanut	Arachis hypogeae	10
Cotton	Gossypium hirsitum	26
Sugar beet	Beta vulgaris	9
Sunflower	Helianthus annus	17

sary for physical support in plants. The turgor pressure of guard cells controls the opening and closing of stomata (pores in the epidermis of leaves).

Plastids   Do genes occur exclusively in the nucleus? *Plastids* are protein-containing structures capable of synthesizing some of their own protein (they are semiautonomous). The genes they contain are, however, not subject to Mendelian laws of inheritance. Plastids are dynamic and occur in different forms. One form that contains **chlorophyll**, the green pigment of green leaves, is called a *chloroplast.* Chloroplasts are sites of photosynthesis. Chloroplasts occur only in plants and are involved in the variegation of leaves (the development of patchy white or purple coloration found in green leaves). Some plastids are colorless and are called *leucoplasts;* others are called *chromoplasts* and produce numerous pigments found in fruits and flowers. Pigments of bright yellow, orange, or red are called *carotenoids.* The various forms of plastids are interconvertible. As such, when plants are grown in darkness, chloroplasts change to another form of cells called *etioplasts,* cells that cause spindly growth (etiolation). The abnormality is corrected upon exposure to normal light.

Chlorophyll.   A complex organic molecule in plants that traps light energy for conversion into chemical energy through photosynthesis.

Mitochondria   **Mitochondria** are double-membraned structures that are sites of the energy-production processes of the cell, called respiration. The food synthesized by photosynthesis is converted into chemical energy by respiration. Like chloroplasts, mitochondria have their own DNA, and hence they are semiautonomous.

Mitochondria. Organelles found in the cytoplasm of eukaryotic cells that is associated with cellular respiration.

Other Organelles   The cell contains other **organelles** with distinct functions. The *ribosomes* are tiny structures consisting of RNA and protein and are the sites of protein synthesis. They occur on the *endoplasmic reticulum* (a network of sacs and tubes). The *Golgi apparatus* consists of flattened sacs that have secretory functions. There are also substances in the cell called *ergastic substances* that are metabolites (e.g., tannins and resins), some of which are toxic (phytotoxins) to animals and insect pests and thus protect plants from herbivores and insect attack.

Organelle.   A membrane-bound region in the cell with specialized function.

2.7: PLANT CELLS AGGREGATE TO FORM DIFFERENT TYPES OF TISSUES

Aggregates of cells produce a **tissue**. There are two types of tissue—*simple tissue* (consisting of one cell type) and *complex tissue* (consisting of a mixture of cell types). There are three basic types of simple tissue—*parenchyma, collenchyma,* and *sclerenchyma.* They differ in their cell wall characteristics.

Tissue.   A group of cells of similar structure that performs a special function.

2.7.1   SIMPLE TISSUE

Parenchyma

Parenchyma cells have a thin cell wall. The tissue they form is called parenchyma tissue. It is found in actively growing parts of the plant called meristems. These cells are undifferentiated (they do not have any particular assigned functions). The fleshy and succulent parts of fruits and other swollen underground structures (e.g., tubers and roots) are comprised of large amounts of this simple tissue. Some parenchyma cells have secretory roles, while others (e.g., chlorenchyma) have synthetic roles in photosynthesis.

Collenchyma

The primary cell wall of collenchyma cells is thicker than those of parenchyma cells. Collenchymatous tissue occurs in plant parts such as the leaf, where it has a mechanical role in the plant support system by strengthening the parts. It is found in the petiole, leaf margins, and veins of leaves. This mechanical role is confined to regions of the plant where active growth occurs. Fruit rinds that are soft and edible contain collenchyma tissue.

Sclerenchyma

This type of cell has the thickest cell wall of the three primary cells. This is caused by the presence of both primary and secondary cell walls. Sclerenchymatous tissue has primarily a strengthening or mechanical role in the plant. It is elastic and resilient and occurs in places where movement is needed—for example, in the leaf petiole. Structurally, there are two basic sclerenchyma cell types—*short cells (sclereids)* and *long cells (fibers)*. Field crops that are grown for fiber, such as cotton, kenaf, and flax, have large amounts of this tissue.

2.7.2 COMPLEX TISSUE

Complex tissues are comprised of combinations of the three basic simple tissues. They are found in the epidermis, secretory structures, and conducting tissue of the plant.

Epidermis

Epidermis. The outer layer of cells on all parts of the primary body of the plant, except meristems.

The outermost layer of the plant is called the **epidermis**. By virtue of its location, this structure is involved in a variety of roles, including structural (protective), physiological, regulation of water and gaseous movement between the external and internal part of the plant, and anatomical variability. In some plants, an additional protective layer of polymerized fatty acids called *cutin* produces a waterproof layer called a *cuticle*. Pores called *stomata* occur in the epidermis for gaseous exchange. In some species, pubescence (hair-like structures) may occur on the epidermis. This plays a role in pest control by interfering with oviposition, or egg deposition.

Secretory Tissue

There are numerous locations on the plant where secondary tissues perform secretory functions (Table 2–6). Some of these tissues are on the surface of the plant, while others occur on the inside.

Conducting Tissue

Xylem. Specialized cells in plants that transport water and minerals from the soil through the plant and constitute the woody tissue of woody plants.

Phloem. Specialized cells in plants through which carbohydrates and other nutritive substances are translocated through the plant.

Vascular plants conduct inorganic materials up to the leaves and photosynthates down through a network of conducting tissues. The tissue for upward transportation of minerals and water is called **xylem**, and that for conducting assimilates is called the **phloem**.

Xylem constitutes the wood of woody plants and consists of two types of conducting cells—*tracheids* and *vessel elements*. Collectively, they are called *tracheary elements*. Water with nutrients moves up the xylem conducting tissue by water potential. The movement is caused by passive transport, since the xylem cells lack protoplasm and hence function essentially as dead cells. The two types of xylem cells have lateral perforations called *pits* that function in the cell-to-cell flow of fluids. The phloem conducting elements are called sieve elements. The sieve elements consist of two cell types—*sieve cells* (primarily parenchyma cells) and *sieve tube members*. Unlike xylem

Table 2-6 Secretory Tissues of Plants

Those Found Outside the Plant

Nectaries: Occur on various parts of the plant. In flowers, they are called floral nectaries and they secrete nectar that attracts insects for pollination.

Hydathodes: Secrete pure water. Droplets of water may form along leaf margins of certain plants due to secretory activities.

Salt glands: Found in plants that grow in desert or brackish areas

Osmophores: Secrete fragrance in flowers. Repulsive odor of aroids is attributed to the amines and ammonia secreted by osmophores.

Digestive glands: Found in insect-eating (insectivorous) plants (e.g., pitcher plant)

Adhesive cells: Secrete materials that aid attachment between host and parasite

Those Found Inside the Plant

Resin ducts: Found commonly in woody species. They secrete sticky resin.

Mucilage cells: Slimy secretions found at the growing tip of roots and believed to aid the passage of roots through the soil.

Oil chambers: Secrete aromatic oils

Gum ducts: Cell wall modification results in the production of gums in certain trees.

Laticifers: Latex-secreting glands

Myrosin cells: Secrete an enzyme called myrosinase, which when mixed with its substrate, thioglucosides, produces a toxic oil called isothiocyanate. This occurs when cells are ruptured by insects or animals during chewing.

---

cells, which are nonliving (lack protoplasm) and hence play a passive role in the movement of materials, phloem cells are living and thus are actively involved in movement of food. These two cells also have associated cells called *companion cells* in angiosperms (flowering plants) and *albuminous cells* in non-angiosperms. These cells aid in phloem loading of newly synthesized sugars into cells for transport to other parts of the plant. Phloem cells are not durable and must be replaced constantly.

Meristems

Do plants grow in the same fashion as animals? **Meristems** are areas of active growth in plants. The cells at these locations divide rapidly. They are also undifferentiated. When meristems occur at the tip (or apex) of the plant, they are called *apical meristems*. Those that occur at the leaf axil are called *axillary meristems*. Unlike animals, in which growth is a diffused process (i.e., growth occurs throughout the entire individual), plants have localized growth (growth is limited to specific areas—the meristems). Meristems occur in other parts of the plant (basal, lateral, intercalary). Localized growth permits juvenile and mature adult cells to coexist in a plant, provided favorable conditions exist. In some species, the plant can continue to grow indefinitely without any limit to final size, while certain organs and tissues are fully mature and functional. This is called an *indeterminate* growth pattern. In certain plant species, the apical meristem dies at some stage. Such plants are described as *determinate* in growth pattern. Even though it appears many plants have predictable sizes that are characteristic of the species, this feature is believed to be a largely environmental and statistical phenomenon. Under optimal and controlled conditions, many annual plants have been known to grow perennially. What are the advantages and disadvantages of determinacy and indeterminacy in crop production?

Meristem.
Undifferentiated tissue whose cells can divide and differentiate to form specialized tissues.

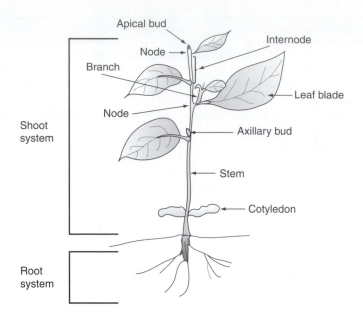

FIGURE 2-19 The shoot system of a dicot plant.

Apical bud

Node

Branch

Shoot system

Node

Internode

Leaf blade

Axillary bud

Stem

Cotyledon

Root system

2.8: THE SHOOT SYSTEM

The shoot system comprises the part of the plant that grows above ground. Shoot growth occurs in two phases—*vegetative* and *reproductive.* When in the vegetative phase, the shoot comprises the stem or cylindrical axis, to which appendages or lateral organs are attached (Figure 2–19). These appendages are leaves and axillary or lateral buds or branches. When a plant enters the reproductive phase, additional appendages arise, depending on whether it is an angiosperm or a gymnosperm. These include cones, sporangia, flowers, and fruits. In terms of external morphology, the shoot has certain features. The leaves attach to the stem at jointlike regions called *nodes.* The space between nodes are *internodes.* A shoot has an apical meristem inside the tip of each terminal bud and axillary bud. This tissue is responsible for forming bud scales, leaves, and, in the reproductive phase, flowers or cone parts. The axillary (lateral) buds give rise to lateral branches of the shoot system. The number of leaves attached to each node determines the leaf arrangement of the shoot system. A single leaf attached to each node produces an *alternate* arrangement that usually forms a spiral pattern up the shoot. When two leaves occur at each node, directly across from each other, the arrangement produced is called *opposite,* while three or more leaves at each node produces a *whorled* arrangement. When a leaf drops off the plant, it leaves a *leaf scar.* In some cases (e.g., grasses), the leaf base is expanded into a *leaf sheath.*

2.9: THE LEAF

Leaf morphology used in plant identification was discussed earlier in this chapter. There are five types of leaves: *foliage leaves, budscales, floral bracts, sepals,* and *cotyledons.* Foliage leaves are the most prominent.

The epidermis on the upper side of the leaf is called the *adaxial side;* the opposite side, the lower side, is called the *abaxial side.* The distribution of stomata is not always equal on both sides. In most plants, very few stomata occur on the upper side. However, in wheat and onions, there are equal numbers of stomata on both sides of the leaf. Both sides of the leaf may be covered with *trichomes* (pubescence). In species such as hemp *(Cannabis),* the trichomes contain certain compounds (e.g., cannabinoid hydrocarbons in *Cannabis*).

Older leaves on trees eventually drop. Some leaves drop during certain seasons of the year (summer, fall). Leaf fall is called *abscission.* In crop production, some producers use artificial methods to induce premature leaf abscission (e.g., in cotton production) to facilitate crop harvesting and to obtain a high product quality. Leaf abscission occurs in a specialized region, called the *abscission zone,* where the petiole detaches from the node, leaving leaf scar.

2.9.1   FUNCTIONS

The functions of leaves include the following:

1. Food synthesis. Leaves manufacture food by the process of photosynthesis. Foliage leaves conduct this function. Are leaves the only plant organs capable of photosynthesis?
2. Protection. This role is performed by non-foliage leaves (bud scale, floral bracts, and sepals) through protection of vegetative and floral buds.
3. Storage. Cotyledons, or seed leaves, store food that is used by seeds during germination.

Can you name additional functions of leaves?

2.9.2   INTERNAL STRUCTURE

The internal structure of the leaf is shown in Figure 2–20. The parts and functions of leaf internal structure are summarized in Table 2–7.

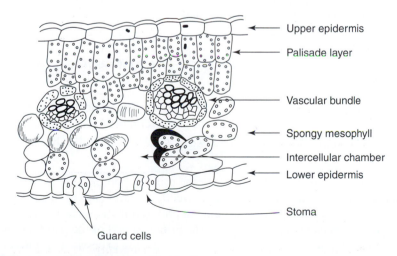

FIGURE 2–20   Internal structure of a dicot leaf.

Table 2–7	Internal Structure of the Leaf and Functions of Parts

Cuticle: A layer of waxlike material called cutin. It protects the leaf and prevents evaporative loss through the epidermis. It is water repellent.

Epidermis: A layer of cells that form the outer protective layer of the leaf. There is an upper epidermis and a lower epidermis.

Mesophyll: Layers of cells beneath the epidermis. Consists of two types of cells:

    Palisade parenchyma: Blocky or rectangular cells that are arranged on the shorter side against the epidermis. They contain chloroplast for photosynthesis.

    Spongy parenchyma: Consists of irregularly shaped cells that are loosely arranged, allowing large intercellular spaces to occur between them. They have chloroplasts. The spaces allow for gases to move through the leaf.

Vascular bundle: Comprised of the conducting tissues, xylem and phloem. Xylem tissue conducts water and minerals from the roots, while phloem conducts photosynthates and other materials from the leaf to other parts of the plant.

Bundle sheath: A ring of cells around the vascular bundles of leaves of certain species. May be involved in storage of photosynthates.

Stomata: Pores in the epidermal layer defined by two special cells called guard cells. The closing and opening of the guard cells regulate the rate of movement of carbon dioxide and water between the leaf and the atmosphere.

2.9.3 MODIFIED LEAVES: WHEN LEAVES DON'T LOOK LIKE LEAVES

Leaves may be modified to perform functions other than photosynthesis or to carry on photosynthesis under unusual environmental conditions. Some leaf modifications are as follows:

1. *Glands.* For secretion.
2. *Spines or thorns.* For protection against herbivores.
3. *Storage tissue.* For food storage, as in bulbs (e.g., onions).
4. *Thickened leaf surface.* To reduce moisture loss under xeric (dry) conditions.
5. *Thin cuticle and gas chamber.* For survival under submerged conditions.
6. *Tendrils.* Stringlike structures for additional support. The terminal leaflet of the trifoliate of pea may be modified into a tendril.

Can you think of other economically useful plant parts that are modified leaves?

2.9.4 SUMMARY OF LEAF-RELATED CROP MANAGEMENT PRACTICES

The management practices summarized here are discussed in detail in appropriate sections of the book. These represent decisions a producer must make in order to have success with crop establishment. The factors discussed here are similar to those presented for stems.

1. *Photosynthetic surface.* The leaves are the primary photosynthetic organs of the plant. Reduced photosynthetic surface directly impacts crop productivity. Many of the practices presented next impact the photosynthetic surface.
2. *Forage management and harvesting.* The principal part of the plant removed during grazing or harvesting is the leaf. Overgrazing excessively removes the herbage, reducing regrowth and promoting soil erosion. Ample amount of leaf

area must be left after grazing or harvesting to allow the plant to photosynthesize enough to meet the carbohydrate needs of the plants for regrowth. Younger leaves are more nutritious and digestible, making higher-quality hay.

3. *Drought management.* Most of the moisture lost through transpiration occurs through the leaves. Cultivars used in dryland production should have drought tolerance traits.

4. *Light interception.* Photosynthesis is the harnessing of light energy and its conversion into chemical energy. Leaves absorb light. The effectiveness of light interception depends on the leaf angle and arrangement. Some cultivars have upright and narrow leaves that reduce leaf shading in the canopy for more effective light interception.

5. *Plant density.* Proper spacing of plants is essential for establishing an ideal leaf area index (total leaf area per unit area) for effective use of the land for high crop productivity.

6. *Planting date.* A crop is seeded at a certain time to take advantage of environmental conditions or to avoid hazards. The management of time of planting in crop production can allow plants to take advantage of a longer growing season. That is, plants can photosynthesize for a longer period.

2.10: THE STEM

The stem is the central axis of the shoot of a plant. It produces various appendages—leaves, axillary buds, floral buds—from the shoot apical meristem. The dermal tissue consists of epidermal cells, stomata, and trichomes. The guard cells of the stomata have chloroplast; hence, the stem, at least in early growth, is green and is capable of photosynthesis. The stem in grasses is called a *culm* and may be hollow (e.g., in rice, wheat) or solid (e.g., in corn). The hollow or central cavity arises when the central pith matures and stops growth earlier than in the peripheral epidermis and cortex. As a result, the central tissue is torn apart as the internodes extend, creating a hollow center.

2.10.1   FUNCTIONS

The functions of the stem include the following:

1. Provision of mechanical support to hold branches, leaves, and reproductive structures (flowers). It is important that the leaves are well displayed to maximize light interception for photosynthesis.
2. Conduct water and minerals up to leaves and assimilates from leaves to other parts of the plant. This function occurs through the vascular system.
3. Usable as material for crop propagation (e.g., in yam and Irish potato).
4. Modified stems for food storage that is of economic value in crops (e.g., Irish potato and yam).

Can you think of additional functions of the stem?

2.10.2   INTERNAL STRUCTURE

The outer layer of the stem is the epidermis and the inner layer is the *cortex*. Dicots and monocots differ in the internal structure of the stem regarding the arrangement of the vascular structures (xylem and phloem). The vascular tissues form a central cylinder called the *stele*. The stele consists of units called *vascular bundles*. The vascular bundles are arranged in a ring in dicots while they are scattered in monocots (Figure 2–21).

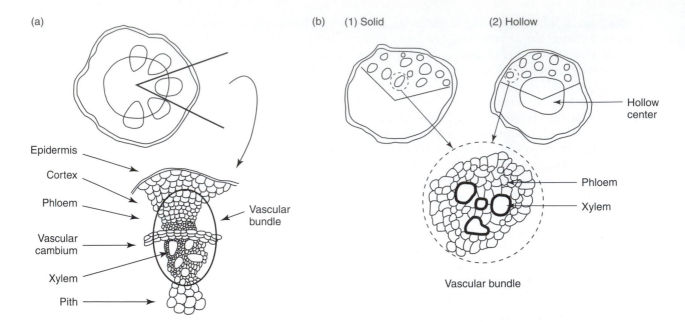

FIGURE 2-21   Internal structure of (a) dicot stem and (b) monocot stem. The key distinguishing feature is the arrangement of the vascular bundles in a ring in dicots, while they are scattered in monocots. Further, the stem may be solid, as in a corn plant, or hollow, as in rice.

Table 2-8   Internal Structure of the Stem and Function of the Parts

Epidermis: Outer protective layer of the stem. It is usually one cell thick and often bears trichomes.

Vascular bundles: Comprised of the conducting tissues, xylem and phloem. Xylem tissue conducts water and minerals from the roots, while phloem conducts photosynthates and other materials from the leaf to other parts of the plant. Vascular bundles are arranged in a single ring in most dicots, while they are scattered throughout the ground tissue of monocots.

Ground tissue: Mostly parenchyma tissue that occurs in two regions in dicots:

  1. **Cortex:** Ground tissue that occurs between the epidermis and the ring of vascular tissue.
  2. **Pith:** Ground tissue in the center of the stem. It is specialized for storage. Not readily discernable in monocots.

In the center of the stem lies a region of purely parenchyma cells called the *pith*. The stem internal structures and their functions are presented in Table 2–8.

2.10.3   STEM MODIFICATION: WHEN STEMS DON'T LOOK LIKE STEMS

Rhizome.   An unusually thickened and horizontally growing underground stem.

Just like the leaf, stem modifications occur, some of which are the economic parts of the plant for which the crops are produced. Most stem modifications occur underground. Some important world food crops are modified stems. These include tubers such as Irish potato and yam *(Dioscorea)*, both of which are swollen ends of stems. A grass plant may have a vertical stem and may have modified stems of one of two kinds: **rhizomes** or

(a)

(b)

(c)

FIGURE 2–22 Stem modifications. Modified stems feature prominently in the list of crops that feed the world. (a) Potatoes are modified stems, (b) onions are stems modified as bulbs, and (c) ginger is a stem modified as a rhizome, or an underground storage organ.

stolons. A rhizome is a lateral stem that grows underground (e.g., in ginger and indian-grass), while a stolon is a modified stem that grows horizontally on the surface of the ground (Figure 2–22). Can you name other examples of economic plant parts that are modified stems?

Stolons are also called *runners*. They develop **adventitious roots** at sites along the stem where their shoot tips turn upward. Stolons and rhizomes are a natural means of

Stolon. An unusually slender and prostrate above-ground stem.

Adventitious roots. Plant structures arising from unusual places (e.g., adventitious root or bud).

propagation. Corms, such as tubers, are fleshy underground stems. They are erect and bear scale leaves (e.g., *Crocus* and *Gladiolus*).

2.10.4   SUMMARY OF STEM-RELATED CROP MANAGEMENT PRACTICES

The management practices summarized here are discussed in detail in appropriate sections of the book. These represent decisions a producer must make in order to have success with crop establishment.

1. *Plant density.* Plant population and spacing are decisions made by the grower to obtain the optimal number of plants for optimal yield under a specific cultural system. Plant spacing depends on several factors, including the planting equipment, the crop species and cultivar, and the moisture available (dryland or irrigated farming). The closer the spacing, the less the branching of some plants or tillering of grasses. Also, closer spacing promotes elongation of the stem (etiolation) and lodging. Consequently, dwarf environmentally responsive cultivars are selected for high-density planting. High-density planting promotes the development of a closed canopy, which reduces weeds in the field. However, extreme density may be counterproductive.

2. *Grazing management and forage harvesting.* Perennial forage species, both grasses and non-grasses, must be carefully removed (grazing or cutting) such that the remainder can sustain regrowth and protect the plant from damage. A rule of thumb is "take half leave half." The time of harvesting determines the quality of the product. Grasses are best cut at or before flowering. Alfalfa is cut in the early bloom for high yield and long-term survival of the plants. The carbohydrate content of the stem and roots is critical in determining the best times for harvesting forage.

3. *Weed control.* Species with horizontal stems, underground stems (rhizomes), or above-ground stems (stolons) (e.g., johnsongrass) are difficult to control by tillage. Herbicides should be applied at the proper stage of stem growth to minimize or avoid injury to the crop. As previously indicated, high density of the crop provides a more effective ground cover for shading out weeds.

4. *Fertilization.* Tall cultivars are generally susceptible to lodging, which is exacerbated by high nitrogen fertilization. Dwarf cultivars are more responsive to high fertilization. Phosphorus promotes rooting, whereas potassium promotes the development of strong stems.

5. *Cultivar morphology.* Grass crop cultivars differ in tillering habits, some being low tillering while others are high tillering. In crops such as soybean, there are thin-line varieties, determinate-stemmed and indeterminate-stemmed, branching and less branching types. In cotton, there are the picker varieties and stripper varieties. In pea, there are bush and pole types, whereas peanut has bunch and runner varieties.

2.11: THE ROOT

Roots are underground vegetative organs of plants. There are two types of roots—*seminal* (derived from seed) and *adventitious* (derived from organs of the shoot system rather than the root system). A germinated seed provides a young root called a *radicle*. It grows to become the primary root from which secondary or lateral roots emerge. The tip of the root is

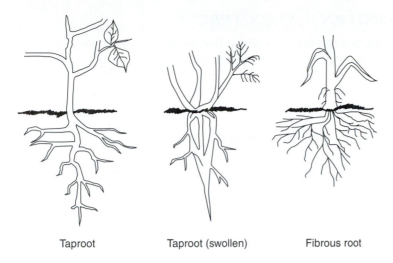

Taproot   Taproot (swollen)   Fibrous root

FIGURE 2–23   Root systems. Dicots have a taproot system in which the central axis may or may not be swollen. Moncots have a fibrous root system that lacks a defined central axis.

protected by a *root cap. Root hairs* are tiny roots that develop from larger ones and serve as structures for absorbing water and minerals from the soil. There are two basic root systems (distribution of roots in the soil)—*taproot* and *fibrous roots* (Figure 2–23).

In the taproot root system, there is a primary root consisting of a large central axis and several lateral roots. The taproot is usually deeply penetrating and is found in dicots and gymnosperms. Sometimes, it is swollen and is harvested as the economic part (e.g., carrot and sugar beet).

Fibrous roots occur in the grass family (e.g., rice, corn, wheat) and other monocots (e.g., onion, banana). This root system lacks a dominant central axis and is shallowly penetrating. The seminal not senesces after completion of seedling growth. Fibrous roots have more soil-binding effect and are used in erosion control in soil conservation practices.

Adventitious roots may occur on rhizomes (underground) or on aerial shoots (above ground). Adventitious roots develop in some species when the nodes and internodes of the stem come into contact with the soil or are buried in it (e.g., in currant, gooseberry). When plants are propagated by cuttings (pieces of shoot—stem, leaf, root—used for propagation), sometimes an application of rooting powder to the cut surface prior to planting is needed to promote rapid rooting. Aerial roots are produced by many vines and epiphytes (plants that grow on other plants), such as orchids and bromeliads. Other forms of aerial roots are *prop roots,* which grow down from the stem to the soil (e.g., in corn).

2.11.1   FUNCTIONS

The functions of the root include the following:

1. Roots anchor plants in the soil, holding stems and leaves upright and preventing toppling by wind.
2. Roots absorb the nutrients and water used by plants in photosynthesis and other physiological functions.
3. Modified roots have storage roles, as occurs in sweet potato where they are the economic part of the plant. Aerial roots occur in certain species, where they provide additional support for the plant through attachment to physical support. Prop roots of corn are aerial roots.

2.11.2 INTERNAL STRUCTURE

The pericycle is a meristematic region that produces, among other structures, lateral or branch roots that grow outwardly through the cortex and the epidermis. The pericycle together with the vascular cylinder form the *stele* (Figure 2–24). The solution of water and nutrients from the soil enters the inner tissue through the permeable endodermal cell walls and the protoplast.

2.11.3 ROOT MODIFICATION

Why is sweet potato a root, while Irish potato is a stem, even though they both occur underground? Do all roots grow underground? As previously mentioned, roots do not always come from the seed but can develop from other parts of the plant. However, true roots can be modified to become swollen, as in sweet potato or cassava, where they store water and foods (Figure 2–25). Some roots provide additional support to the plant stem by becoming modified for attachment, as in aerial roots and prop roots (for such species as corn).

2.11.4 ROOT NODULE

Through symbiotic association between a plant host and bacteria, localized root modifications, called *root nodules,* appear on roots as roundish structures. This symbiotic association occurs between legumes (e.g., soybean, peanut, bean) and a soil bacterium, *Rhizobium.* Nodulation starts when the appropriate strain of bacterium enters a root hair and changes into bacteroids (irregularly shaped bacteria). These bacteroids enter the cortical cell of the root, invading the cytoplasm of the root cells. These invaded cells divide, proliferating into tumorlike structures (nodules). A nodule is hence cortical cells that are filled with bacteroids. The bacteria derive nourishment from the plant and in turn incorporate molecular nitrogen from the air into the plant by the process of biological nitrogen fixation.

2.11.5 SUMMARY OF ROOT-RELATED CROP MANAGEMENT PRACTICES

The management practices summarized here are discussed in detail in appropriate sections of the book. These represent decisions a producer must make in order to have success with crop establishment.

FIGURE 2–24 Internal structure of roots: (a) dicot root, (b) monocot root.

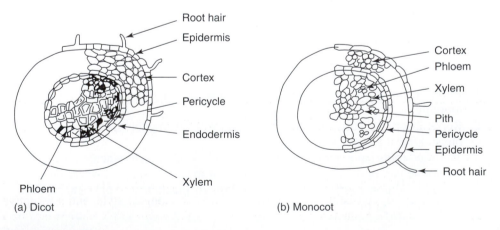

(a) Dicot

(b) Monocot

FIGURE 2–25 Modified root, represented by sugar beet (*Beta vulgaris*).

1. *Depth of irrigation.* The goal of irrigation is to provide water in the root zone of the plant. Plants vary in root depth and spread. Depth of irrigation for grasses is different from depth of irrigation suitable for deep-rooted plants.
2. *Cultivation.* Cultivation of growing crops may be undertaken to control weeds. Shallowly rooted plants are more susceptible to damage from cultivators.
3. *Tillage.* In preparing the land for planting, deep-rooted plants require deep soil for proper growth. Tillage, in this case, should be deep. Using heavy implements in tillage can cause subsoil compaction, which affects deep-rooted species more than shallow-rooted species in terms of soil volume available for exploiting. However, shallow rooting predisposes plants to root lodging.
4. *Fertilization and fertilizer placement.* Deep-rooted plants can benefit from deep placement of fertilizer as well as leached nutrients. Phosphorus tends to promote good rooting. Starter application of fertilizer usually contains a high amount of phosphorus for this reason.
5. *Grazing management and harvesting.* Deep-rooted species are more resistant to damage from uprooting by grazing animals. The carbohydrate content of roots is an important factor in grazing management.

2.12: THE FLOWER

2.12.1   REPRODUCTION

The female reproductive part of the flower is the *gynoecium*. It is made up of one or more *carpels* (or pistils), consisting of an enlarged ovary, an elongate style, and a receptive stigma. The male reproductive part of the flower is the *androecium* and consists of a

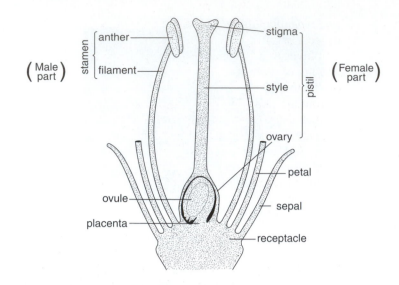

FIGURE 2–26 Reproductive parts of a dicot flower.

filament and a distal *anther.* The anther contains four pollen sacs, or microsporangia. When mature, the anthers dehisce to shed the pollen grains that contain the haploid number of chromosomes for the species (Figure 2–26).

2.12.2 POLLINATION

Pollen grains are transferred to the stigma of a flower by the process of *pollination,* using a variety of agents. Flowers may be *self-pollinated* (receive pollen only from the same plant or genotype) or *cross-pollinated* (receive pollen from a different plant or a different genotype). Agents of pollination include insects (e.g., bees, wasps, flies), birds (e.g., hummingbirds), bats, and wind. Bees tend to pollinate flowers that are showy and that are blue or yellow. Bees are the most important insect pollinators. Growers of certain crops sometimes rent colonies of bees from bee keepers for use during the growing season for effective pollination for a good fruit set. Wind-pollinated species produce large amounts of smooth, small pollen grains. Flowers that are self-pollinated often have some amount of cross-pollination, and vice versa. Hence, it is best to describe plants as *predominantly* either self- or cross-pollinated.

2.12.3 FERTILIZATION

Fertilization is the union of gametes (pollen and egg in plants). It starts with the germination of pollen on a receptive stigma and subsequent growth of the pollen tube through the style to the ovule of the ovary. The pollen tube enters the embryo sac and discharges two sperm nuclei, one of which fuses with the egg to form a zygote ($2n$). The other fuses with two polar nuclei to form the primary endosperm nucleus (a triploid, $3n$). These two fertilization events constitute what is called *double fertilization.* The primary endosperm nucleus builds a nutritive endosperm around the developing embryo. Fertilization does not always follow pollination. The process of embryo formation is called *embryogenesis.*

2.12.4 SUMMARY OF FLOWER-RELATED CROP MANAGEMENT PRACTICES

The management practices summarized here are discussed in detail in appropriate sections of the book. These represent decisions a producer must make in order to have success with crop production.

1. *Effectiveness and type of pollination.* In flowering species where the economic part is the fruit, seed, or grain, crop yield depends on effective pollination. Some growers may rent hives of bees for use during the growing season for effective pollination. Spraying of pesticides at flowering should not be done to avoid killing insect pollinators. Some flowers are predominantly self-pollinated (receive pollen from the same flower or plant), while others are predominantly cross-pollinated (can use pollen from any compatible source). Hybrids of cross-pollinated species are developed such that seed cannot be saved for planting the next season's crop. Self-pollinated cultivars breed true. Hence, seed can be saved for planting the next season's crop.

2. *Fruit set.* The yield of the crops depends on the number of flowers that develop fully to desired harvest maturity. Fruit set is affected by environmental factors, such as drought, frost, and pests. Growth regulators may be sprayed to control fruit set so that only a desirable number of flowers are permitted to develop, resulting in larger fruits.

3. *Plant identification.* Flowers are less environmentally labile. They maintain their properties across a wide range of environments, making them desirable as a means of identifying crop species.

2.13: THE FRUIT

Embryogenesis leads to the production of a fruit. The mature ovary with the associated parts form the *fruit*. In some species, the fruit develops without fertilization, a phenomenon called *parthenocarpy*. Parthenocarpic fruits are seedless—for example, "Cavendish" banana, "Washington navel" orange, and many fig cultivars. The natural function of the fruit is to protect the seed, but the fruit is what is most desired by animals and humans.

A typical fruit has three regions: the *exocarp* (which is the outer covering, or skin), the *endocarp* (which forms a boundary around the seed and may be hard and stony or papery), and the *mesocarp* (the often fleshy tissue that occurs between the exocarp and the endocarp).

A common way of classifying fruits is according to the succulence and texture on maturity and ripening. On this basis, there are two kinds of fruits—fleshy fruits and dry fruits. However, anatomically, fruits are distinguished by the arrangement of the carpels from which they develop. A carpel is sometimes called the pistil (consisting of a stigma, a style, and an ovary), the female reproductive structure, as previously described. A fruit may have one or more carpels. Even though the fruit is basically the mature ovary, some fruits include other parts of the flower in their structure and are called accessory fruits. Combining carpel number, succulence characteristics, and anatomical features, fruits may be classified into three kinds—*simple, multiple,* or *aggregate* (Figure 2–27).

2.13.1   SIMPLE FRUITS

Simple fruits develop from a single carpel or sometimes from the fusing together of several carpels. This group of fruits is very diverse. When mature and ripe, the fruit may be soft and fleshy, may be dry and woody, or may have a papery texture.

Fleshy Fruits

There are three types of fleshy fruits—drupe, berry, and pome.

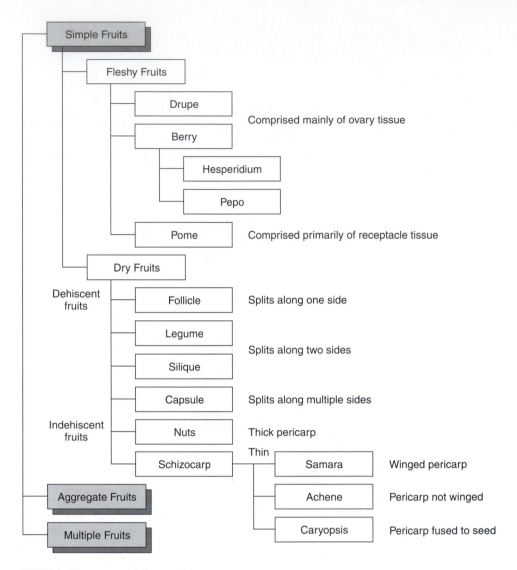

FIGURE 2–27 A classification of fruits.

Drupe Drupes are simple fleshy fruits with a hard, stony endocarp (or pit) that encloses a single seed. The mesocarp may not always be exactly fleshy in the true sense of the word. The coconut (*Cocos nucifera* L.) has a fibrous layer of a combination of mesocarp and exocarp called a *husk*. This is a source of fiber for a variety of crafts. The edible part of the fruit is the seed, which occurs as a thick, white layer and can be scooped with a spoon if the fruit is young. The seed is hollow and contains a liquid that is casually referred to as coconut "milk." The seed is covered by a hard endocarp. Sometimes, this kind of fruit is described as fleshy-dry. Examples include peaches, plums, cherries, apricots, almonds, and olives (Figure 2–28).

Berry A berry originates from a compound ovary. Usually, it contains more than one seed. The pericarp of berries is fleshy. There are distinguishable variations among berries. A true berry has a thin exocarp. At maturity, the pericarp is usually soft. Common examples include the tomato, grape, banana, eggplant, and pepper

(Figure 2–29). Exceptions to this description are date and avocado, which are fleshy fruits with one seed, as opposed to the typical multiple seeds. Just because a fruit has the word *berry* in its common name does not make it one. For example, blackberry, strawberry, and raspberry are not berries at all. In species such as orange, lemon, and lime, the pulp juice sacs are enclosed in a leathery rind (Figure 2–30). The fruit in this case is called a *hisperidium.* Another kind of fleshy fruit called a *pepo* is exemplified by melon, squash, and cucumber (the gourd family) (Figure 2–31). The pericarp forms a much harder rind than in the citrus family.

Pome Pomes originate from the receptacle located around the ovary called the *hypanthium.* In apples, the same adjacent tissue is involved in the development of

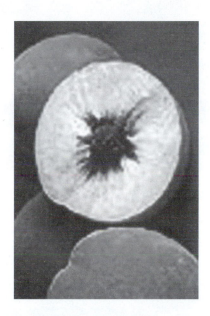

FIGURE 2–28   A drupe, represented by a peach *(Persica spp.).*

FIGURE 2–29   A berry, represented by tomato *(Lycopsersicon esculentum).*

FIGURE 2–30 A hesperidium, represented by Citrus *(Citrus spp.)*. (Source: USDA)

FIGURE 2–31 A pepo, represented by muskmelon. (Source: USDA)

the fruit (Figure 2–32). The edible part of the fruit occurs outside the seed. Quince and pear are examples of pomes.

2.13.2 AGGREGATE FRUITS

Aggregate fruits are formed from a single flower with many pistils. Each of these pistils develops into a fruit (sometimes called *drupelet*). However, these tiny fruits form a cluster on a single receptacle upon maturity. Examples of aggregate fruits are strawberry, blackberry, and raspberry (Figure 2–33).

2.13.3 MULTIPLE FRUITS

Multiple fruits are distinguished from aggregate fruits by the fact that they are produced by many individual flowers occurring in one inflorescence. Like aggregate fruits, the individual flowers produce separate tiny fruits, or fruitlets, which develop together into a larger fruit. Common examples are fig and pineapple.

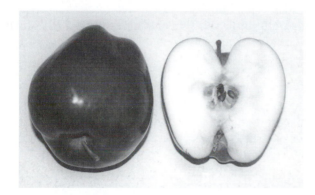

FIGURE 2–32   A pome, represented by an apple *(Pyrus malus).*

FIGURE 2–33   Aggregate fruit, represented by strawberry. (Source: USDA)

Dry Fruits

As the name implies, dry fruits are not juicy. Some have one seed; others have many. The exocarp, mesocarp, and endocarp may be fused into one layer, called the **pericarp,** which surrounds the seeds as a thin layer. When mature and dry, the fruit may split open and discharge the seed (*dehiscent fruits*) or retain it (*indehiscent fruits*). Based on this characteristic, there are two kinds of dry fruits.

Dehiscent Fruits   There are a number of dehiscent fruits, which may be distinguished according to how they split. When the fruit splits along one side only, it is called a follicle. Follicles include peony (*Paeonia* spp.), milkweed (*Asclepias* spp.), and larkspur (*Delphinium* spp.). Certain dehiscent fruits split along both sides of the fruit, as typified by legumes, such as pea, lentil, garbanzo bean, and lima bean. Follicles and legumes are collectively called *pods* (Figure 2–34). Siliques are fruits that split along both seams but whose seed is borne on a structure between the two halves of the split fruit. Sometimes the fruit is long and is called a *silicle.* Plants that bear this kind of fruit include those in the mustard family (Brassicaceae), such as cabbage, radish *(Raphanus sativus),* watercress *(Nasturtium officinale),* and broccoli. Plants such as poppy, lily, snapdragon, and iris are classified as capsules. They split in a variety of ways and consist of two or more carpels.

Indehiscent Fruits   Fruits in this category do not dehisce because the seed is united in some way to the pericarp. One kind is the *achene* (one-seeded dry fruit with firm pericarp), as found in buttercup and nettle. The other kind is called a *caryopsis* and is the seed (or fruit) of grasses (such as corn, wheat, and sorghum). The pericarp is

FIGURE 2–34 A lugume or pod, represented by garden bean *(Phas eolus vulgaris).*

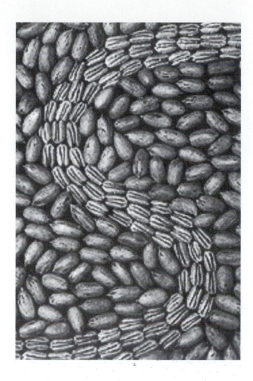

not separable from the seed. The third indehiscent dry fruit is the nut, which is similar to the achene, except that nuts usually are much larger and have a thicker, harder pericarp which must be cracked to reach the edible seed. True nuts include chestnut *(Castanea* spp.) and hazelnut *(Corylus* spp.). Just like berries, there are many fruits that have *nut* in their common names but are not true nuts. For example, walnut and pecan *(Carya illinoensis)* are actually drupes (Figure 2–35).

The seed is a mature, or ripened, ovule. It develops in the ovary portion of the carpel as the ovary is differentiating into a fruit. Seeds are the end products of sexual reproduction in seed plants, the means by which they are propagated. All seeds are covered by a seed coat (testa) comprised of fused inner and outer integuments of the ovule. The seed coat surrounds the endosperm and the embryo. The space occupied by the embryo in the seed is variable among species. In grasses, the embryo is more differentiated.

2.14: THE SEED AND CARYOPSIS

The seed is the propagational unit of flowering species. The economic part of grass crops is not seed but rather the entire fruit, called a *grain* or *caryopsis.*

2.14.1 DICOT SEED

Dicot seeds are so called because they have two *cotyledons,* or seed leaves, the structures that contain the stored food of the seed. The outer covering is called the *testa* (Figure 2–36). Legumes have this seed type. The parts of seed and their functions are summarized in Table 2–9.

Dicot. A subclass of flowering plants with two cotyledons.

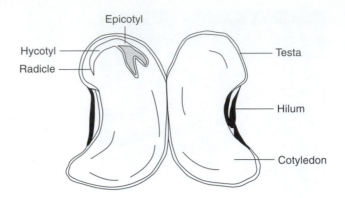

FIGURE 2–36 Structure of a dicot seed. Dicot seed can be split into two halves, exposing a fragile embryo.

Table 2-9 Parts and Functions of a Legume Seed

Raphe: A ridge on seeds formed by the stalk of the ovule (in seeds in which the funiculus is sharply bent at the base of the ovule).

Hilum: Present as a scar, it represents the point of attachment of the seed to the pod and through which nourishment was transferred to the seed during development. In species such as soybean, the hilum occurs in different colors that are used as a tool for seed identification purposes.

Micropyle: Opening representing the point of entry into the ovule for the purpose of fertilization.

Embryo area: A region (on the opposite side of the raphe when it occurs) where the embryonic axis is located.

Testa: The seed coat of a legume seed. It has protective functions.

Hypocotyl: The part of the stem tissue between the epicotyl and the radicle. The elongation of this structure causes the arching characteristic of epigeal seedling emergence.

Radicle: The part of the embryonic axis that becomes the primary root. It is the first part of the embryo to start growth during germination.

Epicotyl: The upper portion of the embryonic axis or seedling, above the cotyledons and below the first true leaves.

Cotyledons: A pair of seed leaves that contain food reserves for use by the embryonic axis and seedlings in early stages of growth.

2.14.2 MONOCOT CARYOPSIS

Monocot. A subclass of flowering plants with a single cotyledon.

Monocots have one cotyledon, also called the *scutellum* (Figure 2–37). The storage tissue is called the *endosperm*. The cereal grain or kernel, is called a *caryopsis*. The fruit cover, or *pericarp,* is not loose like the testa of a legume seed. It is fused to the aleurone. The parts of the cereal caryopsis and their functions are summarized in Table 2–10.

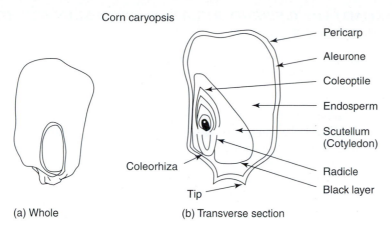

Corn caryopsis

Pericarp

Aleurone

Coleoptile

Endosperm

Scutellum
(Cotyledon)

Coleorhiza

Radicle

Black layer

Tip

(a) Whole   (b) Transverse section

FIGURE 2-37   Structure of a monocot caryopsis. Corn caryopsis and other monocot kernels cannot be split naturally into two-parts.

Table 2–10   Parts and Functions of a Cereal Caryopsis

Pericarp: The outer covering of the kernel. May contain red color in corn. Consists primarily of ovary tissue and hence its characteristics are subject to maternal inheritance.

Aluerone layer: Lines the inside of the pericarp and consists of several layers of the outer part of the endosperm occurring to the inside. Its characteristics are subject to biparental inheritance. Produces enzymes that break down the endosperm to release nutrients during seed germination.

Endosperm: Represents the bulk of the mature cereal kernel and is a source of nourishment during the germination and early seedling stages. May be starchy (soft) or flinty (hard) and variable in color (e.g., reddish, yellow, white). In wheat, rye, and triticale, the endosperm contains *gluten* (a proteinaceous substance that is responsible for the stickiness and elasticity of dough). Soft starch may shrink upon drying to cause a dent to form on top of the kernel in corn (hence *dent corn*).

Scutellum: This is the equivalent of the cotyledon. It contains enzymes that are released during the germination process to digest the endosperm to release energy for the young seedling.

Coleoptile: Occurs at the top of the embryonic axis as a protective sheath for the tender and emerging leaf.

Epicotyl: Or plumule. This is the portion of the embryonic axis above the cotyledon.

Apical meristem: The growing point of the embryo.

Scutellar node: The point of attachment to the cotyledon or scutellum to the embryonic axis.

Radicle: The part of the embryonic axis that becomes the primary root. It is the first part of the embryo to start growth during germination.

Coleorhiza: A protective sheath surrounding the radicle and through which the young developing root emerges.

Tip: Or pedicel. This is the point of attachment of the kernel to the flower stock.

Other parts found in certain cereal grains:

Brush: A patch of pubescence or hairs. Occurs at the upper tip of rice caryopsis, for example.

Black layer: A black layer formed near the tip of corn as an indicator of physiological maturity.

2.14.3 SUMMARY OF SEED-RELATED CROP MANAGEMENT PRACTICES

The management practices summarized here are discussed in detail in appropriate sections of the book. These represent decisions a producer must make in order to have success with crop establishment.

1. *Tillage and seedbed preparation.* Tillage precedes seeding as an operation designed to disturb the soil to provide a suitable environment for seed germination. A key consideration in this operation is seed size—the smaller the seed, the finer and firmer the seedbed needed to provide an effective seed-soil contact for imbibition of moisture for germination.
2. *Seeding method and equipment.* The method of seeding depends on seed size. Larger seeds are easier to plant at predetermined equidistant spacing using row planters (e.g., in cotton, corn, peanuts), whereas smaller seeds (e.g., wheat, rice, barley) are best closely spaced using grain drills or broadcast using broadcast seeders.
3. *Seeding depth.* The depth of planting is relative to seed size. Smaller seeds are shallowly planted, while larger seeds are planted deeper in the soil. Seeds differ in the mode of emergence. Those that are epigeal (cotyledons emerge above the soil) are best planted at a shallower depth than those with hypogeal emergence (cotyledons remain in the ground). Soil texture and soil moisture are important in determining seeding depth. Seeding is deeper in sandy soil, especially, under dry conditions. Seeding deeper allows the seed a better chance of obtaining moisture for initiate germination.
4. *Seed treatment.* When seeding into a soil with pest problems (e.g., nematodes, wireworm, and other fungal problems), the producer should consider seed dressing with appropriate pesticides. Some seed treatments are designed to protect the seed against herbicide injury. Seeding into soil that previously received a herbicide treatment, or to allow the seed to resist certain herbicides in the current season, growers should know about any susceptibility of the seed to the herbicide. For example, Lassso herbicide is injurious to sorghum seedling. However, treating the seed with Screen prior to seeding protects it from such injuries. These chemicals are called safeners or protectants. Inoculation of legume seed with *Rhizobium* bacteria is a common practice in the production of crops such as soybean. This seed treatment enhances biological nitrogen fixation for enhancement of crop performance. Sometimes, seeds need special treatment to break dormancy before germination can occur. Dormancy can be broken by physical or chemical methods.
5. *Date of seeding.* Seeds require appropriate temperature and moisture conditions for germination. Extreme cold or drought may kill the seed. Sometimes, a producer may delay seeding to allow volunteer plants and certain weeds to emerge and be controlled by tilling before seeding.
6. *Seeding rate.* Each crop has an ideal seeding rate to obtain the optimal plant density for optimal yield under a specific cultural practice. The planter used for seeding must be properly adjusted to seed at the desired rate.

SUMMARY

1. Diversity is a fact of nature.
2. The enormous natural biological variability needs to be classified to facilitate its use. There is consensus in classifying and naming plants.

3. Plant taxonomy is the science of classifying and naming plants based on information from various sources, including cytology, anatomy, and ultrastructural properties.

4. Taxonomy is a work in progress. Changes are made as new information becomes available.

5. Taxonomy follows certain universal rules so that a particular crop plant name means the same thing to all users worldwide. The system of classification is the binomial nomenclature, which assigns two names to a plant type: a genus name and a species name.

6. Plants may be operationally classified according to seasonal growth cycle as annual, biennial, or perennial. Plants may also be classified according to stem type, stem growth form, agronomic use, and adaptation, among others.

7. Important crop plant families include Poacea (grass family—e.g., corn, wheat, and sorghum), Fabaceae (legume family—e.g., soybean, peanut), and Solanaceae (e.g., tomato, potato).

8. Crop plants are identified by using morphological factors, including leaf shape, form, margin, arrangement, and attachment in broadleaf plants. In grasses, the leaf characteristics of importance include venation, sheath, ligule, collar, and auricles.

9. Flowers are important in the taxonomy of flowering plants because they are stable in expression over environments.

10. Seeds are important in identifying seed-forming plants. They vary in shape, color, size, and other features.

11. The cell is the fundamental unit of organization of organisms. Some organisms are unicellular; others are multicellular.

12. Plant cells have a cell wall around the plasma membrane. This wall may be lignified in certain situations.

13. The nucleus is the most prominent cell organelle. It contains the primary hereditary material (DNA). Other major organelles are the mitochondria (site of respiration), chloroplast (site of photosynthesis), and plastids.

14. There are two groups of tissues—simple tissues and complex tissues. There are three types of simple tissue—parenchyma (have thin walls), collenchyma (have medium wall thickness), and sclerenchyma (have thick walls). Thin-walled cells occur in fleshy and succulent parts, while thick-walled cells occur in strengthening tissue.

15. Complex tissues have a mixture of basic cell types. The major ones are the epidermis (occurs on the outside of plant parts and provides protection), secretory tissue (secretes various substances), and conducting tissue (or vascular system), consisting of the xylem (conducts raw materials up the plant for photosynthesis) and phloem (conducts photosynthates to plant parts).

16. There are different types of leaves; foliage leaves are the most prominent. They are the primary organs in which photosynthesis occurs. Monocot leaves have a parallel venation, while dicot leaves have a reticulate venation.

17. The stem is the central axis of the shoot of a plant. It provides mechanical support for plant parts and conducts nutrients and photosynthates. It may also be modified for food storage and used for propagation in certain species.

18. Roots are underground plant structures. They provide anchorage to the plant and absorb water and nutrients. There are two basic types of root systems—taproot (in dicots) and fibrous roots (in monocots).

19. Seeds are propagational units for flowering plants. Dicot seeds have two cotyledons and are characteristic of legumes. Monocot seeds (caryopsis) characteristic of grasses have one cotyledon.

20. The flower is the reproductive structure of flowering plants.

REFERENCES AND SUGGESTED READING

Acquaah, G. 2002. *Horticulture, principles and practices.* Upper Saddle River, NJ: Prentice Hall.

Benson, L. 1979. *Plant classification.* Washington, DC: Heath and Co.

Esau, K. 1977. *Anatomy of seed plants.* 2d ed. New York: Wiley.

Hayward, H. E. 1967. *The structure of economic plants.* New York: Lubrect & Crammer.

Kaufman, P. B., T. F. Carlson, P. Dayanandan, M. L. Evans, J. B. Fisher, C. Parks, and J. R. Wells. *Plants, their biology and importance.* New York: Harper & Row.

Mauseth, J. D. 1988. *Plant anatomy.* San Francisco, CA: Benjamin/Cummings.

Moore, R., and W. D. Clark. 1994. *Botany: Form and function.* Dubuque, IA: Wm. C. Brown.

SELECTED INTERNET SITES FOR FURTHER REVIEW

http://www.csdl.tamu.edu/FLORA/gallery.htm

Photo gallery of vascular plants, with good description.

http://www.csdl.tamu.edu/FLORA/Wilson/tfp/ham/history.htm

A great site on plant taxonomy.

http://www.cgiar.org/centers.htm

List, location, and functions of all international agricultural research centers.

OUTCOMES ASSESSMENT

PART A

Answer the following questions true or false.

1. T F   Linnaeus developed the binomial nomenclature for plants.
2. T F   In binomial nomenclature, the genus name must always start with an uppercase letter.
3. T F   The stamen is the male reproductive part of the flower.
4. T F   Flowers occurring in a cluster form an inflorescence.
5. T F   Corn has a spike inflorescence.
6. T F   In taxonomy, a name ending in *aceae* is likely to be genus name.
7. T F   Some perennials are cultivated as annuals.
8. T F   Angiosperms are non-flowering plants.
9. T F   All plant cells have a secondary cell wall.
10. T F   Chloroplasts have DNA.
11. T F   Photosynthesis occurs in the mitochondria.
12. T F   Meristems consist of parenchyma cells.
13. T F   Xylem vessels conduct raw materials for photosynthesis.
14. T F   Seminal roots are derived from seeds.

PART B

Answer the following questions.

1. The science of classifying and naming plants is _____.

2. The bionomial name of a plant consists of a two-part name: _____ and _____.

3. Classify plants according to seasonal growth cycle.

4. Give the four main parts of a typical flower.

_____

_____

_____

_____

5. The collectivity of petals is called the _____.

6. Give four important plant families, in terms of world food production.

_____

_____

_____

_____

7. Give an example of a type of indeterminate inflorescence.

_____

_____

8. Give two examples of annual crops.

_____

_____

9. List, giving two examples each, four operational classes of crops.

_____

_____

10. Leaf shape may be simple or _____.

11. Give the three basic types of plant cells.

_____

_____

_____

12. Give three functions of roots.

_____

_____

_____

13. Give three functions of leaves.

_____

_____

_____

14. Give three functions of stems.

_____

_____

_____

15. The two basic root systems are _____ and _____.

PART C

Write a brief essay on each of the following topics.

1. Distinguish between a botanical variety and a cultivar.
2. Classify plants on the basis of temperature adaptation.
3. Describe a typical grass flower.
4. Discuss the features used in identifying grasses.
5. Compare and contrast the dicot seed and monocot.
6. Compare and contrast the dicot stem and monocot stem.
7. Discuss the structure and function of plastids.
8. Discuss the structure and function of conducting tissues.

PART D

Discuss or explain the following topics in detail.

1. Develop your own operational system of classification of plants and justify its utility. (Do not use any example from the text.)
2. Why is it important for an agronomist to understand plant structure and function?
3. How will knowledge of plant structure and function help someone become a better crop producer?
4. How important are modified plant parts in U.S. agriculture?

3

Fundamental Plant Growth Processes

PURPOSE

The purpose of this chapter is to discuss the processes involved in the manufacture of food, the conversion of photosynthates into economic yield, the harvesting of energy for cellular activities, and the roles of these plant activities in growth and development and ultimately crop productivity.

EXPECTED OUTCOMES

After studying this chapter the student should be able to:

1. Discuss the plant as a metabolic machine.
2. Describe the steps involved in the conversion of light energy to chemical energy by photosynthesis.
3. Discuss the factors that affect photosynthesis.
4. Compare and contrast C_3 and C_4 photosynthetic pathways (or plants).
5. Discuss the role of photosynthesis in crop yield or productivity.
6. Discuss plant respiration for maintaining the plant and supporting its functioning.
7. Discuss plant respiration for impacting biomass synthesis.
8. Discuss the process of fermentation.
9. Discuss the process of transpiration.
10. Discuss the process of absorption.
11. Discuss how water moves in plants.
12. Discuss the process of translocation.
13. Discuss how photosynthesis is managed in crop production.

KEY TERMS

Action spectrum

Aerobic respiration

Anabolism

Anaerobic respiration

C_3 pathway

C_4 pathway

Catabolism

Cellular respiration

Crassulacean acid
  metabolism (CAM)

Fermentation

Glycolysis

Light compensation point

Maintenance respiration

Metabolism

Photophosphorylation

Photorespiration

Photosystem (PS)

TO THE STUDENT

Crop production entails the management of production resources such that the crop grows and develops to express its genetic potential to the highest profitable extent. The ultimate source of energy for life on earth is sunlight. However, humans and other animals cannot utilize radiant energy directly. Only organisms that contain the special pigment chlorophyll are capable of utilizing solar energy directly. They do so by the process of photosynthesis. This makes green plants the ultimate source of all food on earth. This is why plants are called primary producers in the ecosystem. It is therefore said that crop production is about the management of photosynthesis for productivity. Photosynthates are metabolized to release energy and other materials for growth and development as well as for maintenance of plant structure. Respiration, in effect, is the reverse of photosynthesis. You are about to study some of the most important biochemical reactions in the world. As you proceed, pay attention to the factors that affect photosynthesis, for the management of these factors is what crop production is all about. Also pay attention to, and compare and contrast, the inputs and outputs of photosynthesis and respiration. Further, take note of crop production activities and conditions that impact respiration, as well as how to manage them for high productivity.

3.1: THE PLANT IS A METABOLIC MACHINE

Anabolism. The synthesizing reactions of metabolism that occur in a living cell.

Catabolism. The destructive reactions of metabolism that occur in a living cell.

Metabolism. The overall physiological activities (all chemical reactions) occurring in a cell.

Plant productivity depends on the production and use of assimilates. The production aspect is made possible through the chemical reactions of the plant process called *photosynthesis*. Photosynthesis consists of processes by which assimilates (carbohydrates and other substances) are synthesized or manufactured. The term **anabolism** is used to describe the synthesizing reactions that occur in the plant. The energy locked in anabolic products is made available through *respiration,* a destructive process that breaks down photosynthates into simpler compounds to release stored energy in the chemical bonds. The term **catabolism** is used to describe the destructive reactions in the plant's physiology. The two sets of reactions, anabolism and catabolism, are collectively called **metabolism**. The plant is thus a metabolic machine. Respiration produces energy and reducing agents used by plants for constructing new materials and for maintaining existing structure that result in growth and development.

3.2: PHOTOSYNTHESIS: HARNESSING LIGHT ENERGY TO MAKE FOOD

3.2.1 LIFE ON EARTH IS MOSTLY SOLAR-POWERED

Is life possible on earth without the sun? Except for a few bacteria that derive their energy from sulfur and other inorganic compounds, life at any level depends on photosynthesis, the process by which plants use light energy to synthesize food molecules from carbon dioxide and water. The sugars produced may be further converted to structural materials and other cell constituents. Over 90% of dry matter of crop plants depends on photosynthetic activity. Crop yield depends on the rate of duration of photosynthesis. In the ecosystem, plants are called *primary producers* because they are the ultimate source of food for all life on earth.

3.2.2 THE SITE OF PHOTOSYNTHESIS IS THE CHLOROPLAST, WITH CHLOROPHYLL AS CHIEF PIGMENT

Why are leaves characteristically green? Can a plant manufacture food in any other part apart from the leaf? The leaf is the principal plant organ of eukaryotes in which photosynthesis occurs. The assimilation process occurs in the chloroplasts, which are found mainly in leaves. These are plastids that contain the green pigment *chlorophyll*. There are different types of chlorophyll. *Chlorophyll a* occurs in all photosynthesizing eukaryotes. However, trachaeophytes (vascular plants) and bryophytes also require *chlorophyll b* for photosynthesis. Apart from chlorophyll, photosynthesis involves two other pigments— *carotenoids* and *phycobilins,* the latter being water-soluble accessory pigments not found in higher plants.

3.2.3 PHOTOSYNTHESIS OCCURS IN TWO PHASES

What materials are required by the plant for photosynthesis? How do these materials combine to produce food products usable by plants? In what way do humans involuntarily contribute to photosynthesis? The general reaction for photosynthesis is

$$6CO_2 + 12H_2O \rightarrow C_6H_{12}O_6 + 6O_2 + 6H_2O$$

The carbon dioxide is obtained from the air, while plants obtain water from the soil. The ultimate result of photosynthesis is the reduction of carbon dioxide to carbohydrate. In this regard, photosynthesis can be discussed with reference to the fate of carbon dioxide in three critical steps:

1. Entry of carbon dioxide into plant organs (especially leaves) by the process of diffusion. This is a physical activity.
2. Harvesting light by photochemical processes.
3. Reduction of carbon dioxide (assimilation) by biochemical processes.

All organisms contain a group of compounds called *tetrapyrroles*. Certain pigments are derived from tetrapyrroles. These are large rings made of four smaller rings (each a pyrrole ring and consisting of four carbons and one nitrogen), and carbon bridges that sequester a metal atom then link these four rings. This metal may be iron (Fe), in which case the pigment is a heme (as in hemoglobin in blood). The pigment of many feathers of birds consists of copper as the metal, while plant chlorophyll has a magnesium atom.

Chlorophyll is the primary pigment of photosynthesis. There are several types, the most important and the primary photosynthetic pigment being chlorophyll *a*. Two German chemists, Richard Willstatter and Hans Fisher, determined the molecular structure of chlorophyll *a* ($C_{55}H_{72}O_5N_4Mg$) in 1940, for which they were rewarded with a Nobel Prize. Similarly, Robert Woodward was awarded the Nobel Prize in 1960 for synthesizing the chlorophyll *a* molecule in vitro. Whereas the pigment color of hemoglobin is not significant in the molecule's ability to function in oxygen transport, chlorophyll depends on its color to be effective. It absorbs maximally at wavelengths of 400 to 500 nm (violet-blue) and 600 to 700 nm (orange-red) of the electromagnetic spectrum (the range of wavelengths emanating from the sun).

Other useful photosynthetic pigments (called *accessory pigments*) include chlorophyll *b* (bluish-green pigment) and *carotenoids* (e.g., beta-carotene), which produce the colors of fruits (e.g., tomato, carrot, and banana) and leaves (fall colors).

The structure of a cholorphyll molecule. The difference between chlorophyll *a* and chlorophyll *b* lies in the *R* group, being—CH_3 in chlorophyll *a* and—CHO in chlorophyll *b*.

Further, the processes in these steps may be classified into two categories or phases, one set involving light *(light-dependent reactions)* and the other not needing it *(light-independent reactions)*.

3.2.4 CARBON DIOXIDE ENTERS THE PLANT BY DIFFUSION

Diffusion is the movement of molecules (or substances) from a region of higher concentration of those molecules to a region of lower concentration. The entry point is through the stoma. Diffusion occurs because of a carbon dioxide gradient created by a lower concentration of carbon dioxide inside the leaf (because of carbon dioxide uptake by tissues) than in the air on its surface. The diffusion occurs across a series of barriers formed by tissues, by cell walls, and eventually through the protoplasm of the mesophyll cells to the chloroplasts. Carbon dioxide availability for fixation is affected by factors such as these barriers. For example, the opening and closing of these pores controls the amount of carbon dioxide entering the leaf through the stomata. The opening of the aperture of the pores is controlled by water in the guard cells. Dry atmospheric conditions will cause stomata to close. This is a moisture-loss-prevention mechanism. Plants also differ in the structure and organization of leaf internal cells. The passage of carbon dioxide varies from one genotype to another and is further influenced by the internal cellular environment (e.g., the turgor pressure of mesophyll cells).

3.2.5 LIGHT REACTIONS DEPEND ON VISIBLE LIGHT

Harvesting light is a photochemical process. The ultimate source of energy in the ecosystem is sunlight. Visible light constitutes only a small proportion of the electromagnetic spectrum. The plant pigments associated with photosynthesis absorb light energy. Each of these pigments has a specific **action spectrum** (the relative effectiveness of different wavelengths of light for a specific light-dependent process such as photosynthesis or flowering). The various pigments absorb a range of wavelengths of light *(absorption spectrum)* (Figure 3–1). Can plants utilize light from other sources apart from the sun?

The light-absorbing pigments are packed in the thylakoids in discrete units called **photosystems (PS)**. There are two photosystems, *photosystem I (PS I)* and *photosystem II (PS II)*. In a photosystem, only a pair of chlorophyll molecules (called the *reaction center chlorophyll*), out of all the pigments present, is capable of utilizing the light energy absorbed. The other pigments form the antenna pigments. The reaction center chlorophyll in PS I is a special molecule of chlorophyll *a* called P_{700} because it has peak

Action spectrum. A graphical representation of a physiological reaction in which physiological activity is plotted against light wavelength.

Photosystem (PS). An energy-collecting and energy-transferring system that operates in a chloroplast (two interacting photosystems are present in a cell).

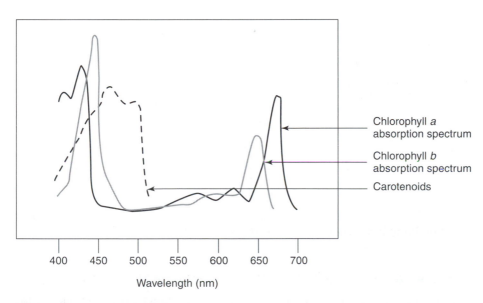

Chlorophyll *a* absorption spectrum

Chlorophyll *b* absorption spectrum

Carotenoids

Wavelength (nm)

FIGURE 3–1 Absorption spectra of various pigments associated with photosynthesis. Chlorophyll *a* absorbs maximally at 430 nm and 662 nm and is the primary photosynthetic pigment. Together, chlorophylls *a* and *b* absorb maximally at wavelengths of 400 to 500 nm (violet-blue) and 600 to 700 nm (orange-red). Carotenoids are accessory pigments for photosynthesis. They absorb maximally at wavelengths between 460 and 550.

absorption at 700 nanometers. Photosystem II, also called P_{680} (for similar reasons) is more efficient in light harvesting than PS I. The high energy absorbed is transferred in a relay fashion from one photosynthetic pigment to another until it reaches the reaction center molecule (Figure 3–2). The light energy excites electrons of the photosystem. The energy differential as the electrons return to a lower energy level is utilized in forming chemical bonds. The process of accomplishing this is called **photophosphorylation**, the reaction by which ATP (adenosine triphosphate) is formed from ADP (adenosine diphosphate) and phosphorus. Electrons are excited simultaneously from P_{700} and P_{680}. As the P_{700} electrons return to lower energy via a photon gradient, they reduce coenzyme $NADP^+$ (nicotinamide adenine dinucleotide phosphate) to NADPH (nicotinamide adenine dinucleotide phosphate hydrogenase).

Photophosphorylation. The synthesis of ATP from ADP and inorganic phosphate during the light phase of photosynthesis.

The light reaction can be summarized by the following chemical equation:

$$\text{Light quanta} + 2H_2O + 2NADP^+ + 3ADP + 3Pi \rightarrow$$
$$2NADPH + 2H + 3ATP + 3H_2O + O_2$$

The ATP is the usable energy. Along with the reducing potential (NADPH), these two products enter the next phase of the photosynthetic cycle—light-independent reaction or dark reaction.

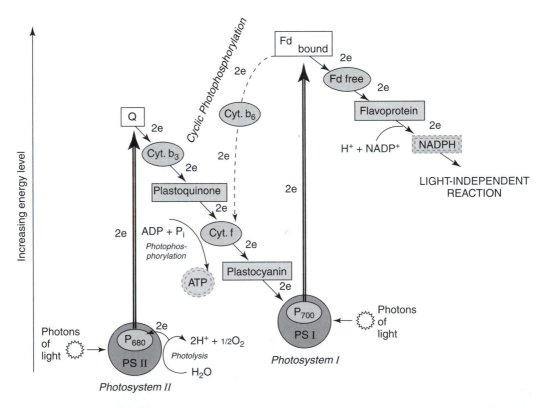

FIGURE 3–2 A summary of the light reaction of photosynthesis showing the role of the two photosystems (PS I and PS II) in photosynthetic energy transport, photophosphorylation, and formation of NADPH. The components are organized within the thylakoid membrane. PS I consists of chlorophyll *a*, chlorophyll *b*, and carotenoids. PS I is excited by radiant energy at wavelengths of 680 and longer.

3.2.6 SUGAR SYNTHESIS USES ENERGY (ATP) AND REDUCING POWER (NADPH) FROM THE LIGHT PHASE

The final step in photosynthesis involves biochemical processes that result in the *fixation* or *assimilation of carbon dioxide,* using the energy (ATP) from the light-dependent reaction (photochemical) and the reducing agent (NADPH). These processes occur by one of three pathways, depending on the plant species. Based upon the first stage product of the pathway, the three processes are called C_3 pathway, C_4 pathway, and Crassulacean acid metabolism (CAM).

C_3 Pathway

The first stable product of the **C_3 pathway** is a three-carbon compound called *phosphoglycerate* (Figure 3–3). The C_3 pathway is called the *Calvin cycle,* after its discoverer, and plants using this pathway for carbon dioxide assimilation are called *C_3 plants.* C_3 plants include both monocots (e.g., barley, wheat, rice, and oat) and dicots (e.g., soybean, pea, peanut, and sunflower). Tubers such as potato also utilize this route for carbon dioxide assimilation. The assimilation process is catalyzed by an enzyme called *ribulose 1,5-bisphosphate carboxylase oxygenase* (or simply *rubisco,* or *RuBP*). The overall reaction is represented by the following chemical equation:

C_3 pathway. The Calvin cycle of photosynthesis, in which the first products after carbon dioxide fixation are three-carbon molecules.

$$6CO_2 + 12NADPH + 12H^+ + 18ATP \rightarrow$$
$$1 \text{ glucose} + 12NADP^+ + 18ADP + 18 \text{ Pi} + 6H_2O$$

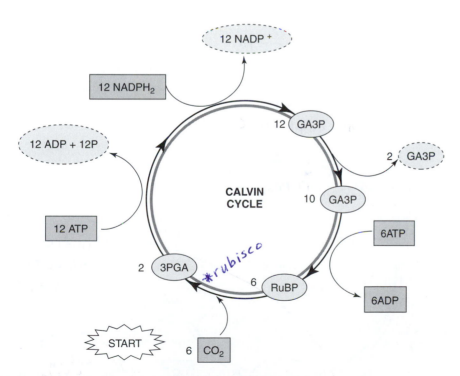

FIGURE 3–3   The Calvin cycle of the C_3 pathway of carbon dioxide fixation. The biochemical reactions involved occur in the stroma of the chloroplast. The products of the reactions in the thylakoids (ATP and reduced $NADP^+$) are used to reduce carbon dioxide to carbohydrates.

C₄ Pathway

C₄ pathway. The Hatch-Slack cycle of photosynthesis, in which the first products after carbon dioxide fixation are four-carbon molecules.

The first stable product resulting from the **C₄ pathway** is a four-carbon molecule called *oxaloacetate* (Figure 3–4). *C₄ plants* include corn, sorghum, sugarcane, and millet. These plants produce two types of chloroplasts. One type occurs in the parenchymatous tissue around the bundle sheath. They are large in size. Small chloroplasts occur in the mesophyll cells where phosphoenol pyruvate (PEP) carboxylase catalyzes the assimilation of carbon dioxide into PEP, which then is transported to the chloroplasts in the bundle sheaths. Malate is converted to pyruvate, releasing carbon dioxide in the process. The carbon dioxide enters the dark reaction cycle.

Is more light always better for photosynthesis? C_3 plants experience a loss in energy through a process called **photorespiration**. It is the light-dependent production of glycolic acid in chloroplasts and its subsequent oxidation in peroxisomes. Unlike respiration, photorespiration produces no ATP. This reduces the overall efficiency of this pathway, making it less efficient than the C_4 pathway. This energy loss occurs under hot, sunny skies. The reduction in energy is increased under conditions of low carbon dioxide, which occurs when stomata close under intense light.

Photorespiration. The oxidation of glycolic acid produced during photosynthesis in C_3 plants.

Crassulacean Acid Metabolism (CAM)

Crassulacean acid metabolism (CAM). The metabolic pathway in plants of the Crassulaceae (Stonecrop) family and certain other families that allows them to fix large amounts of carbon dioxide in the form of organic acids during the night and release carbon dioxide during the day.

A third pathway in carbon dioxide assimilation, **Crassulacean acid metabolism (CAM)**, operates by fixing carbon dioxide in the dark. It depends on carbon dioxide that accumulates in the leaf in the night. This pathway is utilized by many houseplants and field crops in the Agave family such as sisal *(Agave sisalina)*, century plant *(Agave americana)*, and henequen *(A. fourcroydes),* as well as bromeliads such as the pineapple *(Ananas comosus).*

3.2.7 CARBON DIOXIDE ASSIMILATION IS AFFECTED BY ENVIRONMENTAL FACTORS

Can producers influence the rate of photosynthesis? The rate of photosynthesis in higher plants is affected by a number of environmental factors—irradiance, carbon dioxide concentration, temperature, water, and length of day.

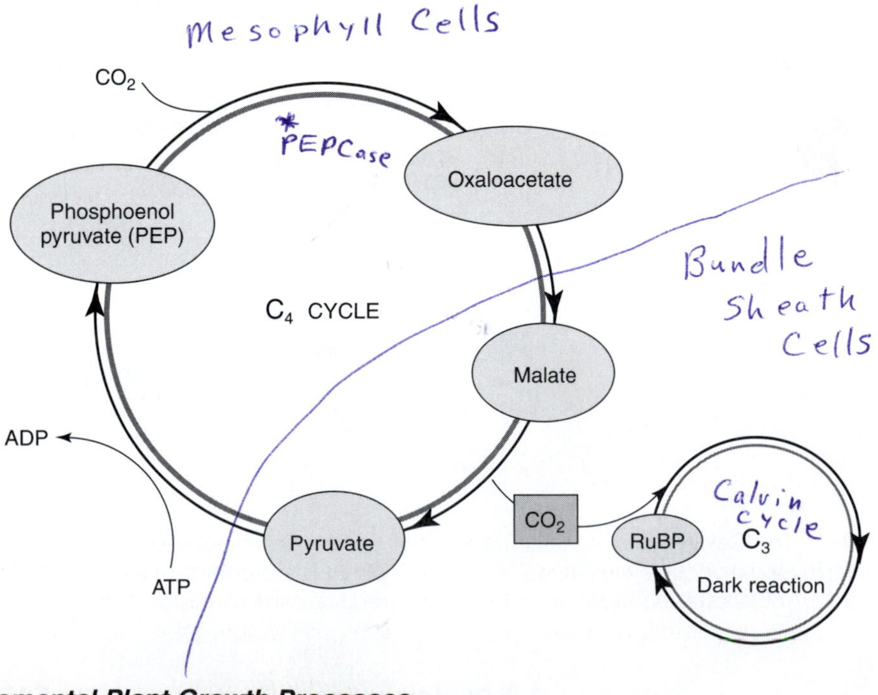

FIGURE 3–4 The C_4 pathway of carbon dioxide fixation. Mesophyll cells fix carbon dioxide into four-carbon acids that move into bundle-sheath cells, where they are carboxylated.

C_4 plants like it bright and hot

It is often said that C_4 plants (e.g., sugarcane, sorghum, and corn) are more efficient than C_3 plants (e.g., soybean, wheat, and peanut). However, this generalization should be made with caution, since it is true only when the conditions are bright and hot. These conditions are naturally available in the tropics and subtropics. When they occur, the stomata of plants close to conserve moisture and in the process reduce the amount of carbon dioxide entering the leaf. These conditions promote the phenomenon of photorespiration, which leads to the undoing of the results of photosynthesis. That is, about 50% of the carbon dioxide fixed in photosynthesis is reoxidized and lost as carbon dioxide, since the photosynthetic enzyme rubisco (capable of both oxidation and reduction) utilizes the oxygen that occurs in high concentration internally in preference to the lower concentration of carbon dioxide. C_4 plants have a mechanism that enables carbon dioxide to be pumped into bundle sheath cells to keep the internal concentration of carbon dioxide at about 20 to 120 times the normal level. This allows the enzyme rubisco to continue to fix carbon dioxide (instead of oxygen), even until the carbon dioxide level reaches zero. It is virtually impossible to light saturate C_4 plants. Consequently, they continue to photosynthesize even in full sunlight. Biochemically, however, C_4 plants are less efficient than C_3 plants because they use two extra ATPs per carbon dioxide for photosynthesis:

C_3 cycle: $6CO_2 + 30ATP + 12NADPH + 12H_2O \rightarrow C_6H_{12}O_6 + 30ADP^+ + 30Pi + 12NADP^+ + 6H_2O + 6O_2$

C_4 cycle: $6CO_2 + 18ATP + 12NADPH + 12H_2O \rightarrow C_6H_{12}O_6 + 18ADP^+ + 18Pi + 12NADP^+ + 6H_2O + 6O_2$

However, when you consider that under hot and bright conditions C_3 plants would photorespire to reduce photosynthetic products, then the additional expense of the C_4 cycle is advantageous in the long run. On the other hand, when conditions change to wet and cool (less than 25°C) or under poorly lit conditions, the additional expense becomes a factor and C_3 plants become more competitive.

Irradiance

Increasing intensity of light increases the rate of photosynthesis to a point. After the system is saturated, increasing irradiance does not cause photosynthesis to increase (Figure 3–5). At high light intensity, carbon dioxide and other factors may become limiting. However, at low light intensity, the rate of carbon dioxide fixation is balanced by the rate of carbon dioxide released through respiration leading to no net fixation. The point of zero net carbon dioxide assimilation rate that starts to level with increasing irradiance is called the **light compensation point**. C_4 plants have a lower rate of carbon dioxide assimilation at low light intensity than C_3 plants. As irradiance increases, the reverse is true; C_4 plants have a higher rate of carbon dioxide assimilation than C_3. This difference is primarily attributable to the effect of photorespiration that occurs in the C_3 pathway.

The intensity of light also affects leaf morphology. When plants grow in low light, their leaves grow longer. Similarly, leaves of plants grown in high light intensity are thicker and broader than when grown in low light. What crop would be better off grown under high light intensity? What about under low light intensity?

Light compensation point. The light intensity at which energy stored by photosynthesis equals the energy released by respiration at a given carbon dioxide concentration in the environment.

Carbon Dioxide

As light intensity increases, increasing carbon dioxide concentration results in an increased rate of carbon dioxide assimilation (Figure 3–6). C_4 plants have a lower carbon dioxide compensation point because they are more efficient in trapping carbon dioxide than C_3 plants. Increasing the carbon dioxide content of the atmosphere (*carbon dioxide*

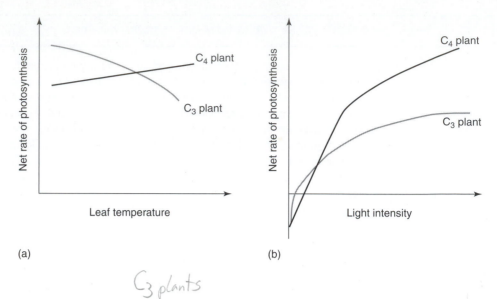

FIGURE 3–5 The relative rates of photosynthesis in C_3 and C_4 plants as influenced by (a) temperature and (b) light intensity. C_4 plants are photosynthetically more efficient than C_3 plants under conditions of hot, dry, and bright sunlight. Under such conditions, C_4 plants do not photorespire like C_3 plants.

FIGURE 3–6 The effect of carbon dioxide concentration on photosynthesis. Increasing the carbon dioxide concentration in the air to 0.10% is known to double the rate of photosynthesis in C_3 plants by reducing photorespiration. The benefits of increased concentrations of carbon dioxide are limited because stomata close and photosynthesis stops at carbon dioxide concentrations higher than 0.15%.

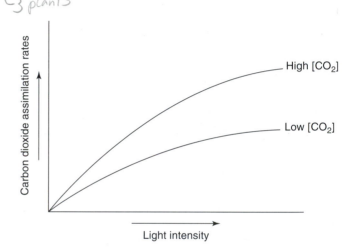

fertilization) is a strategy in greenhouse production for increasing crop productivity. The enclosed atmosphere is enriched to increase the carbon dioxide level from 0.035 to 0.10%. This doubles the rate of carbon dioxide assimilation.

C_3 and C_4 plants differ significantly in the way increasing carbon dioxide concentration increases net carbon dioxide assimilation. The increase is more gradual for C_3 plants than for C_4 plants (Figure 3–7). This difference is explained by the fact that rubisco, the primary carbon dioxide acceptor system in C_3 plants, has a lower affinity for carbon dioxide than PEP carboxylase in C_4 plants. This lower affinity makes rubisco require a higher carbon dioxide concentration to become saturated. The other reason is that the two plant types have different carbon dioxide compensation points, the point being 5 to 10 times greater in C_3 than C_4 plants.

Temperature

How does temperature affect photosynthesis? Carbon dioxide assimilation is a biochemical process, involving enzymes. Enzymes are temperature-sensitive. The photosynthetic rate decreases as temperature decreases. The effect of temperature on photosynthesis is affected by light intensity. Generally, in temperate or C_3 plant species,

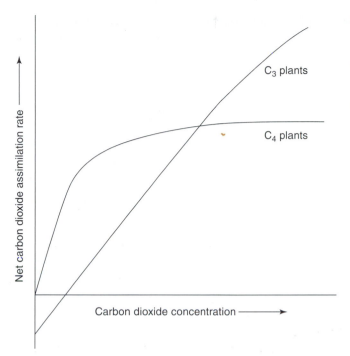

FIGURE 3–7 The relative rates of photosynthesis in C_3 and C_4 plants as influenced by carbon dioxide concentration. The carbon dioxide compensation point (the concentration of CO_2 at which plants show no net fixation of carbon dioxide) is high for C_3 plants: 50 ppm at 25°C and 21% oxygen. C_4 plants have a low CO_2 compensation point that occurs at 0 to 5 ppm.

Tinkering with rubisco?

Is rubisco a candidate for genetic engineering? The enzyme rubisco (ribulose-1,5-bisphosphate carboxylase-oxygenase) has the dubious title of being the world's most abundant protein as well as, perhaps, the world's most incompetent enzyme. Rubisco initiates the chain of biochemical reactions that creates carbohydrates, protein, and fats that feed and sustain the world. However, it does so very inefficiently. Chemically, this enzyme is a giant molecule (16 parts) that is encoded by many genes in both the nucleus and chloroplast. Rubisco captures carbon dioxide and helps turn it into starches, sugars, and other compounds. Photosynthesis is itself not an efficient plant process. General enzymatic rates are of the order of 25,000 reactions per second. It has been suggested that rubisco generates a dismal rate of fewer than 5 reactions per second.

A major negative view of rubisco comes from the fact that it is both a carboxylase and an oxygenase, able to fix carbon dioxide or oxygen. In other words, it can undo what it does through carbon dioxide fixation in photosynthesis by fixing oxygen instead of carbon dioxide under certain conditions. The process of this anti-photosynthesis is called photorespiration. In crop species such as rice and wheat, rubisco is 100 times more likely to take up carbon dioxide than oxygen. Unfortunately, given the high concentration of oxygen in the atmosphere, the natural affinity for carbon dioxide is diminished, resulting in about 20 to 50% of the carbon dioxide fixed by photosynthesis being lost through photorespiration.

Scientists are working toward the possibility of using genetic engineering techniques to enhance the efficiency of rubisco. If successful, this would help plants to increase their biomass. Further, the large amount of nitrogen that is currently needed to produce the enzyme (a protein) would be reduced, thereby reducing the need for nitrogen fertilization of crops in production. In 1992, scientists at Ohio State University, F. R. Tabita and B. R. Read, discovered that some diatoms and red algae have more specific rubisco than higher plants. In 1997, A. Yokota of Japan discovered red algae with rubisco that is 2.5 to 3 times more efficient than that of higher plants. Scientists hope to transfer this more efficient enzyme from algae to higher plants. Another strategy being considered for making photosynthesis more efficient is to introduce the C_4 photosynthesis cycle that is more efficient under high light into C_3 plants.

the rate of carbon dioxide assimilation doubles (approximately) for each 10°C (18°F) increase in temperature, provided light is not limiting. Further, tropical plants generally need a higher temperature for maximum photosynthesis than temperate plants.

Water

The effect of water is imposed through its role in the regulation of stomatal aperture. Under conditions of moisture stress (water deficit), caused by low soil moisture or drying winds, the rate of transpiration loss exceeds root absorption of water. Stomatal aperture is drastically reduced, accompanied by a reduction in enzyme activity. Increased stomatal resistance reduces carbon dioxide and oxygen exchange and consequently reduces carbon dioxide assimilation.

Length of Day

All factors being equal and within certain limits, plants benefit from a longer photoperiod, or daylength, by increasing photosynthesis for faster growth. Greenhouse operators who provide supplemental high intensity of light through artificial sources adopt this strategy.

3.2.8 PHOTOSYNTHETIC RATE IS AFFECTED BY GROWTH AND DEVELOPMENT NEEDS OF PLANTS

Photosynthesis is also regulated by sink activity of the plant. Its rate is lower when leaves are younger and still expanding, and it declines with senescence, eventually ceasing. The sites of rapid growth and metabolism in the plant need and use photosynthates. Certain parts of the plant also accumulate and store photosynthates. These parts of the plant are called *sink tissues* (or simply *sinks*). The fully expanded leaf is the part of the plant where photosynthesis occurs to produce photosynthates and is called the *source leaf* (or simply *source*). The needs of sink tissues place a demand (called *sink tissue demand* or simply *sink demand*) on photosynthetic activity. This demand influences the photosynthetic activities of the source leaf.

3.3: CELLULAR RESPIRATION: HARVESTING CHEMICAL ENERGY FROM FOOD MOLECULES

Cellular respiration. The sequence of enzyme-catalyzed reactions that result in the oxidation of organic substrates (usually carbohydrates) to carbon dioxide and water and production of energy in the form of ATP.

Aerobic respiration. Cellular respiration that occurs in the presence of oxygen.

Anaerobic respiration. Cellular respiration that occurs under oxygen-free conditions.

Do all cells in the plant respire? **Cellular respiration** is the process by which active cells obtain energy from food. Not all cells photosynthesize, but all active cells undergo respiration. Respiration occurs continuously, 24 hours a day, in active cells, regardless of whether photosynthesis occurs simultaneously in the same cell. The energy harvested is stored in the chemical bonds of energy molecules called ATP. Cellular respiration, like photosynthesis, is a reduction-oxidation (redox) reaction.

Photosynthesis occurs in chloroplasts; respiration occurs in mitochondria. Respiration may be described as the reverse of photosynthesis. It usually requires oxygen to occur and thus is called **aerobic respiration**. However, under certain conditions, respiration occurs under an oxygen-deficient environment and is called **anaerobic respiration**. In crop production, the producer should strive to avoid conditions that cause plants to respire anaerobically. However, silage-making depends upon anaerobic respiration. This is discussed further in Chapter 17.

3.3.1 NOT ALL HARVESTED ENERGY DIRECTLY IMPACTS BIOMASS

How does the plant use energy obtained from respiration? Plants use energy obtained from respiration in two general ways: maintenance and growth.

Maintenance Respiration

Some of the energy from respiration is used for maintaining the plant and supporting its functioning. This housekeeping activity, called **maintenance respiration**, may not directly impact plant biomass. The maintenance includes replacement or repair of the plant chemical constituents. Some energy is utilized in maintaining the photosynthetic process. Maintenance respiration may be summarized as follows:

Maintenance respiration. Cellular respiration that does not directly impact biomass production.

Plant maintenance energy required = energy conserved to replace/repair plant constituents + energy consumed to maintain concentration gradients + energy lost through spontaneous chemical hydrolysis

Energy is required in the uptake of nutrients and maintenance of osmotic pressure in cells. The amount of maintenance energy required depends on the amount of biomass and the composition and activities related to biomass. Since these maintenance activities represent energy drain that reduces net biomass, it is important that maintenance energy be minimized in agricultural production.

The amount of energy and materials needed for maintenance is affected by factors including plant size, age, nitrogen content of existing structures, and climate of the production region. Molecular activity increases as temperature increases. As plants grow older, they increase in size and accumulate large amounts of tissues and materials associated with aging or maturity, such as lignin, cellulose, and starch. These materials generally require minimal maintenance. However, older plants have a larger proportion of non-photosynthetic tissues that require maintenance. This causes a larger proportion of new assimilates to go into maintenance. This is the case when crops in production attain full canopy cover. Plants at this stage photosynthesize at a constant rate while biomass and maintenance respiration continue to increase.

The rate of maintenance respiration increases linearly with the nitrogen content of the plant. Proteins and lipids hydrolyze (break down) slowly and, similarly, their resynthesis at the same site consumes energy (ATP). A net change in protein or lipid content occurs. Nitrogen-rich plants generally need larger portions of new assimilates than plants growing in cool environments.

Growth Respiration

For crop productivity, the respiration that directly impacts biomass synthesis, sometimes called *growth respiration,* is of primary importance. Catabolism provides the energy required for the biosynthesis and the constructive reactions leading to formation of cellular products used in growth and crop yield. As indicated previously, the plant product of interest may be the gross plant product (i.e., biomass) or a specific product (e.g., oil or protein content). The amount of assimilates required for biosynthesis and constructive activities depends on the composition of the economic product desired.

3.3.2: AEROBIC RESPIRATION OCCURS IN THREE STAGES

Just how do cells extract the energy locked up in organic molecules? The general reaction involved in aerobic respiration may be summarized by the equation

$$C_6H_{12}O_6 + 6O_2 \rightarrow 6CO_2 + 6H_2O + \text{energy}$$

Glycolysis. The first phase of cellular respiration, which involves the enzymatic breakdown of substrates (glucose or other carbohydrates) to pyruvic acid with the trapping of some of the energy released in the form of NADH and ATP.

There are three stages in aerobic respiration, *glycolysis, Krebs (tricarboxylic acid, or TCA) cycle,* and *electron transport chain* (Figure 3–8).

Glycolysis

Glycolysis, or "sugar splitting," involves breaking glucose down into *pyruvic acid* (Figure 3–9). It is a 10-step process that occurs in the cytosol. This stage in aerobic respiration may be summarized by the equation

$$\text{Glucose} + 2\text{NADH}^+ + 2\text{ADP} + 2\text{Pi} \rightarrow 2 \text{ pyruvic acid} + 2\text{NADH} + 2\text{H}^+ + 2\text{ATP} + 2\text{H}_2\text{O}$$

FIGURE 3–8 The three primary stages of aerobic respiration. Respiration consists of three key sets of reactions, called glycolysis, tricarboxylic acid (TCA) cycle, and electron transport chain. They occur in a sequence, the products of one feeding the next. Glycolysis occurs in the cytosol of the cell, while the next two groups of reaction occur in the mitochondrion.

FIGURE 3–9 A summary of glycolysis. Glycolysis is a 10-step process that occurs in the cytosol and involves the splitting of glucose into two three-carbon compounds. Chemical bond energy from a substrate is used to bond phosphate to ADP to make ATP (called substrate level phosphorylation).

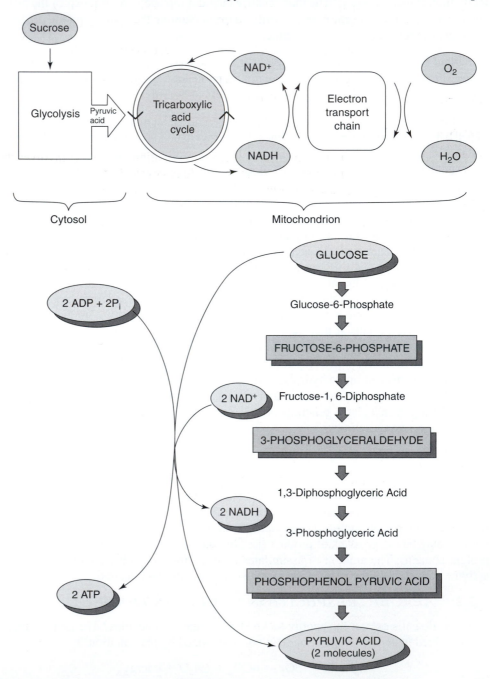

Most steps in the glycolytic pathway are reversible (e.g., glucose can be synthesized from fructose-6-phosphate that has been dephosphorylated). The first half of glycolysis uses ATP. ATP and NADH are made in the second half of glycolysis. Most of the energy of glucose remains locked up in pyruvic acid until the next stage.

Krebs Cycle

The pyruvic acid enters the *Krebs cycle* (or *TCA cycle*) in the mitochondrion (Figure 3–10). The summary of the reactions (called *oxidative decarboxylation*) is as follows:

$$\text{Oxaloacetic acid} + \text{acetyl CoA} + \text{ADP} + \text{Pi} + 3\text{NAD}^+ + \text{FAD} \rightarrow$$
$$\text{oxaloacetic acid} + 2\text{CO}_2 + \text{CoA} + \text{ATP} + 3\text{NADH} + 3\text{H}^+ + \text{FADH}_2$$

where FAD = flavin adenine dinucleotide, and CoA = acetyl coenzyme A.

In both glycolysis and the Krebs cycle, chemical energy from different substrates is used to bond phosphate to ADP to make ATP by the process called *substrate-level phosphorylation,* which does not involve a membrane.

Electron Transport Chain

The *electron transport chain* is the final step in the respiration process. It involves the transfer of electrons generated in glycolysis and the Krebs cycle downhill in an electron gradient. The energy released is used to form ATP from ADP in a process called *oxidative phosphorylation* (Figure 3–11). In oxidative phosphorylation, a series of oxidation-reduction reactions of the electron transport chain makes the cell use the energy in NADPH and ubiquinol to phosphorylate ADP to ATP. Oxygen is required only at the end of the electron transport chain as an electron acceptor. It is the strongest acceptor in the chain. Lack of oxygen inhibits both the electron transport chain and the TCA cycle in glycolysis.

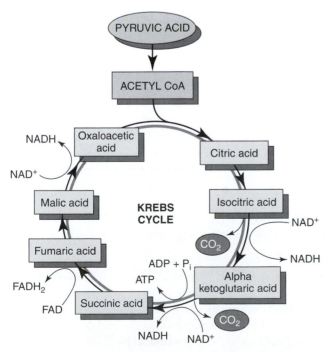

FIGURE 3–10 A summary of the tricarboxylic acid cycle, or Krebs cycle. The cycle always begins with acetyl CoA, which is the only real substrate. Like glycolysis, the Krebs cycle is involved in substrate-level phosphorylation. Seven of its eight steps occur in the mitochondrial matrix. Step six, involving ubiquinol and the oxidation of succinic acid to fumaric acid, occurs in the mitochondrial membrane.

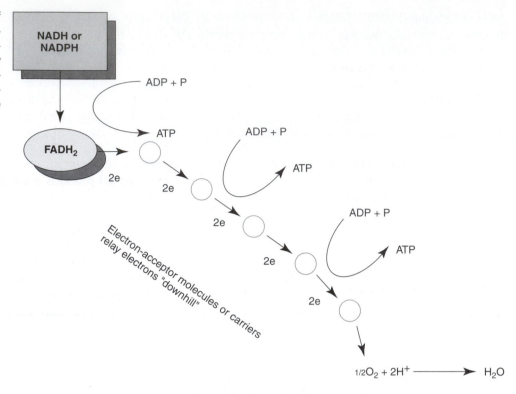

FIGURE 3–11 A summary of oxidative photophosphorylation. The electron transport chain involves the flow of electrons in an energetically downhill fashion, resulting in energy release for the formation of ATP.

Table 3–1 Energy Produced from One Cycle of Aerobic Respiration

Metabolic Reaction	Coenzyme Type Produced	ATP Yield per Coenzyme	Total ATP Yield
Glycolysis	(Direct)	2	2
	2 NADH	2	4
Oxidative decarboxylation of pyruvate to acetyl CoA	2 NADH	3	6
Krebs cycle	(from GTP)	2	2
	6 NADH	3	18
	2 FADH$_2$	2	4
Total yield of ATP			36

3.3.3 CELLS CAN EXTRACT NEARLY HALF OF THE POTENTIAL ENERGY OF ORGANIC MOLECULES USED AS FUEL

How efficient is respiration in extracting energy from organic molecules? Provided oxygen is abundant, cells can extract about 40% of the potential energy of organic molecules used as fuel in respiration. Different kinds of organic molecules can be used as a source of energy by cells (e.g., fats, proteins, and polysaccharides). At the end of one cycle of aerobic respiration, one molecule of glucose yields a net of 36 ATPs (Table 3–1).

Respiration in most organisms requires oxygen (aerobic respiration). However, glycolysis is a part of aerobic respiration that does not require oxygen. Under conditions of low oxygen, pyruvic acid does not enter the Krebs cycle but is converted in a two-step enzymatic reaction to ethyl alcohol and carbon dioxide. The processes involved in the conversion of glucose to alcohol and carbon dioxide constitute fermentation. Certain bacteria and fungi do not utilize oxygen as the terminal acceptor and hence their respiration process is anaerobic (absence of oxygen).

The tissues of higher plants begin to ferment when deprived of oxygen. When seeds are sown in very wet soil or plants grow in poorly drained soil, they are predisposed to fermentation. Seedlings vary in their ability to survive prolonged exposure to anaerobic conditions. Rice especially, and to a lesser extent pea and corn seedlings, are able to survive for some time under conditions of oxygen depletion.

Anaerobic respiration by anaerobic bacteria and fungi is exploited in commercial food processing—for example, in flavoring cheese and yogurt. Similarly, fermentation products are important to the alcoholic beverage and baking industries. The alcohol produced by fermentation is the basis of the wine and beer brewing industry, while the carbon dioxide released during fermentation by baker's yeast *(Sacharomyces cerevisiae)* causes dough to rise. Fermentation is also managed for successful production of high-quality silage.

3.3.4 HARVESTING ENERGY UNDER ANAEROBIC CONDITIONS IS INEFFICIENT

Under conditions of oxygen deficiency, the end product of glycolysis is not pyruvate but *ethyl alcohol* (ethanol), a breakdown product of pyruvate. This process is also called **fermentation** (Figure 3–12). Unless they have aerenchyma cells to adapt them, many plants ferment when grown in mud or water depleted of oxygen. The anaerobic pathway is less efficient than the aerobic, yielding only two ATP molecules. However, fermentation is useful and desirable in certain post-harvest processing of products. Silage-making, for example, is a process whose success depends on proper fermentation.

Fermentation. The metabolic breakdown of a substrate such as glucose in the absence of oxygen or in an oxygen-deficient environment.

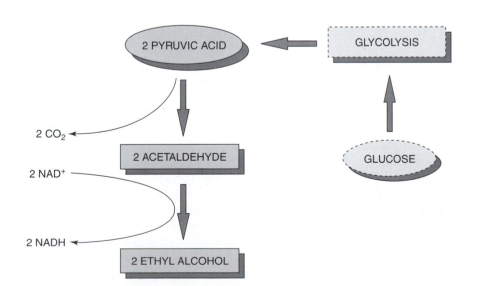

FIGURE 3–12 A summary of the process of fermentation. The reactions of anaerobic respiration are collectively called fermentation. The process produces little or no ATP. However, it is economically important to society in the area of food processing, such as the preparation of cheese and yogurt, and the preparation of various alcoholic beverages.

3.3.5 ABSORPTION

Water and dissolved nutrients in the soil solution enter the plant through the roots by the process of *absorption.* Absorption entails two other natural processes—*diffusion* and *osmosis.* Diffusion is the movement of molecules or ions from a region of higher concentration to a region of lower concentration, along a gradient (diffusion gradient). This movement is affected by factors such as temperature, density of molecules, and the medium in which the movement occurs. Osmosis is the diffusion of water through the differentially permeable (semipermeable) membrane from a region where water is more concentrated to one where it is less concentrated. The cell walls of the root hairs are semipermeable membranes. Water enters the cell by osmotic potential (pressure required to prevent osmosis) and is balanced by the resistance of the cell wall to expansion.

Another force involved in the absorption of water by plants is imbibition. Colloidal materials and other large molecules in the cell (e.g., starch, cellulose) attract water molecules. This process occurs in both living and dead tissues. Imbibition results in the swelling of tissues. It is the initial step in the germination of seeds.

3.3.6 TRANSPIRATION

Transpiration is the loss of moisture through the above-ground parts of the plant, primarily the leaves. Most of the water (about 98%) absorbed by a plant is eventually lost to the atmosphere through the process of transpiration. Two steps are involved in this process. First, the spongy mesophyll region of the leaf provides a tremendous amount of surface area for *evaporation* to occur. This is then followed by *diffusion,* which occurs through the stomatal pores of the leaf epidermis and proceeds from the areas of high moisture saturation (leaf internal part) to the outside environment, where the air is drier.

Transpiration through woody stems is minimal, especially in older plants that have developed thick barks. Younger horticultural woody plants, when transplanted, may need a wrapping of paper around the stem to reduce moisture loss through evaporation. Some plants, such as some grasses, roll their leaves under moisture stress, due to the shrinking of the large vacuolate cells on their surfaces, called *bulliform cells.* This reduces transpiration in such species.

Transpiration is regulated by the closure of the stomata, an event that excludes CO_2 and reduces photosynthetic rate. However, respiration produces some CO_2, which can be trapped and used by plants after the stomata have closed. The stomata open or close according to changes in turgor pressure in the guard cells. Water, in most cases, is the primary factor that controls stomatal movement. However, other factors in the environment also affect stomatal movement. Generally, an increase in CO_2 concentration in the leaf causes stomatal closure in most species. Some species are more sensitive than others to the effects of CO_2. Similarly, the stomata of most species open in light and close in the dark. An exception to this feature is plants with the CAM pathway of photosynthesis. Photosynthesis uses up CO_2 and thus decreases its concentration in the leaf. Evidence suggests that light wavelength affects stomatal movement. Blue light and red light have been shown to stimulate stomatal opening. The effect of temperature on stomatal movements is minimal, except when excessively high temperatures (more than 30°C or 86°F) prevail, as occurs at midday. However, an increase in temperature increases the rate of respiration, which produces CO_2 and thereby increases the CO_2 concentration in the leaf. This increase in CO_2 may be part of the reason stomata close when the temperature increases. Apart from environmental factors, many plants have been known to accumulate high levels of abscisic acid, plant growth hormone. This hormone accumulates in plants under conditions of moisture stress and causes stomata to close.

Transpiration is accelerated by several environmental factors. The rate of transpiration is doubled with a more than 10°C or 50°F rise in temperature. It is slower under high humidity. Also, air currents may accelerate the transpiration rate by preventing water vapor from accumulating on the leaf surface. Certain plants are adapted to dry environments. These plants, called xerophytes, have anatomical and physiological modifications that make them able to reduce transpiration losses.

3.3.7 HOW WATER MOVES IN PLANTS

Water moves in plants via the conducting elements of the xylem. It moves along a water potential gradient from soil to root, root to stem, stem to leaf, and leaf to air, forming a continuum of water movement. The trend is for water to move from the region of highest water potential to the region of lowest water potential, which is how water moves from the soil to the air. Transpiration is implicated in this water movement, because it causes a water gradient to form between the leaves and the soil solution on the root surface. This gradient may also form as a result of the use of water in the leaves. Loss of moisture in the leaves causes water to move out of the xylem vessels and into the mesophyll areas, where it is depleted. The loss of water at the top of the xylem vessels causes water to be pulled up. The movement of water up the xylem according to this mechanism is explained by the cohesion-tension theory. It is the cohesiveness of water molecules that allows water to withstand tension. Water is withheld against gravity by capillarity. The rise is aided by the strong adhesion of water molecules to the walls of the capillary vessel.

Capillary flow is obstructed when air bubbles interrupt the continuity of the water column, called embolism. This event is preceded by cavitation, the rupture of the water column. Once the tracheary elements have become embolized, they are unable to conduct water. In the cut flower industry, the stems of flowers are cut under water to prevent embolism. Water enters the plant from the soil via the root's hairs, which provide a large surface area for absorption. Once inside the root hairs, water moves through the cortex and into the tracheary elements. There are three possible pathways by which this movement occurs, depending on the differentiation that has occurred in the root. Water may move from cell to cell (passing from vacuole to vacuole), passing through the cell wall (the apoplastic pathway), or from protoplast to protoplast (the symplastic pathway) through the pores in the plasmodesmata (minute cytoplasmic threads that extend through openings in the cell wall and connect the protoplast of adjacent living cells).

Root pressure plays a role in water movement, especially at night when transpiration occurs to a negligible degree or not at all. Ions build up in the xylem to a high concentration and initiate osmosis, so that water enters the vascular tissue through the neighboring cells. This pressure is called root pressure and is implicated in another event, guttation, whereby droplets of water form at the tips of the leaves of certain species in the early morning. These drops do not result from condensation of water vapor in the surrounding air but, rather, are formed as a result of root pressure forcing water out of hydathodes (structures that secrete pure water).

The absorption and movement of water in plants occurs via the xylem vessels. Inorganic nutrients are also transported through the vessels. Solutes are moved against a concentration gradient and require an active transport mechanism that is energy-dependent and mediated by carrier protein. Some amount of exchange between xylem and phloem fluids occurs such that inorganic salts are transported along with sucrose, and some photosynthetic products are transferred to the xylem and recirculated in the transpiration stream.

3.3.8 TRANSLOCATION

Translocation in plants is the transport of organic and inorganic solutes through the plant. This activity involves both xylem and phloem tissue. Organic solutes are transported from sites of food manufacture *(source)* to parts of the plants where they are used or stored *(sink)*. Translocation usually occurs in an up-down fashion, but lateral movements also occur. Some sinks are located below the source—for example, roots (sink) are below leaves (source). However, materials may be transported to a developing fruit, flower and seed (sinks) from below. Translocation occurs through the phloem tissue as *sap,* which consists primarily of sugar and nitrogenous substances, such as amino acids. Phloem transport is believed to occur by the mechanism of *pressure flow.* This hypothesis suggests that assimilates are moved from translocation sources to sinks along a gradient of hydrostatic pressure (turgor pressure) of osmotic origin. Sugar is asserted to be transported in the phloem from adjacent cells in the leaf by an energy-dependent active process called *phloem loading.* The effect of this process is a decrease in water potential in the phloem sieve tube, which in turn causes water entering the leaf in the transportation stream to move into the sieve tube under osmotic pressure. The water then acts as a vehicle for the passive transport of sugars to sinks, where they are unloaded, or removed, for storage. The water is recirculated in the transpiration stream because of the increased water potential or the sink resulting from the phloem unloading.

Translocation does not have a fixed pattern but, rather, changes to meet the need of a plant. For example, when a plant is growing and has active terminals—or when reproductive or productive functions, such as the development of fruits and tubers, are occurring—organic solutes are translocated preferentially to such areas of high consumption. As leaves senesce, they cease to be sinks; therefore, food supplies must be redirected to other areas of need. This prioritization in sink-source relationships explains why vegetative growth in fruit plants slows down or even ceases while fruits are developing. In terms of environmental impact on translocation, increasing light intensity tends to increase translocation from shoot to roots, while darkness promotes the reverse.

3.3.9 SUMMARY OF PHOTOSYNTHESIS–RELATED CROP MANAGEMENT PRACTICES

The management practices summarized here are discussed in detail in appropriate sections of the book. These represent decisions a producer must make in order to have success with crop production.

1. *Leaf area.* Leaf area management includes the manipulation of plant density through seeding rate and protection against diseases and pests that destroy leaves.
2. *Light management.* As discussed elsewhere, light interception is critical to photosynthesis. Light management includes proper crop density, plant arrangement, and cultivar selection with advantageous morphology (e.g., vertical leaves, narrow leaves that avoid shading in the canopy).
3. *Nutrient management.* Proper fertilization is needed to keep plants healthy and to provide all the essential nutrients for the basic plant growth processes to proceed efficiently for enhanced productivity. Chlorophyll contains nitrogen and manganese.
4. *Moisture management.* Water is critical for photosynthesis. Plants are unable to grow and develop properly under drought conditions.

SUMMARY

1. Photosynthesis is the process by which plants convert light energy to chemical energy. It is the source of over 90% of dry matter of crop plants.
2. The ultimate result of photosynthesis is to reduce carbon dioxide to carbohydrate by three steps—entry of carbon dioxide into the plant, harvesting light by photochemical processes, and reducing carbon dioxide (assimilation) by biochemical processes.
3. Light absorbing pigments occur in the thylakoids of chloroplasts, photosystems (PSI, PSII), and antenna pigments.
4. Carbon dioxide assimilation occurs by three pathways: C_3 pathway yields 3-phosphoglycerate as its first stable product, while C_4 pathway yields oxaloacetate. The third pathway is CAM.
5. The C_4 pathway is more efficient than the C_3 pathway; C_3 plants experience a loss of energy through the process of photorespiration.
6. Carbon dioxide assimilation is affected by light intensity, light duration, carbon dioxide concentration, and temperature.
7. The rate of photosynthesis is also affected by growth and development needs of the plant.
8. Respiration is the process by which energy and reducing agents are used by plants for constructing new materials and for maintaining existing structures.
9. Respiration is a catabolic process and requires oxygen (aerobic). Under conditions of limited oxygen, the less efficient anaerobic respiration occurs.

REFERENCES AND SUGGESTED READING

Acquaah, G. 2002. *Horticulture, principles and practices.* Upper Saddle River, NJ: Prentice Hall.

Galston, A. W. 1980. *Life of the green plant.* 3d ed. Englewood Cliffs, NJ: Prentice Hall.

Gregory, R. P. 1989. *Photosynthesis.* New York: Routledge, Chapman and Hall.

Kaufman, P. B., T. F. Carlson, P. Dayanandan, M. L. Evans, J. B. Fisher, C. Parks, and J. R. Wells. *Plants, their biology and importance.* New York: Harper & Row.

Moore, R., and W. D. Clark. 1995. *Botany: Plant form and function.* Dubuque, IA: Wm. C. Brown.

Noggle, G., and G. F. Fritz. 1983. *Introductory plant physiology.* 2d ed. Englewood Cliffs, NJ: Prentice Hall.

Salisbury, F. B., and C. W. Ross. 1992. *Plant physiology.* 4th ed. Belmont, CA: Wadsworth.

Stern, K. R. 1997. *Introductory plant biology.* Dubuque, IA: Wm. C. Brown.

SELECTED INTERNET SITES FOR FURTHER REVIEW

http://www.biology.clc.uc.edu/courses/bio104/photosyn.htm

Discussion of photosynthesis.

http://www.hcs.ohio-state.edu/hcs300/photosyn.htm

Discussion of photosynthesis.

http://www.biology-online.org/1/3_respiration.htm

Discussion of respiration with tutorials.

http://staff.jccc.net/pdecell/cellresp/respintro.html

Excellent and complete discussion of cellular respiration.

http://www.botany.hawaii.edu/faculty/webb/BOT311/PPloem/Phloem/phloem_
translocation.htm

Excellent and comprehensive summary of translocation with good figures.

http://users.rcn.com/jkimball.ma.ultranet/BiologyPages/G/GasExchange.html#
Closing_stomata

Excellent discussion of exchanges of gasses in plants.

OUTCOMES ASSESSMENT

PART A

Answer the following questions true or false.

1. T F C_3 plants are more photosynthetically efficient than C_4 plants.
2. T F   Photosynthesis occurs in the mitochondria.
3. T F   Carbon dioxide enters the leaf by osmosis.
4. T F   The parts of the plant that accumulate and store photosynthates are called sources.
5. T F   Respiration is a catabolic process.
6. T F   Aerobic respiration yields more ATPs than anaerobic respiration.
7. T F   The end product of glycolysis is pyruvate.
8. T F   Glycolysis is sugar splitting.
9. T F   Under oxygen deficit conditions, glycolysis produces ethanol as an end product.

PART B

Answer the following questions.

1. Write the chemical equation summarizing photosynthesis.

   _____

2. Write a general equation summarizing the light reaction of photosynthesis.

   _____

3. Give the three critical steps involved in photosynthesis.

_____

_____

_____

4. The C_3 pathway is also called the _____ cycle.

5. What is photorespiration? _____

_____

6. Metabolism consists of two sets of reactions: anabolism and _____.

7. Write the chemical reaction that summarizes aerobic respiration.

_____

8. What is fermentation? _____

_____

9. One molecule of glucose yields a net of _____ ATPs.

10. Give two electron acceptors in the reactions of respiration.

_____

_____

11. What is the role of oxygen in aerobic respiration?

_____

12. Respiration that is associated with housekeeping activities of the cell is called

_____.

13. Substrate-level phosphorylation occurs at what stages of aerobic respiration?

_____

14. What is oxidative phosphorylation? _____

_____

PART C

Write a brief essay on each of the following topics.

1. Discuss the process of photophosphorylation.

2. Describe the C_3 pathway of carbon dioxide assimilation. *fixation*

3. Describe the C_4 pathway of carbon dioxide assimilation.

4. Discuss the effect of carbon dioxide on the rate of photosynthesis.

5. Discuss the effect of light intensity on the rate of photosynthesis.

6. Discuss the Crassulacean acid metabolism pathway of carbon dioxide assimilation.

7. Describe the Krebs cycle.

8. Discuss the energy budget of respiration.

9. Discuss maintenance respiration in plants.

10. Describe the reactions involved in glycolysis.

PART D

Discuss or explain the following topics in detail.

1. If C_3 plants are more photosynthetically efficient than C_4, why don't they dominate the earth?

2. Can the yield potential of crops be raised indefinitely?

3. How important is photosynthesis to life on earth?

Plant Growth and Development

PURPOSE

The purpose of this chapter is to discuss the concepts of growth and development of plants. The fundamental growth processes discussed in Chapter 3 are involved in the growth and development of plants.

EXPECTED OUTCOMES

After studying this chapter, the student should be able to:

1. Distinguish between growth and development.
2. Describe the general pattern of growth.
3. Define yield and the types of yield.
4. Discuss the concept of yield components.
5. Discuss harvest index and factors affecting it.
6. Discuss photoperiod and its effect on plant growth.
7. Discuss the concept and application of growing degree days.
8. Discuss the growth and development–related management practices in crop production.

KEY TERMS

Biological yield
Biomass
Economic yield

Growth
Phenology

Vernalization
Yield

TO THE STUDENT

Producers often target a certain part, or component, of the crop in a production enterprise as the commercial product. This part produces the economic yield for which a crop is cultivated. If a crop grows luxuriantly but the economic part is not enhanced satisfactorily, it is to no avail that the producer invested time and resources. For example, if grain is the goal, it is of little use to have good, healthy-looking corn with tiny ears. Thus, the goal of the crop producer is to maximize economic yield. Just like animals in which some are poor converters of feed, some plant cultivars are similarly less efficient in converting photosynthates into economic yield. Apart from choosing an efficient, high-yielding cultivar, the crop producer should provide the proper cultural environment in which plants can optimize their metabolic activities. By studying plant growth, scientists are able to redesign plant types (through plant breeding procedures) that best utilize production resources and efficiently convert them to economic yield. They can also advise producers on how to produce these crops (e.g., spacing, nutritional management) for best results.

4.1: WHAT IS GROWTH?

Growth. An irreversible increase in cell size or cell number.

How does growth of a crystal or snowball differ from growth of a plant? **Growth** is a progressive and irreversible process that involves three activities—*cellular division, enlargement,* and *differentiation.* Cell division occurs by the process of mitosis followed by cytoplasmic division. Cells enlarge when they take in water by *osmosis* (the diffusion of water or other solvents through a differentially permeable membrane from a region of higher concentration to one of a lower concentration). Water status of a plant is thus critical to its growth and development. Differentiation entails the development and modification of cells to perform specific functions in the plant.

Growth produces an increase in dry matter when the plant is actively photosynthesizing. The rate of growth follows a certain general pattern described by a sigmoid curve (Figure 4–1). The *s*-shaped curve has four distinct parts:

1. *Lag growth phase.* This phase includes activities in preparation for growth. Dormant cells become active; dry tissue imbibes moisoure; cells divide and increase in size; the embryo differentiates.

FIGURE 4–1 A typical sigmoid growth curve (a) lag phase (b) log phase, (c) decreasing phase, (d) steady phase.

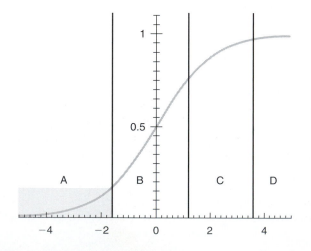

2. *Logarithmic growth phase.* During this phase, the plant experiences an ever-increasing rate of growth. Events that occur in this phase include germination and vegetative growth.
3. *Decreasing growth phase.* The decreasing growth phase is characterized by a slowing down of the rate of growth. The plant activities during this period include flowering, fruiting, and seed filing.
4. *Steady growth phase.* The growth rate either declines or stops during this phase. Plant activities in this phase include maturation.

4.2: VARIATIONS IN GROWTH PATTERN

The specific shape of the curve varies according to, for example, growth cycle and the part of the plant measured. The shoot growth pattern varies among annuals, biennials, and perennials. The growth pattern for plant height and dry weight differ for the same growth cycle. A rapidly growing species has a growth pattern different from that of a slowly growing species. Biennials require two growing seasons to complete their growth cycle. They go dormant in winter and resume growth in spring. The roots of a herbaceous perennial may remain alive indefinitely, but the shoot may be killed each cold winter.

4.3: NOT ALL DRY MATTER ACCUMULATION IS ECONOMIC

Crop productivity or production is the rate at which a crop accumulates organic matter per unit area per unit time. It depends primarily on the rate of photosynthesis, the conversion of light energy to chemical energy by green plants.

Yield is the generic term used by crop producers to describe the amount of the part of a crop plant of interest that is harvested from a given area at the end of the cropping season or within a given period. The plant part of interest is that for which the crop producer grows the crop. This could be grain, seed, leaf, stem, root, flower, or any other morphological part. It could also be the chemical content of the plant, such as oil or sugar. In certain industrial crops, such as cotton, the part of economic interest to the producer is the fiber while, for the producer of tea or tobacco, the part of interest is the leaf. A producer may harvest multiple parts of a plant for use or sale.

Yield. The product of metabolism that may be economic or non-economic.

Biomass is a measure of yield used by scientists such as ecologists to describe the quantity or mass of live organic matter (total individuals present) in a prescribed area at a given point in time. This definition includes both above- and below-ground parts of plants in the area. Crop producers do not use this measure of yield.

Biomass. The quantity of live organic matter in a given area at a given point in time.

Weighing or measuring a product volume quantifies yield, as is the case in liquid products. The weight of the morphological part harvested depends on its moisture content. The moisture content is an important factor in yield measurements, especially in the case of crops that are harvested dry (e.g., grains). The weight of a specified amount of corn taken at 10% moisture content will be higher than one taken at 5% moisture content. For uniformity and fair pricing, industries associated with these kinds of crops observe certain standards with regard to grain or seed moisture content at the time of sale.

Scientists sometimes use a more stable basis for yield measurements that eliminates the moisture factor. Yield is then defined as the amount of dry matter produced per unit area of land. To obtain this value, the plant part is dried in an oven to a stable weight before weighing.

4.4: Types of Yield

Yield may be divided into two types: biological and economic.

4.4.1 BIOLOGICAL YIELD

Biological yield. The amount of dry matter produced per unit area.

Biological yield is the total dry matter produced per plant or per unit area. This measurement takes into account all plant parts, above and below ground levels. Researchers use this measurement of yield in agronomic, physiological, and plant breeding studies to indicate dry matter accumulation.

4.4.2 ECONOMIC YIELD

Economic yield. The amount (volume or weight per unit area) of the part(s) of the plant of usable or marketable value.

Also called *agricultural yield,* **economic yield** represents the total weight per unit area of a specified plant material that is of marketable value or other use to the producer. The particular valuable plant part or product differs from one crop to another and may be located above ground (as in cereal crops, forage crops, cotton, and soybean) or below-ground (as in root and tuber crops such as sugar beet and Irish potato). The producer determines the part of the plant that is of economic value. The part may be used for food, feed, or industrial purposes. A producer of corn for grain is interested in the grain; a producer of corn for silage is interested in the young, fresh stem and leaves.

All economic yield is biological yield. However, not all biological yield is necessarily economic yield. For example, whereas the above ground part of corn (grain, stalk, leaves) may be used for a variety of economic purposes (e.g., food or feed), the roots have no economic value and are not harvested. On the other hand, root crops such as sugar beet produce roots that are harvested for sugar extraction, while the top leaves may be sold to livestock producers for feed. Thus, in the latter case, the total plant (biological yield) is of economic value.

4.5: Photosynthetic Products Are Differentially Partitioned in the Plant

Partitioning describes the pattern of carbon use in a plant. The effect of partitioning of assimilates is manifested as changes in plant morphology (e.g., changes in size, shape, and number of plant organs) during the growing season. Unlike animals, in which growth occurs throughout the organism, growth in plants occurs in specific regions called meristems. There are different locations of these growing points that are named accordingly as apical, intercallary, or axillary meristems. These centers of active cell division continue activity throughout most of the lifetime of the plant. Plasticity occurs in plants because of the presence of a variety of meristematic centers. It is important that these competitive centers of activity be coordinated so that plant resources are properly utilized. Certain centers may need to have accelerated activity, while others need to be slowed down at certain periods in the plant growth cycle.

How does the plant decide where to allocate photosynthates? Partitioning is controlled primarily by the capacity of the meristem to grow. This capacity is influenced by intrinsic and extrinsic factors. It has been proposed that growth susbstances (growth hormones), nutritional substrates, and environment are particularly responsible for controlling partitioning. The control by growth substances is especially manifested through apical dominance. Apical meristems produce auxin, a growth hormone that suppresses axillary meristem activity (i.e., branching). When the shoot apex is removed, this apical dominance is abolished, allowing branching to occur. Certain organs have the capacity to act as sinks (importers of substrates), while others are sources (exporters of substrates). However, an organ may be a source for one substrate at one point and then a sink at another time. For example, leaves are sinks for nutrients (e.g., nitrates) absorbed from the soil, while they serve as sources for newly formed amino acids.

Can crop producers influence the way plants allocate photosynthates? Genotypes differ in patterns of partitioning of dry matter. For example, in legumes, bush or erect cultivars differ from pole or prostrate cultivars. In cereals, tall cultivars differ from dwarf types in dry matter partitioning. Plant breeders strive to breed for cultivars that have a larger harvest index. C. M. Donald developed the *ideotype* concept to describe a model of an ideal plant phenotype that represents optimum partitioning of dry matter. An ideotype is developed for a specific cultural condition—e.g., monoculture, high-density mechanized production, irrigated production, or production under high input (fertilizers, pest control).

4.6: HARVEST INDEX

4.6.1 WHAT IS HARVEST INDEX?

Harvest index is defined as the proportion of the crop that is of economic importance. In early literature, the term used to describe this proportion was *coefficient of effectiveness.* This may have been because one of the uses of this measurement in plant breeding research is as an indicator of the efficiency with which photosynthates are partitioned or distributed to the plant parts that have a bearing on economic yield. Harvest index is calculated as a ratio as follows:

Harvest index = (economic yield/total yield)

Its value theoretically ranges from 0.0 to 1.0. The higher the value, the more efficient the plant is in directing assimilates to the part of the plant of economic use.

Harvest index is influenced by several factors. In some instances, the researcher deliberately or inadvertently excludes certain plant parts from the estimate. For example, even though assimilates are partitioned to roots, this part of the plant is difficult to harvest (except where roots are the economic parts) and hence is routinely excluded from estimations of harvest index. Also, plants may shed some leaves when they reach maturity. These fallen leaves are not usually picked up and included in the measurement of total yield (biomass). Consequently, harvest index is overestimated on such occasions.

The economic parts of certain plants are very small. There are some crop plants that are nearly completely useful in the sense that nearly all parts have known economic value. This notwithstanding, the producer may choose to emphasize only one component in a crop production enterprise. When this happens, the harvest index values may be very small. A classic example is sugar cane, a versatile crop that may be grown for sugar, syrup, or cane. Harvest index values based only on sugar content may be about 0.2. When sugar and syrup are combined, the value is about 0.6. However, if economic

Plant *hormones* are organic molecules that are produced in small amounts in one or several parts of the plant and then transported to other parts called *target sites,* where they regulate plant growth and development. Because of this physiological role, plant hormones are also called *plant growth regulators,* a broad terminology that includes and is often associated with synthetic chemicals that have a similar effect as hormones. The way of classifying plant hormones is on the basis of their origin—*natural* or *synthetic.* Unlike animal hormones that have specificity in site of production and target site, plant hormones tend to be more general with respect to both source and target.

There are five basic groups of natural plant hormones:

1. *Auxins.* Auxins are produced in meristematic tissue such as root tips, shoot tips, apical buds, young leaves, and flowers. Their major functions include regulation of cell division and expansion, stem elongation, leaf expansion and abscission, fruit development, and branching. An example of a natural hormone is indole-3-acetic acid (IAA). Synthetic hormones that are auxins include 2,4-dichlorophenoxyacetic acid (2,4-D), which is actually a herbicide for controlling broadleaf weeds such as dandelion in lawns, α-naphthaleneacetic acid (NAA), and indole-3-butyric acid (IBA). Other uses of auxins in horticulture are
   a. As rooting hormones to induce rooting (adventitious) in cuttings
   b. To prevent fruit drop in fruits trees shortly before harvest
   c. To increase blossom and fruit set in tomatoes
   d. For fruit thinning to reduce excessive fruiting for larger fruits
   e. For defoliation prior to harvesting
   f. To prevent sprouting of stored produce, for example, in potatoes; when applied to certain tree trunks, basal sprouts are suppressed

   The concentration of auxin in plants can be manipulated in horticultural plants in cultivation. Auxins are produced in relatively higher concentrations in the terminal buds than in other parts. This localized high concentration suppresses the growth of lateral buds located below the terminal bud. When terminal buds are removed (for example, by a horticultural operation such as pruning or pinching), lateral buds are induced to grow. This makes a plant fuller in shape and more attractive.
2. *Gibberellins.* Gibberellins are produced in the shoot apex and occur in embryos and cotyledons of immature seeds as well as roots. They occur in seed, flowers, germinating seed, and developing flowers. This class of hormones promotes cell division, stem elongation, seed germination (by breaking dormancy), flowering, and fruit development. Fruit size of seedless grapes is increased through the application of this hormone. An example is gibberellic acid (GA_1).
3. *Cytokinins.* These hormones stimulate cell division and lateral bud development. They occur in embryonic or meristematic organs. Examples of natural cytokinins are isopentenyl adenine (IPA) and zeatin (Z). Synthetic cytokinins included kinetin (K) and benzyl adenine (BA). Cytokinins interact with auxins to affect various plant

value is based on fresh cane delivered to the processor or buyer, the harvest index may be above 0.90

The developmental pathway followed by the plant part or chemical component of economic value affects harvest index. In cereal crops such as corn and wheat, the economic part, the grain, fills in a linear fashion up to a definite point and then ceases. Harvest index in these crops depends on the relative duration of vegetative and reproductive phases of plant lifecycle. In root and tuber crops, such as sugar beet and Irish potato, the economic part follows a protracted developmental pathway. In these crops, harvest index depends more on genetics and environmental factors.

functions. A high cytokinin-auxin ratio (that is, low amounts of auxin, especially IAA) promotes lateral bud development because of reduced apical dominance. The principal role of cytokinins in plant physiology is the promotion of cell division.

4. *Ethylene.* Ethylene is a gas that is found in tissues of ripening fruits and nodes of stems. It promotes fruit ripening and leaf abscission. In the horticultural industry, ethylene is used to aid ripening of apples, pineapples, and bananas and to change the rind color of fruits (as in oranges and grapefruits from green to yellow and tomatoes from green to uniform red). On the other hand, ripening of apples produces this gas in large quantities, which tends to lower storage life of fruits. Ethaphon is commercially used for this purpose. This can be reversed by removing the gas with, for example, activated charcoal. Ethylene in the growing environment may cause accelerated senescence of flowers and leaf abscission. Carnation flowers close while rose buds open prematurely in the presence of ethylene. In cucumbers and pumpkins, ethaphon spray can increase female flowers (disproportionately) and thereby increase fruit set.

5. *Abscisic acid.* Abscisic acid (ABA) is a natural hormone that acts as an inhibitor of growth, promotes fruit and leaf abscission, counteracts the breaking of dormancy, and causes the stomata of leaves to close under moisture stress. ABA has an antagonistic relationship with gibberellins and other growth-stimulating hormones. For example, ABA-induced seed dormancy may be reversed by applying gibberellins. Commercial application of ABA is limited partly because of the high cost of the chemical and scarcity of the substance.

Plant hormones may also be classified based on their effect on plant growth as *stimulants* or *retardants.* Cytokinins and gibberellins have a stimulating effect on growth and development, while ABA is a growth inhibitor. Alfalfa is known to produce the alcohol *triacontanol,* which stimulates growth. Naturally occurring inhibitors include benzoic acid, coumarin, and cinamic acid. A number of synthetic growth retardants are in use in producing certain horticultural plants. Their effect is mainly to slow down cell division and elongation. As such, instead of a plant growing tall with long internodes, it becomes short (dwarf), compact, fuller, and aesthetically more pleasing. Examples of these commercial growth retardants are

1. Daminozide (marketed under such trade names as Alar and B-nine; plants affected include poinsettia, azalea, petunia, and chrysanthemum
2. Chlormequat (CCC, cycocel), which retards plant height in poinsettia, azalea, and geranium
3. Ancymidol (A-Rest), which is effective in reducing height in such bulbs as Easter lilies and tulips, as well as chrysanthemums and poinsettias
4. Paclobutrazol (Bonzi), which is used to reduce plant height in bedding plants, including impatiens, pansy, petunia, and snapdragon

Plant maturity plays a role in harvest index, the values increasing with early maturity in crops such as rice. Increasing plant population density in crops such as corn decreases harvest index. Drought (water stress) and fertilization (e.g., nitrogen application) tend to lower harvest index values.

4.6.2 MANIPULATING HARVEST INDEX

Harvest index in certain crops has been successfully increased through plant breeding. In wheat, this has been generally accomplished by increasing the efficiency with which the plant partitions photosynthates to the grain, while keeping the biomass virtually unchanged. Increase in harvest index in small grain cereals and other species has been achieved by reducing plant height through selection or the use of dwarfing genes (such as the *Rht* genes in wheat breeding). Decreasing plant height in modern cultivars is partially responsible for increased harvest index in certain species. The reduction in the size of plant organs is compensated for in modern production through the application of agronomic inputs such as fertilizers. Such a practice appears to reduce the need for plants to remobilize accumulated reserves in leaves to the grain. This trend, however, is not conclusive in all cases.

The use of early-flowering cultivars, coupled with good crop management (e.g., fertilization, pest control, irrigation) allows the plant to allocate assimilates to seed sooner, thereby leading to reduced accumulation of reserves in the leaves. Early-flowering cultivars in species such as rice may benefit from a high harvest index through this process.

4.7: THE BIOLOGICAL PATHWAY
TO ECONOMIC YIELD IS VARIABLE

4.7.1 YIELD COMPONENTS

In an effort to manipulate crop yield, researchers attempt to construct the path by which the reproductive, developmental, and morphological features of plants in a crop stand contribute to yield of a specified plant product. There are many such pathways to high yield *(yield components)*. For grain yield, a model is as follows:

Yield/unit area = (plants/unit area) $\times$ (heads/plant) $\times$ (mean number of seeds/head) $\times$ (mean weight/seed)

where the plant species produces tillers, the model may be presented as follows:

Yield/unit area = (plants/unit area) $\times$ (mean number of tillers with ears/plant) $\times$ (mean number of grains/ear) $\times$ (mean grain weight)

These plant features describe yield. They all depend on energy in a fixed pool that is furnished through photosynthesis. Plant breeders and agronomists seeking to influence yield manipulate the components to positively affect photosynthesis.

In interpreting correlation between yield and its components, one should not evaluate the components in terms of relative importance. Seasonal sequence of environmental conditions that affect plant development should be considered. Growing conditions may be ideal in the early growth and development of the crop, leading to good initiation of reproductive features. However, if there is an onset of drought, few pods may complete their development and be filled with seed, leading to low correlation between yield and the number of seeds per pod.

Yield components vary from one species to another in terms of optimum value relative to other components. Further, yield components affect each other to varying degrees. For example, if increasing plant density drastically reduces the number of pods per plant, the number of seeds per pod may only be moderately affected, while seed size remains unchanged or only slightly affected.

Whereas a balance among yield components has great adaptive advantage for the crop, the components are environmentally labile. High yield usually results from one component with extreme value. Further, yield components are determined sequentially. As such, they tend to exhibit *yield compensation,* the phenomenon whereby deficiency or low value for the first component in the sequence of developmental events is made up with high values for the subsequent components. The net effect is that yield is maintained at a certain level. However, yield compensation is not a perfect phenomenon. For example, it may occur over a wide range of plant densities in certain species. In beans, reduction in pod number can be compensated for by an increase in seed number per pod and weight per seed.

4.7.2 YIELD POTENTIAL

What determines how much a crop can produce? Crop yield is a complex trait (quantitative trait). A given crop cultivar has an inherent optimum capacity to perform under a given environment. This capacity is described as its *yield potential.* When this potential is attained, attempts to improve yield or crop performance will be difficult even if levels of production inputs are increased. At this point, the genetic potential needs to be increased by genetic manipulation through a plant breeding program.

4.8: PLANT DEVELOPMENT CONCEPTS

Development is the term used to describe the continuing change in plant form and function as the plant responds to environmental factors. It involves the coordination, growth, and longevity of new vegetative and reproductive parts. Newly produced cells from the division of meristematic cells undergo change and specialization through differentiation, a process called *morphogenesis.* The activities of the various meristems produce change in the plant. The shoot meristem supported by intercalary meristems in the internodes produces elongation in the plant. Also, axillary meristems in leaf axils produce branches, while lateral meristems (mainly in the cambium) produce increases in girth.

Since the fruit and seed are commonly the economic yield of most crop plants, crop producers are especially interested in reproductive development. Flowering plants have two phases of development—vegetative and reproductive. At a certain stage in the development, the shoot apices change from the vegetative phase to the reproductive phase. The environment induces this conversion. The yield of the crop depends on the extent to which this conversion occurs, thus making it critical for a crop to complete its reproductive phase during the growing season and without stress. For high productivity, there has to be a good balance between the two phases. A protracted vegetative phase reduces reproductive functions. However, good vegetative development is needed to support good reproductive development.

4.8.1 DEVELOPMENTAL PHASES

Scientists recognize certain distinctive developmental stages, or *phenostages* (phenological stages), in the course of a plant's development. The lifecycle of a seed-producing crop plant may be divided into five general developmental stages as follows:

1. *Seed germination to emergence.* This is the first critical stage in crop production. The success of the crop starts with a good establishment.
2. *Emergence to floral initiation.* This is the most prolonged phase in the development of the plant. It is the vegetative period of development, during

which the plant roots, stems, and leaves grow and accumulate food for the transition from vegetative to flowering, developing flower buds.

3. *Floral initiation to anthesis.* Anthesis is the opening of the flower buds, called the reproductive stage. The plant needs to sustain an adequate number of flowers for producing the economic product.
4. *Anthesis to physiological maturity.* This critical stage includes the filling of the grain. The grain fills to a stage where additional agronomic inputs do not further enhance its size.
5. *Physiological maturity to cessation of growth.* This cessation of growth may be terminal (such as death in annuals) or temporary (such as dormancy in perennials).

Most of these phenostages can be readily recognized by observation only. Others, such as floral initiation, require dissection and microscopic examination of the tissues. The rate of advance within the intervening *phenophases* is called the *developmental rate.* **Phenology** is the study of the progress of crop development in relation to environmental factors. Individual plants in a crop stand vary in developmental rate. This is caused by the prevalence of different microclimates and some genotypic factors. Crop cultivars of narrow genetic base (e.g., purelines) provide opportunity for synchronized development that enables the producer to administer certain management operations (e.g., fertilizer or pesticide application) as close to the best time as possible.

Normally, vegetative growth ceases when the apical meristem converts from vegetative growth to a reproductive structure. In certain crops, flowering occurs within a limited period of a few days. These are called *determinate crops;* examples include cereal crops. In other crops, vegetative growth and reproductive growth overlap. The apical meristem continues to produce new leaves while flowering continues progressively from axillary meristems. These crops are called *indeterminate crops;* examples include certain legumes, such as soybeans. Some crops have both determinate and indeterminate cultivars in use in crop production.

Phenology. The study of the timing of periodic phenomena such as flowering, growth initiation, or growth cessation, especially as related to seasonal changes in temperature or photoperiod.

4.8.2 GROWTH STAGING IN CEREAL GRAINS

As previously discussed, plants have distinct developmental stages in their lifecycles. Over the years, a number of systems (staging systems) have been developed by researchers to quantify development for scientific and crop management purposes. The staging systems for cereal grains have been some of the most widely studied and applied. These systems are applied in the crop production industry for proper identification of the best times to apply various production inputs, as well as other management practices.

Of the staging scales in existence, the most widely used are the Feekes, Haun, and Zadoks scales. The characteristics and strengths of these scales are briefly discussed in this section.

The Major Growth Stages of Cereal Grains

Researchers have identified 10 major growth stages that all cereal grain plants go through during a normal lifecycle. These stages are important to producers because they base many of their agronomic practices on stages of plant growth and development. The stages are

1. Germination—the emergence of the radicle and coleoptile from the seed
2. Seedling—the young, newly emerged plant
3. Tillering—the formation of lateral branches that originate from below-ground nodes on the stem
4. Stem elongation (jointing)—stem nodes appear above the ground level

5. Booting—The inflorescence expands inside the upper leaf sheath.
6. Heading—cluster of florets form.
7. Flowering (anthesis)—Anthers extend from the glumes
8. Milk—early development of kernel; kernel has watery and whitish consistency
9. Dough—kernel has semi-solid consistency
10. Ripening—kernel loses moisture and is more solid

Common Staging Scales

Table 4–1 compares the three major staging scales in terms of physical changes they describe and the various numerical codes used to refer to various growth and development stages.

Feekes Scale This scale is one of the earliest and perhaps the best known and most widely used. The scale describes 11 major developmental stages, starting with the first leaf stage to the grain ripening. The stages in the vegetative phase are assigned numeric values ranging from 0 to 10. The stages in reproductive phase are characterized in more detail, starting from 10.1 (first spikelet of the inflorescence is visible) to 11.4 (hard kernel). A major agronomic application of the Feekes scale is the timing of pesticide application.

Haun Staging System The Haun staging system focuses mainly on leaf production stages of development. The length of an emerging leaf is expressed as a fraction of the length of the preceding fully emerged leaf. A Haun numeric value of 3.2, for example, means that three leaves are fully emerged, the fourth leaf having only emerged to about two-tenths of the length of the third leaf. The scale is less useful when developmental indicators other than leaf numbers are used in decision making.

Zadoks Staging System This two-digit code system is very detailed and more precise than the others. It provides more information during the early developmental stages than the Feekes scale. The Zadoks scale recognizes 10 developmental stages, beginning from germination (0) to ripening of kernels (9). The first of the two digits represents the principal stage of development. Each major developmental stage is subdivided into secondary stages (the second digits). A leaf has to be 50% unfolded or emerged to be counted. A code of 3, for example, is interpreted as a plant seedling with 3 leaves that are at least 50% extended. A second digit of 5 usually indicates that the midpoint of the primary stage has occurred.

4.8.3 GROWTH STAGING IN LEGUMES

Developmental stages of legumes are usually divided into two: vegetative and reproductive. The following example is of growth staging in soybean.

Vegetative Stages (Vn)

VE—Cotyledons emerge.
VC—Cotyledon emerges and the unifoliate leaf unfolds.
V1—First node has leaves.
V2—Second node has leaves.
V3—Third node has leaves.
Vn—*n*th node has leaves.

Table 4.1 A comparison of three cereal growth staging scales.

Zadoks	Feekes	Haun	Description
Germination			
00	–	–	Dry Seed
01	–	–	Start of imbibition
03	–	–	Imbibition complete
05	–	–	Radicle emerged from seed
07	–	–	Coleoptile emerged from seed
09	–	0.0	Leaf just at coleoptile tip
Seedling Growth			
10	1	–	First leaf through coleoptile
11	–	1.0	First leaf extended
12	–	1.+	Second leaf extending
13	–	2.+	Third leaf extending
14	–	3.+	Fourth leaf extending
15	–	4.+	Fifth leaf extending
16	–	5.+	Sixth leaf extending
17	–	6.+	Seventh leaf extending
18	–	7.+	Eighth leaf extending
19	–	–	Nine or more leaves extended
Tillering			
20	–	–	Main shoot only
21	2	–	Main shoot and one tiller
22	–	–	Main shoot and two tillers
23	–	–	Main shoot and three tillers
24	–	–	Main shoot and four tillers
25	–	–	Main shoot and five tillers
26	3	–	Main shoot and six tillers
27	–	–	Main shoot and seven tillers
28	–	–	Main shoot and eight tillers
29	–	–	Main shoot and nine tillers
Stem Elongation			
30	4-5	–	Psuedo stem erection
31	6	–	First node detectable
32	7	–	Second node detectable
33	–	–	Third node detectable
34	–	–	Fourth node detectable
35	–	–	Fifth node detectable
36	–	–	Sixth node detectable
37	8	–	Flag leaf just visible
39	9	–	Flag leaf ligule/collar just visible
Booting			
40	–	–	—
41	–	8-9	Flag leaf sheath extending
45	10	9.2	Boot just swollen
47	–	–	Flag leaf sheath opening
49	–	10.1	First awns visible

Zadoks	Feekes	Haun	Description
Inflorescence Emergence			
50	10.1	10.2	First spikelet of inflorescence visible
53	10.2	–	1/4 of inflorescence emerged
55	10.3	10.5	1/2 of inflorescence emerged
57	10.4	10.7	3/4 of inflorescence emerged
59	10.5	11.0	Emergence of inflorescence completed
Anthesis			
60	10.51	11.4	Beginning of anthesis
65	–	11.5	Anthesis 1/2 completed
69	–	11.6	Anthesis completed
Milk Development			
70	–	–	—
71	10.54	12.1	Kernel watery-ripe
73	–	13.0	Early milk
75	11.1	–	Medium milk
77	–	–	Late milk
Dough Development			
80	–	–	—
83	–	14.0	Early dough
85	11.2	–	Soft dough
87	–	15.0	Hard dough
Ripening			
90	–	–	—
91	11.3	–	Kernel hard (difficult to divide by thumbnail)
92	11.4	16.0	Kernel hard (can not be dented by thumbnail)
93	–	–	Kernel loosening in daytime
94	–	–	Overripe, straw dead and collapsing
95	–	–	Seed dormant
96	–	–	Viable seed giving 50% germination
97	–	–	Seed not dormant
98	–	–	Secondary dormancy induced
99	–	–	Secondary dormancy lost

Reproductive Stages (R)

R1—Beginning bloom
R2—Full bloom
R3—Beginning pod
R4—Full pod
R5—Beginning seed
R6—Full pod
R7—Beginning maturity
R8—Full maturity

Because all plants will not be at the same developmental stage at the same time, staging is sometimes generalized as follows:

Late vegetative—stem is about 12 inches tall but no buds or flowers are visible
Early bud—about 1% of plants in the field have flower buds
Mid bud—about 50% of all plants have at least one flower bud
Late bud—about 75% of all plants have at least one flower bud

These stages can be applied to blooming (i.e., first bloom, one-tenth bloom, midbloom, full bloom).

4.8.4 PHOTOPERIODISM

What makes plants switch from one developmental phase to another? There are several significant developmental switches from the initiation of leaves (vegetative) to flowering (reproductive) that are determined by environmental factors. Some crops are facultative in response to temperature or daylength, in that flowering is not promoted by desirable conditions but occurs nonetheless. In obligate crops, the plants will not flower if the temperature or daylength is above or below a certain critical value. Two developmental switches are *photoperiodism* and *vernalization.*

Duration of length of day is called its *photoperiod* (Figure 4–2). It affects developmental switches, as from vegetative phase to flowering phase. Based upon plant response to light duration, crop plants may be classified into three general groups: long-day, short-day, and day-neutral.

1. *Long-day* (short night) plants. Long-day plants require a light period longer than a certain critical length in order to flower. These plants will flower in continuous light. The response is common among cool-season plants such as wheat, barley, alfalfa, oat, and sugar beet.
2. *Short-day* (long night) plants. Short-day plants will not flower under continuous light. They require a photoperiod of less than a certain critical value within a 24-hour daily cycle. This response is common among warm-season crop plants, including corn, rice, soybean, peanut, and sugarcane.
3. *Day-neutral* (photoperiod insensitive) plants. These plants will flower under any condition of photoperiod. Examples are tomato, cucumber, buckwheat, cotton, and sunflower.

Effects of photoperiod on crop production are time of flowering of plants and time of maturity. This is critical to the production of crops whose economic part is the grain or seed.

Photoperiodism is photomorphogenic response in plants to day length. Photoperiodic plants actually track or measure the duration of darkness or dark period rather than the duration of light. Thus, short-day plants (or long-night plants) flower only if they receive continuous darkness for equal to or more than a critical value

(Figure 4–3). If the dark period is interrupted by light of sufficient intensity for even a minute, flowering will not be induced. Similarly, a long-day plant (or short-night plant) will not flower if the critical duration of darkness is exceeded. However, if a long-night period is interrupted by light, flowering will be induced. Interrupting the long night with such a short period of lighting is called flash lighting, a technique that

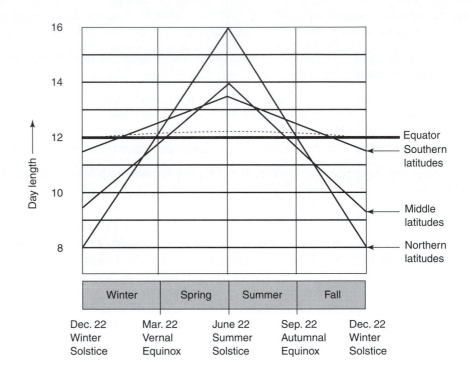

FIGURE 4–2 The average day length varies with the seasons. In the northern latitudes (20° to 50° above the equator), day lengths range between 8 hours in winter and 15.5 hours in summer. The day length at the equator remains at about 12 hours all year long.

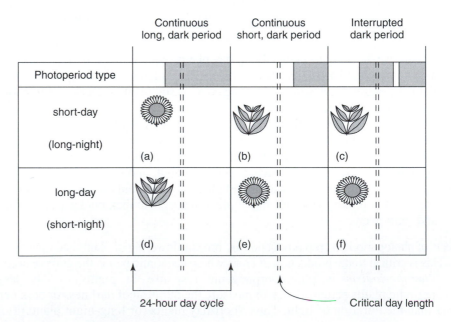

FIGURE 4–3 Photoperiodic response in flowering species. Light interruption of darkness affects short- and long-day plants differently.

Plants respond to the duration and timing of day and night. This seasonal response of plants to the environment is critical to the successful production of certain flowering plants. Work by plant physiologists Karl Hamner and James Bonner revealed that the length of light period is unimportant; rather, plants track or measure the *duration of darkness,* or dark period. The duration of uninterrupted darkness is critical, not the duration of uninterrupted light.

Photoperiod-sensitive plants will flower not according to a certain absolute length of photoperiod but, rather, based on whether the photoperiod is longer or shorter than a certain critical length required by the species. This critical length can be exacting in certain species such that a deviation of even 30 minutes can spell disaster. For example, in the henbane, the critical photoperiod is 10 hours and 20 minutes; 10 hours is ineffective. The environment (e.g., temperature) can modify this critical period. Further, the most sensitive part of the dark period is the middle of the period of exposure, the effect di-minishing before or after the mid-period. Some species require only one exposure to the appropriate photoperiod to be induced to flower, whereas others require several days or even weeks (as in spinach).

Certain light-sensitive plant pigments called *phytochromes* are involved in photoperiod response. Phytochromes exist in two photoreversible forms. The *Pr* form absorbs red light, while the other form, *Pfr,* absorbs far-red light. A molecule of Pr is converted to Pfr when it absorbs a photon of 660 nanometer light. Similarly, when Pfr absorbs a photon of 730 nanometer light, it reconverts to Pr instantaneously. The ratio of Pfr to Pr (or P730 to P660) decreases during the growing season as the days become shorter and nights longer. At a critical level, flowering is induced in short-day plants. In the case of long-day plants, as the day length increases and the nights become shorter during the early growing season, the P730 to P660 ratio increases. Flowering is initiated when a critical level is reached.

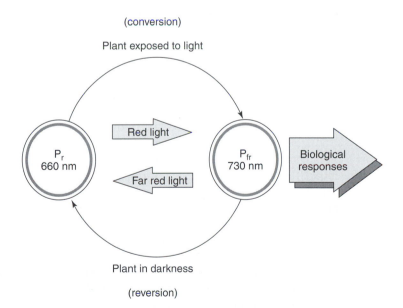

The role of phytochromes in the photoperiodic response in plants. Dark reversion of far red phytochrome to red phytochrome has so far been detected in dicots but not monocots.

is used by producers of ornamental flowering plants to manage the timing of flowering of plants and readiness for the market. Growers also manipulate the photoperiod of certain seasonal and high-income greenhouse plants (e.g., poinsettia) to produce plants in a timely fashion. To do this, growers start the plants under long-day conditions and then finish them under appropriate photoperiods. The required photoperiod is provided by covering the plants with a black cloth between 5 P.M. and 8 A.M. Conversely, the photoperiod may be prolonged during the natural short days by artificial lighting to keep the plants vegetative.

Photoperiod and temperature appear to compensate in their effect to some degree. Most cool-season crops tend to have a long-day response, while most warm-season crops are short-day in their response to photoperiod. The photoperiodic response of some crop plants may be modified by temperature. For example, early-maturing sorghum varieties may grow larger under long photoperiods of the north than they do in the south, where temperatures are more conducive to sorghum production. The time of seeding of crops affects the periods required to reach maturity because of the interaction of temperature and photoperiod.

4.8.5 VERNALIZATION

Vernalization. The process by which floral induction in some plants is promoted by exposing the plants to chilling for a certain length of time.

Vernalization is the cold-temperature induction of flowering that is required in a broad variety of plants. The degree of coldness and duration of exposure required to induce flowering vary from species to species, but temperature requirements usually lie between 0° and 10°C (32° and 50°F). While plants such as sugar beet and kohlrabi can be cold-sensitized as seed, most plants respond to the cold treatment after attaining a certain amount of vegetative growth. Some plants that need it are not treated because they are cultivated not for flower or seed but for other parts, such as roots (in carrot and sugar beet), buds (Brussels sprouts), stems (celery), and leaves (cabbage). Apple, cherry, and pear require vernalization as do winter annuals, such as wheat, barley, oat, and rye. Flowers such as foxglove (*Digitalis* spp.), tulip, crocus (*Crocus* spp.), narcissus (*Narcissus* spp.), and hyacinth (*Hyacinthus* spp.) need the treatment that may be applied to the bulbs, thereby making them flower in warmer climates, at least for that growing season. However, they must be vernalized in order to flower in subsequent years. In some of these bulbs, the cold treatment is needed to promote flower development after induction, but not for induction. Sometimes, vernalization helps plants such as pea and spinach to flower early but not as a requirement for flowering.

In onion, the bulbs are the commercial products harvested in production. Cold storage (near freezing) is used to preserve onion sets during the winter. This condition vernalizes the sets; if planted in spring, they will flower and produce seed. To obtain bulbs (no flowering), the sets should not be vernalized. Fortunately for growers, a phenomenon of *devernalization* occurs, in which exposure to warm temperatures of above 27°C (80°F) for two to three weeks prior to planting will reverse the effect of vernalization. Onion producers are therefore able to store their sets and devernalize them for bulb production during the planting season.

Flowering in certain species is affected by a phenomenon called *thermal periodicity,* in which the degree of flowering is affected by an alternation of warm and cool temperatures during production. For example, tomatoes in the greenhouse can be manipulated for higher productivity by providing a certain cycle of temperature. Plants are exposed to a warm temperature of 27°C (80°F) during the day and cooler night temperatures of about 17° to 20°C (63° to 68°F). This treatment causes increased fruit production over and above what occurs at either temperature alone.

4.9: Growing Degree Days

As an organism develops, its development is closely related to the daily accumulation of heat. Organisms have specific growth stages. In order to move from one stage of development to the next, a certain amount of heat is needed to provide adequate energy for the process (e.g., the emergence of a new leaf in corn or the hatching of insect eggs). Even though the amount of heat required to move from one stage to the next is constant from one year to the next, the amount of time within which the specific developmental transition occurs is dependent on the environment and hence is variable. Researchers have determined through experimentation the minimum base (threshold) temperature below which development in various organisms would not occur. As temperature rises above this minimum temperature, the growth rate increases up to a certain maximum and then declines.

4.9.1 CALCULATING DEGREE DAYS

Growing degree days (GDD) (sometimes called *heat units*) are obtained by calculating the heat accumulations above a minimum threshold temperature. Several methods of calculating degree days accumulations are available, the common ones including the averaging methods, Baskerville-Emin (BE) method, and the electronic real time data collection method. The variety in methodologies is due in part to the fact that these base temperatures are affected by photoperiod. The BE method uses a technique that fits a sine curve. It is advantageous over the averaging method when the minimum daily temperature is below the base temperature. The electronic method uses devices (heat unit accumulators) that record temperatures every few minutes, thus giving the most accurate results of all the methods.

4.9.2 APPLICATIONS

Degree days estimates have many useful applications in crop management. Growing degree days may be used to track the development of many crops and insect pests of interest. For example, in the Central Corn Belt, 2,100–3,200 GDD are needed (depending on the hybrid) to grow corn successfully. In this region, 200 GDD are needed for corn to reach the two-leaf stage, 1,400 to reach silk emergence, and 2,700 to reach physiological maturity (for a 2,700 GDD hybrid). Some alfalfa producers use GDD to determine the best time to cut the crop. GDD is also useful in integrated pest management for scouting. Scouting is more efficient when one knows the number of GDD required for a pest to reach a destructive stage. For example, if a study shows that it takes 300 heat units for alfalfa weevil eggs to hatch, then scouting for the pest may begin once the producer's calculations have accumulated 300 heat units beginning at a certain critical date in the growing season (e.g., from January 1, using 48°F as the developmental threshold). The grower may then use economic thresholds to determine the most appropriate management practices to implement.

4.9.3 CROP HEAT UNITS

Crop heat units (CHU) are based on a similar principle as GDD, but the calculations are different. The maximum and minimum temperatures are calculated by separate formulas. The CHU for a site is calculated as follows:

$$\text{Daily CHU} = (Y_{max} + Y_{min})/2$$
$$\text{where} \quad Y_{max} = (3.33 \times (T_{max} - 50°F)) - (0.084 \times (T_{max} - 50°F))^2$$
$$Y_{min} = (1.8 \times (T_{min} - 4.4))$$

Crop varieties are rated according to CHU accumulated between seeding time to physiological maturity. Corn hybrids are rated according to the CHU required to reach 32% kernel moisture, while soybean varieties are rated according to the CHU required to have 95% of the pods turn brown. CHU are accurate to within approximately 100 heat units. To avoid risk, a producer may select a hybrid or cultivar with less CHU rating than the prescribed.

4.10: Growth and Development–Related Crop Management Practices

The various stages in the lifecycle of flowering plants may be summarized as follows: Seed—seedling—early vegetative—late vegetative—early flowering—late flowering—early seed maturity—late seed maturity. The duration of each of these stages is variable and can be influenced to some degree by manipulating the growth environment. Various agronomic practices are conducted at these stages for optimal plant growth and productivity. Some of these activities are for protection (e.g., seed dressing, pesticide spraying), while others promote vegetative growth (e.g., application of fertilizers, growth regulators). Other practices regulate fruit set and maturity.

The principal plant morphological features are roots, stems, leaves, and flowers. Crop production management practices for each of these organs have been discussed in the corresponding sections.

SUMMARY

1. Crop productivity depends on how dry matter is partitioned.
2. Plant cultivars differ in patterns of partitioning of dry matter. Plants with a larger harvest index are more efficient at partitioning dry matter for higher productivity.
3. Growth is a progressive and irreversible process that involves cellular division, enlargement, and differentiation.
4. The rate of growth follows a general sigmoid curve pattern.
5. Growth may be measured on the basis of individual plants or a community of plants.
6. Yield may be biological or economic. The choice of economic part depends on the crop producer.
7. Harvest index is the proportion of the crop that is of economic importance. It is affected by genetic and environmental factors.
8. The concept of yield components is an attempt by scientists to construct the path by which the reproductive, developmental, and morpohological features of plants in a crop stand contribute to the yield of a specified plant product.
9. Certain developmental switches occur in plants, including photoperiodism and vernalization, that affect the onset of flowering.

REFERENCES AND SUGGESTED READING

Acquaah, G. 2002. *Horticulture: principles and practices,* 2nd ed. Upper Saddle River, NJ: Prentice Hall.

Foskett, D. E. 1994. *Plant growth and development: A molecular approach.* San Diego: Academic Press.

SELECTED INTERNET SITES FOR FURTHER REVIEW

http://www.ag.iastate.edu/departments/agronomy/corngrows.htm #how

Growth of corn; informative.

http://www.plant-hormones.bbsrc.ac.uk/education/kenhp.htm

Great discussion of plant growth hormones.

OUTCOMES ASSESSMENT

PART A

Answer the following questions true or false.

1. T F   All yield is biological.
2. T F   Photoperiod measures the duration of darkness.
3. T F   Photoperiodism is temperature-induced response in plants.
4. T F   Yield components are determined sequentially.
5. T F   Growth is an irreversible process.

PART B

Answer the following questions.

1. Crop yield may be biological or _____.

2. Provide a model for grain yield based on yield components.

   _____.

3. What is phenology? _____.

4. What are growing degree days? _____.

PART C

Write a brief essay on each of these topics.

1. Discuss the sigmoid growth curve.

2. Discuss the concept of yield components.

3. Discuss the concept of yield component compensation.

4. Discuss the occurrence of developmental switches in plants.

5. Discuss the application of growing degree days in crop production.

PART D

Discuss or explain the following topics in detail.

1. Discuss how crop producers can manipulate developmental switches in plants for increased crop productivity.

2. Can the yield potential of crops be raised indefinitely?

3. Discuss the role of growth regulators in plant growth and development.

Crop Improvement

PURPOSE

This chapter is devoted to discussing the genetic basis of crop improvement and the methods employed by plant breeders in the development of new crop cultivars.

EXPECTED OUTCOMES

After studying this chapter, the student should be able to:

1. Discuss basic Mendelian concepts.
2. Discuss the principles of plant breeding.
3. Describe the steps in plant breeding.
4. Discuss the various breeding systems.
5. Discuss the strategies for increased crop productivity (breeding objectives).
6. Discuss the role of biotechnology in crop improvement.

KEY TERMS

Allele	Genotype	Phenotype
Biotechnology	Harvest index	Recombinant DNA
Cross-pollination	Heritability	Self-pollination
Gene	Heterosis	Transgenic plant

TO THE STUDENT

Plant breeders genetically manipulate crop plants for high productivity. They manipulate plants based on their understanding and application of various basic sciences, especially genetics, the science of heredity. They are more concerned about the heritable aspects of crop productivity. A plant cannot be forced to produce a product for which it has no genes. Genes are not expressed in a vacuum but in an environment. Organisms can express their genetic potential only to the extent permitted by their environment. Therefore, a high-yielding cultivar is only as good as its cultural environment. That is, in order for a crop to yield highly, it must be provided adequate nutrients, moisture, and other essential growth factors. Some of the specific goals of plant genetic manipulation are to:

1. Improve the yield of a desired plant product (e.g., grain, oil)
2. Improve the quality of a desirable product (e.g., protein quality)
3. Adapt a crop to different climatic areas
4. Change the morphology of a plant (e.g., change a climber to a bush cultivar)
5. Protect plants from diseases and pests

Crop improvement is an art and a science. As the years go by, it is increasingly becoming science-based. In this chapter, both the traditional and cutting-edge plant breeding methodologies are discussed. The topics discussed include a review of pertinent genetic principles, conventional plant breeding methodologies, common plant breeding objectives, and the role of biotechnology in plant improvement.

5.1: THE SCIENCE OF GENETICS

5.1.1 GENES CONTROL THE EXPRESSION OF HERITABLE TRAITS

Gene. The basic unit of heredity comprised of base pairs in the DNA and functioning to determine the synthesis of a particular polypeptide.

Allele. One of two or more alternate forms of a gene.

The science of genetics is critical to understanding and conducting plant breeding. Gregor Mendel first discovered that plant traits, or characteristics, are passed from parents to offspring according to certain predictable patterns. He discovered that traits are controlled by hereditary factors called **genes.** The alternative expressions of a gene are called **alleles.** A pair of such alleles resides at a location on a chromosome and interacts to produce the observed trait. For example, Mendel observed that the gene for flower color in the pea plant (*Pisum*) had two alleles, one that determined purple color and the other white. When two plants with these contrasting expressions of this trait were crossed, the offspring had purple flowers. When two of the offspring with purple flowers were crossed (selfed), he observed both purple- and white-flowered plants in the progeny. These outcomes were explained by the genetic analysis in Figure 5–1. Purple flower is controlled by a dominant allele that suppresses the expression of the white flower allele (the recessive allele) in the first cross product or *first filial generation* (F_1). Upon selfing of two F_1s, the two alleles reappeared in the F_2. Mendel formulated two laws to explain these events:

1. *Mendel's law I (the law of segregation).* This law states that the pair of alleles of a gene separate (segregate) independently during gamete formation such that only one form of each pair ends up in each gamete (egg or pollen grain).
2. *Mendel's law II (the law of independent assortment).* This law states that pairs of alleles of a gene controlling different traits separate independently of each other during gamete formation and combine randomly to form zygotes.

The pattern is the same for cases involving more than one gene (*monohybrid cross*), only more complex. Try genetic analysis involving two genes (*dihybrid cross*) and then three genes (*trihybrid cross*). A plant breeder needs to know the inheritance of a trait before he or she can develop appropriate strategies to manipulate it. Breeding methods for traits controlled by dominant genes differ from those used for traits controlled by recessive genes.

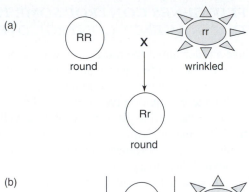

(a)

RR
round

X

rr
wrinkled

Rr
round

(b)

	R	r
R	RR round	Rr round
r	Rr round	rr wrinkled

FIGURE 5–1 Mendel's laws: (a) dominance and recessiveness and (b) segregation in a monohybrid cross with dominance. The expression of the recessive allele is suppressed in the F_1 but is expressed in the F_2.

Landmark genetic–based impact on crop production

Domestication of plants (the invention of agriculture in 10,000 B.C.) was accompanied by thousands of years of primitive and slow efforts at improving crop productivity. Farmers learned through experience and trial-and-error to identify and select better-performing plants with greater adaptation to the cultural environment. The quality and quantity of agricultural products gradually increased over long periods of time.

The first significant and dramatic change in the performance of crop plants came in the 1860s, when Gregor Mendel discovered the laws of inheritance. With a better understanding of plant genetics, scientists were equipped to manipulate plants with more purposefulness and greater success. Plant breeding became more of a science than an art. Plant breeding is viewed as the "old biotech." Crop plant products increased in quality. Further, crop productivity also increased. The most dramatic change in crop productivity came with the development of *hybrids*. Exploiting the phenomenon of hybrid vigor, hybrids

outyield their counterparts. These hybrids were first developed in cereal grains. Using the hybrid technology, the next major impact of genetics on crop productivity came with the Green Revolution of the 1960s. Short-stalked corn, rice, and wheat cultivars that were environmentally responsive (adapted to high-input agriculture) revolutionized crop production in the tropics. Hybrids have been developed for many crops including non-cereal plants.

The next most significant leap in crop productivity was brought about by the introduction of molecular techniques into plant breeding in the 1980s. Molecular plant breeding technology was able to circumvent some of the limitations of conventional breeding. Genes can now be transferred across literally all natural biological barriers, since deoxyribonucleic acid (DNA) is universal. New plant types equipped to exploit the production environment and resist diseases and insect pests have been produced. The "old" and "new" biotechnologies work together.

5.1.2 ONE OR A FEW GENES CONTROL SOME TRAITS; SEVERAL TO MANY GENES CONTROL OTHERS

Plant breeders need to have an idea about the number (few or many) of genes that control a trait. Traits controlled by one or a few genes are classified as *qualitative traits* or simply *inherited traits,* while traits controlled by many genes are called *quantitative traits* or *polygenic traits.* In the F_2, qualitative traits can be categorized into non-overlapping groups (e.g., white vs. purple flowers, green vs. yellow cotyledons) by counting. Quantitative traits (or metrical traits), on the other hand, are measured or weighed. Many traits of agronomic importance, such as yield, are quantitative traits. They are influenced by the environment to a greater extent than are qualitative traits. That is, changing the growth environment (through irrigation, fertilization, or temperature control) can change the degree of expression in production.

5.1.3 CERTAIN GENES MAY INTERFERE WITH THE EXPRESSION OF OTHER GENES

Can scientists always predict the exact outcome of a cross between two parents of known genotypes? There are certain occasions on which the outcomes of a cross cannot be predicted. In the expression of a trait that is controlled by several genes, the phenomenon of *epistasis* may cause one gene to mask the expression of another. When this occurs, the expected Mendelian ratios in the F_2 are not observed (i.e., *non-Mendelian inheritance*). For example, instead of 9:3:3:1, various outcomes such as 15:1 or 9:7 may be observed. Another factor that causes this deviation from expected ratios is *genetic linkage,* the physical association of adjacent genes on a chromosome that prevents independent assortment. It is desirable for certain genes to be linked. However, in certain situations a desirable gene is linked with an undesirable one. This presents a problem for breeders when they attempt to enhance the desirable trait. The process of *crossing over* that occurs during *meiosis* (the process by which gametes, or sex cells, are formed) breaks linkage naturally.

5.1.4 GENES INTERACT WITH THE ENVIRONMENT TO PRODUCE A VISIBLE TRAIT

Genotype. The genetic constitution of a cell or an organism.

Phenotype. The appearance of an organism as a result of the combined influence of its genetic constitution and environmental factors.

The genetic makeup (the sum total of all the genes in a cell) of an individual constitutes an individual's **genotype** (or *genome*). For academic purposes, genes are represented by letters, an uppercase for a dominant allele and a lowercase for a recessive allele (e.g., GG, Gg, gg). When the pair of alleles is identical (GG), the location *(locus)* of the gene is said to be *homozygous;* otherwise it is *heterozygous* (Gg). The observed trait is called a **phenotype.** A homozygous locus and a heterozygous locus have the same phenotype if the locus is under dominance gene action. The environment and modifier genes may also alter the manifestation of a genotype.

5.1.5 DNA IS THE GENETIC MATERIAL

The hereditary material DNA occurs in chromosomes. Is the DNA the sole hereditary material for all organisms? (A few organisms such as certain viruses have RNA and no DNA.) DNA consists of four *nitrogenous bases* (adenine, thymine, cytosine, and guanine, which are represented as A, T, C, and G, respectively), a *sugar* (deoxyribose), and a *phosphate.* When these three components link up, they form a *nucleotide,* the unit of

Mitosis is a process by which the nucleus of a cell divides after the chromosomes have doubled. This division results in two daughter nuclei that are identical to the nucleus of the parent. Mitosis is the foundation for growth and development of eukaryotes. A multicellular organism begins life as a *zygote,* the product of the fusion of an egg and sperm. Genetically, the zygote is a diploid (two sets of chromosomes). Mitotic division results in identical cells that later may differentiate into complex structures in the plant. Increase in size of the plant is largely due to an increase in the number of cells resulting from mitosis. When plants become injured, damaged tissue is repaired by producing new cells through mitosis.

Mitosis usually consists of two kinds of division: *nuclear* (or *karyokinesis*) and *cytoplasmic (cytokinesis).* The latter sometimes fails to occur. The sum of the phases of growth of an individual cell is called the *cell cycle.* Upon proper stimulation, the cells proceed to interphase (S). The S stage is where DNA synthesis begins in preparation for chromosome replication. It is followed by the G2 phase, in which DNA synthesis stops. Mitosis follows G2 and occurs in five stages:

prophase, prometaphase, metaphase, anaphase, and telophase.

The cell cycle is under genetic control. A gene called *cdc2* (cell division cycle 2) that codes for the enzyme cdc kinase is essential for the entry of a cell into mitosis. Another gene, P.53, is responsible for arresting the cell cycle in the G1 stage.

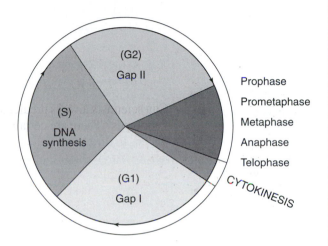

A summary of the cell cycle.

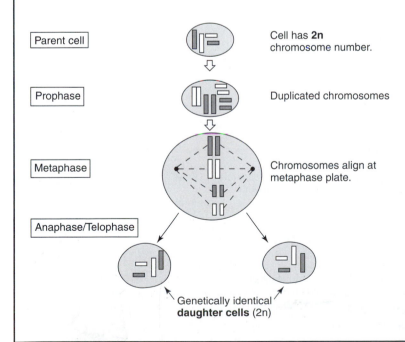

A summary of mitosis.

Meiosis, unlike mitosis, occurs only in sexually reproducing organisms. It involves two successive nuclear divisions, after which the resulting daughter nuclei have reduced amounts of genetic material and are unidentical to the parent nuclei. The resulting cells have half the chromosomes (haploid number) of the parent nucleus. This is the process by which gametes (egg and pollen) or spores are formed. The parental diploid number is restored following fertilization, the process that unites haploid cells. Meiosis ensures genetic variation among members of a species by shuffling the parental chromosomes, and by the process of crossing over (exchange of parts of chromosomes between nonhomologous, or unidentical, chromosomes). This is followed by the independent assortment of chromosomes into gametes, thereby creating biological variation.

The process of meiosis is more involved than mitosis. Just like mitosis, it is preceded by DNA synthesis. The first nuclear division (meiosis I) is described as re-ductional because the homologous chromosomes separate. The homologous chromosomes pair, or synapse (called *synapsis*), forming a structure called a *bivalent*. Bivalent chromosomes replicate to produce a tetrad of four chromatids. Prophase I in meiosis has five stages. During diplonema, non-sister chromatids exchange parts (*crossing over*) at one or a few areas where they intertwine. These regions are called *chiasmata* (singular, *chiasma*). At the end of meiosis I, the two homologs separate into two dyads of two chromosomes each, still joined at the centromere.

In meiosis II, the division is equational, each dyad separating into two *monads* of one chromosome each. Monads that were involved in crossing over are called *recombinant chromosomes* (have undergone genetic recombination to mix maternal and paternal genetic information). After telophase II, four unidentical gametes are produced.

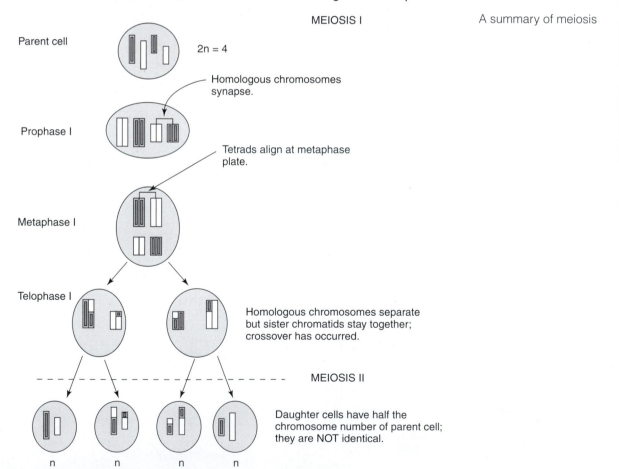

MEIOSIS I

A summary of meiosis

Parent cell   2n = 4

Prophase I

Homologous chromosomes synapse.

Tetrads align at metaphase plate.

Metaphase I

Telophase I

Homologous chromosomes separate but sister chromatids stay together; crossover has occurred.

MEIOSIS II

Daughter cells have half the chromosome number of parent cell; they are NOT identical.

n   n   n   n

DNA (Figure 5–2). Numerous nucleotides join to form a chain called a *polynucleotide chain.* A DNA molecule consists of two polynucleotide chains that are linked in predetermined fashion by hydrogen bonds. In the bonding of the two chains, C always bonds with G, and A always with T. The two strands are complementary.

The message of the polynucleotide chain occurs in a *genetic code.* This code is a *triplet code* (three nucleotides), also called a *codon,* that codes for a specific amino acid. Amino acids then join in various fashions to create long chains called *polypeptide chains.* These chains combine in various ways to form protein. Each gene codes for one polypeptide chain.

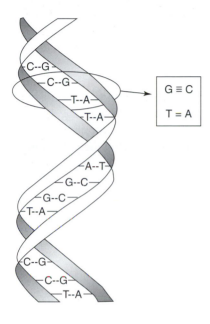

FIGURE 5–2 Structure of a DNA molecule. The double helical structure of the DNA molecule consists of a sugar-phosphate backbone connected by nitrogenous bases that pair in predictable and restrictive pattern.

Central dogma of molecular biology

Genetic information flow occurs in a certain fashion. The genetic message naturally flows in one direction, from DNA to proteins and not the reverse. This is called the *central dogma of molecular biology.* Because of advances in science, the reverse process is now routinely accomplished in the laboratory. Given a protein product, scientists can synthesize the corresponding DNA, called a *complementary DNA* or *cDNA.*

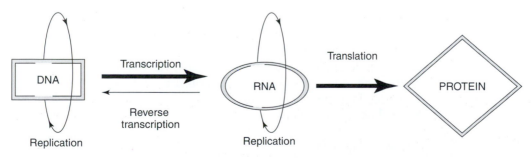

The central dogma of molecular biology describes the path of genetic information travel as normally from DNA to protein.

5.2: Plant Breeders Genetically Manipulate Plants to Produce New Traits

5.2.1 PLANT BREEDERS ARE GUIDED BY GENETIC PRINCIPLES IN THEIR WORK

Plant breeders are not able to genetically manipulate every trait. Therefore, it is important for them to determine whether or not a trait can be genetically manipulated before embarking on an improvement program. An underlying concept in making this decision is embodied in the following equation:

$$P = G + E$$

where P = phenotype, G = genotype, and E = environment. Simply stated, what is seen is a product of the interaction of the genotype with its environment. To change the phenotype, the genes that code for the trait may be changed (e.g., through crossing), the environment may be changed, or both factors may be changed. Changing the environment is done through agronomic practices (e.g., irrigation, fertilization, and pest control). Changing the genotype is permanent; changing the environment is only temporary, meaning that the conditions must always be reintroduced for the trait to be expressed.

Heritability. The degree of phenotypic expression of a trait that is under genetic control.

Another concept of importance to plant breeders is **heritability.** This is especially important when considering the improvement of a quantitative trait. Heritability is the degree of phenotypic expression of a trait that is under genetic control. Mathematically, it is expressed by the following equation:

$$H = Vg/Vp$$

where Vg = genetic variance and Vp = phenotypic variance. In this form, the formula estimates *heritability in the broad sense.* The values of this estimate range between 0.0 and 1.0 (or 0.0 and 100%). If the estimate of heritability is high, it indicates that the success of manipulating for improving a trait through breeding is likely. Otherwise, the trait may be enhanced through improving the cultural environment during crop production. In reality, both genotype and environment are important in production. A genotype is only as good as the environment in which it grows. High-yielding cultivars require good environment to attain their potential yield capacity.

5.2.2 CONVENTIONAL PLANT BREEDING FOLLOWS CERTAIN BASIC STEPS

How is plant breeding done? The general steps in a plant breeding program are as follows:

1. Determine the breeding objective(s).
2. Assemble genetic variability (or heritable variation).
3. Recombine the variation (cross, hybridize).
4. Select desirable recombinants.
5. Evaluate the selections.

Breeding Objectives

A breeding program is initiated for a specific purpose or objective. This may be yield increase, disease resistance, improved quality (e.g., high oil or protein content), and others as determined by the breeder. The method used for breeding depends on the objective (the trait to be manipulated and the direction of manipulation). A breeding objective may arise from the need of producers. For example, producers may desire a cultivar that does not shatter if harvesting time is delayed. They may desire disease resistance in an adapted cultivar or improvement in the architecture of the plant to adapt it to mechanized cultivation. Consumers may also dictate breeding objectives. If the end users of a crop product prefer a certain quality (e.g., sweeter taste, smaller size, high starch content), they will influence the breeding program in that direction. Plant breeders may also initiate objectives on their own. They may survey the producers and users to find problems that they encounter and then develop breeding programs to solve them. Selected objectives are discussed in detail later in this chapter.

Heritable Variation

Is all variation heritable? No, some variation is caused by differences in the environment. Without heritable variation, it is not possible to conduct a breeding program—that is, to genetically manipulate plants. If a breeder desires to increase the protein content of an existing cultivar, somewhere there must exist a genotype with high protein. Otherwise, such a trait must be induced, if possible, by mutation. Variation used in plant breeding can be obtained from many sources:

1. *Adapted cultivars.* These may be old and "retired," or "heirloom," cultivars, landraces (unimproved local variety), or even current cultivars in use. The advantage of using adapted material is that the new cultivar that is bred is already adapted to the environment and can be released to farmers much sooner.
2. *Recombinants.* New combinations or recombinants can be generated through recombination that occurs when genetically unidentical plants are crossed.
3. *Breeder seed.* When plant breeders conduct breeding programs, the newly created genotypes that do not make it to the farmer as cultivars may nonetheless have certain desirable qualities. These breeding materials are kept and may be incorporated into future breeding programs as needed.
4. *Plant introductions.* New plant types may be imported from other regions of the world. To aid researchers, certain facilities throughout the world are devoted to the collection and maintenance of plant variation. Plant breeders may request specific kinds of variation from such collections (called *germplasm banks*).
5. *Wild plants.* Cultivated plants have wild ancestors, or progenitors. These are excellent sources of genes for incorporating into breeding programs.
6. *Mutation.* Mutation is the ultimate source of variation. When all else fails, the breeder may attempt to induce the desired trait by subjecting plant materials to agents of mutation, or *mutagens.*

Recombination

The conventional way of creating variation is through hybridization, or crossing of two different plants. The effect of this action is the reorganization of the genotypes of the two plants into a new genetic matrix to create new recombinants. The ease of crossing varies from one species to another. In species where plants utilize pollen from the same plant, the breeder usually has a more difficult time with crossing. The flower of one plant must be designated as male and the other as female. The female flower is usually rid of all male organs by the often tedious process of *emasculation.* However, plants that are naturally cross-pollinated, and thus utilize pollen from other sources, do not need emasculation. The breeder simply plants the parents next to each other for pollen transfer to occur naturally by agents such as wind and insects.

It is critical that a cross be authenticated before it is used to continue a breeding program. A tag is often used, at a minimum, to identify the emasculated flower that becomes artificially pollinated. The seed from the putative cross can be further evaluated when the F_1 seed is planted, provided a genetic marker is incorporated in the breeding program. A *genetic marker* is a trait that is readily identified or assayed and is linked to another trait the breeder seeks to improve. When the marker is observed, the other trait, which is usually difficult to observe or evaluate, is assumed to be present.

Selection

After crossing, the breeding program proceeds with a series of selections (genetic discrimination) of desirable recombinants. The way selection is conducted depends on the method of breeding.

Evaluation

The evaluation phase of plant breeding entails testing a number of genotypes over several locations and years, in comparison with existing commercial varieties. The most desirable and superior performing genotype is then released as a new variety following existing protocol, and increased for distribution to farmers.

5.2.3 CONVENTIONAL PLANT BREEDING HAS CERTAIN LIMITATIONS

Conventional breeding is beset by the following weaknesses:

1. *Long duration.* The breeding program lasts for several to many years in some cases.
2. *Limited to crossing within species.* To hybridize, the parents must be compatible and belong to the same species (occasionally, crosses between different species is possible though problematic).
3. *Lower selection efficiency.* The methods used to sort among the enormous variations generated from a cross is not precise. This is the reason that markers are used to improve breeding efficiency.
4. *Large segregating population.* In order to have a high chance of identifying the recombinant of interest, plant breeders usually plant large numbers of plants in the segregating population (e.g., F_2). This requires large amounts of space and thus increases breeding expense.

5.3: METHODS OF BREEDING DEPEND ON THE BREEDING SYSTEM OF THE PLANTS

5.3.1 SOME PLANTS ARE PREDOMINANTLY SELF-POLLINATED, WHILE OTHERS ARE PREDOMINANTLY CROSS-POLLINATED SPECIES

Self-Pollination

Self-pollination, or *autogamy* is the mating system in which pollen grains are transferred from the anther of one flower to the stigma of the same flower or that of another flower on the same plant. Progeny from such mating are said to be naturally inbred and are more

Self-pollination. The transfer of pollen from the anthers of a plant to stigmas of flowers of the same plant "or a plant of the same genotype."

uniform genetically and phenotypically. *Cleistogamy* (self-pollination that occurs in a flower before the bud opens) is a mechanism enforcing self-pollination. Examples of crop plants that are normally self-pollinated are presented in Table 5–1. Self-pollinated crops may experience some natural cross-pollination, normally less than 5%.

Cross-Pollination

Cross-pollination, or *allogamy,* is the sexual production in which a stigma may receive pollen (more than 40%) from sources other than the flower itself. Such species are prone to adverse consequences of loss of vigor (inbreeding depression) when they are artificially crossed. Like self-pollinated crop plants, certain natural mechanisms enforce cross-pollination. Dioecy (the occurrence of male and female plants in one species) and *self-incompatibility* (lack of self-fruitfulness) encourage cross-pollination. Examples of crops with normally predominant cross-pollination are given in Table 5–2.

Cross-pollination. The transfer of pollen from the anthers of one flower of a plant to the stigma of a flower of a different plant.

Table 5–1 Selected Predominantly Self-Pollinated Species	
Barley	*Hordeum vulgare*
Clover	*Trifolium* spp.
Common bean	*Phaseolus vulgaris*
Cotton	*Gossypium* spp.
Cowpea	*Vigna unguiculata*
Flax	*Linum usitatissimum*
Oat	*Avena sativa*
Pea	*Pisum sativum*
Peanut	*Arachis hypogea*
Rice	*Oryza sativa*
Sorghum	*Sorghum bicolor*
Soybean	*Glycine max*
Tobacco	*Nicotiana tabacum*
Tomato	*Lycopersicon esculentum*
Wheat	*Triticum aestivum*

Table 5–2 Selected Predominantly Cross-Fertilized Species	
Alfalfa	*Medicago sativa*
Buckwheat	*Fagopyrum esculentum*
Cassava	*Manihot esculentum*
Coconut	*Cocos nucifera*
Cucumber	*Cucumis sativa*
Corn	*Zea mays*
Fescue	*Festuca* spp.
Onion	*Allium cepa*
Potato	*Solanum tuberosum*
Pumpkin	*Cucurbita* spp.
Rye	*Secale cereale*
Sugar beet	*Beta vulgaris*
Sunflower	*Helianthus annus*
Sweet potato	*Ipomea batatas*

5.3.2 COMMON PLANT BREEDING METHODS

1. *Breeding methods for self-pollinated species.* The common methods of breeding self-pollinated species (Table 5–3) include mass selection (Figure 5–3), pedigree selection (Figure 5–4), and backcross (Figure 5–5).
2. *Breeding methods for cross-pollinated species.* Mass selection is applicable to cross-pollinated species. A very widely used method of improving cross-pollinated species, also used for self-pollinated species to a lesser extent (because of practical reasons), is *hybrid breeding* (Table 5–4).

Table 5–3   Selected Methods of Plant Breeding

Key Concepts in Mass Selection (See Figure 5–3)

1. Mass selection is the oldest of all breeding methods.
2. It is an easy and rapid method of breeding; it can be done by growers themselves.
3. It improves the population rather than creating a cultivar from a single plant.
4. It is based on visual selection (phenotypic-based) of the desired trait; hence, heritability is important. Success is limited if heritability is low.
5. It is applicable to both self- and cross-pollinated species.
6. It is suitable for breeding for horizontal resistance to diseases.
7. The product is heterogeneous.
8. Since selection is based on phenotype, the breeder should ensure that selection environment is homogeneous.

Key Concepts in Pedigree Selection (See Figure 5–4)

1. It is an extension of the pure-line breeding method.
2. It is started by crossing parents of a known genotype.
3. Record keeping is the key activity for maintaining an accurate record of relationship (pedigree).
4. Success depends on operator's skill and ability to identify desirable individuals.
5. Method is long if only one growing season per year is available.
6. Products have high genetic purity.
7. Suitable for breeding for vertical resistance to disease.

Key Concepts in Backcross (See Figure 5–5)

1. It is a conservative method of breeding; no recombination is allowed to occur to create new recombination of traits.
2. It is used for improving an existing highly desirable cultivar that is deficient in one or a few genes (e.g., lack of resistance to a particular disease).
3. It is effective for transferring qualitative trait genes.
4. It is easier for transferring monogenic dominant allele.
5. The outcome is predictable from the beginning.
6. The derived cultivar is like its progenitor plus the transferred trait.
7. It is suitable for introgression through wide crosses.
8. It may be combined with other breeding methods for specific purposes.

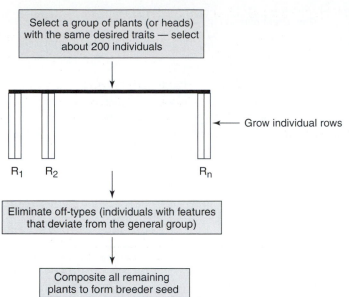

Select a group of plants (or heads) with the same desired traits — select about 200 individuals

Grow individual rows

R_1 R_2 R_n

Eliminate off-types (individuals with features that deviate from the general group)

Composite all remaining plants to form breeder seed

FIGURE 5–3 A summary of the mass selection breeding method. Mass selection is an "inclusive" strategy whereby only the atypical materials are discarded. The result is a product of a wide genetic base.

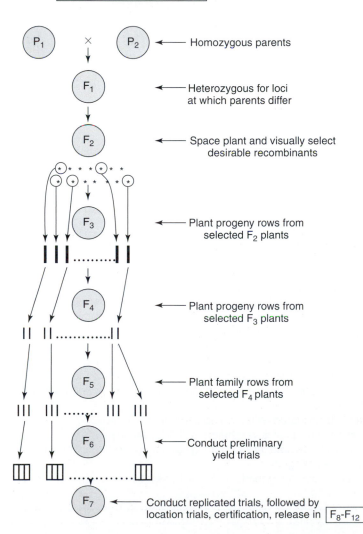

P_1 × P_2 Homozygous parents

F_1 Heterozygous for loci at which parents differ

F_2 Space plant and visually select desirable recombinants

F_3 Plant progeny rows from selected F_2 plants

F_4 Plant progeny rows from selected F_3 plants

F_5 Plant family rows from selected F_4 plants

F_6 Conduct preliminary yield trials

F_7 Conduct replicated trials, followed by location trials, certification, release in F_8-F_{12}

FIGURE 5–4 A summary of the pedigree breeding method. This method starts with a cross that is followed through the breeding program by maintaining records of lineage. This way, the breeder can always go back to reconstitute a particular cultivar. It is an "exclusive" method in which desirable recombinants are retained while all others are discarded.

155

FIGURE 5–5 A summary of the backcross method of breeding. This is a conservative method of breeding in which genetic recombination that produces genetic variability is prevented. The genotype of the original recipient parent is kept intact while incorporating only the donor gene.

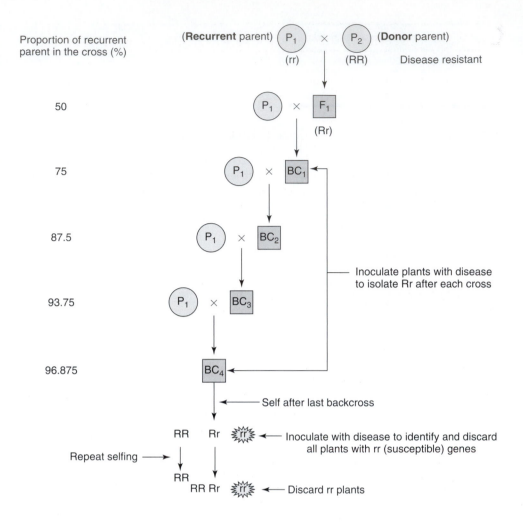

Proportion of recurrent parent in the cross (%)

(**Recurrent** parent) P₁ × P₂ (**Donor** parent)
(rr)   (RR)   Disease resistant

50   P₁ × F₁
                 (Rr)

75   P₁ × BC₁

87.5   P₁ × BC₂

                    Inoculate plants with disease
                    to isolate Rr after each cross

93.75   P₁ × BC₃

96.875   BC₄

                    Self after last backcross

          RR   Rr   rr ← Inoculate with disease to identify and discard
                        all plants with rr (susceptible) genes

Repeat selfing →

          RR
             RR Rr   rr ← Discard rr plants

Table 5–4 Breeding Hybrid Corn: Key Concepts

1. A hybrid is the progeny of a cross between two parents.
2. It depends on the phenomenon of **heterosis** (hybrid vigor) for success.
3. Commercial hybrid production requires the use of inbred lines (because they reproduce true to type and allow hybrids to be produced consistently year after year).
4. There are three basic types of crosses: single cross (A × B); double cross [(A × B) × (C × D)]; triple cross [(A × B) × C]. There are variations of these types.
5. The basic steps in hybrid productions are (a) development of inbred lines, (b) identification of compatible inbreds (combining ability for productivity), and (c) production of commercial seed by crossing and increasing seed from compatible inbred.
6. To cross, one inbred line is designated as male and the other as female (by emasculation—removing male parts of flower before self-fertilization).
7. Emasculation is tedious; hence, hybrids are easier to produce in a crop in which it can be done easily or avoided altogether.
8. To avoid emasculation, a cytoplasmic-genetic male sterility system is used in crops such as corn.

Heterosis. The increased vigor, growth, size, yield, or function of a hybrid progeny over the parents that results from crossing genetically unlike organisms.

5.4: COMMON PLANT BREEDING OBJECTIVES

Plant breeding is conducted to improve a variety of aspects of plants that eventually make for increased productivity and quality of product.

5.4.1 CROP YIELD IS OFTEN THE ULTIMATE GOAL OF CROP IMPROVEMENT

Yield is a product of the interaction of numerous physiological and biochemical plant processes. Breeding for higher yields requires understanding plant physiology, genetics, and agronomy, among other factors. In addition, the plant breeder has to be able to develop a model of that ideal genotype suited to a specified production environment.

Photosynthesis is the basis of crop yield through the production of dry matter. In order to increase crop productivity, the photosynthetic rate needs to be increased. Identifying or producing genotypes with improved morphology and functioning of the photosynthetic apparatus may accomplish this. Further, the dry matter produced should be partitioned or directed to organs of economic importance. This may be accomplished through the manipulation of the growth functions of these organs and the interaction between them.

5.4.2 BREEDING STRATEGIES FOR INCREASED CROP PRODUCTIVITY

There are several strategies for increased crop productivity, including improving photosynthetic efficiency, breeding for photosensitivity, determinate stem habit, dwarf stature, early maturity, yield stability, and improved harvest index.

Improving Photosynthetic Efficiency

Photosynthetic efficiency of a plant may be improved by improving light interception and leaf orientation, among other factors.

Light Interception Photosynthetic efficiency is the primary component of dry matter productivity. It depends on several factors. Canopy light interception is the most important of these factors. Photosynthetic rates (crop growth rate) are higher in genotypes with more erect leaves, lower extinction coefficients, and therefore higher critical leaf area indices. Genetic control of short, stiff, upright leaf habit in rice has been studied. In order for a crop to realize maximum yield, leaf area should expand to reach its optimum rapidly. Further, the leaves should remain photoactive for a long period and during senescence be able to supply assimilates to the reproductive and/or storage organs.

Leaf Orientation (Angle) Open crop canopy enhances the penetration of light into the crop canopy. Erect leaves (smaller leaf angle) are effective in increasing photosynthetic efficiency in small grains. These plants are planted at high densities for significant yield advantage.

Other Strategies Scientists have studied the genetics of stomatal frequency and conductance and found high stomatal frequency to be associated with high photosynthetic rates.

A hybrid is the product of a cross between two parents that differ in one or more inherited traits (i.e., the parents are not identical). Hybrid breeding is often used to breed cross-pollinated species (even though it can be applied to self-pollinated species, but with less ease). Hybrid production exploits the phenomenon of heterosis. The following conditions are necessary to increase the chance of success of a hybridization program:

1. *Heterosis.* The parents must be compatible enough (high combining ability) to manifest a high degree of heterosis in the offspring.
2. *Elimination of fertile pollen from the female parent.* The female parent should not contribute any pollen to the progeny, only the egg. There should be a practical and inexpensive way of excluding the female pollen from the cross.
3. *Adequate pollination and fertility restoration.* The pollen source should supply adequate pollen to pollinate all plants in the breeding program.
4. *Availability and maintenance of parents.* Parents used in hybrid programs are inbred lines. There should be a system in the breeding program to maintain the parental lines in good condition.
5. *Efficient pollen transport.* Hand pollination is laborious, slow, and expensive. Commercial

hybrid production should utilize economical pollen distribution—e.g., by wind or insects.
6. *High economic return.* Hybrid seed production is expensive and hence uneconomical to use in plants with low economic return.

Types of hybrids

There are three basic types of hybrids: single cross, double cross, and three-way cross. A single cross is the most common hybrid produced. Its products have the highest heterosis but are less uniform.

Eliminating emasculation

In corn the female plant is detasseled. The silk is covered with a paper bag to eliminate any stray female pollen until it is time to be pollinated with the desired sources of pollen. This process is slow and tedious. To circumvent this, the female is rendered male sterile through deliberate introduction of sterility genes into the plant. The system of male sterility commonly used is *cytoplasmic male sterility.* A comparison of the two methodologies is shown in the figure on the following page for a single cross and a double cross. The inbred male-sterile lines (female) are called A lines while the maintainer lines are called B lines. The fertility restorer lines are called R lines.

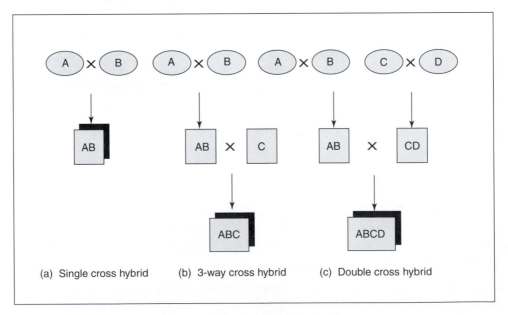

(a) Single cross hybrid   (b) 3-way cross hybrid   (c) Double cross hybrid

The three types of hybrids: single, three-way cross, and double cross.

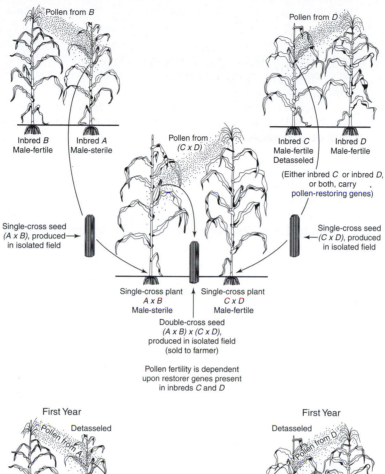

Pollen from B

Inbred B
Male-fertile

Inbred A
Male-sterile

Pollen from
(C x D)

Pollen from D

Inbred C
Male-fertile
Detasseled

Inbred D
Male-fertile

(Either inbred C or inbred D,
or both, carry
pollen-restoring genes)

Single-cross seed
(A x B), produced
in isolated field

Single-cross seed
(C x D), produced
in isolated field

Single-cross plant
A x B
Male-sterile

Single-cross plant
C x D
Male-fertile

Double-cross seed
(A x B) x (C x D),
produced in isolated field
(sold to farmer)

Pollen fertility is dependent
upon restorer genes present
in inbreds C and D

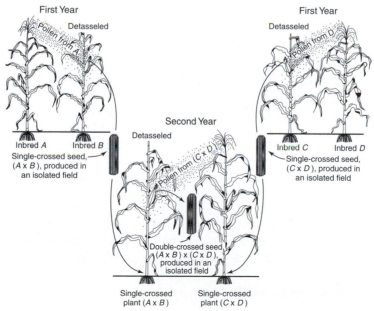

First Year

Detasseled

Pollen from A

First Year

Detasseled

Pollen from D

Second Year

Detasseled

Inbred A

Inbred B

Single-crossed seed,
(A x B), produced in
an isolated field

Pollen from (C x D)

Inbred C

Inbred D

Single-crossed seed,
(C x D), produced in
an isolated field

Double-crossed seed,
(A x B) x (C x D),
produced in an
isolated field

Single-crossed
plant (A x B)

Single-crossed
plant (C x D)

Older methods of conventional breeding of hybrid corn required emasculation or detasseling.
(Source: USDA)

Breeding for Photoinsensitivity

Plants may be categorized as short-day, long-day, or day-neutral. The daylength requirement of plants may limit their adaptation. Day-insensitive plants flower under both long-day and short-day conditions. Seed stocks of such new cultivars can be increased very rapidly. Major world cereal crops (wheat, rice, corn) have wide adaptation because of the development of photoinsensitive cultivars. Photo sensitivity is under genetic control, being single recessive or double recessive in rice. Similar genetic control is reported in other species. In soybean, photoinsensitivity is dominant over photosensitivity.

Both conventional breeding and mutation breeding have produced day-neutral genotypes. Day-neutral cultivars have been bred in cowpea, *Phaseolus* bean, potato, cotton, and pigeon pea.

Breeding for Determinate Stem Habit

Determinacy is a trait that may occur naturally or can be induced. Among other traits, determinate genotypes are short stemmed, and they have fewer and thicker internodes. These plants are less prone to lodging, have a shorter life cycle, and have early pod set. They have high harvest index and are amenable to mechanization. These genotypes in which vegetative growth becomes attenuated at some point in time have been shown to increase yield in certain species. This trait has been most widely studied in soybean. Determinacy is controlled by a single gene, dt_1, and indeterminacy, Dt_1. Semideterminacy was found to be controlled by a single dominant gene, Dt_2. Further, dt_1 is epistatic to Dt_2 or dt_2. Determinate cultivars require less space and can be planted at high density to increase yield.

In terms of grain yield, climbing, or pole, cultivars tend to yield higher than determinate, or bush, cultivars. This deficit is compensated for, however, by planting bush cultivars in higher densities. A recessive gene controls determinate stem habit in *P. vulgaris*.

Breeding for Dwarf Stature

Dwarf genes were pivotal in the success of the Green Revolution. The Norin 10 dwarf genes (Rht_1, Rht_2) used to improve Mexican wheat by Norman Borlaug in 1954 revolutionalized tropical agriculture. These vegetative growth reducing genes produced high harvest index and increased grain yield. The Norin 10 genetic system consists of two partially recessive independent genes that act additively together. Duplicate recessive genes control dwarfness in pigeon pea, while additive and non-additive gene action for height has been recorded in peas. Dwarf genomes have been developed in rice using nonconventional methods such as tissue culture and mutation breeding.

Dwarf cultivars are more responsive to fertilization without lodging and have high grain yield. The root system of dwarf cultivars of wheat were generally deeper penetrating than the non-dwarf cultivars. This is advantageous when water and nutrients are limiting in production. Dwarf cultivars have better partitioning of assimilates to the developing ear. This results in higher grain yields.

Breeding for Early Maturity

Photoinsensitivity is desirable for wide adaptation. Coupled with early maturity, crop producers are able to produce multiple crops per growing season. It also enables crops to be produced in regions where the growing season is short (e.g., arid regions). Early-maturing crops yield lower than late-maturing ones, but their smaller plant size enables the producer to compensate for low yield by adopting high plant populations.

Early-maturing crop plants have shorter vegetative and reproductive phases and hence reduced dry matter accumulation. The challenge in breeding early-maturing and high-yielding cultivars is to reduce the vegetative phase while protracting the grain or seed filling period for maximum partitioning of dry matter into economic parts.

Several genes control earliness. These genes may have dominant, additive *x* additive, or epistatic effects. In wheat, earliness was found to be partially dominant over lateness and of high narrow sense heritability. Early flowering in rice is dominant to late. Flowering duration in cotton is controlled by additive gene action. However, some dominance gene action has also been reported. Induced mutagenesis (mutation breeding) has been employed to develop early-maturing mutants in barley, maize, rice, cotton, and other species.

Improving Harvest Index

All yield is biological, but not all yield is economic. **Harvest index** is the ratio of economic yield to total plant biological yield. It is believed that low grain yield in legumes is attributable primarily to low harvest index. Harvest index is highly correlated with grain yield. Cultivars with high harvest index are found to be more efficient in converting plant growth nutrients into grain production. A high harvest index is a good indication of yield stability. It is less variable under changes of environment than biological or grain yield. There is significant genetic variability in high harvest index in different varieties of most crop plants. Highest harvest index values for cereal range between 0.50 and 0.60.

Harvest index. The proportion of the crop that is of economic importance.

High harvest index is partially dominant over low harvest index. Breeders have several strategies for increasing harvest index. Reducing vegetativeness by, for example, reducing plant height (dwarfing) has increased harvest index. Dwarf rice cultivars, for example, generally have higher harvest index than tall ones. To be economic, high harvest index should be accompanied by early maturity. This decrease in crop duration, however, should affect only the vegetative phase of plant development.

Other plant architectural traits that should complement high harvest index are determinacy, suppressed tillering, and branching (uniculm). Further, with the reduced stature of plants, producers can increase yield by increasing plant population density.

Achieving Yield Stability

Yield stability is a key breeding objective. However, it is difficult to achieve. It is the attribute of a crop to maintain its yield over changing environments. This attribute is especially important in production regions where the growth environment is variable and capital for providing supplemental inputs is limited. For a cultivar to be successful in production, it should perform predictably across the range of environments in which it will be produced. That is, it should have minimum genotype *x* environment interaction.

Yield stability is achieved through three general categories of mechanisms—*genetic heterogeneity, yield component compensation,* and *tolerance to environmental stresses.* A crop cultivar that has genetic heterogeneity (mixture of genotypes) has insurance against changing environments. Yield buffering of multiple cross (double, three-way crosses) genotypes tends to make them more stable than single cross genotypes.

The phenomenon of yield compensation enables one yield component to make up for the reduction in the expression of another. Crop yield is affected by biotic and abiotic factors. The effect of stress depends on the stage of development of the plant.

Breeding for Yield per se

Breeding for yield per se is a daunting task. This is because yield is a genetically complex trait that is quantitatively controlled. It has low heritability. Yield is a product of the interaction among numerous physiological and biochemical processes.

Since yield is a quantitative trait, plant breeders employ quantitative genetics in selecting appropriate parents for crossing. Molecular markers have been developed to tag quantitative trait loci (QTLs) to facilitate breeding of complex traits.

Some breeders attempt to improve yield by targeting certain yield components or physiological processes. These components (e.g., seed weight, number of seeds per pod) tend to be more heritable than yield per se. Various genetic analyses are available for use in predicting the performance of a parent in a cross. One of the oldest and widely used is the Jink's *diallel analysis,* which evaluates parental lines crossed in all possible combinations. This method enables plant breeders to identify parents most likely to produce the highest-yielding segregates.

Finlay-Wilkinson's strategy of breeding for yield per se focuses on selecting for increased adaptability. Good yielders perform well over a wide range of environments.

Breeding for Improved Product Quality

There are numerous quality factors in crop production; some of them are chemical, while others are physical. One nutritional component of importance in food crops is proteins. Proteins may be classified according to their solubilities in various solvents as follows:

1. *Albumins*—proteins that are soluble in salt-free water
2. *Globulins*—proteins that are insoluble in salt-free water but are soluble in dilute neutral salt solutions
3. *Prolamins*—proteins that are soluble in 60 to 80% ethyl alcohol or dilute alkali but not in the solvents mentioned in 1 or in 2
4. *Glutelins*—proteins that are soluble in dilute alkali or acid

Albumins and globulins are involved in regulatory processes, enzymatic reactions, nutrient reserves, and other roles. Prolamins *(zein)* are the nutritionally least useful of the four categories of proteins. They lack lysine and tryptophan. The *opaque*-2 and *floury*-2 genes have the effect of reducing the prolamin protein and increasing albumins and globulins.

Plant breeders breed for quality of the economic product. The goal in this objective differs according to the end use of the product. For food, nutritional quality includes protein content and protein quality. A success story in this regard is the breeding of the *high-lysine corn* that specifically improved the content of the amino acid lysine. Quality improvements may focus on taste, flavor, quantity of a nutritional factor, shape of fruits, and others. For industrial products such as cotton, quality may mean fiber length, strength, color, or some other factor. One of the outstanding modern day successes in breeding product quality is the development of the *Flavr Savr* tomato that has prolonged the shelf-life of vine-ripened tomatoes.

Breeding for Disease Resistance

The biological environment of crop plants includes pathogens that cause yield-reducing diseases. Under pathogenic attack, plants may resist or tolerate the pathogen or, if susceptible, may die. There is a genetic basis of host-pathogen relationship. The genetic basis for disease resistance is described by the *gene-for-gene concept.* For each gene that confers resistance in the plant (host), there is a corresponding gene that confers virulence

Table 5-5 The Genetics of Disease Resistance in Plants: The Gene-for-Gene Concept

| | Resistance or Susceptibility Genes in the Plant | |
Virulent or Avirulent Genes in the Pathogen	R (Resistant) Dominant	r (Susceptible) Recessive
A (avirulent) Dominant	AR (−)	Ar (+)
A (virulent) Recessive	aR (+)	ar (+)

Note: Where − is incompatible (resistant) reaction (no infection) and + is compatible (susceptible) reaction (infection develops). AR is resistant because the plant (host) has a certain gene for resistance (R) against which the pathogen has no specific virulent (A) gene. This does not mean other virulent genes do not occur. Ar is susceptible due to lack of genes for resistance in the host and hence susceptible to other virulent genes from the pathogen. The aR host has the resistance gene but the pathogen has a virulent gene that can attack it; ar is susceptible because the plant is susceptible and the pathogen is virulent.

Table 5-6 Genetics of Disease Resistance in Plants: A Case of Multiple Genes

| | | [Resistance (R) or Susceptibility (r) Genes in Plants] | | | |
		R_2R_2	R_1r_2	r_1R_2	r_1r_2
[Virulence (a) or avirulence (A) in the pathogen]	A_1A_2	−	−	−	+
	A_1a_2	−	−	+	+
	a_1A_2	−	+	−	+
	a_1a_2	−	+	+	+

in the pathogen, and vice versa (Table 5–5). *Virulence* is the capacity of the pathogen to overcome the influence of a resistant gene in the host. *Avirulence* is the inability of the pathogen to infect the host. This relationship is very specific. A fungal pathogen has a range of different types (called *physiological races*) that differ in their capacities to overcome the resistance in the host. Unless a race of a pathogen carries all the virulent genes corresponding to the number of resistance genes in the host, it will be unable to attack the host (Table 5–6). A challenge to plant breeding against fungal diseases is the problem that arises because some pathogens may mutate rapidly to produce the appropriate virulent genes to overcome the resistance in plants. Plant breeders thus have to breed new cultivars regularly using new sources of resistance to the pathogen. Sometimes, they incorporate multiple genes for resistance to pathogens in one cultivar, in order to delay the breakdown by virulent genes.

Breeding for Resistance to Insect Pests

The genetic basis of insect resistance is similar to that of fungal diseases. Plant breeders employ one of three strategies in breeding against insect attack. One strategy involves improving the mechanical or physical resistance to insect access to plant tissue through the strengthening or creation of barriers to feeding by sucking or chewing insects. The epidermal layers of leaves or stems are thickened or strengthened through breeding. Another strategy is to introduce genes in the plant that modify the plant's metabolism to produce toxins *(antibiosis)* that injure insects upon feeding on the plant. The third approach is to modify plant morphology or palatability such that insects avoid the plant. Pubescence on the plant interferes with oviposition in insects.

Breeding for Lodging Resistance

Lodging is the tendency for the stem of a plant to incline. The bending may occur at the base of the stem or the roots (especially in cereals), or the stem may break. Lodging may occur at any stage during plant development. The effect of lodging depends on the stage in which it occurs. Recovery from lodging may be possible if it occurs early in development.

Lodging is caused by several factors:

1. Stem or stalk decay caused by disease which, in turn, is caused by fungi such as *Gibberella* spp
2. Attack from insect pests such as stalk borer, rootworm, and cutworm
3. Weather conditions, including strong wind, excessive rain, hail, and other storms, cause stems to bend or break
4. Cultural damage from improper use of machinery that causes plants to be injured; planting at high density causes etiolation or spindly growth and weak stems; further, overfertilization causes excessive vegetative growth, which results in top-heavy plants that are easily toppled or bent by winds
5. Genetic susceptibility of certain plants
6. Heavy bearing or yielding, which causes plants to become top-heavy

Harvest Losses Lodging may cause yield reduction due to reduction in net assimilation rate from reduced interception of light and absorption of nutrients and water. Apical dominance may be destroyed, leading to increased tillering and branching. The additional vegetative growth is wasteful if it does not contribute to harvestable yield. Severe lodging causes the grain to be too close to the ground such that the combine is unable to harvest it.

Reduced Quality of Product In severely lodged plants, the ears, pods, or heads may come into contact with the soil and cause pathogenic infection. Dry seed may sprout in the pod. Deterioration reduces product quality and the harvestable quantity.

Reduction of lodging can be reduced through the following actions:

1. Space plants properly (proper density) to avoid etiolation.
2. Fertilize plants at appropriate rates and good nutritional balance for proper growth.
3. Use lodging-resistant cultivars for planting.
4. Harvest on time to avoid stalk rot that occurs—for example, in mature corn.
5. Spray against insect pests and diseases.
6. Use dwarf cultivars.

Field crops in which lodging is a problem include small grains, corn, soybean, and sorghum. To breed for lodging resistance, breeders seek to improve plant architecture by improving the culm strength (stiff, sturdy stalks) and reducing plant stature (dwarfing). Resistance to insects and diseases that weaken the stalk and roots is also a goal in breeding against lodging. A major accomplishment in this regard is the discovery and use of the *Bt* gene, developed by Monsanto Company, which protects plants that have the gene (e.g., *Bt* corn and cotton) from insects that contribute to lodging.

Breeding for Shattering Resistance

Shattering is the discharge and loss of seeds from pods or heads prior to harvest. Plants such as soybean and small grain are prone to shattering. A delay in harvesting by a few days can result in significant yield loss. Shattering resistance is a quantitative trait. Cultivars that are resistant to shattering have been developed for major field crops.

Breeding for Heat and Drought Resistance

Moisture stress and excessive heat occur in regions of high temperature and low rainfall. These conditions, coupled with strong winds, cause accelerated desiccation of plant tissue. Respiration also occurs at a higher rate than photosynthesis. Consequently, yield is drastically reduced. The effect on yield depends upon the stage in the growth cycle at which drought occurs. Heat stress is most devastating to crop productivity when it occurs during the flowering period. At this stage, pollen viability, stigma receptivity, and seed formation are reduced by excessive heat.

There are two basic mechanisms for drought resistance in plants. The plant may avoid *(avoidance mechanism)* or tolerate *(tolerance mechanism)* it. Genotypes with deep root systems can exploit moisture from a lower depth in the soil and thereby avoid moisture stress. Some species have avoidance mechanisms such as leaf rolling under moisture stress, the effect being a reduction in transpiration loss. Plant breeders sometimes breed early maturity in genotypes for use in drought-prone areas. This technique allows the genotypes to flower before severe heat stress sets in.

Breeding for Winter Hardiness

Winter hardiness is important in regions where winters are severe. Under such weather conditions, plant tissues may freeze or plants may heave (uplift).

5.5: BIOTECHNOLOGY EXCEEDS CONVENTIONAL BREEDING METHODS

Biotechnology offers new avenues for manipulating plants beyond the capabilities of conventional breeding methods. Conventional plant breeding involves manipulating plants at the whole plant level. The breeder manipulates plants on the basis of phenotype. A plant with purple flowers is assumed to have genes for purple flower. Further, crossing is limited to parents that are compatible by way of pollination.

Molecular plant breeding involves the manipulation of plants at the cellular and subcellular levels. The DNA is manipulated directly, using the methodologies of biotechnology. **Biotechnology** is a collection of tools based on a living system used to manipulate organisms or use them to make products. Conventional plant breeding is thus considered to be biotechnology.

One of the tools of biotechnology is **recombinant DNA** technology (or *genetic engineering*), whereby a piece of DNA can be taken from any organism and transferred into another. The DNA of all organisms obeys the same laws (i.e., DNA is universal). Through advances in science, plant breeders are able to circumvent a major limitation in conventional plant breeding—that is, genes do not have to be mixed—by recombination through meiosis.

When the segment of the polynucleotide chain corresponding to the gene of interest has been identified, it is chemically snipped out using an enzyme called *restriction endonuclease*. The isolated piece of DNA is unable to function independently. It is inserted

Biotechnology. A collection of tools used to manipulate organisms or use them to make products.

Recombinant DNA. The biotechnology whereby genes can be mixed across biological barriers.

Perhaps the most prominent of the biotechnologies is the recombinant DNA (rDNA) technology. Also called *gene cloning* or *molecular cloning,* rDNA involves a number of research protocols employed to transfer DNA from one organism to another. The key activities in rDNA research are as follows: (a) the DNA to be transferred is identified in the donor and cleaved; (b) the target DNA is inserted into a carrier that will deliver it into the DNA of the host; (c) the target DNA is transferred into and maintained in a host cell until the next step; (d) the host cell with the foreign DNA is selected from among numerous others without the foreign gene; and (e) depending upon the goal of the research, the incorporated target DNA can be manipulated to be expressed in the host cell.

Upon identification, the target gene is enzymatically cleaved by using *restriction endonucleases* (bacterial enzymes). These enzymes cleave DNA upon recognizing enzyme-specific base sequences (called

recognition sequences) that are typically a few base pairs long (e.g., 4, 6, or 8). The cleaved piece of DNA is inserted into a carrier molecule called a *cloning vector.* A common cloning vector is a bacterial *plasmid,* a circular DNA molecule that is self-replicating and double-stranded. Plasmid cloning vectors are engineered to have recognition sites embedded in genes for resistance to antibiotics. These are used in a selection scheme to help identify and select cells that have incorporated the target gene. The plasmid is cleaved with the same restriction enzyme used to cleave the target DNA. The recombinant plasmid (containing the target DNA) is inserted into a bacterium by a process called *transformation.* The bacteria are transformed using a laboratory technique called *electroporation,* which involves incorporation of the plasmids in an electric field. Once inside the bacterial host, it can be maintained and replicated clonally.

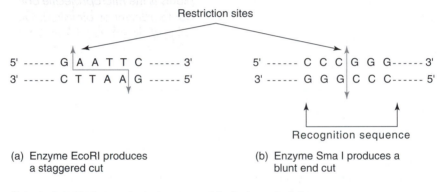

(a) Enzyme EcoRI produces a staggered cut

(b) Enzyme Sma I produces a blunt end cut

Selected restriction endonucleases and their characteristics.

into a carrier molecule capable of replication called a *cloning vector.* This molecule, a circular DNA called a *plasmid,* is found in bacteria. The recombinant plasmid (plasmid with the alien piece of DNA) is reinserted into a bacterium. The bacterium is then said to be *transformed.*

The next stage is to transfer the cloned gene into a plant. Cloned genes are transferred into plants in a variety of ways. The tumor-forming bacterium *Agrobacterium tumifaciens* is used as carrier of the recombinant plasmid. This bacterium is then introduced into plant cells in tissue culture. Some plant cells incorporate the recombinant plasmid and become transformed. They then individually develop into plants with foreign DNA in their genome and are called **transgenic plants**.

Transgenic plant. A plant whose genotype consists of introduced foreign DNA from an unrelated source.

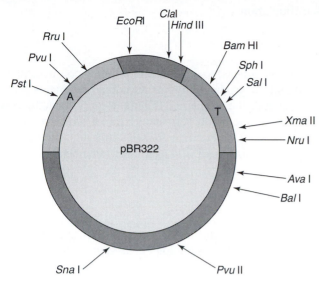

A restriction map of a plasmid.

cell types that grow. Cells from colonies in the first selection step are transferred to a tetracycline-agar plate. Cells with intact pBR322 will grow; transformed cells whose tetracycline resistance gene has been disrupted by the insertion of the target gene will not grow. The corresponding colonies on the first plate carrying the pBR322-cloned DNA constructs are identified and maintained in cultures.

The cloned target gene is transferred into plant host cells in a variety of ways. The "tumor-inducing" (Ti) plasmid found in most strains of the soil bacterium *Agrobacterium tumifaciens* is used as a vector to carry target genes into plants. The Ti plasmid used is incapacitated to be ineffective in causing tumors. Upon inserting the target gene into the Ti plasmid, the bacteria are attached to plant cells for the bacteria to enter through wounds to naturally enter the cell's genome. Target genes can also be transferred by physical methods. A physical DNA delivery system for plants is the *microprojectile bombardment* (or *particle bombardment* or *biolistics*). Gold or tungsten spherical particles of about 1 to 4 mm diameter are coated with DNA and accelerated in an apparatus called a *particle gun* or *gene gun,* using gunpowder or compressed gas, into plant tissue.

After integrating the foreign DNA into the plant genomic DNA, the transformed cells are identified with the aid of selectable markers and genes whose proteins produce a readily detectable response to a specific assay. These are called *reporter genes.* The transformed cells are cultured and nurtured into plants, and the plants are called *transgenic plants.*

After transformation, the successful cells are selected by a two-step process. For example, if the *pBR322 cloning vector* is used along with the *Bam*HI restriction enzyme, the *Bam*HI gene is incapacitated as a result of the insertion of the target gene. The transformation mixture is plated onto a medium containing the antibiotic ampicillin. Bacterial cells with intact pBR322 or recombinant pBR322 will grow; all nontransformed cells will not. The purpose of the second selection step is to distinguish between the two

5.5.1 BIOTECHNOLOGY IS USED TO SOLVE A VARIETY OF PROBLEMS IN CROP PRODUCTION

There are some general ways in which biotechnology is used in agriculture to benefit society:

1. Biopharming
2. In vitro techniques
   - Micropropagation
   - Somatic cell genetics
   - Transgenic plants

3. Artificial seed production
4. Diagnostics and control of diseases and pests
5. Remediation
6. Biofertilization and phytostimulation

Biopharming

Biopharming uses plants (and other organisms) to develop and produce pharmaceuticals for humans. Using transgenic animals to produce proteins and other chemicals for humans is well advanced. Successes include the production of human serum albumin (used to treat emergency blood losses and chronic blood deficiency), factor VIII (required by the body to repair injuries of blood vessels), and human growth hormone. These are produced in goat milk. The use of plants in this regard is in its infancy. Plants under consideration include tobacco, banana, and potato.

Micropropagation

Micropropagation is the propagation of plants in *tissue culture* (nurturing of plant parts in vitro, sometimes to produce full plants). This technology is used in a variety of ways. The advantages of this technology include the following:

1. Quick way to generate new plants
2. Large-scale production of new genotypes
3. Requires only a small amount of plant tissue
4. Raising pathogen-free materials from diseased plants

Micropropagation is conducted in a sterile environment. The tissue used to start the process may be obtained from any part of the plant, such as the leaf, stem, pollen, or root. The starting material is called the *explant*. It is surface-sterilized and placed on a sterile nutrient medium consisting of macronutrients and micronutrients, sugar, vitamins, and growth regulators. A common recipe for a medium is the Murashige and Skoog (MS) medium. The explant changes into an amorphous, undifferentiated mass of tissue called a *callus*. The growth regulators can be manipulated to induce rooting and other vegetative growth into a full plant.

Sometimes, the growth medium is manipulated by scientists to induce mutations through the inclusion of mutagens (mutation-inducing substances). This procedure is called *somatic cell selection*. Heritable variability can, however, arise spontaneously in tissue culture without any intervention by researchers and is called *somaclonal variation*. Some of these variants have agronomic value and have been used by plant breeders for crop improvement.

Disease-free propagules can be obtained from plants infected by systemic pathogens. The older cells have the systemic pathogen (virus or bacterium), but the meristematic tissue is usually free of such pathogens. Scientists are able to carefully extract the disease-free material and culture it in tissue culture. This is a way of purifying an infected parental stock.

In crop improvement, plant breeders cross genetically divergent plants to produce new recombinants. This is usually successful if the parents derive from the same species. Sometimes, plant breeders find it necessary to cross parents from different species, a procedure that is frequently problematic. The hybrid embryo often fails to develop properly. Plant breeders then remove the immature embryo and culture it in tissue culture, a technique called *embryo rescue*.

Somatic Cell Genetics

In addition to propagating plant materials directly in tissue culture, scientists are able to propagate haploid plants from microspores (e.g., pollen grains). Haploid cells can then be hybridized (called *somatic hybridization*) by the process of *protoplast fusion.* The process enables crossing for introgression of agronomically useful traits across genetic barriers.

Transgenic Plants

The genetic engineering of crop plants has revolutionized crop improvement by allowing gene transfer across genetic crossing barriers. This substantially improves the results of the techniques of embryo rescue of offspring from interspecific crosses and cell fusion techniques. This novel technique, in theory, does not only allow gene transfer but allows the plant breeder to select the organ in which to express the gene, as well as the strength of expression. Major accomplishments in transgenic plants include those described in Table 5–7.

Synthetic Seeds

Plant in vitro techniques can be used to induce a high rate of *somatic embryogenesis* (the development of embryos in tissue culture). These structures are then encapsulated in biodegradable protective coating to produce *synthetic seeds,* or *artificial seeds,* for propagation. This coating may be fortified with fertilizers, pesticides, or some other seed dressing. Artificial seeds provide a way of propagating plants clonally.

Diagnostic Tools and Control of Plant Pests

Numerous biotech diagnostic tools have been developed for early monitoring and detection of plant diseases, pests, and chemical residue in the environment. These include nucleic acid probes, dot-blot hybridization, monoclonal antibodies, and enzyme-linked immunosorbent assay (ELISA).

Inorganic pesticides damage the environment. Biopesticides are safer to humans and the environment and are more pest-specific. Microbial sprays in use include the *Bacillus thuringiensis* spray for controlling cutworms, corn borers, and cabbage worm, among others, and the aerial application of spores of the fungus *Collectotrichum gloesporides* for controlling northern jointvetch in rice fields. Biocontrol is employed in the postharvest control

---

Table 5–7 Examples of Plants Genetically Engineered for Specific Traits

Species	Traits Incorporated
Corn	*Bt* gene for resistance to European corn borer; herbicide resistance (Roundup® ready)
Cotton	*Bt* gene as in corn
Rice	Provitamin A (Golden rice)
Tomato	*Flavrsvr* (for delayed postharvest spoilage)
Canola	Lue-enkephalin (a mammalian neuropeptide)
Sweet potato	Protein quality gene
Sugar beet	Herbicide resistance (phosphinoticin or Basta®)
Arabidospsis	Polyhydroxybutyrate (biodegradable plastic)

This list is designed only to sample the immense variation in the use of plant biotechnology to solve various problems. Numerous species have been transformed and successfully regenerated to express a wide variety of genes.

Biotechnology is embroiled in a great deal of controversy. Because biotechnology is truly ubiquitous in its impact, the debate involves the news media, consumers, politicians, government officials, business executives, and scientists, to name a few. Much of the controversy is because the field of biotechnology is developing rather rapidly, and our knowledge about it is incomplete and fragmentary. Such gaps in our knowledge often trigger fears and fuel public apprehension about scientific innovation. The capacity of molecular biotechnologists to transfer genetic material across natural boundaries, especially, is at the heart of the current ethical controversy over the development and application of biotechnology.

Ethics, to a great extent, is concerned with procedures or rules society employs to convert value and value-free knowledge into prescriptions as to "what ought or ought not to be done." Traditionally, many scientists have tended to view ethical issues and questions as matters beyond the realm of objective investigation or research. They generate information that contributes to knowledge that is value-free with respect to the characteristics of conditions, situations, things, and acts. Biotechnology has realized and anticipated beneficial impacts on various aspects of society, but it also has anticipated adverse impacts, both of which need to be addressed objectively. Ethical issues are wide-ranging.

Are bioengineered foods the "silver bullet" they are touted to be, or the "frankenfoods" they are being projected to be by some? In 1999, about 50% of the American soybean crop carried an herbicide-resistant gene, while about 25% of the corn crop was Bt corn (i.e., corn carrying the insect resistance gene from the bacterium *Bacillus thurigiensis [Bt]*). There are other major crops that have been bioengineered for a variety of purposes, so how is bioengineering of food different from conventional plant breeding? Simply, biotechnology provides more precision and control over what new characteristics are introduced into plants. Instead of reorganizing the entire genomes of two parents in a new genetic matrix, resulting in the inheritance of a combination of genes from the two parents in the offspring, biotechnology enables scientists to insert one or more specific genes of interest into the recipient parent.

The obvious question is do the introduced genes or their protein products have any adverse effects on consumers? Since the genetic code is universal, the foreign DNA per se does not have any ill effects on humans. However, it is known that insertion of foreign DNA in genomes can cause defective growth and alter levels of plant compounds. For this reason, breeders conduct extensive research and product testing prior to releasing new cultivars. The protein products are virtually identical to nontoxic enzymes already present in the plant and are present in very low levels that can be easily digested. What about the alleged possibility of allergies? Certain products are known to cause allergies (e.g., eggs, fish, cow's milk, shellfish, tree nuts, wheat, and legumes). Biotechnology products are not inherently different from their conventional counterparts, and no research data are available to show that biotechnology foods have allergens.

Another source of apprehension from the public is about the inadvertent introduction of antibiotic resistance into the environment as a product of the actual research methodology. Certain molecular biotechnology techniques utilize antibiotic resistance marker genes. This is suspected by some to be a source of transfer of health risk, as this could increase the antibiotic resistance of bacteria already in the system. To preclude such a possibility, the Food and Drug Administration (FDA) has advised biotech companies involved in food development to refrain from using clinically important antibiotics in their marker selection procedures. The FDA, the Environmental Protection Agency (EPA), and the United States Department of Agriculture (USDA) are the three federal agencies that monitor and regulate bioengineered foods for safety. This includes labeling of foods and monitoring of environmental impact. Foods containing bioengineered products for industrial purposes, such as canola oil with altered fatty acid composition, are required to be labeled appropriately.

What do you get when you cross a bacterium with a daffodil and then with rice? A few years ago, such a question would have been considered a riddle. Nowadays, thanks to modern biotechnology, science solves such riddles with relative ease.

Thanks to a scientist with a passion to help the poor and malnourished, today the world has a miracle grain that has the potential to make more than just a dent in the worldwide efforts at attaining food security and fighting malnutrition. Professor Ingo Potrykus of the Swiss Federal Institute of Technology in Zurich is responsible for this wonder product, called "Golden rice," which is fast becoming a "postercrop" for what is good about biotechnology, in the face of all the bad press the technology has been receiving.

The successful creation of Golden rice was not a solo effort by Potrykus. In 1990, Gary Toenniessen, the director of food security for the Rockefeller Foundation, identified the specific problem for Potrykus. Toenniessen recommended the use of the more sophisticated tools of genetic engineering in addressing the lack of beta-carotene in the grain of rice, a crop that is a staple worldwide. Potrykus hooked up with Peter Beyer of the University of Freiburg, an expert on the beta-carotene pathway in daffodils. With $100,000 seed money from the Rockefeller Foundation, the duo embarked on what turned out to be a seven-year journey and a $2.6 million tab, with contributions from the Swiss government and the European Union. With 100% noncommercial support, the scientists hoped that they could give the product away without any consequence.

The feat was accomplished by following the steps described in the figure. In all, three organisms unrelated to rice were involved in creating the new rice: daffodils and the bacterium *Erwinia uredovora* provided the genes that encode beta-carotene, while the crown gall bacterium (*Agrobacterium tumifaciens*) provided the plasmids that served as gene couriers into rice tissue.

Potrykus and Beyer wanted to give away their Golden rice free of charge, but one of the genes they used had been patented by a biotech company, AstraZeneca of London. Needless to say, a commercial company is in business to make money. After some negotiations in which AstraZeneca received exclusive commercial marketing rights for Golden rice, the company decided to support the scientists' cause and work toward making the seeds available free of charge to farmers in the poor regions of the world.

The dream of two scientists may have come true in the end, but the manner in which it ended was not anticipated. Anti-biotechnology activists and critics saw the transaction between the scientists and a big biotech company as a pact with the devil and as more ammunition for their campaign against genetically modified foods. This notwithstanding, the world should be thankful that, because of Golden rice, millions of children potentially would not have to suffer from vitamin A deficiency and be predisposed to blindness. However, the finished product is still years away from the plates of the targetted consumers.

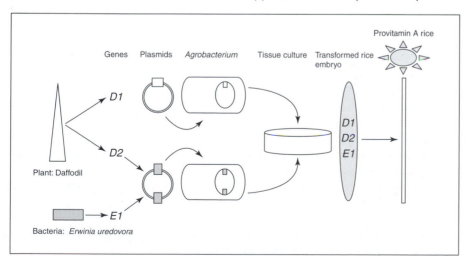

Golden rice was created with genes for beta-carotene from daffodils and the bacterium *Erwinia uredovora*. These were transferred into the embryo of rice by the method of *Agrobacterium* mediated transformation.

of spoilage in horticultural fruits such as citrus and apples. A suspension of the bacterium *Bacillus subtilis* is used to delay brown rot caused by the fungus *Monilinia fruticola.*

Researchers have moved a step further by developing transgenic plants that express endotoxins. Proteins such as protease inhibitors, alpha-amylase, and lectins from *Bacillus* spp. have been shown to have potent insecticidal activity. Plants resistant to certain herbicides such as glyphosate (in corn, soybean, cotton), 2,4-D (in cotton, tobacco), Atrazine (in tobacco), and Dalapon (in tobacco) have been developed.

Phytoremediation

Certain plants have the capacity to absorb heavy metals and salts and are used to clean up sites that need detoxification or desalinization.

Bioremediation

Bioremediation is the removal of toxic pollutants from both terrestrial and aquatic ecosystems using microbes. Like phytoremediation, microbes are used for cleaning sites that have been contaminated by chemicals such as oil. Fungi are used to degrade pollutants such as PCB, TNT, and DDT. These substances are anthropogenic (human-made) and recalcitrant in the soil.

Biofertilization and Phytostimulation

Many microorganisms, in association with plant roots, are able to produce compounds that stimulate plant growth and productivity. Microbes fix nitrogen for plant use by one of several processes. *Rhizobia,* by living in the roots of legumes, undertake biological nitrogen fixation in a symbiotic relationship. Biological nitrogen fixation is responsible for over 60% of soil nitrogen used in crop production. This host-specific process involves the development of root swelling called *nodules. R. leguminosarium* bv. *viciae* nodulates pea, while the bv. *trifolii* nodulates clover. Soybean is nodulated by *Bradyrhizobium* (previously called *R. japonicum*). Biological nitrogen fixation occurs in non-leguminous trees involving bacteria of the genus *Azospirillum.* These bacteria in the rhizosphere affect the development and function of grass and legume. They help to improve mineral uptake (especially NO_3^-, PO_3^{3-}, and K^+) and water uptake. *Azospirillum* promotes root hair formation in the roots of seedlings.

Mycorrhizae (fungi-plant root association) increases the effective root surface area and thereby aids in nutrient uptake.

SUMMARY

1. Plant breeders manipulate plants for high productivity.
2. Plant breeding is based on the principles of genetics.
3. Some traits are controlled by one or a few genes (qualitative traits), while others are controlled by several to many genes (quantitative traits).
4. Plant breeding depends on heritable variation for success.
5. The methods used in breeding depend on the breeding system of the plant—self-pollination or cross-pollination.
6. Crop productivity may be improved by adopting a variety of strategies—such as improving photosynthetic efficiency, breeding dwarf stature, breeding early maturity, breeding for yield stability, and improving harvest index.
7. Breeding may be done by conventional methods or by using genetic engineering. The two are used in a complementary fashion.
8. Genetic engineering allows plant breeders to transfer genes across genetic barriers.

References and Suggested Reading

Acquaah, G. 2002. *Horticulture: Principles and practices.* 2nd ed. Upper Saddle River, NJ: Prentice Hall.

Klug, W. S., and M. R. Cummings. 1996. *Essentials of genetics.* Englewood Cliffs, NJ: Prentice Hall.

Nash, J. M. 2000. Grains of hope. *Time.* 31 July, 39–43.

Poehlman, J. M. 1995. *Breeding field crops.* 4th ed. Westport, CT: AVI.

Simmonds, N. W. 1979. *Principles of crop improvement.* New York: Wiley.

Wallace, R. A. 1997. *Biology: The world of life.* 7th ed. New York: Benjamin/Cummings.

Selected Internet Sites for Further Review

http://www.aba.asn.au/

A discussion of general biotechnology.

Outcomes Assessment

Part A

Answer the following questions true or false.

1. T F   Plant traits are controlled by hereditary factors called genes.
2. T F   One or a few genes control quantitative traits.
3. T F   In dioecy, male and female plants occur in one species.
4. T F   All yield is biological, but not all yield is economic.
5. T F   Traits with low heritabilities are best improved by agronomic strategies.
6. T F   DNA is universal.

Part B

Answer the following questions.

1. Give the chemical composition of DNA.

   _____

   _____

2. Give four specific breeding objectives in crop improvement.

   _____

   _____

   _____

   _____

3. _____ is the ultimate source of biological variation.

4. What is heterosis?

_____

5. Define biotechnology.

_____

6. What is a transgenic plant?

_____

7. What is the central dogma of molecular biology?

_____

PART C

Write a brief essay on each of the following topics.

1. Discuss the concept of heritability in plant improvement.
2. Discuss the sources of variation for crop improvement.
3. Describe how hybrids are developed.
4. Describe how breeding for dwarf stature improves crop productivity.
5. Discuss the importance of yield stability in crop productivity.
6. Discuss the importance of shattering resistance in crop productivity.
7. Discuss the importance of micropropagation in crop production.
8. Discuss the landmark genetic-based impact on crop production.
9. Describe how transgenic plants are developed.

PART D

Discuss or explain the following topics in detail.

1. Do you think the general public should have a direct say in how scientists produce new crop cultivars for the market?
2. Can crop performance (yield) be increased indefinitely through genetic manipulation?
3. In the use of technology, does the end justify the means?
4. In biotechnology and crop production, is the best yet to come?
5. How has crop improvement (plant breeding) impacted crop production?
6. In what specific way would you like to see genetics (crop improvement) used in crop production?

6

Climate and Weather

PURPOSE

The purposes of this chapter are to distinguish between climate and weather and to discuss the roles or factors of these phenomena in crop production.

EXPECTED OUTCOMES

After studying this chapter, the student should be able to:

1. Define the terms *climate* and *weather* and distinguish between them.
2. Discuss the role of climate in crop adaptation.
3. Discuss weather as a chance factor in crop production.
4. Discuss the major climatic types and their distribution.
5. Discuss air as the source of carbon dioxide for photosynthesis and as prone to pollution from human activities that adversely impact crop production.
6. Discuss air as the source of oxygen for respiration.
7. Discuss the role of temperature in biochemical reactions such as photosynthesis and its role in crop adaptation and productivity.
8. Discuss the effect of light intensity, quality, and duration on plant growth and development processes.
9. Discuss selected global issues that affect crop production.

KEY TERMS

Climate	Global warming	La Niña
Cool season plant	Greenhouse gases	Leaf area index
Desertification	Heaving	Meteorologist
El Niño/Southern Oscillation	Hopkins Bioclimatic Law	Microclimate
	Ionosphere	Orographic effect

Ozone layer   Stratosphere   Warm season plant
Relative humidity (RH)   Troposphere   Weather
Soil solarization

TO THE STUDENT

Crop producers produce crops that are adapted to specific regions of production. Climate determines crop adaptation; unadapted cultivars grow and perform poorly. Field crop production is subject to the short-term variations in climatic factors, frequently referred to as the vagaries of the weather. Once an adapted cultivar has been selected, the producer has to contend with the weather in the local area of operation. Even though there are patterns in the weather, it is difficult for meteorologists to predict weather precisely. When producers are surprised by an unpredicted weather episode, crop productivity is jeopardized. Climate and weather are two critical factors in crop production that are not within the control of the producer. Even though patterns occur in these factors, surprises sometimes spring up and may overwhelm the producer and cause loss of a crop. You will also learn strategies that may be used to minimize the adverse effects of these factors of production. Photosynthesis is the single most important plant process that is responsible for crop productivity. This process depends on radiation (light), temperature, and carbon dioxide. Unlike rain, these factors operate on a continuous basis, varying in intensity either on a daily or seasonal basis. As you study this chapter, pay attention to how meteorological factors (air, water, temperature, radiation) interact to impact crop plant development and function. Also, note the cultural strategies that crop producers use to optimize the exploitation of these growth factors for increased crop productivity.

6.1: CROP PRODUCTIVITY IS SUBJECT TO THE VAGARIES OF THE WEATHER

Climate. Patterns of variation in meteorological factors over a large area formed over many years.

What is the difference between climate and weather? Even though the terms **climate** and **weather** are sometimes used as though they were synonymous, there are significant differences between them. Weather is what is presented on daily TV or radio broadcasts by the **meteorologist**. It describes the short-term (e.g., hourly, daily, or weekly) variations in meteorological factors within a local area. It is caused by the reactions of certain environmental factors within a local area, especially factors in the *troposphere* (the part of the atmosphere closest to the earth's surface). Climate, on the other hand, is the meteorological conditions over a broader area (regional or global) and depends on patterns developed over years.

Weather. The environmental condition described by the short-term variations in meteorological factors within a local area.

Meteorologist. One who studies the environmental factors that define weather and climate.

Climate is the basis for crop adaptation. The producer selects a crop that is adapted to the area where it will be grown. However, it is the weather in the locality that will eventually determine the crop development and productivity. The producer should pay attention to weather forecasts in planning field production activities. Of course, when producing under a controlled environment (greenhouse), the vagaries of the weather may be inconsequential to crop production.

6.1.1 THE ENVELOPE OF GASES AROUND THE EARTH IMPACTS CROP PRODUCTION

Troposphere. The layer of atmosphere nearest the earth's surface; stretches from 0 to 10 miles above the earth.

What part of the atmosphere impacts crop production? Layers of gases, called the *atmosphere,* envelope the earth. There are three layers of the atmosphere; the closest and the most important to crop production is the **troposphere** (Figure 6–1). Forces within this

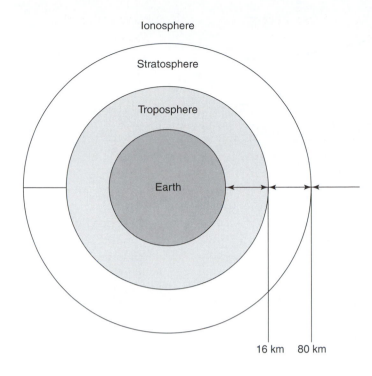

Ionosphere

Stratosphere

Troposphere

Earth

16 km   80 km

FIGURE 6–1 A profile of the atmosphere. The forces operating in the troposphere are largely responsible for generating the weather factors that impact crop production. The stratosphere is a filtering layer that blocks out solar radiation that could be harmful to life on earth. The ionosphere indirectly impacts crop production through radio technology used in production.

layer that may reach 16 km (10 miles) high in certain places determine weather and climate fluctuations. The **stratosphere**, reaching about 80 km (50 miles) above the earth, is important to life on earth because of its capacity for filtering out solar radiation, especially ultraviolet radiation that is harmful to organisms. The **ozone layer** occurs in the stratosphere. The **ionosphere**, the last layer of the atmosphere, reaches as high as 500 miles in certain parts. It is the warmest part of the earth's atmosphere. While it has no direct impact on crop production, it is the most suitable part of the atmosphere for radio wave transmission. Satellite technology plays a significant role in modern crop production. It is used in high-tech production methods such as precision farming.

The fluctuations in the meteorological conditions in the troposphere are caused by changes in air pressure and air masses. These two factors influence temperature, wind, humidity, cloud, and precipitation patterns in a given area. At any given time, one of two pressure systems may prevail in an area with different consequences. A *high-pressure system* involves a downdraft (sinking air). The system is accompanied by decreasing temperature and water vapor. A *low-pressure system,* on the other hand, is characterized by many clouds, much precipitation, and updraft (rising air). The air temperature rises as water vapor decreases. Rising air cools at the rate of 1°C per 100 meters. The reverse is true for descending air.

Stratosphere. The layer of atmosphere spanning 10 to 50 miles above the earth's surface.

Ozone layer. The protective layer of gases in the stratosphere that absorbs short-wave radiation, such as ultraviolet, preventing it from reaching the earth.

Ionosphere. The outermost layer of the earth's atmosphere, occurring between 50 and 500 miles (80 and 1,000 km) above the earth's surface.

6.1.2 GEOGRAPHIC FACTORS CAUSE VARIATIONS IN CLIMATE AND WEATHER PATTERNS

The major climatic regions of the United States are the *Steppe, Continental,* and *Subtropical Wet* (Figure 6–2). Climatic patterns are determined primarily by geographic factors, including landform, latitude, altitude, distance from large bodies of water, ocean currents, and the direction and intensity of winds. *Air masses* (large bodies of air that form in horizontal directions) influence weather patterns and vegetational types. Temperate Continental climates originate over the Arctic lands and produce severe winters

FIGURE 6–2 The major climatic regions of the United States. BS = Steppe (arid); Dc = Temperate Continental; Do = Temperate Oceanic; BW = Desert; Cr = Subtropical Wet; and Cs = Subtropical Winter Rain (or Mediterranean climate). The boxed figures prefixed by the letter *A* represent climate stations monitored by the National Weather Service.

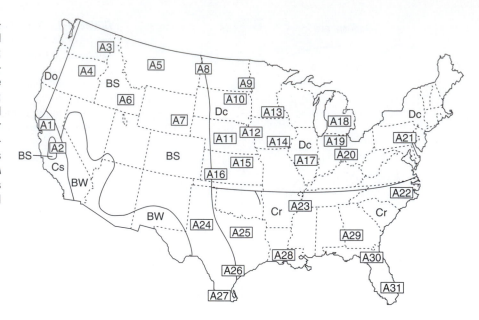

and cool summers. The Steppes occur in the interior regions (e.g., Southwest and Great Plains). Extreme diurnal temperatures and hot and dry summers characterize Steppes. This climatic type is affected by even relatively small bodies of water through thermal buffering, creating subclimatic types in the process. Subtropical wet climate develops over large bodies of water in the tropics. It causes hot and humid summers and occurs in areas including the central and eastern United States. This climate is also more equable.

Hopkins Bioclimatic Law. States that events (such as planting, harvesting, and morphological developments in plants) are delayed by 4 days for 1 degree of latitude, 5 degrees of longitude, and 400 feet of altitude, as one moves northward, eastward, and upward in the United States.

Three geographic factors (altitude, latitude, and longitude) are important for crop production. These are embodied in the **Hopkins Bioclimatic Law**. It states that crop production activities (e.g., planting, harvesting) and specific morphological developments are delayed 4 days for each 1° latitude, 5° of longitude, and 122 meters (400 feet) of altitude, as one moves northward, eastward, and upward, respectively. This indicates that a successful crop producer should have a good understanding of the climate and weather patterns in order to implement sound crop production management strategies.

Orographic effect. The modification of the climate in a region by the presence of a mountain range in the path of prevailing winds, resulting in wet conditions on the windward area and drier conditions on the leeward part of the mountain range.

Geographic surface features such as large bodies of water and mountain ranges modify the meteorological factors in a region. A mountain range in the path of the prevailing wind creates moist conditions on the windward side and dry conditions on the leeward side. The air cools adiabatically (occurs without heat entering or leaving the system) to the dew point as it rises up the mountains. The air that is forced upward by the mountain range loses its moisture and becomes a dry air mass once it goes over the top (Figure 6–3). This is called the **orographic effect** and is different from rainfall caused by weather fronts and thunderstorms in regions without mountains. In the United States, the Westerlies deposit most of their rain (by orographic effect) on the Pacific slopes of the Cascade Mountains and Sierra Nevada Mountains in the Rocky Mountains. Similarly, as the wind blows over a large body of water, it picks up moisture, thus creating a more moderate leeward condition (Figure 6–4). Crop producers tend to allocate more delicate crops to the leeward areas of a large body of water and hardier field crops such as cereal grains on the windward region where the weather is less stable.

Topography influences weather and microclimate through temperature modification. Due to the phenomenon of *adiabatic expansion* of air, temperature declines with altitude. The normal adiabatic lapse rate for an unsaturated atmosphere is $-1°C$ per 100 meters of rise in altitude. Moist air cools at about $0.6°C$ per 100 meters rise. However, once it loses

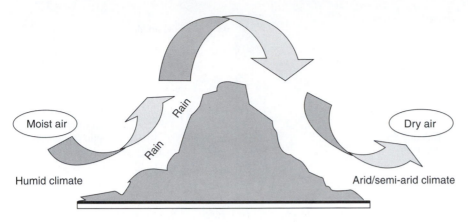

FIGURE 6–3 The effect of relief on the climate of a region. Mountain ranges in the path of a prevailing wind force it to rise, cool, condense, and then fall as rain on the windward side. Upon reaching the leeward side, the wind is dry, thus creating an arid condition in the area.

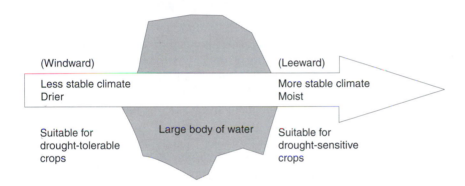

FIGURE 6–4 The effect of large bodies of water on the immediate climate. Water is normally warmer than the surrounding land. When a prevailing wind moves over a large body of water, it becomes modified, thereby moderating the climate on the leeward side of the body of water. The leeward side of the body of water is subject to less temperature fluctuation.

moisture after depositing it through rain, the unsaturated air warms at a rate of 1°C per 100 meters as it descends on the leeward side of the mountain. The descending air on this side is not only drier but also hotter.

North-facing slopes and south-facing slopes may differ significantly in climate and natural vegetation. South- and west-facing slopes receive more sunshine and are warmer than slopes facing north and east. These sites differ in the crops that can be grown. Other topographical influences include the temperature inversion that occurs between valley bottoms and surrounding hillsides (Figure 6–5). In the night, air is cooled by contact with surfaces. It becomes dense and drains down to the valley bottom, forcing warmer air up the hillsides. Crops on the valley bottom are exposed to frost.

World climate is not static but is subject to change caused by factors such as systematic changes in solar activity, sea level, atmospheric carbon dioxide, and continental drift. Continental drift gradually changes the latitudinal position of the landmasses and the sea level. Volcanic explosions discharge large amounts of dust and gases into the atmosphere. This may cause a decline in temperature (reverse of the greenhouse effect).

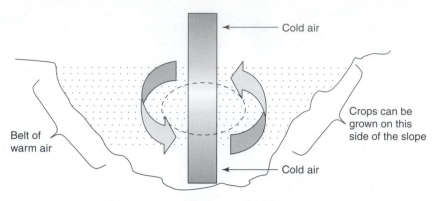

FIGURE 6–5 The phenomenon of temperature inversions. Warm air rises and cools at high altitude. The colder and denser air descends to the bottom of the valley, displacing the warmer air. As the warm air rises, it meets a cold layer at a higher altitude. The colder air descends and the cycle continues. A convectional flow of warmer air is set into motion between the two layers of cold air. Growing crops at the bottom of the valley predisposes the crop to frost. Producers can utilize the middle belt region for frost-free crop production.

Global warming. A general and gradual increasing trend in global average temperature suspected to have a role in the greenhouse effect.

The accumulation of greenhouse gases (especially carbon dioxide) and methane is responsible for the trend in **global warming**. Carbon dioxide increases are attributed to the burning of fossil fuels and the release of trapped carbon dioxide in the soil caused by tillage. Farm animals produce methane through their digestive systems. The impact of the greenhouse effect on world climate includes changes in rainfall pattern and, subsequently, crop productivity.

The early history of the world was characterized by dramatic changes in climate in which cold ice age (glacial ages) and warmer interglacial periods occurred. There have been periods of rise in temperature, abundant rain, drought, and cooling. All these changes impacted crop production positively or adversely and influenced population migration.

6.1.3 UNSEASONABLE WEATHER EFFECTS ON CROP PRODUCTION

In what ways do the uncertainties of the weather impact crop production? In spite of technological advances in modern crop production, crop producers are still subject to the vagaries of the weather that are manifested in three main ways: moisture stress (drought) temperature stress, and natural disasters.

1. *Moisture stress (drought).* Most agricultural production problems associated with water are related to its quantity and quality. Drought is a period when soil moisture is limited to an extent that crops are adversely affected. Drought occurs when precipitation rates are too low, temperatures are too high, or both. High temperatures cause high evaporation rates. Drought can be intensified by poor land management, as was the case in the Great Plains in the 1930s, resulting in the Dust Bowl. For a good yield, moisture should be available to crops in the appropriate amounts and at the appropriate times. Improper timing or distribution of moisture will result in a decrease in crop yield. The effect of these unpredictable short-term changes in moisture regime can be lessened or eliminated if the producer has the resources to install an irrigation system or through the adoption of soil moisture-conserving tillage and crop rotation practices. Also, drought-resistant cultivars are able to survive dry spells.

The greenhouse effect is the warming of the earth's atmosphere through the activities of certain gases that block the re-emittance of solar energy that reaches the earth. About 85% of solar radiation reaching the earth is absorbed by water vapor in the atmosphere. The remainder is reflected back to the earth due to the accumulation of carbon dioxide, methane, chlorofluorocarbons (CFCs), and nitrous oxide, among others. These gases are called **greenhouse gases**. Their concentration in the atmosphere is influenced by human activities such as the burning of fossil fuels (that increases carbon dioxide content of the air), deforestation (which reduces plant population and hence reduces use of carbon dioxide), livestock gases (produce methane), and the use of CFCs in cooling units (e.g., refrigerators). Carbon dioxide is estimated to account for 50% of the greenhouse effect, methane for 20%, CFCs for 14%, and other sources for the remainder.

Greenhouse gases.
Gases, especially carbon dioxide, that create a barrier in the atmosphere against the re-radiation of heat, leading to the greenhouse effect.

These gases act as a shield or glass pane that allows light in but reflects radiation back to the earth. This re-radiance is less energetic, having long wavelength. The atmospheric gases filter the harmful short waves during the entry process.

Without greenhouse gases, the earth's surface temperature would be about −18°C and thus unsuitable for crop production. On the other hand, excessive heat also has consequences by way of climate modification. Some of the expected effects include drought that causes crop production in affected regions to be moisture-stressed and less productive. In regions where ice occurs, the high temperature causes ice to melt, raising the sea level and causing flooding in lowlands. Further, colder regions receive additional warmth and experience a longer crop-growing season. The effect of this change in temperature also has the potential of reducing crop production in regions of the United States where production is currently adequate, while increasing it in Canada, where currently most of the land is covered with ice. World food production is impacted with this reversal in production trends.

To reduce these possible effects of global warming from the greenhouse effect, governments may institute legislation to control the emission of greenhouse gases into the atmosphere. Laws are currently in place in certain countries to regulate the emission of exhaust fumes of automobiles, the burning of forests, and the use of CFCs.

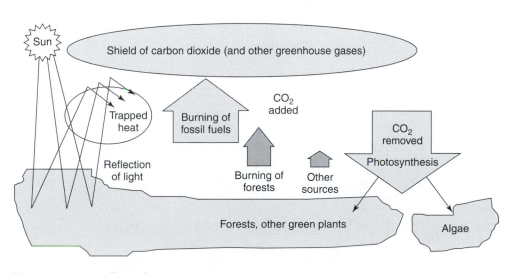

The greenhouse effect of carbon dioxide.

2. *Temperature stress.* For each plant species, there is a minimum temperature that kills plants outright. Low temperatures cause plant damage, which varies from partial tissue damage to the death of the entire plant. The damage depends on the physiological status of the tissue at the time of exposure to the adverse temperature. Hardened or dormant tissue may survive an unseasonably low temperature, while actively growing succulent tissue (e.g., flowering buds) are severely damaged. Freezing temperatures cause ice crystals to form in the tissues, resulting in injury. Early frost or freezes are of concern to producers in the temperate zones, especially producers of warm-season crops. Frost can kill emerging seedlings and reduce crop stand. The best protection against frost damage is to plant the crop after the threat of frost has passed. Frost damage is greatest during the hours just before sunrise. Unseasonable temperature may delay planting of a crop and reduce the growing season. When the crop is established, an unexpected cold or dry spell may affect the reproductive phase of the plants, leading to severe yield reduction. It is more difficult to protect crops from adverse temperature. Horticultural crop producers utilize a variety of strategies to protect field plants from a freeze. These include using hot caps, heaters, or wind machines to circulate the air around the plants. Growers may minimize their production risks by consulting guides such as the average dates of the last spring freeze (Figure 6–6) and the first autumn freeze (Figure 6–7).

3. *Natural disasters.* Modern technology is no match for major weather systems. Excessive monsoon rains and cyclones devastate crops and property in Asia,

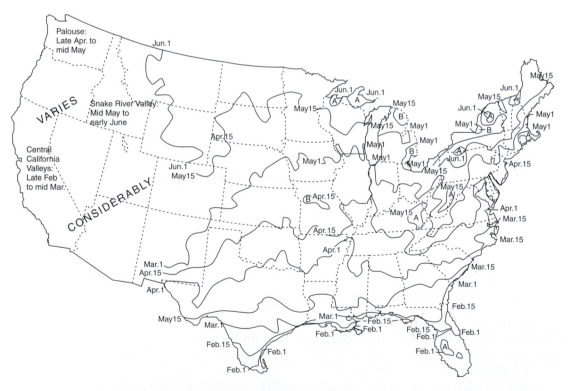

FIGURE 6–6 The average dates of the last killing frost in spring. Because temperature is subject to variation, these dates should be used only as a general guide. It is helpful to listen to and use weather information. (Source: USDA)

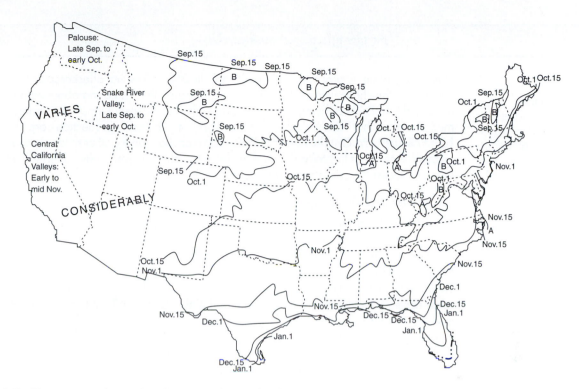

FIGURE 6–7 The average dates of the first killing frost in fall. Because temperature is subject to variation, these dates should be used only as a general guide. It is helpful to listen to and use weather information. (Source: USDA)

while hurricanes, frost, floods, drought, and hail destroy crops and other natural resources in other parts of the world. Prolonged drought in places such as Africa and the Middle East can lead to severe famine. Seeds fail to germinate; established crop cannot be productive. If plants do not grow, water is not likely to infiltrate the soil and will run off the soil surface when it becomes available. Floods cause soil erosion and soil degradation, reducing soil productivity.

6.1.4 CULTURAL PRACTICES IN CROP PRODUCTION CAN CREATE AN ARTIFICIAL ENVIRONMENT AROUND PLANTS IN THE FIELD

When plants grow together in close proximity to each other, as occurs in field crops, they are able to significantly modify the meteorological characteristics of the vicinity, creating a **microclimate**, or *microenvironment*. The nature of the microclimate depends on plant features, such as height and architecture, and planting density. The crop canopy created by the collectivity of the plants in the stand modifies their immediate environment in a way that has implications for crop production and crop productivity. The wind speed, humidity, light penetration and interception by leaves, and temperature are subject to alteration within the canopy. Slow wind speeds and high humidity may create conditions that favor plant pathogens. The carbon dioxide level within the canopy is less than that above it. Shading of leaves (a factor of leaf area) reduces light interception and the effective photosynthetic leaf surface. These conditions reduce photosynthesis and hence crop productivity.

Microclimate. The atmospheric environmental conditions in the immediate vicinity of the plant.

6.1: Crop Productivity Is Subject to the Vagaries of the Weather **183**

El Niño/Southern Oscillation. A weather system responsible for the warming of the tropical Pacific surface waters.

There are certain natural and human-induced climatic factors that affect crop production. Weather is the chance element in field crop production that can bring about crop failure in spite of the efforts of the producer to provide all other necessary inputs. Two unusual climatic events that dramatically impact world agricultural production are the **El Niño/Southern Oscillation** and the *Indian monsoon.* Unlike other mild unseasonable weather events that last for a short while, these two climatic events have aftermath effects that can persist in various forms for a long period.

El Niño/Southern Oscillation

The El Niño phenomenon is caused by a reversal of the normal high air pressure over Tahiti and the low pressure over Darwin. Meteorologists have devised a measure called the *Southern Oscillation Index (SOI)* to indicate pressure difference between these two locations. When the surface pressure is high at Darwin and low at Tahiti, the SOI is negative. The unusual climatic system that this condition brings is called the El Niño. When the converse occurs (i.e., low pressure at Darwin and high pressure at Tahiti), the SOI is positive and the climatic event called **La Niña** occurs. El Niño is Spanish for "little boy" and describes a warm weather system that periodically occurs across the tropical pacific. This warm southern current appears periodically around Christmas time on the Peruvian coast. It is characterized by temperature oscillation, the warm episode called the El Niño and the cold La Niña ("little girl").

La Niña. A cold episode that periodically occurs across the tropical Pacific and impacts surface atmosphere and weather patterns (opposite of El Niño).

For most of the year, the northeast and southeast trade winds (Easterlies) dominate the tropical pacific region around the *International Date Line.* During this period, a cool ocean current occurs, producing nutrient-rich water to support fish life. This condition changes in late December, when the upwelling of the Easterlies relaxes. The

Fluctuations in the Southern Oscillation Index (SOI) over a 60-year period and the incidence of cold and warm episodes. Cold episodes are called La Niña, while warm episodes are called El Niño. (Source: NOAA)

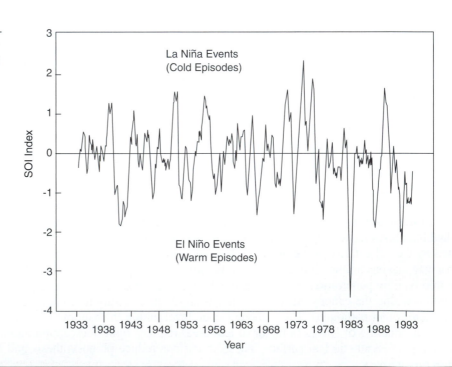

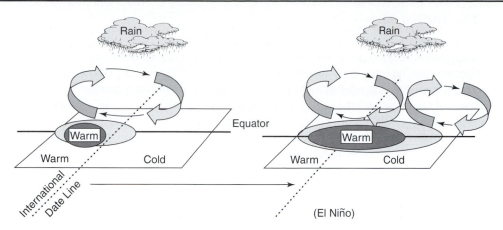

Warm water from the west moves eastward during late December, causing unusual warming and excessive rain near the Equadorian Peruvian region. (Source: Modified after NOAA)

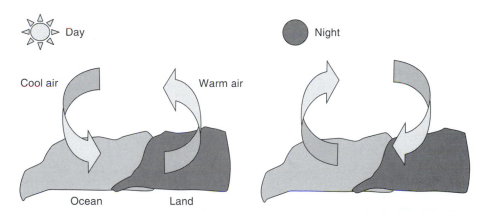

Warm air rises from land during a hot afternoon and moves to the ocean where it displaces the cooler air. The cool air moves toward land to provide a cooling effect called the sea breeze. The cycle reverses at night.

effect is that the water becomes warmer and consequently nutrient-poor. The fishing season comes to an end. The western warm water moves eastward. The *Tropical Convergence Zone* moves farther south than normal at that time. The El Niño brings an unusual warming and excessive rainfall to the Ecuadorian Peruvian (north) region. The warm episode prevails for a period, but as the surface pressure changes, cooler than normal equatorial weather arises in the central and eastern equatorial Pacific. This is the La Niña. The effect of this cool weather system is drought in some regions. The two opposing unseasonable weather phenomena generally reduce crop productivity through floods or droughts.

(continued)

The Indian Monsoon

A monsoon is a seasonal shift in wind direction that can produce a drought or excessive precipitation. The monsoon weather pattern is widely distributed, occurring in low-latitude climates ranging from West Africa to the western Pacific Ocean. The mechanism for creating a monsoon is similar to that for the phenomenon of land-sea breeze that occurs in areas where large bodies of water (e.g., oceans) are capable of absorbing and retaining radiant energy from the sun for a long time. This is due to the fact that such bodies of water are able to absorb heat at different depths and reflect less back to the atmosphere. Landmasses, on the other hand, lack depth in their absorbing surfaces. As a result, they lose heat much more quickly. Heat is transferred from the region of high amount to one of low amount. Coastal regions experience daily shifts in the wind. On a sunny day, the land absorbs heat and warms more quickly than the ocean. The hot air on land rises, allowing the cooler air over the ocean to move inland to replace it. The land enjoys a cool sea breeze during the day. The cycle is reversed during the night, when the land cools more quickly than the sea. This causes the warmer air on the sea to be displaced by the cooler air on the land. The land thus experiences a warm breeze during the night.

The monsoon is a land-sea breeze phenomenon on a massive scale. Energy imbalances occur on a large scale involving the continent and the surrounding oceans. Heat is accumulated during the summer months, creating areas of low pressure or low-density air masses. Crop producers in the semiarid tropics depend on monsoon rains for production. These tropical rains come about when moisture-laden air from the south meets falling cooler air from the north in an area called the *Intertropical Convergence Zone*. This zone is pushed northward in summer, where it causes the monsoon rains to fall in the semiarid tropics. The monsoon penetrates the northern region differentially, sometimes reaching the Sahel and northern India. Inadequate penetration leaves the northern regions with insufficient rain for crop production, leading to low productivity and famine in the Sahel.

6.2: WATER IS CRITICAL TO CROP PRODUCTIVITY

Precipitation reaches the earth in various forms, including rain, snow, and ice. Rainfall is the most important form of precipitation in crop production. The average annual precipitation in the United States is shown in Figure 6–8. Agricultural crop production regions may be described on the basis of annual precipitation (Table 6–1). Regions described as semiarid have annual total rainfall ranging between 25 and 50 centimeters and are suited to small grain production, while subhumid regions (50 to 75 centimeters of precipitation per annum) are suited to row crops and agronomic forage crops. In both regions, supplemental irrigation may be required (more so in semiarid regions) for high crop productivity. However, good management can eliminate or minimize the need for this activity.

6.2.1 WATER DISTRIBUTION

Water is the limiting resource for crop production in each of the most populous countries on earth—China, India, Russia, the United States and Indonesia. Agriculture is the world's largest user of fresh water resources, consuming (mostly through irrigation) about 80 to 85%

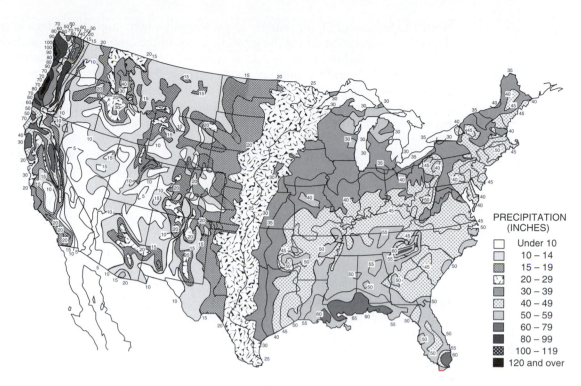

FIGURE 6–8 The average precipitation in the United States. The eastern half of the country receives precipitation ranging from about 30 inches to over 120 inches. The western half is drier, with parts receiving under 10 inches of annual precipitation. The heaviest rainfall occurs on the northern parts of the West Coast. (Source: USDA)

PRECIPITATION (INCHES)

☐	Under 10
☐	10 – 14
▨	15 – 19
▨	20 – 29
▨	30 – 39
▨	40 – 49
▨	50 – 59
▨	60 – 79
▨	80 – 99
▨	100 – 119
■	120 and over

Table 6–1 Classification of U.S. Crop Production Regions Based on Rainfall

Annual Rainfall (Approximate Range)	Crop Region	Cultural Methods for Success
10 in. (0–25 cm) or less	Arid	Irrigation is needed; grazing lands
10 in.–20 in. (25–50 cm)	Semiarid	Tillage methods that conserve moisture; moisture-conserving practices; irrigation often needed; grow small grains
20 in.–30 in. (50–75 cm)	Subhumid	Good moisture management; conservation practices; supplemental irrigation sometimes needed; row crops
30 in. (75 cm) or more	Humid	Moisture not a problem; forest crops can be grown; row crops

of fresh water in the United States. An estimated one-third of the world's food is grown on about 18% of the cropland that is irrigated. The availability of high-quality fresh water is critical to the sustainability, productivity, and dependability of crop production in the future.

Factors that determine the effectiveness of a given quantity of rainfall in crop production are the time of year it falls, the rapidity, the intensity, and seasonal evaporation. Rainfall is most useful for production when it falls during the growing season, which is

April 1 to September 30 for summer crops. Regions where rain falls in the winter experience low evaporation during the growing period of the crop, and hence producers can conserve more soil moisture by fallow cropping than can producers in other regions. Furthermore, corn and sorghum are poorly adapted to regions of winter rainfall because the period of their greatest moisture needs occur in the dry, hot season.

The total rainfall fluctuates widely from year to year. For example, over a 10-year period, the annual precipitation at North Platte, Nebraska, in the Great Plains region, experiences a variation from 10 to 40 inches.

Generally, the monthly distribution of precipitation in the region east of the Mississippi River tends to be relatively more uniform throughout the year. This is more so in the area north of the Cotton Belt. On the other hand, the period from November to May in Florida is dry, most of the rainfall occurring in the summer months. In the Great Plains region, about 70 to 80% of the annual rainfall occurs from April to September. On the Pacific Coast, rainfall occurs in the winter.

Rainfall in the Great Plains and states to the east often come in torrential showers, much of which is lost to surface runoff. On the other hand, light showers of less than 1/2 inch may cool the plants and reduce evaporation but is scarcely enough to be stored in the soil for plant use.

In corn production, the critical time (when moisture is needed) is 10 days after tasseling and silking. For most crops, this time occurs around flowering. Rainfall is of most value to crop productivity when it falls at these critical periods (flowering, tasseling, silking) in the plant growth and development.

6.2.2 SUMMARY OF WATER-RELATED CROP MANAGEMENT PRACTICES

The management practices summarized here are discussed in detail in appropriate sections of the book. These represent decisions a producer must make in order to have success with crop production.

1. *Moisture conservation.* Dryland (rainfed) crop production depends on moisture-conserving practices for successful production. These include practices such as fallow cropping, mulching, and cover cropping.
2. *Cultivar selection.* Some plant cultivars are drought-tolerant and useful for production under conditions of moisture stress.
3. *Supplemental moisture.* Crop production in dry regions often needs supplemental moisture during the periods of drought. There are various sources of irrigation water and various methods of application of irrigation water.
4. *Drainage.* Drainage removes excess moisture from the field for proper soil aeration and nutrient utilization.

6.3: ALL PHYSIOLOGICAL PROCESSES ARE AFFECTED BY TEMPERATURE

Temperature is the intensity factor of heat energy. Temperature affects chemical reactions, increasing reaction rate when high and decreasing it when low. Plants grow slowly under cold temperature. Temperature extremes can lead to death of plants. Temperature decreases as one moves high above the soil surface and lower below the surface. The phenomenon of temperature inversion sometimes occurs (at night) when the air temperature at the soil surface is lower than that of the air above. For each crop there are three generalized and important temperatures of which the producer needs to be aware. These are:

1. *Minimum growth temperature.* This is the lowest temperature needed for growth to start or seed to start.
2. *Optimum growth temperature.* Each crop has a most favorable temperature at which growth occurs at the highest rate.
3. *Maximum growth temperature.* Above this maximum, denaturation of essential proteins occurs, causing physiological processes to fail.

6.3.1 TEMPERATURE-BASED LAWS IN CROP PRODUCTION

Temperature, as previously indicated, is influenced by latitude and altitude. Most crops in the temperate region are grown where the monthly temperature for 4 to 12 months averages between 50° and 68°F. As previously stated, each crop has its minimum, optimum, and maximum temperature range for growth, but most crops grow and develop well between 60° and 90°F. Most crop plants cannot tolerate temperatures of 110°–130°F. Similarly, annual crop plants cannot grow and develop at temperatures below 32°F. Whereas some crops can show some signs of growth at 40°–43°F, crops such as sorghum will cease growth at 60°F, while sugarcane will cease growth at 70°F.

The normal daily temperature for spring planting of a given crop is relatively uniform throughout the United States. For example, the temperature at the time of seeding for wheat (37°F), spring oats (43°F), corn (55°F), and cotton (62°F) is the cultural practice adopted by producers in all producing areas in the country. Apart from the Hopkins Bioclimatic Law, two other temperature-based laws are applicable to crop production. Generally, crop growth doubles for each 18°F rise in temperature, a relationship that corresponds to the *Vant Hoff-Arrhenius Law* of molecular chemical reactions. Crops differ in the number of heat units (as sum of degree days for a growing season) needed to mature them. According to the *Linsser Law,* in different regions the ratio of total heat units to the heat units necessary to mature a given crop is nearly constant.

6.3.2 SOME CROPS LIKE IT WARM; OTHERS LIKE IT COOL

Can any plant be successfully grown anywhere? Temperature influences crop plant adaptation and determines the length of the growing season. Temperature affects photosynthetic rate and thus crop plant growth rate and productivity. The USDA *plant hardiness zone* map shows the temperature-based adaptation of crop plants (Figure 6–9). In Chapter 2, plants were classified on the basis of temperature adaptation in two groups— warm-season and cool-season crops.

Crops differ in length of growing season needed for maximum productivity. Within the same species are cultivars with varying durations of lifecycle; some are early-maturing, while others are late-maturing. For example, corn requires an average of about 120 days, while cotton requires about 180 days to complete its lifecycle.

Adaptation

Plants may be classified on the basis of temperature adaptation as either **cool season** or **warm season plants**.

Cool Season or Temperate Plants   These plants, such as wheat, sugar beet, and tall fescue, prefer a monthly temperature of between 15° and 18°C (59° to 64°F) for growth and development.

Cool season plant. A plant that grows best at daytime temperatures ranging from 59°–64°F (15°–18°C) and is frost tolerant.

Warm season plant. A plant that grows best at daytime temperatures ranging from 64°–80°F (18°–27°C).

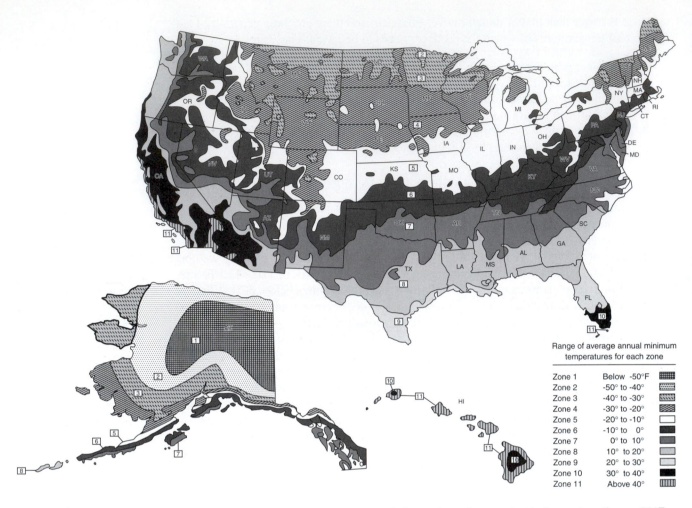

Range of average annual minimum temperatures for each zone	
Zone 1	Below -50°F
Zone 2	-50° to -40°
Zone 3	-40° to -30°
Zone 4	-30° to -20°
Zone 5	-20° to -10°
Zone 6	-10° to 0°
Zone 7	0° to 10°
Zone 8	10° to 20°
Zone 9	20° to 30°
Zone 10	30° to 40°
Zone 11	Above 40°

FIGURE 6–9 The USDA plant hardiness zone map. There are 11 zones of plant adaptation ranging between less than −50°F in the north and above 40°F in the south.

Warm Season or Tropical Plants These plants, such as corn, sorghum, and buffalograss, require warm temperatures of between 18° and 27°C (64° to 80°F) during the growing season.

6.3.3 SOILS ABSORB HEAT FROM THE SUN

Solar radiation is the primary source of soil heat. The temperature a soil eventually attains depends on *heat supply* (the amount of heat reaching the soil surface from the source) and the *dissipation* of heat (what happens to the heat once in the soil). Some of the solar radiation is reflected back to the sky, while the energy absorbed is mainly used in evaporating water from the soil and other surfaces. It is estimated that only 10% of the total radiant energy absorbed by the soil is used to warm it. The amount of energy absorbed is also influenced by factors including soil color, vegetative cover, and slope of the land. Darker soils absorb more heat than do light-colored soils. Darker soils have high organic matter and hold more moisture. They require a higher amount of energy to become warm. Soils that are bare receive more solar radiation and warm up more rapidly than soil with vegetation cover. Once absorbed, soil heat moves primarily by conduction.

Traditionally, the lifecycle of crops is reckoned in terms of time (duration of the cycle). However, it is known that physiological processes such as growth and development are affected by temperature. If seeds are planted in a cold soil, germination and growth will be slow, prolonging the lifecycle of the plant. It is hence desirable to take temperature into account when describing the lifecycle of a crop. This is the basis for the calculation and use of heat units in crop production. The correlation between temperature and growth is used by producers to predict the harvest dates for crops, and also to determine the adaptability of plant species and cultivars to an area. Plant breeders may also use this estimation to stagger the planting of parents to be used in a hybrid production program for best results in crossing.

The calculations of heat units should be considered only as approximations. Heat units is calculated by the equation

Heat units = [(daily minimum temperature + daily maximum temperature)/2] − base temperature (°F)

The base temperature is the minimum temperature required for growth and varies from one species to another (e.g., 40°F for small grains and 60°F for cotton). The number of heat units accumulated in a week for corn, for example, with minimum temperature of 50°F and maximum of 86°F, may be calculated as follows:

| | Temperature (°F) | | Mean Temperature | |
Day	Max	Min	(°F)	Heat units
1	68	60	64	14 DD
2	86	60	73	23 DD
3	90	80	85	35 DD
4	99	85	92 (86)	0 DD
5	60	52	56	6 DD
6	50	44	47	0 DD
7	72	60	64	14 DD

Total DD_{10} = 146 DD_{10}

Degree days cannot be negative. Similarly, heat units cannot be accumulated when the temperature is above the maximum desired for the plant. For a more accurate estimation, heat accumulation may be calculated on an hourly basis (i.e., degree hour). On day 4 in the table the mean temperature was 92°F, 6° higher than the maximum for corn (86°F). Thus, no heat units were accumulated. Similarly, the average for day 6 was below 50°F and hence no heat units were accumulated. Certain horticultural crops (e.g., apples, pears, and cherries) require an exposure to a period of cold temperatures before they will fruit. In this case, an opposite of heat units, called *chilling days*, is calculated for these crops.

The movement is affected by soil moisture. Water conducts heat more efficiently than air; thus, heat moves faster in moist soil than in dry soil.

Soil microbes involved in nutrient recycling work more efficiently at higher temperatures (27°–32°C or 80°–90°F). At temperatures below 10°C (50°F), microbial oxidation of ammonium ions to nitrates is minimal. Crop producers may take advantage of this fact and apply ammonia fertilizers to cold soils in spring. This way, the NH_4^+ ions will not be oxidized to NO_3^- (that are readily leached) until soil temperature rises. By the time the nitrates are available, the plants will have grown enough to absorb them.

Soil microbes differ in temperature sensitivity. In summer, spreading transparent plastic sheeting over the soil can increase its temperature to about 50°C (122°F) or higher. This strategy, called **soil solarization**, is used to control pathogens such as wilt-causing fungal diseases.

Alternative freezing and thawing of the upper layers of the soil affects soil structure. This temperature effect causes **heaving**, or movement of plants (and stones) from lower depths to the top of the soil, eventually. Heaving can kill perennial forage crops such as alfalfa.

Soil solarization. Sterilization of the soil using solar radiation as the source of heat.

Heaving. The partial lifting of plants (and other structures such as roads and buildings) out of the ground, as result of freezing and thawing of the surface soil in winter.

6.3.4 CROP PRODUCERS CAN MODIFY SOIL TEMPERATURE

What practices can be employed by crop producers to modify soil temperature? The crop producer can modify soil temperature to a reasonable extent. The two practices used are *mulching* and *tillage:*

Mulching

The purposes of mulching include the following:

1. *Temperature moderation.* A mulch keeps the soil warm for seed germination and increased activity of soil organisms. It also tends to limit soil temperature fluctuations by protecting the soil from excessive heat in the daytime and the retention of more heat at night. Mulched soils warm up slowly in spring, an event that may be detrimental to crops. On the other hand, mulched soils cool slowly in fall, allowing plant root activity to continue for a longer time.
2. *Improved soil fertility.* Organic mulches are biodegradable and hence eventually break down and may become incorporated into the soil as a soil amendment to improve soil fertility and structure.
3. *Moisture retention.* Mulches are effective in reducing soil moisture loss from evaporation. Impermeable plastic mulches are the most effective for retaining moisture.
4. *Weed suppression.* A layer of mulch suppresses weeds by preventing them from receiving light. It is best to apply the mulch before weeds grow. A cheaper material, such as straw, is applied as a thick layer (6–12 inches) to be effective.

Mulches can be organic or synthetic:

1. *Organic mulches.* A common organic mulch used in crop production is straw. It is slow to decay. Application of straw mulch is not uniform over a producer's field.
2. *Synthetic mulches.* Synthetic mulches include fabric or plastic materials (Figure 6.10). Plastic mulches are most commonly used in commercial crop production. They are not

FIGURE 6–10 Plastic mulching is a production practice of crops such as strawberry, where, in addition to controlling weeds and other benefits, the fruit is protected from contamination from soil. (Source: USDA)

FIGURE 6–11 Conservation tillage is a practice that leaves some plant material on the soil surface. (Source: USDA)

biodegradable. Synthetic mulches may be clear or opaque. Opaque mulches come in white, gray, black, or other colors. They can absorb some solar heat, but the soil underneath them does not reach as high a temperature as achievable under clear plastic. Plastic mulch is relatively uniform and adapted to mechanized application. Plastic mulch suppresses some weeds, but nutsedge may puncture and grow through it, and the disposal of old plastic mulch may be problematic.

Tillage

Conservation tillage is a type of tillage that leaves crop residues on the soil surface or in the top layer (Figure 6–11). This tillage practice conserves soil moisture, among other effects, thus influencing soil heat movement. Planting on raised beds is recommended as a practice for increasing soil temperature. Raising the seedbed improves soil drainage and improves aeration, which has a warming effect on the soil.

6.4: AIR IS A SOURCE OF ESSENTIAL PLANT NUTRIENTS

The air above ground environment consists mostly of nitrogen, carbon dioxide, hydrogen, and oxygen. Carbon dioxide is critical to photosynthesis, while oxygen is required for aerobic respiration. Photosynthesis and respiration are the two metabolic processes needed for plant growth and development. Though nitrogen is the most abundant, it is not directly usable by plants. However, through symbiotic association between legume roots and certain bacteria (*Rhizobium* spp.), molecular nitrogen (N_2) can be directly fixed in the roots for use by plants.

Air is also important for crop productivity through its role in *humidity*, the water content of the air, which is measured in units of **relative humidity (RH)**. Relative humidity is the ratio of the weight of water vapor in a given quantity of air to the total weight

Relative humidity (RH). The amount of water vapor in the air expressed as a percentage of the amount of water vapor the air can hold at the same temperature.

of water vapor that quantity of air can hold at a given temperature, expressed as a percentage. Relative humidity decreases if temperature increases while water vapor remains the same. Plants lose water through their stomata. This loss, called transpiration, is rapid under high temperature, wind, and low RH. High humidity in the microclimate predisposes plants to diseases. Air movement reduces the adverse effect of high humidity of the microclimate. However, strong winds may damage crop plants by uprooting them, causing them to lodge, snapping the branches, or causing fruit drop.

How does atmospheric air differ from soil air? The nitrogen content of soil is relatively constant. Soil air is usually lower in oxygen, the concentration decreasing with depth. Soil carbon dioxide, on the other hand, is higher and similarly increases with depth. The relative humidity of soil air is generally above 95% and increases with depth.

One of the best measurements of soil air is the *oxygen diffusion rate (ODR)*. This is the rate at which oxygen in the soil exchanges with oxygen in the atmosphere. Root growth ceases when the oxygen diffusion rate is less than 20×10^{-8} g/m^2 per minute. Plants vary in their ODR requirements. For example, sugar beet taproots are stunted if ODR is low at even 30 centimeters depth. Apart from gaseous air and ODR, a third measure of soil air is the oxygen reduction (redox) potential. The reduction and oxidation states of soil chemical elements vary under different oxygen levels. When soils are well aerated, oxidized states of common soil elements are Fe^{3+}, Mn^{4+}, NO^{3-}, and SO_4^{2-}. When soils are poorly drained and aerated, the reduced forms of ions include Fe^{2+}, Mn^{2+}, NH_4^+, and S^{2-}.

The oxygen needed by most plants is taken from the soil. A few species, such as paddy rice, can take in adequate oxygen through their above-ground parts. This explains why this species of rice can grow under stagnant water. Conditions that promote poor soil aeration include waterlogging and compaction, as well as the dominance of expanding clays. Further, in summer, the warm temperatures accelerate decomposition of organic matter and the buildup of carbon dioxide, all byproducts of the activities of aerobic bacteria.

The effects of poor soil aeration on crop production include the following:

1. It slows the decay of organic matter.
2. Anaerobic organisms dominate, leading to the production of methane (swamp gas).
3. Root growth is curtailed. Wheat and barley are less tolerant of poor aeration, while crops such as ladino clover and fescue can grow in poorly drained soils.
4. Nutrient and water absorption are hindered when oxygen levels are low.

6.4.1 SOIL AIR CAN BE IMPROVED WITH CERTAIN CULTURAL PRACTICES

Most field crops perform best under good soil aeration. Thus, one of the important soil management activities in crop production is improving soil aeration. Air and water occur in the spaces in the soil (pore spaces). If the pore spaces are filled with water, air is automatically excluded. Since both air and water are critical to plant growth, crop producers must maintain a good balance between the factors. Specific strategies in this regard include the following:

1. *Drainage.* Certain soil types such as those high in clay and organic matter (e.g., histosols) may need drainage to remove excess moisture.
2. *Tillage.* Tillage is used to mechanically disturb the soil and to prepare the seed bed for planting. Tillage can also be used to break up compacted soil to improve drainage.
3. *Raised beds.* Raising the bed improves drainage in the seedbed.

There are two major categories of air pollutants—primary and secondary. Primary air pollutants enter the atmosphere directly, while secondary pollutants form when primary pollutants combine with other substances already present in the atmosphere to form new pollutants. In terms of origin, air pollutants may be classified as either of human origin (anthropogenic) or of non-human origin. Human causes of air pollution include the combustion of fuels, petroleum refinery processes, mining activities, and farming. About 47% of air pollution is caused by transportation activities, with the combustion of fossil fuels and industries activities contributing 27% and 15%, respectively.

The key air pollutants are carbon monoxide, nitrous oxides, hydrocarbons, sulfur oxides, and suspended particles. The principal kinds of nitrous oxides are nitric oxide (NO), nitrogen dioxide (NO_2), and nitrous oxide (N_2O). Nitrous oxide is the primary catalyst for smog formation, while NO_2 is involved in acid rain formation. N_2O is implicated in global warming. The major hydrocarbons include methane (CH_4) and octane, or gasoline (C_8H_{18}). Hydrocarbons are common in nature, arising from the decaying of plants and animal materials, among other sources. They are also released when humans burn fossil fuels or have accidents (spills) with fuels.

The primary sulfur oxide pollutant is sulfur oxide (SO_2). This compound is subsequently converted to sulfur trioxide (SO_3) in the atmosphere, which reacts with water to produce sulfuric acid (H_2SO_4), a component of acid rain. The sulfur oxides are primarily generated by human activities (the burning of fossil fuels, especially coal, and from the refinery of petroleum).

Suspended particles (or suspended particulate matter) include solids (e.g., pollen, ash), liquids (e.g., droplets of liquids, such as H_2SO_4), and gaseous materials, such as smoke, that carry particles. They cloud the atmosphere, thereby reducing visibility.

Air pollutants adversely impact crop production in various ways:

1. They destroy the photosynthetic surface of plants, resulting in reduced photosynthesis and consequently reduced productivity.
2. Ozone is injurious to plants (it reduces photosynthetic surface).
3. Acid rain injures plants, causing defoliation and reduced photosynthetic surface.
4. Acid rain also increases soil pH, affecting plant adaptation and soil microorganisms.

6.4.2 SUMMARY OF AIR-RELATED CROP MANAGEMENT PRACTICES

The management practices summarized here are discussed in detail in appropriate sections of the book. These represent decisions a producer must make in order to have success with crop production.

1. *Soil drainage.* A typical mineral soil consists of about 25% air. The key to management of soil is to recognize that both soil air and moisture occur in the same pore spaces. If all pores are filled with water, there is no soil air, and vice versa. Excessive soil moisture must be drained to have ample air for root respiration.
2. *Subsoiling.* Subsoiling is an operation conducted to break compaction in the subsoil that impedes soil drainage.
3. *Variation of tillage depth.* Maintaining the same plow depth may eventually create an impervious layer, the plow pan, that impedes soil drainage.
4. *Cultivation of soil.* Cultivation of the soil improves soil infiltration and reduces water-logging.
5. *Tillage conditions.* Tilling the soil when wet destroys soil structure and causes compaction, which impedes water infiltration into the soil and encourages water-logging.

6.5: SUNLIGHT IS ELECTROMAGNETIC ENERGY

Solar radiation is the portion of the electromagnetic spectrum that reaches the earth. *Visible radiation,* or simply *light,* constitutes only a small part of the radiant energy spectrum (Figure 6–12). Radiant energy is described by its wavelength and frequency; the shorter the wavelength and the higher the frequency, the greater the energy of the radiation. The effect of solar radiation on crop productivity depends on the angle at which it strikes the earth, the cloud cover, and the interception by plants. The curvature of the earth makes the polar regions of the earth receive radiation at an oblique angle. The effect of this is reduced energy and hence colder temperatures in those regions. Light exerts its effect in different ways through three properties: *quality, quantity,* and *duration.*

6.5.1 LIGHT INTENSITY MEASURES ITS BRIGHTNESS

How much light can plants use? Light intensity measures its quantity or brightness. At midday, light intensity may be as high as 10,000 foot candles. Species differ in the maximum intensity of light they can use (Figure 6–13). It is best to have two leaves receiving 50% each of full light than one leaf receiving 100% light it cannot convert to dry matter. Corn and sorghum increase their photosynthetic rate as light intensity increases more so than soybean and alfalfa. Most plants more efficiently utilize low light intensity than high light intensity.

Another key factor in light interception and utilization is plant architecture, especially regarding leaf angle and shape. Horizontal leaf display and broadleaf form cause overlapping (shading) of leaves and hence reduced penetration of light through the crop canopy. Upright leaf angle in corn is known to have increased yield of grain by 40% while reducing bareness by 50%.

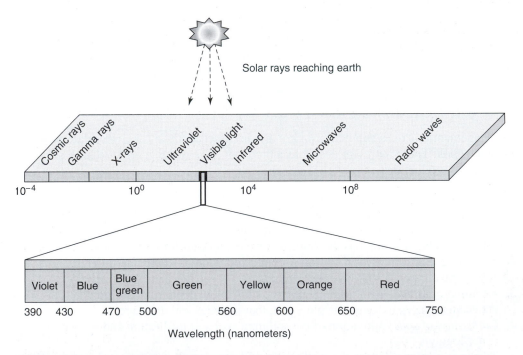

FIGURE 6–12 The radiant energy spectrum. Visible radiation or light constitutes a small proportion of the range of wavelengths of electromagnetic radiation, occurring between about 400 and 735 nanometers.

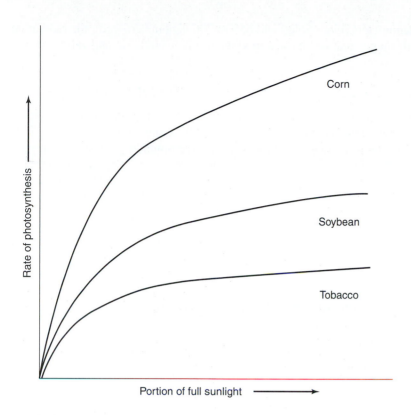

FIGURE 6-13 Plant species differ in the amount of light intensity they can use for photosynthesis. Species such as corn and sorghum can utilize full sunlight for photosynthesis, while tobacco can tolerate only partial sunlight.

Rate of photosynthesis →

Corn

Soybean

Tobacco

Portion of full sunlight ⟶

Optimal **leaf area index** is about 3.5 to 4.0. Plant density (spacing) influences crop canopy characteristics. The grower should utilize the best density that allows optimal light utilization. Shaded leaves may contribute less to accumulation of assimilates by consuming more of it through respiration.

Leaf area index. The ratio of the total leaf area per unit area.

Light intensity conditions several plant responses that have consequence in crop productivity, the key ones being etiolation, tillering, branching, and bareness.

1. *Etiolation.* Etiolation or spindly growth is induced by low light intensity or prolonged darkness. In field crop production, close spacing (high density) causes plants to overshade each other. In an attempt to receive light, these plants grow spindly. Shading increases the levels of auxins (IAA, especially), while high light intensity destroys IAA. Shaded plants thus experience excessive stem elongation (long internodes). Spindly growing plants are prone to lodging.
2. *Tillering.* Tillering is the production of secondary stems that arise from the crown area of the plant. Species that tiller include wheat and barley. Even though tillering is enhanced by cultural environment (cool temperature, high fertility, and good soil moisture), it is triggered by high IAA concentration.
3. *Bareness.* Bareness reduces grain yield by reducing seed filling. Ineffective penetration of light into the canopy to reach lower leaves contributes to this condition.

6.5.2   THE WAVELENGTH OF THE RADIATION DETERMINES ITS QUALITY

Plant pigments associated with photosynthesis have different absorption spectra. That is, they absorb different wavelengths of light. Leaf chlorophyll absorbs radiation efficiently at 400 to 500 nanometers (blue-violet) and 650 to 700 nanometers (red), the radiant energy wavelength ranges that are most important to photosynthesis and reflect mainly green (its visible color).

6.5.3 IMPROPER LAND USE MAY LEAD TO DENUDATION

Desertification. The human-induced spreading of desert conditions that disrupts semiarid and arid ecosystems and agroecosystems.

Desertification is a human-induced deterioration of vegetation that adversely affects climate. There is a general chain of events leading to the denudation of the land. In the Sahalian region, wells were dug to provide water for the nomadic people of the region. This facility changed their lifestyle from nomadic to sedentary. Population increases around these watering holes placed enormous pressure on the land. The land became overgrazed, and trees were cut for firewood, leading to denudation. The bare land reflected more heat and less water vapor into the atmosphere. Consequently, fewer clouds formed and less rain fell in the locality. The vegetation loss increased progressively. Desertification has occurred in various parts of the world. To forestall the advance of the desert, nations affected have embarked on reforestation (tree planting), among other strategies.

6.5.4 SUMMARY OF LIGHT-RELATED CROP MANAGEMENT PRACTICES

The management practices summarized here are discussed in detail in appropriate sections of the book. These represent decisions a producer must make in order to have success with crop production. Light management practices in crop production include the following:

1. *Plant density.* Plant population or density affects the amount of light trapped by plants in the field. Low plant population means that plant leaves will not intercept much of the light from the sun.
2. *Plant arrangement.* Plant arrangement in the field affects how effectively the plants can intercept light. The crop producer decides the pattern of distribution and the spacing between and within rows.
3. *Photoperiod response.* Plants differ in photoperiod response, the basis for placing plants into maturity groups. Some prefer short days, while others are long-day species, flowering only when the daylength exceeds a certain critical minimum. Photoperiod response is critical to crop production in plants whose economic parts are flower-based because flowering will occur only under the appropriate photoperiod.
4. *Weed control.* Shading can smother weeds. The producer can use the proper plant density, planting at the right time, cultivation, and other practices to create an effective plant canopy to shade out weeds.

SUMMARY

1. Climate pertains to a region and determines crop adaptation.
2. Weather describes meteorological factors pertaining to a locality and varies over short periods of time. It determines crop productivity. It is the chance factor in crop productivity.
3. Geographic surface factors such as large bodies of water and mountains modify meteorological factors in a region.
4. The major climatic regions in the United States are the Steppe, Continental, and Subtropical Wet.
5. Climate is not static. It keeps changing as factors such as volcanic discharge, continental drift, and industrial activities impact it.

6. The El Niño/Southern Oscillation and La Niña, as well as the Indian Monsoon, are climatic systems that have devastating effects on crop production when they occur.
7. The greenhouse effect may be partially responsible for global warming and is a potential threat to crop production.
8. Atmospheric air consists of only 0.035% carbon dioxide, which is needed for photosynthesis.
9. Most of the atmospheric air (79%) consists of nitrogen in a form that plants cannot use directly, except to a limited extent by legumes with the aid of bacteria in symbiotic association.
10. Air pollutants inhibit photosynthetic processes, cause acid rain, and reduce crop productivity.
11. Solar radiation is the primary source of soil heat. Temperature regulates all biochemical reactions, colder temperature slowing them while higher temperature accelerates reactions.
12. Planting and duration of growing season are influenced by temperature.
13. Crop production and productivity are affected by light intensity, quality, and duration. Intensity impacts plant structure (etiolation, tillering, and branching) as well as photosynthetic efficiency. Light duration impacts flowering and crop maturity.
14. Light affects photosynthesis and plant developmental switch (photoperiod) involving changes from vegetative phase to reproductive phase.

REFERENCES AND SUGGESTED READING

Acquaah, G. 2002. *Horticulture, principles and practices,* 2nd ed. Upper Saddle River, NJ: Prentice Hall.

Arshad, A., and R. H. Azooz. 1996. Tillage effects on soil thermal properties in a semi-arid cold region. *Soil Sci. Soc. Am. J.* 60:561–67.

Critchfield, H. J. 1974. *General climatology.* Englewood Cliffs, NJ: Prentice Hall.

Green, D. E., D. G. Wooley, and R. E. Mullen. 1981. *Agronomy: Principles and practices.* Edina, MN: Burgess.

Kellogg, W. W., and R. Schware. 1981. *Climate change and society.* Boulder, CO: Westview Press.

Martin, J. H., and W. H. Leonard. 1967. *Principles of field crop production.* New York: Macmillan.

Merrill, S. D., A. L. Black, and A. Bauer. 1996. Conservation tillage affects root growth of dryland spring wheat under drought. *Soil. Sci. Soc. Am. J.* 60:575–83.

Pruitt, W. O., F. J. Lourence, and S. Von Oettingen. 1972. Water use by crops as affected by climate and plant factors. *California Agriculture* 26:10–14.

Rosenberg, N. J., and B. L. Blad. 1983. *Microclimate, the biological environment.* 2nd ed. New York: Wiley.

Rykbost, K. A., L. Boersma, J. J. Mack, and W. E. Schmisseur. 1975. Yield response to soil warming: Vegetable crops. *Agron. J.* 67:738–43.

Stefanski, R. J. 1994. El Niño: Background, mechanisms, and impacts. In *Major world crop areas and climate profiles.* Washington, DC: U.S. Department of Agriculture.

Unger, P. W. 1978. Straw mulch effects on soil temperatures and sorghum germination and growth. *Agron. J.* 70:858–64.

SELECTED INTERNET SITES FOR FURTHER REVIEW

http://ceres.ca.gov/elnino/background.html

A discussion of various aspects of El Niño.

http://www.nhq.nrcs.usda.gov/land/index/basemaps.html

Relief of United States; other U.S. base maps.

OUTCOMES ASSESSMENT

PART A

Answer the following questions true or false.

1. T F   Climate determines crop adaptation.
2. T F   Oxygen is a greenhouse gas.
3. T F   El Niño is characterized by a warm episode around the Peruvian coast.
4. T F   The windward side of a mountain range is drier than the leeward side.
5. T F   Most of the atmospheric air consists of carbon dioxide.
6. T F   The pH of acid rain is between 6.0 and 6.9.
7. T F   Heaving is a temperature effect that causes plants to be moved upward from lower depths in the soil.
8. T F   Light intensity measures its brightness or quantity.
9. T F   Etiolation is caused by low light intensity.

PART B

Answer the following questions.

1. Give the three major climatic types of the United States.

   _____

   _____

   _____

2. Give three specific gases involved in the greenhouse effect.

   _____

   _____

   _____

3. Most irrigated lands in the U.S. are located in the _____ and _____.

4. What is a microclimate?

_____

_____

5. Give four common air pollutants.

_____   _____

_____   _____

6. State the Hopkins Bioclimatic Law.

_____

7. _____ is the strategy of spreading a plastic sheet over the soil to trap heat to control pathogens.

8. Leaf chlorophyll absorbs radiation efficiently between _____ nanometers and _____ nanometers.

9. Light quality is measured by its _____.

10. _____ is the growth hormone involved in etiolation and tillering.

PART C

Write a brief essay on each of the following topics.

1. Distinguish between climate and weather.

2. Discuss the Hopkins Bioclimatic Law.

3. What is the orographic effect?

4. Discuss the greenhouse effect.

5. Discuss the factors that influence global climatic change.

6. Discuss how the correlation between temperature and plant growth is used to predict harvesting time and crop adaptation.

7. Describe how acid rain is formed.

8. Discuss the air pollutants and their effect on crop productivity.

9. Discuss how soil temperature may be managed for crop production.

10. Discuss how the soil is warmed.

11. Discuss the effects of poor aeration of soil on crop growth and development.

12. Compare and contrast soil chemistry under aerobic and anaerobic conditions.

Discuss or explain the following topics in detail.

1. What impact do you think the widening of the ozone hole will have on world crop production?

2. Predict and describe the performance of a soybean variety that is adapted to the Midwest if it is grown in the southern United States.

3. Discuss how a crop producer can minimize chance effects due to weather in crop production.

4. Discuss government and other programs available to U.S. crop producers, prior to crop loss and after crop loss, that are geared toward reducing the unpredictable effects of the weather.

7

Soil and Land

PURPOSE

The purpose of this chapter is to discuss soil physical, chemical, and biological properties and how they influence crop production. The distribution of arable lands and classification of land based on land capability are also discussed. Also, soil fertility and its management are discussed.

EXPECTED OUTCOMES

After studying this chapter, the student should be able to:

1. List and briefly describe the twelve soil orders in relation to crop production.
2. Discuss the role physical characteristics (texture, structure, soil bulk density, porosity, permeability, and aggregation) play in crop production.
3. Define and discuss CEC and its importance.
4. Define and discuss soil pH and its importance.
5. Discuss the distribution of arable land in the world.
6. Describe and discuss soil organisms and their importance.

KEY TERMS

Arable land
Bulk density
Cation exchange capacity
  (CEC)
Denitrification
Expanding clay
Flocculation

Humus
Igneous rock
Liming
Loess
Metamorphic rock
Mycorrhizae
Non-expanding clay

Parent material
Ped
Sedimentary rock
Soil aggregate
Soil colloid
Soil horizon
Soil permeability

Soil profile   Soil structure   Topsoil
Soil reaction   Soil texture   Weathering
Soil separates

TO THE STUDENT

"Dirt" is what you sweep off the floor. "Soil" is the medium in which field crop production is carried out. Nutrients, water, and air, key factors that are needed for plant growth and development, are provided through the soil. The soil also provides physical support to plants and serves as a recycling center for nutrients and organic waste. As a natural resource, soil can be degraded and depleted of plant nutrients through natural factors as well as human activities. Certain soils are more suitable for crop production than others. However, provided any weaknesses can be corrected or managed in crop production, most soils can be used to produce certain crops.

7.1: WHAT IS SOIL?

The Soil Science Society of America defines soil as the unconsolidated mineral or organic material on the immediate surface of the earth that serves as a natural medium for the growth of land plants. It is also defined as the unconsolidated mineral or organic matter on the surface of the earth that has been subjected to and shows the effects of genetic and environmental factors of climate (including water and temperature effects) and macro- and microorganisms, conditioned by relief, acting on parent material over a period of time. Further, a product-soil differs from the material from which it was derived in many physical, chemical, biological, and morphological properties and characteristics.

The society also defines soil science as the science that deals with soils as a natural resource on the surface of the earth, including soil formation; classification and mapping; the physical, chemical, biological, and fertility properties of soils per se; and these properties in relation to the use and management of soils.

7.2: SOIL FORMATION

How is soil formed? Soil formation is a slow process; for all practical purposes, soil is not a renewable resource. It is important, therefore, that soils be managed properly because it is difficult to reclaim damaged soil. Soil is weathered rock. The process of weathering, by which soil is formed, is slow and sequential, involving physical, chemical, and biological factors. It involves the alteration of rocks and their minerals, resulting in their disintegration, decomposition, and modification. Soil is a dynamic system, changing as it is impacted by environmental factors and by the plant and other organisms that live in it.

The five factors of soil formation are *parent material, climate, organisms* (organic matter), *topography* (relief), and *time.* These factors of soil formation are interrelated, each one affecting and being affected by the others.

7.2.1 SOIL ORIGINATES FROM MINERAL AND ORGANIC MATERIALS

The starting point in soil formation is the parent material, the rock material on which agents of weathering act. This material is classified into two types, based upon origin. Some parent material is formed in place *(sedentary)*, while some is *transported* by various agents and deposited in certain places. A soil formed from parent material that was transported by ice is called *glacial,* while a soil formed from parent material transported by wind is called **loess**. Most soils in the United States are formed from non-glaciated parent material. Glacial soils occur in the central parts of the United States, especially the midwestern states.

Parent materials are formed from parent rocks (these consist of primary minerals such as augite, feldspar, hornblende, mica, olivine, and quartz), which are of three basic types: **igneous, sedimentary**, and **metamorphic**.

1. Igneous rocks (e.g., granite) are consolidated, hard rocks that consist of minerals, including quartz and feldspar.
2. Sedimentary rocks are unconsolidated and consist of pieces of rock transported by various agents. Upon deposition, various products are formed, such as limestone, sandstone, and shale.
3. Metamorphic rock is produced when other rock types (igneous and sedimentary) are subjected to intense heat and pressure. The resulting products (metamorphic rocks) include slate, gneiss, marble, and schist.

Some soils originate from organic material. Peat and marsh (bog) soils are high in organic matter.

7.2.2 WEATHERING PROCESSES CAN BE PHYSICAL OR CHEMICAL

Weathering is the process by which parent materials are broken down into the finer particles that form the soil. Agents of weathering act upon parent materials. These agents can be physical or chemical.

Physical Weathering

Physical weathering occurs primarily through disintegration of rock material. This occurs in several ways. Rocks consist of different chemical materials with different rates of expansion and contraction. Under temperature fluctuation, rocks crack and peel off (exfoliation) as the chemical components expand and cool at different rates. Cracks in rocks collect water, which upon freezing expands to further widen the crack. Glacial movement involves grinding and scraping of rocks as the ice moves on the earth's surface. Similarly, moving water carries rocks that crack and crush as they tumble with the currents. Sometimes, plant roots grow into cracks in rocks and widen them, as the roots grow larger.

Chemical Weathering

Chemical weathering occurs by one of several distinct chemical processes. These reactions increase the disintegration rates of minerals.

1. *Carbonation.* Carbon dioxide dissolves in water to form a weak acid [carbonic acid (H_2CO_3)], which then reacts with carbonates and other minerals in rock

Loess. Soil particles, predominantly silt-sized, transported and deposited by wind.

Igneous rock. Rock formed by the cooling and solidification of molten rock (magma).

Sedimentary rock. Rock formed from material originally deposited as a sediment, then physically or chemically changed by compression and hardening.

Metamorphic rock. A rock (igneous or sedimentary) that has been greatly altered from its previous condition by a combination of high temperature and pressure.

Weathering. The processes by which parent material changes in character, disintegrates, decomposes, and synthesizes new compounds and clay minerals.

materials that are easier to decompose, such as calcium bicarbonate. This reaction is called *carbonation:*

$$CaCO_3 + H_2O + CO_2 \rightarrow Ca(HCO_3)$$

2. *Hydration.* Hydration entails adding water molecules to materials to form hydrated products that are easier to break down—for example,

$$CaSO_4 + 2H_2O \rightarrow CaSO_4 \cdot 2H_2O \text{ (hydrated calcium sulfate, or gypsum)}$$

3. *Hydrolysis.* Hydrolysis is a reaction involving water in which ions in parent materials are made more readily available through the formation of more soluble products such as KOH.

$$KAlSi_3O_8 \rightarrow HAlSi_3O_8 + KOH$$

$$2HAlSi_3O_8 + H_2O \rightarrow Al_2O_3 \cdot H_2O + 6H_2SiO_3$$

4. *Oxidation.* Minerals in rocks may react with oxygen (oxidation) to produce less stable materials. This occurs especially where minerals contain iron. The oxides formed are visible as reddish-yellow coloration (rust).

$$4FeO + O_2 \rightarrow 2Fe_2O_3$$

5. *Reduction.* Under conditions of low oxygen, the oxidation reaction may proceed in the reverse direction to produce products that are less oxidized.
6. *Solution.* Certain rock minerals dissolve readily in water, creating weaknesses in the rock material and promoting physical disintegration.

Weathered parent materials produce a large variety of metallic ions that may be recrystallized into new minerals, called *secondary minerals,* such as clay.

7.2.3 SEVERAL FACTORS ARE RESPONSIBLE FOR SOIL FORMATION

Parent material. The material from which soil is developed by the process of weathering.

Soil development occurs at a rate that is dependent on **parent material**, *time, climate, biota,* and *relief.*

Parent Material

What kinds of materials can be converted to soil? Most soils are formed from unconsolidated materials produced by the action of erosive forces (wind, water, air) or sediments that are glacial in origin. These deposits vary in particle size as well as chemical composition. Whether the parent material will play a dominant role in determining the physical and chemical properties of the soil formed will depend on the extent of weathering of the material. Slightly weathered parent material plays a more dominant role than well-weathered parent material in this regard. Parent materials differ in their rate of weathering, quartz being among the most resistant.

Climate

Climate affects the rate of soil formation and the type of soil that is ultimately formed. The role of climate is both direct and indirect:

1. *Direct.* What are the general differences in soils formed under high rainfall conditions and those formed under drier conditions? High rainfall and temperature provide conditions that promote rapid weathering, high leaching (downward loss of nutrients through the soil profile), and high oxidation. The soils formed tend to have red or yellow colors, which are characteristic of oxidized soil material. Soils formed under drier conditions, on the other hand, experience salt (e.g., calcium and magnesium) accumulation (not leaching). Such soils are thus high in salt and may require leaching *(desalination)* to render them useful for crop production.

2. *Indirect.* The indirect effect of climate is in the type of vegetation it supports. Semiarid climates support shrubs and grasses, while heavy rains encourage trees and forest conditions. Organisms and organic matter affect soil formation as described next.

Biota (Organisms)

Do soils formed under forest vegetation tend to have more organic matter than those formed under grassland? The *biota* (living things) affects soil formation. The vegetation determines the type and amount of organic matter in the soil. Grasses differ from pines in the type of organic matter they produce. Soils developed under grasses have high organic matter. Much of the organic matter in forested regions lies on the soil surface and decomposes slowly. It is not incorporated into the soil as effectively as organic matter from grasses. Further, soils under forested areas are heavily leached, depleting the top horizon of organic matter and clay minerals, which accumulate in the lower horizon. Soils formed under conditions where soil-burrowing organisms (e.g., moles, gophers, earthworms, and termites) abound experience constant mixing of the soil within the profile. This causes fewer but deeper horizons to develop. Soil microbes are also beneficial to soil formation by decomposing organic matter and causing weak acids to form to aid in dissolution of minerals.

Topography

Topography, or *relief,* determines the rate of soil erosion and hence the depth of the soil formed through its effect on water and temperature. Erosion is more pronounced on steep slopes. Rapid soil movement does not allow the soil enough time to develop. The soil profile on gentle slopes is deeper and supports more luxuriant vegetation. The soil profile under such conditions is not well defined. Topography influences soil drainage and thus the amount of moisture available for weathering processes. In low-lying regions, drainage may become a problem, thereby causing anaerobic conditions to occur. Organic matter decomposition is slow and hence it accumulates, leading to the formation of high organic matter soils such as peat and muck.

Time

Soil formation is a continuous process. Rock materials differ in rates of decomposition. Limestone decomposes rapidly, while granite is very resistant to weathering. Profile development is dependent on time. Old soils have more defined profiles, because they have been exposed extensively to weathering agents. This allows decomposition and modification of component chemicals. The rate of soil development depends on the effect of time on soil forming factors. Are older soils necessarily better for crop production?

Soil profile. A vertical cross section of a soil through all of its horizons.

Soil horizon. A layer of soil, approximately parallel to the soil surface, with distinct characteristics produced by soil-forming processes.

Topsoil. The layer of soil that is disturbed during cultivation.

A vertical cross section of the soil reveals its vertical distribution, called the **soil profile**. A profile is characteristic of the soil genesis, or origin. Depending upon the age of the soil, the conditions under which it was formed, and parent material, among other factors, a soil profile will show different and distinguishable layers called the **soil horizon** (Figure 7–1). Figure 7–2 presents an actual soil profile. There are six master horizons designated by uppercase letters as O, A, E, B, C, and R. The R horizon is consolidated rock. Lowercase letters may also be added to further distinguish the horizons. For example, a Bt horizon indicates that the layer has clay deposits. Further, there are transitional horizons that are indicated by two uppercase letters—for example, AE. For crop production purposes, a simplified and typical soil profile will have three horizons. In agricultural soils, the O horizon is often absent because of tillage operations that mix it up with the A horizon material.

1. *A horizon.* This is also called the **topsoil** and consists of the uppermost layer of the soil. It has the most organic matter and is the most leached layer of the

FIGURE 7–1 Soil profile. There are six master horizons in a typical soil profile. There are subdivisions and intermediate layers between the layers. For crop production purposes, a simplified profile has three parts: topsoil, subsoil, and the underlying parent material. Most crop roots occur in the topsoil. (Source: Modified after USDA)

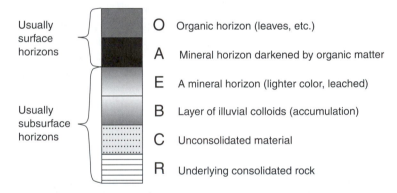

Usually surface horizons

Usually subsurface horizons

O Organic horizon (leaves, etc.)

A Mineral horizon darkened by organic matter

E A mineral horizon (lighter color, leached)

B Layer of illuvial colloids (accumulation)

C Unconsolidated material

R Underlying consolidated rock

FIGURE 7–2 (a) A soil profile, showing variation in soil color between the top and low horizons. (b) A technician extracts a soil core using a truck-mounted powered soil auger. (Source: USDA)

(a)

(b)

profile. Thus, it is sometimes referred to as the *zone of leaching (zone of eluviation).* This is the portion of the soil that is tilled for crop production. Does it mean we should not care about the lower layers?

2. *B horizon.* Materials leached from the A horizon accumulate in the B horizon, also called the *zone of accumulation (zone of illuviation).* This includes leached salts and clay minerals. The soil in this layer is less weathered. Roots of deep-rooted plants may reach this layer.

3. *C horizon.* This zone consists of unconsolidated material. There is little or no profile development. The underlying consolidated rock material is called the *R horizon.*

The soil profile can be modified by agricultural activity. As previously indicated, plowing mixes soil in the O and A horizons to form the *plow layer.* Further, the use of heavy tillage machinery can create compacted layers, called *hard pans,* that are impermeable to root penetration, and impede drainage.

7.4: SOIL GROUPS IN U.S. CROP PRODUCTION

The United States may be divided into two broad regions according to soil characteristics by dropping a line from western Minnesota through central Texas (Figure 7–3). The soils to the east of this line are called *Pedalfers* and are generally leached (low or deficient in basic materials—Na, K, Ca, Mg). The absence of these basic materials makes the soils on the eastern region acidic in reaction. Similarly, the soils of the humid Pacific Northwest are acidic. The soils on the west side of the United States are called *Pedocals.* Contrary to the eastern soils, these soils have an accumulation of soluble salts in the subsurface or subsoil. This accumulation of soluble salts occurs because the rainfall in the region is inadequate to leach the salts out of the soil. These leached soils are naturally fertile. However, low rainfall limits plant growth, making it necessary to irrigate crops in order to realize optimal productivity unless the season is wet.

The soils in most of the humid parts of the drier region of the west were formed under grassland, called *Chernozem soils;* they are black in color and very high in organic matter. They are very fertile and suited to the production of grasses and small grains, such as wheat. Where the climate is drier, the vegetation is sparse and consequently soil organic matter is low, making these soils lighter in color in the A and B horizons. Other soil groups in the west are the *Chestnut soils, Brown soils, Sierozem (Gray) soils,* and *Desert soils.* The Chestnut soils are suited to wheat production but growers should be aware of unseasonable weather. Similarly, the Brown soils are suited to wheat and sorghum (with the same warning of adverse weather) in the moist season, and for grazing in the driest areas. The Sierozem soils of the semidesert regions are generally low in organic matter, high in soluble salts, and relatively shallow. These soils are used for crops only when irrigation is possible; otherwise, they are used for grazing.

The Corn Belt region has soils that are very dark brown in color, high in organic matter, and naturally fertile. Called *Prairie soils,* they are suited to producing grains and a variety of other field crops. The climate of this region is very conducive to crop production. The combination of good natural fertility and good climate makes the Corn Belt one of the most productive soils in the world.

The northern humid regions, including the Great Lakes regions, are also acidic because of the leaching under the prevailing climate. Called *Podzol soils,* they are not naturally fertile and are suited to hay and pastures. Next to these soils to the south lie the Gray-Brown Podzolic soils of the Great Lakes and eastern Corn Belt states. They were

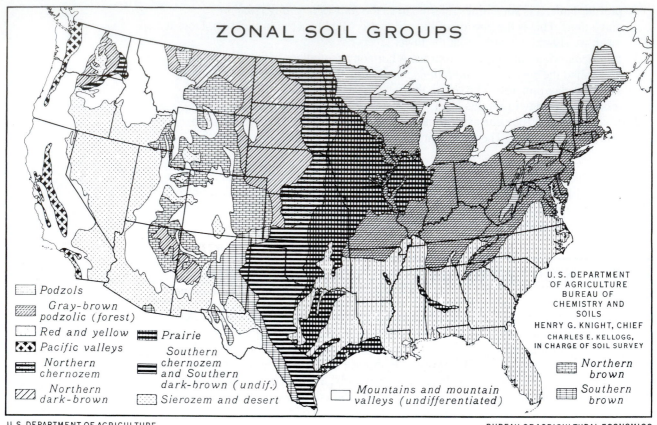

FIGURE 7–3 The United States is divided into two broad regions of soil groups, the Pedalfers (leached with no accumulation of lime) to the east and the Pedocals (accumulation of calcium carbonates) to the west. (Source: USDA)

developed under deciduous forests and a moist temperate climate. These soils are naturally relatively more fertile than the Podzols (but less fertile than the Prairie soils) and are suited to producing various crop plants. The Gray-Brown Podzolic soils are suited to producing a variety of crops, especially when limed and fertilized.

There are pockets of peat and marsh soils that have unusually high organic matter. Similarly, these are regions with excessive accumulation of bases such that they are unsuitable for cropping until leached by heavy irrigation. These alkaline soils are called *Solonchack soils;* they occur in arid regions.

7.5: THE USE OF SOIL FOR AGRICULTURAL PRODUCTION DEPENDS ON ITS PROPERTIES

What are the key soil physical properties that impact crop production? To what extent can a producer modify these properties for higher crop productivity? Some soils consist primarily of mineral matter (e.g., sand in the desert areas), whereas others consist predominantly of organic matter (e.g., peat bogs). A good soil for crop production, however, should have a good balance of certain basic components. These are mineral matter, organic matter, air, and water (Figure 7–4).

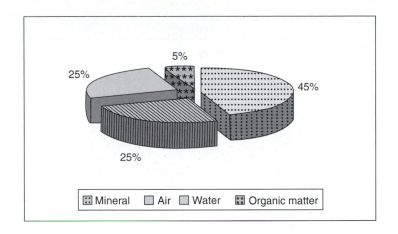

5%

25%

45%

25%

☐ Mineral ☐ Air ☐ Water ☐ Organic matter

7.5.1 SOIL CONSISTS OF PARTICLES THAT ARE ARRANGED IN A CERTAIN PATTERN

Understanding soil physical properties is important to crop productivity. These properties impact how soil functions in an ecosystem and how it can be best managed. Soil physical properties affect how soil water and solutes move through and over the soil, as well as the suitability for agricultural use. Two important soil physical properties of mineral soils are its **soil texture** and **soil structure**. They help determine the nutrient-supplying ability of soil solids and the ability of the soil to hold and conduct the water and air required for proper root activity for proper plant growth.

Soil texture. The relative proportions of sand, silt, and clay in a soil.

Soil Texture

The soil textural class may be determined in the field by the "feel" method. A moist soil sample is rubbed between the thumb and forefingers and squeezed out to make a "ribbon." A non-cohesive appearance plus a short ribbon indicates a sandy loam, while a smooth appearance plus a crumbly ribbon indicates a silt loam. Clay makes a smooth, shiny appearance plus a flexible ribbon.

Soil structure. The arrangement of primary soil particles into secondary particles, units, or peds.

Soil may be physically separated into three particle size groups called **soil separates**. These are *sand, silt,* and *clay* (Table 7–1). Soil texture is defined as the proportions (or percentages) of sand, silt, and clay particles in a soil. Agricultural soils typically consist of a combination of all three particle sizes (called a *loam*). To determine the textural grade or textural class of a soil after mechanical analysis, the USDA textural triangle may be used (Figure 7–5). Can a soil that is predominantly clay, silt, or sand be successfully used for crop production?

Soil separates. The individual particle size groups (sand, silt, and clay) of a mineral soil.

Soil texture is very important in crop production. Fine-textured soils (clay soils) generally have poor drainage and are prone to waterlogging. However, they have high water-holding capacity. Clay soils are further described as heavy soils and are difficult to till. They impede root development and thus are not suitable for root crop production. Texture affects soil porosity. Clay soils have a preponderance of *micropores,* or capillary pores (small pores) that impede drainage. Sandy soils have more *macropores,* or non-capillary pores (large pores). Clay soils are also described as cold soils and tend to require warming (e.g., by drainage) for use in early cropping. Poorly drained soils have low microbial activity, but clay soils have high **cation exchange capacity (CEC)**, the ability of soil to attract and hold cations. Thus, clay soils have high nutritional status, while sandy soils, which are coarse-textured, have low CEC. On the other hand, sandy soils are light soils and are easier to till. They also have greater infiltration rates, drain better, and

Soil permeability. The amount of water that moves downward through the saturated soil.

Cation exchange capacity (CEC). The sum total of exchangeable cations that a soil can adsorb.

Table 7–1 Selected General Properties of the Three Soil Separates of Inorganic Soil

Property	Sand	Soil Separate Silt	Clay
Range of particle diameter (mm)	2.0–0.05	0.05–0.002	Smaller than 0.002
Visible with/under	Naked eye	Microscope	Electron microscope
Dominant minerals	Primary	Primary	Secondary
Aeration	Excellent	Good	Poor
Cation exchange capacity	Low	Moderate	High
Permeability by water	Fast	Moderate	Slow
Water-holding capacity	Low	Moderate	High
Drainage	Excellent	Moderate	Poor
Consistency when wet	Loose, gritty	Smooth	Sticky, malleable
Consistency when dry	Very loose and gritty	Powdery with few clods	Hard clods
General ease of tillage	Easy	Moderate	Difficult
General spring temperature	Warms fast	Warms moderately	Warms slowly
General erodibility by water	Easy	Moderate	Difficult

FIGURE 7–5 The USDA soil textural triangle. To use this guide, first obtain a physical analysis of the soil (i.e., the proportion of sand, silt, and clay, such as 40, 35, and 25). Locate 40% on the axis labeled "sand" and draw a line parallel to the axis labeled "silt." Next, locate either the clay or silt. Locate 25% on the clay axis and draw a line parallel to the sand axis to intersect the previous line from the sand axis. The section of the triangle in which the intersection occurs indicates the soil textural class. (Source: USDA)

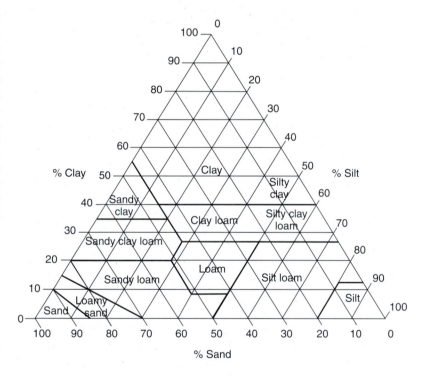

are warm. However, crop production on sandy soil requires frequent irrigation to make up for its poor water-holding capacity. Soil texture has implications in irrigation of field crops. The method and frequency of irrigation depend on soil texture.

Soil Structure

Ped. A unit of soil structure, such as an aggregate or a crumb, formed by natural processes.

Soil structure is the arrangement of soil primary particles (soil separates) into secondary particles, units, or **peds**.

Soil particles are arranged in different shapes and sizes. The common arrangements are

1. *Prismlike.* Particles are arranged in the vertical plane.
2. *Platelike.* Arrangement of particles is in the horizontal plane.
3. *Blocky.* This soil structure consists of rectangular shapes.
4. *Granular-crumb.* For agricultural purposes, the granular-crumb structure is ideal, and occurs in the furrow slice or plow layer. This structure consists of spheroidal units or circular arrangement of peds. However, this desirable structure is influenced by cultural practices and is subject to alteration.

Soil structure can be destroyed through compaction from traffic from farm animals, vehicles, and raindrops. Tilling the soil while it is wet, and applying certain soil amendments (e.g. liming), can destroy soil structure. Poor soil aggregation is partly the cause of poor drainage and poor water-holding capacity. Soil structure can be improved through, for example, the addition of organic matter.

Certain factors force these particles to make contact and to form secondary units called **soil aggregates**. These factors include soil tillage, organic matter, cations, and physical processes. The kind of cations adsorbed by soil colloids affects aggregation. Na^+ ions cause dispersion, while Ca^{2+}, Mg^{2+}, and Al^{3+} ions encourage aggregation by the process of **flocculation**. In crop production, it is important to maintain soil aggregate stability. Soils vary in their resistance to alteration from factors such as rain and tillage implements. Heavy rains can cause crusting of the soil surface, thus impeding seedling emergence and water infiltration. Clay soils are naturally prone to crusting. Artificial soil conditioners such as *synthetic polymers* can aid the stabilizing of soil aggregation.

Soil aggregate. A clump of many primary soil particles.

Flocculation. The aggregation of especially clay particles into clumps.

7.5.2 SOIL PARTICLES MAY BE LOOSELY OR DENSELY PACKED

How important is the degree of packing of soil particles to crop production? **Bulk density** of soil is defined as the mass (weight) of a unit volume of dry soil. This is the volume of soil as it exists naturally and thus includes air space and organic matter. *Particle density* is the density of the solid soil particles only. Bulk density is calculated as follows:

$$\text{Bulk density} = [\text{oven dry soil mass}]/[\text{soil volume}]$$

Bulk density. The mass of dry soil per unit of bulk volume, including the air space.

Bulk density is useful in estimating the differences in compaction of a given soil. Compaction on the farm is caused by the use of heavy farm machinery and implements. Trampling of soil by humans and farm animals also causes soil compaction. A compacted soil has an increased bulk density. Finely textured soils such as clays are more packed and are called *heavy soils.* They have high bulk densities and are more difficult to till. Heavy soils also drain poorly and have high water-holding capacity. Sandy soils are loose and are called light soils. They have lower bulk densities and are thus much easier to till. Root growth is impeded when bulk density is high. Bulk density is generally higher in the lower layers of the soil profile.

For good plant growth, a clay soil should have a bulk density of less than 1,400 kg/m^3. For cultivating sands, a bulk density of less than 1,600 kg/m^3 is desirable. Greenhouse soils contain a variety of soilless components such as vermiculite and perlite. These materials make them light and of very low bulk density, about 100 to 400 kg/m^3. To improve bulk density, organic matter may be incorporated into the soil. The system of crop and soil management adopted also affects soil bulk density.

A typical mineral soil consists of about 25% pore spaces, or voids. There are two types of pore size: *micropores* and *macropores.* Clay soils have fine particles and a preponderance of micropores (small pores). Sand and coarse-textured soils have more macropores (large pores) than micropores. Total pore spaces in sandy soils may range between 35 and 50%. Cropping of soil tends to reduce soil organic matter and lower total space. Macropores play a role in drainage and aeration. Compaction reduces aeration and drainage. Is soil compaction more of a problem under mechanized farming than non-mechanized farming?

7.6: CLAY AND HUMUS HAVE COLLOIDAL PROPERTIES THAT ARE CRITICAL TO SOIL FERTILITY

7.6.1 SOIL COLLOIDS

How do soils hold and make available the nutrients essential to plants? The chemical properties of soils are determined by **soil colloids**. Colloids are minute particles that remain in suspension for a long time. They have large surface areas that are charged. The predominant soil colloids of importance to crop production are clay and humus.

Clay colloids have a net negative charge and hence attract cations. They are secondary minerals (reformed from other partially dissolved minerals). Most clays are crystalline. A clay particle is called a *micelle* and consists of layers of atoms of oxygen, silicon, and aluminum. The layers may be spread apart to allow access to cation exchange sites on the internal surface. Such clays are called **expanding**, or *swelling*, **clays**, as opposed to **non-expanding**, or *non-swelling*, **clays** that have tightly bonding layers. There are several arrangements of the layers in the micelles in crystalline clays, which form a basis of their classification (Table 7–2).

Humus is an amorphous, largely water-insoluble temporary remainder of the decomposition of organic matter. It is called an *organic colloid* and has a net negative charge arising solely from mineralization of hydrogens from R–OH groups.

7.6.2 SOIL SOLUTION AND LEACHING

Soil water is not pure (drinking water) but, rather, contains hundreds of dissolved organic and inorganic substances. Consequently, soil water is best described as *soil solution*. This dilute solution supplies dissolved nutrient elements to plant roots. Some of these nutrients are critical to plant growth and development and are called *essential elements*. These elements are discussed later in the chapter. The soil solution has a reaction (acidic or alkaline), depending on the preponderance of certain dissolved ions (H^+, OH^-), a property called soil pH.

Soil nutrients move in the soil solution. The desired effect of soil water movement is to deliver nutrients to the root zone. Under some conditions, the nutrients are moved out of the root zone, a process called leaching. Leaching has consequences: (1) soil nutrients

Soil colloid. Organic and inorganic matter with very small particle sizes and a correspondingly large surface area per unit of mass that remain in suspension for a long time

Expanding clay. Clay mineral with a 2:1 lattice structure that is subject to significant swelling upon wetting and shrinking upon drying.

Non-expanding clay. Clay mineral with a 1:1 lattice structure that resists swelling upon wetting and shrinking upon drying.

Humus. The usually darker fraction of soil organic matter remaining after the major portion of added residues have decomposed.

Table 7–2   Selected Features of Major Categories of Clay Minerals

1. *Noncrystalline (amorphous) clays:* e.g., glass and opal
2. *Crystalline layer silicate clays:*
   a. *Kandite group:* e.g., kaolinite
      Has one sheet of tetrahedral: one octahedral sheet per layer (i.e., 1:1 crystal lattice). It is non-expanding (non-swelling) clay.
   b. *Smectite group:* e.g., talc and montmorillonite
      Has two tetrahedral sheets: one octahedral sheet per layer (i.e., 2:1 crystal lattice). It is expanding (swelling) clay (expands when wet and shrinks and cracks upon drying).
   c. *Hydrous mica group:* e.g., vermiculite (used in horticultural potting soils). It is 2:1 clay.

---

are not accessible to plants, leading to reduced yield and economic loss; (2) leached nutrients may contaminate groundwater and pose a health hazard to humans. A high CEC shows cation leaching because cations are bound to the soil colloids. Coarse soils with more macropores are more susceptible to leaching because of high gravitational water flow. On the other hand, compact soils have poor internal drainage but are susceptible to surface runoff, which also moves dissolved nutrients into surface water. Nitrates are most susceptible to leaching to contaminate groundwater.

Leaching losses are more likely to occur when high rates of fertilizer are applied at a stage in the crop growth cycle when nutrient uptake is slow. Loss of nutrients to leaching is likely to be greater in spring because high rainfall and low evapotranspiration occur at about the same time that mineralization and nitrification are just beginning and plant growth is slow.

7.6.3 SOILS DIFFER IN THE AMOUNT OF CATIONS THEY CAN HOLD

The cation exchange capacity (CEC) is an index of soil fertility that is based on the base-exchange capacity or measure of the total exchangeable cations a soil can hold. Cations (positively charged ions) useful to plant nutrition include Mg^{2+}, Ca^{2+}, K^+, and Mn^+. These ions are adsorbed by colloids. Adsorbed cations can be replaced (exchanged) by other cations in solution by *mass action* (competition for the negative sites because of a large number of ions present). The strength of adsorption of the ions on the surface depends on the valence of the cation, the charge density, and the strength of the site's negative charge. Protons (H^+) are small and have high charge density. They are very tightly ionic-bonded to clay. Na^+, on the other hand, has a small charge density and hence is weakly adsorbed and highly prone to leaching. Ranking of ions in decreasing charge density (lyotrophic series) and decreasing adsorption to colloids is as follows:

$$H^+ > Al^{3+} > Ca^{2+} > Mg^{2+} > K^+ = NH_4^+ > Na^+$$

The kinds of mineral ions present in the soil depends on the chemical nature of the parent material and the effect of climate. CEC is measured in meq/100 g of soil. A small CEC means the soil can hold small amounts of essential nutrients. It will be dominated by H^+ and Al^{3+} ions, making it acidic and of low fertility. Organic matter is more capable of cation exchange than clay, having 100 to 300 meq/100 g as compared with only 80 to 150 meq/100 g for clay. In terms of CEC, the order for soil colloid is generally humus > vermiculite > montmorillonite > illite > kaolinite > sesquioxides. Since clay and organic matter are the principal colloidal materials in the soil, can it be said that, without some clay and organic matter, a soil is useless for crop production?

7.6.4 SOIL PH

What makes one soil acidic and another alkaline? How can producers correct soil acidity or reclaim alkaline soils? **Soil reaction**, or *pH*, defines the soil acidity or alkalinity based on the hydrogen ion concentration. On the pH scale, a value of 7.0 indicates a neutral reaction. All values below 7.0 indicate acidity, while all values above 7.0 indicate alkalinity (Figure 7–6). The pH scale is logarithmic (i.e., pH = $-\log_{10}[H^+]$). This means that a pH of 4.0 is 10 times as acidic as a pH of 5.0.

Total acidity in the soil can be divided into three categories:

1. Active acidity—caused by H^+, Al^{3+} ions in the soil solution
2. Salt-replaceable acidity—due to H^+ and Al^{3+} ions that are easily exchangeable by other cations in the simple, unbuffered soil solution

Soil reaction. The degree of soil acidity or alkalinity, usually expressed as a pH value.

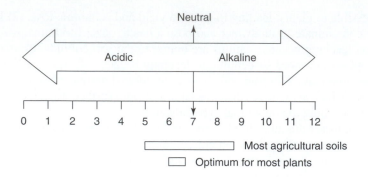

FIGURE 7–6 The soil pH scale is logarithmic and divided into units ranging from 0 to 14. Car battery acid has a pH of less than 1, while household bleach has a pH of about 12.5. Pure water is neutral; lemon juice has a pH of about 2; coffee has a pH of about 4.2.

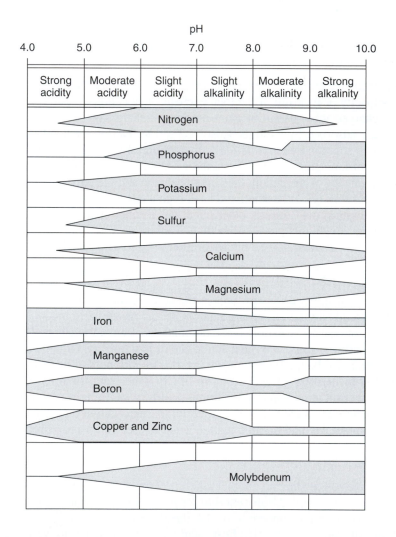

FIGURE 7–7 An illustration of the soil nutrient availability as affected by pH. The range of availability for each nutrient element depends on whether the soil is mineral, organic, or in between.

3. Residual acidity—can be neutralized by limestone but not susceptible to salt-replaceable techniques

Even though only a small fraction of the total soil acidity, active acidity is very important because plant roots and soil microorganisms are exposed to the soil solution.

Each nutrient element has a pH range within which it is available to plants (Figure 7–7). Similarly, plants have pH adaptation, some being sensitive to acidity,

Table 7–3 Relative Crop Yields Affected by Soil pH

| Crop | Relative Yield at Indicated pH | | | | |
	4.7	5.0	5.7	6.8	7.5
Alfalfa	2	9	42	100	100
Barley	0	23	80	95	100
Corn	34	73	83	100	85
Oats	7	93	99	98	100
Red clover	12	21	53	98	100
Soybeans	65	79	80	100	93
Sweet clover	0	2	49	98	100
Wheat	68	76	89	100	99

Source: Ohio Agricultural Experimental Station, Special Circulation number 53, 1938.

while others are sensitive to alkalinity (Table 7–3). The optimum pH range for most plants is 6.5 to 7.0. Most agricultural soils have a pH of between 5.0 and 8.5. Not only is soil pH important for knowing the nutrient availability, but certain pH values cause the release into the soil of toxic amounts of certain elements (e.g., Al^{3+} at highly acidic pH) that are required by the plant in small amounts.

7.6.5 SALT-AFFECTED SOILS

Most soils in arid and semiarid regions are *alkaline* and *salt-affected* because of the accumulation of high amounts of salts in the soil. Even though most are not used for cropping, these soils cannot be totally ignored in crop production. In the United States, 15% of the cropland is irrigated and accounts for 40% of the total crop value in such dry areas. The high alkalinity results from the lack of adequate precipitation to leach these base-forming cations (Ca^{2+}, Mg^{2+}, K^+, Na^+, etc.) salts out of the plant root zone. These soils have a high base saturation and high pH values, often above 7.0.

High salt concentration usually reduces plant growth by the osmotic effect. As soil salt concentration increases, the force with which water is held in the soil increases, requiring plant roots to expend more energy to extract water from the soil solution. Sometimes, the salt concentration is so high that, instead of roots absorbing water, they lose water by exosmosis into the soil solution. Such a reverse trend in moisture flow can result in the death of some plants, especially young plants.

Salt accumulation in the soil inhibits the germination of seeds, reduces crop growth and development, and decreases crop yield. Salt-loving plants are called *halophytes*. Plants used in crop production vary in salt tolerance (Table 7–4). Plant breeders have developed salt-tolerant cultivars in certain crops. Barley and cotton are both relatively salt tolerant. Salt-affected soils have reduced permeability to water, encouraging surface erosion.

Soil salinity may be measured by electrical conductivity, whereby an electrode is used to obtain an indirect measurement of the salt content. A soil sample is saturated with distilled water and mixed to a paste (saturation paste extract method). Soils that contain large quantities of soluble salts (Na^+, Ca^{2+}, Mg^{2+}, Cl^-, SO_4^{2-}, etc.) are called *saline (salty) soils*. When the Na^+ saturation of the soil is 15% or higher, the soil pH may rise above 8.5, creating a *sodic soil* that is unproductive and very hard to manage.

Table 7-4 Relative Salt Tolerance of Selected Crops

Tolerant	Moderately Tolerant	Moderatively Sensitive	Sensitive
Barley (grain)	Barley (forage)	Alfalfa	Strawberry
Cotton	Sugar beet	Corn	Bean
Bermudagrass	Wheat	Peanut	Potato
	Oats	Rice (paddy)	Tomato
	Fescue (tall)	Grape	Pineapple
	Soybean	Sweet clover	Onion
	Sorghum	Broad bean	Raspberry
	Sudangrass	Sweet potato	
	Cowpea	Timothy	
	Rye (hay)	Clover (red, ladino)	

Another measurement of soil alkalinity is the *sodium adsorption ratio (SAR),* calculated as

$$SAR = Na^{2+}/\sqrt{1/2[Ca^{2+}] + Mg^{2+}}$$

SAR is used to evaluate the suitability of water for irrigation.

A saline soil has an electrical conductivity of 4.0 decisiemens/meter (dS/m) or SAR of 13 or less. A sodic soil has SAR of 13 or greater. Salty soils can be reclaimed by (1) establishing an internal drainage in the soil to remove excess water, (2) replacing the excess exchangeable sodium, or (3) leaching out most of the soluble salts by heavy irrigation. Plants may also be planted on the slope of a ridge to avoid the salt that has been drawn up to the top of the ridge.

7.6.6 CORRECTING SOIL ACIDITY

Liming. The application of agricultural lime or other materials containing the carbonates, oxides, and/or hydroxides of calcium and/or magnesium to neutralize soil acidity.

Low soil acidity can be corrected by **liming**, the procedure of applying limestone ($CaCO_3$) or gypsum ($CaSO_4$) to raise soil pH. On the other hand, to lower soil pH, sulfur may be applied. Certain fertilizers tend to increase soil acidity because they leave acid residues. Liming is beneficial to soil microbial growth and activity, since bacterial activity is reduced under low pH. Liming is discussed in more detail in Chapter 8.

7.7: THE SOIL IS HOME TO ORGANISMS THAT ARE EITHER BENEFICIAL OR HARMFUL TO PLANTS

What role do soil organisms play in plant production? Soil organisms from all of the five kingdoms in both living and dead forms influence soils. Some of the activities of these organisms are beneficial to crop production, while others are detrimental, causing diseases and loss of productivity.

7.7.1 SOIL ORGANISMS

The soil is teeming with a wide diversity of organisms. The animals (fauna) of the soil range in size from macrofauna (e.g., earthworms, moles), through mesofauna (e.g., springtails, mites), to microfauna (e.g., nematodes). The plants (flora) in the soil include the roots of higher plants (e.g., weeds, crops) and microscopic algae. In addition to these, some of which are visible to the naked eye, there are other microorganisms (e.g., bacteria, fungi, actinomycetes) that are visible under the microscope. The classification system given in Chapter 2 applies also to soil organisms.

1. *Animalia*
   a. *Burrowing animals.* These include moles, prairie dogs, gophers, mice, badgers, and rabbits. They bore large holes into the soil and incorporate plant materials, thereby improving aeration and soil fertility. Unfortunately, these animals often eat the crop plants of economic interest to the producer. Hence, overall, burrowing animals are destructive to crop production.
   b. *Earthworms.* Earthworms are perhaps the most important macroanimals in the soil. They ingest about 2 to 30 times their weight of both plant and animal residues, as well as soil, each day. They excrete large amounts of partially digested organic and soil material as small granular aggregates called *casts.* Their activities in the soil improve aeration and drainage, stirring the soil for enhanced water infiltration and root penetration, as well as enhancing soil fertility. However, they can also spread diseases. Earthworms work best under moist soil conditions where organic matter is available as a source of food and the pH is about 5.5 to 8.5. The optimum temperature for worms is about 10°C; hence, their peak activities in the temperate regions are in spring and autumn. They function less in compacted soil (from heavy machinery use), sandy soils, salty soils, drought conditions, acidic soils, extreme temperature, and rodent-infested soil.
   c. *Arthropods and gastropods.* The important arthropods (invertebrates with jointed foot) include termites, mites, millipedes, beetles, and ants. Termites (or white ants) are most prominent in the grasslands (savannas) and forests of tropical and subtropical regions. They are known for constructing mounds of varying sizes. They are not as useful as earthworms in positively affecting soil fertility and physical qualities. Their activities are more localized (in termite mounds). Important gastropods (belly-footed organisms) include slugs and snails. They feed on decaying vegetation and can feed on crop plants.
   d. *Nematodes.* Nematodes (eelworms, threadworms) may be omnivorous (living mainly on decaying organic matter), predaceous (preying on soil bacteria, fungi, algae), or parasitic (infecting plant roots). The most common nematodes are omnivorous. Parasitic nematodes are known for the characteristic and conspicuous knots they form on the roots of susceptible plants. Many food crops and vegetables are susceptible to nematodes, but more so are soybean, sugar beet, and corn, in which they cause yield losses exceeding 50%. Nematodes are beneficial on golf courses and sod farms, where they control white grub.
2. *Plantae.* Living roots of plants impact soil physical properties as they push through it. They secrete a variety of compounds into the soil. When plants die or are harvested, a large mass of roots comprising about 10 to 40% of the plant vegetative body is left in the soil, contributing to soil organic matter. The area in the soil immediately around the living roots is called the *rhizosphere.* Some of the root exudates exert growth-regulating effects on other plants and soil microorganisms, a phenomenon called *allelopathy.* Algae growth occurs in production systems that use surface irrigation (e.g., in flood-irrigated rice fields).
3. *Soil fungi.* Fungi may be unicellular (e.g., yeast) or multicellular (e.g., molds, mildews, rusts, mushrooms). They are responsible for most of the economic

diseases of crop plants, as described elsewhere in this book. Some of the toxins (mycotoxins) they produce are deadly to animals (e.g., aflatoxin). Fungi have beneficial roles in the ecosystem. They decompose organic matter and thereby enhance soil fertility. They help some plants absorb soil nutrients through a symbiotic association called **mycorrhiza**.

Mycorrhiza. The association, usually symbiotic, of fungi with roots of some seed plants.

4. *Protista.* This group includes protozoa and slime molds. They cause a few plant diseases but cause many animal diseases (e.g., malaria, amoebic dysentery, Texas cattle fever).

5. *Monera.* Actinomycetes are known for producing antibiotic compounds (actinomycin, neomycin, streptomycin). They also decompose organic matter. Soil bacteria are important in their symbiotic relationships with plants. They may be autotrophic (self-nutritive, making their own food) or heterotrophic (deriving nutrition from organic substances). Soil bacteria are engaged in several important processes that help soil fertility including biological nitrogen fixatron, nitrification, and mycorrhiza

7.7.2 CULTURAL PRACTICES IN CROP PRODUCTION MAY BE USED TO MANAGE SOIL MICROBES

Since the soil contains both beneficial and harmful microbes, the crop producer should adopt practices that encourage the beneficial microorganisms while controlling harmful ones. Just as a good growth environment favors both weeds and crop plants, good conditions for beneficial organisms may also encourage the growth of harmful ones. Thus, how can a grower selectively promote beneficial organisms while suppressing harmful ones? There are certain practices that can give an advantage to beneficial microbes:

1. *Inoculation.* There are native bacteria in the soil. Since symbiotic activities involve specificity for host bacteria, it is advantageous to inoculate the soil with the *Rhizobium* of interest.

2. *Application of lime.* Acidic soils should be limed to raise the pH to about 6.0. High acidity destroys bacteria.

3. *Reduction of fumigation.* Soil fumigation or sterilization kills both harmful and beneficial microbes and should not be conducted too frequently. In the greenhouse, however, sterilization of soil and other materials is routine and desirable.

4. *Sanitation.* Remove and destroy infected and diseased plants and residue to avoid spreading harmful microbes.

5. *Maintenance of good soil environment.* The soil should be well aerated, should be drained (no waterlogging), should have good moisture (no drought), and should not contain abnormal levels of salts.

6. *Organic matter content.* Soil microbes depend on soil organic matter for energy. A good, healthy soil should have adequate organic matter.

The following strategies are helpful in reducing the populations of harmful microbes:

1. Establish crops by using healthy, disease-free planting material.

2. Observe good sanitation: clean tools after use. Remove and destroy infected plants and their remains.

3. Eliminate vectors: insect vectors are carriers of diseases and can be eliminated to reduce the spreading of a pest.

4. Maintain good microclimate: high humidity, still air, and high temperature encourage growth of fungi and other microbes. Pruning, a horticultural practice

of removing branches and other plant parts, can be used to open the plant canopy for aeration. Irrigation should be done in the morning, so that plant leaves dry during the day. This helps to reduce disease.

5. Maintain soil at a slightly acidic pH: the spread of some pathogens such as fungal rots of sweet potato are reduced when soil pH is slightly acidic. Potato scab is effectively controlled at a pH of below about 5.2.
6. **Protect plants from mechanical injury:** pathogens enter plants through natural pores and wounds.
7. Do not allow infestations to build up.

7.7.3 SELECTED ROLES OF SOIL MICROORGANISMS

Soil microbes are involved in activities that enhance soil fertility, one of the basic roles being in decomposition of organic matter to recycle immobilized nutrients. Some additional specific roles of soil microbes are as follows.

Nitrification

Nitrification is the process by which ammonia ions are enzymatically oxidized by autotrophic nitrifying bacteria—*Nitrosomonas* and *Nitrobacter* in the following reactions:

$$2NH_4^+ + 3O_2 \xrightarrow{\textit{Nitrosomonas}} 2NO_2^- + 4H^+ + 2H_2O + \text{energy}$$

$$2NO_2^- + O_2 \xrightarrow{\textit{Nitrobacter}} 2NO_3^- + \text{energy}$$

This transformation process is slowed under cold temperature, strong acidity, and waterlogging (anaerobic conditions). NO_3^- ions are most rapidly absorbed by plants.

Denitrification

Denitrification, a biological process involving bacteria, is the source of the most extensive gaseous nitrogen loss. Under anaerobic conditions, bacteria convert NO_3^- into nitrogen gas:

$$2NO_3^- \rightarrow 2NO_2^- \rightarrow 2NO \rightarrow N_2O \rightarrow N_2$$

Consequently, it is undesirable to fertilize crops with nitrate fertilizers if the soil will be flooded in a production practice (e.g., paddy rice production).

Denitrification. The biological reduction of nitrate or nitrite to gaseous nitrogen or nitrogen oxides.

Biological Nitrogen Fixation

There are two kinds of biological nitrogen fixation (BNF) processes—*symbiotic* and *nonsymbiotic*. Nonsymbiotic BNF is undertaken by free-living bacteria and cynobacteria that are not associated with plants. The bacteria involved include *Clostridium* and *Azotobacter.* Symbiotic nitrogen fixation involves bacteria that live mostly in the roots of legumes and non-legumes. Infected plants produce root nodules (root modifications that are the site of nitrogen fixation) by the following general reaction:

$$N_2 + 8H^+ + 6e^- \xrightarrow{\textit{nitrogenase}} 2NH_3 + H_2$$

$$NH_3 + \text{organic acids} \rightarrow \text{amino acids} \rightarrow \text{protein}$$

The reduction of N_2 involves the enzyme nitrogenase. The best known symbiotic bacteria belong to the genus *Rhizobium*. The amount of nitrogen fixed depends on the species. Symbiotic BNF in alfalfa yields an average of about 200 kg/ha of nitrogen per season versus 45 kg/ha of nitrogen per season under ideal conditions. In the commercial production of legumes, such as peanut and soybean, producers treat seeds with commercial inoculants containing the appropriate bacterium to augment the native bacteria population.

Mycorrhizae

Mycorrhizae (or fungus roots) are of two kinds, based on growth habit—*ectomycorrhiza* and *endomycorrhiza*. Ectomycorrhizae penetrate only the outer cell layers of the root walls, forming a fungus mat on the root surface. Endomycorrhizae penetrate the host cell. Some hyphae (filaments) are from vesicular-arbuscular mycorrhizae (VAM). They form highly branched structures called *arbuscules*. VAM are the most important form of mycorrhizae found on plants and are very helpful in phosphate absorption. In addition to phosphate absorption, they help reduce stress due to drought. Mycorrhizae function best in highly weathered tropical soils that are low in base ions, are acidic, and are low in levels of phosphorus and aluminum. Most agronomic crops (e.g., cotton, potato, soybean, rice, wheat), vegetables, and tree crops benefit from VAM. Mycorrhizae are absent in Cruciferae (e.g., cabbage, canola) and Chenopodiaceae (e.g., sugar beet, spinach).

7.8: SOIL ORGANIC MATTER AFFECTS BOTH PHYSICAL AND CHEMICAL PROPERTIES OF SOIL

Soil organic matter is the decomposed remains of organisms that have been incorporated into the soil. Soil organic matter consists of about 45 to 50% carbon. The skeletal residue remaining after decomposition of organic matter is called *soil humus*. This is a very complex organic material. Based upon solubilities, there are different classes of soil humic substances. *Humin* is the part of humus that is insoluble in sodium hydroxide. *Fluvic acid* and *humic acid* are soluble in dilute sodium hydroxide solution but the latter precipitates out at acid pH.

Decomposition of organic matter is an enzymatic process. Since organic matter consists of different chemical bonds, there are different enzymes for breaking each of these bonds. The products of decomposition depend on the conditions under which the biological reaction occurred. Under aerobic conditions, products of decomposition include CO_2, NH_4^+, NO_3^-, SO_4^{2-}, $H_2PO_4^-$, and H_2O. When decomposition occurs under anaerobic conditions, some toxic products, such as H_2S and dimethylsulfide, are produced. The characteristic foul odor associated with swamps is caused by the evolution of CH_4 (methane), also called *swamp gas*.

The greater the population of decay microbes, the faster the decomposition. However, the rate of decomposition of organic matter usually depends on the amount of nitrogen present. Soil microbes use nitrogen and carbon for growth and reproduction. The nitrogen content of an organic matter is measured by its *carbon to nitrogen ratio (C:N)*. Straw has a C:N of 80:1, while sawdust has a C:N of 400 to 800:1 (Table 7–5). Alfalfa crop residue has a C:N of only 13:1. The lower the C:N ratio the faster the crop residue de-

Nutrient cycling is critical to the success of organic farming. Inorganic nutrients are absorbed by plants and then metabolized and converted into organic matter (cells and tissues). Upon dying, organic matter is subject to decomposition, which results in the conversion of the organic form of nutrients into inorganic components. De-composition is simply the breakdown of organic matter. Either way, the processes of decomposition involve chemical or biochemical events that often require oxygen.

There are two categories of decomposition. *Abiotic decomposition* (without living organisms) occurs primarily through burning in which oxygen reacts with the organic matter to produce carbon dioxide and other substances in the ash. The ash provides nutrients for cropping in a slash-and-burn production system. However, most of the carbon is lost as carbon dioxide.

Biotic decomposition (with living organisms, especially bacteria and fungi) may involve aerobic or anaerobic microbes. These microbes utilize organic matter as a source of energy. Unlike abiotic decomposition, biotic decomposition takes place in a series of steps in which organic matter is sequentially broken down into inorganic compounds through a series of intermediate organic compounds.

The decomposition of importance to cropping occurs in the soil or the soil surface. *Litter* consists of fragments of the source of organic matter that makes it easy to identify the organism. This may be leaves, stems, and other parts of the plant. When litter is broken down into smaller unrecognizable organic material, the product is called *detritus*. The next stage of decomposition is *humus*, a dark brown mixture of various compounds and substances. Inorganic compounds are freed from the organic compounds by the process of *mineralization*.

Mineralization: setting minerals free

Inorganic nitrogen (e.g., NH_4^+, NO_3^-, NO_2^-, and N_2 gas) is used by plants and soil bacteria and converted

into organic nitrogen compounds in the tissues and other structures. The nitrogen in this organic form is said to be immobilized and the process is called *immobilization*. The organic form of nitrogen is unavailable to plants. Upon dying, the organic matter undergoes decomposition. A stage in the process of decomposition called *mineralization* releases the immobilized nitrogen in inorganic form for use by plants and other organisms.

Mineralization is an enzymatic process that may be summarized as follows

$$\longrightarrow \textbf{Mineralization}$$

$$+ H_2O^+ \qquad\qquad + O_2 \qquad\qquad + [O]$$

$$RNH_2 \rightleftharpoons ROH + NH_4^+ \rightleftharpoons NO_2^- + 4H^+ \rightleftharpoons NO_3^-$$

$$- H_2O^+ \qquad\qquad + O_2 \qquad\qquad + [O]$$

$$\textbf{Immobilization} \longleftarrow$$

Through hydrolysis, the hydrogen in the amine group (NH_2) is removed in the form of ammonia (NH_3), which reacts with water or acids in the soil solution to produce the ammonium ion (NH_4^+). Plants can utilize this form of nitrogen. However, it may also be further transformed by enzymatic oxidation by the process of *nitrification* to produce nitrates. Nitrification is a two-step process involving two different kinds of microorganisms called *nitrifying bacteria*:

Step 1: $2NH_4^+ + 3O_2^- \rightarrow 2NO_2^- + 2H_2O + 4H^+ +$ energy

This step is undertaken by bacteria called *Nitrosomonas*.

Step 2: $2NO_2^- + O_2 \rightarrow 2NO_3^- +$ energy

This step is carried out by bacteria called *Nitrobacter*.

Nitrates are highly soluble in water and hence prone to leaching. Further, nitrogen may be lost through *denitrification*, which occurs under conditions of poor drainage and aeration.

composes. Sawdust, therefore, is highly resistant to decomposition. The C:N ratio narrows as decomposition progresses and carbon is released as carbon dioxide.

The conditions required for decomposition are warm temperature and good moisture. Waterlogged soils (anaerobic conditions) slow decomposition. Similarly, dry soils inhibit microbial action. Rate of decomposition is most rapid during the first 2 weeks. A

Biological nitrogen fixation is the process by which elemental nitrogen is combined into organic forms. An estimated 175 million metric tons of nitrogen is fixed by bacteria either independently or in association with plants, the latter being called *symbiotic nitrogen fixation*. Symbiosis is the cohabitation between two organisms for mutual benefit. Certain bacteria, especially those of the genus *Rhizobium*, invade the roots of legumes, causing the production of swellings called *nodules*. The legume host provides the bacteria with carbohydrates for energy, and in return the bacteria fix elemental nitrogen in organic form directly into the plant roots. This legume-bacterial symbiosis has significant specificity—that is, one *Rhizobium* species will effectively inoculate a certain legume species but not others. Based on symbiotic associations, legumes may be classified into seven cross-inoculation groups:

Group	Rhizobium *Species*	Legumes
Alfalfa	*R. meliloti*	Alfalfa (*Medicago*), certain clovers (*Melilotus*)
Clover	*R. trifolii*	Clovers (*Trifolium* spp.)
Soybean	*R. japonicum*	Soybean (*Glycine* max).
Lupine	*R. lupini*	Lupines (*Lupinus*), serradella (*Ornithopus* spp.)
Bean	*R. phaseoli*	Dry bean (*Phaseolus vulgaris*), runner bean (*P. coccineus*)
Peas and vetch	*R. leguminosarum*	Pea (*Pisum*), vetch (*Vicia*), sweet pea (*Lathyrus*), lentil (*Lens* spp.)
Cowpea miscellany	Various species	Cowpea (*Vigna*), lespedeza (*Lespedeza*), peanut (*Arachis*), pigeon pea (*Cajanus*), crotalaria (*Crotolaria*), kudzu (*Pueraria*), stylo (*Stylosanthes*), Acacia, desmodium (*Desmodium*)

The general mechanism by which biological nitrogen fixation occurs is the reduction of nitrogen gas to ammonia, a reaction that is catalyzed by *nitrogenase* and involves iron and molybdenum. The ammonia is subsequently combined with other organic acids to form amino acids and eventually proteins.

Table 7–5 Carbon–Nitrogen (C:N) Ratios of Selected Organic Materials

Material	C:N ratio
Wheat straw	80:1
Corn stover	57:1
Alfalfa (mature)	25:1
Alfalfa (young)	13:1
Sawdust	400:1 or more

crop residue with a narrow C:N ratio will make nitrogen available sooner than one with a wide C:N ratio (Figure 7–8). The producer can plant immediately after plowing under a crop such as alfalfa. On the other hand, after plowing under a material such as straw, a waiting period of about 6 weeks is required before planting. The effect of soil organic matter is summarized in Table 7–6.

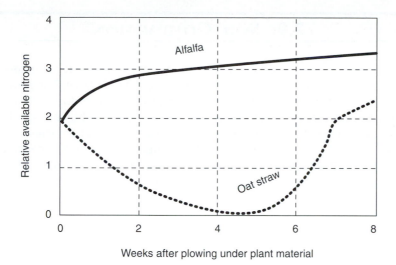

FIGURE 7–8 The carbon–nitrogen ratio (C:N) for alfalfa is lower than that for oat straw. When plowed under, alfalfa decomposes rapidly to release nitrogen into the soil. However, oat has a high C:N ratio and decomposes slowly. Initially, it causes a decrease in available nitrogen until after about 4 weeks, when microbial population decreases; then more nitrogen becomes available for plant use.

Table 7–6 Effects of Soil Organic Matter

Benefits
1. Provides 90 to 95% of all the nitrogen in unfertilized soils (acts as a "slow-release" fertilizer)
2. Supplies soil-binding factors (polysaccharides) for soil aggregation
3. Has colloidal properties and accounts for about 30 to 70% of cation exchange in the soil
4. Increases soil water content at field capacity
5. Improves soil aeration and water flow in fine-textured soils
6. Acts as chelate
7. Provides carbon for microorganism in the soil
8. Can be used as mulch to reduce soil surface erosion and conserve moisture
9. Can modify soil temperature (insulation), increasing it in cold weather and cooling it in hot weather
10. Acts as a soil buffer against rapid pH alteration and other changes in soil chemistry

Adverse Effects
Organic plants may release plant toxins (phytotoxins) upon decomposition. This effect is called allelopathy. Known allelochemicals include alkaloids, benzoxainones, coumrins, cyanogenic compounds, quinines, flavanoids, and terpenes. Crops with such properties include sorghum, johnsongrass, black walnut, and peach.

Crop residues contain chemicals that may be toxic *(phytotoxins)*. *Allelopathy* is the term used to describe the effect of harmful or beneficial chemicals *(allelochemicals)* produced by one plant on another plant. These chemicals are leached out of crop residues. Many perennial weeds such as johnsongrass, quackgrass, and nutsedge are suspected to have allelopathic effects.

Soil compaction is a form of soil degradation that is characterized by an increase in soil density by the packing of soil particles when pressure is applied to it. The packing of the primary soil particles (sand, silt, clay) and soil aggregates closer together dramatically changes the balance between soil solids and voids (pore spaces) (Figure 7–9). The degradation starts with the collapse and elimination of macropores (large pores), thereby severely restricting water movement (especially gravitational water), root penetration, and air movement. Soils of uniform texture (e.g., sandy soil) are less susceptible to compaction than are those with a wide range of textures (e.g., fine sandy loam). Under pressure, the finer particles readily fill the larger pores created by the coarse components of the textural class, increasing soil density.

7.9.1 CAUSES OF SOIL COMPACTION

The causes of soil compaction occurring on cropping land may be categorized according to farm activities, location of the problem, or origin.

1. *Farm activities.* According to farm activities, compaction may be tillage-induced or traffic-induced.

   Tillage operations exert pressure on the soil from the tractor or the implements it draws. Primary tillage implements (e.g., disc plows) tend to compact the soil at the plow depth, creating an impervious layer over time if the plow depth is not varied periodically. Secondary tillage also affects soil structure by destroying soil aggregation and increasing soil density. The tillage-induced compaction is worsened when soils are tilled under wet conditions. Also, this compaction type affects mainly the plow layer (about 6–12 inches) through contact pressure with implements.

   Traffic-induced compaction is caused mainly by the wheels of farm vehicles and, in pastures, by the trampling of livestock. This compaction affects the subsoil layers in proportion to the axle load of the vehicles. Modern mechanized crop production predisposes the field to compaction because of the use of machinery at all stages of production (seeding, chemical treatment, harvesting).

FIGURE 7–9 Soil compaction hinders normal root growth in the soil. Areas in the field where there has been no traffic shows more root development, whereas interrow traffic and plowpans severely restrict the penetration of the roots into the soil. Roots that find cracks in compacted layers are able to penetrate into the subsoil. (Source: Drawn from USDA photo)

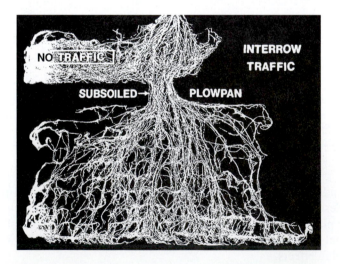

2. *Location of Problem.* Compaction may also be classified, according to the location, as surface or subsoil compaction. The classification essentially corresponds to that based on farm activities, the surface being equivalent to tillage-induced compaction, and the subsoil being equivalent to traffic-induced compaction. Surface compaction is caused mainly by the contact pressure from vehicle tires, while compaction from below the plow layer is related to the total axle load of the vehicles.
3. *Origin.* Compaction may also be classified as being human-induced (e.g., from farming activities) or as originating from natural causes (e.g., the splashing of rain drops).

7.9.2 DIAGNOSING COMPACTION IN THE FIELD

A producer may suspect that a field is compacted when he or she observes the presence of tell-tale signs in the soil or plants or when he or she has made some investigations.

Soil Observations

The major soil changes that suggest compaction include the following:

1. *Ponding of water on the soil surface.* When water collects in a pool in relatively level parts of the field, it may suggest drainage problems.
2. *Presence of dark streaks on the soil surface.* As a result of the prolonged ponding of water on the field in tire tracks, the soil in those spots becomes blackened as the water dries up.
3. *Increased power needed for tillage.* When the operator notices that more machine power than usual is needed for a tillage operation, the soil may be more dense than normal.
4. *Increased runoff.* When gently sloping land has excessive surface runoff following a moderate amount of rain or irrigation, there may be soil compaction.
5. *Surface soil crusts.* Crusting of the surface soil can be the result of surface compaction.

Plant Observations

Soil compaction is manifested in plant-related symptoms:

1. *Incomplete crop stand.* Crusting and other compaction-related structures may hinder seed germination, leading to an incomplete stand.
2. *Uneven crop stand.* Plants may germinate but, because of compaction in various parts of the field plant growth may be uneven. Plants in compacted sections may be stunted, while those in normal soil grow properly, leading to variation in height and plant size.
3. *Changes in plant color.* Compacted areas of the field may be waterlogged, causing moisture stress to plants that may manifest as chlorosis (yellowing) or purpling of the stem in early growth.
4. *Restricted root development.* Instead of plant roots growing deep into the soil, an obstruction may promote more horizontal root growth or balling of the roots.
5. *Unexpected wilting.* Compaction may limit plants' roots to the top, preventing the extraction of moisture from lower depths and causing earlier than normal wilting.
6. *Reduced yields.* Soil compaction reduces plant roots and hence nutrient and water absorption, consequently reducing crop yield.

Soil Investigations

The soil's resistance to penetration by an object indicates the soil density. Various instruments and tools (e.g., soil probe, penetrometer, small-diameter soil sampling tube) may be used to physically test the penetrability of the soil. A soil probe or shovel is pushed through the soil to determine resistance to penetration. A penetrometer provides electronic readings to indicate soil strength and resistance to root penetration. However, it is very sensitive to soil moisture; hence, the reading should be interpreted with caution. One may also dig a small pit to expose the root zone of the plant to examine the growth pattern.

7.9.3 EFFECTS OF SOIL COMPACTION

Soil compaction may have desirable effects in crop production, depending on the degree of compaction, the season of occurrence, and the crop species.

Desirable Effects

Light soil compaction is needed at seeding for good contact between seed and soil for seed germination. Some planters are designed to cover seeds and press down the soil for those purposes. Sometimes, special implements (soil packers) are mounted on tractors and used at seeding times to compact the soil over the seed. Studies have shown that a medium soil texture of bulk density of 1.2 gm per cubic centimeter (74 lb/cubic foot) is favorable for root growth. Where this bulk density is less, moderate amount of compaction may be beneficial for optimal root development. This will bring the roots into more effective contact with the soil to absorb more nutrients, especially those that are immobile (e.g., phosphorus).

Undesirable Effects

As previously indicated, excessive soil compaction hinders proper root growth and consequently reduces the amount of soil moisture and nutrients that can be absorbed. In the dry season, moderate compaction is desirable, increasing yields because the roots are in better contact with the soil to extract whatever moisture is available. As compaction increases, yield decreases as other factors come into play. However, in the wet season, increasing compaction reduces soil aeration, promoting denitrification and reduced root metabolism, leading to decreased yield.

7.9.4 MANAGING SOIL COMPACTION

Soil compaction is largely caused by the activities of producers through the use of machinery. Because mechanization in crop production is here to stay and likely to increase, and because some compaction is desirable, the producer should find ways of minimizing its adverse effects in production. The approaches used in managing soil compaction include prevention and correction.

It is best to prevent soil compaction because, once it occurs, it is usually very expensive to correct.

1. *Proper timing of field operations.* Tillage and other field operations should not be undertaken when the soil is wet. This may not always be prudent because delaying an operation (e.g., planting, harvesting) may have serious economic consequences.
2. *Proper choice of machinery.* Use vehicles with larger-diameter tires, reduced tire pressure, and reduced axle loads. Use all-wheel-drive vehicles when possible. Also, use only enough ballast to reduce slippage.

3. *Traffic patterns.* Reduce the number of passes over the field with equipment. Use larger-capacity equipment with wider working width for fewer passes. Fewer passes means that a smaller portion of the field actually comes into contact with the wheels of farm machinery and equipment.
4. *Drainage.* Improve drainage of the field to reduce soil wetness.
5. *Crop rotation.* Plant the field to crops with different root depths, including deep-rooted species, such as alfalfa.
6. *Tillage depth.* Instead of tilling the field to the same depth season after season, it is best to vary the depth regularly (e.g., reduce the depth in a wet year and increase it in a dry year to break the compacted layer).

When compaction has occurred, it may be corrected by one of several means. Some of these measures may also be used as preventive measures.

1. Subsoiling is the practice of tilling the soil to a very low depth to break compaction at the subsoil level. This is a very expensive activity that is done when the soil is dry. It should be followed with preventive measures.
2. Variation in the tillage depth may be used to alleviate light to moderate compaction caused by tillage operations.

7.10: SOIL EROSION

Even though modern advances in science enable some crop production to occur in soilless (lacking natural rock mineral materials) media (e.g., in greenhouses), most crops are produced in the soil in the field. *Soil erosion* is the removal of soil by water and wind. Soil is a non-renewable resource and hence should be properly managed, so that it is able to sustain crop production indefinitely. Most crop roots occur in the topsoil, the part that is most prone to erosion. The displaced soil must be deposited elsewhere, often with adverse consequences. Soil erosion may also be defined as a process that transforms soil into sediment. There are two main processes of soil erosion—by *water* and by *wind*. The Great Plains region and the Corn Belt have significant amounts of soil erosion (Figure 7–10). Excessive cropland erosion occurs on farms, especially in these regions of the United States.

7.10.1 WATER EROSION

Water erosion occurs in three basic steps—detachment of soil particles, transportation of the dislodged particles downhill, and deposition at a lower elevation (Figure 7–11). The direct impact of raindrops may disintegrate aggregated soils by the splashing effect. The particles are transported by runoff water. The soil surface may be depleted in more or less a uniform manner as water flows gently over the surface. This soil removal is called *sheet erosion.* When the sheet flow concentrates into tiny channels, soil removal is concentrated in those channels (called *rill erosion*), as commonly occurs on bare land, newly planted land, and fallows. When the volume of runoff is further concentrated, it moves with more turbulence and cuts deeper as it rushes through the field, creating larger channels (called *gulley erosion*).

The amount of soil loss by water erosion by rill and sheet can be estimated by the Universal Soil Loss Equation (USLE) or by the more recent computerized version, the Revised Universal Soil Loss Equation (RUSLE) as follows:

$$A = RKLSCP$$

FIGURE 7-10 The heavily farmed croplands of the Great Plains and the Midwest are susceptible to soil erosion. (Source: USDA)

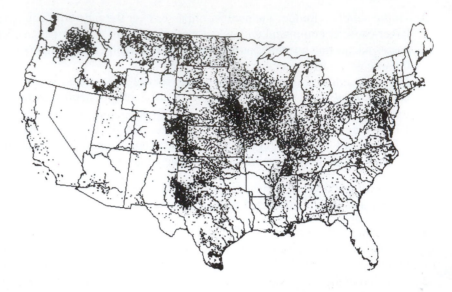

FIGURE 7-11 Water erosion starts as gentle and light soil removal, as in rill erosion (left), and soon develops into massive soil removal, as in gulley erosion (right), if it is unchecked. (Source: USDA)

where A = erosion (soil loss in tons/acre/year)
        R = rainfall factor
        K = soil erodibility factor (ranges from 1 = most easily eroded to 0.01 = least easily eroded)
      LS = field length and slope factor
       C = vegetative cover and management factor
       P = practice used for erosion control

R and K are rain-related factors; K and LS are soil-related factors; C and P are management-related factors.

Water erosion can be controlled by controlling sediment detachment or sediment transport:

1. Sediment detachment is controlled by maintaining a vegetative cover on the soil (e.g., planting cover crops, stubble mulching, or keeping crop residues on the soil). These and other preventive practices are discussed in detail under conservation tillage (Chapter 13).
2. Controlling sediment transport is achieved by slowing the speed of water flow (e.g., by the reduction of the slope in the land, with terraces, or by contour strip cropping). These practices are also discussed under conservation tillage.

7.10.2 WIND EROSION

Wind erosion entails the removal, transport, and deposition of fine soil particles by winds (Figure 7–12). The mechanics are similar to those of water erosion. The damage is greatest when the wind is strong, and the soil is dry, bare, or weakly aggregated. One of the ecological devastations on record is the Dust Bowl of the 1930s, during which wind erosion precipitated by drought devastated the agriculture of the Great Plains region (see boxed reading). Wind erosion occurs in three ways. *Suspension* is the movement of small soil particles over long distances in dust clouds. *Saltation* is the mode of transport of soil particles by short successions of bounces. Coarser soil particles are too heavy to be airborne and, instead, are moved by rolling along the soil surface, called surface creep. The wind erosion equation (WEQ) and the revised equation (RWEQ) are as follows:

$$E = f(ICKLV)$$

where erosion is a function (f) of

I = soil erodibility factor
C = climate factor
K = soil-ridge-roughness factor
L = width of field factor
V = vegetation cover factor

Wind erosion can be controlled by:

1. Maintaining good soil moisture with irrigation
2. Using windbreaks to slow down wind speed

FIGURE 7–12 An unprotected and recently plowed field shows a cloud of dust as the wind blows, whereas the cropped field in the foreground is unaffected (left). A roadside ditch adjoining a farm is gradually becoming clogged with wind-deposited soil (right).

Dryland farming is crop production without irrigation (i.e., rainfed production) in semiarid regions. Although some dryland farming occurs in parts of the Pacific Northwest and the Pacific Southwest, this production system is concentrated in the Great Plains region, an expansive semi-arid region that includes parts of 10 states in the U.S. (Montana, North Dakota, South Dakota, Wyoming, Nebraska, Colorado, Kansas, New Mexico, Oklahoma, and Texas) and the prairie dryland region of Canada. It is the largest dryland farming region in the U.S.

Once called the "Great American Desert," the history of the Great Plains is dominated by two weather-related events. The "Dust Bowl" of the 1930s was, without a doubt, one of the most devastating ecological disasters the world had ever experienced. Nevertheless, in the 1950s, the region experienced the driest years on record. The Great Plains region generally experiences scarce precipitation that severely hampers crop production. However, the region is no stranger to drought. In fact, severe droughts occur about every 20 years, and minor ones occur every 3–4 years. However, there is no other historical account similar to the peculiar winds and the erosion that accompanied the drought of the 1930s, which are often referred to as if they were a single event. In actuality, there were at least 4 distinct drought events during that period—1930–1931, 1934, 1936, and 1939–1940. During this period, numerous dust storms were recorded (e.g., 14 storms in 1932, 38 in 1933, 22 in 1934, etc.).

Drought is often cited as the cause of the Dust Bowl, but it also points out that the poor land management practices of that period's farmers made the Great Plains totally vulnerable to, and exacerbated the consequences of, the drought. The period between 1925–1930 is described as the boom years of the Great Plains. The region was first settled in 1875–1876. These settlers had some success with agricultural production in the wet cycles, which encouraged them to remain in the region in spite of the periodic drought episodes. Proponents of dryland agriculture encouraged the rapid population of the Great Plains through misleading information about the region's agricultural potential. The Great Depression (economic decline), which occurred during that period, and the economic over-expansion contributed to

3. Tillage to provide clods on the surface of the soil, vegetative materials on the soil, and adoption of conservation practices such as strip cropping, cropping perpendicular to the wind direction, and stubble mulching.

7.10.3 IMPORTANCE OF RESIDUE MANAGEMENT IN EROSION CONTROL

Residue management through conservation tillage is one of the most effective and least expensive methods of reducing soil erosion. Maintaining a 20% residue cover can reduce soil erosion by 50%. No-till planting systems leaves the most residue on the soil surface. In such cases, soil erosion can be reduced by 90 to 95% of what occurs in conventional tillage.

Farmhouse in a dust storm.

struction through the pulverizing impact of their trampling on the soil exposed by cultivation.

When the drought hit in the early 30s, the conditions were ideal for massive wind erosion to occur. Dust storms characterized the events of this period. The more dramatic dust storms called "black blizzards," rose to heights of 7,000–8,000 feet. Crops and property were destroyed. One of the storms in 1935 was recorded to have lasted 908 hours. The Soil Conservation Service estimated the Dust Bowl to have covered about 100 million acres in 1935, declined to 22 million acres by 1940, and disappeared in the late 40s.

The most visible impact of the 30s drought was the devastation to agricultural production in the Great Plains. The most severely affected areas were west Texas, eastern New Mexico, the Oklahoma Panhandle, western Kansas, and east Colorado. The Federal Government intervened to provide economic relief assistance to the affected states. About 68 percent of the relief went to farmers. In 1934 alone, the Federal Government provided assistance to the tune of $525 million. The Natural Resources Conservation Service (NRCS), formed from the Soil Conservation Service, developed proactive programs to prevent the recurrence of the catastrophe of the 1930s. Currently, the NRCS continues to emphasize the implementation of soil conservation measures. Other practical measures developed after the Dust Bowl included increased irrigation, crop diversification, federal crop insurance, regional economic diversification, and removal of sensitive agricultural lands from production.

the adverse effects of the drought that followed. New crop production technologies were introduced, including new varieties, machinery, row-cropping, and disc plowing. Most of the newly plowed land was cultivated to wheat, increasing the production of the crop by 300 percent and creating a glut in the process in 1931. In the early 20s, the expansion was necessary to pay for the new farming equipment and to offset the low crop prices that occurred after World War I. Farmers were compelled to cultivate more land to meet their financial obligations, resulting in the cultivation of submarginal lands. Soil conservation practices were abandoned to reduce production costs. Livestock was introduced into the agriculture of the regions, adding to the soil's de-

7.11: SOIL DRAINAGE

Drainage is the removal of excess gravitational water from soils by natural or artificial methods. Artificial drainage is used on over 10% of all cropland in the world and on about 30% of the cropland in North America. Poor soil drainage has adverse consequences, including the following:

1. Reduced soil aeration, leading to reduced root respiration and growth
2. Reduced absorption of nutrients, especially phosphorus and potassium
3. The formation and accumulation of toxic substances

7.11.1 BENEFITS

The benefits of drainage include the following:

1. Wet soils are naturally fertile because of high clay and organic matter. Drainage makes them suitable for cropping.
2. Drainage improves soil aeration for quick warming in spring.
3. Wet spots are eliminated to give a uniform field soil moisture.
4. Microbial activity is promoted for more efficient decomposition of organic matter.
5. Denitrification that occurs under anaerobic conditions is reduced.
6. Root growth and development are promoted for more effective exploitation of the soil for improved crop productivity.
7. A drained soil is adaptable to the production of a wider variety of crops, not just those that tolerate excessive soil moisture.

7.11.2 CAUSES

The factors that cause poor soil drainage include the following:

1. High water table
2. Shallow depth of the bedrock
3. Presence of a low permeable clay layer
4. Accumulation of water at a rate faster than natural drainage occurs

7.11.3 METHODS

Soil drainage systems are of two main types—*surface* and *subsurface.*

Surface Drainage

Surface drainage systems are used for removing excess irrigation water from the field, lowering the water table, and removing water trapped in surface depressions. Drainage is accomplished with open ditches, ridge-tillage, and the grading or smoothing of a field to remove minor depressions that collect water (Figure 7–13).

FIGURE 7–13 Excess water from a field can be drained through grassed open ditches.

FIGURE 7–14 Engineers install a perforated tube for subsurface drainage (left); drained water from the subsurface is discharged into an open drainage line (right).

Subsurface Drainage

Subsurface drainage systems consist of submerged units that include tiles and porous plastic tubes (Figure 7–14). An advantage of this system is that the area above the drainage units can still be cultivated, unlike surface systems that take significant amounts of land out of agricultural production. Subirrigation systems are more expensive to install, those using tiles being more so than those using plastic tubes.

7.12: LAND SUITABLE FOR CROP PRODUCTION

Land varies in its use for agricultural production. Certain pieces of land are naturally fertile and most preferred for crop farming. Others need various levels of management to become usable for cropping. What soil orders tend to have natural fertility for crop production? The USDA-NRCS has developed certain criteria for classifying lands according to the level of soil management needed. This is called the *land capability classification*. There are eight classes identified in this system (Table 7–7). A Class I land is one that can be used continuously for intensive crop production with good farming practice. This land also has few limitations. Prime farmlands have good average temperature that favors most crops. They are deep, are well-drained, have a desirable pH (4.5 to 8.4), and have a reliable source of water for crop production (either from rain or irrigation). Class VIII land, on the other hand, has severe limitations (e.g., wet land, rocks, and steep slopes) and is usually unfit for crop production (Figure 7–15).

A *unique farmland* has characteristics that favor the production of specific crops. This type of land is used for producing specific high-value food and fiber crops with unique soil and/or climatic requirements (e.g., citrus, tree nuts, and vegetables).

By definition, *prime farmland* is the best land available for a specific enterprise. Prime farmland is land whose soils are best suited to producing food, feed, fiber, and oilseed crops. These lands produce the highest yields with minimal production inputs and least damage to the environment.

Table 7-7	Land Capability Classes and Their General Properties
Class	**Properties**
Class I	The land has few limitations on its use. It is nearly level and has deep and well-drained soils. It can be used continuously for cropping under normal management. The land is suitable for cultivation of row crops and others.
Class II	The land has significant limitations to its use for intensive crop production. It usually has a gentle slope (2 to 5%) and requires the implementation of some conservation practices as part of the management practice. Minimum tillage is desirable. Row crops can be cultivated with proper management.
Class III	The land has severe limitations on its use for intensive cropping. The slope is more pronounced (6 to 10%). Special conservation and management practices (e.g., drainage) are necessary prior to use. Crops that protect the soil from erosion (grasses and legumes) may be cultivated.
Class IV	The land has more severe limitations than Class III. Row crops are not suited to this land (unless they are grasses and densely populated species that protect the soil from erosion). It is best to grow perennial species such as pastures. No-till cropping systems, terracing, or other special conservation practices should be implemented.
Class V	The land has very severe limitations that are impractical to rectify by conventional conservation practices. It is prone to excessive moisture from flooding and is too rocky to be tilled. The land can be developed into a pastureland but is unsuited for row cropping.
Class VI	The limitations of this land are extremely severe. The slopes are very steep. It is best left as a woodland or wildlife reserve. Pasture may also be developed on this land.
Class VII	The land has extreme limitations. Pasture production is not practical. It is best developed as a woodland or wildlife reserve.
Class VIII	The land is so severely limited that the only recommended uses are for wild-life, recreation, esthetic purposes, and watershed protection. The land has steep slopes and is rocky, is swampy, or has very sensitive vegetative cover.

7.12.1 DISTRIBUTION OF ARABLE LAND IS NOT EQUITABLE

Arable land. Land suitable for production of crops.

Arable land, or *cropland*, is land that can be utilized for crop production. Whether a piece of land can be utilized for crop production or not depends on the climate (temperature, rainfall) and soil type. Of the over 13 billion hectares of land in the world, about 50% is completely unusable for crop production. Only about 11% of the total land area of the world is arable. About 24% of the land is in pasture, while another 31% is in forests. Most of the arable land is located in the United States, Europe, Russia, India, China, and Southeast Asia. Most productive soils were formed under grassland (e.g., the prairies of the midwestern United States) or hardwood forests (e.g., Europe and India).

There are twelve soil orders in the United States (Figure 7–16). The most dominant soil orders in the world are aridisols (19%) and alfisols (13%); the least dominant include histosols (1%) and vertisols (2%). Of the 9% of oxisols and 6% of ultisols in the world, about 90% and 45%, respectively, are located in Latin America and Africa. The character-

FIGURE 7–15 An expanse of land in San Mateo County, California, is used to demonstrate a variety of soil capability classes. (Source: USDA)

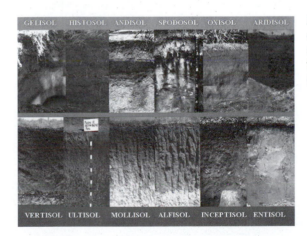

FIGURE 7–16 The 12 common soil orders recognized by the Natural Resources Conservation Service, with the typical soil profile of each order. (Source: USDA)

istics of these soil orders are summarized in Table 7–8. What are the challenges of crop production in tropical regions such as Latin American and Africa? The soils in these regions are highly leached, acidic, and low in nutrients. However, they can be made productive through the application of soil amendments and fertilizer. Unfortunately, producers in these regions can ill-afford these production inputs. The practice of slash-and-burn agriculture and shifting cultivation are prevalent in these regions. Whereas the slash-and-burn agriculture has the short-term benefit of providing nutrients from the ashes for crop use, the high rainfall in these regions quickly leaches out nutrients. The exposed soil is also prone to soil erosion, increased temperature, and rapid organic matter decay. The consequences are that

Histosols

Histosols are organic soils. They occur in wet, cold areas. Organic soils have high water retention, high CEC, low bulk densities, and usually deficiencies in nutrients, especially nitrogen, potassium, and copper. Properly drained, the organic matter can decompose rapidly. The soil is used for vegetable production.

Entisols

Entisols are slightly developed soils that lack defined horizons. They may derive from recent alluvial deposits or recent volcanic ash deposits. They occur in floodplains and rocky mountain regions. Those in floodplains are very good agricultural soils.

Inceptisols

Inceptisols are more weathered than entisols. Many soils used for paddy rice production are of this soil order. Some are too wet or exist in cold regions.

Andisols

Andisols are generally weakly developed soils. They are high in organic matter and have high amounts of amorphous aluminum and iron clays. Andisols are among the most productive soils in the world when well managed. They include the volcanic soils of Hawaii.

Aridisols

Aridisols occur in regions dominated by long dry periods. They have low organic matter content and high basic cation saturation of 100%. They are among the most productive soils under irrigation and fertilization.

Mollisols

Mollisols are dark-colored soils formed mainly under grasslands but also under some hardwood forests. They have deep, dark color and more than 50% base saturation. They are very fertile soils, with high nitrogen and humus.

Vertisols

Vertisols have more than 30% swelling clays (montmorillonite) and go through swelling and shrinking cycles with moisture and drying. They are very sticky when wet and very hard when dry with cracks. They have high CEC and relatively high humus content.

Alfisols

Alfisols are characterized by the translocation of clay downward to accumulate in the Bt layer. They are the most naturally productive soils without fertilization or irrigation. The top soil may be moderately acidic. Application of fertilizers usually produces high yield.

Spodsols

Spodsols have high salt content. They usually occur in cold, wet climates under acidic conifer forests or other vegetation. They have a white leached E horizon and a very low basic cation saturation percentage.

Ultisols

Ultisols are warm and low in basic cation saturation (acidic), occurring in humid regions. Without fertilization, they become worn out with time. However, with high-level management (fertilizers, liming, etc.), ultisols can be among the world's most productive soils.

Oxisols

Oxisols are the most widely weathered soils. They occur on old landforms in humid, tropical, and subtropical climates. They usually have bright yellowish to red colors. They are rich in residual irons and aluminum hydrous oxide residues. Oxisols are very low in nutrients. They are used for producing carbohydrate and oil crops. With adequate nitrogen phosphorus potassium, they are productive for bananas, sugarcane, coffee, rice, and pineapple.

Gelisols

Gelisols have permafrost (perennially frozen soil horizon under the upper soil) within the upper 1 to 2 meters. They are young soils with little profile development. Gelisols mainly support tundra vegetation of lichens, grasses, and low shrubs during the brief summers. Very few areas of Gelisols are used for agriculture.

the deterioration of the soil is accelerated. It is estimated that about 5 to 7 million acres of land are lost to degradation from various sources.

Even though most of the land in semiarid regions consists of alfisols that are less leached than ultisols and oxisols, production in these regions is constrained by erratic rainfall. Periodic drought devastates crop production and causes famines. Mollisols are base-rich and very productive. Unfortunately, the distribution of this soil order in the world is not widespread.

Aridisols occur in areas where sunlight is abundant. They are productive soils if supplemental irrigation and fertilizers are provided. This is the case in Pakistan, India, and the Middle East, where crop production is very successful with high inputs.

The rate of cropping of potentially arable land varies from one region to another and is influenced by socioeconomic factors, among other factors. Europe and Asia have cropped most (80% and 65%, respectively) of their potentially arable land. This is partly due to intense and prolonged population pressures in these places. On the other hand, only about 20 to 30% of the arable land in Africa, South America, and Oceania is in crop production. Whereas the European countries and some in Asia can afford to import additional food, many economically poor countries are being overwhelmed by population growth and the inability to increase food production.

To increase crop production, there are three strategies countries may adopt:

1. Clear new (virgin) arable land for cropping.
2. Increase cropping intensity (number of crops per year).
3. Intensify the use of existing cropped arable lands.

The third strategy is applicable to regions such as Europe where arable land is scarce and population is high but resources are available to increase productivity on existing cropped land. The situation is much different in regions of poor economies such as Africa and South America. Much of the unutilized arable land is not readily accessible. The management of these new lands for sustainable production is yet to be determined. Clearing of virgin forests and steep slopes only results in degradation of the land. Watersheds are destroyed, while rivers and streams are polluted. Also, the domestic water supply is adversely impacted while biodiversity is jeopardized.

7.12.2 THE DECISION TO USE LAND FOR A PARTICULAR PURPOSE IS INFLUENCED BY SEVERAL FACTORS

How does a producer decide the kind of production enterprise to carry out on a piece of land? Land is the most basic requirement for crop production. The decision to use land for a particular purpose is influenced by several factors, including the following:

1. *Location.* This pertains to the climatic zone as well as the distance from major transportation links, distance from urban center, and distance from parks and recreational facilities. Land located near the city center is attractive and valuable for residential construction and for commercial development. The potential use of land and its location determine its economic value.
2. *Topography.* Not only are steep slopes difficult to farm, but they also predispose the soil to erosion. Undulating land may require some leveling to make it amenable to certain practices such as irrigation.
3. *Productivity.* Prime farmlands and unique farmlands are preferred for crop production. Unproductive lands are utilized for construction (industrial, residential, etc.).
4. *Erodibility.* Highly erodible land is not suitable for crop production. Certain government programs are implemented to prevent the cultivation of erodible lands.

Properties in urban areas have exact addresses that permit them to be located without much difficulty. The system used to show the exact location of property in rural areas is called the *legal description*. This system dates back to 1784, when a congressional committee was charged with preparing a survey ordinance to address the boundary litigations arising from indiscriminant settlement. Since many of the old states had been well settled prior to the survey ordinance, they were left to continue their system of property demarcation—the *metes and bounds system*. Most states west of the Mississippi (excluding parts of Texas) and many east of the Mississippi use the product of the 1784 ordinance called the *rectangular survey system*.

The rectangular survey system begins with the establishment of a *principal meridian* and a *base line*. The principal meridian runs in a true north and south direction, while the base line runs east and west at right angles to the principal meridian. The point of intersection of these lines is the *initial point*, or *starting point*. Beginning at the initial point, surveyors mark out lines in the north-south direction and on either side of the meridian, every 6 miles. Similarly, lines spaced 6 miles apart are marked parallel to the base line. The

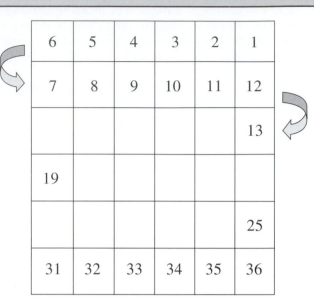

A township is divided into 36 smaller areas measuring 1×1 miles square. Each smaller area is called a "section"; sections are numbered from the north right towards the left and down in a serpentine fashion.

lines parallel to the base line are called *township lines*, and those parallel to the principal meridian *range lines*. The result is a grid of squares.

Each of these 6-square-mile areas constitutes a *congressional township* (not to be confused with *civil township*). Each congressional township is further divided into 36 squares, each square called a *section*. The sections are numbered from the top right-hand corner to the bottom right corner in a serpentine fashion. A section is assumed to be 640 acres. This size is not constant for all sections, due to the curvature of the earth's surface, which causes range lines to converge. Surveyors correct for convergence at specific intervals by measuring from the principal meridian. This explains why jogs occur in the north-south section line roads at regular intervals. Consequently, sections on the north (1-6) and those on the west side (6-31) are not full sections. A section may be subdivided into quarters. The partial sections on the north and west sides will yield parcels of land that are less than 40 acres and are called *correction lots*. These lots also occur along rivers (*river lots*).

The description of land is based on a quarter section as the basic unit, while using the north pole as ref-

Each square is called a "township" and has an area of 6×6 miles square. Area "X" is located in "Township 2 North Range 3 East" while area "Y" is called "Township 2 South Range 4 East."

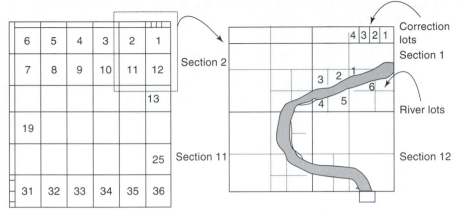

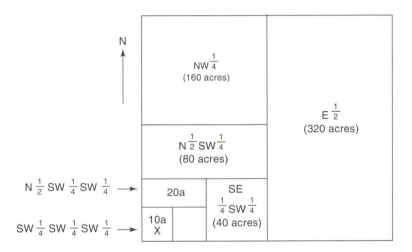

Each section is divided into quarters. Some tracts may be less than a quarter and called correction lots or, when they border a river, river lots. River lots are numbered downstream in a sequence until they reach a section line, then the numbering continues upstream.

Each section measures 640 acres in area. A 40-acre tract of land is called a lot or correction lot. Correction lots are less than 40 acres.

erence. A tract of land that is smaller than a quarter is described as part of the quarter in which it occurs. Uppercase letters and fractions are used in writing the legal description of a tract of land (backwards—starting from the right to the left). The plot labeled "X" occurs in the $SW\frac{1}{4}$ section. Further, it occurs in the southwest corner of the southwest sections (i.e., $SW\frac{1}{4} SW\frac{1}{4}$). Again, it occurs in the southwest corner of the $SW\frac{1}{4} SW\frac{1}{4}$ subsection (thus $SW\frac{1}{4} SW\frac{1}{4} SW\frac{1}{4}$). The area represented by this description is 10 acres. The legal description is completed by tagging on the section number (e.g., $SW\frac{1}{4} SW\frac{1}{4} SW\frac{1}{4}$ section 10). A clue to this nomenclature is that, when two letters appear together (e.g., SW), it always means $\frac{1}{4}$ of what follows (e.g., SW section 5 means $\frac{1}{4}$ of section 5). Similarly, when a letter appears by itself, it indicates $\frac{1}{2}$ of whatever follows (e.g., N sec-

tion 10 means $\frac{1}{2}$ of section 10). If a tract to be described consists of more than one parcel (40 acres), it is described in more than one part. For example, a land of area 120 acres consists of $80 + 40$ acres tracts. The two are described separately and linked (e.g., $N\frac{1}{2} SW\frac{1}{4}$ and $SE\frac{1}{4} SW\frac{1}{4}$). Half of a section (e.g., $N\frac{1}{2}$) may also be designated as N^2. Since a section consists of 640 acres of land, one can determine the acreage of a tract of land from the legal description. For example, $SE\frac{1}{4} SE\frac{1}{4} = 40$ acres, while $S\frac{1}{2} SE = 80$ acres.

To get to an address by reading a legal land description, one needs to have the general highway map for the county in which the property is located. The townships are also designated with reference to the starting point (e.g., *T 16N R 14W* indicates township 16 north and range 14 west).

5. *Government policies and programs.* The government intervenes to protect lands and soil from devastation through the creation and implementation of conservation programs.
6. *Management skills.* Prime and unique farmlands require minimal management for use in crop production. Marginal lands require more challenging management operations to make them productive.
7. *Age of landowner.* Older people tend to be more conservative, while younger ones may be more adventurous and have longer-term goals.
8. *Landowner's expectations.* When all is said and done, the landowner will use the land for whatever he or she deems profitable or rewarding.

SUMMARY

1. Mineral soil is weathered rock. Some soils are derived from organic material.
2. Weathering agents may be physical, chemical, or biological.
3. Soil formation depends on parent material, climate, biota, topography, and time.
4. As soils form, they develop a vertical cross section called a profile, consisting of layers called horizons. The profile is dependent upon the age of the soil and conditions under which formation occurs, among others.
5. In soil taxonomy, soils are classified into 13 orders in the United States.
6. A typical mineral soil has some mineral matter, air, water, and organic matter.
7. Soil texture is dependent on the proportions of soil separates—sand, silt, and clay. It affects drainage, water-holding capacity, tillage, and other factors.
8. Soil structure depends on the arrangement of soil particles. It can be destroyed by compaction from farm machinery, animals, and others.
9. Bulk density and soil porosity are important soil physical properties in crop production. They affect drainage, tillage, and other production activities.
10. Soil colloids (clay and humus) are able to attract cations, the source of fertility for crop production.
11. Soil reaction (pH) determines the availability of nutrients.
12. The soil has organisms, some of which are beneficial, while others are pests.
13. The soil can be managed to control the population of harmful microbes.
14. When organisms die, they decompose and the minerals are recycled.
15. Only about 11% of the total land area of the world is arable.
16. Certain lands are suited for farming of a specific crop (unique farmlands), while others are suited for general food, feed, and fiber production (prime farmland).
17. The use to which land can be put depends on certain factors, including topography, zoning restrictions, land capability, management skills, and age of the operator.

REFERENCES AND SUGGESTED READING

Brady, C. N., and R. R. Weil. 1999. *The nature and properties of soils.* 12th ed. Upper Saddle River, NJ: Prentice Hall.

Burns, R. C., and R. W. F. Hardy. 1975. *Nitrogen fixation in bacteria and higher plants.* Berlin: Springer-Verlag.

Hauck, R. D., ed. 1984. *Nitrogen in crop production.* Madison, WI: American Society of Agronomy.

Kletke, D. (no date). *Legal land descriptions in Oklahoma.* Oklahoma State University, Extension facts, No. 9407.

Mengel, K., and E. A. Kirby. 1987. *Principles of plant nutrition*. 4th ed. Bern, Switzerland: International Potash Institute.

Miller, W. R., and D. T. Gardner. 1999. *Soils in our environment*. 9th ed. Upper Saddle River, NJ: Prentice Hall.

Natural Resources Conservation Service. 2000. Soil Taxonomy USDA, Washington, DC.

Penny, D. C., S. C. Nolan, R. C. McKenzie, T. W. Goddard, and L. Kryzanowski. 1996. Yield and nutrient mapping for site specific fertilizer managment. *Communications in Soil Science and Plant Science Analysis* 27:1265–79.

Stevenson, F. J. 1986. *Cycles of soil carbon, nitrogen, phosphorus, sulfur, and macronutrients*. New York: Wiley.

Usery, E. L., S. Pocknee, and B. Boydell. 1995. Precision farming data managment using geographic information systems. *Photogrammetric Engineering and Remote Sensing* 61:1383–91.

SELECTED INTERNET SITES FOR FURTHER REVIEW

http://www.nhq.nrcs.usda.gov/land/index/soils.html

Soils and soil types.

http://www.statlab.iastate.edu/soils/photogal/orders/soiord.htm

Photos of soil orders.

http://www.nhq.nrcs.usda.gov/land/index/lcc.html

Land capability classification.

http://www.statlab.iastate.edu/survey/SQI/sqiinfo.html

Soil quality factors—available water, erosion, leaching, pesticide, compaction, organic matter.

http://www.fertilizer.org/PUBLISH/PUBMAN/introdc.htm

OUTCOMES ASSESSMENT

PART A

Answer the following questions true or false.

1. T F Soil is weathered rock.
2. T F A pH of 7.8 is acidic.
3. T F Clay soils have a finer texture than sandy soils.
4. T F Incorporating organic matter into the soil increases its bulk density.
5. T F Kaolinite clays have 1:1 clay lattice structure.
6. T F Hydrogen ions are more tightly bound to soil colloids than potassium ions.

Answer the following questions.

1. Give the five major factors of soil formation.

   _____

   _____

   _____

   _____

   _____

2. Give the three basic types of rocks. _____

   _____ _____

3. What is weathering? _____

   _____

4. Give 5 of the 12 soil orders in the United States.

   _____

   _____

   _____

   _____

   _____

5. The porter's clay is a silicate clay and is also called _____.

6. What does the acronym CEC stand for? _____

   _____

PART C

Write a brief essay on each of the following topics.

1. Discuss physical weathering.
2. Discuss the role of climate in soil formation.
3. Describe the composition of a typical mineral soil.
4. Discuss the role of soil texture in crop production.
5. Discuss the role of soil structure in crop production.
6. Discuss the nature and role of humus in soil structure.
7. Discuss the role of soil organisms in crop productivity.
8. Discuss the distribution of arable land in the world.

PART D

Discuss or explain the following topics in detail.

1. Discuss the impact of leached fertilizer on the environment.
2. Discuss the role of fertilizers in boosting crop productivity in the United States.
3. Is soil a renewable resource? Explain.

8

Plant Nutrients and Fertilizers

PURPOSE

The purpose of this chapter is to discuss crop nutrition and its management. The discussion includes the source of essential nutrients, their roles in plant growth and development, and how to provide supplemental nutrition for enhancing crop productivity.

EXPECTED OUTCOMES

After studying this chapter, the student should be able to:

1. List the 18 essential nutrients for plant growth and development.
2. Discuss the criteria for essentiality of an element in plant nutrition.
3. Discuss the role of essential nutrients in plant growth and development.
4. Discuss how nutrients are lost from the soil.
5. Describe plant nutrient deficiency symptoms for all essential elements.
6. Discuss the methods of diagnosing soil nutrient status.
7. Discuss fertilizers and their commercial sources.
8. Discuss methods of application of fertilizers.
9. Discuss fertilizer use in U.S. crop production.

KEY TERMS

Cholorosis	Mineralization	Phosphorus fixation
Eutrophication	Nutrient cycling	Soil test
Luxury consumption	Nutrient deficiency symptom	Necrosis
Macronutrients		
Micronutrients		

TO THE STUDENT

Plants use particular nutrient elements more than others. Thus, soil nutrients are not depleted proportionally. There are natural nutrient cycles for replenishing soil nutrients. However, these processes are often slow. In modern crop production, crop producers often supplement soil nutrients by applying inorganic fertilizers. The soil should be well managed in crop production. The goal of soil fertility management is to sustain the soil's ability to supply essential nutrients to crops. To accomplish this, there is a need to understand

1. How soil nutrients become depleted
2. How to detect nutrient deficiencies
3. How to correct nutrient deficiencies

8.1: ESSENTIAL ELEMENTS

Do plants absorb and utilize all of the 90 elements available? For optimal growth and development, crop plants need certain *essential elements* or *nutrients* from the soil. These elements are deemed essential because of the following reasons:

1. Plants cannot grow and develop properly without them.
2. They play critical roles in plant metabolism.
3. Their roles cannot be replaced by another element.
4. Deficiency symptoms can be corrected only by supplying that deficient element.

There are 18 essential elements that are required by most plants. Even though silicon is not universally accepted as essential, it occurs in the tissues of most plants. Selenium is referred by some range plants. Essential plant nutrients may be placed into two general categories according to source as either *non-mineral* or *mineral,* of which there are 3 and 15, respectively. The non-mineral elements are carbon (C), hydrogen (H), and oxygen (O). Carbon, hydrogen, and oxygen are present in the largest amounts in the plant. Carbon is fixed in the plant by the process of photosynthesis from CO_2 derived from the atmosphere to produce organic compounds. Hydrogen and oxygen enter the roots as water through root hairs. The mineral elements may be subdivided into three groups according to quantities used by the plant, as **macronutrients** (or major elements), secondary nutrients, and **micronutrients** (or trace elements) (Table 8–1).

Macronutrients are used by plants in large amounts, while micronutrients are needed in minute amounts. Further, calcium, magnesium, and sulfur are sometimes

Macronutrients.
Chemical elements needed in large amounts for plant growth.

Micronutrients. Chemical elements required in small amounts for plant growth.

Table 8–1 The 18 Soil Mineral Elements Essential for Plant Growth and Development

Macronutrients	Secondary nutrients	Micronutrients
Nitrogen	Calcium	Boron
Phosphorus	Magnesium	Iron
Potassium	Sulfur	Molybdenum
		Manganese
		Zinc
		Copper
		Chlorine
		Cobalt
		Nickel

classified as *secondary nutrients* because they are utilized in larger quantities than the other micronutrients.

A plant requires all these essential elements in proper amounts and proper ratios to each other for development and crop productivity. Deficiency of one element may result in improper development of the plant. The severity of response to deficiency differs among plant species. Certain characteristic symptoms accompany deficiency of each of these elements. Plant species differ in their need and use of these essential elements. Each element has a specific role in plant growth and development. Certain elements promote vegetative growth and development, while others promote reproductive functions and root development, among other roles.

Do plants know when to stop absorbing nutrients? When the nutrients are present in non-limiting amounts, plants have a tendency to use more than they need. Up to a certain point, uptake of nutrients does not translate into increased biomass (or productivity). At this level, the use of nutrients is described as **luxury consumption.** If consumption continues, it might reach a toxic level (Figure 8–1).

Luxury consumption.
The intake by a plant of an essential nutrient in amounts in excess of what it needs.

Nutrient deficiency symptom. A visible change in plant morphology or appearance associated with the deficiency of a specific plant nutrient.

FIGURE 8–1 Plants vary in their need for and use of nutrients. Nutrients accumulate in plants and may be economically used up to a certain concentration. At low concentrations, plants start to display **nutrient deficiency symptoms**. This stage is followed by the hidden hunger stage, at which time the nutrient concentrations are less than optimal but not low enough to cause deficiency symptoms to be observed. The critical nutrient range or the threshold level is the level below which an essential element is deemed to be deficient in the plant. Sufficiency occurs when the nutrient is present in the plant in an adequate amount and proper balance with other nutrient elements. After this stage, accumulation of nutrient elements may become toxic to the plant. This curve may be interpreted in another light, as in Figure 8–1b. Soil nutrients may accumulate in plants in four main categories. At the deficiency level, plant growth is hampered. When nutrients are present in adequate amounts, the plant can use them for economic production. After this level, additional amounts of nutrients are wasteful and will not result in economic use. Finally, higher levels may become toxic to plants, causing a decline in growth and yield of the economic product.

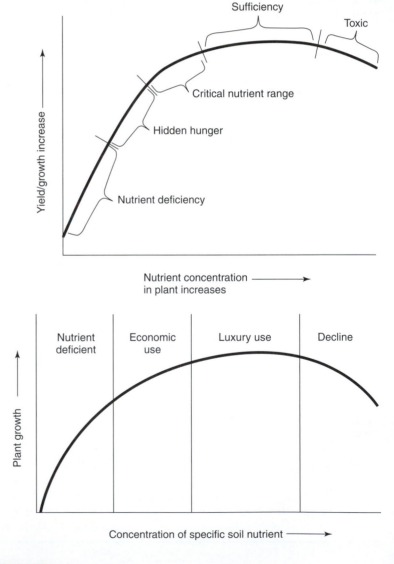

In crop production, the grower may increase certain nutrients in the general nutrition for enhanced productivity according to the yield objective. This may be accomplished in a variety of ways, using inorganic (artificial) or organic (natural) fertilizers.

8.2: NITROGEN

Nitrogen is the key nutrient in plant growth and productivity. It is often the limiting nutrient in plant growth and the soil nutrient required in the greatest amount. It is a constituent of nucleic acids and therefore plays a role in plant heredity. Nitrogen is also a constituent of proteins and chlorophyll, the primary pigment in photosynthesis. It promotes vegetative growth, making tissues more tender and succulent and plants larger. Crop yields are adversely affected when nitrogen is deficient.

Nitrogen uptake by plants is in the form of NO_3^- and NO_4^+ ions. The dinitrogen (N_2) form, though most abundant in the atmosphere, is unusable directly by plants. Nitrogen is very mobile in the plant. As such, when deficient in nutrition, it is translocated from older leaves to younger ones, where it is most needed. The older leaves lose color (**chlorosis**) and become yellowish.

Soil nitrogen derives from mineral, atmospheric, and organic sources. Weathering of minerals adds to soil nitrogen as rocks decompose and interact with other chemicals in the soil environment. The decomposition of organic remains of plants and animals by soil microbes releases the organic nitrogen (*immobilized nitrogen*) into inorganic forms. This conversion process is called **mineralization.** Soil microbes also convert atmospheric dinitrogen into usable forms through the process of nitrogen fixation. This can happen independently when bacteria fix nitrogen in the soil (non-symbiotic) or in association with plant roots (symbiotic).

Nitrogen is released into the soil when organic matter decomposes by the process of mineralization. Organic matter decomposition is accelerated by increased temperature. Thus, it occurs fastest in moist, sandy soils (especially in summer), which are well aerated and warmer, and slower in clay and silts in cool spring.

Even though plants can absorb nitrogen as either NO_3^- or NO_4^+, the latter ion is subject to microbial transformation into the former. *Nitrosomonas* oxidizes NO_4^+ to nitrite (NO_2^-) before *Nitrobacter* converts the product to nitrates (NO_3^-). Nitrates are readily leached from the soil. The positively charged NO_4^+, though soluble in water, is more resistant to leaching. Instead, it is readily rendered unexchangeable and unavailable to plants by the process of *ammonium fixation*. Nitrogen is also lost in gaseous form through the process of denitrification, another bacterial process. This rapid process occurs under warm anaerobic conditions when bacteria are compelled to use an alternative source (NO_3^-) as a source of oxygen. Nitrate ions are changed to N_2 and N_2O and lost through volatilization as gas into the atmosphere. For this reason, it is best to use the ammonium form of fertilizer as a source of nitrogen under paddy rice cultivation so as to eliminate waste due to denitrification. It may also be lost through ammonium volatilization when ammonium or urea fertilizer is applied to the soil surface. To reduce ammonium volatilization, the fertilizer should be covered immediately after application or watered. Broadcast application should be avoided when using these materials for fertilization. When deficiency occurs in the soil, leaves become yellow (chlorosis). Chlorosis starts with older leaves, since nitrogen is a mobile element in the plant.

Chlorosis. A condition in which a plant or a plant part is light green or greenish yellow because of poor chlorophyll development or the destruction of chlorophyll resulting from a pathogen or a mineral deficiency.

Mineralization. The conversion of an element from an organic form to an inorganic form as a result of microbial decomposition.

8.3: PHOSPHORUS

Phosphorus is second to nitrogen in importance in plant nutrition. Plants utilize about 1/10 as much phosphorus as nitrogen. It is part of plant nucleoprotein and hence important in plant heredity. Phosphorus also plays a role in cell division, stimulates root growth, and hastens plant maturity. One of the notable physiological roles of phosphorus is in the energy storage and transfer bonds of ATP (adenosine triophosphate) and ADP (adenosine diphosphate). Phosphorus deficiency shows up as purpling of plant parts.

Soil phosphorus, unfortunately, is largely of very low solubility and therefore not readily available to plants. It is very reactive to both the soil solution and solids, and hence very immobile, except in organic soils or soils with very high CEC. The original source of phosphorus is rock phosphate (apatite). Plants absorb phosphorus in the soluble ions of $H_2PO_4^-$ and HPO_4^{2+}. The orthophosphate form, however, is readily precipitated and adsorbed to soil particles (a process called **phosphorus fixation**). At low pH, iron and aluminum ions are very soluble and react with phosphorus to form precipitates. At high pH, calcium precipitates phosphorus. Phosphorus is most available at pH of 6.5. It is more soluble under anaerobic conditions than aerobic. Phosphorus is lost through surface runoff into surface water, where it is implicated in **eutrophication** (nutrient enrichment of surface water, leading to excessive plant growth and oxygen deficit). Clay soils fix more phosphorus, fixation being higher in expanding clays than kaolinites.

When deficient, leaves become dark green and plants stunted. This deepening in color is attributed to increase in nitrates in the leaves. Purplish color may also develop, especially in older leaves. Phosphorus is mobile and thus symptoms appear first in older leaves.

Phosphorus fixation. The rendering of phosphorus unavailable to plants through the formation of insoluble complexes in the soil.

Eutrophication. Nutrient enrichment of surface waters such as lakes and ponds resulting in the stimulation of growth of aquatic organisms, and leading eventually to oxygen deficiency in the water.

8.4: POTASSIUM

Potassium is a very soluble cation. However, its common natural forms (micas and orthoclase feldspar) are very slowly soluble. However, it is not an integral part of any specific cell compound. It is required by many enzymes for activation and plays a role in cell division, formation of carbohydrates, and translocation of sugars. Potassium is also known to enhance the resistance of certain plants to certain diseases. It is a key factor in the regulation of osmosis (water control) in plants.

Potassium can be adsorbed between clay layers *(potassium fixation)*. Further, both soluble and exchangeable potassium can be adsorbed in excessive amounts by plants with little or no consequence (luxury consumption). Sometimes, excess potassium can hinder the adsorption of magnesium. Potassium is also immobile in the soil, even though less so than phosphorus. It is lost through erosion.

Potassium deficiency symptoms include marginal leaf burn *(marginal necrosis)*, speckled or mottled leaf, and lodging.

8.5: CALCIUM, MAGNESIUM, AND SULFUR

8.5.1 CALCIUM

Calcium is important for cell growth, cell division, cell wall formation, and nitrogen accumulation. Apart from being directly used by plants, calcium is used to make other nutrients more available to plants. Calcium is used in soil amendment to correct soil acidity (liming). Calcium also forms organic salts (e.g., calcium oxalate) with certain organic acids in the plant that cause irritations to humans when ingested. Calcium is absorbed as Ca^+ ions by plants. Its deficiency in plant nutrition causes a variety of symptoms such as strap leaf, poor root development, and defective terminal bud development (blunt end).

8.5.2 MAGNESIUM

Magnesium is absorbed by plants as Mg^{2+} ions. It is the central atom in the structure of a chlorophyll molecule. It is essential in the formation of fats and sugars. Large amounts of potassium may interfere with the uptake of magnesium. Magnesium is mobile in plants and, as such, deficiency symptoms first appear in older leaves as chlorosis.

8.5.3 SULFUR

Sulfur is a constituent of many amino acids and many vitamins. The oils of plants in the mustard families contain sulfur, which is responsible for the characteristic flavors of onion, cabbage, and other cruciferous plants. When organic matter is incinerated, the sulfur oxides are washed down in rain. This rainwater can be acidic (acid rain). It is absorbed by plants as SO_4^{2+} ions. When deficient, plants become chlorotic.

8.6: MICRONUTRIENTS (TRACE ELEMENTS)

Plants require micronutrients in minute amounts but are by no means less important. They are predominantly involved in plant physiology as activators of many enzyme systems.

8.6.1 BORON

Boron is one of the most commonly deficient of the trace elements. Boron is absorbed by plants BO_4^{2+}. It affects flowering, fruiting, cell division, and water relations (translocation of sugars) in the plant. When deficient, symptoms appear first at the top. The terminal bud produces growth described as *witches' broom*. Lateral branches form rosettes while young leaves thicken.

8.6.2 IRON

Iron is abundant in the soil. It can be absorbed through the leaves or roots as Fe^{2+} and sometimes as Fe^{3+} ions. It is a compound in many enzymes and a catalyst in the synthesis of chlorophyll. Its deficiency shows up as *interveinal chlorosis* (yellowing between the veins) of young leaves. The symptoms appear first in younger leaves, since iron is immobile.

8.6.3 MOLYBDENUM

Molybdenum is involved in protein synthesis. It is also required by certain enzymes that reduce nitrogen. Vegetables, cereals, and forage crops are among a number of plant species that are known to show very visible symptoms when the element is deficient in the soil. A classic symptom in cauliflower and other crucifers is narrowing of leaves *(whiptail)*. Plant leaves may also become pale green and roll up.

8.6.4 MANGANESE

Manganese has an antagonism reaction with iron (i.e., as one increases the other decreases). Deficiencies are common on sandy soils. Manganese is absorbed as Mn^{2+} ions.

It is crucial to photosynthesis because of its role in chlorophyll synthesis. It is also important in phosphorylation, activation of enzymes, and carbohydrate metabolism. When deficient, plants develop interveinal chlorosis in younger leaves.

8.6.5 ZINC

Zinc is absorbed as Zn^{2+} ions. It is an enzyme activator that tends to be deficient in calcareous soil with high phosphorus. When deficient, plant leaves become drastically reduced in size, while internodes shorten to give a rosette appearance. Leaves may also become mottled.

8.6.6 COPPER

Copper is absorbed mainly as Cu^{2+} ions. Copper deficiency is common in soils that are high in organic matter. The element is important in chlorophyll synthesis and acts as a catalyst for respiration and carbohydrate and protein metabolism. When copper is deficient, terminal buds die and the plant becomes stunted. Younger leaves may show interveinal chlorosis, while the leaf tip remains green.

8.6.7 CHLORINE

Chlorine deficiency is rare. Excessive levels of chlorine are a more common problem. It is absorbed at Cl^- ions. When deficient, plants may be stunted and appear chlorotic with some necrosis.

8.6.8 COBALT

Cobalt is essential for the symbiotic fixation of nitrogen. However, some plant species have non-symbiotic nitrogen fixation-associated need of cobalt. It is found in vitamin B_{12}.

8.6.9 NICKEL

Nickel is essential for the functioning of enzymes, such as urease, hydrogenase, and methyl reductase. It is essential for grain filling, seed viability, iron absorption, and urea and ureide metabolism. When nickel is deficient, legumes accumulate toxic amounts of urea in their leaves. Also, the seeds of cereal plants that are deficient in nickel are not viable and fail to germinate.

8.7: NEED FOR NUTRIENT BALANCE

Plants need all of the 18 essential nutrients in the proper quantities and balance in order to grow and develop properly. An imbalance may result in the improper use of nutrients or even toxicity of the one(s) present in excessive amounts. According to the concept first proposed by Liebig, plant production can be no greater than the level allowed by the growth factor present in the lowest amount (limiting factor) relative to the optimum amount for that factor. In other words, if a factor (e.g., temperature, nitrogen, water) is not the limiting factor, increasing it will hardly enhance plant growth. On the contrary, increasing the non-limiting factor may cause an imbalance in the growth factors in the plant environment. For example, increasing nitrogen in a situation where phosphorus is the limiting factor may actually intensify the deficiency of phosphorus. On the other hand, increasing phosphorus (the limiting factor) may cause the plant to respond better to an increase in nitrogen. This synergistic effect is desirable in fertilization. By the same token, antagonistic effects occur when certain nutrient elements compete with or cause a reduction in the uptake of other nutrients. Antagonism may be used to reduce the toxicity of certain elements in the soil. For example, adding sulfur to calcareous soils that contain toxic quantities of soluble molybdenum may reduce the availability and hence the toxicity of molybdenum.

The most modern precision application of fertilizers for proper balance is *precision agriculture*, the application of fertilizers and other agronomic inputs using geospatial technology. This technology is discussed later in the chapter.

8.8: SOIL NUTRIENT STATUS DETERMINATION

How can a producer tell if the soil is deficient in an essential element for plant nutrition? Soil nutrient status can be diagnosed visually or by chemical analysis. In order to determine the fertilizer needs during crop production, the producer should first determine what nutrients are deficient. Two general methods are used for this purpose: *visual examination* of plants and *chemical analysis.* Dark color of soils indicates high organic matter content. Color is used to identify the presence of certain elements in the soil. Just as plants respond to fertilizers, they also respond adversely, and sometimes readily visibly, to nutrient deficiency.

8.8.1 VISUAL EXAMINATION

Plants growing in soils that are deficient in essential nutrients may display certain telltale symptoms (called nutrient deficiency symptoms). How easy are these symptoms to spot and how reliable an indicator of nutrient deficiency are they? These visible signs are frequently ambiguous; thus, their use requires skill and experience. The general symptoms are change in growth (e.g., stunting), change in color (e.g., chlorosis, purpling, deepening of green color, a whitish appearance), and tissue death.

The symptoms manifest to different intensities, depending on the severity of the deficiency and the age of the plant part. As stated previously, certain nutrient elements such as potassium, magnesium, sodium, and sulfur are mobile and hence translocated from older tissues to younger ones. When these elements are deficient, the symptoms for these elements appear first in older leaves. On the other hand, relatively immobile elements such as iron, manganese, zinc, and boron, exhibit deficiency symptoms first in younger leaves. Further, nutrients that affect chlorophyll formation produce color abnormalities (especially yellowing) when they are deficient.

The major visual symptoms associated with nutrient deficiency are

1. *Chlorosis.* When older leaves are affected, nitrogen deficiency is suspected. In this case, the yellowing is also uniform over the entire leaf. When both young and older leaves are chlorotic, sulfur deficiency is more likely. Sometimes, the yellowing does not occur over the entire leaf but is restricted to between the veins. When only older or recently mature leaves are involved, magnesium deficiency is suspected. However, if younger leaves are affected, iron, magnesium, copper, and zinc may be implicated.
2. *Purpling.* When phosphorus is deficient in plant nutrition, leaves may appear dark green with a tint of purple (or sometimes blue or red). This purpling is caused by an accumulation of anthocyanin pigments.
3. *Necrosis.* **Necrosis,** or tissue death, may occur as patches or spots on leaves, or as marginal or leaf tip necrosis (also called leaf scorch or tip burn). This damage is also symptomatic of severe weather (e.g., drought or frost) damage.

Necrosis. Tissue death associated with discoloration and dehydration of all or some parts of plant organs.

4. *Stunted growth and other growth abnormalities.* Generally, deficiency symptoms include some growth abnormality, especially stunting of growth. Zinc deficiency causes young leaves to be severely reduced in size. Short and thick roots and the cessation of growing points of plants indicate calcium deficiency. If boron is deficient, the terminal buds usually die, while the leaves become thickened and chlorotic.

8.8.2   CONCEPTS IN SOIL TESTING

Purpose of a Soil Test

Soil test. A chemical, physical, or microbial operation that estimates a property of the soil.

A **soil test** is key to the development of an effective nutrient management plan for crop production. Its purpose is to allow the crop producer to assess the nutritional status of the soil in order to develop and implement a cost-effective and environmentally sound lime and fertilizer practice. Producers may use soil testing kits to determine the general fertility status of their field. However, commercial crop producers normally farm large acreages and usually submit soil samples to professional labs for analysis and recommendations.

Philosophies of Soil Testing

Soil fertility testing is not an exact science. There are general steps that are followed by soil testing labs: analysis, calibration, interpretation, and recommendation. However, there is no consensus on that approach to use to make recommendations from lab results. Three of the common approaches are basic cation saturation ratio, nutrient maintenance, and sufficiency level concept:

1. *Basic cation saturation ratio.* This approach assumes that a certain ideal ratio of exchangeable bases exists in the soil that will optimize plant nutrient use for optimal crop yields. The benchmark used is 60% Ca, 10% Mg, and 5% saturation of 80%), corresponding to Ca/Mg ratio of 6.5, Ca/K ratio of 13, and Mg/K ratio of 2.0. A departure from these ideal ratios indicates the deficiency of one or the other elements. The approach does not address P, S, and trace elements. The assumption is valid in soils with fairly high CEC and high natural pH. Some experts argue that cation balance is not useful for estimating the nutrient needs of most crops.
2. *Nutrient maintenance.* This is a more liberal philosophy in the sense that it assumes that a level of nutrient sufficiency to replace what was removed during previous crop production should be added, notwithstanding the soil levels of nutrients. In other words, it does not matter if nutrient levels are high and capable of supporting one or two production cycles without fertilization. Nutrients removed by plants must be replaced, nonetheless, even if it may be at the risk of toxicity to plants.
3. *Sufficiency level concept.* This conservative philosophy simply states that when a soil test calibration indicates that the nutrient levels are adequately high, there is no need to add fertilizers to the soil. This is common with many state-run testing labs. It has the greatest potential for producing the highest economic yield in the least environmentally intrusive way. An underlying principle of this philosophy is that the goal of fertilization is to fertilize the crop, not the soil.

8.8.3 SOIL SAMPLING

How useful is a soil test? The usefulness of a soil test depends on the soil material submitted for testing. The key to success is to sample the field such that the composite sample represents what it is supposed to represent. Since soil is heterogeneous, sampling for soil analysis is a more critical step than the method used for the analysis. Soils vary in two dimensions (vertically and horizontally). The depth of sampling is important to a soil test. Most testing labs recommend a depth of 30 centimeters (13.3 inches). The nutrients to be tested also affect the depth of sampling, being deeper (at least 60 centimeters, or 23.6 inches) for nitrogen, and shallower (15 to 20 centimeters, or 5.9 to 7.9 inches) for phosphorus and potassium. Even though it is recommended to sample 15 to 20 cores that are then mixed to obtain a composite sample, fewer samples are taken in practice. More samples (about 25) are needed for a phosphorus test and fewer (about 5) for potassium.

Depth of sampling

For mobile elements, it may be helpful to sample at different depths to determine their relative positions in the soil profile. Sampling depth is deeper (24–48 inches) for sugar beet and malting barley than for most crops. Once collected, soil samples should be immediately air-dried to avoid changes in NO_3^-N as a result of microbial activity.

The best time to sample a soil for analysis is as close as possible to the planting time. This way, the nutrients that the soil can supply on its own to the crop are best estimated. An established producer may sample his or her field every 2 to 3 years. Samples should be obtained from the areas between rows, making sure to avoid sampling soils from previous fertilizer bands. Unique spots should also be avoided.

Another key consideration in sampling is uniformity. Areas that differ in topography (flat, slope, valley), soil texture (clay, sandy), soil structure, depth, color, productivity, and previous management (e.g., liming, fertilizing), should be sampled separately. The land user should delineate these areas prior to sampling. Factors to consider in sampling are depth, number of samples, frequency, time, location, and uniformity.

Time to sample

Whereas a soil may be sampled at any time of the year for the determination of pH, salt content, Zn, and P levels, sampling when the soil is frozen may give higher than actual K value. Similarly, because NO_3^-N, S, and Cl are mobile, they are best tested by sampling the soil in fall or spring. In addition to the method of composite soil sampling described so far, there are other sophisticated procedures that provide more accurate assessment of total nutrient levels that are more compatible with the needs of precision or site-specific farming. These other methods, grid sampling and directed sampling, may involve the use of GPS technology or prior knowledge about the field. Satellite imagery and aerial photography, as well as data from yield monitoring, may be used to guide sampling for within-field nutrient levels determination.

8.8.4 SOIL TEST COMPONENTS AND THEIR IMPORTANCE

Soil analysis labs differ in the specific tests they conduct as standard and the methods used, as well as the units in which results are reported. Some of the important soil test components are as follows:

1. *Humic matter content.* The importance of soil organic matter is discussed elsewhere in this book. The relatively stable fraction of organic matter is called the humic matter (humus), which comprises 60 to 80% of the actual organic

matter status of the soil. A conversion formula is used to convert humic matter to organic matter percent. The soil component is used to provide guidance for adjusting lime, P, and micronutrient recommendations and for safe use of pesticides. Manufacturers of pesticides may restrict application to soil with a minimum humic matter percentage to prevent groundwater contamination. On the other hand, a high humic matter may prevent the effectiveness of certain pesticides because the latter becomes trapped in the former through strong binding.

2. *Weight per volume.* Measured in gm/cm^3, this determination is used for assigning the soil testing class of the soil. A value of 1.5 or higher indicates a soil that is high in sand, while silt and clay loams have a weight/volume ratio of about 1.0. Soils that are high in organism matter have values of about 0.4. Soil textural class helps to make proper recommendations for liming and application of fertilizers, especially regarding P$_2$O$_5$, Cu, and Zn rates.

3. *Cation exchange capacity.* Soils that are rich in colloids (organic matter, clay) have a high CEC (hold more cations, especially, Ca^{2+}, K$^+$, Mg^{2+}). Soils with low CEC are susceptible to leaching. Split application of fertilizers is recommended to minimize leaching losses.

4. *Base saturation percent.* Calcium is usually the most dominant cation in the CEC determination. This ion is also the principal component of aglime. When a crop production enterprise requires high availability of CA, the producer may use gypsum (CaSO$_4 \cdot$ 2H$_2$O).

5. *Mg saturation.* This determination of the proportion of total CEC represented by Mg is used to determine the type of lime or the need to include Mg in fertilizers. If a test indicates a deficiency of Mg (extractable Mg CEC of < 0.5) and liming is recommended, dolomitic lime should be used.

6. *pH.* A soil pH test is used mainly to determine the lime requirements.

Nutrients Reported as Index Values

For trace elements and sometimes for P and K, the index system of reporting soil nutrient levels is used for ease of interpretation of the results. Specific critical values are not reported for these elements. Rather, a range is provided to indicate the likelihood of the crop's response to supplemental application of an element. A soil test may be reported as follows: a range of 0–25 = low, 26–50 = medium, 51–1,000 = high, and 1,000$^+$ = very high presence of the element. A rating of low normally means a crop would benefit from a supplemental application of the element. A medium rating indicates a likelihood of response for P and K. However, crops would seldom respond to added nutrients if the soil test rating were high or very high.

Soluble Salt Index

The soil accumulates soluble salts from fertilizers, manures, and irrigation water. A moderate amount of soluble salts is desirable. The soluble salt index indicates the amount of fertilizer elements that are soluble in the soil. An excessive amount of soluble salts is injurious to sensitive plants. The injury level depends on the soil type, the soil moisture content, and the species.

8.8.5 CALIBRATION OF A SOIL TEST

The ultimate goal of soil testing is to provide actual fertilizer recommendations from the results and interpretations made (Figure 8–2). As a service to growers, state extension services and soil analysis labs usually develop practical ratings to indicate the likelihood

Agricultural Testing Services

The Samuel Roberts Noble Foundation, Inc.

2510 Sam Noble Parkway
Ardmore, Oklahoma 73402

P.O. Box 2180
Telephone (580) 223-5810

Field Name:	10 Very S Field			

Field Name: 10 Very S Field

Lab Number: 55572 Date: 8/18/03

County: OK-Jefferson

Howard Sheep & Cattle Ranch
Route 2 Box 129A
Waurika, OK 73573

Intended Crop: Wheat

Yield Goal:

Test Run: Test A

Sample Depth	pH	Buffer Index	Pounds per Acre Extractable Nutrient						Soluble Salts (ppm)	Organic Matter %	Cation Exchange Capacity Meg/100 gms	Parts Per Million Extractable Nutrient				
			Nitrogen NO3	Phosphorus P	Potassium K	Calcium Ca	Magnesium Mg	Sodium Na	Sulfur SO4				Iron Fe	Zinc Zn	Manganese Mn	Copper Cu	Boron B
0-6" TOP	4.30 Strongly Acid	6.30	80	146 Sufficient	398 Sufficient	706 Adequate	162 Adequate	32 Normal			1.5	10.0					

Fertilizer Recommendations
(Pounds per Acre Actual Nutrient)

N Nitrogen	P2O5 Phosphorus	K2O Potassium
70	0	0

Tons Lime per Acre   0.0
100% ECCE

Recommendations and Comments:

pH is below the optimum for wheat. You may wish to re-sample this field this year or next to make sure you need to lime. If you want to go ahead and lime, you would need to apply 2.0 ton ECCE lime/ac to raise pH to 6.4.

Phosphorus is adequate.

Potassium is adequate.

Nitrogen is based on yield goal. It takes about 60 lb N/ac to produce one ton of winter forage minus soil carryover, so if you want to produce 2 ton/ac forage, you need to apply 40 lb N/ac. If you plan to combine wheat as well, you should add another 30 lb/ac Nitrogen. I recommend splitting this application with some at planting and some as a spring topdress.

Laboratory Analysis by Ward Laboratory Inc., Kearney, NE
Recommendations by Noble Foundation Specialist

Wade Thomason
Wade Thomason

FIGURE 8–2   Soil test results of a farmer's field. (Source: Courtesy of Dennis Howard)

of crop response with reference to specific soil test outcomes. These ratings are not only crop-specific but also cultivar-specific, and hence the labs must constantly calibrate new cultivars as they are produced by plant breeders. Calibrations are based on replicated field trials over time and locations.

8.8.6   NUTRIENT AVAILABILITY

Plant nutrient availability depends on the pH of the soil solution (Figure 7–7). As pH decreases (increasing acidity), elements such as Mn, Zn, Cu, and Fe become more available. When pH decreases below 5.5., toxicity of Mn, Zn, and Al start to be a problem in crop production. On the other hand, increasing soil acidity (low pH) inhibits the availability of N, K, Ca, Mg, and S. The availability of P and B is decreased at both very low and very high pH, being most available at between pH 5.5 and 7.0.

8.8.7   FACTORS AFFECTING ACCURACY OF SOIL TEST

A soil test is conducted for the purpose of recommending appropriate fertilizer application. A good recommendation depends on several factors.

1. *Crop producer.* What can the producer do to improve the usefulness of a soil test? The role of the producer is to provide the following information and materials to the testing laboratory:
   a. Accurate and representative soil samples
   b. Accurate cropping history of the land
   c. Accurate projection of expected yield; a recommendation is based on the yield goals of the producer
2. *Laboratory analysis.* The technicians at the laboratory should have the skills and expertise to conduct a correct soil analysis. Further, the laboratory should have an up-to-date good correlation to field plot data. Without correlation of results to crop responses to different fertilizer levels in field plot trials, the laboratory results are meaningless. The laboratory should continually test new cultivars and fertilizers as well as new crop improvement systems.
3. *Evaluator.* The scientists who evaluate soil test results should be well trained in the science and art of evaluation and should be familiar with the production area and the plants for which the predictions or recommendations would be applied. The evaluator should know the plant and its nutritional needs, the rainfall regime of the area, and other natural nutrient cycling activities in the soils of the area.

   After a recommendation has been made, the successful implementation depends on several factors:
   a. *The producer.* Crop production depends to a large extent on the producer's ability as a decision maker. The soil test recommendation should be properly implemented. The evaluators usually recommend that the crop producer amend the recommendations according to his or her experience and management skills, rather than following them to the letter. In addition, the producer should select the best planting materials, plant on time and at proper density, control pests, and implement other appropriate management practices.
   b. *Weather.* Crop productivity is weather-dependent. Inclement weather, such as prolonged drought, excessive rain, and hail, is destructive to crops and will erode the benefits of implementing the soil test recommendations.

8.8.8 TISSUE TEST

An indication that nutrients are available in adequate amounts in the soil does not necessarily mean that the plants will absorb nutrients. Analyzing plant tissue (*tissue test)* is the only way to ascertain that the soil nutrients are available to plants. However, plant analysis has shortcomings:

1. Plants cannot be tested until they are growing.
2. The purpose of a soil test is to obtain information for application to crop production. Unfortunately, plant analysis results are not available for preplanting decision making.
3. Sometimes, when plants show deficiency symptoms, it is too late to salvage the crop.

Plant analysis, however, is useful for predicting fertilizer needs of perennial crops such as sugarcane, fruit trees, permanent pastures, and forest trees that remain in the field for long periods in the growing season.

Plant tissue analysis is performed on samples obtained with care, taking into account the physiological status, age, health, and representativeness of the plant population. Differences in results have been obtained between different portions of even a single leaf. Like soil testing, the laboratory performing the tests should have a database of results from analysis using tissue from the same plant parts sampled.

Plant Tissue Sampling

The time of tissue sampling varies with the crop. For example, tobacco is best sampled before bloom, taking the uppermost leaf. In corn, the ear leaf should be sampled when about 50% of plants are at the stage where the silk is just beginning to appear. Soybeans and other legumes should be sampled when 10% of the plants are in bloom. The uppermost leaf of soybean is picked, while in the case of leguminous forage species (e.g., alfalfa, red clover), the upper one-third of the plant is picked. In wheat and barley, the flag leaf is sampled when about 25% of the plants have headed. The top 6 inches of forage grasses constitute the best sample to take for plant tissue analysis.

The parts of the plant sampled for effectiveness depends on the species. For crops such as cassava, the bark of the plant is most reliable for analysis, while the whole seedling is most effective for annual crops such as corn and beans.

These plant samples should be washed to remove contaminants such as soil and chemicals sprayed by the producer or accumulated from pollutants in the environment. The producer should provide the name of the plant cultivar, the planting density, and other meteorological pieces of information.

In interpreting plant analysis, there is a threshold (critical) nutrient level against which the results are compared. An element is declared deficient when it is present in the tissue at a level lower than the threshold value. Unfortunately, the critical nutrient levels vary readily with growth environment (e.g., pH, moisture, and soil nutrients), cultural conditions, plant health, and other factors. It is difficult to prescribe threshold values, thus making interpretation of plant tissue analyses problematic.

8.9: ANTAGONISM AND SYNERGISM

Plant nutrients, especially micronutrients or trace elements, interact in the soil either to enhance or to hinder their absorption by plants. This emphasizes the need for nutrient balance, so that no particular one is present in excess amounts. Nutrient interaction may be *antagonistic* (negative) or *synergistic* (positive). Antagonism is when one substance reduces the detrimental effects of a second substance. Synergism, on the other hand, is when two factors interact to cause a greater effect than the sum of the two substances separately. Antagonistic effects may be used to reduce the toxicities of certain trace elements in the soil. For example, adding sulfur to calcareous soils containing toxic quantities of Mo may reduce the availability of Mo and thereby avoid toxicity of the element. Similarly, adding Mn or Zn reduces the availability of Fe. Examples of synergistic effects include the increase in Mo utilization as a result of increasing P in the soil and the enhanced utilization of Zn by increasing N application.

8.10: LIMING

High soil acidity is corrected (decreased) by liming to raise soil pH. However, the main goal of liming is to neutralize the toxic elements that occur under high acidity (especially Al and Mn) while improving the availability of desired elements. The effect of liming is to add hydroxide (OH^-) ions to the soil solution to decrease the solubility of Al^{3+}, Mn^{2+}, Fe^{3+}, Zn^{2+}, and Cu^{2+} ions. Liming causes these ions to precipitate out of the soil solution while supplying basic elements (Ca, Mg, depending on the type of liming material). Liming also increases the availability of P, Mo, and B and promotes microbial growth, development, and processes (e.g., biological nitrogen fixation, nitrification). Overliming is undesirable, since it can result in the deficiency of some elements—for example, Zn, Fe, Mn, and Ca.

8.10.1 LIMING REACTION

Liming materials contain mainly carbonates of Ca and Mg that are generally sparingly soluble. The oxides and hydroxides of Ca are more soluble in water. Soil acidity is neutralized as follows:

1. $CaCO_3 + H_2O \rightarrow Ca^{2+} + HCO^{3-} + OH^-$
   $CaO + H_2O \rightarrow Ca^{2+} + 2OH^-$
2. $H^+ + OH^- \rightarrow H_2O$

8.10.2 SOURCES OF LIMING MATERIALS

Limestone varies in purity. Limestone consists of sedimentary rocks that are rich in calcite ($CaCO_3$) or dolomite [$CaMg(CO_3)_2$]. When the rock is relatively pure in calcite, it is called *calcitic limestone*. Similarly, when magnesium predominates in the rock, it is called *dolomitic limestone*. The CEC percentage (or total neutralizing power) is a measure of the acid neutralizing value of the liming materials. This value is calculated by using pure calcite ($CaCO_3$) as a standard. Values greater than 100 (e.g., 110) indicate that the source is higher (10%) than pure calcite in neutralizing power (Table 8–2).

Slaked or hydrated lime [$Ca(OH)_2$] is difficult to handle because it is very reactive and caustic. It produces quick results and may be applied when there is a need to rapidly raise soil pH.

8.10.3 CALCULATING LIME REQUIREMENT

Soil pH measures the relative intensity or the ratio of the base cation saturation of the exchange complex. It does not provide information on the total amount of acidity that must be neutralized. To be able to calculate the amount of lime to apply, the pH requirement of the crop, the current actual pH of the untreated soil, and the CEC should be known (i.e., the soluble and exchangeable acidity). The amount of lime required depends on (1) the change in pH required, (2) the CEC, (3) the chemical composition of the liming material, and (4) the fineness of the lime. Calculations and recommendations are best done by the soil-testing lab than by producers because of the chemistry involved. However, as previously indicated, field testing kits may be used by producers for estimating soil nutritional status.

Particle size is determined by passing the ground material through a series of mesh screens. The sieve number is the number of holes per square inch (e.g., a 20 mesh has $20 \times 20 = 400$ holes). States have laws regarding the fineness of liming materials. The beneficial effects of liming are achieved only when the material is in contact with the soil.

Table 8–2 Sources of Lime Materials, Their Chemical Formulas, and Their Calcium Carbonate Neutralizing Equivalents (CCE)

Source	Formula	CCE%
Calcitic limestone	$CaCO_3$	85–100
Dolomitic lime	$Ca/Mg(CO_3)_2$	95–108
Hydrated lime	$Ca(OH)_2$	110–135
Burnt lime	CaO	150–175
Marl	$CaCO_3$	50–90
Basic slag	$CaSiO_3, CaO$	50–70

Liming materials are relatively insoluble in water and immobile in the soil. When applied, they affect only the top 2 to 3 inches. It is desirable to mix the liming material into the soil in order to have soil pH modified in the root zone.

The amount of lime needed to increase the pH of a sandy soil by a certain degree is generally less than the amount needed to cause the same change in a clay soil. This is because the sandy soil has less residual or reserve acidy (buffering capacity).

8.11: CONCEPT OF REALISTIC YIELD

The crop producer should manage his or her production enterprise to obtain an optimum economic yield rather than maximizing yield. Optimizing economic yield implies that a producer is careful not to waste resources (i.e., uses cost-effective management practices). As management and inputs increase, the cost of production also increases. At a certain critical point, profits are maximized. Adding inputs after this point only increases production cost without increasing profits, since the added yield gain is not cost-effective. A producer can eliminate unnecessary costs from production inputs by practices such as periodic soil testing to determine if added fertilizers would be economically beneficial.

Application of fertilizers should be based on the realistic yield expectation (RYE) of the producer's field. Determining a RYE is not an exact science. Averaging the crop yield on the producer's farm over 5 years, and using the yield from the three most productive years, is generally the common method of estimating RYE for agronomic crops. The challenge is finding such historic records for the farm. Once a RYE for a farm has been determined, the appropriate rate of nitrogen to be applied is calculated by multiplying the RYE by the rate of nitrogen suggested by soil analysts from the soil test.

RYE is influenced by a host of soil-related factors, including depth of subsoil, soil texture and structure, organic matter content, permeability, drainage, slope, and local climate. The amount of nutrients needed to optimize crop yield and economic return while minimizing undesirable effects on the environment is called the agronomic rate. Excessive amounts of nutrients can adversely impact both yield and the environment, while insufficient amounts can cause yield reducing nutrient deficiency symptoms in plants.

Commonly, agronomic rates are determined by the RYE of a site and the amount of nitrogen (often the priority element—the nutrient element most likely to have adverse effects on crop yield or the environment) and the amount of nitrogen required to produce a unit of a specific crop. For example, if it takes 1.0 to 1.20 lb of nitrogen to produce a bushel of corn, and the RYE is determined to be 100 bushels per acre, the agronomic rate is between 100 and 120 lb/acre of nitrogen (obtained as 100 bu/acre × 1.0 lb/acre = 100 lb/acre, and 100 bu/acre × 1.2 lb/acre = 120 lb/acre).

8.12: SOIL NUTRIENTS CAN BE SUPPLEMENTED THROUGH NATURAL PROCESSES AND ARTIFICIAL METHODS

8.12.1 CERTAIN ESSENTIAL NUTRIENTS MAINTAIN NATURAL CYCLES

Through a combination of physical, chemical, and microbial interactions, certain essential nutrients maintain natural cycles. The major cycles include the nitrogen, phosphorus,

FIGURE 8–3 Nutrient cycling in nature is characterized by changes, additions, and losses in nutrients. Enrichment of the soil nutrients comes through application of fertilizers as well as naturally through microbial decomposition of organic matter and weathering of minerals. Depletion of soil nutrients may be wasteful and occurs through avenues such as leaching, soil erosion, and microbial activities. Crop production is a means of economic depletion of soil nutrients.

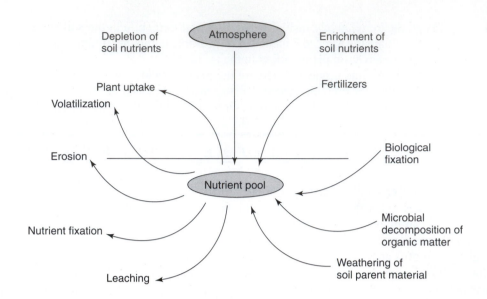

Nutrient cycling. The absorption of inorganic forms of elements, their conversion into organic forms, and their subsequent conversion back to inorganic forms through microbial decomposition.

and potassium cycles. These elements change state from organic to inorganic, existing for varying periods of time in various nutrient pools. Natural **nutrient cycling** plays a significant role in the sustainability of the ecosystem. The cycles differ in complexity. Nutrient cycling consists of processes that can be categorized into three: *addition, fixation,* and *loss* (Figure 8–3).

In nutrient cycling, certain factors increase the status of various nutrient elements in the soil. Some of these are natural processes, while others are human-made. Certain atmospheric phenomena (e.g., lightning) cause chemical reactions to occur. In the nitrogen cycle, dinitrogen gas is converted to NO_4^+, while fires produce oxides of nitrogen such as NO_2. Similarly, atmospheric sulfur in various compounds [SO_2, SO_3, H_2S, $(CH_3)_2S$, etc.] are washed down to the soil as acid rain. Atmospheric sources of phosphorus are minimal. A large amount of dinitrogen gas is fixed by microbial activity into the soil (e.g., symbiotic and non-symbiotic nitrogen fixation).

These natural processes are untargeted in the sense that they occur in both agricultural and non-agricultural areas. Crop producers, however, deliberately add large amounts of selected nutrients in desired amounts to specific cropping lands to boost crop productivity. These are in the forms of organic manure or industrial chemicals (inorganic fertilizers). Nutrients removed by plants (immobilized) are reconverted by microbial actions into inorganic nutrients by mineralization, when plants are deliberately incorporated into soil.

8.12.2 SOIL NUTRIENTS ARE DEPLETED IN A VARIETY OF WAYS SOME BENEFICIAL, OTHERS WASTEFUL

Nutrient cycles are also leaky, and hence the nutrients that enter the soil can be lost. In crop production, the producer should adopt practices that minimize the losses. Losses occur through the following main avenues: *soil erosion, volatilization of gases*, *leaching, removal by crops,* and *fixation.* These mechanisms vary in importance from one nutrient element to another.

1. Soil erosion is a process of removal and translocation of soil. Soil is the medium that supports agricultural crop production. Soil erosion depletes the soil, especially of nitrogen and phosphorus. The nutrients may end up in bodies of water and cause eutrophication.

2. Volatilization entails the loss of nutrients in gaseous forms into the atmosphere. The most affected elements are nitrogen and sulfur. The forms of these nutrients, such as N_2O, NO_2, and SO_2, occur when plant material is burned. Some nitrogen can be lost directly from applied fertilizer under improper environment. For example, ammonia gas applied as a source of nitrogen can be lost if the soil is dry or alkaline. Soil microbes can also cause loss of nitrogen through the process of denitrification that involves the reduction of NO_3 to N_2 and N_2O. Sulfur is also volatilized through reduction to SO_4 or H_2S.

3. Leaching is the loss of nutrients in soluble form. The mechanism is important in the loss of nitrogen, potassium, and sulfur in the nitrate or sulfate forms.

4. Crop use of nutrients through absorption from the soil is the fourth mechanism. Crops are cultivated to exploit the soil nutrients for economic yield. Crops vary in amounts of nutrients removed. Alfalfa and corn are heavy feeders of all macronutrients, while rice and cotton remove small amounts of nutrients.

5. Ammonium ions may become highly bonded to clay mineral lattices such that they are unavailable to plants *(ammonium fixation)*. Similarly, soluble phosphate ions ($H_2PO_4^-$) may react with other soil chemicals to become insoluble *(phosphate fixation)*. Some of these fixed nutrients can become available at the appropriate pH.

8.12.3 FERTILIZER TERMINOLOGY

If a fertilizer recommendation is made following a soil test, what does a producer use to supply the needed supplement? Commercial fertilizers are formulated as either liquids or solids. The nutrient composition of a fertilizer is displayed on the container or bag in a certain standard format. The label indicates the *fertilizer grade* and the minimum guaranteed percentage of nitrogen, phosphorus, and potassium (NPK), in that order. The remainder of the content consists of an inert material called a *filler* (used to bring the formulation up to the desired weight or volume) that is used as a carrier of the active ingredients. The fertilizer grade information is presented as

1. Percent total nitrogen (measured as elemental nitrogen, N)
2. Percent available phosphorus (measured as phosphorus that is soluble in ammonium citrate solution and calculated in practice as phosphorus pentoxide, P_2O_5)
3. Percent water-soluble potassium (calculated as potassium oxide, K_2O)

A fertilizer grade of 10-40-5 means the material contains 10% nitrogen, 40% phosphorus, and 5% potassium. Why not revise the convention to say exactly what is in the bag? Efforts have been made to change the nomenclature to a simplified elemental format of N-P-K, instead of N-P_2O_5-K_2O. It is believed that resistance to change by the fertilizer industry is because an elemental format will produce smaller equivalent numbers. This may seem to indicate that the fertilizer has a lower grade than it really does. For example, a conventional fertilizer grade of 0-45-0 is equivalent to 0-19-0 in elemental form (P). A fertilizer may also be described according to the concentration of the nutrients present. For example, a fertilizer grade of 10-10-10 is described as *low analysis* as compared with a grade of 30-25-25 (high analysis). In applying fertilizer, the producer will need more bags of a low analysis fertilizer than a high analysis one to provide the appropriate rate.

Certain fertilizers leave residual acidity in the soil when applied. Most nitrogen fertilizers, all ammonium materials, and many organic nitrogen fertilizers are acid forming. On the other hand, fertilizers such as urea, diammonium phosphate, and anhydrous ammonia, initially produce soil alkalinity. All potassium fertilizers, except potassium nitrate, produce neutral residual soil reaction.

8.12.4 FERTILIZER SOURCES AND THEIR USE

Nitrogen Fertilizers

The major fertilizer sources of nitrogen and their characteristics are summarized in Table 8–3.

Anhydrous ammonia has the highest available nitrogen per unit source (82%). It is applied by special equipment used to inject it into the soil. To reduce losses (by volatilization), it should be sealed into the soil immediately after application. The equipment usually has an attachment for this purpose. Sandy and dry soils are susceptible to high losses. This source of nitrogen is applied between rows. Urea is also susceptible to volatilization losses and hence is not suitable for surface applications (e.g., in pastures, no-till systems). Ammonium nitrate is easy to apply but can cause fertilizer burn. UAN solutions are commonly applied to winter wheat. These may be injected into the soil in no-till production systems or surface-applied with the aid of streamer bars to reduce fertilizer burn. Ammonium sulfate is recommended where sulfur deficiency occurs along with low nitrogen.

Phosphorus

The major commercial sources of phosphorus and their characteristics are summarized in Table 8–4. Diammonium phosphate (DAP) produces alkalinity in the soil and hence is suitable for application when soil acidity needs to be reduced. It is best applied in concentrated bands. Triple superphosphate (TSP), on the other hand, produces acid residues and hence is suitable where soil alkalinity needs to be reduced. It is more readily available than single superphosphate (or ordinary superphosphate) (SSP). SSP contains gyp-

---

Table 8–3 Commonly Used Nitrogen Fertilizer Materials

Fertilizer	Formula/Symbol	Percent by Weight Nitrogen
Anhydrous ammonia	NH_3	82
Urea	$CO(NH_2)_2$	45
Ammonium nitrate	NH_4NO_3	33
Sulfur-coated urea		30–40 (plus 13–16 sulfur)
Ureaformaldehyde	UF	30–40
UAN solution		30
Isobutylidene diurea	IBDU	30
Ammonium sulfate	$(NH_4)_2SO_4$	21 (plus 24 sulfur)
Sodium nitrate	$NaNO_3$	16
Potassium nitrate	KNO_3	13 (plus 36 phosphorus)

Table 8–4 Commonly Used Phosphorus Fertilizer Materials

Fertilizer	Formula	Percent by Weight P
Monoammonium phosphate	$NH_4H_2PO_4$	21–23 (plus 11 N)
Diammonium phosphate	$(NH_4)_2HPO_4$	20–23 (plus 18–21 N)
Triple superphosphate		19–22 (plus 1–3 S)
Phosphate rock	$Ca_3(PO_4)_2 \cdot CaX$	8–18 (plus 30 Ca)
Single superphosphate		7–9 (plus 11 S, 20 Mg)
Bone meal		

Table 8–5 Commonly Used Potassium Fertilizer Materials

Fertilizer	Fertilizer	Percent by Weight K
Potassium chloride (muriate of potash)	KCl	56
Potassium sulfate	K_2SO_4	42
Magnesium sulfate	$MgSO_4$	18 (plus 9 Mg, 18 S)
Potassium nitrate	KNO_3	37 (plus 11 N)

sum. Ammonium polyphosphate is used primarily in fertigation and for weed-and-feed and seed placement applications. It can be surface applied without incorporation into the soil. Surface application of phosphorus is the least efficient method. Hence, it is best to build up soil P before initiating a no-till system.

Even though nitrates are more soluble than phosphates, eutrophication is attributed to P, the element that is often the limiting factor in surface water because of its relatively high insolubility. Hence, an entry of relatively small amounts of P into the body of water is enough to stimulate accelerated growth of algae and other water plants, leading to reduced oxygen.

Potassium

Some of the most commonly used commercial sources of K are presented in Table 8–5. KCl is a commonly used source of K. The sources presented in the table usually have a high salt index and hence should not be applied close to crop seeds. Further, each of these sources provides more than one of the essential plant nutrients. Crops that are responsive to K fertility include soybean, alfalfa, and corn. The chlorine in KCl makes it unsuitable for fertilizing tobacco because it leads to an elevated Cl level in the leaves and poor product quality.

Secondary Nutrients

Numerous materials are used as sources of secondary nutrients (Ca, Mg, S) in crop production. Calcium and magnesium are relatively low in mobility in the soil. When fertilizers such as SSP are applied, they supply significant amounts of S. Fertilizer materials containing sulfate ($SO_4{}^{2-}$) do not impact soil acidity in the short run, but over time the pH may decrease with use. Elemental sulfur needs to be converted to sulfate through biological oxidation involving the bacterium *Thiobacillus.*

8.12.5 METHODS OF FERTILIZER APPLICATION

How is fertilizer applied in the field? There are several techniques of applying fertilizers in crop production. The common methods are *starter, broadcast, deep banding, split, side dress, top dress, injection, fertigation,* and *foliar.*

Starter Application

Also called *pop-up application, starter application* is the application of fertilizer at the time of seeding. Depending upon the implement used, the fertilizer may be dribbled in the soil near the seed or placed in a band near it. This application is generally at a low rate and is more suited to phosphorus and potassium than nitrogen. It is desirable for plants that grow rapidly, or when a soil test indicates a very low nutritional level.

Some species are sensitive to salt damage and hence no more than 10 kg/ha nitrogen and potassium should be applied, and contact with seed should be avoided. Phosphorus is

Following a soil test, recommendations are made as to the type and amount of fertilizer the producer may apply. An amount of a certain nutrient may be prescribed. However, the weight of fertilizer applied depends on the source, since fertilizers come in all kinds of grades. The three common calculations involving dry fertilizer formulations are nutrient percentage, the amount or weight of source (commercial) fertilizer to apply, and the amount (weight) of component materials to use in preparing a bulk of mixed fertilizer.

a. Nutrient percentage

Problem: What is the percentage of nitrogen in the fertilizer urea?

Solution: Urea has a formula of $(NH_2)_2CO$ and a molecular weight of 60.056 g
Molecular weight of nitrogen (N_2) = 28.014 g
Percentage of nitrogen = $[28.014/60.056] \times 100$ = 46.6, or 46%
approximately (round down to nearest whole number)

b. Simple fertilizer mixture

Problem: Given ammonium nitrate (34-0-0) and treble superphosphate (0-45-0), prepare 1,000 kg of fertilizer of grade 15-10-0.

Solution: Final mixture will contain 150 kg of nitrogen (i.e., 15% of 1,000 kg).
It will also contain 100 kg P_2O_5 (i.e., 10% of 1,000 kg).
Amount of ammonium nitrate needed (note: it contains 34 kg of N per 100 kg of ammonium nitrate):

$[100 \times 150]/34$
= 441 kg of 34-0-0

Similarly, for phosphorus (contains 45 kg of P_2O_5 per 100 kg)

$[100 \times 100]/45$
= 222 kg of 0-45-0

less mobile and less likely to cause salt damage. As the rates of application of fertilizer increase, the distance between seed and fertilizer should be increased.

Broadcast Application

In broadcast application, the fertilizer is spread over the soil surface in no particular pattern, making this method the one with the lowest *fertilizer efficiency* (the percentage of fertilizer added that is actually used by the plant). It may be left on the soil surface or later incorporated by, for example, disking. This method is desirable for paddy rice fields and established pastures. It is easy to apply. Further, large amounts (a high rate) of fertilizer may be applied by this method. Certain fertilizers may volatilize unless disked. Watering after application also helps to move nutrients down more quickly to the root zone.

Sometimes, as in pastures, broadcasting is the only method for applying fertilizer. Broadcast efficiency is increased if the fertilizer is incorporated into the soil or if the soil remains moist (e.g., from shading).

Deep Band Application

The goal of deep band application is to place most of the fertilizer where plant roots have most access. Phosphorus fertilizers have low mobility and hence are best placed near the roots. Further, by concentrating phosphorus this way, the danger of fixation is

Total nutrients = 441 + 222 = 663 kg (leaving 337 balance of the desired 1,000 kg)
The balance is satisfied by adding a filler (inert material) or lime. The procedure is the same for compounding a mixture of three components (i.e., N-P-K).

What if calculated proportions of nutrient components add up to more than total desired weight? In this example, what if the two amounts exceeded 1,000 kg? The component amounts cannot exceed the total desired weight. If that happens, the desired grade should be lowered or a source with higher analysis (e.g., for nitrogen, use urea with 46% N) should be used.

c. Amount (weight) of sources (fertilizers) to apply

Problem: It has been recommended that a producer apply 80 kg of nitrogen and 40 kg of phosphorus per acre to the field. How much of ammonium nitrate and treble superphosphate should be applied to achieve the recommended rate?

Solution: Ammonium nitrate contains 34% N; treble superphosphate contains 45% P_2O_5.
From example (b), the amount of source to be added to provide 80 kg of N

$$= [80 \times 100]/34$$
$$= 235 \text{ kg of } 34\text{-}0\text{-}0 \text{ per hectare}$$

Similarly, for 40 kg of P_2O_5

$$= [40 \times 100]/45$$
$$= 88.9 \text{ kg of } 0\text{-}45\text{-}0 \text{ kg per hectare}$$

Total of mixture = 235 + 89 = 324 kg per hectare

To convert from kg per hectare to pounds per acre, multiply by 0.89.

minimized. Anhydrous ammonia gas is applied by this technique, using equipment under high pressure.

Deep banding is expensive to conduct because of the specialized stronger equipment and the energy needed to apply at lower soil depth.

Split Application

Split application is the technique of dividing the recommended fertilizer rate into several portions and applying at various stages in the growth and development of the crop. Split application may be accomplished by more than one technique of fertilizer application (e.g., starter application, followed by side dressing). The advantage of this technique is that waste of especially highly mobile nitrogen fertilizers is reduced. By timing the application, the producer is able to provide the appropriate amount of nutrients at times when the plant needs it the most. Split application helps to control vegetative growth of the plant. Fertilizer efficiency is increased especially for nitrogen. However, multiple applications add additional cost to crop production operations.

Side Dressing

Side dressing is undertaken after the crop is growing in the field. Usually, it is done as part of a split application. The fertilizer may be broadcast to the soil surface or applied as a shallow band application in rows along the row of plants or around individual plants. This method is not very effective for phosphorus and potassium fertilizers. This method is applied to row crops.

Top Dressing

This post emergence application is usually applied to small grains and pastures.

Fertigation

Liquid fertilizer may be applied through fertigation. It is very commonly used in greenhouse production. It is most effective where soils have low nutrient retention and for mobile nutrients such as nitrates and sulfates. It allows a quick and convenient way of applying fertilizer, especially where nutrients are needed immediately.

Foliar Application

Foliar application is the application of fertilizers to the leaf as a liquid spray. The nutrients are absorbed through stomata. The technique is adapted to very low rates of fertilizer application. Further, it is commonly used for applying micronutrients. For example, iron chelates are readily mobilized in the soil and thus are best applied to leaves.

8.13: SOIL FERTILITY VERSUS SOIL PRODUCTIVITY

Soil fertility is the capacity of a soil to provide the essential plant nutrients in adequate amounts and proportions for plant growth and development. The productivity of soil, on the other hand, is the capacity of a fertile soil to perform under a set of environmental conditions. A fertile soil is not necessarily productive, but a productive soil must be fertile. This is so because a soil can have all the essential nutrients in the right amounts and proportions but is useless for crop production unless there is adequate moisture and appropriate temperature as well as other essential growth factors. For practical purposes, two kinds of fertility may be identified—general fertility and enterprise-specific fertility. Some soils are naturally fertile and suitable for producing a wide variety of crops. Other soils lack such all-around fertility but are nonetheless ideal for specific crop production enterprises.

8.14: NOT ALL APPLIED FERTILIZER IS USED BY PLANTS

How much of the fertilizer applied to crops is actually used by plants? The percentage of added fertilizer that is actually used by the plants is called the *fertilizer efficiency*. The efficiency differs for various fertilizer elements. It is estimated that about 30 to 70% of added nitrogen is used, while 5 to 30% of phosphorus and 50 to 80% of potassium are utilized by plants. Fertilizer efficiency is influenced by several factors, including the operator's technique of application, the weather conditions, the soil type, and the crop. Plants differ in the ionic form of nutrients preferred. For example, legumes are known to prefer divalent cations such as Ca^{2+}, while grasses prefer monovalent ones such as K^+. Soil impediments such as rocks and hard pans restrict root growth and development and subsequently nutrient absorption.

Fields are seldom homogeneous, or uniform. They vary in their physical and chemical characteristics. A farmer growing a large field is likely to have several soil areas in the field that react differently with respect to their ability to grow plants under a particular management. In such a case, it is not prudent to treat them all alike and apply the same rate of fertilizer to the whole area. Certain portions may be overfertilized and wasteful in terms of resources, whereas other areas would be underfertilized.

Precision agriculture (or *high-tech agriculture, spatial variability management*) is a *site-specific management* approach in modern crop production for applying agrochemicals to the field in an economic and environmentally sound fashion.

Precision farming entails the collection and management of a wide variety of agronomic information with the purpose of identifying and meeting the *real* needs of variable parts of a field, through supplying what is *actually* needed rather than the *average* needs of the whole field, as is the case in conventional management. The key objective of a variable rate technology (VRT) strategy is to develop an accurate application map for the field to be used as the blueprint for determining the level and location of agronomic inputs applied to the field. This technology is applicable to a wide variety of agronomic production operations, including fertilizer application, liming, seeding rate, pesticide application, irrigation, and tillage operation. To be effective, each application requires a clearly developed and accurate guide.

Conventional ways of variable application of agronomic inputs are based on operator intuition, soil survey maps, soil data from sparsely spaced grid samples, and visual observation of variations in the field by the operator while traversing it. Grid soil sampling is unguided and has serious limitations. Conventional applications are not usually automated and often subjective. Modern strategies used in site-specific applications involve dividing the field into smaller, homogeneous *management zones*. A precision farming management zone is a subregion of a field that expresses a functionally homogenous combination of yield-limiting factors for which a single rate of a specific cropping input is appropriate. A good management zone strategy will maximize economic return by optimizing rates of yield-limiting inputs and controlling the adverse effects of weeds and other crop pests.

Accurate delineation and classification of spatial variability should take into account four categories of site characteristics: (1) quantitative and stable (e.g., relief, pH, and soil organic matter), (2) quantitative and dynamic (e.g., crop yield and weed density and distribution), (3) qualitative and stable (e.g., soil color, immobile nutrients like phosphorus and potassium, and soil drainage), and (4) intuitive or historical (e.g., grower past experience of field characteristics, history of cultural practices, soil tilth and quality, and crop rotations).

This modern production method depends on two technologies—*geographic information systems* and *global positioning systems*.

Geographic information systems

A geographic information system (GIS) is a computer-based system for storing very large amounts of data (collected based on spatial location) and retrieving, manipulating, and displaying them for easy interpretation. The term *geographic* should be interpreted to mean "space" or "spatial." For crop production, some of the data of importance are soils, land use, vegetation, fertility, hydrology, and rainfall averages. Spatial analysis is concerned with analyzing data involved with changes with space or location within an area. GIS has the capability of linking multiples sets of data (in layers) to study relationships among various attributes and creating new relationships.

GIS is an aid for decision making. It can be used to answer the location question "What is there?" or "Where is it?" It can also be used to study trends—that is, to answer the question "What has changed since a certain point in time?" Another application of GIS is in the area of prediction of change or modeling.

The success of GIS analysis depends on the availability of accurate and reliable data. These databases are created by various private and public entities and are available for a fee or freely accessible via the Internet or other means. One of the leading suppliers of GIS products is the Environmental Systems Research Institute (ESRI), makers of ArcView® and ArcInfo® software and various accessories. In view of the enormity of data routinely involved in GIS applications, one

continued

needs to have access to a computer with appropriate capabilities and speed, in order not to make use of the technology cumbersome to apply.

Global positioning systems

A global positioning system (GPS) is another of the cutting edge technologies of the information age. Whereas GIS basically asks the question "What is it?" GPS asks the basic question "Where is it?" It is a versatile navigational aid. Also computer-based, the key components in this technology are satellites. GPS depends on 24 satellites (courtesy of the U.S. Department of Defense) that are strategically positioned in orbits such that, at any given time and place on earth, one can have a line of sight to at least four of these satellites. Using the technique of *trilateration*, a user is able to access at least three satellite signals, each producing a surface that will overlap each other to locate the site in question, providing both longitude and latitude. A fourth satellite signal allows the elevation of the site to be calculated.

Satellite signals can be clearly received even under inclement weather conditions. However, tall building, dense forests, mountains and hills, and other such structures can obstruct clear signal communications. The GPS technology is used in everyday life as navigational aids in automobiles, in tracking sites of breakdown, in routing emergency response crew, and in other applications. Individuals can purchase handheld units of varying resolution.

Implementing a precision agricultural production operation

There are two basic methods for implementing a precision farming strategy.

1. *Map-based technologies.* This is currently the most widely used method. It may involve a GIS-based method of pre-sampling and mapping of the field. This is effective for quantitative stable characteristics of the field. The computer-

generated maps are then converted into a form that can be used by the variable rate applicator. The applicator's controller then calculates the desired amount of an agronomic input to apply at each moment in time as the equipment traverses the field. The farmer knows how much of the agronomic input is needed before starting the application. A DGPS (differential global positioning system) must be used to constantly evaluate the location in the field with a coordinate on the map and the desired application rate for that coordinate.

Nutrients are not absorbed effectively unless there is adequate soil moisture because mineral nutrients must go into solution (for/solution) to be absorbed. However, excessive moisture causes leaching of nutrients. Nutrient efficiency is also reduced when timing is improper. Nutrients should be applied when the plant needs them the most and can absorb them effectively. Ammonia fertilizers are prone to volatilization and should be applied

2. *Sensor-based technologies.* This method offers real-time sensing and variable rate control. The strategies provide on-the-go sensing of field characteristics, thereby eliminating the need for a positioning system. The sensors must be mounted strategically (e.g., at the front of the tractor) to allow the variable rate applicator's controller adequate time to adjust the rate of the agronomic input accordingly before it passes the sensed location. More sensing packages need to be developed for the various agronomic applications.

Sensed yield data should be interpreted carefully. Variable yield data by themselves are not informative. If the monitor records that grain yield varies 50 bushels per acre from one location in the field to another, there is no clue to the causes of this variation. A field is influenced by many factors such that, if the sensor records identical low yields for two different locations, the causes could be very different. To increase utility of yield monitoring, it should be coupled with walking the field to record site-specific variations.

The area to be cultivated is first delineated into portions according to soil differences (i.e., the soil special variation is delineated). This may be done by inspection or use of computer programs. A map is made to show the variations in the levels of fertility. The soil is then sampled on a grid pattern. Plant health data from previous cropping may also be used to assist in delineating the field. Record keeping is critical to success in precision farming. All data must be tied to locations in the field (i.e., *georeferenced*). *Ground control points* should be set up using the DGPS. All sampling is done with reference to the ground control points and entered into a GIS for processing. The data are then processed to determine the variable rates of application to use. Precision farming requires specialized equipment with sensors. The tractor or other vehicle being used should have an onboard computer and GPS receiver. The GIS information is coupled to the guidance system of the

GPS receivers to dispense variable treatment as the tractor or other vehicle moves over the field.

Cost and profitability

Information gathering and analysis for precision farming is expensive. Precision farming requires a significant initial investment into equipment. Variable rate management does not necessarily increase yield all over the field. In some cases, yield increase is observed in lower-yielding parts of the field. The value of yield gains is more important than savings from reduced application of the agronomic input. Generally, in cases of variable rate fertilization, high-value crops that are responsive to the technology tend to be more profitable than low-value crops, because the yield gains are worth more.

Farmers should adopt this technology only after careful consideration. It is not the "golden egg" to financial security, as some think. Producers with weaker financial standing should be more cautious about adoption of this technology.

with care. If the soil pH is not at the proper reaction, certain nutrients may be rendered immobile and unavailable through fixation.

To improve fertilizer efficiency, the producer should avoid adding a large amount of fertilizers, especially nitrogen and potassium, at one time. It should be borne in mind that Liebig's law of the minimum (i.e., a growth factor in the least relative amount will limit the growth of the plant) is operational in the case of fertilizer application to plants. Regular soil tests will aid in discovering any improper pH levels and lack of nutrient balance that can cause nutrient deficiency symptoms in plants.

8.15: U.S. Agriculture Is a Big User of Fertilizers, Especially in Areas Such as the Corn Belt

The application of fertilizers played a significant role in the dramatic rise in yield per unit land area for major crops in the United States. For example, corn yield rose from 55 bushels per acre in 1960 to 139 bushels per acre in 1994. Nutrient sources used in U.S. crop production are primarily commercial fertilizer and to a lesser extent animal waste.

Of the major nutrients, nitrogen is by far the most widely used nutrient supplement in crop production (Figure 8–4). In 1960, nitrogen use constituted about 37% of total commercial nutrient use. By 1981, the proportion had reached over 50%. In 1995, 55.2% of total commercial nutrients used in crop production in the U.S. was nitrogen, a total of 11.7 million tons of the element. On the other hand, the use of phosphate commercial fertilizers declined from its share of 34.5% of total nutrients used in 1960 to 20.8% in 1995. Potash use exceeded that of phosphorus in 1977. In 1995, it accounted for 24.0% of total commercial fertilizer used in the United States. Most (63%) of the fertilizers used in 1995 were mixed (i.e., supplied more than one major nutrient).

Most of the commercial fertilizers in the United States are used by producers in the Corn Belt (i.e., Ohio, Indiana, Illinois, Iowa, and Missouri). This is because corn is the most fertilizer-using crop. The Northern Plains region (North Dakota, South Dakota, Nebraska, and Kansas) is second after the Corn Belt in the use of commercial fertilizers. Though the most frequently fertilized crop in the United States is corn, the rate of application of nitrogen, the most frequently used fertilizer for corn, is normally highest for fall potatoes, averaging 221 pounds per acre in 1995 (Figure 8–5). The average application rates of nitrogen

FIGURE 8–4 Primary fertilizer nutrients applied in U.S. crop production. Nitrogen accounts for more than 50% of fertilizer consumption. In 1997, it was applied to 71% of cropland. Only 29% of cropland was not treated with nitrogen. Similarly, 60% and 56% of cropland were treated with phosphorus and potassium fertilizers, respectively. (Source: USDA)

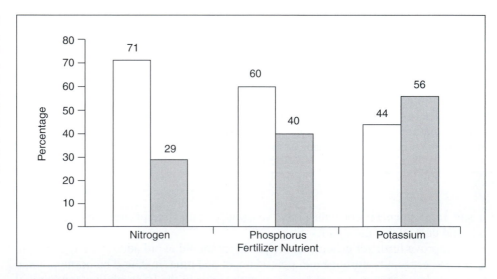

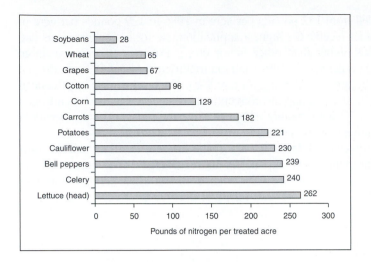

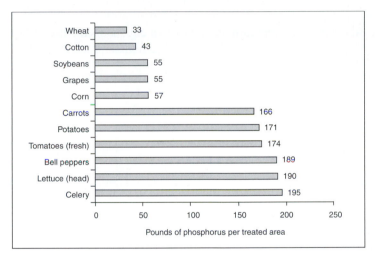

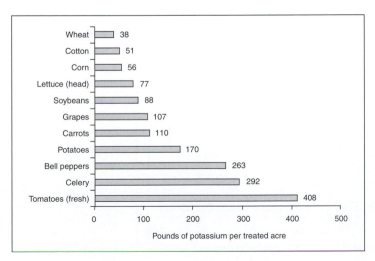

FIGURE 8–5 Nutrient application rates are higher for vegetables than field crops. In 1997, potatoes received the highest rates of all the three primary nutrients applied to field crops. After potatoes, corn received the next highest rates of fertilizer nutrients, receiving 129 lb/acre of nitrogen, 57 lb/acre of phosphorus, and 51 lb/acre of potassium. Wheat generally receives moderate amounts of NPK. (Source: USDA)

in corn production dropped from 132 pounds per acre in 1986 to 129 pounds per acre in 1995. Similarly, fall potatoes receive the highest application rate for both phosphorus and potash, two to three times higher than other major crops. Potatoes also receive large amounts of sulfur (82 pounds per acre in 1994) and micronutrients. About 75% of nitrogen fertilizers used is applied before or at planting (Figure 8–6). The fertilizer use is influenced by the cropping practice. Producers adopting conventional tillage tend to use less nitrogen fertilizer but often supplement with manure. However, livestock manures are not widely used in crop production in the United States (Figure 8–7).

The amount of fertilizer used depends on acreage farmed, fertilizer prices, commodity programs, and nutrient management. As more acreages of major crops are farmed, the use of fertilizer increases, especially when corn and wheat acreages increase (since they account for 45% and 16% of total fertilizer used, respectively). Larger crop acreages will be planted if crop prices are favorable. Fertilizers constitute about 6% of crop production cost. Generally, fertilizer use is unresponsive to changes in fertilizer prices, at least in the short term. Events such as the oil embargo brought about sharp rises in fertilizer prices. Also, increased demand for fertilizers (e.g., after the floods of 1993 and export demand) increases fertilizer prices. Government programs such as the commodity programs tend to create a stable farm economy that engenders the use of fertilizers. As producers incorporate more effective nutrient management practices such as

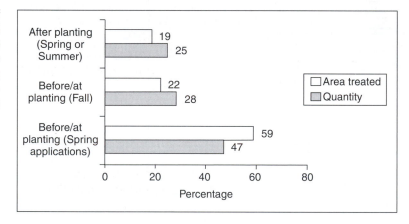

FIGURE 8–6 About 75% of nitrogen fertilizer is applied before or at planting in the production of most of the major field crops, including corn, soybeans, wheat, cotton, and potatoes. (Source: USDA)

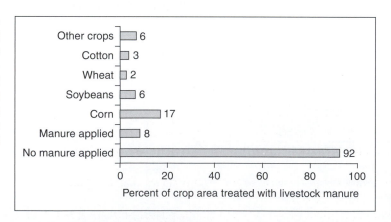

FIGURE 8–7 Livestock manures are not widely used in crop production in the United States. Corn received the most amount of livestock manure. About 17% of all corn acreage was treated with livestock manure between 1994 and 1996. (Source: USDA)

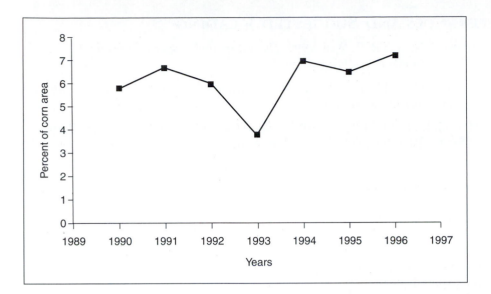

FIGURE 8–8 The use of nitrogen stabilizers reduces the leaching loss of applied nitrogen fertilizers. In 1993, soils in the Corn Belt were extremely wet from the heavy rains, thus delaying the normal application of fertilizers, resulting in a reduction in the use of nitrogen stabilizers. (Source: USDA)

split application, nitrogen stabilizers, and timing of application, loss of nutrients through leaching and other avenues is decreased (Figure 8–8). Lower rates of fertilizer use can then be practiced. Modern technologies such as precision farming further reduce the amounts of fertilizers used in crop production through more effective application.

Soil fertility management in crop production should be conducted as part of the general soil management practice that includes managing soil physical properties (through tillage, drainage, and control of erosion).

SUMMARY

1. The goal of soil fertility management is to sustain the soil's ability to supply essential nutrients to crops.
2. The nutrients required by plants are called essential nutrients. There are 18 essential nutrients. These are categorized into three according to quantities absorbed. Major or macronutrients (required in large amounts—nitrogen, phosphorus, and potassium), secondary nutrients (required in medium amounts), and minor nutrients, or micronutrients (required in minute or trace amounts).
3. Soils lose nutrients through leaching, erosion, volatilization, and crop removal.
4. A soil test is used to determine the nutritional status of the soil. Based on the tests, fertilizer recommendations for amendment are made for application by the farmer.
5. Plant tissue may also be analyzed to detect soil nutritional deficiency.
6. Nutrient availability is influenced by the soil pH.
7. There are natural cycles (e.g., N, P, S) in nature that return nutrients absorbed and used by plants into the soil upon decay.
8. Fertilizers are used to supply supplemental nutrition to the soil for crop production.
9. There are several ways of applying fertilizer to crop plants: starter, split, top, broadcast, banding, foliar, and fertigation.
10. Not all of the applied fertilizer is actually utilized by plants. Some of it is lost to leaching, fixation, and other causes.

REFERENCES AND SUGGESTED READING

Brady, C. N., and R. R. Weil. 1999. *The nature and properties of soils.* 12th ed. Upper Saddle River, NJ: Prentice Hall.

Burns, R. C., and R. W. F. Hardy. 1975. *Nitrogen fixation in bacteria and higher plants.* Berlin: Springer-Verlag.

Mengel, K., and E. A. Kirby. 1987. *Principles of plant nutrition.* 4th ed. Bern, Switzerland: International Potash Institute.

SELECTED INTERNET SITES FOR FURTHER REVIEW

http://www.fertilizer.org/PUBLISH/PUBMAN/introdc.htm

Fertilizer use/abuse, types, pollution, pH, diagnosis of deficiency.

http://www.fertilizer.org/PUBLISH/PUBMAN/manual.htm

Fertilizer recommendations for various crops.

OUTCOMES ASSESSMENT

PART A

Answer the following questions true or false.

1. T F   Nitrogen is a micronutrient in plant nutrition.
2. T F   Nitrogen is associated with vegetative plant growth.
3. T F   Purpling is associated with potassium deficiency.
4. T F   Necrosis is tissue death.
5. T F   An established farmer should conduct a soil test every year.

PART B

Answer the following questions.

1. Give the three macronutrients or major nutrients in plant nutrition.

   _____

   _____

   _____

2. Soil nutrient diagnosis is also called _____.

3. Give a function of nitrogen in plant nutrition.

   _____

4. Give a list of the secondary nutrients in plant nutrition.

_____

_____

_____

_____

_____

5. Give two examples each of the commercial fertilizers used to supply
   a. Phosphorus

_____

_____

   b. Nitrogen

_____

_____

PART C

Write a brief essay on each of the following topics.

1. Discuss chlorosis as a symptom of nutrient deficiency in plant nutrition.

2. Discuss the criteria for declaring a nutrient as essential in plant nutrition.

3. What is phosphorus fixation?

4. Discuss nutrient cycling.

5. What is eutrophication?

6. Discuss the methods of fertilizer application.

7. Discuss the factors that affect a soil test.

8. Discuss the factors that affect the implementation of recommendations from a soil test.

PART D

Discuss or explain the following topics in detail.

1. A fertile soil is not necessarily productive. Explain.

2. What can be done to improve fertilizer efficiency?

3. Discuss the role of fertilizers in boosting crop productivity in U.S. agriculture.

4. Discuss the impact of leached fertilizer on the environment.

9

Plant and Soil Water

PURPOSE

The purpose of this chapter is to discuss the properties of soil water, its movement in the soil and the plant, and its management for crop production under rainfed and irrigated conditions.

EXPECTED OUTCOMES

After studying this chapter, the student should be able to:

1. Discuss the hydrologic cycle.
2. Classify soil water according to availability to plants.
3. Discuss how water moves in the soil.
4. List the primary goals of soil water management.
5. Discuss water use efficiency and factors that affect it.
6. Discuss water management in rainfed production in humid conditions.
7. Discuss water management in rainfed production in dry conditions.
8. Discuss water management in production under irrigated conditions.
9. Discuss the methods of irrigation and the role of water quality.

KEY TERMS

Aquifer	Cohesion	Furrow irrigation
Available water	Embolism	Gravitational water
Capillary rise	Evapotranspiration	Hydraulic conductivity
Capillary water	Fallow	Hydrologic cycle
Cavitation	Field capacity	Hydrophytes

Hygroscopic water
Infiltration
Irrigation
Matric potential
Mesophytes
Microirrigation

Percolation
Permanent wilting point
Salinization
Saturated flow
Sprinkler irrigation

Subirrigation
Surface irrigation
Unsaturated flow
Water-use efficiency
Xerophytes

To the Student

Water for crop production is stored in the soil. The capacity of soil to store water depends on several factors, including its depth, texture, structure, and organic matter content. The amount of water stored in the soil at some point in time depends on how much was received (through rain or irrigation) and how much was lost (transpiration, evaporation, percolation). Certain regions of crop production have climatic conditions that support production solely on water received as rain. In other areas, no crop production can occur without irrigation. Because of the uncertainty of the weather, drought periodically devastates crop production when it is conducted under solely rainfed conditions. Whenever possible, provisions should be made for supplemental irrigation, in case of droughty days. Since water for crop production is stored in the soil, the producer should adopt cultural practices to manage the soil so that it does not lose moisture.

Regarding the crop being produced, the grower needs to know certain factors, including the root system and stage of crop development (affect the time when the plant needs water the most). Apart from soil and plant factors, effective management should take into account the effects of local weather. Weather modifies crop needs for water. In this chapter, you will learn how water is managed under various crop production systems and the various methods used to provide supplemental moisture in crop production. Pay attention to the soil-plant-water relationship and how the atmosphere impacts it.

Water can exist in three states in nature: *solid, liquid,* and *gaseous.* The liquid state is the most useful to plants. Soil water is the solvent in which all the soil nutrients needed for plant growth and development are dissolved. A soil may have high concentrations of nutrients but will be unproductive if there is no soil water to dissolve them for root absorption. Water needed for photosynthesis is absorbed from the soil (or growing medium) by plant roots. Other physiological processes such as transpiration and translocation depend on water.

9.1: Hydrologic Cycle

Water in the soil, plants, and the atmosphere exists in a continuum. Just like soil nutrients, soil water is a dynamic system influenced by factors that deplete it or replenish it. There is a relationship between water in the atmosphere and soil water that is described by the **hydrologic cycle** (Figure 9–1). This cycle is a complex phenomenon of movement of water that includes factors that add water to the system and factors that remove water from it. Plants act as conduits for the flow of water from the soil through plant organs (e.g., roots, stem, leaves) into the atmosphere. This pathway of water movement is called the *soil-plant-atmosphere continuum.* The movement of water through plants into the atmosphere constitutes transpiration, while the direct loss of water into the atmosphere from the soil is evaporation.

Hydrologic cycle. The circuit of water movement from the atmosphere to the earth and back to the atmosphere through various processes, such as precipitation, runoff, percolation, storage, evaporation, and transpiration.

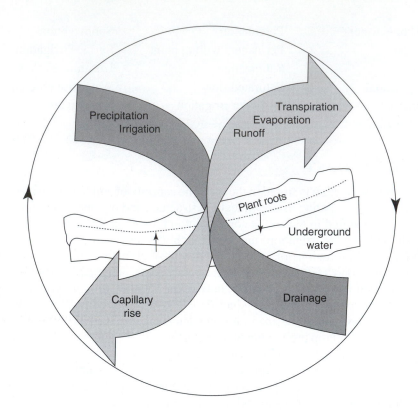

FIGURE 9–1 The hydrologic cycle operates on the principle that water evaporates from the earth's surface, including water bodies, and returns in the form of precipitation. Life on earth depends on the hydrologic cycle. The cycle described in this figure is in particular reference to plants. Precipitation and irrigation are the main avenues for replenishing soil water from above ground. Underground, water can reach plant roots from the underground water pool through capillary rise. Similarly, water is lost through transpiration, evaporation, surface runoff, and drainage from the root zone into the underground water pool.

Soil water is stored at different depths (levels) of the soil. Plant roots draw water from different depths of the soil and store it in their organs. For crop production, the water stored in the root zone is of importance. The level or amount of this water depends on a balance between factors that increase it and those that deplete it, as already stated.

9.1.1 THREE FORCES ARE INVOLVED IN SOIL WATER MOVEMENT AND RETENTION

How does water move in the soil? The rate of water flow depends on the *potential gradient* (the driving force) and the **hydraulic conductivity** (the ease of flow through pore spaces in a particular medium). Water can perform work when it moves. *Soil water potential* is the work performed when a quantity of water moves from its present state to a pool of pure, free water at the same location and at normal atmospheric pressure. This potential has several components. The effect of the surface area of soil particles and micropores, called **matric potential**, and the effect of dissolved substances (*osmotic potential*) are key components.

Other key components are the *pressure potential* (effects of atmospheric gas) and the effect of the water's elevation as compared with a specified position (*gravitational*

Hydraulic conductivity. An expression of the soil's ability to transmit flowing water through a solid, such as soil, in response to a given potential gradient.

Matric potential. The amount of work an infinitesimal quantity of water in the soil can do as it moves from the soil to a pool of free water of the same composition and at the same location.

potential), the latter not related to soil properties. Water flows from an area of high water potential (usually wetter soil) to low water potential (usually drier soil). Soil water potential is mostly matric potential (work done when a quantity of water in the soil moves from the soil to a pool of free water of the same composition and at the same location). Matric potential nearly equals water potential in non-salty soils.

Three forces are involved in soil water movement and retention. Water enters the soil by infiltration through the pore spaces. The infiltration capacity of a soil depends on its physical characteristics (e.g., texture, structure, and the presence of impediments such as impervious layers, like pans). When downward movement of water encounters an obstruction, or the rate of infiltration is overwhelmed by the rate at which water is supplied, water pools or ponds on the soil surface. Where there is a slope, surface runoff occurs as a consequence.

Water flow may be classified into two categories: **saturated flow** and **unsaturated flow**. Soil is said to be saturated when all the pores (both macropores and micropores) are filled with water. Under this state, water movement depends on *hydraulic force* (usually gravitational pull) and hydraulic conductivity. Sandy and loamy soils have higher hydraulic conductivity than clay soils. Why is this so?

The process by which water (rain or irrigation) enters the soil body is called water **infiltration**. Infiltration wets the soil profile. More water then moves through the wetted soil by the process called **percolation** (the process responsible for leaching of dissolved salts).

Water infiltration is influenced by various soil factors:

1. Soil texture (the percentage of sand, silt, and clay in the soil). Infiltration is rapid in coarse sands and slow in clay soils.
2. Soil structure (particle arrangement). Infiltration is high in a soil with granular structure and poor in structureless soil such as clay.
3. Soil organic matter. Organic matter improves infiltration. Mulching also protects soil structure by protecting against destruction of soil aggregation. Infiltration is facilitated by good soil aggregation.
4. Presence of impervious layers, such as pans (clay pan, plow pan) and other obstructions such as bedrock, which impedes infiltration. The problem is more significant if the depth of the impervious layer is shallow. Can you explain why?
5. Compaction. Soil compaction reduces pore spaces and slows water infiltration.
6. Soil temperature. Cold soils absorb water slowly.
7. Soil moisture status. Wet soils have low infiltration rates.

Unsaturated flow of soil water occurs when its potential is low (less than −20 to −33 kPa). Under such conditions, water moves in any direction but from regions of high potential (relatively wet) to those of low potential (relatively dry). The rate of this flow is fastest when the water potential gradient between the wet and the dry areas is high. This flow is also faster when soil moisture is near field capacity.

Under unsaturated soil conditions, the matric potential gradient is more important than hydraulic conductivity and hydraulic force. Hydraulic conductivity is higher in sand than clay when matrix potential is high. However, under drier soil conditions, water movement through clay has the advantage of higher capillarity.

Capillary rise (water movement in micropores) is responsible for the loss of water from the soil by evaporation (Figure 9–2). While gravity moves water downward, the retention of water by soil depends on two surface forces, namely **cohesion** (the force of attraction between water molecules) and *adhesion* (the force of attraction between soil particles and water molecules). The ability of the soil to retain water is called its *water-holding capacity*.

Saturated flow. The movement of water through the soil by gravity flow, as in irrigation or during a rainstorm (i.e., saturated soil).

Unsaturated flow. The movement of water in soil that is not filled to capacity with water (i.e., soil is not water-saturated).

Infiltration. Entry of water downward through the soil surface.

Percolation. The downward movement of water through the soil, especially the downward flow of water in saturated or nearly saturated soil.

Capillary rise. Rise of water in small or capillary pores against gravity.

Cohesion. Force holding a liquid or solid together because of attraction between like molecules.

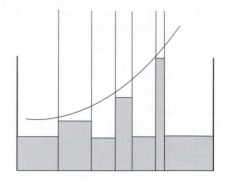

FIGURE 9–2 The upward capillary rise of water through tubes depends on the diameter of the bore. The rise is higher in tubes with finer bores than those with wider bores. Similarly, the capillary rise in the soil is higher in finer-textured soils (more micropores) than coarse-textured soils (more macropores).

9.1.2 WATER UPTAKE BY PLANTS OCCURS LARGELY WITHOUT ENERGY EXPENDITURE

How does soil water enter plant roots? More than 90% of water entering plants is absorbed by *passive absorption* (no energy involved). As plants transpire, a force is generated that pulls a continuous column of water through plant cells. The trend is for water to move from the region of highest water potential to one of lowest water potential in the soil. Water is moved through xylem conducting vessels. This movement is possible because of the strong cohesive bond among water molecules. This movement of water is called the *cohesion-tension theory* of water movement. The cohesiveness of water molecules allows water to withstand tension and retain the continuous column. Water in the xylem vessels is held against gravity by capillarity. Capillary flow can be obstructed when air bubbles interrupt the continuity of the water in the water column (**cavitation**). The condition in which capillary flow is cavitated is called **embolism**. An embolized tracheary element is incapacitated and unable to conduct water.

Water enters the plant from the soil through the root hairs. Once inside, water moves through the root cortex into the tracheary elements. Transpiration is negligible at night. As such, root pressure is known to play a role in water movement during this period. Ions build up in the xylem to a high concentration to initiate osmosis and cause water to enter the vascular tissue through neighboring cells. This *active absorption* requires energy expenditure.

Water occurs in the soil in the pore spaces and is adsorbed to soil particles. Not all soil water is accessible to plants for use. The portion of stored soil water that can be readily absorbed by plants is called plant-**available water**. This water is held within a potential of between −33 and −1,500 kPa. At water potential greater than −1,500 kPa, water absorption is slower than moisture loss through transpiration. This results in wilting of plants. The water content of soil at a water potential of −1,500 kPa is called the *wilting point* (or **permanent wilting point**). Plants such as corn show signs of wilting even at −100 to −200 kPa, as occurs on a hot, dry day. Plants usually recover during the cool of the night. This is because the wilting is only temporary.

Soil water at a soil water potential of −33 kPa is called **field capacity**. This is the maximum amount of water a soil can hold after wetting and free drainage. The difference between the soil water content at field capacity and the soil water content at permanent wilting point is the plant-available water. The water available depends on soil texture (Figure 9–3). Explain the trends in the figure based on your knowledge of the physical properties of soil.

The difference between the total water potential between two locations in the soil is called *water-potential gradient*. Water in micropores (**capillary water**) moves in all directions when a water-potential gradient exists.

Cavitation. An interruption in the continuity of flow in a pore by air bubbles.

Embolism. The cavitation of a capillary flow.

Available water. The portion of water in a soil that can be readily absorbed by plant roots.

Permanent wilting point. The largest amount of water in the soil at which plants will wilt and not recover when placed in a dark, humid atmosphere.

Field capacity. The amount of water remaining in a soil after being saturated and then freely drained.

Capillary water. The water held in capillary or small pores of a soil that is available to plants.

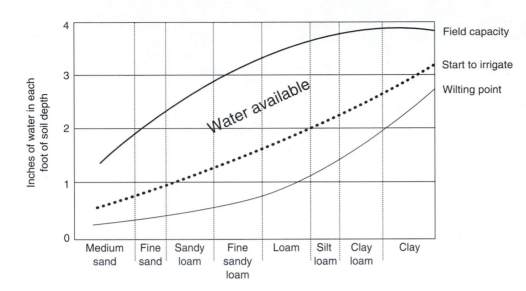

FIGURE 9–3 The water available for plant use depends on the soil texture. Field capacity and wilting coefficient are also influenced by soil texture. Field capacity increases as soil texture increases, leveling off as soil becomes finer in texture. Loamy soil drains well, has good water-holding capacity, and is recommended for cropping.

9.1.3 NOT ALL FORMS OF SOIL WATER ARE AVAILABLE TO PLANTS

How useful is a heavy downpour of rain to crop in the field? Based on the degree of wetness (which depends on the tightness with which the moisture film is held around the soil particles), soil water may be classified into three forms: gravitational, capillary, and hygroscopic (Figure 9–4).

1. **Gravitational water**. Gravitational water occupies the macropores as well as the micropores after a heavy rain or excessive irrigation. It drains under gravity and is therefore not available to plants for growth and development.
2. *Capillary water*. Capillary water is held in the micropores, after the macropores have been drained by gravitation pull. This is the only important source of water for plants.
3. **Hygroscopic water**. Hygroscopic water is held too tightly around soil particles and therefore not available to plants. It can be removed by air-drying of the soil.

Plants differ in their capacity to grow and produce under varying moisture regimes. In this regard, there are three general categories of plants:

1. **Hydrophytes** are plants that can live and can be productive under waterlogged conditions (e.g., paddy rice).
2. **Mesophytes** are plants that prefer normal rainfall conditions. Most field crops are mesophytes (e.g., corn, wheat, cotton).
3. **Xerophytes** are plants adapted to conditions of moisture deficit (drought). Drought tolerance is an important breeding objective for areas that are prone to erratic rainfall (e.g., *Agave* spp., such as sisal).

9.1.4 SOME PLANTS USE WATER MORE EFFICIENTLY THAN OTHERS

What makes certain species use water more efficiently that others? The water needs of plants change as they grow and develop. Most of the water that is used by most of the

Gravitational water. Water that moves into, through, or out of the soil under gravitational force.

Hygroscopic water. The component of soil water that is held by adsorption to surface of soil particles and is not available to plants.

Hydrophytes. Plants that grow in an exceptionally moist habitat, usually partly or entirely submerged in water.

Mesophytes. Plants requiring moderate or intermediate amounts of water for optimum growth.

Xerophytes. Plants that tolerate and grow in or on extremely dry soils.

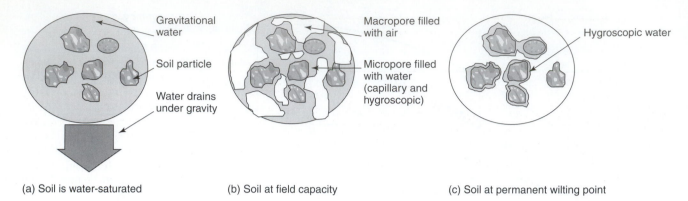

Gravitational water

Soil particle

Water drains under gravity

(a) Soil is water-saturated

Macropore filled with air

Micropore filled with water (capillary and hygroscopic)

(b) Soil at field capacity

Hygroscopic water

(c) Soil at permanent wilting point

FIGURE 9–4 Forms of water. Water in a saturated soil drains freely under gravitational force (gravitational water). This water is not available for plant use. When water is drained out of the macropores, drainage ceases, the remainder of the water being held against gravity in micropores. After plants use up this water, or after it is lost to evaporation, there is some water held too tightly to soil particles to be accessible to plant roots. This is the hygroscopic water.

Water use efficiency. The dry matter or harvested portion of a crop produced per unit of water consumed.

cultivated crop plant species is drawn from the top 5 feet (1.5 meters) of the soil profile. Some crops, such as alfalfa, have a deep root system and thus are capable of extracting more water from lower depths of the soil than other species. Similarly, drought-tolerant species (such as sorghum with its extensive root system) are able to extract more water from the soil than other grasses.

Plants differ in the water needed for development and productivity. Corn, for example, requires between 30 and 100 cm (11.8 and 39.4 inches) of water per month in cultivation for proper development and productivity. As previously indicated, it is important that moisture be available during the critical periods of the crop plant's lifecycle. Crops may be able to grow and produce well using the cropping season's rainfall or stored soil moisture. However, under certain cultural conditions, irrigation may be necessary to supplement soil moisture for maximum productivity. Irrigation may be needed because of seasonal variation in soil moisture.

The role of water in photosynthesis has been previously discussed. Crop plants differ in their **water-use efficiency**. Moisture stress causes stomata to close and, consequently, carbon dioxide assimilation is reduced. Certain species are more efficient at fixing carbon dioxide under low water conditions by keeping stomatal aperture open for a longer time. This, of course, means opportunity for water loss is extended. *Transpiration ratio* is a measure of the effectiveness of a crop plant in photosynthesis under moisture stress through moderating moisture loss while allowing sufficient carbon dioxide to be assimilated. This is calculated as follows:

Transpiration ratio = amount of water transpired/amount of carbon dioxide fixed

Water-use efficiency is a measure of the total amount of water required to produce a unit of dry matter. This total amount of water includes usable water (for plant growth) and water lost (through drainage, transpiration, evaporation, and surface runoff). In order for water to be used efficiently, the producer should experience the most advantage from the least amount of water.

There are several ways of measuring water-use efficiency. The transpiration ratio is a measure of water-use efficiency that is obtained by determining the effectiveness in moderating water loss while allowing sufficient carbon dioxide uptake for photosynthesis.

The amount of dry matter produced depends on the amount of carbon dioxide fixed. It is clear that transpiration ratio and water use efficiency are closely linked. In fact the transpiration ratio may be defined as the amount of water transpired per unit of dry matter produced. The reciprocal of transpiration ratio is the water-use efficiency (i.e., WUE = 1/transpiration ratio). Water use efficiency ranges from 1/200 to 1/1,000, the common range being 1/300 to 1/700. As the climate becomes hot and dry, values decline sharply. Insufficient water causes stomata to close and photosynthesis to cease. Crops with high water-use efficiencies are able to produce high yields under drought conditions by keeping their stomata open for longer periods of time for carbon dioxide fixation. Drought-tolerant species such as sorghum (or C_4 plants) have higher WUE than others such as wheat (or C_3 plants).

In crop production, certain biotic and abiotic factors will influence the ability of plants to utilize the water available. Weeds compete with crop plants for water. They disrupt the patterns of water use and cause shortages at critical times in the crop cycle, leading to reduced yield. Weed control is important under rainfed production.

Pathogens cause diseases that reduce physiological activities and deprive plants of energy for growth and development. Those that affect leaves reduce photosynthetic surface; those that are soil borne may destroy roots and reduce their ability to absorb water and nutrients, resulting in reduced photosynthesis, growth, and development.

Plants that receive balanced nutrition develop properly and exploit water and other resources effectively. However, excessive nitrogen causes luxuriant and early growth and greater leaf area that may place a greater demand on water. Plants growing under low fertility and moisture stress have reduced plant growth. However, water is better distributed between vegetative and reproductive functions for maximizing yield under the given environmental conditions even though the yield is low.

The loss of moisture from leaves (transpiration) and from any exposed surface (evaporation) together constitutes the combined effect called **evapotranspiration**. Evapotranspiration is influenced by solar radiation, increasing on a clear, sunny day. It is also affected by the leaf area index (LAI). As such, evapotranspiration is high when plants are younger and the ground in the field is more exposed (because of low LAI). Evaporation occurs when the atmospheric vapor pressure is less than that at the immediate leaf or soil surface. Drier and windy atmospheric conditions accelerate evapotranspiration. Evapotranspiration is also higher when the soil is near field capacity than when it is drier.

Evapotranspiration. The sum of water transpired by vegetation and evaporated from soil in a given area within a given period of time.

Crop producers may reduce evapotranspiration through mulching, weed control, or **irrigation**. Certain irrigation methods, such as drip, are more effective in this regard. In corn under irrigation, narrower spacing and higher plant population density (28,000 plants per acre or 69,188 plants per hectare) produce higher WUE than spacing at 14,000 plants per acres (34,594 plants per hectare).

Irrigation. The intentional application of water to the soil.

9.1.5 SOIL WATER CONTENT CAN BE MEASURED IN A VARIETY OF WAYS

Soil water content is measured by several methods, the most classic and standard being the *gravimetric method*. The soil sample is weighed (wet) and then dried at 105° to 110°C (221° to 230°F) to constant weight. The following calculations are then made:

Mass water ratio = mass of water/mass of oven-dry soil

Mass water percentage = mass water ratio $\times$ 100

Soil water is frequently expressed in terms of volume:

Volume water ratio = volume water/volume of soil

= (weight of water/density of water)/(weight of oven-dry soil/soil bulk density)

Volume water % = volume water ratio × 100

For the purpose of irrigation, it is important to know the reservoir of water in a soil volume or the amount of water required to wet the soil by irrigation or rain. Plant roots explore a volume of soil for water.

A variety of instruments are available for quick estimates of soil water status. One such instrument is the *tensiometer* that measures the matric potential of soil moisture. Some growers of crops such as sugarcane and potatoes use tensiometers to schedule irrigation. One instrument is placed such that it monitors moisture at the depth of maximum root density and activity, and another near the bottom of the active root zone. When the tensiometer readings at the top and lower soil levels are, for example, -50 kPa and -40 kPa, respectively, an automatic irrigation system is triggered into action.

Tensiometers are, however, unable to measure soil matric potential at levels low enough to warn against the danger of wilting. Other instruments for measuring soil water are the *neutron probe* and the *gamma ray absorption unit*.

9.2: GOALS OF SOIL WATER MANAGEMENT

The goal of water management in crop production is to optimize water use. Water should be supplied when plants need it so it is not wasted. The general goals of managing soil water in crop production are

1. To optimize the use of soil water while it is available
2. To minimize nonproductive soil water losses. Irrigation is prone to evaporative and percolative losses.
3. To optimize the use of supplemental moisture when it is applied. Supplemental moisture is supplied at additional costs to crop production. It is therefore uneconomical to irrigate crops at the wrong time and in the wrong amounts.

9.2.1 WHAT IS DRYLAND AGRICULTURE?

Dryland agriculture comprises systems of production that rely exclusively on rainfall as the source of moisture for crops and pasture production. Consequently, moisture is often the limiting factor in production during at least part of the cropping season. Dryland areas receive less than 20 inches of annual rainfall. Crop rotations are commonly practiced on dryland farms. The principal environmental problems that dryland farmers face include susceptibility to soil erosion (by both wind and water), increased soil salinity and acidification, soil structure deterioration, and soil and nutrient depletion.

Drylands are called by different names all over the world—e.g., plains, grasslands, savannas, steppes, and pampas. They are among the most productive ecosystems in the world, accounting for about 75% of the world's food supply. The major crops grown include potato, cassava, wheat, corn, and rice. Dryland farming occurs in arid and semiarid regions of the world, which make up about 40% of the world's total land area, and is home to over 700 million people. Most (about two-thirds) of this land area is in developing countries. The natural resource base of these areas is fragile. The soils generally are low in fertility and organic matter, as well as low in water-holding capacity. In the United States, dryland farming is common in the eastern and central states.

As energy costs rise and water resources for irrigation decline, dryland production will be increasingly important in producing food to meet the needs of the growing populations of the world.

9.2.2 CROP PRODUCTION UNDER RAINFED CONDITIONS DEPENDS ON THE HYDROLOGIC CYCLE

There are two basic soil water management practices, depending on whether the production will be rainfed or irrigated. In managing soil water for production under rainfed conditions, the features of the hydrologic cycle of importance are

1. *Rainfall.* The aspects of rainfall of importance are total rainfall per year, time of year of rainfall, and reliability of the rain.
2. *Crop.* The water requirement of the crop is important.
3. *Atmosphere.* Water is lost to the atmosphere through evaporation and transpiration.

The role of the weather in rainfed (non-irrigated) crop production is critical. Precipitation is depended upon for all the water used in production. The schedule of watering is dictated by nature. Since the weather is unpredictable, how can rainfed crop producers ensure reasonable success in their operations? The crop producer operating under rainfed conditions should adopt a cropping system that exploits soil water while it is available. Production decision making under such circumstances is based largely on years of experience. It is important to know when the rainfall season starts and, more important, the reliability of the rainfall, timing, frequency, and intensity. Crop cultivars selected for production should mature within the growing season. Early-maturing cultivars are suited to regions of a short wet season. Crop production decisions regarding planting density and fertilization should be made with caution to avoid premature exhaustion of stored soil moisture before the cropping season is over. Cultivars with drought resistance are important in rainfed crop production systems because they provide insurance against unexpected drought occurrences.

Is crop production in the world predominantly rainfed or irrigated? Most crops are produced under rainfed conditions. Rain may be supplemented with irrigation to varying degrees. Rainfed production strategies may be categorized into two, according to rainfall characteristics—practices for *wet conditions* and those for *dry conditions.* Humid regions of the world experience large quantities of rainfall. Unfortunately, the rain does not fall in amounts that are ideal for cropping. Occasional drought is possible even under humid and wet conditions. The more common problem, however, is excessive rainfall that causes a variety of problems, including water erosion, surface flooding, waterlogging, and leaching. Low-lying lands may require drainage to be productive as agricultural lands. Further, these lands usually have cold soils.

Excessive rainfall delays crop production activities such as tillage, planting, and harvesting of produce. Grain dryers may be needed to keep harvested grains in good condition. Humid conditions provide opportunities for diseases and pests to thrive. Producing hay in wet regions is problematic. Instead, ensilaging may replace haying. Weeds are also problematic in wet agriculture. Cloud cover in wet regions is greater than in dry regions. This reduces production potential. However, the abundant moisture compensates for the deficiency.

9.2.3 IN RAINFED CROP PRODUCTION, DISTRIBUTION OR VARIABILITY IN RAINFALL PATTERNS IS MORE IMPORTANT THAN THE TOTAL AMOUNT OF RAINFALL

The rainfed producer should select crops and cultivars very judiciously. The crops and cultivars should have developmental cycles that avoid or tolerate periods in the growing period when water shortage occurs. Further, they should be able to optimize the use of available water for high productivity.

Plants whose economic yield is vegetative (non-grain or seed) are more successful under conditions of unpredictable water supply. Such species include grasses and other forage species. Roots and tubers are adapted to rainfed agriculture because they are able to halt the growth cycle temporarily but resume with rainfall to add more dry matter. Cassava *(Manihot esculenta)* is exceptionally suited to drought-prone regions.

Plants with determinate stem types are very susceptible to fluctuations in rainfall during the growing season. Crop yield is significantly reduced when drought occurs near flowering time (anthesis). Under such unpredictable conditions, short-duration cultivars, even though less productive, are best to use to avoid water stress.

To use water efficiently under rainfed conditions, the grower should select cultivars with high water-use efficiency. All the water that is available under the production system should be fully utilized for crop production. The crop should be planted such that stress is avoided; otherwise, the crop should be able to tolerate transient water shortages. Further, the soil water should be used such that it supports vegetative and reproductive growths.

9.2.4 DRYLAND PRACTICES

Fallow. Cropland left idle and free of weeds for a period of time to restore productivity through accumulation of water, nutrients, or both.

Irrigation is usually not the practical solution to drought in dryland production. Rather, farmers adopt practices such as **fallow**, mulching, alternative plant arrangement in the field, use of catch crops, and xerophytic crops to conserve or efficiently use moisture in production. Sometimes, it is also prudent to plan for crop failure by purchasing crop insurance policies.

Crop producers use fallow under rainfed conditions to accumulate moisture for crop production. The land is idled for a period of time weed-free. The success of this practice depends on the climate, soil, and management skill of the producer. The soil should have good water-holding capacity. Part of fallow management includes improving soil infiltration or penetration, minimizing losses by evaporation and drainage. Sandy soils have high penetration but low water-storage capacity. The efficiency of a fallow (the proportion of the total rainfall that contributes to crop production during the subsequent cropping) depends on the soil type, rainfall amount and distribution, and evapotranspiration. The efficiency of a fallow can be enhanced by implementing water conservation practices that increase the irregularities on the soil surface (e.g., ridges, plant residues on soil). These structures trap surface water by pooling and allowing more time for infiltration to occur, while reducing surface runoff.

Sometimes, soil structure must be improved through incorporation of organic matter, addition of soil amendments, and tillage. Tillage may be used to break up pans and control weeds. However, frequent tillage to control weeds during a fallow is expensive. Application of a mulch or maintenance of surface residue will control weeds, increase infiltration, protect the soil surface from destruction, and help the soil to retain water for a longer time. In spite of management practices adopted, a fallow is subject to losses by evaporation, transpiration by weeds, and drainage below the root zone.

There are two basic kinds of fallow systems in crop production—conventional and conservation. These systems are associated with certain tillage practices.

Conventional Fallow System

The chief characteristic of the conventional system is the lack of plant residues on the surface of the soil. The bare soil is vulnerable to both wind and soil erosion. Extensive tillage is used in conventional tillage systems to control weeds and conserve moisture. Tillage raises soil temperature and increases aeration, thereby accelerating the decomposition of soil organic matter and the recently incorporated crop residues. In addition to a decline in soil organic matter, the nutrients resulting from the mineralization of soil organic matter are susceptible to leaching or volatilization, as the case may be. The reduced soil organic matter also adversely impacts soil structure, reducing the infiltration of precipitation and thereby increasing surface round off.

Conservation Fallow Systems

Conservation fallow systems aim at minimizing soil erosion during the fallow period by leaving plant residue on the soil surface. An estimated minimum residue cover of about 1,300 lb/acre is required to protect the soil sufficiently from soil erosion. Residue cover is reduced by natural processes (e.g., microbial activity, oxidation) as well as by producer activities, such as tillage. Various conservation fallow systems are in use:

1. *Chemical summer fallow.* Chemical fallow provides an alternative to tillage for controlling weeds. Herbicides are used to control weeds. Vegetative growth during the fall or summer months is controlled by applying one or two consecutive applications of suitable herbicides. The herbicides kill weeds and leave plant residue on the soil surface for protection against soil erosion. The potential risk of using herbicides is the risk of contaminating groundwater and other water resources. Such potential risks can be minimized by the timely application of herbicides with short residual and limited mobility in the soil. The chemicals should be applied at the right rates and under the proper environmental conditions. One practice involves an early spring application of 24-D to control winter annual weeds and a late spring application of glyphosate or paraquat mixed with an appropriate phenoxy herbicide for a broad spectrum control of all broad weeds and volunteer grains until July. This approach eliminates three or four tillage operations.

2. *Eco-fallow.* This is similar to chemical fallow that is solely dependent on herbicides. In eco-fallow, a tillage operation (e.g., subsurface sweep) is performed to spread the plant residue. The tillage application follows a previous application of a herbicide, such as atrazine. This application into plant residue may miss weeds that were covered by the debris. The tillage controls volunteer crop plants and weeds that were hidden under the windrow or some other debris during the first chemical application.

3. *Stubble mulch fallow.* Stubble is the basal parts of plants that are left after the straw has been harvested. Stubble mulch fallow is designed to leave the field covered with straw residue during the non-cropping period. Wide-sweep blades undercut the stubble and weeds without pulverizing the soil surface. The mulch cover prevents soil erosion and conserves soil moisture. The amount of residue left on the soil surface depends on the implement used for the tillage operation. A wide blade cultivator or offset disc leaves only about 35 to 65% residue cover after each pass.

Cover Crops

Cover crops are critical to the success of any dryland annual cropping system that strives for sustainability. Cover crops are grown primarily to prevent soil erosion, to suppress weeds, and to conserve moisture. To be effective, cover crops should provide a high percentage of ground cover as quickly as possible. Leguminous cover crops enrich the soil with nitrogen. There are different types of cover crops. When cover crops are incorporated as green manures, they recycle other essential nutrients (nitrogen, phosphorus, potassium) and improve soil organic matter for improved soil physical conditions.

1. *Winter cover crop.* This crop is sown in late summer or fall to provide soil cover during the winter. In northern climates, hardy species, such as hairy vetch and rye, are desirable cover crops. In the southern region, adapted varieties of clovers, vetches, or field peas may be used. Other cereals are also grown as cover crops.
2. *Living mulch.* A living mulch is a cover crop that is inter-planted with an annual or a perennial cash crop. These living mulches suppress weeds, reduce soil erosion, enhance soil fertility, and enhance water infiltration. Examples of living mulches are overseeding hairy vetch into corn at the last cultivation or sweet clover drilled into small grains. In perennial crop fields (e.g., vineyards, orchards), grasses or legumes may be planted in the alleyways between crop rows.
3. *Catch crop.* A catch crop is one that is planted to reduce nutrient leaching following a main crop. These nutrients would have been lost to future crops. The term is also applied to a short-term cover crop that occupies a niche in a crop rotation. An example is planting cereal rye following corn harvest to scavenge the residual nitrogen from fertilizing corn. Some catch crops are winter-killed (e.g., buckwheat, oats, spring wheat, crimson clover), while others overwinter (hairy vetch, fall rye).

Skip-Row Planting

Skip-row planting is the practice of eliminating certain rows in the field to allow the producer to move through the field with implements without damaging crops or to reduce the effective crop area to reduce irrigation needs. This tactic can buy time while the farmer waits for rain and reduces the risk of crop failure.

Crop-Rotation

When crop rotation is practiced in dryland agriculture, the primary goal is to optimize the use of soil moisture for optimal crop yields. Rotations may or may not include a fallow (e.g., wheat-corn-millet-fallow or wheat-corn-millet-no fallow) as used by growers in Colorado. Producers using crop rotations have recorded 10 to 30% increase in income. Crop rotations are discussed in detail later in this book.

9.3: IRRIGATED CROP PRODUCTION

Whenever feasible and affordable, irrigated production can dramatically boost crop yield. Irrigation is one factor responsible for the great productivity of crops in places such as the Pacific states of the United States, especially California. About 50 million acres of the over 300 million acres of cropland in the United States are under irrigated production.

Irrigated crop production depends on a reliable source of clean water. Irrigation may be used to supplement rainfall during crop production at times when drought spells occur. However, in arid and semiarid conditions, irrigation may be used as the sole source of water for crop production.

Water for irrigation may be derived from a variety of sources:

1. *Surface water.* Surface water for crop irrigation includes rivers, lakes, streams, and ponds. To increase the volume of water available, the river or stream may be dammed.
2. *Groundwater.* Water may be pumped through wells from an **aquifer** (water stored underground) for irrigation. The aquifer can be overdrawn. Thus, management of irrigation water from this source includes methods for replenishment as well as safe withdrawal. There has to be a way of monitoring the withdrawal of the aquifer by various producers operating in the same general regions and drawing from a common pool. Not all aquifers can be recharged, leading some to question if irrigated production is sustainable.
3. *Surface runoff.* Rainwater may be collected and stored in, for example, ponds for use in irrigation.

Aquifer. A geologic formation, usually a permeable layer such as sands, gravel, and vesicular rock, that transmits water underground under ordinary water pressures.

9.3.1 IRRIGATION WATER SHOULD BE FREE FROM POLLUTANTS

What are the sources of irrigation water? How do they differ in terms of water quality for irrigation? The quality of irrigation water depends on its source. Rainfall contains salts and dissolved pollutants. In certain regions, the pH may be very low (acid rain) and have corrosive effects on plants. Direct surface runoff is relatively salt-free. However, running water (streams, rivers), by virtue of the fact that it contains water that has percolated through soil and underlying rock minerals, contains high concentrations of dissolved salts. Groundwater is created from accumulation of deeply percolated water and hence has the highest concentration of dissolved salts. The most common ions of dissolved salts found in irrigation water are Na^+, Cl^-, Ca^{2+}, Mg^{2+}, K^+, HCO_3^-, CO_3^{2-}, NO_3^-, SO_4^{2-}, and BO_3^{3-}. The kinds of ions and their concentrations vary from one place to another and depend on the geology, leaching, and hydrologic cycles, among others. Na^+ and Cl^- ions are more common in arid regions. When they combine with weak anions such as HCO_3^- or CO_3^{2-}, they produce alkaline soil (pH greater than 8.5). Low rainfall produces less leaching. As such, the salts are trapped in the top layers (B horizon).

The suitability of water for irrigation is determined by its salt content (total dissolved solids) and the sodium adsorption ratio:

$$SAR = [Na^+]/\sqrt{[Ca^{2+}][Mg^{2+}]}$$

The accumulation of salts in the soil is called **salinization**. All water used for irrigation has some soluble salts. As irrigation becomes more efficient, less water is applied to the crops, reducing the amount available to leach salts out of the soil profile. Consequently, salt accumulates as the added water evaporates and is transpired. Irrigation-induced salinization is a problem with irrigation, especially in arid climates (Figure 9–5).

Salinization. The accumulation of soluble salts in the soil.

Irrigation itself is a significant contributor to accumulation of salts in the soil through the use of water with high salt content. The salt accumulates over a period and may reach toxic proportions in the crop root zone. At this stage, leaching is needed to wash excess salts away to avoid reduction in the crop productivity.

Crops differ in tolerance to salinity. Grasses generally are more salt-tolerant than legumes. Highly sensitive crops include common bean, chickpea, strawberry, white clover, and red clover. Corn, alfalfa, rice, and oats are moderately sensitive to salts, while wheat, tomato, cotton, soybean, and sorghum are moderately salt-tolerant. Crops such as barley, cotton, sugar beet, and bermudagrass are tolerant of salts.

FIGURE 9–5 Irrigation-induced salinization, showing a field covered with salts that have been drawn in the evaporative stream and deposited on the soil surface. (Source: USDA)

The two most important factors to consider in irrigation water quality analysis are the total dissolved solids (TDS) and the sodium adsorption ratio (SAR). The TDS is basically a measure of the concentration of soluble salts in the water sample (i.e., salinity), while the TDS is measured in terms of electrical conductivity (dS/m). The U.S. salinity laboratory has developed a suitability classification system for irrigation water that combines salinity and sodicity. A classification of C2-S2 indicates water with high sodicity and medium salinity. Generally, irrigation water with electrical conductivity of 2 dS/m or a SAR of more than 6 could be harmful for irrigation. The SAR value can be lowered by adding calcium using a source that has high solubility (e.g., $CaCl_2$).

9.3.2 IRRIGATION WATER MAY BE APPLIED IN ONE OF SEVERAL WAYS

How is irrigation water applied in crop production? What determines the method selected for application? Which methods are most water-efficient? Techniques of irrigation may be categorized in several ways, one of which is the degree of control they offer the producer over application rate and distribution. Technologies of irrigation also differ in cost of installation and maintenance. Generally, the higher the producer's control over the amount and placement of the water applied, the higher the cost of the system. Another useful way of categorizing irrigation application systems is according to the energy or force needed to distribute water—by *gravity* or *pressurized*. Surface irrigation systems are gravity application systems, while sprinkler systems are pressurized application systems that depend on pumps to move water.

The choice of an irrigation system depends upon several factors, including

1. *Cost.* Irrigation is a capital-intensive undertaking. It requires a high initial capital investment for installation and then maintenance.
2. *Profitability of the enterprise.* Irrigation is justifiable if the returns on investment are high.
3. *Topography.* Terrain affects the method of distribution of water. Sometimes, land leveling is needed for successful use of certain techniques.
4. *Soil physical properties.* Soil infiltration is a key factor in the choice of a method of irrigation. Coarse-textured soils (sandy) have high infiltration. In surface techniques (e.g., flood), high soil infiltration is undesirable.

5. *Salinity.* Crops differ in tolerance of salinity. The source of water differs in salinity.
6. *Severity of frost.* In cold regions, water in irrigation pipes buried in the ground may freeze during certain times.
7. *Crop.* Certain crops may not tolerate the pool of water needed for certain methods of irrigation. Also, certain methods are expensive to operate and thus suited to crops with high economic value.
8. *Rainfall.* Some producers need irrigation to supplement rainfall in certain regions. In other regions, production may totally depend on irrigation.

9.3.3 METHODS OF IRRIGATION

The methods of irrigation may be classified operationally as follows:

1. Surface irrigation
   a. Gravity flow
      • Flood
      • Furrow
      • Border/basin
   b. Sprinkler systems
      • Center pivot
      • Mechanical move
      • Hand move
      • Solid set and permanent
   c. Low-flow irrigation (drip/trickle)
2. Subirrigation

Surface Irrigation

Surface irrigation is a technique of irrigation in which water is spread over the soil surface. Sometimes, the entire land area is covered such that the plants stand in water, as in the case of *flood irrigation.* In one variation of flooding, called *border strip,* small levees are built around strips of leveled land and then flooded (Figure 9–6). In other practices, the water flows between close rows (called *corrugations*), as in **furrow irrigation** (Figure 9–7). Furrow irrigation is the oldest method of crop irrigation in agriculture. About 40% of all crop land is furrow irrigated. It is applicable to row crops such as corn and cotton. Since water moves by gravity in surface irrigation, the land requires grading to create a gentle slope. A steep slope promotes soil erosion. Further, smooth furrows allow faster flow of water and reduced percolation. The ideal slope should be less than 0.25% (no more than 2%). Some producers plant Kentucky bluegrass in the furrows to reduce erosion on steep slopes. This method is not suitable for soils of high infiltration (e.g., sandy soil). Water would have to be run rapidly, but such high rates would cause erosion. In some production areas, such as California, the furrows are deliberately smoothed by dragging a steel cylinder filled with concrete with a cone cap (called a *torpedo*) down the furrow. Up to 30% faster water flow has been achieved in some cases as a result of this treatment.

Another method of flooding, called *basin irrigation,* is used on impermeable soil for growing crops that tolerate flooding (Figure 9–8). In paddy rice cultivation, clay soils are deliberately tilled in such a way (by puddling) that it further decreases the soil impermeability. Water is thus held in the basin so that plants grow in water for a long period. Infiltration under this condition is uniform.

Surface irrigation. The intentional application of water to plants from aboveground.

Furrow irrigation. A method of surface irrigation in which water is delivered to the field through shallow ditches between ridges on which plants are growing.

FIGURE 9–6 Flood irrigation is suited to soils that are not readily permeable. The land surface should be level to permit unimpeded and even water flow without surface erosion. (Source: USDA)

FIGURE 9–7 In furrow irrigation, water flows through furrows or corrugations by gravity. Crops are planted on the ridges before water is applied. About 40% of irrigated land in the United States is furrow irrigated. (Source: USDA)

The disadvantages of surface irrigation include the cost of grading the land, the use of excessive water, and water loss to evaporation. One advantage is no need for high-pressure pumps to move water. Surface systems are not amenable to automation, making them more labor-intensive. A technique called *surge flow surface irrigation* delivers water intermittently and improves water efficiency of surface systems. The flow rates are larger, causing wetting to occur farther down the irrigation row but accompanied by reduced percolation at the upper end (head) of the furrow (Figure 9–9).

9.3.4 SPRINKLER IRRIGATION

Sprinkler irrigation is a technique in which water is moved under pressure through pipes and distributed through nozzles as a spray. Since water is pumped under high pressure, no grading of the land is necessary prior to installation of the system. Rolling terrain can be irrigated with this system. Further, a wide variety of soil textures (light to

Sprinkler irrigation. The method of aerial application of water through pipes fitted with sprinkling units.

FIGURE 9–8 Basin irrigation is a modified form of flood irrigation in which the field is divided into checks or basins by using ridges. Each tree or several trees are enclosed by a small dike. (Source: USDA)

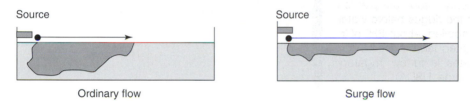

Ordinary flow   Surge flow

FIGURE 9–9 Surge flow improves surface irrigation efficiency by decreasing water through unnecessary depth of soil wetting at the origin. A surge flow controller may be programmed to deliver water intermittently at surge intervals of between 1 and 10 minutes or more. The smooth bursts of water allow the entire furrow to be wetted before large amounts infiltrate at the upper part of the field.

heavy) can be irrigated. The user has control of distribution and rate of delivery. The system is readily amenable to automation.

Sprinkler designs are variable. Some systems are fixed permanently in the plot where the irrigation is required (Figure 9–10). This involves laying pipes in the ground permanently with risers to distribute water. This design is used for high premium horticultural production. Other systems are moved periodically, either by hand (e.g., water guns and supply lines) or by automation (Figures 9–11 and 9–12). A widely used automated sprinkler system is the *center pivot* (Figure 9–13), which moves in a circle. Properly installed and operated, sprinklers wet the soil evenly. The water used for sprinklers should be clean to avoid clogging the sprinkler nozzles. An advantage of sprinklers is their adaptability to *chemigation* (application of chemicals such as pesticides and fertilizers through irrigation water).

There are several methods for increasing the uniformity of sprinkling. There are several causes of lack of uniformity of sprinkling, the key ones being improper irrigation pressure regulation and obstruction to water flow through the nozzles. There is new technology called *low-energy precision applicators (LEPA)* that are able to improve uniformity of distribution of irrigation water through sprinklers. Uniformity can be reduced by

(a)   (b)   (c)

FIGURE 9–10 Center pivot mobile sprinklers can cover areas as large as 53 hectares (130 acres) or larger in one circular sweep. (a) Water is supplied from a well located at the center pivot. The sprinkler is programmed to rotate at a specific speed. (b) Aerial photograph of irrigation by center pivot reveals the characteristic circle of green cropped land. (c) The wheel line sprinkler system does not rotate but moves in one direction on wheels.

FIGURE 9–11 Cable tow irrigation involves the use of a powerful spray gun connected to a long cable and mounted on wheels. The rig is moved with the cable in tow from one part of the field to another.

drift (movement of water away from intended area). This can be reduced by proper equipment adjustment and applying at wind speeds of less than 10. The boom height should be lowered and nozzles that spray larger droplets selected. High pressure of the spray causes fine droplets and increases drift.

Microirrigation

Microirrigation. A method of intentional application of water to plants in drips.

Microirrigation systems are designed to deliver water in small amounts, either in drips or sprays (hence *drip* or *trickle irrigation*). This technique offers the user the most control over the amount and placement of irrigation water. Usually, individual plants are targeted and hence this method is most suited to orchard crops. Water is delivered through

FIGURE 9–12 Solid set irrigation requires that irrigation pipes be permanently set in the field (buried underground). Sprinkler heads are mounted on risers of height according to the crop being irrigated.

FIGURE 9–13 Movable pipe irrigation is labor-intensive; it requires the irrigation pipes be moved and set up for use and dismantled after use. It is practical for use where small acreages are to be irrigated.

tubes to *emitters* (drippers or sprayers) located near individual plants (Figure 9–14). Only a small area (the root zone of the plant) is wetted with practically no surface flow. This system is the most water-efficient, each emitter delivering water at less than 3.7 liters per hour. The water flow is slow and at low water pressure. It is adaptable to steeper slopes. However, it requires the use of very clean water to avoid clogging the emitters. The water should not have debris, microbes, or any substance that will precipitate. Initial investment and operating costs are usually higher than for other systems. It is widely used for trees and other perennial crops, especially high-premium crops such as grapes.

Subirrigation

In both surface and sprinkler irrigation systems, water is applied over the soil surface. In **subirrigation** systems, the water table is raised to the rhizosphere. Water is pumped into a drainage system (e.g., tile or tube) laid underground. This method is effective on light soils (e.g., sandy, muck).

Subirrigation. The method of intentional application of water to plants through conduits located underground.

FIGURE 9–14 Drip irrigation enables water to be applied to individual plants in steady drips and minimizes evaporative losses. It is used for high-value field crops, such as grapes.

9.3.5 FURROW IRRIGATION IS THE MOST COMMON GRAVITY IRRIGATION METHOD USED IN U.S. CROP PRODUCTION

How important is irrigation in U.S. crop production? Where is irrigation practiced the most in the United States? What irrigation techniques are most widely used? Irrigated production areas are shown in Figure 9–15. The use of gravity irrigation application systems has declined by 20% since 1979 (Table 9–1). However, more than 50% of all irrigated cropland in the United States utilized gravity-flow systems in 1994. The practice is widely used in the arid regions of the western United States, especially in California, Nevada, Arizona, New Mexico in the Southwest; Wyoming, Colorado, and Utah in the Central Rockies; Texas and Oklahoma in the Southern Plains; and Arkansas, Louisiana, and Mississippi in the Delta Region. The most (60%) widely practiced method of irrigation is the furrow system.

Conventional systems have been improved in various states. Terrain modification and land management measures that have been implemented include improved on-farm water conveyance systems, precision field leveling, shortened water runs, alternative furrow irrigation, surge flow and cablegation, and tailwater reuse. About 60% of farms implementing the gravity-flow system utilize open-ditch systems as the main method of on-farm water conveyance. Improvements have come by way of ditch lining, ditch reorganization, and pipeline installation. Pipelines may be installed aboveground or underground. These improvements reduce percolation losses and minimize surface runoff. Producers in the Northern and Southern Plains and Delta regions lead the nation in the use of gated-pipes for water conveyance. Surge-flow and cablegation systems are concentrated in the Delta states.

The Farm and Ranch Irrigation Survey of 1994 also indicates that 20% of gravity flow users adopted alternative furrow strategies. Shortened rows were used widely in the Southwest (in Arizona and California) and in the Southern Plains regions. In 1994, 12% of gravity-irrigated farms in the Southwest, Delta, and Southeast farming regions were precision laser-leveled. Other improved water management practices adopted included

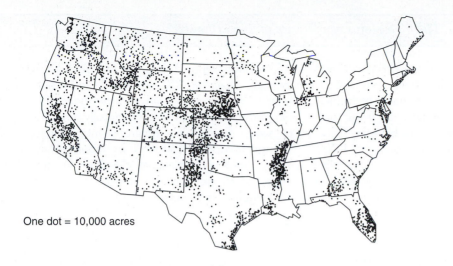

One dot = 10,000 acres

FIGURE 9–15 The state with the most irrigated areas is California, followed by Nebraska and Texas. The Ogallala aquifer supplies water for irrigation of lands in Colorado, Kansas, Nebraska, New Mexico, Oklahoma, South Dakota, Texas, and Wyoming. (Source: USDA)

Table 9–1 Changes in U.S. Irrigated Acreage According to Type of Irrigation System

System	Million Acres		Change (Percent)
	1979	1994	
All systems	50.1	46.4	−7
Gravity-flow systems	31.2	25.1	−20
Sprinkler systems	18.4	21.5	17
Center pivot	8.6	14.8	72
Mechanical move	5.1	3.7	−27
Hand move	3.7	1.9	−48
Solid set and permanent	1.0	1.0	2
Low-flow irrigation (drip/trickle)	0.3	1.8	445
Subirrigation	0.2	0.4	49

Source: USDA, ERS.

deficit irrigation techniques, such as reduced irrigation set-times, partial field irrigation, and reduced irrigation. These strategies were practiced on 10% of farms concentrated in the Northwest regions (e.g., Idaho, Oregon, and Washington). Further, 20% of farms, especially those in California, installed tailwater reuse systems.

9.3.6 CENTER-PIVOT IRRIGATION SYSTEMS ARE THE MOST WIDELY USED PRESSURIZED IRRIGATION SYSTEMS IN U.S. CROP PRODUCTION

Pressurized irrigation systems, including sprinkler and low-flow systems, accounted for about 46% of irrigated acreage in 1994 (Figure 9–16). Sprinkler irrigation is concentrated in the Northern Pacific, Northern Plains, and Northern Mountain states. They are also used for supplementary moisture supply in production of specialty crops in the humid eastern states. Self-propelled *center-pivot* systems were developed in the 1960s. Cropland on which center-pivot systems are used increased by 6.2 million acres from 1979 to

The High Plains regional aquifer system occupies a total of 174,000 square miles of underground area beneath parts of eight states—Colorado, Kansas, Nebraska, New Mexico, Oklahoma, South Dakota, Texas, and Wyoming. Elevations in this area vary between 7,800 feet on the west near the Rocky Mountains and 1,000 feet in the east near the Central Lowlands. The High Plains have gently sloping and smooth plains that are suitable for agricultural production. However, the region is dry, receiving only about 16 inches of rain in the west and about 28 inches in the east. About 54% of the land in this region is devoted to dryland and irrigated cropping, accounting for 19% of wheat, 19% of cotton, 15% of corn, and 3% of sorghum produced in the United States. Further, the High Plains region accounts for 18% of U.S. cattle production and a significant amount of swine production. These agricultural uses place a great demand on the aquifer for irrigation purposes. It is estimated that 27% of irrigated land in the United States occurs in the High Plains region.

The High Plains aquifer yields about 30% of the groundwater used for irrigation in the United States. It also provides about 82% of the domestic water needs of the over 2 million residents in the region. The High Plains aquifer is sometimes referred to as the Ogallala aquifer because the geologic unit called the Ogallala Formation forms about 80% of the aquifer. The aquifer has a saturated thickness ranging from 0 feet to more than 1,000 feet in Nebraska, with an average of about 200 feet. Depth to water ranges from 0 to 500 feet, with an average of 200 feet. Groundwater in this aquifer generally flows from east to west, discharging naturally into streams and springs. Evapotranspiration is important in areas where the water table is very shallow. In addition to these natural discharges, active pumping from wells for irrigation is a major mechanism of groundwater discharge. The primary source of recharge to the Ogallala aquifer is precipitation. Several river systems, including the Platte, Republican, Arkansas, Cimarron, and Canadian rivers, cross this aquifer and may be sources of recharge through leaks.

The High Plains aquifer is experiencing water level declines resulting from withdrawals in great excess of recharge. Drawing water for irrigation began in the 1940s. Since then, water levels in this aquifer have declined more than 100 feet in parts of Kansas, New Mexico, Oklahoma, and Texas. The declines are greater in the southern areas of the region than in the north. In certain cases, declines are severe enough to make the use of the aquifer for irrigation either impossible or cost-ineffective. Whereas most of this management has been exploitation of natural resources, the Ogallala is fossil (connate) water that cannot be recharged, at least not at rates economically feasible for irrigation. This is an example of mining groundwater.

Water-Level Changes in the Ogallala Aquifer Since 1980

State	Years: 80–88	Change in Feet (Area-Weighted)							
		80–89	80–90	80–91	80–92	80–93	80–94	80–95	80–96
Colorado	—	−1.45	−3.08	−3.15	−3.04	−3.25	−3.39	−4.20	−4.70
Kansas	—	−4.61	−4.91	−6.21	−7.39	−7.26	−6.12	−7.50	−7.00
Nebraska	—	+1.34	+0.67	+0.23	−0.27	+0.02	+1.88	+1.80	+1.60
New Mexico	—	−0.09	−0.66	−2.27	−1.86	−3.42	−2.31	−3.10	−6.20
Oklahoma	—	−0.36	−0.32	−0.11	−1.80	−0.41	−1.81	−2.80	−3.70
South Dakota	—	−2.42	−1.03	−0.74	+0.08	−0.90	+1.47	−0.60	−0.80
Texas	—	+0.55	−1.17	−1.65	−2.46	−1.96	−3.02	−4.80	−6.10
Wyoming	—	+3.12	+3.14	+2.92	+1.30	+0.63	−2.52	−3.40	−1.90
High Plains	+0.80	−0.23	−1.04	−1.41	−2.24	−2.09	−1.54	−2.40	−2.80

Source: United States Geological Service.

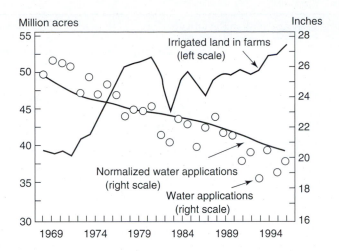

FIGURE 9–16 Trends in irrigation in the United States between 1969 and 1996 indicated a general increase in the acreage irrigated. Currently, over 50 million acres of cropland are irrigated.

1994. They account for about 70% of sprinkler acreage in 1994 (Table 9–2). Most users were in the Northern Plains, Southern Plains, and Delta farming regions. Corn for grain and alfalfa were the two most irrigated crops in production in 1996 (Figure 9–17).

Center-pivot irrigation is the core technology in other new and often more efficient technologies, such as *low-pressure center pivot, low-energy precision application (LEPA),* and *linear move* systems. In 1994, 40% of center-pivot irrigated cropland was irrigated with LEPA systems. LEPA systems operate at a pressure of 30 psi, as opposed to 60 psi in conventional systems. LEPA adoption is highest in the Southern Plains. Low-flow irrigation systems (drip, trickle, or microirrigation) are used for vegetable and orchard production and occur mainly in California and Florida.

Irrigation technology continues to evolve. Advances in irrigation practices for higher efficiency are summarized in Table 9–3.

9.3.7 IRRIGATION WATER SHOULD BE WELL MANAGED FOR PROFITABILITY OF CROP PRODUCTION

The cost of water for irrigation is often high. How do irrigated crop producers use water more efficiently to increase the profitability of their enterprises? To maintain farm profitability, it is imperative that producers of irrigated crops increase water-use efficiency through proper management. Irrigation water management involves the management of water allocation and related inputs in irrigated crop production for enhanced economic returns. Whereas individuals may own separate reservoirs, in areas where irrigation is the primary source of water for crop production, producers usually draw from a common source of water (e.g., river or aquifer).

Further, the needs of farmers in a production region are similar since they tend to grow the same crops. There is the need to manage the distribution of water through a network that will ensure adequate water for all growers. This management includes drainage by which irrigation water is recycled. Producers with on-farm storage of water are able to recycle tailwater from surface irrigation. Further, the performance of irrigators is generally enhanced when the farm has its own water supply or on-farm storage to replenish it. Generally, irrigation drainage gradually declines in quality as salts and other farm chemicals leach into the water.

Table 9–2 Irrigation Application Systems by Type Used in the United States		
System	**Acres (Million)**	**Share of All Systems (Percent)**
All systems	46.4	100.0
Gravity-flow systems	25.1	54.0
Row/furrow application	14.2	31.0
Open ditches	5.0	11.0
Aboveground pipe	7.4	16.0
Underground pipe	1.8	4.0
Border/basin application	7.5	16.0
Open ditches	5.1	11.0
Aboveground pipe	0.9	2.0
Underground pipe	1.5	3.0
Uncontrolled flooding	2.3	5.0
Application		
Open ditches	2.3	5.0
Aboveground pipe	0.0	0.0
Underground pipe	0.0	0.0
Sprinkler systems	21.5	46.0
Center-pivot	14.8	32.0
High-pressure	3.2	7.0
Medium-pressure	5.9	13.0
Low-pressure	5.7	12.0
Mechanical move	3.7	8.0
Linear/wheel move	3.0	7.0
All other	0.6	1.0
Hand move	1.9	4.0
Solid set and permanent	1.0	2.0
Low-flow irrigation (drip/trickle)	1.8	4.0
Subirrigation	0.4	1.0

Source: USDA, ERS, 1994. (Because of rounding and multiple systems on some irrigated acres, percents may not add up to totals.)

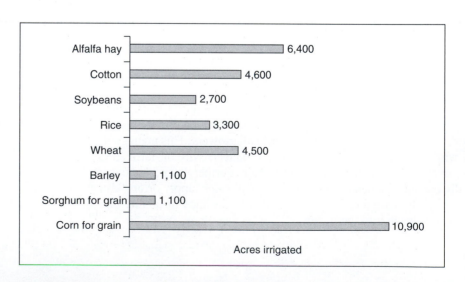

FIGURE 9–17 Corn for grain and alfalfa were the two most irrigated crops in production in 1996, with 10.9 million and 6.4 million acres, respectively. (Source: USDA)

Table 9–3 Advances in Irrigation Technology and Water Management for Higher Water-Use Efficiency

System and Aspect	Conventional Technology or Management Practice	Improved Technology or Management Practice
On-farm conveyance	Open earthen ditches	Concrete or other ditch linings; above or below-ground pipe
Gravity-flow systems		
1. Release of water	Dirt or canvas checks with siphon tubes	Ditch portals or gates; gated pipe with surge flow or cablegation
2. Field runoff	Water allowed to move off field	Applications controlled to avoid runoff; tailwater return systems
3. Furrow management	Full furrow wetting; furrow bottoms uneven	Alternate furrow wetting; furrow bottoms smoothed and consistent
4. Field gradient	Natural field slope, often substantial; uneven surface	Land leveled to reduce and smooth field surface gradient
5. Length of irrigation run	Field length often .5 mile or more	Shorter runs of .25 mile or less
Pressurized systems		
1. Pressure required	High; typically above 60 psi	Reduced; 10–30 psi
2. Water distribution	Large water dispersal pattern	Narrower dispersal through droptubes, improved emitter spacing, and low-flow systems
3. Automation	Handmove; manually operated systems	Self-propelled systems; computer control of water applications
4. Versatility	Limited to specific crops; used only to apply irrigation water	Multiple crops; various uses—chemigation, frost protection
Water management		
1. Assessing crop needs	Judgment estimates	Soil moisture monitoring; plant tissue monitoring; weather-based computations
2. Timing of applied water	Fixed calendar schedule	Water applied as needed by crop; managed for profit (not yield); managed for improved effectiveness of rainfall
3. Measurement of water	Not metered	Measured using canal flumes, weirs, and meters; external inpipe flow meters
Drainage	Runoff to surface-water system or evaporation ponds; percolation to aquifers	Applications managed to limit drainage; reuse through tailwater pumpback; dual-use systems with subirrigation

Source: USDA, ERS.

The primary goals of irrigation water management are (1) improved farm profitability, (2) water conservation, and (3) reduction in water-quality impacts:

1. *Improved farm profitability.* Certain crop production systems depend entirely on irrigation. With the rising cost of fresh water for irrigation, the producer should avoid waste by optimizing water use.
2. *Water conservation.* Over 90% of fresh water withdrawals used in certain western states of the United States are accounted for by irrigation use. Communities have other projects (e.g., industrial use, municipal water, and recreational use) that compete with farm use for fresh water. Prospects of a new large-scale water supply are very limited, thus making the need to conserve irrigation water very compelling.

Irrigation must be conducted judiciously to be effective. Plants need a continuous supply of water during production. When this flow is interrupted for a significant period of time, plant growth and development are adversely affected, as is yield of economic product. Some plants are able to store water against periods of drought. Crop producers producing irrigated crops must know the water needs of their crops and select the most effective method to apply the supplemental moisture. This water should be delivered to the root zone of the crop at the time of need and in amounts that can be used without waste.

The amount and frequency of irrigation depend on several factors, including the root system of the plant (depth and distribution), the water retentive capacity of the soil, the rate of water use by the plant (consumptive use), the availability and timeliness of irrigation water, and the minimum water potential that should be maintained in the root zone to avoid moisture stress. The moisture in the soil should not be allowed to decrease to near permanent wilting point, except when the crop is in the ripening stage.

In terms of the plant root system, the depth of penetration is affected by soil physical structure. Where the soil is heavy (high clay content) or has impervious layers (pans), roots become restricted in growth. Most plant roots occur in the upper layer of the soil, and hence this zone must be kept moist. All things being equal, the recommended irrigation depth for field crops ranges from 1.5 feet for grasses to as deep as 8 feet for deeply penetrating species, such as alfalfa.

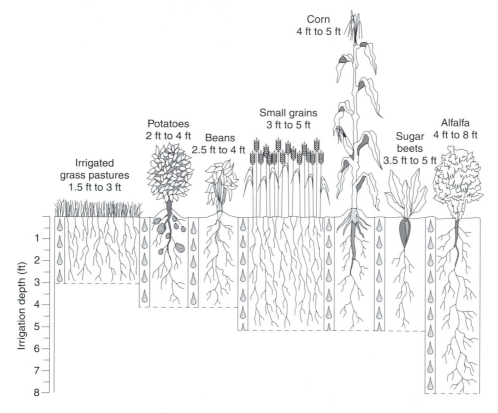

Root depth varies among crop species. Grasses are shallow-rooted, while species such as alfalfa are deeply penetrating. (Source: modified after USDA)

3. *Reduction in water-quality impacts.* Irrigated agriculture impacts off-site water quality through pollution from agrochemicals (pesticides and fertilizers) applied during crop production. Irrigation also increases field salinity and sedimentation of streams, stemming from soil erosion associated with irrigation. Irrigation-induced salinization is an important means by which cropland is lost.

Crop producers may adopt certain practices to effectively manage irrigation water, the key ones being irrigation scheduling, water-flow measurement, and irrigation drainage.

Irrigation Scheduling

Irrigation water is expensive, so farmers must determine the correct timing and quantity to avoid waste. This scheduling depends on knowledge of many factors, including evapotranspiration, the water-holding capacity of the soil, the water needs of the crop, the consequence of insufficient water on crop performance (yield), the probability of rainfall during the growing season, the available water for irrigation, and the quality of irrigation water.

There are three general categories of methods of determining irrigation schedules: (1) producer experience with the crop and production region, (2) use of simple moisture-measuring devices, and (3) detailed calculation (sometimes computerized) called the water budget method. The third method is rather involved and complex and is used mostly by irrigation professionals and consultants. Farmers may use various devices such as moisture tensiometers to determine soil moisture status. Crop water status (manifested as, for example, signs of wilting) and canopy temperature may be used as indicators of time to irrigate. It is wasteful to irrigate and have rainfall soon thereafter. As such, using fixed irrigation intervals is not a prudent irrigation strategy.

In crops such as cotton, favorable growth conditions tend to promote vegetative growth at the expense of reproductive yield. Reducing watering by the strategy of regulated deficit irrigation is known to be successful in creating a balance between the two phases of plant growth that favors economic yield.

Water–Flow Measurement

Crop producers may use water-flow measuring devices such as weirs and flumes that are installed in water conveyance systems to ensure optimal water deliveries to the field, according to the methods of irrigation scheduling.

Irrigation Drainage

Instead of allowing water to drain away, tailwater systems may be installed to recover irrigation drainage flows below the field and recycled. It may be necessary to install drainage systems to remove excess water during periods of heavy precipitation and use it during periods of drought. Methods of irrigation were discussed in Chapter 7.

SUMMARY

1. Soil water is critical to crop production. It provides the solvent for nutrients and is needed for photosynthesis and respiration.
2. Soil water is dynamic and is influenced by factors that deplete (evaporation, runoff, transpiration) or replenish (precipitation, capillary rise, irrigation) it. This is called the hydrologic cycle.
3. Water enters the soil by infiltration and moves under forces depending on whether it is saturated or unsaturated.

4. Soils differ in water-holding capacity.
5. The availability of soil moisture depends on how tightly it is held to soil particles. Too loosely held water (gravitational water) or tightly held water (hygroscopic water) is useless to crops. Capillary water is most useful for crop production.
6. Most of the water entering the plant is absorbed by passive absorption.
7. Water in the atmosphere and soil water exist in a continuum (soil-plant-atmosphere continuum).
8. Plants differ in water needs and water-use efficiency.
9. The standard method of measuring soil water is the gravimetric method. Soil water estimates are made by devices such as tensiometers.
10. Water for field crop production is stored in the soil
11. The goals of soil water management are to optimize the use of soil water while it is available, to minimize nonproductive soil water losses, and to optimize the use of supplemental moisture when applied.
12. In rainfed crop production, precipitation depended upon for all water used.
13. In terms of water depletion patterns, the variability in rainfall is more important to crop productivity than low rainfall per se.
14. In rainfed production in humid conditions, excessive rainfall may be a production problem. In dry conditions, it may be necessary to provide supplemental irrigation for the occasions in which protracted drought occurs. However, growers in dryland regions usually adopt moisture-conserving practices (e.g., catch crop, fallow) to manage soil moisture for production.
15. Water for irrigation is obtained from three primary sources: surface water, groundwater, and surface runoff.
16. Soil moisture, moisture in the atmosphere, and plant moisture occur in a soil-plant-atmosphere continuum.
17. Water quality is critical to irrigation. Sources of water may become polluted with ions, including calcium, magnesium, nitrates, sulfates, and carbonates.
18. Irrigation methods belong to one of three groups: surface (e.g., basin, furrow) sprinkler (e.g., center-pivot), or subsurface.
19. Microirrigation methods are the most water-efficient. They are not used for field crops.
20. Agricultural chemicals may be applied through sprinkler irrigation systems.

References and Suggested Reading

Brady, C. N., and R. R. Weil. 1999. *The nature and properties of soils.* 12th ed. Upper Saddle River, NJ: Prentice Hall.

CAST. 1988. *Effective use of water in irrigated agriculture.* Task Force Report no. 113. Ames, IA: Council for Agricultural Science and Technology.

Economic Research Service/USDA. 1997. *Agricultural resources and environmental indicators, 1996–97.* Agricultural handbook No. 712. Washington, DC/USDA.

Hudson, B. D. 1994. Soil organic matter and available water capacity. *J. Soil and Water Cons.* 49:189–194.

Miller, W. R., and D. T. Gardner. 1999. *Soils in our environment.* 9th ed. Upper Saddle River, NJ: Prentice Hall.

Pruitt, W. O., F. J. Lourence, and S. Von Oettingen. 1972. Water use by crops as affected by climate and plant factors. *California Agric.* 26:10–14.

Waddell, J., and R. Weil. 1996. Water distribution in soil under ridge-till and no-till corn. *Soil Sci. Soc. Amer. J.* 60:230–37.

SELECTED INTERNET SITES FOR FURTHER REVIEW

http://www.encarta.msn.com/find/MediaMax.asp?pg=3&ti=761558496&idx=461565881

Sprinkler irrigation.

http://www.encarta.msn.com/find/MediaMax.asp?pg=3&ti=761558496&idx=461530531

Furrow irrigation.

http://www.encarta.msn.com/find/MediaMax.asp?pg=3&ti=761558496&idx=461530522

Irrigation canal.

http://www.farmphoto.com/album2/html/noframe/m_P01907.asp

Aerial photo of center-pivot irrigation.

OUTCOMES ASSESSMENT

PART A

Answer the following questions true or false.

1. T F   Water enters the soil by infiltration.
2. T F   Under saturated conditions, soil water moves primarily by gravity.
3. T F   Capillary rise occurs through macropores.
4. T F   Sandy soils have higher soil infiltration rates than clays.
5. T F   In crop production, evapotranspiration is higher when plants are younger.
6. T F   Basin irrigation is the most water-efficient of all methods of irrigation.
7. T F   Water for irrigation obtained from a well is drawn from an aquifer.
8. T F   Chlorine and sodium ions are the most common sources of irrigation water pollution in humid regions.
9. T F   Application of chemicals through irrigation water is called eutrophication.
10. T F   Microirrigation is applicable to wheat crop production.

PART B

Answer the following questions.

1. Give the three forms of soil water.

   _____ _____ _____

2. What is field capacity? _____

3. What is the permanent wilting point? _____

4. The ability of the soil to retain water is called its _____.

5. Give three specific factors that influence water infiltration into the soil.

   _____ _____ _____

6. What is water-use efficiency?_____

7. Give the three sources of water for irrigation.

   _____ _____ _____

8. The amount of water required to produce a unit of dry matter is called

   _____.

9. What is an aquifer? _____

10. Give five specific ions commonly associated with water quality for irrigation.

    _____

    _____

    _____

    _____

    _____

11. Give five specific factors that affect the choice of irrigation system.

    _____

    _____

    _____

    _____

    _____

PART C

Write a brief essay on each of the following topics.

1. Describe the hydrologic cycle.

2. Describe the movement of soil water under saturated conditions.

3. Discuss evapotranspiration and its role in crop productivity.

4. Discuss the movement of water through plants.

5. Describe methods of estimating soil water content.

6. Discuss the concept of the soil-plant-atmosphere continuum.

7. Describe the use of fallow in water management for crop production.

8. What is microirrigation?

9. Describe the method of furrow irrigation.

10. Describe the method of sprinkler irrigation.

11. Discuss the importance of water quality in crop irrigation.

12. What are the general goals of managing soil water in crop production?

PART D

Discuss or explain the following topics in detail.

1. Discuss the importance of irrigation in U.S. crop production.

2. Discuss the role of the Ogallala aquifer in U.S. crop production.

3. Why is the quality of irrigation water important?

4. Discuss the importance of water-use efficiency in crop production.

10

Pests in Crop Production

PURPOSE

This chapter explores the living organisms in the environment and the non-living environmental factors that cause diseases and other forms of damage to crop plants, as well as how they are managed in crop production.

EXPECTED OUTCOMES

After studying this chapter, the student should be able to:

1. List and describe the categories of organisms that cause disease.
2. Discuss the factors that cause disease.
3. Describe the disease cycle.
4. Discuss the strategies and mechanisms plants adopt to resist disease.
5. Discuss the genetic basis of disease resistance.
6. Discuss how pathogens affect crop productivity.
7. Describe insect pests, their classification, and their economic importance.
8. Describe weeds, their classification, and their economic importance.
9. Discuss abiotic disease agents.
10. List and describe important crop pests.
11. Discuss the principles of pest management.
12. Describe how pest attacks can be prevented.
13. List and discuss the methods of pest management.
14. Describe the steps in selecting and applying pesticides safely.
15. Discuss the impact of pesticides on the environment.

KEY TERMS

Active ingredient
Aerosol

Annual weeds
Biennial weeds

Biological pest
management (control)

Complete metamorphosis

Cultural pest management (control)

Disease triangle

Economic injury

Formulation

Horizontal resistance

Host range

Hypersensitive reaction

Incomplete metamorphosis

Integrated pest management (IPM)

Legislative pest management (control)

Lethal dose (LD_{50})

Lifecycle

Mechanical pest management (control)

Obligate parasites

Organochlorines

Organophosphates

Overwintering

Pathogenicity

Perennial weeds

Pesticides

Phytoallexins

Plant quarantine

Postemergence herbicides

Preemergence herbicides

Surfactants

Toxicity

Vertical resistance

TO THE STUDENT

Plants, as primary producers, are the main sources of energy that support life on earth. Plants are hence preyed upon by a wide variety of consumers, from microbes to mammals. Obtaining the crop potential or maximum yield in production is elusive for the reason that crop production is subject to physical, chemical, and biological factors in the environment that impact crop productivity. Yield reduction from biological sources alone is estimated at about 30% or more of the potential yield. These yield losses occur at various stages in the production operation, from seedling establishment, to crop growth and development, to harvesting and storage. Some of the organisms in the plant's environment are beneficial to the plant, while others are destructive. In order to manage the plant biotic environment effectively, there is the need to understand how these organisms fit into the ecosystem, their nature, and their behavior. It is important to know the lifecycle of an organism in order to know the stage in which it is most vulnerable to be effectively controlled. This chapter focuses on the biology of the organisms. Chapter 11 focuses on how to manage pests and reduce their destructive effects on crop plants. The term *pest* is all-inclusive for organisms that are bothersome or annoying and undesirable to humans. In this chapter, emphasis is placed on economic loss caused to crops by pests. Pests and adverse environmental conditions cause diseases.

Agricultural production occurs within an ecosystem (i.e., an agroecosystem) in which biotic and abiotic components interact. Agriculture is a "managed ecosystem," as opposed to a natural ecosystem. The producer manages resources, making decisions that are influenced by nature, economic factors, and political factors. Since crop production is goal-oriented, the producer deliberately restricts interaction among organisms in the agroecosystem in order to favor the crop being grown for maximum productivity. Agricultural pests reduce crop productivity. The crop producer must therefore adopt appropriate strategies to exclude harmful associations without endangering the environment. The methods used should also be economical. You will learn that the mere sight of a pest does not warrant the implementation of a control measure. Since pest management can be expensive, the pest population should be at a critical threshold, beyond which economic yield is jeopardized, before pest control is essential. You will also learn that, whereas modern crop production technology enables producers to use chemicals in pest management, there are non-chemical strategies that can be used. Chemicals pollute the environment and must be used judiciously.

Some prefer the term *management* rather than *control* of pests because, in practice, producers rarely control or eliminate any pest. Rather, pest populations are reduced to levels at which they cause no economic loss.

10.1: CONCEPT OF DISEASE

Pathogens and improper physiological environment may cause plants to malfunction. The biotic factors that cause yield reduction are of both plant and animal origins. They may be placed into four general categories:

1. Disease-causing organisms
2. Plants as pests (weeds)
3. Insect pests
4. Non-insect invertebrate or vertebrate pests

A plant is said to be diseased when it deviates from one or more of its normal physiological functions (e.g., cell division, absorption of water and minerals, photosynthesis, and reproduction). The deviation is caused by an agent of disease, which may be either a living pathogenic organism (called a *pathogen*) or some physical environmental factor. The affected plant is unable to perform certain physiological functions to the best of its genetic potential. A disease may be defined as the malfunctioning of afflicted cells and tissues attributed to the presence of a causal agent and which produces a symptom (e.g., change in form, physiology, integrity, or behavior).

The extent of incapacitation is dependent upon the causal agent as well as the plant. Further, the kind of cells and tissues infected will determine the kind of physiological function to be affected first. The first effect will then trigger other effects that may eventually result in the death of the plant. For example, if the root becomes afflicted by root rot, the conducting vessels may be damaged (as is the case in vascular wilt), translocation of water and minerals is interrupted, photosynthesis is adversely impacted, plant growth is reduced, and yield is reduced. Infections involving the foliage leaf (e.g., blight and leaf spot) reduce photosynthetic area and hence photosynthetic output which, in turn, reduces growth and yield. Diseases do not always weaken or destroy afflicted cells. On the contrary, certain diseases stimulate cell division *(hyperplasia)* or cell enlargement *(hypertrophy)*. These abnormal events lead to tissue proliferation, resulting in amorphous overgrowths (e.g., tumors).

There are tens of thousands of crop plant diseases that must be classified to facilitate their study, identification, and management. Just like classification of plants, there are several ways to classify plant diseases, operationally or scientifically.

Operational categories of plant disease include the following:

1. *Symptoms caused.* Diseases cause numerous symptoms, such as rots (root rot), canker, blights, rusts, smut, mosaic, and yellows.
2. *Plant organ afflicted.* General categories of plant disease by this classification include root diseases, foliage diseases, fruit diseases, and stem diseases.
3. *Plant category.* Certain diseases are associated with field crops, ornamentals, trees, vegetables, fruit trees, and turf.
4. *Causal organism or factor.* This system of classification is one of the most useful and widely used. The specific mechanisms by which diseases are produced depend on the causal agent. By knowing and studying the causal agent, as well as its nature, functioning, and behavior, scientists are more readily able to develop effective strategies to control the agents. On the basis of causal agent, plant diseases may be classified into two broad groups: *biotic* (infectious) and *abiotic* (non-infectious):
   a. *Biotic (infectious) diseases.* These diseases are caused by pathogens and can be transmitted from one victim plant to another. The organisms are fungi,

viruses and viroids, parasitic higher plants, prokaryotes (bacteria and mycoplasmas), nematodes, and protozoa.

b. *Abiotic (non-infectious) diseases.* These diseases are caused by environmental
factors and thus not infectious. They include abnormal levels of growth requirements (e.g., extreme high or low temperature, moisture excess or deficit,
low or intense light, nutrient deficiency, nutrient toxicity, lack of oxygen), improper cultural practices, pollution, and improper pH.

10.2: CLASSIFICATION OF PLANT DISEASES

What microorganisms are responsible for plant diseases? What is the relative economic
importance of the diseases they cause? A variety of microorganisms are pathogenic on
plants. Plant pathogens may be grouped into the following main categories:

1. Viruses and viroids
2. Bacteria
3. Fungi
4. Mycoplasma-like organisms
5. Protozoa
6. Nematodes
7. Parasitic higher plants

10.2.1 VIRUSES ARE OBLIGATE PARASITES

Viruses need a host tissue to stay alive (i.e., obligate parasites). They are not cells and do
not consist of cells. They are not true animals and are incapable of digestion and respiration. They consist of a core of nucleic acid that may be RNA or DNA encased in a protein or lipoprotein coat (called a capsid). They are variable in shape and size. Most plant
viruses are RNA viruses.

Viroids are small, naked, single-stranded, circular RNAs. Plant viruses enter plants
through mechanical wounds or vectors (carriers). Sometimes, an infected pollen grain
carries a virus into an ovule. Viruses are primarily systemic, occurring in the host's vascular (phloem) fluids. Once in the phloem, the virus moves towards the apical meristem.
As such, sucking insects (especially aphids) and chewing insects (e.g., thrips) transfer viral diseases from infected plants to healthy ones through feeding. Viruses may be seed-
transmitted (though the ovule of infected plants). Certain nematodes, mites, and fungi
also transmit viruses.

Viral disease symptoms include *mosaics* (light-green, yellow, or white patches
mingled with normal green) and especially the stunting of growth. The pattern of discoloration is variable and may be described accordingly as mottling, streak, ring, line,
vein clearing, vein banding, or chlorosis. Viruses seldom kill their plant hosts. Why do
you think this is advantageous to the pathogen? Viruses instead diminish plant growth
and development, resulting in stunted growth and consequently reduced productivity.
Some infected plants can show no symptoms *(asymptomatic).* In fact, these pathogens
(latent viruses) can remain in the host for as long as the host remains alive. Stunted
growth is caused by a reduction in growth regulatory substances.

Viral infections are customarily described in a variety of ways, but especially according to the host (host specificity) (e.g., tobacco mosaic virus, or TMV; maize dwarf mosaic
virus, or MDMV) and the most visible symptoms caused. They are also described by the
type of nucleic acid (e.g., DNA or RNA), morphology, mode(s) of transmission, and others.

Viral diseases are not controlled by the use of pesticides (chemicals that are used to control plant pests). The most effective control is prevention from entry through quarantine, inspection, and certification. Resistant cultivars are commonly used to control viral diseases. Another method of control is to expose infected tissue to high heat (38°C or 100°F) for 2 to 4 weeks (called *heat therapy*) in order to inactivate the virus. Growing tips of infected plants are free of viruses. These tips may be retrieved and propagated by tissue culture.

10.2.2 ALMOST ALL PATHOGENIC PLANT BACTERIA ARE ROD-SHAPED (BACILLI)

There are two kinds of prokaryotes that cause disease in plants: *bacteria* and *mycoplasma-like organisms.* Ubiquitous in the environment, these unicellular organisms occur in one of four forms: spherical *(cocci),* rod-shaped *(bacilli),* spiral *(spirilli),* and filament *(filamentous).* They may also be classified according to their reaction to the Gram's stain as either *Gram positive* (violet) or *Gram negative* (pink-red). Bacteria multiply by *binary fission* (divide into two parts). Most bacteria are facultative saprophytes and hence useful in decomposing organic wastes. Many species have numerous pathovars (strains that differ in the plant species they infect). Plant pathogenic bacteria occur in places where it is moist and warm. Bacteria contain small, circular DNA that is used in biotechnological research. One genus of importance to biotechnology is the *Agrobacterium.* Some bacteria thrive in plant hosts but many (e.g., common scab of potato or *Streptomyces scabies*) are soil inhabitors.

Bacteria are spread by a variety of means, including wind, splashing from irrigation or rain, and use of infected seed or other propagation material. They gain entry into the host through natural pores (e.g., stomata) or wounds. Bacterial diseases include soft rots, bacterial cankers, wilts, overgrowth, scabs, and rots. These usually affect the stem and roots of the plant.

Bacteria are generally intolerant of temperatures above 125°F (51.7°C). Since they spread through wounds, pruning tools should be properly cleaned between plants. Sanitation should be observed during crop production so as to destroy sources of disease and protect wounds. Like viral diseases, the best control is to use resistant cultivars.

Not all bacteria are pathogenic. Leguminous crop plants (e.g., soybean) benefit from symbiosis (the mutually beneficial plant-bacteria association that fixes atmospheric nitrogen for plant use). In commercial production, soybean seeds may be treated with artificially cultured *Rhizobia* inoculant.

10.2.3 MYCOPLASMA-LIKE ORGANISMS

Mycoplasma-like organisms are wall-less microbes that occur in the phloem tissue. They are capable of reproducing themselves, have both DNA and RNA, but lack a cell wall. Mycoplasma-like diseases have been identified in vegetables and field crops.

Mycoplasma-like organisms are systemic and transmitted by vectors such as grasshoppers and aphids. Spraying is used to control the insect vectors. Tetracycline application has proved successful in the control of pear decline disease.

10.2.4 FUNGI CAUSE MOST OF THE IMPORTANT INFECTIOUS DISEASES OF CROP PLANTS

Over 100,000 fungal species have been identified, most of which are saprophytic (live on dead organic matter). About 8,000 of these cause diseases in plants. They have been responsible for a number of crop destruction episodes that have led to famines.

Fungi attack flowers, seeds, leaves, stems, and roots. Most are multicellular (even though some are unicellular) and have no chlorophyll. Unable to photosynthesize, fungi live parasitically. They may be classified into four categories. Usually, each plant fungal disease is caused by only one fungus.

1. *Obligate saprophytes.* These live only on dead plant and animal tissue.
2. **Obligate parasites**. These live only on living plant tissue.
3. *Facultative saprophytes.* These usually live on living tissue but can also live on dead tissues under certain conditions.
4. *Facultative parasites.* These are normally saprophytic but occasionally parasitic.

Obligate parasites.
Organisms that must live as a parasite and cannot otherwise survive.

Fungi are spread primarily by bodies called *spores* (with a few exceptions) that vary in shape, size, and color. Upon germination, spores produce structures called *hyphae* (singular is *hypha*) that grow and branch to produce other structures called *mycellia* (singular is *mycellium)* or fungal bodies. Spores spread by wind, water, insects, and other agents. They infect through wounds, through natural openings, or by direct penetration of the epidermis. Nearly all the pathogenic fungi spend part of their lifecycles on the plant host and part on plant debris or soil. Their survival depends on the temperature conditions that prevail.

Fungal diseases include leaf spots, blights, mildews, rusts, and wilts. Some spores grow on the leaf surface as molds (e.g., *Helminthosporium* diseases). Other fungi, such as *Septoria* and *Ascochyta,* live embedded in plant tissue. Key symptoms of fungal diseases are *necrosis* (tissue death), *hypotrophy* (stunted growth), and *hypoplasia* (excessive growth):

1. *Necrotic symptoms.* The common necrotic symptoms include leaf spot (localized lesions), blight (rapid browning leading to death of organ), die back (necrosis starting at the tip of the organ), canker (localized wound that is usually sunken), anthracnose (necrotic and sunken ulcerlike lesions), damping off (rapid collapse and death of young seedlings), and scab (localized usually raised lesions).
2. *Symptoms involving growth.* The key symptoms involving hypertrophy or hyperplasia include clubfoot (enlarged wilts), gall (enlarged plant part), witches' broom (profuse upward branching), and leaf curl (curling of leaves).
3. *Other symptoms.* Other common symptoms associated with fungal diseases are wilts (vascular system collapse), rust (numerous small lesions that appear as rusty color), and mildew (covering of mycelium and fructification of the fungus).

Fungal diseases are relatively easy to control. Control measures include the use of chemicals, diseased seed, and resistant cultivars. There are protective pesticides that are commonly used as seed treatment or are applied to the surface of plants. Chemical control is the most effective control measure of the disease. A new generation of systemic fungicides is in use. Sanitation also helps to keep fungi under control. Plant remains should be destroyed, especially those infested.

Like bacteria, some fungi are beneficial to humans. These include *Penicillium* (from which the antibiotic penicillin is produced), edible mushrooms, and yeast (used in fermentation of alcoholic beverages such as beer and for leavening bread and other foods). Like symbiosis in legumes, fungi-plant associations called michorrhizae occur in species such as corn, cotton, soybean, and tobacco, aiding in the absorption of phosphorus.

10.2.5 NEMATODES

Nematodes or eelworms may live parasitically on living plants and animals. Plant-parasitic nematodes are microscopic and mostly inhabit the soil. Some of them are ectoparasites (remain in the soil), while others are endoparasites (enter the plant roots, tubers, bulbs, etc.). A few nematodes are able to infest the leaf (foliar nematodes). Infected roots appear knotted or with galls. Tubers and roots of root crops are malformed. Root knotting interrupts the uptake of soil nutrients and water. Infested plants may, therefore, show symptoms of moisture stress or nutrient deficiency as well as reduced crop yield. Using chemicals or resistant cultivars controls nematodes.

Important nematodes of economic importance in crop production include the root knot and cyst nematodes.

Root Knot Nematodes

Almost all cultivated plants are susceptible to root knot nematodes (*Meloidogyne* spp.). Affected plants develop knots or galls at the points of infection. These swellings are irregular and easily distinguished from the roundish nodules developed from *Rhizobia* infection of legume roots. When root or tuber crops are infected, the roots or tubers are deformed. The development of infection by other soil pathogens such as *Pythium* and *Rhizoctonia* are found to be accelerated through galls formed by nematodes.

Cyst Nematodes

Cyst nematodes cause cysts on the roots and cause root proliferation. The cyst nematodes of importance in crop production are the *Glabodera* and *Heterodera*. *G. rostochiensis* affects potato especially, as well as tomato and egg plant. *H. glycines* affects soybean, while *H. avenae* and *H. trifolii* affect cereals and clover, respectively.

10.2.6 PARASITIC PLANTS

Of the more than 2,500 species of higher plants known to be parasitic on other plants, few are economic pests of cultivated crops.

Dodders

Dodders (*Cuscuta* spp.) infest alfalfa, sugar beet, potato, and other species. They are composed of orange or yellow vine strands that smother above-ground parts of plants. The plant produces seed in summer that overwinters in the field or in harvested produce. Chemical control is effective.

Witchweed

Witchweed (*Striga* spp.) is a parasite of corn, tobacco, rice, sugarcane, and some small grain. Infected plants are stunted and chlorotic, followed by wilting and death in cases of severe infestation. The parasite has attractive features. It grows at the base of the infected plant, where it parasitizes the roots of cultivated plants.

Broomrapes

The causal parasite is *Orobanche* spp. It is a whitish-to-yellow annual plant. It infests herbaceous crop plants, including tobacco, potato, clover, and alfalfa. It produces seeds that overwinter and can persist in the soil for many years. When a susceptible root grows near the seed, germination is induced, followed by penetration of the host roots.

10.3: NATURE OF DISEASE

Disease is the product of the interaction among the causal organism, host, and certain factors within the environment. The capacity of a pathogen to cause disease is called its **pathogenicity**. The variety of plants a pathogen can grow on is its **host range**. Obligate parasites tend to be limited to one host *(host-specific),* while non-obligate parasites can cause diseases in various plant species. What are the essential requirements for disease conditions to develop? In order for disease to occur, three factors are essential:

1. Pathogen (causal organism)
2. Susceptible host (e.g., plant)
3. Favorable environment

These three factors form what is called the **disease triangle** (Figure 10–1). The degree of disease occurrence depends on the nature of these factors. A little amount of disease agent may multiply rapidly and become established if the host is very susceptible and the environment ideal. On the other hand, a large amount of pathogen inoculant on a plant in an unfavorable environment may not cause any pathogenic effect at all, or only mildly so.

Each pathogen has its biological lifecycle. The disease cycle is the series of events that leads to disease development and perpetuation in a host. There are several general stages in disease development:

1. Inoculation
2. Penetration
3. Infection
4. Dissemination
5. Overwintering/oversummering

10.3.1 INOCULATION

The contact between host and pathogen is called *inoculation.* The pathogen or part of it that is involved in the inoculation is called the *inoculum.* This may be spores, viruses, bacteria, or other microorganisms. The initial unit of inoculum that comes into contact with the host is called the *primary inoculum.* It produces the *primary infection.* As indicated previously, the inoculum may be spread by vectors or other agents such as water and air.

Pathogenicity. The capacity of a pathogen to cause disease in a host.

Host range. The variety of plant species a pathogen can successfully invade and grow on.

Disease triangle. The concept that disease occurs only when three factors, namely pathogen, susceptible host, and favorable environment, occur at the same time.

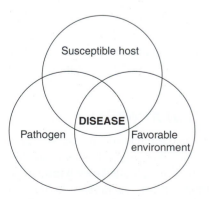

FIGURE 10–1 The disease triangle. Disease will occur only when the pathogen interacts with a susceptible host under favorable conditions. The presence of a pathogen alone is not sufficient to cause disease.

10.3.2 PENETRATION

Penetration is the stage at which the pathogen enters the host tissue. This occurs as germination of spores in fungi. Some pathogens (e.g., fungi and nematodes) and parasitic plants exert mechanical pressure to gain entry into plants. Other pathogens secrete enzymes and other chemical compounds that soften the cell wall to facilitate entry. Penetration may also occur through natural pores (e.g., stomata), wounds (caused by equipment damage during cultivation, hail damage, and other/causes), or directly through the cuticle by a growth called *penetration peg*. Penetration may trigger a defense mechanism to resist the entry of the pathogen, thereby curtailing disease development.

10.3.3 INFECTION

Infection is the process whereby the pathogen establishes contact with host cells or tissues to be nourished for growth and multiplication. For infection to occur, the host must be susceptible to the race of the pathogen (i.e., the pathogen must be virulent on the host). Once entry into the host has been gained, the pathogen starts to invade the cells and tissues to colonize them. This may result in visible signs or symptoms. Colonization (growth and reproduction) may occur primarily on the surface, as in powdery mildew, or the vascular system, as in wilts. Sometimes, only one or a few cells are colonized (called *localized infection*). Viruses, on the other hand, invade the entire plant (called *systemic infection*). Symptoms may appear almost immediately after entry of the pathogen. However, some infections may remain *latent* and undetectable for a period. The time between inoculation and symptom development is called the *incubation period*.

How do plants react to invasion by pathogens? There are three general responses:

1. *Overdevelopment of tissue.* Affected tissue overdevelops, resulting in symptoms such as curling or galls.
2. *Underdevelopment of tissue.* Plant organs and the whole plant may become stunted, as occurs in viral infections.
3. **Hypersensitive reaction**. The pathogen is contained through drastic measures that result in tissue death around the site of invasion (necrosis), as occurs in leaf spots and blights, or rots, as in canker and soft rot.

Hypersensitive reaction. A strategy adopted by a host plant to contain the invasion of a pathogen by self-induced death of the tissue around the site of invasion.

10.3.4 DISSEMINATION

Disease-causing organisms are spread by vectors and other agents to new hosts. Nematodes, fungal zoospores, and bacteria move only short distances. Consequently, they spread slowly from one plant to another. Spores, on the other hand, can be carried long distances by air. Bacteria, spores, and nematodes are also disseminated by water (e.g., rain, splashing of irrigation water). Insect vectors include aphids and leafhoppers (for viral infections). Activities of humans (handling of plants) also contribute to the spread of disease. Humans are blamed for introducing pathogens such as powdery mildews and downy mildews of grapes.

10.3.5 OVERWINTERING OR OVERSUMMERING

Pathogens persist in the environment, in the ground, or on plant material in a certain stage in the lifecycle until the next crop growing season arrives. When a host dies in winter, the pathogen must be able to survive the harsh winter conditions (called **overwintering**). On the other hand, if the host dies in summer, the pathogen must *oversummer* in order to survive. Some pathogens are soil inhabitants (e.g., *Pithium* and *Fusarium*), while others

Overwintering. Strategy by which field pests survive the winter.

are soil transients, living most of their lifecycles on the host. Viruses and mycoplasmas survive only in living tissue.

10.4: PLANTS HAVE CERTAIN DEFENSE MECHANISMS TO WARD OFF PATHOGENIC INVASION

How do plants defend themselves against attack from pathogens? Upon inoculation, the host plant may resist further development of the disease cycle by either *active* (preexisting) or *passive* (induced) resistance strategies. These strategies involve *structural* and *biochemical defense mechanisms:*

1. *Structural defense.* Plants have structural features such as water-repellent waxy deposition on the epidermis or a thick cuticle that hinders penetration by pathogens. Some plants also have pubescence (hairlike structures) on the leaf and other parts that interfere with the pest organism's lifecycle.

   Should the pathogen succeed in penetrating the host tissue, structural degradation of tissue (e.g., necrosis in hypersensitive reaction) restricts the spread of the disease. In some species, a cork layer forms to contain the invasion.

2. *Biochemical defense.* Certain species produce fungitoxic exudates when invaded by pathogens. Red onions resist onion smudge fungus by exuding toxins called protocatechaic acid and catechol. White onions do not produce these toxins and are thus susceptible to the fungal attack.

   Similarly, potato cultivars that are low in reducing sugars are less susceptible to bacterial soft rot caused by *Erwinia carotova* var *atroseptica* than those that are high in reducing sugars.

   Injured plants exude certain chemicals called **phytoallexins**. For example, injured bean plants exude phaseolin, while potato and pepper exude rishitin and capsidol, respectively.

Phytoallexins. Chemicals exuded by certain plants upon injury to ward off pest attack.

10.4.1 SOME PLANTS ARE GENETICALLY EQUIPPED TO RESIST CERTAIN DISEASES

Crop plants may resist disease through several types of resistance:

1. *Non-host resistance.* Plants that are taxonomically outside the host range of the pathogen will not be infected by the pathogen. Certain diseases occur only in certain taxonomic groups. As such, a certain corn disease may not affect potatoes even if the most favorable conditions prevail.

2. *Genetic or true resistance.* The plant with true genetic resistance possesses a certain gene (or genes) that is able to resist the virulence gene(s) of the pathogen.

3. *Apparent resistance.* The plant, through other strategies or mechanisms, is able to avoid the disease.

Vertical resistance. The genetic resistance in a host that is conditioned by one or a few genes.

Genetic (True) Resistance

True resistance to pathogens is genetically controlled, enabling plant breeders to breed cultivars with resistance to certain diseases. According to the origin of control, there are two forms of genetic resistance: *nuclear* and *cytoplasmic*. Further, there are two kinds of nuclear resistance: **vertical resistance** and **horizontal resistance**. Is one of these more advantageous in crop production than the other?

Horizontal resistance. Genetic resistance of a plant host to disease that is conditioned by several to numerous genes.

Nuclear Resistance   This form can be either vertical or horizontal.

1. *Vertical Resistance*   Also called *major gene, oligogenic, monogenic, qualitative, race-specific,* or *differential resistance,* vertical resistance is controlled by one or a few genes. Thus, it is simply inherited and relatively easy to breed into cultivars. Cultivars protected by major gene resistance usually have race-specific resistance. They show complete resistance to a particular pathogen under a wide variety of environmental conditions. However, a simple mutation in the gene (new race) abolishes the host resistance to the disease. This resistance is also described as not durable. The host and pathogen are not compatible. Consequently, the host exhibits a hypersensitive response to invasion. Even though a crop cultivar may have one or two resistance genes, a single species may have over 20 resistance genes against a single pathogen (e.g., wheat has 20 to 40 genes for resistance against Puccinia recondita).

2. *Horizontal Resistance*   Horizontal resistance is controlled by several to numerous genes, each contributing to the total resistance. It is also described as *minor gene, non-specific, general, quantitative, adult plant, field, polygenic, multigenic,* or *non-differential resistance.* This resistance is usually not complete. It reduces certain aspects of the disease (e.g., sporulation, infection frequency, spread of the disease in the field). It is more difficult to incorporate into a breeding program but is more durable than vertical resistance (i.e., mutation effects do not easily overcome it). It is environmentally labile, and hence resistance is variable from one environment to another. All plants have some degree of horizontal resistance. It does not protect the crop plant from infection but, rather, slows down the development and spread of the disease.

Cytoplasmic Resistance   The genes involved in vertical resistance and horizontal resistance are nuclear in origin and hence subject to Mendelian laws of inheritance. However, certain diseases are controlled by genes that are extranuclear or extra-chromosomal, occurring in organelles in the cytoplasm. Examples of diseases under cytoplasmic control are the southern corn leaf blight *(Bipolaris Helminthosprium maydis)* and yellow leaf blight *(Phyllosticta maydis).* The normal cytoplasm has genes for resistance while the genes are suppressed in hybrids with the Texas (T) male-sterile cytoplasm.

Apparent Resistance

Certain plants without resistance genes are able to avoid diseases through two kinds of mechanisms: *escape* and *tolerance.*

1. *Escape*   As described previously, the balance among three factors (pathogen, host, and environment) determines the severity of a potential outbreak of disease. Even though a pathogen may be present on the crop plant, absence of the appropriate environment may cause the plants to escape the infection. Some plants are susceptible to a disease only at a certain stage in growth. For example, younger tissues are more susceptible to certain diseases such as powdery mildew *(Phytium)* and viral infection than older tissues. The producer may also vary the cultural conditions (e.g., spacing, rate of planting, fertilization, sanitation, rouging, use of vigorous seeds) to prevent or reduce disease incidence in the crop. Monoculture favors the buildup of disease inoculum, while mixed cropping or crop rotation lowers the pathogen population. Many plants escape diseases due to lack of moisture and low humidity. In potato production, a low soil pH prevents potato scab *(Streptomyces scabies).*

2. *Tolerance*   Certain plants can be productive while harboring pathogens and are said to be tolerant to the disease. Viral infections usually do not kill the host plants but cause reduced productivity through stunting of plants. Tolerant plants have certain specific heritable characteristics that enable them to allow the pathogen to multiply without any adverse effects. The host plants are either able to inactivate toxins or are able to compensate for any dysfunctional effects and still be productive.

Genetic Basis of Disease Incidence

Disease infection depends on the host and pathogen (and environment), each with its separate genetic properties. The *degree of susceptibility* or *resistance* of the host to infection is called the *host reaction*. The degree of pathogenicity of a pathogen is called its *virulence.* There are genes for resistance (in the host) and virulence (in the pathogen). These two sets of genes are thought to be closely related to the extent that, for each resistance gene in the host, there is a corresponding virulence gene in the pathogen. This is called the *gene-for-gene hypothesis.* The genes for resistance are dominant *(R),* while susceptibility is controlled by a recessive gene *(r).* In the pathogen, *avirulence* (inability to infect) is under dominant gene control *(A),* while virulence is recessive *(a).* A host is resistant to a disease only when it has the resistant gene *(R),* while the pathogen lacks the gene for virulence *(A)* (that is, a host × pathogen gene combination of *AR*). Disease will occur if the host is susceptible *(r)* or if the pathogen is virulent *(a).* In the *Ar* combination, disease occurs because the pathogen, though lacking the specific virulent gene, has other genes for virulence.

10.5: HOW PATHOGENS AFFECT CROP PRODUCTIVITY

What is the economic importance of plant disease? In what specific ways are crop plants affected by disease? Disease adversely affects plant physiology and hence metabolism. Disease may result in the death of plants or their parts. The major physiological processes affected are photosynthesis, respiration, translocation, growth, death, and economic product.

10.5.1   PHOTOSYNTHESIS

Foliar pathogens (pathogens that attack plant leaves, such as blight) reduce the photosynthetic surface, or leaf area. The effect of photosynthesis on plant growth and subsequently on yield depends on the severity of reduction in leaf area. It has been found that certain species (e.g., broad bean) are able to grow normally while suffering about 30 to 40% leaf loss. The remaining leaves somehow increase photosynthetic efficiency to compensate for decreased leaf area. In addition to leaf loss, pathogenic infections may degrade chlorophyll, producing chlorosis (yellowing of leaves). The total amount and the efficiency of chlorophyll may both be reduced as a result of viral infections that produce chlorosis. Starch metabolism is adversely affected by disease. Carbon dioxide fixation is also decreased. Thus, there are less assimilates for plant use and biological and economic yield. Leaf spots are particularly known for reducing leaf area. Shoot blights also cause such an effect.

10.5.2   RESPIRATION

Plants under pathogenic attack have been known to have increased respiration. In fact, where the infection is caused by biotrophic pathogens or obligate parasites (parasites that depend solely on living tissue for nutrients), host plant respiration has been observed to double. The host's metabolism is generally stimulated. Synthesis (or anabolism) is increased to meet the demand for nutrients under infection. Diseased plants also appear to have a shift in respiratory balance from glycolysis to the pentose phosphate pathway. Infected tissue frequently has low ATP.

10.5.3 TRANSLOCATION

Plants infected by obligate parasites tend to experience translocation of photosynthates to regions of infection. Pathogens called *vascular wilts* (caused by *Fusarium, Verticilium,* and *Ceratocytis*) invade the vascular system of the host, blocking the transpiration stream and producing wilting of the plant. *Necrotrophs* (parasites that cause immediate death of the tissue through which they pass) are also capable of directly affecting vascular transport by killing the plant tissue. Root and stem rots affect uptake of nutrients.

10.5.4 GROWTH

The adverse effect of disease on plant physiological processes (photosynthesis, respiration, and translocation) results in reduced biomass accumulation. Growth hormone levels, and hence growth regulation, adversely affect plant growth. Viral infections are noted for stunting plants rather than killing them.

10.5.5 PLANT DEATH

While certain diseases reduce growth, others are capable of completely destroying the entire plant or the part of economic importance to the producer. Damping off disease kills plant seedlings (reduces biomass), leading to an incomplete crop plant stand and reduced economic yield. In the case of grain pathogens, an important disease of field crops that destroys grain, thus leading to reduced economic yield, is the ergot fungus *(Claviceps purpurea).*

10.5.6 ECONOMIC PRODUCT

Whereas all the previously discussed effects of pathogens eventually adversely impact crop yield, some diseases directly affect the marketable plant part (e.g., fruit rot).

10.6: INSECT PESTS

Insect pests are widely adapted. Insects belong to the phylum Arthropoda (have jointed legs, exoskeleton, segmentation, bilateral symmetry) and class Insecta (true insects). Class Arachnida (spiders and mites) along with true insects constitute the source of most pests of plants in production. A majority (about 80%) of animal life consists of insects. Insects are widely adapted and distributed all over the world. They may cause direct damage to crop plants or be carriers (vectors) of pathogens. About 600 species of insects are considered crop pests.

The insect orders that are important to crop production are Lepidoptera, Coleoptera, Hymenoptera, Diptera, Thysanoptera, Aphididae, Pseudococcidae, and Cicadellidae. Examples of important pests in these orders are presented in Table 10–1.

Insects may be classified in various ways, for example, according to their lifecycle or feeding habit.

Table 10-1 Insect Orders of Importance to Crop Production

Order	Examples
Thysanura	Silverfish
Collembola	Springtails
	Several families damage seedlings and succulent stems.
Othoptera	Crickets, grasshoppers, and locusts
	Destroy roots or leaves
Phasmida	Stick insects
	Defoliate plants
Isoptera	Termites
	Destroy roots
Suborder Homoptera	Leafhoppers, whiteflies, aphids, mealy bugs, soft scale, and armored scale
	All are sap-sucking insects.
Suborder Heteroptera	Capsid/mosquito bugs, stink bugs, and shield bugs
	All are sap-sucking insects.
Thysanoptera	Thrips
	Cause leaf rolling and folding
Coleoptera	Beetles
	Biting and chewing mouthparts
Diptera	Flies, gall midges, shoot flies, fruit flies
	Larvae (never adults) are pests, destroying plant parts.
Lepidoptera	Leaf miner, stem borer, stinging/slug caterpillars, loopers, moths, armyworms, cutworms, butterflies, swallowtails
	Only larvae are destructive to plants, destroying leaves, stems, and other parts.
Hymenoptera	Sawflies, ants
	Larvae destroy leaves.
Acarina	Mites
	Piercing and sucking insects

Source: Extracted from list by Dennis, S. H., and J. D. Hill. 1994. *Agricultural Entomology*. Portland, OR: Timber Press.

10.6.1 LIFECYCLE

Based upon metamorphosis (the process of change an insect passes through), insects may be classified into four categories (Figure 10–2):

1. *No metamorphosis.* These insects are hatched as miniature adults. The young look like small versions of adult insects. This type of change occurs in Thysanura and Collembola.
2. *Gradual metamorphosis.* The young of such insects are called nymphs, and do not have all adult features. Insect orders with this type of metamorphosis include Homoptera, Isoptera, and Orthoptera (e.g., grasshopper, aphid).
3. **Incomplete metamorphosis**. The young change shape gradually from a naiad to an adult, as found in Odonata.
4. **Complete metamorphosis**. The **lifecycle** of insects with this type of metamorphosis consists of four distinct stages, namely egg, larva, pupa, and adult. Examples of insect orders with this type of metamorphosis include Hymenoptera, Lepidoptera, and Diptera (e.g., butterfly, housefly, bee). The most destructive stages are the larva (caterpillar) and adult.

Lifecycle. The set of stages an organism goes through from birth or germination, through maturity, and eventually death.

Incomplete metamorphosis. The insect lifecycle in which the young changes shape gradually to adult.

Complete metamorphosis. An insect lifecycle that is characterized by stages that are morphologically different from each other.

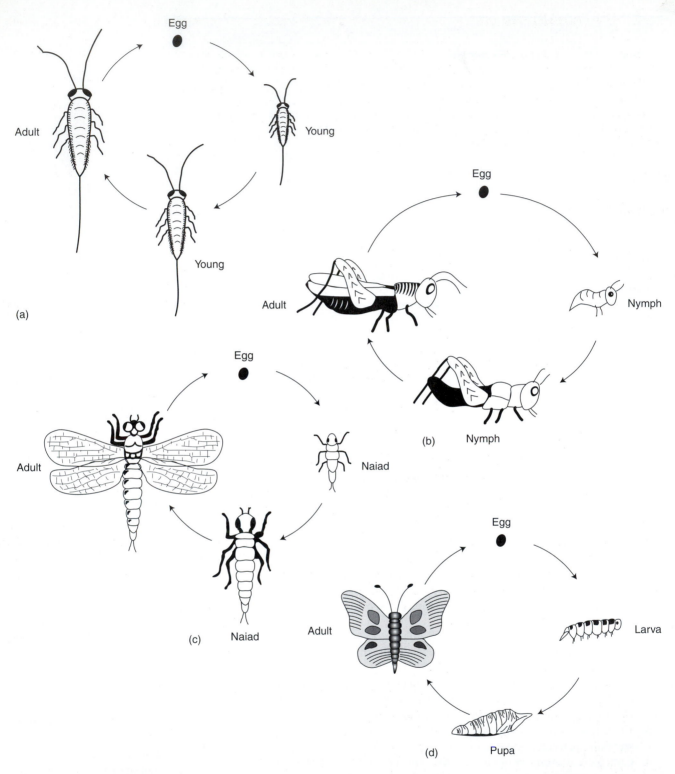

FIGURE 10–2 The lifecycle of insects consists of several stages of growth and development, starting from egg to adult. (a) Some insects have no metamorphosis, their young being miniature adults. (b) Some insects have gradual metamorphosis, their young lacking adult features. (c) In insects with incomplete metamorphosis, the young change shape gradually through the maturation process. (d) Some insects have complete metamorphosis, whereby there are four distinct stages.

The lifecycle of an organism plays a major role in pest management. A pest with one generation per year (e.g., alfalfa weevil) is relatively much easier to manage than a pest with several generations per year (e.g., tobacco aphids). Similarly, an insect pest with complete metamorphosis presents different management challenges than one with no metamorphosis. In the case of complete metamorphosing insects, the grower should target the appropriate management intervention to the most vulnerable stage in the lifecycle, and before the stage in which it causes economic damage. For example, the alfalfa weevil is more vulnerable in the larval stage than in the adult stage. As previously indicated, the larvae are most destructive to plants.

10.6.2   MOUTHPARTS AND FEEDING HABITS

Based on mouth parts and feeding habits, insect pests may be classified as *chewing* or *sucking and piercing*:

1. *Chewing insects.* Chewing insects have chewing mouthparts and chew during feeding. Larval stage of insects, adult beetles, and boring insects is chewing insects. The symptoms of chewing insect attack include defoliation, boring, leaf mining, and root feeding.
2. *Sucking and piercing insects.* These insects puncture plant tissue and suck fluids during feeding. Examples include aphids, scales, mealybugs, thrips, and leaf hoppers. These insects are usually small, sometimes even microscopic. During the feeding process, they may inject poisons into the plant. Sucking and piercing insect attack produces symptoms such as leaf curling or puckering and abnormal growth (gall).

Chewing insects are susceptible to foliar sprays that have either contact (effective on physical contact) or stomach (effective upon ingestion) mode of action. In cases where the pest starts its destructive activities from within the whorl of leaves (e.g., in corn borer), it is important that the pest be controlled before it gains access to the whorl. Insects with sucking and piercing mouthparts suck the food from the plant vessels and hence are best controlled by systemic insecticides (that are distributed throughout the entire plant in its vessels). Some insects have nocturnal feeding habits (e.g., moths). They hide during the daytime and evade direct contact with pesticides. The grower should know where such insects hide in order to be able to spray effectively to kill them.

10.6.3   INSECTS DAMAGE VARIES WITH PLANT GROWTH STAGE

Insects indirectly affect photosynthesis by reducing leaf area through defoliation. Photosynthates are reduced as are, subsequently, growth and yield. Chewing insects such as cutworms reduce crop stand by killing plants and thus reducing the number of established plants and yield. Insects may bore through fruits and tubers in the field, reducing harvestable yield. In the storehouse, insects continue their devastation through the damage they inflict on grain and other crop products. Root-feeding insects (e.g., wire worms and white grub) can kill plants from below the ground level. Insects act as vectors of disease.

The damage caused depends on the crop growth cycle. Seed maggots and cutworms are effective in the very early stages of plant growth but cannot damage larger plants. The economic loss depends on the plant part attacked by the pest. An attack at the vegetative stage is less economically injurious than at the reproductive stage. For example, whereas soybean can tolerate a relatively large amount of leaf damage, the presence of a small number of stinkbugs is a greater cause for concern.

10.6.4 TYPES OF INSECT DAMAGE AND THEIR IMPACT ON PLANT ECONOMIC PRODUCTS

The type of insect damage determines the justification for control and when and how to implement pest management intervention. There are four general types of economic insect damage:

1. *Direct damage.* The insect pest may damage the economic product for which the crop is grown (e.g., an attack on the grain in cereals or leaves of salad crops or greens). The primary effect is reduction in product quality. For example, the earworm destroys the corn kernels, while the cabbage worm destroys the cabbage head by boring holes through the wrapped leaves.
2. *Indirect damage.* Insects that cause indirect damage to crops attack either vegetative or reproductive parts of plants that are not the economic plant parts. Their primary effect is overall yield loss.
3. *Product contamination.* Insects may not attack plant parts, but they decrease crop product value by contaminating the harvested product with their presence as living organisms, dead bodies or parts, or the byproducts of their metabolism (e.g., wastes). These foreign materials in economic products diminish their quality and consequently economic value.
4. *Insects as vectors for diseases.* Some insects may not attack the economic plants (i.e., they are not pests of plants). However, they may serve as vectors, acquiring a pathogen from an infected host and transmitting it to another host. Leafhoppers and aphids are vectors for various viral diseases (e.g., maize dwarf mosaic virus [MDMV] and barley yellow dwarf virus [BYDV]).

10.7: OTHER CROP PESTS

Other pests that plague crops during production include a variety of mites, vertebrates, and storehouse pests.

10.7.1 MITES

Mites belong to the order Acrina. They reproduce very rapidly and frequently. Spider mites *(Bryobia praetiosa, Tetranychus urticae,* and *Panomychus ulmi)* are important pests of both field crops and indoor-produced crops.

10.7.2 VERTEBRATES

Two groups of vertebrates, birds and mammals, are important pests of crop plants. Birds are important pests of cereal grains, wreaking havoc toward the end of the crop production cycle. Birds may eat both mature and immature grains. Small mammals (rodents) destroy field crops by either removing planted seeds or destroying mature grain or herbage and other products.

10.7.3 STORAGE PESTS

Storage pests of importance are discussed in Chapter 15.

10.8: WEEDS

When a plant grows where it should not, it is deemed a pest. A *weed* is a plant growing where it is not wanted (i.e., a plant out of place). This broad definition notwithstanding, the term is normally reserved for certain specific plant species that perpetually are unwanted in crop production. Weeds are more aggressive than cultivated plants and outcompete them in production. They can thrive on marginal soils and have characteristics associated with the wild. Weed management is a major concern in crop production. More than 50% of all agricultural chemicals consists of those designed to control weeds (called herbicides).

10.8.1   ECONOMIC IMPORTANCE OF WEEDS

What is the economic importance of weeds in crop production? Weeds are undesirable in crop production for several reasons:

1. Weeds compete with cultivated plants for growth factors (light, water, nutrients). They are usually better competitors and thus cause reduced crop productivity.
2. They harbor pests such as rodents, snakes, insects, and pathogens.
3. Weeds increase crop production costs. This is so because the producer controls weeds at additional cost through weeding, mechanical cultivation, or use of herbicides.
4. When harvested grain becomes contaminated with weed seeds, cleaning the grain poses another additional production expenditure. Further, the seed quality, and subsequently the market value, is reduced.
5. Some weed species are poisonous or injurious to humans and animals (e.g., ragweed causes hay fever, locoweed causes "blind staggers" in animals).
6. Weeds reduce the aesthetic value of the area. A weed-infested farm is an eyesore.

Weeds may be classified according to plant lifecycle or leaf characteristics.

10.8.2   USEFUL ROLE OF WEEDS IN AN AGROECOSYSTEM

Weeds may be plants out of place, but they arise where they are most at home. They are the most adapted to the conditions under which they grow to maturity. Weeds are hardy plants and a part of the ecosystem. Weeds can have useful roles in crop production. They can serve as indicators of soil fertility, and also protect the soil from erosion.

1. Bare land is susceptible to soil erosion. When crops are not being grown on the land, it is best to have some plants growing on it rather than leaving it bare, especially where the land is erosion-prone.
2. Weeds can be plowed under the soil to improve soil organic matter.
3. A piece of land on which a good population of goosegrass, thistles, chickweed, and yarrow is found usually indicates that the soil is fertile and nutritionally balanced.
4. When dandelions, poppy, bramble, shepherd's purse, bulbous buttercup, and stinging nettle occur in dense populations, the soil is likely to be light and dry.
5. Sedge, buttercup, primrose, thistle, dock, comfrey, and cuckoo flower are found in wet soils.

6. Acid soils support acid-loving plants, such as cinquefoil, cornflower, pansy, daisy, foxglove, and black bindweed.
7. White mustard, bellflower, wild carrot, goatsbeard, pennycress, and horseshoe vetch are found in alkaline soil.
8. Clay or heavy soils hold moisture and favor crops such as plantain, goosegrass, annual meadow grass, and creeping buttercup.

Large populations of mixtures of several of the associated species listed should occur in order for the diagnosis to be reliable.

10.8.3 CLASSIFICATION OF WEEDS

Weeds may be described, according to lifecycle, in three ways:

Annual weeds. Weeds that are seasonal, appearing and completing their lifecycles either in summer or winter.

Biennial weeds. Weeds that complete their lifecycles in two seasons, the first involving vegetative growth and the second flowering and death.

Perennial weeds. Weeds with perpetual lifecycles that reappear without reseeding.

1. **Annual weeds** complete their lifecycle in one year or growing season. Most weeds are annuals. They are relatively easy to control. Many annual weed seeds can remain dormant in the soil for many years. There are two types of annual weeds, *summer annuals* and *winter annuals*. Summer annuals (warm season) germinate in spring and grow through summer. They include foxtail, crab grass, and lamb's quarter. Winter annuals (cool season) germinate in fall, live through winter, and produce seed in spring. Examples are chickweed, broadleaves like shepherd's purse, and grasses such as cheat and ryegrass.
2. **Biennial weeds** germinate in the spring of one year, live vegetatively through winter, and flower the next spring. Examples are pigweed and musk thistle.
3. **Perennial weeds** are difficult to control once they are established. They consist of many grass and non-grass species such as johnsongrass, nutsedge, and bindweed. They may be warm season (e.g., johnsongrass, yellow milksedge, morning glory, milkweed) or cool season species (e.g. wild garlic, curly dock).

Grass Versus Broadleaf

Weeds may be generally classified as either *broadleaf* or *grass weeds*. Grass weeds are very difficult to control. Most noxious weeds are grasses. Grass weeds include nutgrass, quackgrass, and bermudagrass. Broadleaf weeds include milkweed and dandelion.

Other Operational Classifications of Weeds

Weeds differ also in their impact on crop productivity and how difficult they are to control. In seed purity analysis (which provides information on the physical condition of the seed and the presence of unwanted material), evaluators check for the presence of weed seed. Seeds of weeds classified as *common restricted* or *secondary* are relatively less important to crop production in a particular production region and hence can be allowed in minimal amounts in the crop seed. However, weeds classified as *primary, prohibited,* or *noxious* are very difficult to control and are more undesirable. Consequently, crop seed cannot be contaminated by certain noxious weeds and cannot be marketed in a production region where such weeds are of economic importance.

10.8.4 WEED SPECIES ARE PERSISTENT IN THE LANDSCAPE

Adaptive Properties of Weeds

Why are weeds so competitive in the field and difficult to control? Weeds have specific properties that increase their survivability and make it difficult to control them in production. They produce numerous seeds (e.g., corn produces 400 seeds per plant, while a weed such as pigweed produces about 20,000 seeds per plant).

1. Weed seeds have dormancy mechanisms that enable them to avoid adverse weather conditions.
2. Weeds are more resistant to adverse environmental conditions (heat, drought, low light, pests) than are cultivated plants.
3. Certain weeds have features that are similar to the cultivated plants with which they are associated. For example, wheat is similar in growth habit to cheat and both are winter annuals, while sorghum and johnsongrass are both warm season grasses. Similarly, dodder is similar in size to alfalfa seeds and blends in readily.
4. Weeds are very competitive and spread quickly in multiple ways. When clipped, they regrow quickly and produce seed rapidly.
5. Many weeds have adaptive features that make them persist in the soil. Most noxious weeds have rhizomes or stolons.

Succession is an ecological event that describes the natural and directional changes in plant community structure over time. The phenomenon occurs when land is disturbed. The disturbance modifies the environment, destroying existing ecological niches. As new niches form, the first species to colonize the disturbed field are short-lived, have high reproductive rates, and are generally controlled by density-independent factors (these are the so-called r-selected species and differ from the k-selected species, which follow r and have opposite characteristics). In an agroecosystem, the field is often disturbed (by tillage) for planting. This activity promotes the growth of weeds (from the stirring up of weed seeds stored in the ground). Weeds are r-selected species. The frequency of tillage will determine the kinds of weeds that will arise in the field. However, cultivated crops are mostly r-selected species. Since both weeds and crop plants are r-selected species, they compete to the detriment of crop plants.

Weed–Crop Interaction

Weeds and crop plants interact in a competitive fashion, the outcome of the competition depending on such factors as weed density, time of weed appearance in the crop field, weed plant biology, and prevailing environment.

Weed Density   Generally, the more weeds there are in a cultivated field, the greater the competition for growth factors and, consequently, the greater the yield reduction. However, a few large weeds can have a greater adverse effect than many small weeds. Similarly, a high crop density is more effective in suppressing weeds.

Timing of Weeds' Appearance   There are certain times in the lifecycle of cultivated crops at which they are most adversely impacted by competition from weeds. Generally, the period of crop establishment should be weed-free, since the seedlings are most vulnerable. Fast-growing species such as corn and soybean are able to establish sooner and compete well with weeds, while slow-growing species can be devastated when weeds appear early in the growing season.

Plant Biology   The photosynthetic pathway employed by a plant to fix carbon dioxide plays a significant role in how some cultivated crops and weeds compete. C_4 plants have a higher photosynthetic rate than C_3 plants and can continue to photosynthesize under high temperature and high light intensity conditions. C_3 plants, under these conditions, photorespire. Consequently, C_4 weeds are more aggressive and difficult to control in summer when they occur in a field of C_3 crop plants. Corn is a C_4 plant and thus can compete well with many weeds in summer. In cooler periods of early spring, C_3 weeds may be better competitors.

Most weed species reproduce mainly by seed. Hence, it is an effective weed management practice to prevent weeds from flowering and producing seed (i.e., prevent buildup of weed bank in the soil).

Environment   Indigenous weeds are better adapted to the locality and hence more competitive than introduced crop plants.

10.8.5   TROUBLESOME WEEDS

Some of the most troublesome weeds in the world are purple nutsedge *(Cyperus rotundus)*, yellow nutsedge *(Cyperus esculentus)*, bermudagrass *(Cynodon dactylon)*, barnyardgrass *(Echinochloa crusgalli)*, jungle rice *(Echinolchloa colonum)*, goosegrass *(Elensine indica)*, johnsongrass *(Sorghum halepense)*, congograss *(Imperata cylindrical)*, common purselane *(Portulaca oleracea)*, lamb's quarter *(Chenopodium album)*, large crabgrass *(Digitaria sanguinalis)*, and field bindweed *(Convolvulus arvensis)*.

Other common weeds are described in Appendix G.

10.8.6   MANAGING WEEDS IN CROP PRODUCTION

Weeds always occur in field cultivation. Producers can manage weeds for economic crop production by exploiting crop-weed competition to the advantage of crops, using various tactics:

1. *Mechanical/physical weed control.* Weeds may be controlled by weeding with hand tools (e.g., hoes) or inter-row cultivation (tillage) with tractor-drawn implements. Other methods include burning, flaming, mulching, mowing, solarization (heating the soil in the sun under clear plastic), and flooding (e.g., in rice fields to control red rice).
2. *Biological control.* One of the oldest methods of biological weed control is the selective grazing of plants using domestic animals (e.g., goats, sheep). The use of mycoherbicides (herbicides consisting of specially formulated disease-causing fungi) have been tried in some crop production systems.
3. *Cultural control.* Various crop production practices that improve crop compaction or reduce weed numbers are useful in controlling weeds. Tactics for increasing crop competition against weeds include
   a. Using narrow row spacing and high seeding rates to reduce open space in the field after crop establishment.
   b. Using transplants, where feasible, for a quick establishment of the crop to provide a quick ground cover.
   c. Selecting a cultivar that is able to establish an effective canopy quickly to shade out weeds.
   d. Timing of planting such that it quickly follows land preparation to avoid giving weeds a head start over the crop. Seeding at a time when germination conditions are not ideal (e.g., low temperature, low moisture) would allow the more adapted weed seeds to have a competitive advantage over the crop seed to become established ahead of the crop.
   e. Using a high enough crop density to provide an effective ground cover to shade out weeds. However, it is important that the crop density selected is not extreme, since this can promote competition within the crop itself.

Tactics for reducing weed number include the following:
   a. Planting cover crops and using crop rotations to break the cycle of weeds with the competitive crop
   b. Intercropping to fill open spaces in the field
   c. Mulching to suppress weeds
   d. Planting a living mulch to provide a good ground cover
4. *Chemical control.* Herbicides are pesticides used for controlling weeds. They are the most heavily used pesticides in U.S. agriculture. These are discussed in detail later in the chapter. Herbicides may be applied as preplant (before seeding), preemergence (before seedling emergence), or postemergence (after seedling emergence). In the production of crops such as cotton and potato, producers may facilitate harvesting by applying desiccants (to kill the foliage).
5. *Preventive measures.* Quarantine measures may be used to reduce the spread of weeds. States have their own laws restricting the spread of weed seeds through seed inspection that forbid the sale of seeds with certain weed seeds. Seed analysis includes a test for the presence of weed seeds. The producer can also reduce the spread of weed seeds by adopting certain sanitary observances on the farm (e.g., cleaning equipment), purchasing quality seeds, and composting manure before using it (fresh cow manure may contain weed seeds that are killed during composting).

10.9: NON-PATHOGENIC CAUSES OF PLANT DISEASE

How does an adverse environment impact the crop physiology and consequently crop production? Some disease symptoms are not caused by pathogens but by physical factors in the environment. Plants in the wild grow in regions to which they are best adapted. Modern crop production entails growing crops under cultural conditions that are significantly artificial. Because growers are able to manipulate the growing environment through the adoption of various kinds of technology (e.g., irrigation, fertilization, mulching, tillage, pesticides), crops are often cultivated in areas that are less than ideal. Consequently, modern crops are more prone to the vagaries of the weather than wild species. Improper levels of any of the environmental factors required for plant growth and development (temperature, water, light, nutrients, etc.) can produce disease symptoms in plants. Unlike pathogenic diseases, diseases caused by environmental factors are non-infectious and not transmissible.

10.9.1 TEMPERATURE

The best temperature range for plant production is 15° to 30°C. Perennial plants and overwintering or oversummering materials can, however, endure much lower or higher temperatures than indicated. Plant species differ in the temperature extremes they can tolerate. Further, the age of the plant (seedling or adult) also affects plant response to adverse temperature. In addition, plant parts or organs differ in their sensitivity to adverse environment. Buds and flowers are more affected than other plant parts.

High temperature induces plant diseases such as *sunscald* injuries, in which the plant parts (e.g., leaves, fruits) exposed to high sunlight intensity may become discolored, become desiccated, or develop a water-soaked appearance and blisters. Blackheart of potato is associated with excessive temperature. Warm-season crops (e.g., corn, beans) are susceptible to low temperatures. In potato, low temperature causes hydrolysis of starch to sugars that leads to caramelization and excessive sweetening of the root product. Freezing temperatures cause various kinds of frost damage. Young growth, flowers, buds, and meristematic tips are prone to frost-killing.

10.9.2 MOISTURE

Inadequate soil moisture, as well as atmospheric moisture, is detrimental to crop plant growth. Low soil moisture predisposes plant leaves to wilting. Nutrient uptake is also reduced under drought. Plant leaves may drop eventually or die back, reducing the photosynthetic surface. Filling of grain or fruits is adversely affected. Plants grow slowly and become stunted.

Low atmospheric moisture (i.e., low relative humidity) encourages increased transpiration and moisture loss from fruits. Plants may wilt, while fruits dehydrate and shrivel from a dry atmosphere. Low humidity is often a transient event, just as much as excessive soil moisture is infrequent. When flooding occurs, often the whole crop is lost. Poor drainage may cause flooding more frequently even from normal amounts of rain. Waterlogging creates anaerobic conditions in the soil, causing root decay in some cases.

10.9.3 AIR

Waterlogging, as indicated previously, reduces soil air. Blackheart of potato is a storage disease caused by excessive respiration at high temperature (leading to oxygen deficit and abnormal physiology). Air pollution from industrial sources, automobiles, and other sources is injurious to plants. Ozone injury is found in potato, soybean, tobacco, wheat, and many others. Common symptoms of pollution damage include bleaching of leaves, chlorosis, mottling, and bronzing.

10.9.4 NUTRITIONAL DEFICIENCY AND TOXICITY

The subject of the effects of nutrient deficiency on plant growth and development is discussed in more detail in Chapter 8.

10.9.5 HERBICIDE INJURY

Herbicide injury results from improper use of herbicides (e.g., improper choices, improper rate and timing of application) or collateral damage caused by drift. Crops may be accidentally killed or scorched by unintended application of herbicides.

10.10: PRINCIPLES OF PEST MANAGEMENT

How can crop producers manage pests in crop production so their enterprise becomes profitable? Agricultural pest management uses various methods either separately or in combination (integration). These methods may be biological, chemical, physical/mechanical, or cultural. A pest management plan or method includes one or more of these methods.

The methods of pest management in an agroecosystem may be described as *exclusion, eradication, protection, resistance,* or *no action.*

1. *Exclusion.* This is the strategy of preventing the causal organism (pathogen, insect, etc.) from being introduced in the area in the first place. If already present, the

organism should be prevented from establishing itself. One method of exclusion is *quarantine,* the use of laws to restrict import-export of living materials. Observance of certain sanitary regulations also helps prevent entry of pests into the field. The grower should use healthy, clean seed for planting the crop.

2. *Eradication.* An established causal organism can be prevented from spreading and its population reduced, eventually eliminating it from the locality. Previous success of significance to crop production includes the eradication of the screwworm from the southern United States and the medfly from Florida.

3. *Protection.* It takes a pathogen and a susceptible host (plus a favorable environment) for disease to occur. One approach of pest management is to prevent pathogen-host contact. This can be done by physical methods or chemicals (pesticides). This is a preemptive approach to reduce a pest population.

4. *Resistance.* Since disease resistance has a genetic basis, plant breeders can breed cultivars with resistance to important diseases and insect pests. This is also a preemptive approach for reducing pest population.

5. *No action.* Just because a pest is observed does not warrant the implementation of a management approach. Based on knowledge about the biology of the pest and other factors, the producer may be better off ignoring the pest for the time being. This is usually the case for minor pests (i.e., pests that are not economic pests of the crop being produced).

10.10.1   PEST ATTACK MAY BE PREVENTED BY EMPLOYING CERTAIN METHODS

Prevention is better than cure. Since pest control increases production cost, it is better to prevent pest attack from occurring in the first place. Some of the effective preventive methods are as follows:

1. *Use adapted cultivars.* Climate determines crop plant adaptation. Adapted plants have less likelihood of succumbing to disease and insect pests.

2. *Use resistant cultivars.* Where a disease or insect pest is a problem, using cultivars that are resistant to the pest will eliminate the effect of the pest on crop production.

3. *Plant high-quality seed.* The producer should always purchase seed from reputable suppliers. Good-quality seed promotes rapid seed establishment. It should have no noxious weed contamination or diseased seeds.

4. *Prepare seedbed or growing medium properly.* In field production, tillage operations should remove weeds that can jeopardize seedlings in early stages of establishment. In greenhouse production, the soil or growing medium should be sterilized against soil-borne pests.

5. *Plant at the best time.* The soil temperature should be right for seeding in order to avoid seed rot. The producer can plant early or late to make the crop escape disease.

6. *Provide adequate nutrition.* Healthy plants are able to resist pest attack much better than malnourished plants.

7. *Observe good sanitation.* Good sanitation during production reduces plant debris left on the soil after harvest. In the greenhouse, sterilizing soil, cleaning and disinfecting tools, hosing the floor, and hanging the watering hose after use are strategies for reducing disease incidence.

8. *Remove weeds.* Weeds compete directly with crop plants for crop growth factors and may harbor pests.

9. *Avoid conditions that create microclimates that are conducive to disease.* Such conditions include high humidity, improper lighting, poor aeration, and high temperature.

10.10.2 DESIGNING A PEST MANAGEMENT STRATEGY INVOLVES CERTAIN CONSIDERATIONS

There are certain considerations and management decisions involved in designing a management strategy. A good pest management strategy should be effective, inexpensive, and safe to the plant, environment, applicator, and consumers of the product. It should be kept in mind that disease occurrence depends on three factors—causal organism, environment, and susceptible host. The producer should exploit the lifecycle of the causal organism in pest control. The pest management strategy should attack the pest at its most vulnerable stage in the lifecycle and before it is destructive to the crop plant. The management strategy should also take into account the feeding habits of the organism (e.g., sucking or chewing).

From the plant's perspective, the management strategy should consider the stage (young or adult) when it is most vulnerable to the pest. The plant product of economic importance should also be considered. The pathogen or insect may or may not affect the product of interest directly.

The environment under which a management method measure is to be implemented is important. Certain methods of management work best under enclosed conditions. Weather conditions (wind, rain, and temperature) affect the success and effectiveness of the method of management. The pest management decision-making process is a dynamic one. In the management cycle, the farmer or producer makes a decision based on various pertinent information, implements the appropriate management strategy, monitors the crop and the pest status, and then revisits the original decision. The producer must be familiar with pest management strategies. He or she must decide when to manage pests. To do this, the following pieces of information are needed.

Density of Pest Population

Pest population density is determined by sampling methods. Sampling of pests is challenging because their dispersal pattern is not uniform but clumped. This is so because pests tend to lay eggs in clumps or clusters and tend to gather in microhabitats of ideal conditions in the field. In microbes, bioassays may be needed to identify specific strains. Sometimes (in the case of large insects), the pests can be counted. On other occasions, the surface of the plant covered or damaged is the basis of estimating pest population density.

Estimate of Expected Crop Damage

Generally, yield decreases as pest population density increases. There is a level of pest population below which yield loss is negligible. This is the crop *damage threshold.* The amount of damage that can be tolerated differs among crops. Where the economic part is the leaf (e.g., tobacco), leaf damage and defoliation are intolerable. However, in a grain crop such as soybean, even 20% defoliation or more may be inconsequential to crop yield.

Economic injury. The level of pest incidence at which the producer would experience an economic loss of product.

Economics of Management Strategy Being Considered

The producer needs to estimate the cost of implementing the management strategy against the expected return (i.e., price expected from the yield). Cost/benefit analysis is critical in pest management. The **economic injury** level is an estimate of the pest popu-

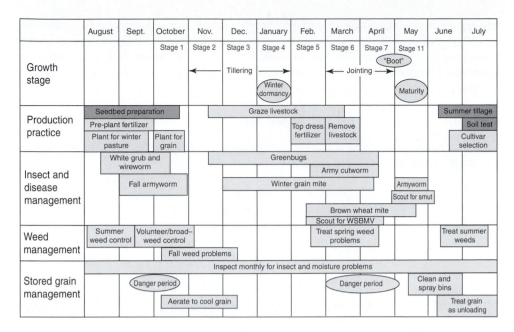

FIGURE 10–3 The different stages in the growth of a crop plant may be plagued by different kinds of pests. A crop producer should have a detailed schedule of pest control for a particular crop. The strategy should include preventive measures as well as control measures. (Source: Adapted and modified from Oklahoma State University Fact Sheet.)

lation density at which the value of the crop yield loss prevented is equal to the cost of implementing a treatment. The *action threshold* is defined as the pest population density at which treatment is necessary to prevent economic injury or prevent the pest population from reaching the economic injury level. It is best to implement a treatment program at a pest density that is slightly below the economic injury level, so that the possibility of exceeding this level is prevented. Further, pests develop at different rates such that in certain cases the window of opportunity between sampling and the attainment of economic injury level by the pest is very narrow.

Frequently, the crop producer is confronted with more than one pest that may occur simultaneously or in succession. These pests may have different biological and ecological characteristics. The management schedule of pests in wheat production in Oklahoma is summarized in Figure 10–3.

10.11: METHODS OF PEST MANAGEMENT

Pests in crop production are managed by one of five major methods: *biological, cultural, legislative, physical/mechanical,* and *chemical.*

10.11.1 BIOLOGICAL PEST MANAGEMENT

Biological pest management (or *biological pest*) **control** is the use of one organism to manage the population of another organism. This is the oldest method of pest management. Biological pest management is based on the ecological principles of parasitism and predation. Every organism has its natural enemies. This design of nature ensures that no

Biological pest management control. The use of a natural enemy to control a pest.

one organism dominates nature through overpopulation. Natural disasters (e.g., drought) can offset this equilibrium, jeopardizing certain organisms. The goal of ecological pest management is to manipulate various biotic factors in the production environment to maintain pest populations at levels below the economic threshold (above which the crop being produced suffers economic injury).

Strategies of Biological Pest Management

Crop producers may exploit and manage certain natural defense mechanisms of plants to control pests. Some of these strategies are

1. *Parasitism.* The alfalfa weevil is a host for the stingless wasp *(Microstomus aethipoides)*, which hatches inside and eventually destroys the weevil. The cyst nematode *(Heterodera)* is parasitized by fungi (e.g., *Catenaria auxilianis*).
2. *Prey-predator relationships.* Predators feed on other living organisms (prey). Certain birds prey on insects and rodents. *Carabid* beetles prey on aphids and caterpillars, while lacewings *(Chryosopa)* prey on spiders and other insects.
3. *Structural.* Certain plant anatomical features (e.g., thick cuticle, pubescence) interfere with feeding and spread of pathogens and insect pests. Sucking insects are unable to penetrate thick cuticles, while pubescence interferes with oviposition of certain insects.
4. *Chemicals.* Certain plant extracts have insecticidal action—for example, pyrethrum from chrysanthemum and nicotine from tobacco. These toxins discourage feeding by susceptible insects.
5. *Phytoallexins.* Phytoallexins are chemicals exuded by certain plants when they become injured. These chemicals are toxic to the invading insects or pathogen.
6. *Repellents.* Plants such as onion and garlic have strong scents that repel aphids. Similarly, marigold repels nematodes.
7. *Trap plants (alternative host).* Pests have preferences for host plants. For example, slugs prefer lettuce to chrysanthemum. Thus, in the production of chrysanthemum, lettuce may be planted in the field as "decoy" plants, or trap plants. In bean and squash production, planting of corn around the field attracts aphids away from these crops. By the time the aphids reach the bean plants, they would have lost much of the viral infection that they are carrying. Trap crops are also used to control nematodes. For example, *Clotalaria* plants trap the larvae of root-knot nematodes.
8. *Biocontrol.* Prior to storage, fruits may be treated with a suspension of the bacterium *Bacillus subtilis* to delay brown rot caused by the fungus *Monilinia fruiticola*.
9. *Microbial sprays (biopesticides).* Artificially cultured *Bacillus thuringiensis* is prepared as a spray application for controlling cutworms, corn borers, and other caterpillars.
10. *Resistant cultivars.* Crop producers may use cultivars with resistance against diseases and other pests of agronomic importance.

Advantages of Biological Pest Management

The advantages of biological pest management include the following:

1. They are safer to apply.
2. Resistant cultivars are cheaper than pesticides.
3. Chemicals are harmful to the user and the general environment.

Disadvantages of Biological Pest Management

The major disadvantages of biological pest management include the following:

1. Handling of living organisms is more difficult than chemicals.
2. Availability and application are limited to relatively few crop plants.

10.11.2 CULTURAL PEST MANAGEMENT

Crop producers may manipulate the way in which plants are cultivated to manage pests in the field. This can be accomplished in several ways:

1. *Crop rotation.* Crop rotation is the production plan of growing different crops on sections of the same piece of land or plot in a predetermined cycle, such that the same crop is not planted in the same section in successive seasons. Specific pests attack crops. Planting the same crop in the same location season after season leads to a buildup of certain pests, especially soil-borne pests such as nematodes. Crop rotation reduces pest buildup.

2. *Sanitation.* Pathogens and other pests overwinter or oversummer in plant remains from harvesting and other debris. Diseased plant materials should be incinerated.

3. *Resistant cultivars (genetics).* Plant breeders have bred cultivars that are resistant to many crop pests. One of the early successes of agbiotechnology is the development of pest-resistant crop varieties—specifically, herbicide-resistant and insect-resistant crop varieties. Genetically, modified crops are discussed in detail in Chapter 18. The use of genetic resistance in crop production eliminates or reduces the use of pesticides in crop production. There are limitations to the use of pest-resistant varieties in production. Disease-resistant varieties are not available in most crops. Furthermore, the resistance is not indefinite because pathogens may adapt to overcome the resistance. Sometimes, having disease resistance in a crop variety causes a reduction in overall productivity because of a possible linkage of the resistance genes to undesirable agronomic genes.

4. *Mulching.* The application of a plastic mulch traps heat in the soil. This may be used to sterilize field soil (solarization). The heat kills soil pathogens and weeds. Certain mulches may leach out phytoallexins that inhibit the growth and development of plants.

5. *Host eradication.* This strategy is employed to preempt the spread of disease. On a small scale, infected plants may be rogued (removed) from the field. On a more drastic scale, a disease outbreak is curtailed by destroying all the crops in the region that are susceptible.

6. *Planting date.* Certain pests appear late in the growing season. Early planting enables a crop to escape infection. On the other hand, aphids that transmit barley yellow dwarf virus (BYDV) are more active before the first hard frost. Hence, planting wheat before the tessian fly free date makes it more susceptible to BYDY than wheat planted later.

7. *Soil reaction.* Liming to increase pH helps to control certain fungal pathogens, while lower pH controls potato scab.

8. *Drainage.* Certain pests (e.g., nematode) prefer moist conditions. Drainage improves aeration and prevents non-pathogenic diseases induced by oxygen deficiency.

9. *Wider spacing (crop density).* Wider spacing improves air circulation around the plants and reduces the occurrence of humid microclimates that support pathogens.

Cultural pest management (control). The manipulation of crop growth environment and cultural practices to control a pest.

10. *Tillage.* No-till methods decrease soil temperature and increase soil moisture. This condition favors certain pests. Soybean cyst nematode may be spread in the direction of tillage.

10.11.3 LEGISLATIVE PEST MANAGEMENT

Plant quarantine is the use of legislation to control the movement (import-export) of plants across certain designated borders. This preemptive and preventive action is aimed at reducing the spread of disease. In the United States, government intervention in movement of biological materials was formally introduced with the enactment of the *Plant Quarantine Act of 1912.* These laws are developed for local, regional, and international scenarios. For example, if a country's primary agricultural crop is wheat, quarantine laws will be very strictly enforced to avoid the introduction of wheat germplasm from parts of the world that are known to be infested with pests. Whereas the laws may be enforced, the success and effectiveness of plant quarantine depend on the experience of the inspector in detecting pathogens when they occur. Certification programs (e.g., seed certification) are also used to ensure that disease-free seeds and plant materials are sold to growers. What other factors can you suggest that are critical in the success of implementing a quarantine measure?

10.11.4 PHYSICAL/MECHANICAL PEST MANAGEMENT

A variety of **physical and mechanical management methods** may be implemented to effect this method of pest control:

1. *Mechanical traps.* Traps may be set to catch large vertebrates such as rodents, while flycatchers are used in the greenhouse to trap insects.
2. *Handpicking.* In small-scale crop production operations, caterpillars and larger bugs may be physically picked up from plants and destroyed.
3. *Barriers.* Mechanical barriers such as fences and tree wraps are used to keep certain pests away from plants.
4. *Tillage.* Tillage operations may be used to reduce weed populations to some extent. The operation brings soil pests to the soil surface, where they are subjected to desiccation. However, disturbance of soil may also favor new weeds that otherwise were buried and lay dormant.
5. *Heat treatment.* Solarization (in the field) and sterilization (in the greenhouse) use heat to control pathogens in the soil.
6. *Radiation.* Ultraviolet radiation is used to sterilize the greenhouse and other enclosures. Gamma radiation is used to prolong the shelf life of produce in storage.

10.11.5 CHEMICAL PEST MANAGEMENT

Chemicals used to manage pests are called **pesticides**. They are designed to kill, hence the suffix *cide* in their names.

General Classification of Pesticides

Pesticides may be classified in several ways: target pest, type of material, chemical structure, mode of action, and formulation.

1. *Target pest.* On the basis of the type of organisms on which they are used, pesticides may be grouped into two broad groups: those used to manage unwanted plants

Table 10–2 General Groups of Pesticides

Pesticides Used to Manage Unwanted Plants
1. Herbicides: used to manage weeds (unwanted plants)
   a. Preemergence: applied before the appearance of a specified weed or crop
   b. Postemergence: applied after the appearance of a specified weed or crop
2. Defoliants: used to cause premature leafdrop to facilitate harvesting (e.g., of soybean, cotton, and tomatoes)
3. Dessicants: used to cause preharvest drying of plants that do not normally shed their leaves (e.g., rice, corn, small grains) or drying of actively growing plant parts when seed or other plant parts are developed but only partially mature

Pesticides Used to Manage Unwanted Animals
1. Those used to manage insects: insecticides
2. Those used to manage vertebrate pests
   a. Small animals
   b. Large animals
   c. Birds
   d. Reptiles
3. Pathogens: cause diseases
   a. Fungi
   b. Bacteria
   c. Viruses
   d. Mycoplasmas
   e. Nematodes

(weeds) and those used to manage unwanted animals (Table 10–2). Pesticides used to manage unwanted plants are generally called *herbicides*. Pesticides used to manage animal pests are usually classified to reflect the category of animal targeted:

Insecticides: for insects

Fungicides: for fungi

Nematicides: for nematodes

Rodenticides: for rodents

Molluscides: for mollusks

Miticides: for mites

Aviacides: for birds

Bacteriacides: for bacteria

2. *Type of material.* Pesticides may be developed from *natural products* isolated from plants (e.g., pyrethrum and nicotine) or other living organisms, or they may be developed from *artificial compounds* or *synthetic products.*
3. *Chemical structure of the compound.* Based on the *active ingredient* (the chemical compound responsible for the killing action of the pesticide), pesticides may be classified as either *inorganic* or *organic* (Table 10–3). Inorganic pesticides (e.g., Bordeaux mixture) are on the decline. Natural organic compounds, also called *botanicals,* are derived from plants. Examples are pyrethrum, nicotine, and rotenone. *Synthetic organic compounds* are effective against a wide variety of insects and other pests. **Organochlorines** (or

Organochlorines.
Pesticides containing chlorinated hydrocarbons as the active ingredient.

Inorganic compounds (inorganics)
Organic compounds (organics)
1. Natural (botanicals)
2. Synthetics
   a. Organochlorines (chlorinated hydrocarbons)
   b. Organophosphates
   c. Carbamates
   d. Pyrethroids
3. Fumigants
4. Spray oils
5. Biologicals (microbial insecticides)

Organophosphates.
Pesticides containing
phosphorus in organic
compounds.

chlorinated hydrocarbons) are not readily biodegradable and persist in the food chain. Examples are DDT (dichlorodiphenyl trichloroethane), chlordane, and lindane. **Organophosphates** have shorter residual action in the environment and are safer to use. Examples are malathion and diazinon. Organic pesticides are more selective and pose less environmental danger. Other classes of pesticides are *carbamates* (e.g., carbaryl), *formamidines* (e.g., amitraz), *organotins* (e.g., fenbutatin), *biologicals* (living organisms, such as *Bacillus thuringiensis* spores) and *pyrethroids* (or synthetic pyrethrins, such as permethrin).

4. *Mode of entry and action.* Pesticides kill pests in a variety of ways:
   a. *Contact action.* They are applied before the infection occurs. Hence, they are less efficacious than systemic pesticides. They are easily washed by rain or sprinkler irrigation. Contact insecticides kill upon making physical contact with the target organism. Examples are malathion, manconeb, captan, and thiram.
   b. *Stomach action (stomach poison).* This kind of insecticide kills upon ingestion. It is effective against chewing insects such as caterpillars and beetles.
   c. *Systemic action.* Systemic insecticides are effective at controlling sucking and chewing insects. Once ingested, the poison moves through the entire organism. They can be applied after the infection. They are absorbed and translocated throughout the plant. Examples are benomyl and metalaxyl.
   d. *Repellent action.* Some pesticides do not kill pests but repel them with strong odors.
   e. *Fumigants.* These pesticides attack the respiratory system of the target organism.

Formulation. The
chemical state (solid,
liquid, or gas) in which a
pesticide is manufactured
for use.

5. **Formulation.** Pesticides contain *active ingredients* (a.i.) that are responsible for the killing of target pests. These ingredients are mixed with inert substances to create preparations that facilitate the application or use of pesticides. The inert substances also reduce the toxicity of the pesticides and make them safer to handle. Pesticides may be formulated as *liquid* or *dry* formulations.
   a. *Liquid formulations.* The common liquid formulations of pesticides include the following:

Aerosol. A chemical that
is applied by propelling it
through pressurized gas.

   1. **Aerosols.** These have low concentrations of active ingredients and are propelled by an inert pressurized gas. They are commonly used indoors.
   2. *Emulsifiable concentrates.* The pesticide contains an active ingredient and an emulsifier that are dissolved in petroleum solvents and prepared in high concentration. They are diluted (e.g., by mixing with water) before use and applied with a sprayer.
   3. *Flowables.* These chemicals are fine suspensions of active ingredients.

4. *Fumigants.* The active ingredient is carried in a volatile liquid. Fumigants are applied to the soil or used in storehouses.
5. *Solutions and ultra low volume solutions.* These chemicals are completely dissolved in water.

b. *Dry formulations.* The common dry formulations include the following:
1. *Baits.* The active ingredient is mixed with food or feed. Baits are commonly used in the storehouse or to trap rodents and other mammals in the field.
2. *Dusts.* The active ingredient is carried in fine talc, clay, or some other material. Dusts are prone to drift (blown by wind).
3. *Granules.* Granules are formulations that are in the form of granular particles. Herbicides, nematicides, and pesticides used in managing soil pests are often formulated as granules.
4. *Wettable powders.* These are formulations in the form of fine powders that require mixing with water before application.

Adjuvants

Pesticide formulations may include compounds that enhance the biological activity. These compounds are called activator *adjuvants* and include **surfactants**, vegetable oils, crop oils, and crop oil concentrate. These additives improve the emulsifying, dispersing, spreading, wetting, and other surface modifying properties of liquids. Sometimes, certain compounds called *spray modifiers* are added to the spray solution for various purposes during field application (e.g., drift control).

Surfactants. Substances that can modify the nature of surfaces (expecially reduction in surface tension).

General Classification of Herbicides

Herbicides are commonly classified by selectivity, mode of action, timing or application, and chemistry (Table 10–4). *Selective herbicides* (narrow spectrum) and kill only certain plant species. Commonly, there are broadleaf and narrowleaf (grass) herbicides. *Non-selective herbicides* are broad action (broad spectrum) and kill indiscriminately, killing both grasses and broadleaf plants (e.g., Roundup®). They are best applied where all plants are not desirable, as is the case around railway tracks.

Herbicides may also kill by *contact* or *translocation* through the plant (systemic). Some herbicides are designed to be applied before sowing crop seed (called *preplant herbicides*), before weeds or crop seedlings emerge (called **preemergence herbicides**), or after the crop has established (called **postemergence herbicides**). Most herbicides in use are organic-based. They include organic arsenicals and phenoxy herbicides, the latter group being hormonal in action (e.g., 2,4-D).

Preemergence herbicides. Herbicides applied before weeds or crop seedlings emerge.

Postemergence herbicides. Herbicides applied to control weeds after crop establishment.

Herbicide Mode of Action Groups

Herbicides have been designed to interfere with various physiological pathways and growth processes. The major modes of action of herbicide groups are as follows:

1. *Contact herbicides.* This group of herbicides kills plant tissue upon contact. Hence, they are applied directly to the target plants. Examples include diquat and paraquat. They are translocated from the point of contact to the point of action.
2. *Growth regulators.* Some herbicides are growth regulators and sometimes are used in research studies in appropriate doses to perform need functions (e.g., 2,4-D is used in tissue culture experiments). When absorbed, these herbicides disrupt plant growth. Examples are phenoxy acids (e.g., 2,4-D), benzoic acids (e.g., dicamba), and pyridines (e.g., picloram). Growth regulators are mainly effective on dicot plants, in which they cause malformation of the leaves and stems.

Table 10–4 A Classification of Herbicides

Based on Selectivity
1. Selective herbicides
2. Non-selective herbicides

Based on Site of Action
1. Contact herbicides
2. Translocated herbicides

Based on Timing of Application
1. Preplant
2. Preemergence
3. postemergence

Based on Chemistry of Active Ingredient
1. Inorganic herbicides
2. Organic herbicides
   a. Organic arsenical
   b. Phenoxy herbicides
   c. Diphenyl ethers
   d. Substituted amide
   e. Substituted ureas
   f. Carbamates
   g. Triazines
   h. Aliphatic acids
   i. Arylaliphatic acid
   j. Substituted nitrites
   k. Bipyridyliums

3. *Photosynthetic inhibitors.* This group of herbicides interferes with the function of the photosystems (PS) of the electron transport system in photosynthesis. Inhibitors of PS I are contact herbicides (e.g., diquat and paraquat). The treated weeds become chlorotic and necrotic. Herbicides that inhibit PS II site A include the triazines (e.g., atrazine) and triazinones (e.g., metbuzin). Others affect the PS II site B (e.g., phenylureas, such as diuron, and nitriles, such as bromoxynil).

4. *Pigment inhibitors.* Pigment inhibitors block the formation of carotene in chloroplasts, causing the leaves to bleach and consequently to be unable to perform photosynthetic functions. Examples include triazoles (e.g., amitrole) and pyridazinones (e.g., norplurazon).

5. *Meristematic inhibitors.* Meristems are the growing points in the plants. When these are destroyed, plants exhibit various malformations of the growing points. Some herbicides are applied to the soil for various reasons (e.g., they are sensitive to light). Dinitroanilines (e.g., trifluralin) are such herbicides. They inhibit root growth and development. Similarly, amides (e.g., alachlor, acetachlor) and carbamothioates (e.g., EPTC) are applied to the soil. They cause malformations in the shoots.

6. *Enzyme pathway inhibitors.* These herbicides block key steps in the metabolic pathways in plants by targeting specific enzymes. Examples include acetyl CoA carboxylase (ACCase) inhibitors (e.g., sethoxydim) and acetolactate synthase inhibitors, of which there are sulfonylureas (e.g., chlorimuron), imidazolinones (e.g., imazethapyr and imazaquin), and sulfonamides (e.g., flumetsulam). There are

also inhibitors of 5-enolpyruvylshikimate-3-phosphate (ESPS) (e.g., glyphosate), glutamine synthetase inhibitors (e.g., glufosinate), and protoporphyrinogen oxidase inhibitors (e.g., lactofen and fomesafen). These inhibitors cause a wide variety of symptoms (e.g., stunting, chlorosis, necrosis, death, purpling or reddening of plant parts) in treated plants, depending on the species.

Pesticide Toxicity

Pesticides are designed to kill pests; thus, users must handle them with great caution. They are assigned a *hazard rating,* which is a function of toxicity and exposure. **Toxicity** is a measure of the degree to which a chemical is poisonous to an organism. Toxicity depends on several factors. A pesticide that is toxic as a concentrate may not be as hazardous as a granule. Further, a pesticide of low toxicity may become very hazardous when used at a high concentration. The frequency of use and the experience of the operator may increase or decrease the hazard level. Toxicity is normally measured by the **lethal dose (LD_{50})** of the chemical. The LD_{50} is the milligrams of a toxicant per kilogram of body weight of an organism that is capable of killing 50% of the organisms under the test conditions. The higher the LD_{50}, the less poisonous or toxic the chemical (Table 10–5). The active ingredient may be an inorganic compound, a natural organic compound, or a synthetic compound.

Toxicity. A measure of the degree to which a chemical is poisonous to an organism.

Lethal dose (LD_{50}). A measure of toxicity that indicates the milligrams of a toxicant per kilogram of body weight of an organism that is capable of killing 50% of the organism under test conditions.

Active Ingredient. The ingredient in a pesticide that determines its effectiveness.

Choosing and Using Pesticides Safely

There are certain general steps that may be followed by a crop producer in the decision-making process of pest management (Figure 10–4).

1. *Detection.* The problem (pest) must first be detected. This may be by visual observation or other tests. Early detection is the key to success in pest control.

Table 10–5 LD$_{50}$ Values of Selected Pesticides	
Pesticide	**LD$_{50}$**
Fungicides	
Captan	9,000–15,000
Maneb	6,750–7,500
Thiram	780
Zineb	8,000
PCNB	1,500–2,000
Insecticides	
Carbaryl	500–850
Dursban	97–279
Malathion	1,000–1,375
Pyrethrum	820–1,870
Rotenone	50–75
Herbicides	
DCPA	3,000
EPCT	1,600
Simazine	5,000
Oxyzalin	10,000

Source: Extracted (and modified) from extension bulletin B-751, Farm Science Series, Michigan State University, University Cooperative Extension Service.

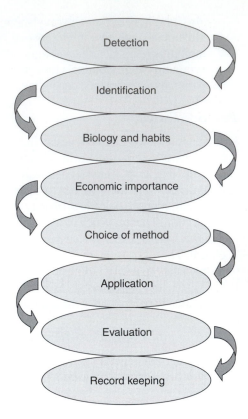

FIGURE 10-4 Pest management involves a number of steps, starting with detection through choice and application of a desirable method. Record keeping is an integral part of a good and effective pest management program.

Detection

Identification

Biology and habits

Economic importance

Choice of method

Application

Evaluation

Record keeping

2. *Identification.* After observing a problem, the producer must make a positive identification of the insect or pathogen. This may be done with the help of an expert from, for example, the Cooperative Extension Service.

3. *Biology and habits.* As indicated previously, an organism has a lifecycle with stages in which it is destructive to plants. There are also stages in the lifecycle in which the pest is most vulnerable. It is important to know the feeding habit of the pest (nocturnal, piercing, chewing, etc.) For example, once the European corn borer has penetrated the stalk of the plant, it is ineffective to spray against it.

4. *Economic importance.* It is important to know the economic injury the pest can cause to crop production. If the threshold population does not occur, it may not be economical to implement pest control measures.

5. *Choice of method of control.* The method that is most effective, most economical, safest (least toxic), and easiest to apply, with minimal environmental consequences, should be selected. Persistent herbicides control weeds for a long time. Atrazine is widely used because it is effective and relatively inexpensive.

6. *Application.* When using a chemical, the best formulation, correct rate, proper equipment, best environmental conditions, and timeliness of application are critical considerations for success.

7. *Evaluation.* The producer should evaluate the effectiveness of an application to determine if there is a need to repeat the application. If repeat application is needed, there may be a need to modify the rates, timing, or some other aspect of the application.

8. *Record keeping.* It is important to keep good records of all pesticide applications. This will provide records for computing production costs and other future needs.

A pesticide label displays certain specific pieces of information for the effective and safe use of the pesticide. The key pieces of information are

1. Name of product (includes trademark name and chemical name)
2. Company name, address, and logo (where applicable)
3. Type of pesticide (e.g., fungicide or insecticide)
4. Product chemical analysis [active ingredient(s), proportions of component compounds, product common name, chemical name(s) of ingredient(s), formulation]
5. Target pest(s)
6. Directions for proper and effective use, plus any restrictions
7. Hazard statements (appear as *caution, warning, danger* messages)
8. Storage and disposal directions
9. Government administrative stipulations (e.g., EPA approval and number)
10. Net content

Of the names of the product, the chemical name is most unwieldy and technical—for example,

Common name: Cyanazine

Chemical name: 2-((4-chloro-6-(ethylamino)-s-triazin-2-yl)amino)-2-methylpropionitrile

Trade name: Bladex® 4L (the "4" indicates that the concentration of the active ingredient is 4%; the "L" indicates the formulation is liquid)

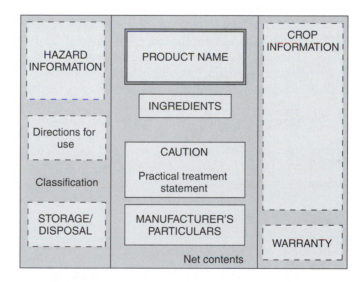

Pesticides are designed to kill and should therefore be handled very carefully all the time. The following are some measures to be observed for the safe application of pesticides.

1. Select the correct pesticide for the job; choose the safer alternative all the time.
2. Purchase only the quantity needed for a job to avoid the need to store leftovers.
3. Read the label on the container and follow directions carefully.
4. Wear protective gear.
5. Do not eat while handling chemicals.
6. Apply under the best environmental conditions.
7. Keep the pesticide out of the reach of children and pets.

8. Apply with care and caution.
9. Know what to do in case of an accident.
10. Clean the applicator after application.
11. Store the pesticide properly, if needed.
12. Be careful about applying near the time of crop harvest.

Methods of Application of Pesticides

In terms of the target to which the pesticide is applied, pesticides may be applied in the following general ways:

1. *Foliar application.* The pesticide is directed at the leaf so that it penetrates the plant through pores (stomata) directly into the plant system. Foliar application is effective against fungal diseases that attack the leaf tissue.
2. *Soil treatment.* In soil treatment, the pesticide is applied to the soil, not the plant directly. Plants absorb the poison through their roots during the process of extraction of water and nutrients. Formulations used for soil application are usually granules. Soil treatments may be fumigants (e.g., chloropicrin), non-fumigant fungicides (e.g., PCNB), or non-fumigant nematicides (e.g., carbofuran).
3. *Seed treatment.* Seed treatment usually takes the form of coating the seed with a pesticide prior to planting. This is sometimes called *seed dressing* and is commonly used for the control of fungal diseases. Seed treatment may be either contact (e.g., captan, thiram) or systemic (e.g., metalaxyl). The protection of contact treatment lasts through the period shortly after emergence of the seedling.
4. *Fumigation.* Fumigation involves using gases and is best done in an enclosed environment. Nematodes and soil fungal pathogens such as *Pythium* may be managed by using soil fumigants. Soil fumigants include chloropicrin and dichloropropene. They work best in moist and loose soil at temperatures between 50° and 75°F. Wet soil impedes diffusion. Dry soil encourages loss. Movement is also impeded by soil compaction and cold soil temperature.

The most common method of pesticide application in the United States is ground application (Figure 10–5).

Methods of Herbicide Application

In terms of the methods of placement, herbicides may be applied in one of several ways, taking into account the location, the nature of the weed problem, the formulation, equipment, and other factors. Granules are suitable for *broadcast application.* When narrow strips of weeds are to be controlled, *band application* is suitable. Single plants may break through cracks in concrete walkways, or weeds may be located in hard-to-reach areas. These weeds are controlled by *spot application* or treatment.

The following are general guidelines for effective weed control:

1. Identify the weed species correctly.
2. Assess the level of weed infestation.
3. Select the appropriate herbicide for the weed species present.
4. Apply at the correct time (e.g., preemergence, postemergence).
5. Apply at the correct rate (prepare correctly and use well-calibrated equipment).
6. Treat the area adequately.

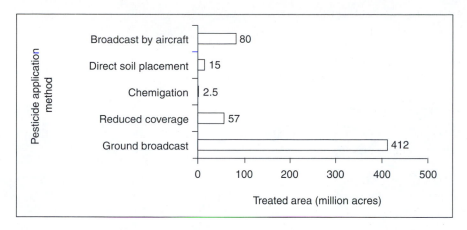

FIGURE 10-5 Ground broadcast is the most common method of pesticide application, accounting for 412 million acres of treated cropland in the 1994 to 1995 cropping season. Corn and soybeans received the most broadcast application. A total of 356 million acres (representing 87% of the total treated area) were treated by ground broadcast application. Chemigation is not a widely used method of pesticide application. (Source: USDA)

7. Consider the weather factors (no rain in forecast, no strong winds).
8. Consider the age of the weed plants (younger plants are easier to control).
9. Consider the ground cover and soil type (organic soils and clay soils absorb pesticides).

Role of Plant Growth Stage and Stress on Herbicide Efficacy

Weeds are best controlled when they are young. They are more responsive to herbicide treatment at a tender stage than when they are grown. At a more advanced age, weeds require more of the herbicide (especially if a contact herbicide) to control them. A more serious problem, however, is that generally a crop is more likely to be injured when herbicides are applied when the plants are bigger. Weed control is more effective when herbicides are applied to actively growing plants. Weeds are less responsive under stress (e.g., drought). On the other hand, it appears that crop plants are more susceptible to herbicide injury when under stress (e.g., drought, disease, nutritional).

Factors Affecting Pollution of Surface Water and Groundwater

Pollution of surface water and groundwater by agrochemicals is affected by factors such as soil CEC, depth of water table, erosion, precipitation leaching, soil texture, plant residue on the soil surface, and rate and timing of application of chemicals. A high CEC indicates the presence of a large amount of soil colloids (e.g., clay, organic matter). The pollutants in the soil solution can be temporarily removed by CEC. Pollutants will have a higher water table sooner than a lower water table. High rainfall promotes high surface runoff and erosion into surface water. Plant residue on the soil surface can slow soil erosion and surface runoff. Coarse-textured soils allow quicker water infiltration and thereby reduce surface runoff that carries pollutants into surface water. Applying agrochemicals at rates higher than recommended, or when the pest pressure is low, leaves excess chemicals to be washed into the soil and surface water. Rainfall after an application of pesticide will wash more of the chemicals into the soil than when there is a dry period after application.

FIGURE 10–6 Application of pesticides in the field requires the use of various kinds of equipment, depending on the acreage to be treated, type of crop, type of chemical to be applied, and the economic situation of the producer. The major methods of pesticide application are ground broadcast, reduced coverage, chemigation, direct soil placement, and aircraft.

10.11.6 PESTICIDE APPLICATION EQUIPMENT

Application equipment may be designed for small areas or large ones. Some field applicators are shown in Figure 10–6. Consult the Internet reference provided at the end of the chapter for more examples.

10.11.7 PESTICIDE INJURY TO CROPS

The goal of pesticide application is to kill the target pest without injuring the crop. However, crop plants may be injured as a result of the operator's inexperience or carelessness but also as a result of the weather conditions at the time of application, the genotype, the characteristics of the chemicals, and the method of application.

1. *Genotype sensitivity.* Certain plants are naturally sensitive to certain chemicals. For example, certain plants are intolerant of fluoride in domestic water and are injured when domestic water is used for irrigation. Plant sensitivity may also manifest at only a certain stage of the lifecycle. For example, some herbicides (e.g., Accent) must be applied before corn plants reach 20 inches tall.
2. *Weather factors.* Pesticides (especially sprays and dusts) are best applied when the air is relatively still, in order to reduce the chance of drift that can damage unintended targets. However, an intended target may be damaged under proper weather conditions. For example, rainfall following an application of preemergence herbicide may cause the chemical to move down to reach the crop seed and possibly injure it.
3. *Residual effect (persistence).* Generally, herbicides that persist longer in the soil are more effective in controlling weeds. However, the residual effect of an herbicide applied to a previous crop may injure the germinating seedlings of the current crop (e.g., Scepter applied to soybean may injure corn in a rotation). Persistence of herbicides in the soil is favored by cool temperatures and a dry environment. Herbicides used in crop rotation should be applied with care, making sure to select with the crop sequence in mind.

4. *Pesticide formulation.* Pesticides may be formulated in various forms (e.g., granules, dust, liquid). Some formulations are safer to handle and are less injurious in certain situations.

5. *Rate of application.* Producers should pay attention to the recommended rates of application of herbicides for effective, safe, and economic application. A higher than normal rate will increase the persistence of the herbicide in the soil and will have a higher chance of crop injury.

6. *Method of application.* Herbicides such as Roundup® may be safely applied as a preplant without injury to crop plants. It may be used in the lawn while the turfgrass is dormant. Some herbicides are more injurious to crop plants when applied as granules to the soil than when applied as foliar sprays and vice versa.

7. *Pesticide interaction.* Applying different pesticides together may result in severe crop injury if the pesticides are incompatible (e.g., applying Asure II and Blazer together is injurious to soybean). Accidental mixtures or contamination may happen when equipment is not properly cleaned between applications.

10.11.8   FACTORS AFFECTING PERSISTENCE OF SOIL-RESIDUAL HERBICIDES

Herbicides that are persistent in the soil are desirable for their long action. However, the residual effect may have potential adverse effect in a crop rotation system. Factors that influence herbicide persistence in the soil and its effect on plants include the following:

1. *Soil moisture.* High soil moisture promotes leaching of herbicides and other soil water-soluble nutrients. Further, it also enhances the rate of biological degradation of herbicides. Consequently, the persistence of herbicides in the soil is less in wetter soils than in dry soils.

2. *Soil temperature.* Temperature accelerates the microbial degradation of soil organic matter, a material that binds and holds herbicides in the soil. Further, the herbicides themselves are prone to more rapid decomposition under high temperatures. Hence, herbicides will persist longer in the soil under cool soil temperatures.

3. *Soil pH.* Soil pH may increase or decrease soil herbicide persistence. Herbicides belonging to the classes of triazines and sulfonylureas tend to persist longer in alkaline soils, whereas clomazone persists longer at acidic pH of 5.9 or less.

4. *Soil microbes.* Soil microbes may directly decompose the herbicide or decompose soil organic matter that has high affinity for some herbicides. Microbial activity is enhanced by warm temperatures.

5. *Rate of application.* Pesticides applied at higher rates persist longer in the soil than those applied at lower rates. It is important to apply herbicides at recommended rates.

6. *Timing of application.* Late application of a herbicide as a result of late planting of the crop means that the gap between the current crop and the next crop in the rotation will be shorter. The persistent herbicide is more likely to be present in the soil at a higher concentration at the time of seeding the next crop.

7. *Type of crop cultivar.* Crop species differ in sensitivity to residual herbicides. Similarly, cultivars of the same crop differ in their sensitivity to residual herbicides in the soil. The producer should be aware of crop and cultivar sensitivity and develop the appropriate crop sequence in the rotation cycle, as well as the proper choice of herbicides to apply. Oat is intolerant of residual atrazine. Similarly, metribuz (Sencor) is injurious to some wheat and barley cultivars.

The producer should consult the manufacturer's instructions for application of herbicides in order to make the correct decisions. The length of delay in the planting date of the next crop following an application of a herbicide depends on the crop and the herbicide. For example, when Raptor (an imidazolinone) is used on soybean, a 3-month delay of seeding is recommended for wheat following soybean in the rotation, a 4-month delay for barley or rye, and a 9-month delay for corn or tobacco.

10.12: INTEGRATED PEST MANAGEMENT

Integrated pest management (IPM).
A pest management strategy that combines the principles of other pest managmenet systems, but with reduced emphasis on the use of chemicals, to reduce pest populations below levels that can cause economic loss.

Pests may be managed by an interdisciplinary approach. **Integrated pest management (IPM)** is a strategy of pest management whose goal is not eradication but, rather, to keep pest population below that which can cause economic loss to crop production. This is accomplished by an interdisciplinary approach (Figure 10–7). The principles of ecology underlie the philosophy of IPM. Several definitions of IPM are in use. The National Coalition on IPM defines IPM as a sustainable approach to managing pests by combining biological, cultural, physical, legislative, and chemical tools in a way that minimizes economic, health, and environment risks. It is implemented not in isolation but as one of the components of the total crop production system, which integrates physical, biological, and management factors for successful crop production. Pesticide use is reduced to the barest minimum and is not a first but a last resort. An insect trap used to trap boll weevils in cotton fields is shown in Figure 10–8. IPM programs strive to achieve safe (to user and environment) and economic management of agricultural pests.

Biological pest management and natural resources conservation are promoted by this method of pest management.

10.12.1 THE UNDERLYING PRINCIPLES OF IPM

The principles underlying IPM may be summarized as follows:

1. *Ecological bases.* IPM considers the agroecosystem as the management unit in crop production. Producers are manipulators of the agroecosystem. Because a

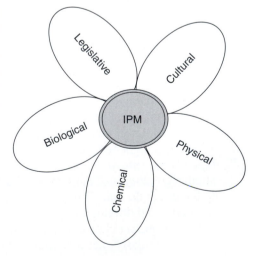

FIGURE 10–7 Integrated pest management employs an interdisciplinary approach to pest management. The goal is not eradication but, rather, to keep pest population at or below a threshold above which it causes economic loss. It combines all the five main methods of pest management: legislative, cultural, physical, biological, and chemical.

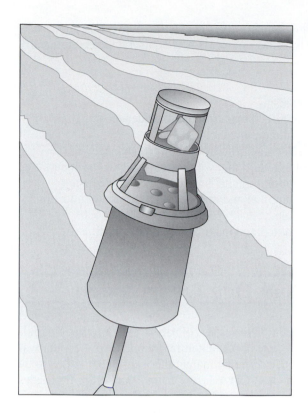

FIGURE 10–8 An IPM insect trap in use on a cotton farm to trap boll weevils.

management action implemented by the producer may yield an unpredicted adverse effect, IPM embraces a holistic approach to pest management, responding to a pest situation through a broad, interdisciplinary approach.

2. *Threshold natural population.* In proper balance, a pest can be present in the agroecosystem at a level that can be tolerated without economic loss. In chemical pest control, as previously indicated, the producer should not spray at the first sight of a pest. There is a pest population threshold beyond which economic injury is imminent. The economic threshold is the pest population level at which an appropriate pest management intervention must be implemented to prevent the pest from reaching the economic injury level. Hence, IPM embraces the fact that there is a natural control mechanism in the ecosystem that can regulate pest populations and keep them below the injury threshold.

3. *Pest suppression, not pest eradication.* All organisms have natural enemies in the ecosystem. Completely eradicating one organism may cause shifts in the balance of the ecosystem such that new pests arise. The concept of a "refuge" for susceptible pests, as promoted in the use of genetically modified crop varieties in production, creates a "buffer system" that maintains a low level of the pest and reduces the chance that pest resistance will occur.

10.12.2 JUSTIFICATION OF IPM

A unilateral approach to pest control involving pesticides has limitations because an agroecological system deals with a complex and dynamic biological system. Sooner or later, target pests develop resistance to pesticides, requiring more potent pesticides to be developed. Pesticides disrupt the natural mechanisms of control in the ecosystem and are environmentally intrusive, endangering humans, livestock, and wildlife. Disrupting natural systems encourages pest outbreaks. IPM optimizes pest control in an economically and ecologically prudent manner.

Steps in an IPM program are

1. Identify pests and beneficial organisms.
2. Know the biology of the pest and how the environment influences it.
3. Determine the tolerable pest population threshold.
4. If economic injury is possible, select the best method of pest management.
5. Select a pest control management method that will destroy pests without harming beneficial organisms. Consider cultural methods first—for example, using resistant cultivars, altering planting time, and providing supplemental nutrition.
6. Develop a pest monitoring schedule.
7. Evaluate the pest management method and make appropriate adjustments.

10.13: PESTICIDES IN OUR ENVIRONMENT

How do pesticides impact the environment? Pesticides contribute to environmental pollution. Improper use of pesticides adversely impacts non-target organisms, groundwater, and food safety. Pesticides are classified among many other substances called toxic chemicals (e.g., polychlorinated biphenyls [PCDs], heavy metals, and oils). Pesticides kill pests that reduce crop yield. However, their effects frequently linger in the environment and threaten human and animal life through contamination of groundwater, food, and feeds. It is estimated that agriculture accounts for approximately 70% of all pesticides used in the United States. Further, only about 5% of the pesticides are estimated to leach into groundwater. Unfortunately, this is enough to endanger human health and other life on earth. The Environmental Protection Agency (EPA) and the Food and Drug Administration (FDA) are federal agencies in the United States that set safe standards for the use of pesticides and monitor their presence in the food chain and the general environment.

The goal of industry and regulatory agencies is to produce that elusive or safe pesticide. The properties of such an ideal pesticide are

1. *Short-lived (non-persistent) in the environment.* A safe pesticide should be able to destroy the target pest within a short period (1 to 2 weeks) and dissipate rapidly. This reduces the chance of its entering and thriving in the food chain.
2. *Effective and safe to operators.* Pesticides are harmful to operators; thus, the pesticide should be readily amenable to safe application. If accidentally spilled or touched, it should be easy to wash away.
3. *Not carcinogenic, mutagenic, or teratogenic.* It is difficult to determine this attribute, which is often the source of much debate and discussion. There has been a shift from the use of persistent organochlorides such as DDT to organophosphates that are less persistent in the environment.

Minimizing Spray Drift

Pesticides are targeted at crop plants, weeds, or the soil, depending on the chemistry and other properties. Application of pesticides should be made under favorable environmental conditions, using appropriate and properly calibrated equipment, as well as according to the guidelines to apply it at the proper rate. The movement away of spray droplets from the intended target is called *spray drift.* Sometimes, a high vapor pesticide is converted to gas during application (called *spray volatilization*), polluting the air in the process. These two major sources of environmental pollution from pesticide application are enhanced by various sprayer characteristics and weather conditions. Spray drift is increased under windy conditions. Certain sprayer additives (called drift control agents) may be included in the tank mixture to reduce spray drift. Drift is higher when the spray volume is low (i.e., smaller droplets) and when high pressure is applied. Off-site drift is increased when the sprayer nozzle is raised high above the target. High wind speed and high temperature tend to increase drift.

10.14: PESTICIDE USE IN U.S. CROP PRODUCTION

How important is pesticide use in U.S. crop production? What types of pesticides are most commonly used? In 1996, the USDA estimated that agricultural pesticides cost U.S. producers a total of about $7.5 billion, of which two-thirds consisted of the cost of herbicides (Figure 10–9). Further, agricultural production and storage use accounted for 75% of all pesticides used in the United States. The cost of pesticides constitutes about 4% of total production expenses.

The application of synthetic pesticides to major crops (corn, wheat, cotton, soybean, fall potato, vegetables, citrus, apples, and other fruits) first peaked in 1982. The conventional way of measuring pesticide use is the amount in pounds of active ingredient. In 1964, pesticides used amounted to 215 billion pounds, while in 1982, producers used a total of 572 billion pounds of pesticides. The 1995 total used was 565 billion pounds. This increase was due to an increase in planted acreage. Application rates during this period were also higher. As commodity prices fell after 1982 and large amounts of farmland were idled, pesticide use also dropped. Most of the pesticide use was accounted for by corn production, being more than double that used in the production of any other U.S. crop (Table 10–6 and Figure 10–10). Further, fall potato production received the most intensive pesticide use on a per acre basis.

The most widely used microbial pesticide in 1997 was the Bt spray (Figure 10–11). Herbicides are the largest pesticide class and accounted for 57% of pounds of active ingredients in 1995. The most commonly used active ingredients of herbicides are atrazine, 2,4-D, and glyphosate (Figure 10–12). Atrazine is used to control weeds in corn and sorghum. It is very persistent in the soil. Trifluralin is most commonly used in cotton, soybean, and vegetable weed management programs. However, with the advent of imazethapyr and other imidazalinone and sulfonylurea herbicides, the use of trifluralin in especially soybean production has declined.

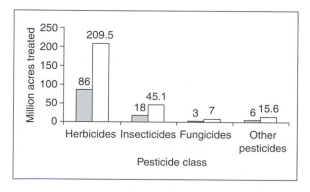

FIGURE 10–9 Crop producers in the United States use large amounts of pesticides, especially herbicides. The figure shows application of pesticides to 244 million acres of corn, soybeans, wheat, cotton, potatoes, fruits, and vegetables in 1997. (Source: USDA)

Table 10–6 Amount of Pesticide Applied and Acres Treated in U.S. Production of Major Crops

Crop	Active Ingredient (Million Pounds)	Acres Planted (Millions)
Corn	201	64
Potatoes	87	1
Cotton	84	17
Soybeans	69	61
Other vegetables	67	3
Citrus and apples	36	3
Wheat	21	49

Source: USDA, ERS 1995 estimates.

FIGURE 10–10 Most of the pesticides used on crops in 1997 were applied to corn. Corn and soybeans accounted for most of the herbicides used. Cotton was the largest user of insecticides, accounting for 32% of the total quantity of insecticides. Pesticides used on potatoes were mostly fumigants and vine killers. (Source: USDA)

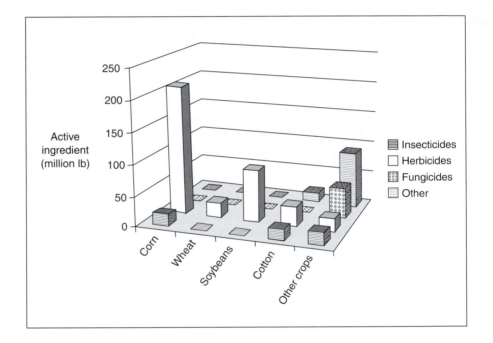

FIGURE 10–11 The most widely used microbial pesticide in 1997 was the Bt (*Bacillus thuringiensis)* spray. It is used to treat the Colorado potato beetle, cotton budworm, and several other fruit and vegetable crop pests. In 1997, it was used to a lesser extent (4%) on field crops and mostly on vegetables (16%) and fruits (11%). (Source: USDA)

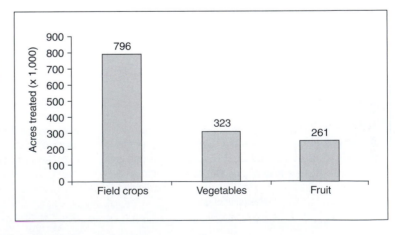

Insecticides made up 13% of the total quantity of pesticides used in crop production in the United States in 1995. The most preferred active ingredients were chlorpyrifos, methyl parathion, and terbufos, accounting for 43% of pesticides used on the five major field crops (Figure 10–12). Chlorpyrifos was the most widely used insecticide in corn fields to combat rootworm larvae, cutworms, and Russian wheat aphid, while methyl parathion was used to manage boll weevils and other cotton insects. The most common insect pests targeted for management in crop production were the cotton boll weevil and bollworm (Figure 10–13).

Fungicides are less used in crop production. They are used to manage pests in vegetable production and potato production. In 1995, 44.6 million pounds of fungicides were used in crop production in the United States. Fumigants are also used to manage soil pathogens and other organisms.

The use of pesticides in crop production is affected by factors including pesticide prices, federal programs, pesticide legislation, pesticide resistance, pesticide registrations, and development of alternative pest management strategies. Pesticide prices are not a major factor affecting pesticide use, since price changes have generally not been dramatic in recent times. Federal programs such as those that encourage land idling tend to reduce use of pesticides. Also, federal legislation through the Environmental Protection Agency (EPA) and laws such as the Clean Air Act of 1970 and Clean Water Act of 1992 tend to decrease the use of pesticides. Various states have various laws regulating the use of pesticides. Further, pesticide manufacturers are compelled to spend more in developing more environmentally safe pesticides, thus driving costs up. The Food Quality Protection Act of 1996 requires that approved pesticides be periodically reevaluated for safety. Those found to have potential safety concerns are withdrawn from use. For example, the use of propargite on fruits such as apples, peaches, and plums was banned, due to the concern about

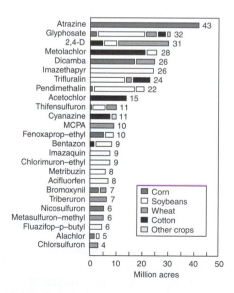

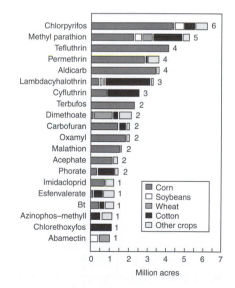

FIGURE 10–12 Atrazine was the most common herbicide ingredient applied to crops in 1999. It was used almost exclusively on corn and grain sorghum. Glyphosate was the second most widely used herbicide ingredient. It was used commonly in orchards and vineyards as well as in no-till systems in corn, wheat, and soybeans. Chlorpyrifos was the leading insecticide ingredient applied to crops. Chlorpyrifos and methyl parathion were the two most widely used insecticide ingredients. They are both organophosphates. Most of the insecticides used were restricted use and applied only by licensed applicators. (Source: USDA)

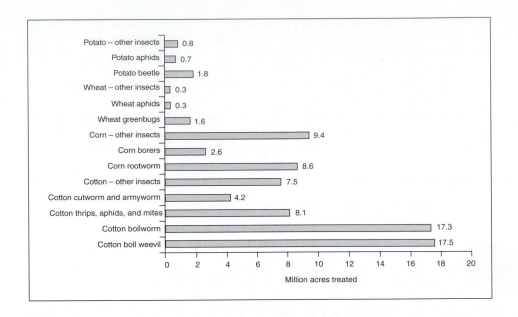

FIGURE 10–13 The most common insect pests targeted for management in crop production in 1994 and 1995 were the cotton boll weevil, bollworm, and corn rootworm. About 17% of all treated acreage was targeted for cotton bollworm or cotton boll weevil. (Source: USDA)

residues that posed potential hazards to infants. The continual use of pesticides has produced pests that are resistant to certain products. For example, weeds resistant to triazine herbicides (atrazine, cyanizine, and simazine) have been reported. To reduce the incidence of pesticide resistance, using lower doses of pesticides is recommended, along with alternating the use of pesticide families. When new and more effective pesticides are produced to replace less efficient ones, there is a reduction in the use of pesticides. For example, Imazethapyr, introduced in 1989, has become the most preferred herbicide in soybean weed management because it is used in lesser concentration than others such as trifluralin. Similarly, the introduction of transgenic corn and cotton seeds (Bt corn and cotton and other crops) has reduced the need for pesticides in the production of these crops.

SUMMARY

1. Yield losses from biological sources occurs at various stages in plant development.
2. Biotic factors that cause yield reduction are of both plant and animal origin. The five categories are disease-causing organisms, weeds, insect pests, non-insect invertebrate, or vertebrate pests.
3. The four disease-causing organisms are viruses, bacteria, fungi, and mycoplasma-like organisms.
4. Some microorganisms are beneficial—for example, certain fungi (mushroom) and bacteria *(Rhizobia)*.
5. In order for disease to occur, there must be a pathogen (causal organism), susceptible host (plant), and favorable environment.
6. A disease cycle consists of inoculation, penetration, infection, dissemination, and overwintering or oversummering.
7. Plants have physical and chemical means of defense against diseases.
8. True resistance to disease is genetically controlled.
9. Insect pests cause a variety of damage to crops in the field and storage house.
10. Weeds are important pests in crop production.
11. The four basic strategies of pest management in an agroecosystem are exclusion, eradication, protection, and resistance.

12. It is better to prevent a pest attack than to control it, since the latter is usually expensive.
13. Pest management strategies should be implemented only when economic injury is imminent.
14. The methods of pest management are biological, cultural, legislative, physical, and chemical.
15. Biological pest management is the oldest method of pest management and is based on the rationale that every organism has its natural enemies.
16. Cultural methods of pest management involve the manipulation of the crop production environment and production practices to control pest populations.
17. Legislative pest management (or quarantine) is the use of laws to control the movement of plant materials across specified borders.
18. Physical pest management involves the physical removal of pests or the use of agents to suppress them.
19. Chemical methods of pest management use pesticides that may be natural or synthetic, as well as organic or inorganic.
20. Pesticides are common in modern crop production. They have an adverse environmental impact.
21. Pesticides should be used judiciously and with caution.
22. Integrated pest management is an interdisciplinary approach to pest management whereby all methods are utilized strategically. Chemicals are especially used only as a last resort, and then very cautiously.

REFERENCES AND SUGGESTED READING

Agrios, G. N. 1988. *Plant pathology.* 3d ed. New York: Academic Press.

Bohmont, B. L. 1997. *The standard pesticide user's guide.* 4th ed. Englewood Cliffs, NJ: Prentice Hall.

Duke, S. O., ed. 1986. *Weed physiology,* vols. 1 and 2. Boca Raton, FL: CRC Press.

Economic Research Service/USDA. 1997. *Agricultural resources and environmental indicators, 1996–97.* Agricultural handbook No. 712. Washington DC: USDA.

Elzinga, R. J. 1987. *Fundamentals of entomology.* 3d ed. Englewood Cliffs, NJ: Prentice Hall.

Klingman, G. C., F. M. Ashton, and L. J. Noordhoff. 1982. *Weed science: Principles and practices.* 2d ed. New York: Wiley.

Lucas, G. B., C. L. Campbell, and L. T. Lucas. 1985. *Introduction to plant diseases.* Westport, CT: AVI.

Powers, E. R., and R. McSorley. 2000. *Ecological principles of agriculture.* Albany, NY: Delmar.

SELECTED INTERNET SITES FOR FURTHER REVIEW

http://www.ipm.ucdavis.edu/PMG/crops-agriculture.html

Pests of agricultural crops, floriculture, and turfgrass; their management guides; great photos of insects, mites, nematodes, and diseases.

http://www.ipm.ucdavis.edu/

Discussion of IPM; photos, methods of management, etc.

http://www./ipm.ucdavis.edu/PMG/WEEDS/california_arrowhead.html

Weed identification; photos.

http://www.farmphoto.com/album2/html/noframe/fld00068.asp

IPM photos.

http://www.farmequip.com/cihfield.jpg

Photos of various farm equipment.

OUTCOMES ASSESSMENT

PART A

Answer the following questions true or false.

1. T F Viruses are obligate parasites.
2. T F Fungi cause most of the infectious diseases of importance in crop production.
3. T F Mushroom is a fungus.
4. T F In insects with no metamorphosis, the young look like small versions of adults.
5. T F The degree of pathogenicity of a pathogen is called its toxicity.
6. T F A pesticide used to control fungi is called a miticide.
7. T F The relative capacity of a substance to be poisonous to a living organism is called its lethal dose.
8. T F The higher the LD_{50} of a pesticide, the more toxic or poisonous it is.
9. T F Pesticides may be formulated as "dusts."
10. T F The use of traps is an example of a cultural method of pest management.
11. T F Aerosols are best used in open air situations.
12. T F Organochlorines are more biodegradable than organophosphates.
13. T F Most herbicides in use are organic-based.

PART B

Answer the following questions.

1. Diseases are caused by organisms called _____.

2. List the categories of microorganisms that cause diseases.

3. The capability of a pathogen to cause disease is called its _____.

4. Give the steps in a disease cycle.

5. What is complete metamorphosis in the lifecycle of insects?

6. Give four of the important insect orders in crop production.

7. Classify weeds according to lifecycle.

8. Give the four basic strategies of pest control.

9. What is economic injury?

10. The method of using laws to manage pests is called _____.

11. What is crop rotation?

12. Give the corresponding names of pesticides used to manage each of the following pests in crop production: rodents, fungi, nematodes, insects.

13. Aviacides are pesticides used to manage _____.

14. What is the toxicity of a pesticide?

15. Give three specific methods of pesticide application.

16. Discuss the classification of herbicides according to (a) mode of action, (b) timing of application, and (c) selectivity.

17. The compound responsible for the killing action of a pesticide is called the _____.

18. What does the acronym IPM stand for?

PART C

Write a brief essay on each of the following topics.

1. Discuss the disease triangle.

2. Discuss the infection stage in a disease cycle.

3. Discuss vertical resistance.

4. Discuss horizontal resistance.

5. Describe the genetic basis of disease incidence in plants.

6. Describe how pathogens affect crop productivity.

7. Discuss the preventive strategies in pest management.

8. Discuss the biological method of pest management.

9. Discuss a specific method of cultural pest management.

10. Discuss the lethal dose of a pesticide.

11. Describe the safe use of pesticides in crop production.

12. Discuss the strategy of integrated pest management.

13. Discuss the impact of pesticides on the environment.

14. Discuss the factors affecting herbicide effectiveness.

PART D

Discuss or explain each of the following topics in detail.

1. Why are viral infections not controlled by pesticides?

2. Discuss the trends in the use of pesticides in crop production (emphasize active ingredients).

3. Discuss the issue of resistance of pests to pesticides.

4. Predict and discuss the status of pesticides in crop production in the next decade.

5. Can modern agricultural production be totally pesticide-free?

11

Agricultural Production Systems

PURPOSE

Crops are produced by farmers in various parts of the world according to certain production packages suited to the region and the socioeconomic situation. In this chapter, basic production systems in use around the world are discussed. Organic crop production is also discussed.

EXPECTED OUTCOMES

After studying this chapter, the student should be able to:

1. Define and describe a natural ecosystem.
2. Compare a natural ecosystem with an agroecosystem.
3. Discuss the components of a crop production system.
4. Discuss plant communities in crop production and their response to competition.
5. Discuss the nature, advantages, and disadvantages of crops grown in monocultures and polycultures.
6. Discuss the role of agroforestry practices in crop production.

KEY TERMS

Agroecosystem	Biodiversity	Monoculture
Agroforestry	Crop rotation	Polyculture
Allelopathy	Ecosystem	Relay cropping
Antagonism	Intercropping	

TO THE STUDENT

Crop production involves managing production resources (e.g., crop cultivars, land, and water) within a given socioeconomic system. Crops are grown as populations of plants. These populations are organized in two basic ways: only one crop at a time on the same land season after season or different crops in one season on the same land. Each approach has pros and cons. As you study this chapter, pay attention to the role of crop diversity and technology in crop production systems. Farmers grow crops as plant stands or populations in various production regions using certain production packages. Plants interact with each other and with the environment during production. Farmers use various technologies along with natural resources to increase crop productivity. Modern crop production depends heavily on agrochemicals to protect plants and to increase soil fertility. Excess agrochemicals find their way into the general environment where they pollute groundwater. Pesticides enter the food chain through animals feeding on contaminated plants and the use of toxic products to protect food crops in the field and the storehouse. As you study this chapter, visualize the farm as a giant organism comprised of biotic components (animals, plants) and abiotic components (soil, environment) that are managed or manipulated by humans (farmers) in an economically sustainable and environmentally sound fashion.

11.1: WHAT IS A CROP PRODUCTION SYSTEM?

A *production system* is a mix of crop(s), natural resources, and socioeconomic factors employed in the production of an agricultural product. Crop production entails the management of (1) inputs, (2) biological processes, and (3) sources of depletion of production resources for productivity. The importance of each of these three basic elements of crop production depends on the farm and the farming system.

11.1.1   INPUTS

Primary production inputs in crop production are seed (or appropriate propagating material), water, fertilizers, labor, pesticides, and energy. The amount of these external inputs depends on the nature of the operation. Some production systems are labor-intensive; others are mechanized and chemical-intensive.

11.1.2   BIOLOGICAL PROCESSES

Photosynthesis is the single most important plant physiological process in crop production. Other biological processes that interact in crop production include natural nutrient cycling, biological nitrogen fixation, biological control, and mycorrhizae effects on phosphorus uptake by plants. Cultivars differ in their capacities to conduct these physiological activities. Genetic factors are thus important in crop production. They determine climatic adaptation, resistance to pests, and efficiency in using nutrients and water.

11.1.3   DEPLETION OF INPUTS

Modern conventional crop production requires repeated inputs for high productivity because the inputs are depleted during production. Natural nutrient cycling, important in sustainable agriculture, is "leaky" and thus experiences losses. Some of the depletion is desirable and planned. Crops are planted to remove soil nutrients and convert them into

A *farming practice* is the way a farmer conducts a specific production activity, such as pesticide application or fertilizer application. A *method of farming,* on the other hand, is the systematic way in which the farmer achieves a particular production function, such as crop stand establishment. A *farming system* is an approach to farming that integrates farming methods, practices, technologies, knowledge, and expertise, coupled with the specific goals and values of the producer.

products. Other factors, such as leaching, fixation, and erosion, deplete soil of its nutrients in ways that are uneconomical. The producer as a manager should implement practices that eliminate or reduce negative and wasteful activities.

11.2: NATURAL ECOSYSTEM VERSUS AGROECOSYSTEM

How is crop production like the operation of an ecosystem? In crop production, living organisms interact with their environment under human supervision. *Ecology* is the study of how living things relate to their environment. In terms of the kinds of groupings of organisms in a given area, there are two terms that need defining. A group of individuals of one species occurring at one location constitutes a *population*. A population may be described based on a simple count (number) of individuals, the number of individuals per unit area (density), and the total mass of individuals (biomass) at the location.

The occurrence of living organisms (biotic factors) and nonliving organisms (abiotic factors) interacting with each other in a specific location constitutes an **ecosystem**. The abiotic factors in an ecosystem are light, temperature, water, soil, and air. These factors determine how plant species are distributed in the ecosystem. Ecosystems are naturally self-sustaining (Figure 11–1). This is possible through three mechanisms: photosynthesis, energy flow through food chains, and nutrient recycling. Photosynthesis is the process by which plants capture and convert light energy into chemical energy. Plants in the ecosystem are called *producers*. They provide food for the herbivorous animals in the system, which are called *primary consumers*. They are, in turn, preyed on by other organisms such as flies, ticks, and leopards (called *secondary consumers*). When producers and consumers die, their organic matter is broken down by microorganisms (bacteria and fungi) called *decomposers*. They cause the organic minerals to be converted into inorganic forms for recycling for plant use, a process called *mineralization*. In an **agroecosystem**, farmers and ranchers constitute the fourth group of players. They are manipulators (or managers) of the ecosystem for producing agricultural products.

In natural ecosystems, a biological balance is created by a diversity of networks of interacting plants and animals in a dynamic equilibrium. The component species are adapted to the prevailing environment. Further, each is a source of food for others. The species survive or escape their competitors and natural enemies. Each ecological niche in the natural ecosystem is occupied in both time and space. Disease epidemics are uncommon under such conditions where there is environmental stability.

On the other hand, a typical agricultural system (agroecosystem) operates in a manner that is counter to the mechanics of biological balance. In modern crop production, especially in developed economies, farmers frequently grow one crop at a time, a practice

Ecosystem. A living community and all the factors in its nonliving environment.

Agroecosystem. The population of selected plants and/or animals growing at a location and interacting with biotic and abiotic environmental factors under the management and supervision of humans.

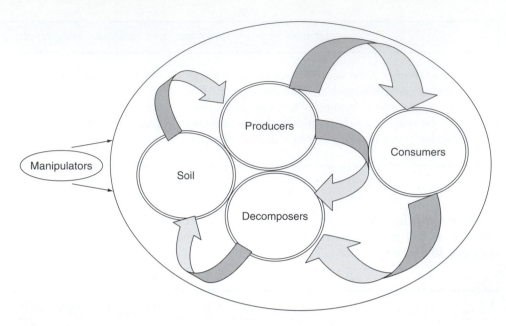

FIGURE 11–1 A conceptual model of a natural ecosystem. Natural nutrient cycling and other natural cycles sustain the ecosystem. Producers (plants) extract nutrients from the soil to make products. When they die, their organic matter is decomposed and chemicals mineralized into inorganic forms that can be reabsorbed by plants. Some products may be used by consumers who, upon dying, release the nutrients back into the soil through the processes of decomposition. These interactions occur within an environment that involves a hydrologic cycle and various gaseous exchanges between plants and animals and the larger environment. The ultimate source of energy in the ecosystem is sunlight, which is used in making food by producers through photosynthesis. The activities of manipulators (humans) impact both natural ecosystems as well as agroecosystems. The thinning of forests by logging, forest fires, and dam construction impact ecosystems.

called *monoculture*. This severely restricts the advantage of diversity. Further, this single genotype is not adapted to the environment well enough to survive without human intervention. Crop producers have to protect their crops from competitors and supplement the nutrients in the environment as needed.

The use of modern production inputs (pesticides, tillage, etc.) and other production practices creates empty niches, resulting in increased frequency of diseases. Modern agricultural production is filled with shocks (e.g., changes in production practices such as irrigation, seeding rate, and tillage) that can upset the balance in the agroecosystem.

To restore biological balance in an agroecosystem, the crop producer may use adapted and disease-resistant cultivars and practices that increase soil organic matter. Changing planting dates and introducing antagonists of pathogens are helpful strategies.

A critical aspect of an ecosystem is natural cycling and interdependency of all its components. On the contrary, modern production systems either bypass or exclude certain key components of natural ecosystems. The consequence is lack of, or limited cycling of, nutrients (Figure 11–2). Modern producers are hence compelled to resort to expensive maintenance of soil fertility through the continuous addition of artificial chemicals. The consequence of the intensive use of chemicals is pollution of the environment. In order to limit the inputs into modern agriculture, production practices should encourage natural cycling of nutrients, as advocated by the concept of sustainable agriculture.

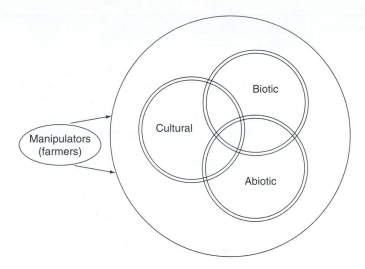

FIGURE 11–2 A conceptual model of an agroecosystem. There is limited nutrient cycling as the crop is harvested for use elsewhere, and consumers do not recycle the nutrients from where they were extracted. Agroecosystems, hence, depend on humans for additional nutrients and other cultural inputs. Further, participants in an agroecosystem are not selected by nature and hence are not necessarily compatible. Through the use of technology, agroecosystems can be designed to overcome natural barriers to growth and development. For example, through irrigation, crops can be produced in arid regions. In crop production, farmers manipulate the production environment through agronomic practices (irrigation, pesticide use, fertilization) to produce crops and raise farm animals.

Like natural ecology, the concept of agroecology (or crop ecology) describes attempts by humans (manipulators) to nudge nature to their advantage, rather than having a balance that favors all components in an ecosystem. Agroecologists utilize technologies to enhance and work with nature to increase productivity of the crop of their choice.

11.3: CROP FARMS AND FIELDS ARE AGROECOSYSTEMS

An agroecosystem is complex in its structure, embracing many agricultural technologies. Agricultural production started as a largely mechanical-based operation in which machines were developed to replace human labor. This reduced the tedium in field production. Modern crop production depends heavily on the use of agrochemicals to manage weeds, insect pests, and diseases and to improve soil fertility. The next step in the evolutionary trend in agriculture was to emphasize the biological aspects of crop production. The goal of this was to reduce the chemical component of modern crop production, which is the chief source of criticism of modern agriculture.

In terms of a scientific basis, an agroecosystem should have three basic technologies: *mechanical, biological,* and *chemical.* However, for sustainability, an agroecosystem should consider the additional dimensions of social, economic, and environmental factors. A sustainable crop production system should be environmentally responsible, being operable with minimal environmental consequence. It should be socially acceptable.

An ecosystem is constituted by living organisms interacting with one another and with factors in their non-living environment. The non-living components of an ecosystem include physical and chemical factors such as light, water, air, soil, and temperature. Plants vary in their requirement for these factors, which consequently affect their distribution in the ecosystem. Plant distribution is also influenced by their interaction with the biotic community. Natural ecosystems are self-sustaining through food manufacture by plants (by photosynthesis), energy flow (by the food chain), and nutrient recycling (through microbial activities in natural cycles).

Light is the ultimate source of energy in the ecosystem. This energy is harnessed by plants (called *primary producers* or *autotrophs*) and converted to energy-storing molecules. Some animals (e.g., cows and sheep) feed directly on plants to obtain this energy. These animals are called *primary consumers* and are food for *secondary consumers* such as lions and other flesh-eaters. Consumers are *heterotrophs* (unable to synthesize their own food and dependent on others). When these three categories of ecosystem components die, the nutrients in their bodies are recycled primarily by the action of bacteria and fungi *(decomposers)* that decompose dead organic matter. Producers and consumers interact to form food chains in which producers are at the bottom. Organisms in the community feed on those below and are in turn preyed on by those above in the chain. The length and complexity of a food chain are variable, since most organisms have multiple sources of food and are in turn food for a number of consumers. Different food chains may link together to produce a *food web*. Energy moves in the ecosystem in the form of a pyramid. The farther away one is from the source (producer), the less energy obtained from the food.

Once an ecosystem has stabilized (attained a natural balance), it becomes self-sustaining. Population explosion is eliminated because every organism has its natural enemy.

Humans are capable of profoundly impacting ecosystems. Their role is both as consumers and manipulators. Through technology and general knowledge of the operation of an ecosystem, humans are able to manipulate the system to their advantage.

Unacceptable social costs associated with modern production include soil erosion and its attendant problems of silting of rivers and reservoirs, deterioration of wildlife habitats, and loss of soil fertility. Silting of water systems is corrected at a cost to society. Municipal water is treated at extra cost if it becomes polluted. Further, changes in crop production practices and farming systems affect farm resource use and eventually farm populations. In this regard, rural communities are most vulnerable.

Production practices and farming systems should be economically viable. Farm profitability is key to a crop production enterprise. Producers will embrace new methods only if they are profitable. Some aspects of profitability stem from government policies and decisions that producers have to comply with in order to benefit from various economic assistance programs. Programs such as the Conservation Compliance Provision and the Conservation Reserve Program of the 1985 Food Security Act were designed to protect U.S. natural resources from deterioration from field production activities.

Crop production is a high-risk operation. Field production is subject to the vagaries of the weather. Profitability is affected by prices of agricultural products, consumption patterns, and global economic patterns. Producers, naturally, make choices that they deem less risky and in their interest. Managerial skill is essential to success in production. Sustainable agriculture does not focus on removing or overcoming natural obstacles to production. Rather, through skillful management, the producer uses innovation to determine the best system to implement, considering the natural resources available. In this regard, no one best system is suitable for all farms.

In effect, an agroecosystem is a purposive (goal-oriented) system. It responds differentially under different environments. However, the producer is able to change goals under a constant environment, or the producer can pursue the same goals in different environments by adopting different behaviors. An agroecosystem is a creation of the knowledge, skills, attitudes, and values of the creator.

A key differentiating factor between the natural ecosystem and an agroecosystem is the human (management) element. An agroecosystem exists because someone willed it into existence. As a purposeful system, it contains human elements (or subsystems) that interact with the other components (physical, biological) within the context of an environment. Physical, biological, socioeconomic, and cultural factors characterize this environment.

Agroecosystems differ also in productivity, stability, and sustainability. Management or manipulators are the key factor in determining the productivity (or yield) of an agroecosystem. Stability (usually of yield of a crop) of an agroecosystem is the amount of variation that occurs around the dynamic equilibrium of this characteristic. There are many factors that cause the variability (e.g., pests, weather factors). A third attribute of an agroecosystem is sustainability. Field crop production is prone to stress in the production environment. Stress may be in the form of, for example, erratic rainfall. The crop producer as a manager of production resources may intervene by providing supplemental moisture (through irrigation). If unseasonable weather brings about excessive moisture, surface runoff or erosion might be a problem. A crop ecosystem in which groundcover occurs would be more resistant to soil erosion forces. The sustainability of a field crop agroecosystem is a measure of the difficulty with which the producer manages production resources to minimize stresses for high crop productivity. A *sustainable system* is one that requires minimal and continuous economical and management inputs to manage productivity constraints.

11.4: PLANT POPULATIONS (AND MIXTURES) AND COMPETITION

How important is spacing among plants in crop production? Plants growing together in a crowded space compete for growth factors. Crop production entails the planting of crop plant populations, of like or dislike plants, depending on the cropping system. When there is crowding, plants in populations tend to encroach upon each other's demands for growth resources. *Competition,* or *interference,* is a crowding phenomenon of plant populations whereby component plants interact in response to physical growth factors—namely, water, nutrients, light, oxygen, and carbon dioxide. These factors all affect photosynthesis, the key to crop productivity. Space is a factor of competition especially in root crops. Competition arises because the supply of resources at a production site is less than the collective ability of closely spaced plants to use them.

Competition, or interference, among plants may also arise as a result of the phenomenon of **allelopathy**, the release of chemical compounds from a plant or its residue. This interference suppresses the growth of other plants in the immediate vicinity.

Crowding response in plants is mostly indirect and manifested through changes in the crop environment. One plant, for example, may aggressively deplete the soil of water, causing the nearest plant to find insufficient amounts of water for growth. Plant response to competitive stress may be classified into three phenomena: *density-dependent mortality, plastic response,* and *hierarchy of exploitation.* Density-dependent response results in reduction in plant density (number of plants per unit area). Plastic response occurs when plants under competitive stress reduce in size of vegetative parts or number of yield components. This results in decreased crop yield. Hierarchy of exploitation occurs between

Allelopathy. The process by which one plant species affects other plant species through biologically active substances introduced into the soil, either directly by leaching or exudation from the source plant, or as a result of the decay of the plant residues.

The primary goal of agriculture is to manage photosynthesis to produce food and fiber for humans. The manipulation of photosynthesis is done according to one of two general models: *industrial* and *sustainable*.

Industrial model

The *industrial model* of agriculture (also called the *conventional model*) considers agriculture as an industrial enterprise in which farms are factories and fields the production plants. The products that are made (production units) may be animals or grain. The producer acquires and uses production inputs to make specific products. Fresh inputs are acquired for each production cycle. The prevailing philosophy is one of a *component approach* to crop production, whereby the producer focuses on individual farming practices and methods. An industrial plant is designed for a specific production practice, where components are assembled to make a product. In the industrial model of crop production, obstacles to production are removed through the use of innovative technologies. The natural environment is modified or controlled (e.g., controlled environment agriculture in a greenhouse). Control also makes it potentially possible for steady progress to occur.

Sustainable ("holistic") model

The philosophy of the *sustainable model* of agriculture is that agricultural production should be viewed from a *whole systems* perspective. It emphasizes the need for producers to conduct their agricultural activities in harmony with the biosphere. This holistic system model promotes *working with* rather than *controlling* or subjugating nature. The natural resource base of agricultural production should be conserved and protected; the environment should also be protected. In effect, the producer, in this approach, focuses more on managing the internal resources (natural resources) of the farm (or agroecosystem) rather than on production inputs that are purchased and introduced into production.

Conventional versus sustainable agriculture

Conventional agriculture is the term for the predominant farming practices, methods, and systems of crop production adopted by producers in a region of production. In technologically advanced regions, these production systems are generally capital-intensive and chemical-dependent. Crop productivity is significantly increased but at a cost to the environment and human health. Agrochemicals build up in the environment and pollute groundwater and the air. Chemical residues in food are health hazards to humans and animals.

Sustainable agriculture seeks to increase crop productivity without the adverse effects to the environment and society as a whole. Since this is a concept that emphasizes a goal rather than a set of practices, producers adopt various farm-based innovations to accomplish the general goal. The innovative aspect of sustainable agriculture comes about because there is no one correct way to arrive at the general goal. This is because each producer's farming situation is different, regarding soils, climate, cropping system, method of production, and market needs.

Sustainable agriculture

The traditional purpose of agriculture is production of food and fiber for society. To this end, traditional agron-

Monoculture. The cropping system in which a single crop is repeatedly cropped on the same piece of land season after season.

Polyculture. The cultivation of more than one crop species on the same piece of land in various temporal and spatial patterns.

unidentical genotypes in competition when one is better able to exploit growth factors than the other(s).

As previously indicated, field crops are cultivated populations of plants. Agronomists manage these populations. There are two basic types of cultivated **monocultures** and *mixed cropping,* or **polycultures**. These types differ in genetic content and structure (in terms of density, spacing pattern, plant size, and stage of development). Agroecosystems are dynamic, being subjected to change from intervention of human managers (farmers) and changes in weather factors. They intensify and change because managers perceive opportunities or simply change their perception of the current status of the crop

omy focuses on removing physical and biological constraints to production. The natural environment is modified or controlled so that crop production is orchestrated according to a schedule that theoretically makes unlimited production possible. Agriculture is evolving in a direction that views crop production within an ecological context, as already stated. The trend is not to control but to farm in harmony with the biosphere. There is a holistic approach in which crop production is conducted in an environmentally responsible manner. Production practices are selected to complement and accommodate the factors in the production environment.

This systems approach to production is called *sustainable agriculture.* This has given rise to different kinds of terminology that generally describe the same concept. These include *organic farming, alternative agriculture, biological agriculture, regenerative agriculture, reduced input agriculture, ecological farming,* and *environmentally sound agriculture.* Notwithstanding the name, the concept describes strategies of achieving synergy for production by integrating several practices. These practices are not new; they are combined strategically such that crop production is enhanced while the natural resource base is protected for posterity.

The goals of sustainable agriculture are

1. Increased profitability of crop production
2. Natural resource conservation in crop production
3. Use of environmentally prudent farming systems in crop production.

Sustainable agriculture is a dynamic concept that emphasizes a goal rather than a set of production practices. *Alternative agriculture,* on the other hand, describes the process of on-farm innovation adopted by crop producers toward achieving the goal of sustainable agriculture.

Sustainable agriculture is also called *limited (or low) input sustainable agriculture (LISA),* a terminology that is sometimes incorrectly thought of as low-technology production. On the contrary, sustainable agriculture calls for the use of the best production technology in a productive, cost-effective, and environmentally responsible manner. Crop producers as managers combine scientific know-how with on-farm resources for highest possible productivity and without adverse environmental consequence or depletion of natural resources. An important objective of sustainable agriculture, however, is to limit the intervention of the crop producer through the use of agrochemicals. Instead, it encourages the production environment to be self-sustaining like an ecosystem. The sustainable agricultural model is founded on the concept of agroecology. Natural processes and production technologies are integrated to develop a particular production system. The sustainable farmer must be knowledgeable to be able to develop a site-specific, integrated, and sustainable system of production.

In conventional crop production, there is intensive crop management and specialization. On the other hand, sustainable agriculture depends on biological interactions and diversification for success. On-farm cycles that are managed in sustainable agriculture include crop rotations, nitrogen fixation, genetic resistance in crops, and several others. This dichotomy is not always as clear cut as presented. There are transitions between these models.

ecosystem. The market dynamics may bring about changing profitability of a production enterprise. The agricultural producer then makes decisions to change production inputs (e.g., fertilizer, cultivars) to respond to prices of products.

Two important attributes of agroecosystems of economic importance to the producer or manager are stability and sustainability of yield. Yield stability is a measure of homeostasis. Sustainability, on the other hand, measures the difficulty the crop producer faces as he or she attempts to manage production resources in response to production constraints. If the management required is not economical or practical, the agroecosystem is deemed unsustainable.

11.5: MONOCULTURE

Why is monoculture a popular crop production system in industrialized economies? Monoculture is at one extreme of the cropping systems spectrum. Crop production in industrialized societies is primarily monoculture. This method of production is characterized by the planting of only one cultivar on a large acreage (Figure 11–3). The land is often flat and readily amenable to mechanization, which is used at all stages of crop production. Monocultures are input-intensive, depending on agrochemicals (fertilizers and pesticides) for high productivity. Plants in this system feed at the same level in the soil and draw the same nutrients. Pests associated with the crop tend to build up, necessitating the intensive use of pesticides to manage them.

Monocultures are cultivated populations that consist of only one species. Monocultures may experience interspecific competition. Biomass accumulation in monocultures is exponential in pattern. This pattern is modified by plant density; the lag-phase shortens as density increases. Competition among plants in a population sets in after a certain period of no interaction, when seedlings have adequate growth resources. The onset of interference occurs as plants increase in size. In the early growth, plants have equal mass. The mass then begins to vary according to intensity and duration of interference. Spacing of field crops is selected such that the plants are ideally crowded to maximize the environmental resources. Under this ideal plant density, no plant dies or becomes unproductive. Individual plants experience less than the optimal yield possible if they had unlimited resources. However, as a population, the yield is optimized.

In crop production, plants that have large adult size are widely spaced at planting. This low plant density means initially the ground cover is low, encouraging weed infestation while crop cover develops. Plant competition sometimes has casualties. Where plant spacing is close (high density), as occurs in the seeding of grasses and forage legumes, smaller plants become crowded out of the population. This competition-induced mortality is called *self-thinning*. This phenomenon is useful in nature but undesirable in crop production, since it wastes soil nutrients; plants use some of it, only to die eventually.

FIGURE 11–3 Monoculture of cotton. (Source: USDA)

Another plant response to competition is *morphological plasticity*. Certain genotypes have the capacity to adjust their size at different plant densities and still be productive. Plants that tiller, like small grains (wheat, barley, oats), are able to attain the same number of heads and final yield over a wide period of planting densities. Modern corn cultivars, on the other hand, have been bred to be single-stalked (uniculmn) or with minimum tillering. As such, corn is unable to maintain final yield over a wide range of planting densities, as opposed to wheat. Corn loses plasticity above a certain maximum seeding rate.

Morphological plasticity is one of the reasons that seeding rates vary widely among crops. The seeding rates, row widths, and plant spatial arrangements are determined based on the degree of plant plasticity. Several patterns of spatial arrangement are used in crop production. The most efficient arrangement (that reduces overshading) depends on plant structure and morphology. Hexagonal arrangement favors plants such as sugar beet that display their foliage in a circle around the axis. Row cropping (rectangular arrangement) suits crops such as corn, cotton, sorghum, and sunflower.

11.5.1 ADVANTAGES OF MONOCULTURE

The advantages of monoculture include the following:

1. Industrialized nations have the technology and know-how to grow crops in monoculture.
2. Monocultures are responsible for producing very large quantities of food that feed the world.
3. Diversity can be introduced into monoculture by practices such as crop rotations and cover cropping.
4. Monocultures are easier to manage. The producer has to contend with one set of practices, rather than different sets for different crops.

11.5.2 DISADVANTAGES OF MONOCULTURE

The disadvantages of monoculture include the following:

1. The production activity is susceptible to pests and disaster. Pest outbreak can wipe out the entire enterprise, since all plants are equally susceptible to a particular pest.
2. There is no insurance against adversity (weather-related or pests).
3. Lack of diversity means the presence of a fewer number of natural enemies of the pests that plague the crop. Further, the crop enterprise has less capacity to rebound from a temporary environmental stress.

11.6: POLYCULTURE

Crop production, especially in developing economies, involves different kinds of crop combinations, simultaneously cultivated on the same site. Growing different kinds of crops on the same piece of land is called *polyculture, mixed cropping,* or simply *mixtures* (Figure 11–4). Some of these mixtures involve unlike genotypes of the same species or different species.

Polycultures are most common in the tropics (Africa, Asia, and Latin America). Why is this so? Polycultures are associated with small-scale, subsistence agriculture. For example, corn and cowpea are often grown in association. There is diversity in crop cultivars

FIGURE 11–4 Polyculture involving corn and soybean. (Source: USDA)

planted. Operations are not mechanized, depending on draft and human labor as sources of farm energy. Subsistence agriculture is generally low-input, with minimal or no use of agrochemicals. Natural methods are depended upon for improving soil fertility. Most production under this system is rain-fed.

Planting crop mixtures is common in forage production where legume-grass combinations are frequently grown. Interspecies mixtures experience interspecific competition that is brought about by mechanisms similar to those that prevail in monocultures. The severity of the competition depends upon the differences between the species in the mixture regarding plant size, growth habit, and response of plants to weather factors.

Donald is credited with developing the principles of competition among plants in mixtures. His four basic principles governing plant association in mixtures are

1. Mixtures generally yield less than the higher-yielding pure culture.
2. Mixtures generally yield more than the lower-yielding pure culture.
3. Mixtures generally yield less than the average of the two pure cultures but may yield more.
4. Evidence of cooperation (or mutual benefit) among plants in mixtures is not widespread; there is little evidence that mixtures exploit the environment better than pure cultures.

Exceptions to these general principles exist. Oat-barley mixtures are found to be highly productive. Mixed grain production strategy results in yield stability, especially in a variable land environment.

Three basic interactions occur among unlike plants cohabiting in time on the same plot as the proportions of the species are varied. As one species is progressively introduced into a pure culture of a second species, one of two types of competitive interaction, *complementation* or competition, may develop progressively.

11.6.1 COMPLEMENTATION

Two or more species planted as mixed cultures are said to complement each other if they perform better in mixtures than in pure stand. This phenomenon may arise by one of two mechanisms. The two species may acquire resources from different spaces in the growing environment. This cohabitation is called *niche differentiation* (different space). Species with different root morphologies may feed at different depths of the soil and have niche differential for water. Legume-cereal mixtures have niche differentiation for nutrients, especially nitrogen. Legumes can utilize atmospheric nitrogen while non-legumes depend on soil nitrogen. Apart from spatial difference, niche differentiation may occur through temporal difference. Producers may maximize the use of production resources through planting practices whereby multiple crops are produced in one growing season. For example, planting cool season and warm season forages together extends the production season.

11.6.2 COMPETITION

If the total yield of the two species is less than either pure culture, the interaction is described as antagonistic or competitive. Mixtures or blends of cultivars of cereal crops, grain crops, and forage crops have been developed to improve crop yield stability. When genetically very similar lines derive from a series of crosses with disease resistance lines and backcross to a common parent, the product is called *multi-line*. Jenkins proposed the multi-line concept. The product has yield stability derived from different genes for disease resistance introduced from the crosses. Hierarchy of exploitation is the response to competition in which mixtures of two species or two cultivars of the same species display an *aggressor-suppressed* relationship. One component in the mixture, the aggressor, is more successful in exploiting the environment to the detriment of the other component. The aggressor uses up a disproportionate amount of the nutrients, light, water, and other growth factors in the environment. In forage mixtures, the producer can manipulate the hierarchy of exploitation by adopting certain management practices that favor one component of the mixture. For example, the application of nitrogen fertilizers to a legume-grass mixture may enhance the vegetative growth of a non-aggressive grass to the extent that it outcompetes the legume for light.

Polycultures have several distinct advantages:

1. The genetic heterogeneity may slow the dispersal and spread of diseases and insect pests if susceptible host plants are interspersed with resistant plants. One component species may have the capacity to serve as a trap crop for insects.
2. Mixtures have yield stability because of the diversity in genotypes.
3. Multiple cropping systems provide insurance against total crop failure, in case of environmental adversity.
4. For subsistence producers, multiple cropping enables the farmer to plant a variety of crops to provide a good balance in diet.
5. Resource-use efficiency can be maximized. The production is integrated.

The disadvantages of polycultures include

1. The genetic diversity that may provide defense against diseases and insect pests may also increase these problems. The greater diversity in hosts may also mean greater diversity of pests and diseases, especially soil-borne problems.
2. Crop rotations are used to control soil-borne diseases in monocultures. This strategy is not practical in mixtures.
3. Planting and harvesting operations are complicated by the diversity of plants. Mechanization is restrictive.
4. Agronomic management is complex; fertilization, irrigation, and pest control are difficult to implement.

There are different ways of combining crops and sequence of planting crops in polyculture. Examples of specific polycultures that are used in crop production include mixtures, **intercropping**, strip cropping, *overlap planting (e.g.* **relay cropping**, *double cropping, nurse crop),* and **crop rotation**. **Agroforestry** practices are polycultures that include trees and shrubs.

11.7.1 MIXTURES

Growers may plant different varieties of crops in mixtures or a mixture of different species. Put another way, crop mixtures range from growing multi-lines (diversity at single genotype level) to agroforestry (diversity at the plant order level). Multi-lines represent a mixture of genotypes that are identical except at a specific trait locus. Diversification is the key characteristic of mixtures. When cultivar mixtures are grown, the population enjoys diversification in desired resistance genes as well as a combination of other mutually complementary traits that are not found in any one genotype. Producers may mix agronomically compatible fodder crop species (e.g., different cereal species or legume-cereal combinations). Mixture components may be selected such that they mature at the same time and can be harvested together (e.g., a combination of winter wheat and winter field beans or spring oats and peas). In a legume plus cereal mixture, the harvested crop can be separated into the components during processing.

Crop monocultures are successful in achieving maximum yield in high-input production at near-optimal environmental conditions. Monocultures are suited for risk-averse, highly profitable, non-sustainable agricultural production systems in which environmental considerations are not a high priority. However, when producing crops under more sustainable conditions that are suboptimal for maximum yield, crop mixtures have proven to be superior to monocultures. Under these less optimal environmental conditions, mixtures are able to better exploit all the resources through enhanced crop plasticity to provide greater yield and quality stability. Crop plasticity enables mixtures to perform appreciably in more marginal and higher-risk environments. The producer with good management can increase profitability and lower the risk of the enterprise by reducing production inputs. Mixtures can also be produced under the high input, near-optimal production environments.

11.7.2 INTERCROPPING

Intercropping is the crop production practice in which one crop is planted in the open areas of another. The goal of this planting format is to increase *land equivalent ratios* (the amount of monoculture land area needed to produce the same amount of polyculture yield). This simultaneous cropping system may not have any competitive interaction at all. It may be used in the early years of establishing an orchard when the tree cover is low. The open spaces can be cropped to annuals until the tree cover is fully established.

Intercropping systems are common in developing countries, especially in the tropics. There are different versions of intercropping systems. Crops may be planted in organized, alternating rows; grouped in different sections of the field; or planted in haphazard fashion. For best success, the plants to be intercropped must be judiciously selected to reduce interplant competition. It is best if the crops fill different niches so they can better use resources in the production environment (e.g., include annuals and perennials, shallow and deep-rooted plants, legumes and non-legumes). These combinations represent complementary associations.

Intercropping. The crop production system in which one crop is planted in the open space of another.

Relay cropping. The cropping system in which one crop is seeded into another standing crop that is near harvesting.

Crop rotation. A planned sequence of crops growing in a regularly recurring succession on the same piece of land.

Agroforestry. The simultaneous cultivation and management of trees and agricultural crops or livestock in various spatial and temporal patterns to optimize productivity of the land and protect natural resources, among other benefits.

FIGURE 11–5 Contour strip cropping. (Source: USDA)

11.7.3 STRIP CROPPING

Strip cropping is used by some farmers in the United States and other regions of the world. This is a practice in which strips of different crops are planted at intervals within the crop in the same field (Figure 11–5). It is an effective and inexpensive method of controlling soil erosion. Three main types of strip cropping are in use—contour, field, and buffer strip cropping. In *contour strip cropping,* alternative strips of row crops and soil-conserving crops (e.g., sods) are grown on the same slope or elevation perpendicular to the wind direction or water flow. If soil is dislodged from the row crops by erosive forces, some of it becomes trapped in the dense soil-conserving strip. It is applicable to short slopes of up to about 8% steepness. In *field strip cropping,* strips of a uniform width are located across the general slope of the land, while in *buffer strips,* strips of grass or legumes are laid between contour strips of crops in irregular rotations.

11.7.4 OVERLAP PLANTING SYSTEMS

In *overlap planting systems,* two or more species overlap for portions of their lifecycles. The producer accomplishes this through selecting and planting crops that differ in maturity in strategic temporal sequence. In one growing season, a monoculture may be used to start crop production. This is intercropped with another species at some point in the growing season, the two crops maturing at different times. With careful choice of specific and proper planting times, interspecific competition is minimal under this system. For example, wide-spaced and long-term crops such as cassava may be interplanted with an early crop such as beans or okra.

The period of overlap is variable. When the overlap period is very short such that the seasonal crop is planted just before the first is harvested, the plant culture is called *relay cropping* (e.g., seeding of winter wheat into a standing crop of soybean).

The advantage of relay cropping is that competition is further reduced between two species (e.g., bean-corn relay). In some production areas, producers are able to maximize the growing season by growing two crops in one season, one after the other, without any cohabitation. This is called *double-cropping* and involves no interspecific competition.

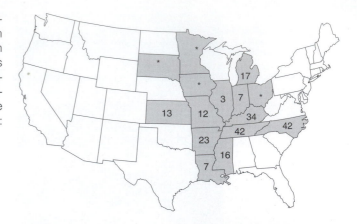

FIGURE 11-6 Double-cropping is possible in some areas in the eastern half of the United States where a longer growing season occurs. In 1997, 67 million acres of soybean were double-cropped. (Source: USDA)

FIGURE 11-7 Double-cropping soybean into corn. (Source: USDA)

Double-cropping is adopted by some producers in the eastern half of the United States where the growing season is long (Figure 11–6). An example of double-cropping is cropping soybean after wheat (Figure 11–7).

11.7.5   NURSE CROP

The concept of a *nurse crop* is incorporated into the practice of overlap planting. It is implemented in various forms. A nurse crop may be an annual, fast-growing species that is planted with the economic or desired crop to suppress weed growth while the economic crop establishes in the field. The nurse crop may then be controlled with chemicals once the desired crop is established, or it may die out eventually as the desired crop gains dominance in the field. For example, to establish alfalfa without the use of herbicides, a grower may seed it along with oats as a nurse crop. The oat plants compete with the early weeds until the alfalfa crop stand is well established. The term *nurse crop* is sometimes used synonymously with *companion crop*. Onions can be planted as a companion crop with carrots, the role of the former being to mask the smell of carrots from the devastating carrot fly. Similarly, cabbage root fly in Brussels sprouts may be controlled by using clover as companion crops in the production of brassica and other crops.

11.7.6   CROP ROTATION

Also called *sequence cropping,* crop rotation is the growing of multiple crops on the same piece of land, one after the other (Figure 11–8). The different crops do not interact

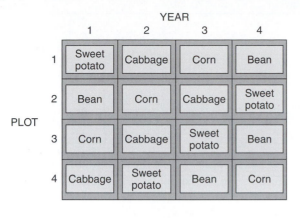

YEAR

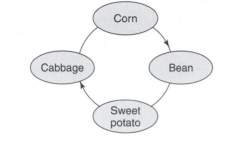

FIGURE 11–8 Crop rotation involves strategic sequencing of crops in production. Species with similar demands on growth factors should not follow each other directly.

but are separated temporally. Crop rotation strategies differ in design. The land may be divided into sections and the crops assigned to specific sites. The sites are changed in subsequent growing seasons. Alternatively, the entire field is planted with one crop species in one season, followed by a different cultivar of the same species or a different species the next season.

The success of a crop-rotation enterprise depends on careful choice and order of cropping. A legume fixes nitrogen for the next crop. However, sorghum has a high C:N ratio; thus, its residue decomposes slowly. Following sorghum with a crop that requires heavy use of nitrogen will cause yield reduction. Another consideration is disease infestation. Crops attacked by the same pest should not follow each other; otherwise, pest buildup will occur.

Crop rotation is a cropping practice in which a set of crops is cultivated in a predetermined sequence, avoiding the same crop being cultivated continuously at the same location. In organic farming, the importance of crop rotation is in the management of soil fertility, diseases, and pests. Continuous cropping of the same piece of land leads to nutrient deficiency and nutritional imbalance.

Crop rotation promotes a buildup of high natural resistance to soil pests and a high level of biological activity. Beneficial microbes are required for mineralization of organic molecules and for suppression of harmful microbes. Crop rotation is the primary strategy for controlling pests in an organic farming system.

The underlying principle in the use of crop rotation is that of creation and maintenance of **biodiversity** in the agroecosystem as a means of establishing equilibrium in the system. Such an equilibrium would prevent the population explosion of any particular pest in the agroecosystem, thereby creating a natural means of pest management. Organic farmers may use cropping systems such as polycultures (e.g., mixed cropping, intercropping) to create biodiversity. Most field crops are commonly rotated in production (Figure 11–9).

There are certain principles applied in the design and effective use of crop rotation systems:

Biodiversity. The occurrence of a variety of biological species.

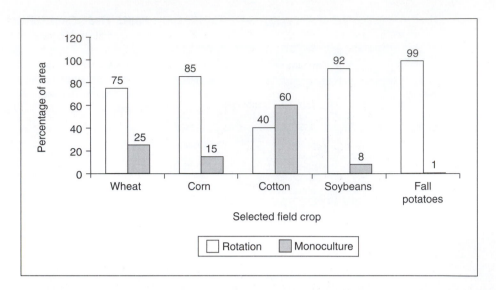

FIGURE 11–9 Cotton is mainly grown in monoculture, especially in the Delta Region. However, most field crops are commonly rotated in production. Common crop rotation systems include corn—soybean, fallow—wheat, and row crops—small grains. From 1994 to 1995, 99% of all fall potatoes was planted in rotation; 82% (162 million acres) of crops was in some form of rotation in this period. Cotton in 1995 was cultivated primarily in monoculture. (Source: USDA)

1. Follow a deep-rooted crop with a shallow-rooted crop for good soil structure maintenance.
2. Alternate between crops with high root biomass and those with low root biomass. This is because a high root biomass provides food for soil microorganisms.
3. Include green manures and catch crops to protect soil erosion and nutrient loss through leaching and to accumulate nitrogen.
4. Include nitrogen-fixing crops and alternate with crops with high nitrogen demand.

11.7.7   FERTILITY RESTORATIVE CROPPING

Regardless of the cropping system (monoculture or polyculture), farmers may interrupt their normal production sequence by leaving the land uncropped *(fallow)* or cropping it to a noncash crop *(cover crop)*. In another practice, the farmer abandons an exhausted soil for a fertile one *(shifting cultivation)*. A farmer may grow certain species for the express purpose of incorporating them into the soil (called green manure). Some of these practices were discussed previously in this chapter.

Cover Crop

As previously discussed, cover crops are used to improve soil fertility and protect an uncropped soil from erosion. They may be planted when normal cropping would be uneconomical. The crop can be grazed by livestock or plowed under as a green manure to fertilize the soil for normal cropping. Cover crops are normally legumes (e.g., alfalfa, pea, red clover, crimson clover, cowpea, and pigeon pea). These crops fix nitrogen. Cover crops also suppress weeds and reduce soil compaction. Grasses (e.g., rye) may be

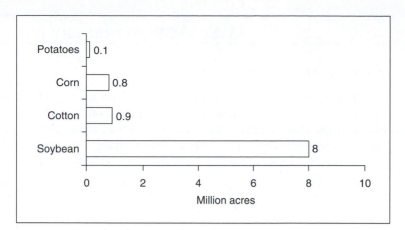

FIGURE 11–10 Cover crops are used most significantly with soybeans in the southern states for winter soil protection. Eight million acres of soybean crop area were planted to winter cover crops. The cover crop in 1995 was usually winter wheat or rye that was harvested as double-crop.

used as cover crops. Cover crops may be grown in association with cash crops to act as living mulches. They are used to a large extent with soybean production in the southern states (Figure 11–10).

Green Manures

Green manures are leguminous plants that are planted and plowed under the soil while still green. These crops can be planted as cover crops and then later incorporated into the soil. All cover crops are not usable as green manures because they are not all leguminous species. Examples of green manure species are hairy vetch, crimson clover, alfalfa, cowpea, and peanut. As an organic amendment (a biodegradable material incorporated into the soil), a green manure crop can improve soil structure, increase soil organic matter, and enhance soil fertility. As previously discussed, the utility of a soil amendment as a source of soil nutrients depends on the C:N (carbon to nitrogen) ratio, which impacts the rate of nitrogen released through decomposition and mineralization. Most green manures have a C:N ratio of 20:1 or less and are a good source of organic nitrogen. If incorporated while green, green manure plants have high nitrogen content, accumulated from biological nitrogen fixation.

Fallow

A fallow occurs when normal cropping of the land is suspended for a period of time. Fallow cropping was discussed previously.

Shifting Cultivation

Also called *slash-and-burn agriculture,* shifting cultivation is a cropping system found in many developing countries in the tropics and subtropics. In this cropping system, the land is cleared (by slashing and clearing with various hand-held tools and burning). The land is then immediately cropped, the ashes fertilizing the soil. After several years of repeated cropping of the land, yield decline sets in as the soil becomes exhausted of nutrients. The producer abandons the land and clears a new site. This trend is repeated in subsequent years. The farmer returns to the beginning land after a fallow period of no cropping. One criticism of this system of cropping is that new land is continually cleared, leading to deforestation (or removal of other vegetation cover) and its attendant problems. Population pressure means shifting cultivation is currently not sustainable because competition for land does not allow the long fallow periods necessary for the land to rejuvenate.

11.8: AGROFORESTRY

11.8.1 WHAT IS AGROFORESTRY?

As previously mentioned, *agroforestry* is the generic name used to describe the land use system in which trees are intentionally combined spatially and/or temporally with agricultural crops and/or animals. As a farming practice, agroforestry is very old and widely practiced, especially in the tropical and subtropical regions of the world. The concept was introduced to the United States in the early 1900s. It was revived, in its present form as agroforestry, in the 1970s.

A modern definition of agroforestry is the integration of agronomy and forestry conservation and production practices into land use systems that can conserve and develop natural resources while increasing economic diversity at both the farm and community levels. The trees in an agroforestry production enterprise are managed as an independent farm enterprise. Existing cropping land is not converted into a forest but is managed as an agroforestry system by integrating trees into the existing operation. Likewise, crop farms with existing woodland may be managed as agroforestry systems for special forest products. The trees in an agroforestry system provide forest products such as timber, nuts, and firewood, as well as shelter for livestock, control of soil erosion, habitat for wildlife, and improved soil fertility (e.g., through nitrogen fixation when leguminous species are used).

Generally, agroforestry practices tend to be more ecologically complex, compared with lands used only for annual crops. This is largely because of the complex interactions that occur among the components of an integrated system. However, with careful consideration, agroforestry practices can be effectively integrated into sustainable agricultural crop production systems to provide numerous environmental and financial benefits to the producer and the rural community.

An agroforestry system may be classified in several ways, depending on the purpose for which it is intended. Two useful and common ways of classifying agroforestry systems are according to structure and function. *Structure* refers to the composition and arrangement of the components of the agroforestry system that may be separated in time and space. Three basic structural classifications are as follows:

1. *Agrisilviculture*—consists of crops and trees
2. *Silvopastoral*—consists of pasture/animals and trees
3. *Agrosilvopastoral*—consists of crops, animals, and trees

The *function of an agroforestry system* refers to the main output and role of, especially, the woody components. The two generalized functional roles are productive and protective functions:

1. *Productive function.* The woody components of an agroforestry system may be used to provide food, fodder, and fuelwood, among others.
2. *Protective function.* The woody components may be used for crop protection (e.g., acting as windbreaks or shelterbelts) or for soil conservation and fertility management.

11.8.2 AGROFORESTRY PRACTICES

The design and technologies involved in an agroforestry system are variable. The distribution of agroforestry practices in the United States is presented in Figure 11–11. There are 5 major types of agroforestry practices: alley cropping, windbreaks or shelterbelts,

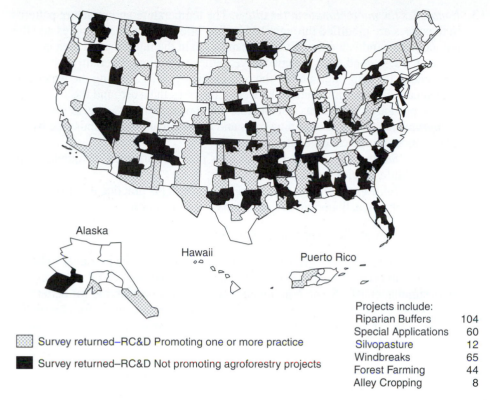

Projects include:
Riparian Buffers	104
Special Applications	60
Silvopasture	12
Windbreaks	65
Forest Farming	44
Alley Cropping	8

▨ Survey returned—RC&D Promoting one or more practice

■ Survey returned—RC&D Not promoting agroforestry projects

FIGURE 11–11 Agroforestry practices occur widely throughout the United States, according to a survey conducted in spring 2000 by the National Association of Resource Conservation and Development Councils (NARC&DC). Among the practices, 81% of respondents indicated the observation of riparian buffers, 68% observed the use of windbreaks, and 13% used alley cropping. (Source: USDA—National Agroforestry Center.)

riparian forest buffers, forest farming, and silvopasture. In addition, agroforestry has special applications.

Alley Cropping

Alley cropping is the growing of an annual or perennial crop between rows of high-value trees. The agricultural crop generates annual income, while the longer-term tree crop matures. Examples include growing soybeans between rows of black walnut or hay between rows of fast-growing pine or poplar. The type of annual crop grown varies as the trees grow larger and produce more shade.

Crop producers adopt alley cropping for several reasons.

1. *It improves farm economics.* The farmer can derive income from both the short-term annual as well as the long-term tree crops, thus increasing productivity of the land, instead of leaving the alley unutilized. The farmer also diversifies his or her enterprise for insurance against adversity. Trees may provide fruits, wood, nuts, or foliage for additional income.
2. *It protects soil from erosion and enhances its quality.* Cropping the alleys provides ground cover that protects the soil from erosion. The litter from the trees adds organic matter to the soil to enrich soil fertility. Nitrogen-fixing trees and shrubs add additional nitrogen to the soil.

3. *It modifies the microclimate in the alleys.* The temperature and moisture patterns in the alleys are modified through the shading effect of the tree canopies and the reduction in wind velocities. Evapotranspiration in the intercropped area is reduced. These conditions are conducive to the cultivation of certain crops. The trees and grass strip filter modify the hydrologic cycle by filtering the water flow and enhancing water infiltration. However, the trees that surround the intercropped area sometimes also deplete soil moisture in these areas.
4. *It improves aesthetics.* Combining trees and crops enhances the landscape by adding more biological diversity.
5. *It provides habitat for wildlife.* Rows of trees provide enhanced surroundings in which wildlife species dwell.
6. *It provides for a gradual transition to agroforestry.* This practice initially removes only a small amount of land from crop production.

The limitations of alley cropping include the following:

1. *It requires intensive management.* Alley cropping requires the producer to be familiar with both trees and annual crop plant and to be capable of managing them simultaneously. Some pruning and thinning may be required for better wood production and to enhance the light conditions in the alleys for cropping.
2. *Trees occupy cropland.* Including trees in a cropping system takes up the space of land that could otherwise be cropped to field crops, as well as some of the soil moisture.
3. *It increases marketing needs.* The producer has to seek marketing outlets for multiple crops instead of just one.

In terms of design, the trees are planted first and in rows. Generally, single rows are used, but two- and three-row sets have been successfully installed. Spacing between the row sets is designed to create alleys that can accommodate the mature trees as well as the crops and the farm machinery to be used. The crop may be an annual or a perennial, a row crop or forage crop. As trees grow bigger, they occupy more space and cast wider shadows. Light management is hence critical to the success of alley cropping. Rows are usually oriented east to west to maximize light in the alleys. One practice in light management is to grow sun-loving crops (e.g., corn and soybean) while the trees are young and small, then switch to shade-tolerant crops (e.g., mushroom and forages) after trees grow older and bigger. If such a switch is not desirable, the row spacing should be wide from the start.

Planting trees in a row facilitates other post-planting activities. Single or multiple rows of either softwood or hardwood may be used. Further, a single species or a mixture of species may be used (Figure 11–12). In certain designs, short rotation species are grown (e.g., Christmas trees), which are then harvested at an appropriate time and replanted. In single-row planting, the trees take up less space but often require pruning to produce high-value wood products. Sometimes, a row of hardwood species is flanked by conifers (called *trainer trees*). If unflanked, hardwood species tend to bend toward light in the alleys, thus reducing the wood quality. If timber or wood is the primary value to the producer, narrower spacing should be used. Nut trees may be widely spaced. Row spacing of 40 feet is desirable for alley cropping cereals, soybean, and corn for 5 to 10 years. If a longer period of cropping is desired, the row spacing should be doubled to 80 feet. Nut trees such as black walnut and pecan are often planted since, in addition to having high-value wood, commercial quantities of nuts can be harvested in as little as 7 to 10 years.

The number-one priority in alley cropping is the proper selection of woody species. Most commonly, the long-term intent is to produce high-quality, knot-free saw timber of

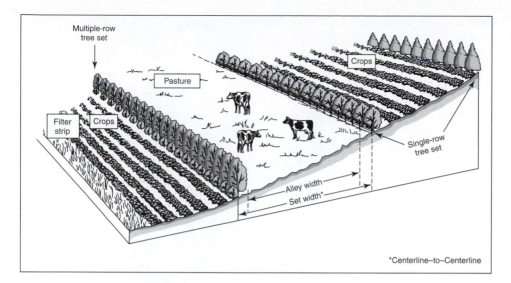

FIGURE 11–12 Alley cropping involves the use of the vacant space between established crops for additional crop production or other beneficial agricultural use, such as pasture for livestock. (Source: USDA—National Agroforestry Center)

commercially valuable species. Tree species should be adapted to the site, mesh with economic markets, and be compatible with companion crops. If trees are deciduous, early foliage loss would enhance the lighting in the alleys for the production of spring-ripening crops such as wheat and barley. Some species have nitrogen-fixing capacity [e.g., black locust (*Robinia psuedoacacia L.*) and leucaena (*Leucaena lecocephala* Lam)]. However, some desirable species such as black walnut have allelopathic effects on some crops. Other species of high commercial value for alley cropping include pecan, chestnut, oak, ash, and conifers. These trees may be intercropped with companion species such as row crops (e.g., corn, wheat, peas, and potato) and forage crops (e.g., white clover, alfalfa, Kentucky bluegrass, big bluestem, tall fescue, and ryegrass). Horticultural species (specialty crops) such as Christmas trees and small fruits may also be alley cropped. Biomass crops (e.g., poplar and birches) may be alley cropped to produce pulp for the paper industry. An advantage of pulp markets is that there is no need to prune the trees; however, if saw lumber markets exist, the price for pruned saw logs may be three to four times that of pulp.

The adoption of alley cropping is highest in the Midwest, where Missouri, Illinois, Indiana, Iowa, and Ohio have a combined total of more than 7.5 million hectares of alley cropped land. The Plains states (Nebraska and Kansas) have 6.5 million hectares of alley cropped land. The eastern half of the United States has more alley cropped land than the western half. Other states with significant alley cropped land include Oregon, Washington, Texas, and Minnesota.

Windbreaks (or Shelterbelts)

Windbreaks can prevent soil erosion and protect crops, livestock, buildings, work areas, roads, or communities (Figure 11–13). Living snowfences primarily protect roads but can also harvest snow to replenish soil moisture or fill ponds. The four basic types of windbreaks are farmstead/community, field, livestock, and living snowfences.

Windbreaks may be designed to enhance the general productivity of the crop farm enterprise, enhance wildlife habitat, or enhance the environment of the general community:

1. *Enhancing farm productivity.* The general farm productivity is enhanced through various functional roles of agroforestry practices in a cropping system. In many instances, windbreaks profoundly increase wildlife populations of game birds and

FIGURE 11–13 Windbreak protects a field crop (Source: USDA)

mammals. Many landowners are able to generate significant revenue from hunting fees for deer and pheasant.

2. *Effect on crop production and produce quality.* It has been well documented that, although windbreak trees do displace some cropland and will compete with adjacent rows of crops for moisture, in areas such as the Central Plains, where hot, dry summer winds prevail, windbreaks almost always increase crop yield on a per-field basis. In essence, the benefits of windbreaks for reducing evapotranspiration in these cropping systems are so pronounced that the farmers can grow the trees for free.

Crops such as cereals, vegetables, and orchard crops are wind-sensitive. Strong winds may cause severe lodging or fruit-drop, resulting in reduced crop yield. Windbreaks modify the microclimate of the cropped area by reducing wind velocity, which in turn modifies soil temperature, evapotranspiration rates, and relative humidities within the cropped area. These climate modifications are caused by the sheltering effects of the canopies of trees. The trees may trap snow that will improve the moisture content of the soil upon thawing. Reducing these meteorological parameters favorably alters the hydrologic cycle to enhance crop growth and development. The reduction in evaporative losses improves irrigation efficiency in sheltered cropping areas.

Windbreaks also protect soils from wind erosion by interrupting the saltation process. In addition to reducing soil fertility depletion, they enhance soil fertility by adding organic matter to the soil through the decomposition of leaves that drop and the effect of tree roots in the soil. In this regard, windbreaks aid in soil nutrient cycling. Trees are deeper feeders than annual crops and hence do not compete for nutrients with the target crop being produced (except when deep-rooted crops such as alfalfa are involved). Another way in which windbreaks enhance farm productivity is through economic diversity. If properly established and managed, the trees in the windbreak may be sources of additional income through the provision of secondary farm products such as wood, nuts, fruits, and foliage. Windbreaks may interrupt the spread of diseases from one section of the field to another, especially where the disease spreads by airborne structures. Several studies have documented the presence of beneficial insects in windbreaks.

3. *General community enhancement.* By reducing wind velocity, windbreaks reduce air infiltration into buildings and consequently reduce heat loss. They also reduce snowdrifts that block roads and pile up in residential areas, necessitating expensive snow removal. Windbreaks increase the biological diversity of the agricultural landscape. They can be designed to screen out unsightly areas on the farm, especially where animal production is involved. Windbreaks also filter the air and improve air quality by capturing airborne particles such as odor, dust, smoke, and drift products from pesticide applications.

The limitations of windbreaks are similar to those of alley cropping:

1. They require a more intensive management system involving annuals and perennials and trees or shrubs.
2. Cropping land is reduced.
3. Reduced farm sizes (because of division) will limit size of farm machinery that can be used.
4. Introduced trees may create an environment that harbors crop pests.
5. Crop performance is reduced in areas adjoining trees because of reduced light and reduced moisture.

Normally, wind speed needs to be at least 20 kilometers per hour, and blowing 30 centimeters above the ground, in order to begin to dislodge soil particles. The design of a windbreak depends on the purpose for which it is intended. For crop production purposes, the key factors the producer needs to consider include the wind direction (especially the direction from which most wind damage results), the crops being cultivated, the tillage practice and conservation measures being implemented, the irrigation practices (if any), the farm machinery and equipment being used, and any specific experiences with wind-related damage (e.g., wind erosion or crop damage). The producer needs to anticipate any future changes in cropping strategies and use of equipment.

The effectiveness of a windbreak depends on its height, density, orientation, number of tree rows, and length. A windbreak must be at least 75 centimeters high to impact wind speed. Height (H) is the most significant determinant of the downward area of protection provided by the windbreak. Windbreaks reduce wind speed for two to five times the height of the windbreak (2 H to 5 H) on the windward side and up to 30 H on the leeward side of the barrier. The area protected is directly proportional to H.

Windbreaks are porous structures. Windbreak density is measured as the ratio of the solid part of the barrier to the total area of the barrier. Barrier density can be adjusted to provide various levels of wind protection and modification of the microclimate. The denser the barrier, the more restrictive the wind flow (Figure 11–14). Deciduous species are more porous (25 to 35% density), while conifers provide a denser barrier (40 to 60% density).

Winds rarely blow exclusively from one direction. The orientation of a windbreak is critical to its effectiveness. The most effective orientation is at right angles to the direction of the prevailing wind. It may be desirable to establish a windbreak with multiple "legs" to provide more effective protection against uncertain wind direction.

The total amount of area protected by a windbreak depends on the length of the barrier. It is recommended that the length of a windbreak be at least 10 times its height. This recommendation will reduce the effect of end-turbulence on the protected area.

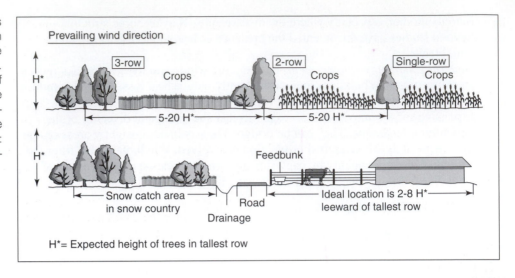

FIGURE 11–14 Windbreaks reduce wind speeds in cropped land. They are used for other purposes. Single or multiple rows of trees may be used in the design of a windbreak, depending on the purpose of the windbreak. (Source: USDA—National Agroforestry Center)

Designing a windbreak system for crop and soil protection starts with the identification of the structural or physical features on the area. A conservation plan map or photo will help locate the areas needing protection, property lines, roads, wind direction, utilities, and other factors. The recommended goal for crop and soil protection is to establish a primary windbreak of 40 to 60% density. This may be accomplished by using a combination of conifers and deciduous species in multiple rows. Further, a field may have a network of barriers spaced throughout the area. For this layout, a break-to-break interval of 15 H to 20 H is adequate for most field crops.

Certain crops are more sensitive to wind-related damage than others. A break-to-break spacing of 6 H to 10 H is recommended for sensitive crops. The farmer may implement additional conventional buffer practices, such as crosswind trap strips, to supplement the effects of the primary windbreaks.

Like all agroforestry practices, trees and shrubs selected should be adapted to the region and suited to the purpose for which they are being established. Tree seedlings should be planted at appropriate spacing to provide the desired density. The trees may be protected from rodents and other pests during establishment. Replanting of trees may be needed to fill in the spots where initial planting failed. Once established, the trees require pruning of the limbs and stems to maintain the proper density.

Windbreaks are often routinely incorporated into cropping systems in the Great Plains and Central Plains farming regions, where strong winds are common. In other regions, windbreaks are used to protect sensitive crops, such as tomato and apple, from bruising.

Riparian Forest Buffers

Riparian forest buffers are natural or planted woodlands adjacent to streams or water bodies and are comprised of trees, shrubs, and grasses. They provide a buffer against non-point source pollution, such as excess nutrient and pesticide runoff generated from agricultural activities. Riparian forest buffers also reduce stream bank erosion, enhance aquatic environments, augment wildlife habitat, and provide aesthetic value.

Modern crop production converts a variety of natural ecosystems (e.g., native prairie, wetlands, forests) into agroecosystems. In search of productive farmlands, producers sometimes utilize land that is sensitive or unstable and prone to rapid deterioration. Sometimes, stream channels are modified to increase the riparian area for crop production. In a typical watershed in places such as central Iowa, about 50% of the total length of a stream channel may be cultivated to the bank edge, while another 30% may be in pasture. Grazing has significantly changed the riparian zones in the rangelands of the West. Livestock adversely impact the stream channel, the stream banks, and the riparian zone through trampling and increasing of the sediment and nutrient load. In addition to this, general agricultural production is responsible for depositing large amounts of pollutants (sediments, fertilizers, pesticides) from uplands into streams through surface runoff.

It is clear from the foregoing that modern agriculture has placed the agroecosystem at risk through its product-oriented activities. It is imperative, therefore, that measures be implemented to restore the rapidly deteriorating riparian regions and the pollution of groundwater aquifers and streams. One effective strategy is through the installation of *riparian forest buffer strips,* consisting of trees, shrubs, or grasses (or a combination of these species), that are planted along streams or water bodies for the purpose of buffering non-point source pollution and sediments from waterways. These strips stabilize the banks and channels of water bodies and forestall the loss of adjacent cropping land. They also provide habitat for terrestrial wildlife and improve aquatic ecosystems.

A riparian zone by nature is a link between the aquatic ecosystem of the stream and the adjacent terrestrial ecosystem. Thus, it plays a critical role in the hydrology of watersheds. Further, the riparian zone is very productive for cropping and should be managed effectively for sustainable use.

Depending on the location of the riparian zone, a riparian forest buffer may serve one or more of the following purposes:

1. *Filter and retain sediments.* Large amounts of sediments can flow toward the stream from the upland region. This may consist of eroded soil particles carried in floodwaters or agrochemicals (pesticides and fertilizers) from upland agroecosystems. Trees and other plants established in the riparian buffer strip can filter and trap these sediments before they reach the stream, especially when the runoff occurs as sheet flow (as opposed to channelized flow). The standing trees and organic debris on the ground provide frictional surfaces that impede the flow of the runoff, thereby causing sedimentation to occur.

2. *Process nutrients and other chemicals.* Plants in the riparian buffer strip remove some of the dissolved nutrients (e.g., nitrates and phosphates) in the runoff, thereby reducing the amounts that reach the stream. Similarly, the activities of soil microorganisms in the buffer immobilize and transform some of the dissolved chemicals from the upland agricultural activities.

3. *Control stream environments and morphology.* Forest buffers stabilize the banks of streams through the soil-binding effects of plant roots. The canopies of trees modify the microclimate (especially light intensity and temperature) of the stream channel through shading. Further, the plants add organic matter to the stream. These events positively alter the aquatic biology of the stream.

4. *Enhance local hydrology.* Forested regions capture and absorb more rainfall than denudated sites. Riparian forest buffers impede the flow of floodwaters, thereby enhancing infiltration for recharging local groundwater.

5. *Protect cropland from flood damage.* Forested buffer zones can reduce the effect of flooding by reducing the erosive force of the running water and out-of-bank flow, thereby saving cropland and crops.

6. *Provide wildlife habitat.* Riparian buffers provide an enhanced habitat for wildlife. They also serve as travel corridors connecting different upland and aquatic habitats. Riparian buffers are rich in plant species diversity.

7. *Supplement income.* Riparian forests can be managed to provide wood, fruits, nuts, and fiber products.

8. *Create recreational activities.* The community may develop the riparian strip for various recreational uses, such as fishing, hunting, and camping.
9. *Enhance carbon storage.* Their favorable hydraulic position in the landscape means riparian zones often have ample moisture and nutrients, which lead to rapid plant growth. This translates to high rates of carbon accumulation.

The limitations of a riparian buffer include the following:

1. A sizable amount of productive cropland will be taken out of production to install a forested buffer strip.
2. By providing an enhanced wildlife habitat, a buffer strip may encourage some of the animals to become pests of crops being produced in the area.
3. The buffer requires management to keep it effective. For example, the riparian vegetation will continue to be an effective nutrient sink only as long as the plants are accumulating biomass. Likewise, grass strips at the interface of the cropped field and the forest buffer need to be periodically disked and reseeded to incorporate trapped sediments into the soil and maintain the ability of the buffer to trap new sediment.

Design   There are certain basic guidelines to follow in the design and implementation of a forested riparian buffer for agricultural lands:

1. *Problem identification and needs of the farmer.* The first step is to identify the problems at the site for which the intervention is intended. This should be viewed in relation to agricultural activities being conducted by the landowner, those that are endangering the riparian region, and the potential benefits that the agroforestry practice would bring. It is important to know the objective of the farmer in his or her desire to install a forested buffer (e.g., need to check erosion that threatens to reduce cropland).
2. *Plant species selection.* Plant species differ in their effectiveness at correcting various problems associated with the riparian region. For example, trees are effective for stabilizing stream banks and absorbing nutrients but are less effective in filtering sediments, a function better performed by grasses.
3. *Determination of minimum acceptable width.* The minimum acceptable width depends on the problem to be tackled with the buffer, the benefits desired by the landowner, and the cost of the project.
4. *Development of a plan.* The plan should include installation and maintenance strategies.

A General-Purpose Riparian Buffer for a Cropland   The USDA has developed guidelines for a three-zone design concept for riparian forest buffers (Figure 11–15). Zone 1 begins at the edge of the stream and is designed to provide bank stability and an undisturbed ecosystem. It provides the final filter of materials entering the stream. Logging and grazing are not permitted in this zone. Zone 2 is a managed forest area that provides maximum infiltration of surface runoff and nutrient uptake, as well as organic matter for microbial processing of agrochemicals. Zone 3 is a non-woody (grasses and herbs) strip that converts concentrated flow to sheet flow for enhanced infiltration of agrochemicals.

A general-purpose riparian buffer may be 50 feet wide, with trees occupying the first 20 feet in zone 1, while shrubs and grasses occupy the next 10 feet and 20 feet, respectively. The trees should be spaced 6 to 10 feet apart, while the shrubs are spaced 3 to 6 feet apart.

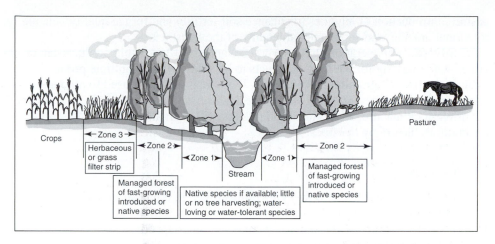

FIGURE 11–15 Riparian buffer strips are used to protect stream banks to prevent the erosion of adjacent cropland, among other objectives. (Source: USDA—National Agroforestry Center)

This design will utilize 6 acres of land per each mile of stream bank. The composition of plants and space allocations may be modified as needed to meet unique needs.

Forest Farming

Forest farming cultivates high-value specialty crops under a forest canopy that has been modified to provide the correct shade and microenvironment for the crop. These specialty crops usually fall into four categories: foods, botanicals, ornamentals, and handicrafts.

1. Foods—e.g., mushrooms and nuts
2. Botanicals—e.g., herbs and medicinals such as ginseng
3. Ornamental—e.g., floral greenery and dyes
4. Handicrafts—e.g., baskets and wood products

Silvopasture

Silvopasture may be defined as the purposeful integration of trees into pastures for the purpose of making a more productive system and gaining improved financial returns for the producer. As an agroforestry practice, silvopasture is a practice specifically designed and managed for the production of trees, tree products, forage, and livestock. The management of the silvopasture system emphasizes the production of high-value timber component, while providing short-term cash flow from the livestock component. To be successful, the producer should be aware of the environmental requirements in the area.

The trees and forage species to be combined should be selected for compatibility, as well as for adaptation to the soil and climate of the area. The timber component should be of high quality, marketable, fast-growing, and capable of providing desired products and environmental services. The forage component should be a perennial species that is suitable for livestock grazing and tolerant of shade. It should also be adapted to the region and amenable to intensive use and management. Both wildlife and livestock may be included in a silvopastoral system. The selected animals should be compatible with the tree and forage species in terms of grazing habits.

Once established, a variety of management tools are used for the proper management of the system. These include proper tree harvesting and thinning, fertilization of the pasture, introduction of legumes to enhance soil fertility and forage quality, fencing for rotational grazing, and supplemental feeding of the livestock.

Silvopastural operations have several advantages. Marketable products can be obtained while waiting for the forest products to be ready. Such a multiple product system reduces the risk of agricultural production. The land is also well used to produce a variety of products, while protecting the environment from degradation. Trees can benefit from livestock manure, while the shade trees provide can also stabilize the climate for the comfortable use by livestock.

SUMMARY

1. Ecology is the study of how living things relate to their environment and to each other.
2. In an ecosystem, there is natural cycling and interdependency of all components.
3. In an agroecosystem, farmers nudge nature to their advantage instead of allowing a natural balance to occur where all components are favored.
4. Crop production entails the management of inputs and biological processes for productivity, as well as sources of depletion of production resources.
5. In crop production, plants are planted in population, either of like plants (monoculture) or unlike plants (polyculture).
6. Examples of polycultures are intercropping, relay cropping, double-cropping, crop rotation, and alley cropping.
7. Mixtures or polycultures are common in forage production where legume-grass mixtures are planted.
8. Modern crop producers who mechanize their production frequently adopt monocultures.
9. Plants in a population interact in a competitive, complementary, or antagonistic fashion.
10. Crop producers employ a variety of techniques to restore soil fertility on site. These include the use of cover crops, green manures, fallow, and shifting cultivation.
11. Crop producers in various parts of the world have adopted production systems that have evolved over the years. Some of these production systems are low-input and less intensive, while others are high-input and intensive. The level of technology is determined largely by socioeconomic factors.
12. Agroforestry systems involve combining trees spatially and/or temporally with agricultural crops and/or animals.
13. In terms of structure, there are three agroforestry categories: agrisilviculture, silvopastoral, and agrosilvopastoral.
14. Agroforestry systems serve two basic purposes—productive and protective functions.
15. There are six agroforestry practices: alley cropping, windbreaks, riparian forest buffers, forest farming, silvopasture, and special applications.

REFERENCES AND SUGGESTED READING

Altieri, M. A. 1987. *Agroecology, the scientific basis of alternative agriculture.* Boulder, CO: Westview Press.

Beetz, A. 1999. *Agroforestry overview.* Appropriate Technology Transfer for Rural Areas (ATTRA).

Garett, H. E., Rietveld, W. J., and Fisher, R. F. 1999 (eds.). *North American agroforestry: An integrative science and practice.* American Society of Agronomy, Madison, Wisconsin.

Hearn, A. B. and G. P. Fitt. 1992. Cotton cropping systems. In Pearson, C. J. (ed.). *Ecosystems of the world: Field crop ecosystems.* New York: Elsevier.

Johnson, L. A. (ed). 1992. *Sustainable agriculture: Enhancing the environmental quality of the Tennessee Valley region through alternative farming practices.* University of Tennessee Agricultural Extension Service.

Juo, A. S. R. and H. C. Ezumah. 1992. Mixed root-crop systems in wet sub-saharan Africa. In Pearson, C. J. (ed.). *Ecosystems of the world: Field crop ecosystems.* New York: Elsevier.

Kormondy, E. J. 1984. *Concepts of ecology,* 3rd ed. Englewood Cliffs, NJ: Prentice Hall.

National Research Council. 1991. *Sustainable agriculture research and education in the field. A proceedings.* Washington, DC: National Academy Press.

Powers, E. R. and R. McSorley. 2000. *Ecological principles of agriculture.* Albany, NY: Delmar.

Rietveld, B., and Irwin, K. 1996. *Agroforestry in the United States.* Agroforestry notes, USDA Forest Service. Rocky Mountain Station.

Ruark, G. A. 1999. Agroforestry and sustainability: Making a patchwork of quilt. *Journal of Forestry.* August 1999.

Smika, D. E. 1992. Cereal systems of the North American Central Great Plains. In Pearson, C. J. (ed.). *Ecosystems of the world: Field crop ecosystems.* New York: Elsevier.

SELECTED INTERNET SITES FOR FURTHER REVIEW

http://www.nal.usda.gov/afsic/afslinks.htm

Alternative farming system information center; good links.

OUTCOMES ASSESSMENT

PART A

Answer the following questions true or false.

1. T F   In a natural ecosystem, bacteria are producers.
2. T F   An agroecosystem has a socioeconomic component.
3. T F   Plants are producers in an ecosystem.
4. T F   Hexagonal plant arrangement favors plants with foliage that is circular around the axis.
5. T F   Trees can be incorporated into field crop production.

PART B

Answer the following questions.

1. The study of how living things relate to their environment is _____.

2. Cultivated communities consisting of one crop species are called _____.

3. Define relay cropping.

4. Define intercropping.

5. What is allelopathy?

6. What is morphological plasticity?

7. Competition-induced plant mortality is called _____.

8. _____ is the agroforestry practice of growing crops, trees, and pasture with animals simultaneously.

9. Define the term agroforestry.

PART C

Write a brief essay on each of the following topics.

1. Describe a natural ecosystem.

2. Compare and contrast a natural ecosystem with an agroecosystem.

3. Discuss the competition among plants in a monoculture.

4. Discuss the competition among plants in a polyculture.

5. Discuss the advantages of polyculture.

6. Discuss the advantages and disadvantages of monoculture.

7. Discuss the phenomenon of self-thinning in a plant population.

8. What is agroforestry?

PART D

Discuss or explain the following topics in detail.

1. How can biotechnology support organic farming?

2. Will sustainable agriculture ever dominate U.S. crop production?

3. What are the benefits of agroforestry?

12

Organic Crop Production

PURPOSE

The purpose of this chapter is to discuss the principles and practices of organic crop production.

EXPECTED OUTCOMES

After studying this chapter, the student should be able to:

1. Distinguish between organic farming and conventional farming.
2. Discuss the principles of organic farming.
3. Describe the standards for organic crop production.
4. Discuss the importance of organic crop production in U.S. agriculture.

KEY TERMS

Compost   Farmyard manure   Organic farming

TO THE STUDENT

Modern crop production depends heavily on agrochemicals to protect plants and to increase soil fertility. Excess agrochemicals find their way into the general environment, where they pollute groundwater. Pesticides enter the food chain through animals feeding on contaminated plants and the use of toxic products to protect food crops in the field and the storehouse. Increasingly, the public is demanding a cleaner environment and a safer food chain. There is a demand for food produced without the use of agrochemicals. This may be likened to a return to the second era of agriculture, the era of resource conservation and regeneration. Organic agriculture is an approach to production in which the producer relies on ecosystem management rather than external inputs. As you study this

393

chapter, visualize the farm as a giant organism comprised of biotic components (animals and plants) and abiotic components (soil, environment) that are managed or manipulated by humans (farmers) in an economically sustainable and environmentally sound fashion.

12.1: WHAT IS ORGANIC FARMING?

12.1.1 DEFINITION OF ORGANIC FARMING

The word *organic* has become part of the vocabulary of participants in the food marketplace. However, it appears the word is interpreted a little differently by different sectors of the food marketplace. To the consumer, a product labeled *organic* means it was produced by methods that exclude agrochemicals. To the crop producer, it means the crop was produced by following the guidelines set forth by certifying agencies and using practices based on ecological principles.

Organic farming. The system of agricultural production in which synthetic agricultural inputs are excluded.

The Organic Producers Association of Manitoba Cooperative Ltd. defines organic products as those raised, grown, stored, and/or processed without the use of synthetically produced chemicals or fertilizers, herbicides, fungicides, or any other pesticides, growth hormones, or growth regulators. A simple definition of **organic farming** is difficult to arrive at, because organic farming systems are very complex. The National Organic Standards Board of the USDA defines organic agriculture as "an ecological production management system that promotes and enhances biodiversity, biological cycles, and soil biological activity. It is based on minimal use of off-farm inputs and on management practices that restore, maintain, and enhance ecological harmony. 'Organic' is a labeling term that denotes products produced under the authority of the Organic Foods Production Act. The principal guidelines for organic production are to use materials and practices that enhance the ecological whole. Organic agricultural practices cannot ensure that products are completely free of food residues; organic growing methods are used to minimize pollution from air, soil, and water. Organic food handlers, processors, and retailers adhere to standards that maintain the integrity of organic agriculture products. The primary goal of organic agriculture is to optimize the health and productivity of interdependent communities of soil life, plants, animals, and people."

The Consumer and Corporate Affairs Canada defines organic farming this way: "Organic farming seeks to create ecosystems that achieve sustainable productivity and provide weed and pest control through a diverse mix of mutually-dependent life forms, through recycling of plant and animal residues, and through crop selection and rotation, water management, tillage, and cultivation. Soil fertility is maintained and enhanced by a system which optimizes soil biological activity as the means to provide nutrients for plant and animal life as well as to conserve soil resources."

12.1.2 "ORGANIC" VERSUS "NATURAL"

Various terms are used to varying extents in the food marketplace to draw attention to the manner in which the product was developed. Newer terms include *eco* or *green* labeling as well as *natural*. Some consumers tend to use the terms *organic* and *natural* synonymously, but this is not correct. Natural foods are generally products that do not contain any "artificial" (without natural counterparts) additives. The U.S. Food Safety and Inspection Service defines natural as a product that contains no artificial ingredients and is no more than minimally processed in accordance with the rules of the agency. *Minimal processes* is interpreted to mean activities that are ordinarily undertaken in a household kitchen (e.g., washing, peeling of fruits, homogenizing of milk, freezing, canning, bottling, grinding of nuts, baking of bread, aging of meats). Just because a product is labeled *natural* does not mean it is *organic*. It does not guarantee synthetic inputs were not used in its production.

12.1.3 GOALS OF ORGANIC FARMING

The International Federation of Organic Agricultural Movements (IFOAM) summarizes the principles and practices of organic farming in its standards manual as follows:

1. To produce food of high nutritional quality in sufficient quantity
2. To work with natural systems rather than seeking to dominate them
3. To encourage and enhance biological cycles within the farming system, involving microorganisms, soil flora and fauna, plants, and animals
4. To maintain and increase the long-term fertility of soils
5. To use, as far as possible, renewable resources in locally organized agricultural systems
6. To work as much as possible within a closed system with regard to organic matter and nutrient elements
7. To give all livestock conditions of life that allow them to perform all aspects of their innate behavior
8. To avoid all forms of pollution that may result from agricultural techniques
9. To maintain the genetic diversity of the agricultural system and its surroundings, including the protection of plant and wildlife habitats
10. To allow agricultural producers an adequate return and satisfaction from their work, including a safe working environment
11. To consider the wider social and ecological impact of the farming system

12.2: PRINCIPLES OF ORGANIC FARMING SYSTEMS

Organic farming is an agricultural production system based on ecological principles. Producers strive to incorporate laws of natural ecosystems into the choice of production practices to use in their enterprises. Generally, organic farming is a farming approach that aims to create an integrated, humane, and environmentally and economically sustainable agricultural production system. In such systems, maximum reliance is placed on locally or farm-derived renewable resources and the management of self-regulating ecological and biological processes and interactions in agricultural production. As previously discussed under the topic "models of agriculture," organic farming is a sustainable model in which producers conduct their activities in harmony with nature (i.e., working with nature rather than controlling or subjugating it). The organic farm is viewed in this regard as an organism in which all its constituent parts (soil, nutrients, water, organic matter, microbes, animals, plants) together with humans (as manipulators) interact to create a holistic, coherent, and stable whole.

The functions of an organic farm within an ecological framework may be summed up by three basic ecological principles—interdependency, diversity, and recycling:

1. *Interdependency.* An ecosystem is a complex entity that consists of mutually dependent life forms that also interact with abiotic systems. The ecological balance can be upset by changing one of its components. For example, overfertilization may cause soluble nitrates to leach into groundwater, and phosphates into streams and lakes, causing eutrophication to occur.

2. *Diversity.* An ecosystem thrives on diversity in life forms, which work to establish the checks and balances that are necessary to suppress the outbreak of pest species (each has natural enemies). Organic crop production mimics such diversity by adopting practices that include the production of different crops on the farm.

3. *Recycling.* Natural ecosystems maintain various nutrient cycles (nitrogen, phosphorus, sulfur, carbon). Green plants (primary producers) harvest solar energy and convert it into photosynthates for use by consumers. These nutrients are returned to the soil upon decomposition. Organic producers exploit nutrient cycling systems to reduce the need for infusion of soil amendments in production.

These three principles will be discussed further in the chapter to describe specific practices employed in the production of crops in an organic farming system.

12.3: ADOPTION OF CERTIFIED ORGANIC FARMING

The worldwide adoption of organic products is increasing. In 2000, there were an estimated 600 certified organic programs in 70 countries. In the United Kingdom, it is estimated that the market for organic products is growing at the rate of over 20% annually. Germany is one of the largest producers and consumers of organic products in the world. The United States is ranked fourth in land area managed under organic farming systems, behind Australia (with 19 million acres), Argentina (6.9 million acres), and Italy (2.6 million acres). In terms of the percentage of total farmland under organic farming, the United States is not among the top 10% in the world. The leaders in 2002 were Switzerland (9%), Australia (8.64%), Italy (6.7%), Sweden (5.2%), Czech Republic (3.86%) and United Kingdom (3.31%).

In the United States, 2.3 million acres were devoted to organic production, representing 0.3 percent of total cropland and pastures in 2001. Of this, 1.3 million acres were certified organic farming cropland and 1.0 million acres as pasture and rangeland. All states, except Mississippi and Delaware, had some certified cropland. The top states in certified organic cropland in 2001 were Minnesota, Wisconsin, Iowa, Montana, Colorado, Idaho, South Dakota, and Michigan (Figure 12–1). In terms of pasture and rangeland, the top states in 2001 were Colorado (514,000 acres), Texas (221,000 acres), and Montana (137,000 acres). Organic acreage declined overall in Georgia, Louisiana, South Carolina, Indiana, West Virginia, Florida, and Idaho between 1997 and 2001. The Southeast generally has less certified organic farmland than any other state in the United States. Some of this decline was due to severe drought (e.g., in Idaho) or decertification of crops (e.g., St. John's wort in Florida and Idaho).

12.4: CROPS IN ORGANIC CULTURE

It is typical for certified organic growers to produce several crops (diversity principle) because of the use of crop rotations and green manures as essential practices in organic production. The major grain crops grown in the United States include wheat, corn, rice, oats, and barley. A total of 457,415 acres were cropped in 42 states in 2001, North Dakota leading with 64,000 acres. Of the total U.S. acreage, 194,000 acres were devoted to

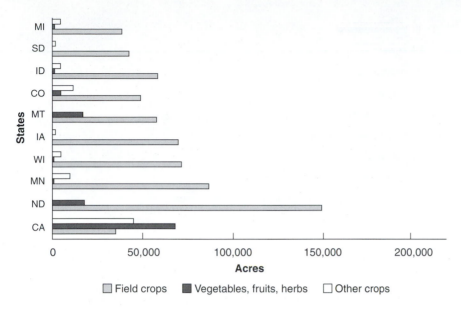

FIGURE 12-1 Top states in certified organic production in the United States in 2001. Vegetables are the most important organic crop in California, whereas other states produce organic field crops. (Source: USDA)

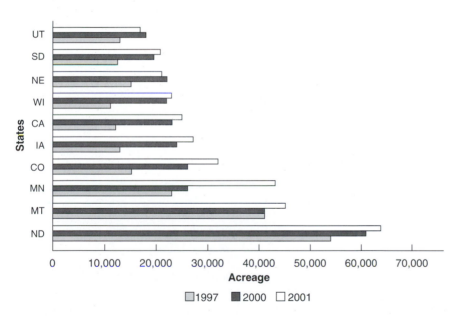

FIGURE 12-2 Organic crop acreage in the top 10 production states in the United States in 2001. North Dakota leads in organic crop acreage. (Source: USDA)

wheat, 93,000 acres to corn, and 30,000 acres each to oats and barley. About 31,800 acres were under rice production in 2001. Montana had the most wheat acreage, while Minnesota had the most corn and rye acreage (Figure 12–2).

Organic soybean was produced in 32 states on 174,400 acres in 2001. This represented a 28% increase from the previous year (Figure 12–3). It was determined that certified organic soybean growers received two times the conventional price for their

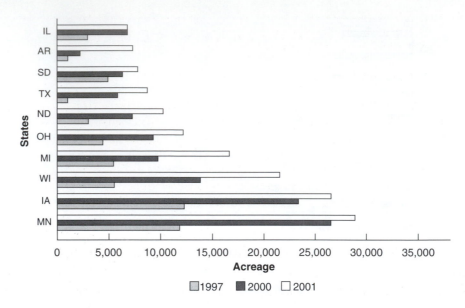

FIGURE 12–3 Top 10 organic soybean production states in 2001. (Source: USDA)

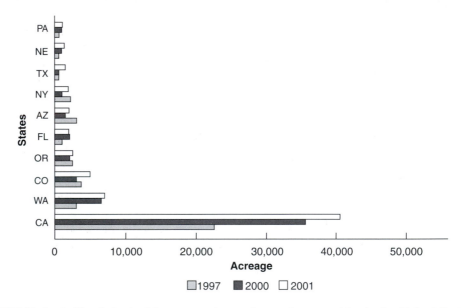

FIGURE 12–4 California is the biggest producer of organic vegetables in the United States. (Source: USDA)

product. Other legumes organically produced include dry beans, lentils, and peas. Twenty-one states produced organic flax and sunflower on 44,000 acres in 2001.

Many farms produce organic hay and silage. In 2001, 40 states produced 253,600 acres of hay and silage, the leading states being Idaho, Wisconsin, New York, North Dakota, and Iowa. States experiencing dramatic increases in production between 2000 and 2001 were New York (35%), Minnesota (43%), and South Dakota (45%). California is the biggest producer of both conventional (57%) and certified organic vegetables (Figure 12–4).

12.5: ORGANIC FARMING PRACTICES

Organic crop producers adopt a variety of practices to eliminate the need for agrochemicals and to conserve soil physical and chemical properties. These practices have been discussed in more detail in previous chapters of the book and hence will only be briefly presented here. The needs of crops in organic production are identical to those of conventional production systems (e.g., temperature adaptation, sunlight), disease-free environment, weed-free environment). The difference lies in how the needs are met. The two areas of need that are supplied in different ways are soil fertility and plant protection (weeds, diseases, insects). In these two areas of need, certified organic production forbids the use of synthetic chemicals. The four categories of practices adopted by organic crop producers are as follows.

12.5.1 PRACTICES FOR INCREASING SOIL FERTILITY

The ecological principle applied here is nutrient cycling. Plant materials are incorporated into the soil to be decomposed to release nutrients into the soil. As previously indicated, organic farming typically grows different crops. This has the advantage of exploiting soil nutrients at different levels in the soil. The different species have diverse needs of soil nutrients and feed at different levels in the soil. Some crops (e.g., legumes) are capable of biological nitrogen fixation and hence enhance soil fertility when they are harvested. Some practices (use of catch crops) prevent soil nutrients from been leached out of the soil during the period when the field is not in an economic crop. The key practices that enhance soil nutritional status in organic crop farming are

a. Green manuring/cover crops—this is the most common method of fertilizing in organic crop production in the United States.
b. Use of organic manures—animal byproducts (e.g., farmyard manure)
c. Crop rotation involving legumes for biological nitrogen fixation
d. Catch crops—to absorb nutrients that otherwise would be lost in off-season
e. Fallow—resting the land to accumulate soil moisture needed for the critical soil solution

Some organic producers sometimes apply aglime to correct soil pH to make nutrients more available.

A key to nutrient cycling is maintaining a biologically active soil that provides microbes for mineralization of organic matter. It is also important to know what crops remove nutrients the most (e.g., grasses extract potassium vigorously) and how to minimize losses while maximizing returns. Losses can be minimized by promoting natural recycling processes (e.g., nitrogen cycle) and biological nitrogen fixation. The availability of nutrients to plants depends on the rate at which nutrients are cycled by these biological processes.

To ensure that the soil remains biologically active, new sources of organic matter must be added at regular intervals. The rate of organic matter decomposition depends on its C:N ratio. Green manuring is used by organic farmers to provide nitrogen to the soil. They are generally readily degradable if plowed under at the appropriate age (i.e., when the plants are young and green). Crop residue from harvesting crops can also be incorporated into the soil. Green manure crops contribute little to soil organic matter due to their rapid rate of decomposition. However, crop residues that add to soil organic matter content decompose slowly. Farmyard manure is more effective at increasing soil organic matter than green manures. Organic farmers should know these characteristics of crop residues and green manure in order to make the most use of the nutrients by planting at the appropriate times. When leguminous crops are used as green manures, they contribute to soil nitrogen through symbiotic nitrogen fixation. Symbiotic nitrogen fixation is a primary source of nitrogen in organic farming.

The major sources of organic matter for organic farming are compost, farmyard manure, slurry, and liquid manures.

Compost

Compost. A mixture of organic residues and soil that has been piled, moistened, and allowed to decompose biologically.

Composting is often considered the essential practice in organic farming. **Compost** is desirable because the decay process is optimized to produce a product that is stable and comparable to what would occur in an undisturbed ecosystem. Farmyard manure is more useful after being composted to a more stabilized product. Composting involves piling suitable materials in a compost heap. The materials are placed in layers. Compostable materials should have a C:N ratio of between 25:1 and 35:1. If the ratio is too narrow, there will be insufficient carbon for microbes. On the other hand, a wide range (e.g., as in straw with 80–100:1) will have low nitrogen. Materials include **farmyard manure**, fallen leaves, grass, and straw from small grains. The material should have good moisture content (55 to 70%). The compost heap should be watered if it is too dry. The third critical requirement is that the heap be adequately aerated. The compost pile may be layered out in windrows to improve aeration.

Farmyard manure. Organic wastes collected from livestock barns for use as organic fertilizer.

The compost site should be within or close to the farm to facilitate transport of the bulky product to the field. A permanent concrete base may be constructed for composting windrows. This base prevents nutrient losses into the ground. The compost heap is covered with plastic and left to decompose. It takes 4 to 6 months for compost to become mature, or ripe, for use. During this time, the compost heap undergoes changes in pH and temperature. The mature compost is slightly acidic in reaction. High temperatures develop in the compost heap, reaching about 70°C. This high temperature helps kill weed seeds and disease pathogens. Further, pesticide residues are decomposed in the heap. The compost heap may be inverted after several months.

Farmyard (barnyard) Manure

Farmyard manure may be applied fresh or partially rotted, the latter being preferred. To minimize leaching losses and environmental pollution, farmyard manure should be applied in the late winter or early spring. It should be applied in thin layers by using spreaders. It is then harrowed into the topsoil. The producer should allow at least 6 weeks after application of farmyard manure before cropping the land, to allow time for the manure to be properly incorporated into the soil.

Slurry and Liquid Manures

Slurry and liquid manures are stored in containers such as tanks and lagoons. These materials are difficult to manage and expensive to store. The source of slurry and liquid manures are livestock operations where the animal waste accumulates in that form. Above-ground storage in tanks is environmentally safer and preferred to ground storage in lagoons. Slurry may be aerated to reduce odors caused by anaerobic decomposition of the materials that produce noxious chemical compounds (e.g., butyric acid and ammonia). Aeration is, however, expensive. Slurry and liquid manures are spread by tankers or by injectors, the latter technique being more effective for reducing losses from volatilization.

12.5.2 PRACTICES FOR DISEASES AND INSECT PEST CONTROL

The applicable ecological principle is diversity (to avoid disease buildup). The key practices for plant protection are

a. Crop rotation—this is the most common practice for pest control in organic production
b. Biological pest control—deployment of beneficial organisms (e.g., Bt)
c. The use of disease/insect resistant cultivars—this is also widely used
d. Sanitation—burying and burning of diseased plants to avoid the spread of disease

Ecological pest and disease management was discussed in Chapter 10. In this approach, the producer seeks to enhance the activities of natural enemies (beneficials) of crop plants. Beneficials include lacewings, spiders, aphids, and parasitic fungi. Enhancing biological diversity over time (e.g., using crop rotation, manipulating planting and harvesting dates, and discontinuing monocultures) and diversity in space (using varietal mixtures) is effective for pest management. Biological diversity helps prevent insect pest outbreaks through a variety of mechanisms such as interference with colonization, repelling, trapping, and increasing the population of beneficials. Trap crops and green manures are techniques that can be used to alter the behavior of pests.

Green manuring has been used to manage cyst nematodes using sugar beet, rape, and brassicas. Species such as corn, peas, beans, and clovers deter the beet cyst nematode. Wheat and oats act as hosts for cereal cyst nematodes, while potatoes and tomatoes act as hosts for potato cyst.

When peanut is intercropped with corn, the corn borer *(Ostrinia furnaculus)* is controlled, while intercropping corn with beans regulates leaf hoppers *(Empoasca kraemeri),* leaf beetle *(Diabrotica balteata),* and fall army worm *(Spodoptera fragiperda).*

12.5.3 PRACTICES USED FOR WEED CONTROL

To control weeds, the applicable ecological principle is diversity in plant species and population density. Planting at an appropriate time provides a rapid ground cover to suppress weeds. The key practices adopted include the following:

a. Use of cover crops—control weeds, protect soil during fallow, improve soil organic matter
b. Cultivation—mechanical tillage to destroy weeds
c. Mulching—to suppress weeds
d. Crop rotation—to break weed cycle and suppress
e. Crop density—to suppress weeds by shading them out

Weed management is one of the major challenges of organic farming. Weed problems can be successfully managed without the use of herbicides, by adopting

1. Appropriate husbandry practices
2. Biological weed management
3. Direct mechanical and physical intervention

Weed species, like other plants, differ in preference for soil factors—physical and chemical. Some species (e.g., horsetail) prefer well-drained sandy or loamy soils, whereas others, such as field bindweed, prefer well-lit, heavy soils that are fertile. Some species are adapted to dry soils, while others prefer soils with good moisture. Drainage of soils can be used to manage horsetail and rushes that prefer good moisture. Liming of acidic soils or increasing soil acidity by applying sulfur can be used to manage pH-sensitive species.

Crop rotation may also be used to manage weeds. Crop species have weed complements. Thus, alternating between contrasting types (e.g., between annuals and perennials, or between autumn crops and spring-germinating crops) helps prevent certain weed species from becoming dominant in the field. Cultivation of the soil buries seeds and kills them, while other seeds are brought to the soil surface where they germinate. Through proper timing between seedbed preparation and cultivation activities, reserves in weed seed bank can be diminished. A "false seedbed" may be created for seeds to germinate, then subsequently harrowed prior to planting.

Another practice in weed management in organic farming is to give the crop a competitive edge over weeds by early planting and quick establishment. Techniques to accomplish this include pre-germination of seeds and transplanting, as well as seeding for high plant density. These practices help in certain cases to establish a ground cover that shades out weeds. Broadleaf species more effectively smother weeds than narrowleaf species (e.g., cereals). The spread of weeds may be reduced by using clean, weed-free seed, cleaning equipment after use, and using manures after they have been composted to kill weed seeds.

Weeds may also be managed by biological pest management techniques. Weeds are commonly controlled by direct mechanical intervention through mechanical cultivation, using various implements such as hoes and harrows. Harrowing should be done carefully to avoid damaging plants. It should be done after the three-leaf stage. Harrowing is not effective against well-established weeds. As a last resort, some producers use flame weeding to control weeds. The technique involves exposing weeds to flame temperatures in excess of 90 to 100°C to dehydrate and kill weeds.

12.5.4   PRACTICES FOR CONSERVING SOIL AND ITS PROPERTIES

To preserve soil structure and nutrients, the ecological principle of nutrient cycling and soil conservation principles to prevent soil loss are applied. The key practices applied include

- **a.** Use of cover crops—to control weeds, protect soil during fallow, improve organic matter when plowed under
- **b.** Use of catch crops—to absorb nutrients that could be lost to leaching in off-season
- **c.** Use of conservation practices—to protect soil from erosion

12.6: CERTIFICATION AND STANDARDS

Products marketed as "certified organic" command premium prices. To reduce consumer fraud, certification and standards were developed for the organic farming industry in the early 1970s by private and nonprofit organizations. State certification was undertaken by several states in the 1980s. The federal Organic Foods Production Act was passed by Congress in 1990 to provide nationwide standards for the industry.

In 2000–2001, a total of 14 state and 39 private organizations provided certification services to organic producers, some of the oldest being the California Certified Organic Farmers, the Northeast Organic Farmers Association of Vermont, and the Maine Organic Farmers and Gardeners Association.

The U.S. Congress passed the Organic Foods Production Act (OFPA) of 1990 to establish national standards for organically produced commodities in order to facilitate the domestic marketing of such products and to assure consumers that these products conform to uniform standards. This legislation is implemented by the USDA's National Organic Program, which was authorized under the OFPA.

The USDA proposed new rules for organic farming and handlers on March 7, 2000:

1. Land would have no prohibited substances applied to it for at least 3 years before the harvest of an organic crop.
2. Crop rotation would be implemented.
3. Use of genetic engineering (included in excluded methods), irradiation, and sewage sludge would be prohibited.
4. Soil fertility and crop nutrients would be managed through tillage practices, supplemented with animal and crop waste materials and allowed synthetic materials.
5. Preference would be given to use of organic seeds and other planting stock, but a farmer could use non-organic seeds and planting stock under certain specified conditions.
6. Crop pests, weeds, and diseases would be controlled primarily through non-chemical management practices, including physical, mechanical, and biological controls; when these practices were not sufficient, a biological, a botanical, or an allowed synthetic chemical substance could be used.

The USDA introduced the Final Rule on October 21, 2002, to require all organic certifiers to be accredited under the USDA's national organic standards. All producers selling more than $5,000 a year must comply by seeking certification by the state or a private agency. The program established the following:

1. National production and handling standards for organically produced products, including a national list of substances that can and cannot be used
2. A national-level accreditation program for state and private organizations, which must be accredited as certifying agents under the USDA national standards for organic certifiers
3. Requirements for labeling products as organic and containing organic ingredients
4. Rules for the importation of organic agricultural products from foreign programs
5. Civil penalties for violations of these regulations

The USDA's requirements are the minimum; additional labeling of organic products is permissible by the certifying agencies. Certain synthetic substances such as insecticidal soaps and horticultural oils are permitted in organic farming as a last resort. However, genetically engineered products and irradiation are not permitted. Any use of materials to control pests must be documented.

12.7: MARKETS AND MARKETING

In 2001, the U.S. organic sales totaled $9 billion to $9.5 billion, up from $7.8 billion, and only $1 billion in 1990. The major world markets are the United States, Japan, Denmark, France, Germany, the Netherlands, Sweden, Switzerland, and the United Kingdom, together receiving $21 billion in sales. Organic products were once rich people's products, but they are gradually becoming mainstream products, sold in a variety of venues—farmers' markets, natural foods supermarkets, club stores, and conventional supermarkets. In 2000, about 49% of all organic products were sold in conventional supermarkets versus 48% in natural food stores.

About 80% of producers in the United States market mainly by wholesale. The prices of organic products are significantly higher than those for conventional products.

In 1999, organic corn sold for about $5.20 per bushel, while soybean sold for $11 to $22 per bushel, depending on the variety. The price premiums were more than 50% for corn, soybean, wheat, and oats between 1993 and 1999. Fresh produce is the top-selling organic category, followed by non-dairy beverages, bread grains, and packaged food. Organic food sales account for 1 to 2% of the total food sales in the United States and other countries. In terms of field crops, organic soyfoods (from soybean) are among the major organic products on the market.

The organic market is largely underdeveloped but is rapidly expanding. It is driven by the fact that more consumers are seeking what they perceive to be healthful and safe foods that are produced in an environmentally responsible manner. In the United States, research conducted in the mid-1990s indicated that about 25% of all adults make purchasing decisions that are influenced by their social and environmental values.

It is projected that organic foods will constitute about 10% of the total retail food market by 2008 in the United States, up from its current proportion of 3 to 5%. Although consumers are willing to pay premium prices for certified organic products, they also demand the quality of conventional products (e.g., appearance, size, etc.).

12.8: BASIC CONVENTIONAL FARMING VERSUS BASIC ORGANIC FARMING

Basic conventional farming and basic organic farming systems have many things in common. The key areas in which they differ are in practices that require the use of agrochemicals. Most agrochemicals are prohibited in organic farming. Some practices, though applicable to both systems, are standard or dominant to one system and optional or less important in the other. The similarities and differences between the two systems are summarized in Table 12–1.

12.9: CONSTRAINTS ON ORGANIC FARMING TECHNOLOGY

Though more energy-efficient than conventional production technology, organic farming has constraints:

1. *High labor input.* Labor input in organic farming is substantially higher than in conventional farming.
2. *Limitation to supply of organic fertilizer.* Organic farmers may utilize green manuring, composting, and other strategies to provide fertility in crop production. However, in order to utilize farmyard manure, slurry, and liquid manures economically, the farm should be located near sources of supply of these bulky manures (i.e., near livestock farms).
3. *Difficulty with certification standards.* There has been a general lack of uniform and nationally enforced standards for defining the term *organic.* The proposed standards by the USDA would help to resolve this constraint.
4. *Other constraints include markets, credit, and lack of research.* These constraints will be resolved with increased acceptability of organic products.

Table 12–1 Key Comparisons Between Organic and Conventional Crop Production

Factors	Organic Production	Conventional Production
Cultivars used	Conventional cultivars only	Conventional cultivars Transgenic cultivars
Certification	Required for production	Not required
Fertilizers	Organic sources Cover crops	Organic sources Cover crops In-organic sources
Pest control	Crop rotation Use of resistant cultivars (conventional only) Biological pest control Cultural control Limited use of chemicals	Crop rotation Use of resistant cultivars (conventional and genetically modified) Biological pest control Cultural control Liberal use of pesticides
Weed control	Mechanical tillage Mulching Crop rotation	Mechanical tillage Mulching Crop rotation Use of herbicides
Environmental impact	Limited	Extensive pollution may occur
Marketing	Special label used	Labels not required
Principles	Sustainable production Ecological basis	Not sustainable

SUMMARY

1. Organic farming is a production practice that avoids or largely excludes the use of synthetically compounded agrochemicals in agricultural production.
2. Organic farming is known by various synonymous names such as alternative farming, regenerative farming, and sustainable agriculture.
3. Organic farming depends on site-specific natural resources and those developed on-site through techniques such as green manuring and composting.
4. In organic farming, the crop producer manages self-regulating ecological and biological processes for sustainable and economic production of crops and products.
5. The soil is cultivated to a shallow depth with light implements, mixing the residue with the topsoil. This promotes water infiltration and prevents soil erosion.
6. A key goal of organic farming is to operate the farm as a closed system as much as possible.
7. Nutrient cycling is a key source of fertility for crop production. It is therefore critical to maintain soil health and thereby promote biological activity in the soil atmosphere.
8. Biological nitrogen fixation is important in organic farming.
9. The major sources of organic matter for organic farming are compost, farmyard manure, slurry, and liquid manures.

10. Crop rotation is used to create biodiversity on the farm as a means of establishing equilibrium in the agroecosystem.
11. Diseases and pests in organic farming are managed by means of biological management techniques such as use of beneficials, crop rotation, and disease-resistant cultivars.
12. With organic farming, cultural practices such as drainage are used to control certain weeds and diseases and pests.
13. The USDA has proposed nationwide rules for organic farming and handling activities.

REFERENCES AND SUGGESTED READING

Kitto, D. 1988. *Composting: The organic, natural way.* New York: Sterling.

Lampkin, N. 1990. *Organic farming.* Cambridge, UK: Farming Press.

Poincelot, R. P. 1972. *The biochemistry and methodology of composting.* Connecticut Agricultural Experimental Station Bulletin 727.

SELECTED INTERNET SITES FOR FURTHER REVIEW

http://www.ofrf.org/about_organic/index.html

Organic Farming Research Foundation site; various related topics; good links.

http://www.ers.usda.gov/whatsnew/issues/organic/

Issues of organic farming; crop acreages from states in the United States; certification information; links.

OUTCOMES ASSESSMENT

PART A

Answer the following questions true or false.

1. T F Organic farmers do not use any synthetic fertilizers.
2. T F Farmyard manure is more stabilized after composting.
3. T F Organic farming is not sustainable.
4. T F Roundup® herbicide can be used in organic farming.
5. T F Vegetables are the most commonly produced crops in organic farming.

PART B

Answer the following questions.

1. Give the two specific reasons organic farming is attractive to some producers.

_____

_____

2. What act was passed in 1990 by the U.S. Congress to regulate organic farming?

_____

3. Give four field crops that are grown by organic farmers in the United States.

_____

4. Give 5 of the top 10 states that are leading producers of organic crops.

_____

5. Give two specific constraints on organic farming.

_____

_____

PART C

Discuss or explain the following topics in detail.

1. Define the term _organic farming_.

2. Briefly discuss the role of nutrient cycling in organic farming.

3. Briefly describe how weeds are managed in organic farming.

4. Discuss the role of crop rotation in organic farming.

5. Explain why organic farmers adopt shallow tillage.

PART D

Discuss or explain the following topics in detail.

1. Organic farming is sustainable crop production. Discuss.

2. Organic crop production is on the rise in the United States. Discuss.

13

Transgenics in Crop Production

PURPOSE

The science of biotechnology and its applications in food agriculture were discussed in Chapter 5. This chapter is devoted to discussing the importance of *genetically modified* (GM) *transgenic* varieties in crop production. GM varieties have become a standard practice in the production of certain crops.

EXPECTED OUTCOMES

After studying this chapter, the student should be able to:

1. Discuss the adoption of transgenic crops in United State agriculture.
2. Discuss the adoption of transgenic crops in world agriculture.
3. Discuss the issues impacting transgenic crop adoption.
4. Give the major traits incorporated into transgenic crop varieties.
5. Discuss the rationale and science of engineering herbicide tolerance in crops.
6. Discuss the rationale and science of engineering insect (*Bt*) resistance in crops.
7. Discuss the rationale and science of engineering fruit ripening.
8. Discuss the benefits of transgenic varieties in crop production.
9. Discuss the risks of transgenic varieties in crop production.

KEY TERMS

Acetohydroxyacid
  synthase (AHAS)
Acetolactate synthase (ALS)
aro A

Climacteric fruits
Coat protein-mediated
  resistance
Cross-protection

Endotoxins
Flavr savr tomato
Genetically modified
  (GM) varieties

Glufosinate	Imidazolinones	Transgenic varieties
Glutathione-S-transferase (GST)	L-Phosphinothricin (PPT)	Triazopyrimidines
	Polygalacturonase (PG)	

TO THE STUDENT

The development and application of biotechnologies are steadily on the rise. However, the applications are currently concentrated in a few major crops and involve a few traits. Some of these crops are food crops, whereas others are fiber crops. Further, the technology is available for traits that are simply inherited and are most readily adaptable to manipulation by genetic engineering. As you study this chapter, it is instructive to know how their scientific basis, as well as the precautions adopted in their safe and practical deployment in the field.

Genetically modified (GM) varieties Plant varieities developed by incorporating desired gene(s) from a genetically unrelated source.

13.1: WORLD TRANSGENIC CROP PRODUCTION

13.1.1 ACREAGE

The adoption of GM crop varieties is on the rise. In 1998, more than 27 million hectares of GM crops were grown worldwide, representing a 10-fold increase from their production in 1996. In 1999, GM crops were grown on 40 million hectares. About 15% of the land devoted to GM production was located in developing countries. Revenues from GM crops totaled $75 million in 1995 and reached about $1.64 billion in 1998.

The major growers of transgenic crops in 2000 were the United States, Argentina, Canada, China, South Africa, and Australia. The United States leads world production, with 74.8 million hectares, followed by Argentina and Canada, with 24.7 and 7.4 million hectares, respectively (Figure 13–1). Other producers of GM crops include Mexico, Bulgaria, Romania, Spain, Germany, France, and Uruguay.

Transgenic varieties Also called genetically modified (GM) varieties, these are plant varieties developed by incorporating desired gene(s) from a genetically unrelated source.

13.1.2 CROPS

More than 20 crop species were known to have **transgenic varieties** suitable for commercial production in 2000. Of these, the most important crop species were soybean, corn, cotton, canola, potato, squash, and papaya (Figure 13–2). Soybean GM acreage was

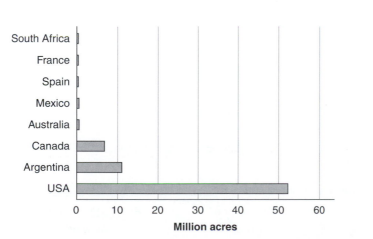

FIGURE 13–1 The top transgenic crop producing nations. The U.S. leads the world in the production of transgenic crop production. (Source: Drawn with data from ISSA)

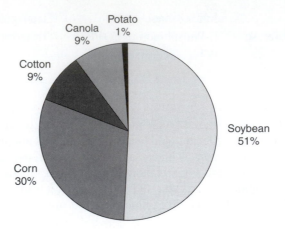

FIGURE 13–2 The top crops in transgenic crop production. Soybean is the single most important transgenic crop in the world. (Source: drawn with data from ISSA)

the highest, with 53.4 million acres, followed by corn and cotton, with 27.4 and 9.1 million acres, respectively. Canada leads the world in canola production, whereas Romania leads in potato production. However, the United States leads in the production of GM corn, soybean, and cotton.

13.2: U.S. TRANSGENIC CROP PRODUCTION

The adoption of transgenic crops in U.S. agriculture has been on the rise since 1996. In 1999, about 25% of U.S. corn and 60 to 70% of the soybean crop area was planted to transgenics. Most of the transgenics were soybean, corn, and cotton, as previously indicated.

13.3: TRAITS INCORPORATED INTO TRANSGENIC CROP VARIETIES

Three traits currently dominate all GM crops in commercial production worldwide. These are herbicide tolerance, *Bt* insect resistance, and virus resistance (Figure 13–3).

FIGURE 13–3 The top traits in transgenic crop development. Herbicide-resistant soybean is the most commonly grown transgenic product.

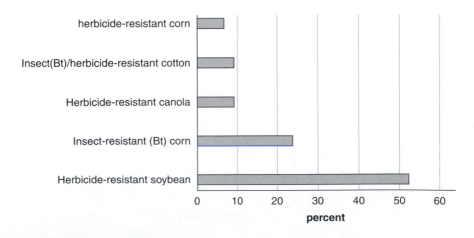

There are also varieties that combine these traits (e.g., *Bt* and herbicide tolerance). Of these traits, about 70% represented crop varieties with herbicide tolerance genes in 1999.

13.3.1   HERBICIDE TOLERANCE

Why Engineer Herbicide–Resistant Crops?

A successful herbicide should destroy only weeds, leaving the economic plant unharmed. Broad-spectrum herbicides (non-selective) are attractive, but their use in crop production can be problematic, especially in the production of broadleaf crops such as soybean and cotton. There is a general lack of herbicides that will discriminate between dicot weeds and crop plants. Preplant applications may be practical to implement; however, once the crop is established and too tall for the safe use of machinery, chemical pest management is impractical. Grass crops (e.g., wheat, corn) may tolerate broadleaf herbicides better than the reverse situation. Consequently, when cereal crops and broadleaf crops are grown in rotation or adjacent fields, the latter are prone to damage from residual herbicides in the soil or drift from herbicides applied to grasses. When a crop field is infested by weed species that are closely related to the crop (e.g., red rice in a rice crop or nightshade in a potato crop), herbicides lack the sensitivity to distinguish between the plants.

To address these problems, one of two approaches may be pursued: (1) the development of new selective postemergent herbicides or (2) the genetic development of herbicide resistance in crops to existing broad-spectrum herbicides. The latter strategy would be advantageous to the agrochemical industry (increased market) and to farmers (safer alternative to pesticides that are already in use). New herbicides are expensive to develop and take time.

Modes of Action and Herbicide Resistance Mechanisms

Most herbicides are designed to kill target plants by interrupting a metabolic stage in photosynthesis. Because all higher plants photosynthesize, most herbicides kill both weeds and desirable plants. Plants resist phytotoxic compounds via one of five mechanisms:

1. The plant or cell does not take up toxic molecules because of external barriers, such as cuticles.
2. Toxic molecules are taken up but sequestered in a subcellular compartment away from the target (e.g., protein) compounds the toxin is designed to attack.
3. The plant or cell detoxifies the toxic compound by enzymatic processes into harmless compounds.
4. The plant or cell equipped with resistance genes against the toxin may produce a modified target compound that is insensitive to the herbicide.
5. The plant or cell overproduces the target compound for the phytotoxin in large amounts such that it would take a high concentration of the herbicide to overcome it.

Molecular Methods of Weed Control

The molecular genetic strategies for engineering herbicide resistance in plants can be grouped into two broad categories:

1. Modification of target enzyme of the herbicide
2. Development of herbicide-tolerant genes

Modification of the Target of the Herbicide   The common mechanisms by which genetically engineered plants resist herbicides are by

1. Inhibiting a pathway in photosynthesis
2. Inhibiting amino acid biosynthesis

Genetic analysis of weeds that had developed resistance to triazine herbicides, and their progeny, revealed that genetic resistance is cytoplasmic in origin. It was determined that a mutant gene, *psbA,* located in the chloroplast, encodes a 32-kDa protein in the thylakoid membrane that is involved in the photosystem-II electron transport system. Triazine herbicides (and others, including triazinones, urea derivatives, and uracils) bind to this protein in susceptible plants. This binding blocks quinone/plastoquinone oxidoreductase activity and incapacitates or interferes with electron transport. It was also discovered that resistant weeds have about a 1,000-fold decrease in affinity for atrazine. Further analysis showed that the amino acid composition of the 32-kDa chloroplast proteins has minor modifications in amino acid composition. Gene sequencing has revealed that more than 60% of the mutations are the result of a substitution of Ser-264 (the number indicates amino acid position in the polypeptide) by Gly or Ala.

Using the *psbA* modification as a genetic engineering strategy for developing herbicide resistance is problematic. In the first place, techniques for transforming chloroplasts are lacking. However, selectable markers (streptomycin phosphototransferase gene) have been developed to facilitate chloroplast transformation. Other researchers have converted the *psbA* mutant gene from plastid to nuclear gene. In *Amaranthus hybridus,* this transfer was achieved by fusing the coding region of the mutant gene to the transcription regulation and peptide-encoding sequences of a nuclear gene for a small subunit of rubisco. The chimeric construct was transferred to herbicide-sensitive tobacco plants via *Agrobacterium* transformation protocols. Some transgenic plants exhibited atrazine resistance and produced the protein product of the nuclear *psbA* gene in the chloroplasts.

Herbicide–Tolerant Genes Certain herbicides inhibit amino acid biosynthesis (e.g., glyphosate [Roundup®]), whereas others are photosynthetic inhibitors (e.g., S-triazines, such as atrazine).

Inhibition of Amino Acid Biosynthesis

Acetolactate Synthase (ALS) Genes

Only plants and microorganisms can synthesize about 50% of the amino acids found in proteins. Examples of these are phenylalanine, tyrosine, tryptophan, lysine, and methionine. Enzymes that are involved in the biosynthesis of these amino acids by pathways that are exclusive to plants provide a unique opportunity for targeting herbicides. Somatic selection techniques have been utilized to isolate mutants with resistance to sulphonylureas. Genetic analysis indicated that resistance is conferred by a single dominant or semidominant nuclear gene. Further studies with the bacterium *Salmonella typhimurium* showed that growth inhibition by sulphonylureas on minimal media can be prevented by the inclusion of certain branched-chain amino acids (leucine, isoleucine, and valine) in the medium. Consequently, the enzyme **acetolactate synthase (ALS)**, also called **acetohydroxyacid synthase (AHAS)**, was implicated as the target for the herbicide (the enzyme is required for the biosynthesis of these amino acids). This has been confirmed by other tests. ALS/AHAS is also the protein target for the **imidazolinone** (e.g., Impazapyr) and the **triazopyrimidine** (or **sulphonanilides**) herbicides. These classes of herbicides are structurally distinct. This makes ALS a particularly susceptible target for herbicides.

Attempts to genetically engineer herbicide resistance in plants by targeting ALS are also problematic. Like the *psbA* gene, ALS occurs in the chloroplast. The metabolic pathways for amino acid biosynthesis in plants and microorganisms are identical and therefore attract opportunities for microbial information to be applied to the understanding of plant processes. However, because bacterial ALS comprises two subunits, the generation of herbicide resistance in plants by using microbial genes is complicated. It would require the expression of two protein subunits of the enzyme. Further, like *psbA,* chloroplasts would have to be transformed. ALS genes have been isolated from several species, including *Nicotiana tabaccum* and *Arabidopsis thaliana.* Site-directed mutagenesis was used to induce mutations in plants with ALS genes. Tobacco plants were successfully transformed with the mutant genes.

Engineering glyphosate resistance; the aro A gene

Gylphosate (*N*-[phosphonomethyl-glycine]) is the active ingredient in the broad-spectrum herbicide developed by Monsanto called Roundup®. This herbicide is toxic to plants, fungi, and bacteria. It is phloem-mobile and has little residual effect, as well as being less toxic to all major crops. Because of these desirable properties, engineering glyphosate resistance has received tremendous attention in the scientific community.

The primary target of glyphosate is an enzyme in the aromatic amino acid biosynthetic pathway called 5-enolpyruvyl-shikimate-3-phosphate synthase (EPSPS, or EPSP synthesis). It is also called 3-phoshoshikimate-1-carboxyvinyl transferase. The role of EPSPS in aromatic amino acid biosynthesis is the catalysis of the condensation of phosphoenolpyruvate and shikimate-3-phosphate to form 5-enolpyruvyl-shikimate-3-phosphate, a precursor of several amino acids (e.g., tryptophan, phenylalanine) as well as a number of aromatic secondary metabolites. Glyphosate binds to prevent the binding of phosphoenol-pyruvate to EPSP synthase, thereby incapacitating the enzyme.

It has been discovered that the gene ***aro* A** encodes EPSP synthase in bacteria. Mutants of this gene have conferred glyphosate resistance upon *Salmonella typhimurium* and *E. coli.* The mechanism of resistance is by overproduction of the enzyme or overexpression of the gene. Mutant *aro* A genes have been transferred to plants under the control of a variety of gene promoters (e.g., the CaMV 35S RNA gene). The mutant bacterial genes have also been fused to plant EPSPS chloroplast transit peptide-encoding sequences of the rubisco small subunit gene and transferred to plants, resulting in a 1,000-fold higher level of resistance.

Whereas the aforementioned strategies produced significant "laboratory resistance" to glyphosate, "field resistance" that enabled commercial application of the technology came as a result of isolating a bacterial strain, *Agrobacterium* CP4, with a suitable EPSP synthase. Likewise, the gene for this EPSP was fused to the coding sequence for the plant EPSPS chloroplast transist peptide and coupled to a powerful form of CaMV 35S. The Monsanto Company used this strategy to develop its Roundup Ready® soybean.

Engineering herbicide resistance using herbicide–detoxifying genes

Herbicide-detoxifying genes may be derived from plants or bacteria.

Plant origin

A strategy for engineering herbicide resistance in plants is introducing herbicide-detoxifying enzymes into plants. This strategy is considered superior to those that modify the target for the herbicide in plants. Target modification may alter the enzyme to a degree that may adversely affect its physiological function to the detriment of the plant. On the other hand, suitable herbicide-detoxifying enzymes are limited. Those enzymes occur in both plants and microorganisms. Plant herbicide-detoxifying enzymes include conjugation enzymes and mixed function oxidases. Researchers have isolated genes for the detoxifying enzyme

glutathione-S-transferase (GST) from maize. These genes encode detoxifying enzymes for alachlor and atrazine. In tomato, another conjugation enzyme *N*-glucosyltransferase provides resistance to metribuzin through increased activity of the enzyme. Mixed function oxidases have been found to detoxify 2,4-D in pea and dicamba in barley.

Bacterial origin

Soil bacteria have the capacity to detoxify many herbicides. Consequently, soil bacteria are a potential source for herbicide-tolerant genes. Several bacterial genes have been isolated and expressed in plants. For example, the plasmid encoded nitralase gene *bxn* converts the herbicide 3,5-dibromo-4-hydroxy benzonitrite to its inactive metabolite, 3,5-dibromo-4-hydroxy benzoic acid (bromoxynill). BXN cotton is available on the market, produced by using a gene obtained from a strain of *Klebsiella ozaenae*. This bacterium was isolated from a contaminated field.

L-Phosphinothricin (PPT) is an antibiotic produced by *Streptomyces* spp. It is a competitive inhibitor of glutamine synthetase, the enzyme that catalyzes the conversion of glutamate to glutamine. This is the only enzyme in plants capable of detoxifying the ammonia that is generated as a product of nitrogen metabolism (e.g., amino acid degradation or nitrate reduction). In the absence of this enzyme, ammonia accumulates to lethal levels in the cell. Another antibiotic used as an herbicide is bialaphos, a tripeptide produced by *Streptomyces hygroscopicus*. Resistance to bialaphos is conditioned by a gene, *bar*, which codes for the enzyme phosphinothricin acetyltranferase (PAT). This enzyme acetylates the free amino acid group of PPT. The *bar* gene has been transferred into plants such as tobacco, tomato, and potato, in which the gene is under the transcriptional control of the CaMV 35S promoter. Resistant cultivars have been produced.

Derived from the *Streptomycete* fungus, **glufosinate** blocks the synthesis of glutamine, causing an accumulation of toxic levels of ammonia in the plant. The fungus produces an enzyme that detoxifies the antibiotic it produces. The gene encoding the detoxifying enzyme has been isolated, cloned, and genetically engineered for optimal expression in plants. The technology was used in the development of resistance in corn to the commercial herbicide Liberty® (also called Ignite or Basta). The transgenic corn hybrid is marketed as Liberty Linky®.

Adoption of herbicide tolerant GM crops

Weed control in crop production is important to producers because weeds cause significant economic loss by reducing crop yield and produce quality. As previously discussed, herbicides may be selective (kill only certain species) or non-selective, or broad-spectrum (kill nearly all kinds of weeds—both broad- and narrowleaf). Transgenic varieties of crops with transgenic genes for herbicide tolerance are able to resist broad-spectrum herbicides. Two major herbicides have corresponding transgenic crop varieties. These are Roundup® (glyphosate) and Liberty® (glufosinate). The corresponding transgenic varieties are called Roundup Ready® and Liberty Link®, respectively. Roundup Ready® products are developed by Monsanto. A grower of this variety will spray Roundup® herbicide to control all other plants except the Roundup Ready® crop. In 1999, 12 to 14 million hectares of soybean with glyphosate resistance were grown in the United States. In Iowa alone, glyphosate-resistant soybean accounted for 40% of the crop acreage in 1998.

Glyphosate-resistant corn adoption lags behind the adoption of glyphosate-resistant soybean, with an estimated 1.6 to 2 million hectares in production in 1999. This slow adoption is attributed in part to the availability of good alternative weed management programs in corn production, a lack of consistent efficacy, and questionable yield potential.

In the case of cotton, Roundup Ready® cotton became available for commercial use in 1997. Prior to that, in 1995, the BXN system for resistance to Buctril® (bromoxynil) herbicide was introduced. The herbicide is effective against major weeds, such as morning glory, cocklebur, and velvet leaf. Companies offering this technology are Monsanto, Calgene, and Rhone Poulenc.

Weed management practices in transgenic production

There are specific guidelines provided by the company to the growers as to how to grow and manage the GM crop for best results. However, some claims by the producing company have been disputed by some farmers. For example, claims that early season weed interference was not a major problem for soybean, and thus glyphosate applications could be delayed without a risk of yield loss, were not always the observation of farmers. This is partly due to differences in cultural practices and the vagaries of the weather that affect planting dates. A single application of glyphosate in conjunction with the Roundup Ready® variety is suggested by the company to be adequate to control weeds. Planting dates are earlier in the Midwest, and hence a single application of a postemergence herbicide is not likely to be adequate. Another advantage of using an herbicide-tolerant transgenic is that, should weeds emerge later during the production season, they can be controlled effectively with multiple applications of the herbicide without injuring the soybean crop.

The management of weeds in the transgenic production of corn is more problematic in early stages of adoption of herbicide-tolerant corn, both for glyphosate- and glufosinate-resistant varieties. Most GM corn development for herbicide tolerance has been for glufosinate tolerance. Glufosinate does not readily translocate in plants; thus, effective weed coverage is critical during application for success to be achieved in weed control. Where applications are delayed, the weeds grow larger, thus making coverage a more significant factor in applications of the herbicide. Growers are reluctant to adopt multiple applications because glufosinate is expensive. Some problems with phytotoxicity to corn was initially reported by growers.

Early glyphosate-resistant corn hybrids that were introduced were not favorably adopted by growers. Corn is more sensitive to weeds early in crop establishment. Consequently, herbicide applications are needed earlier in corn production than in soybean production. A residual soil-applied herbicide used in conjunction with glyphosate-resistant corn would provide growers a greater opportunity for effective weed management in corn fields. Such a practice has been proposed by Monsanto for the future.

Roundup® should be applied after the four-leaf stage to avoid injury to the Roundup Ready® cotton. When only one treatment was made at the four-leaf stage, cockleburs appeared later and caused more than 80% reduction in lint yield.

The DuPont Company is offering sulfonylurea-tolerant GM crop varieties. Crop producers should be aware of the significant difference between the management of Roundup Ready® soybeans and Roundup Ready® cotton. Soybean fields need to be kept weed-free for 2–4 weeks after planting. Narrow-row spacing can create an early crop canopy to control weeds. Weeds in wider rows can be controlled with an application of Roundup®, over-the-top, without adverse consequences to yield. In the case of cotton, the field needs to be kept weed-free for 6–10 weeks after planting. Should there be a need to spray over the top after the four-leaf stage, it is often done at a significant yield reduction.

13.3.2   ENGINEERING INSECT RESISTANCE IN CROPS

The Science

There are two basic approaches to the genetic engineering of insect resistance in plants: the use of protein toxins of bacterial origin and the use of insecticidal proteins of plant origin.

Protein Toxins from Bacillus Thuringiensis (Bt)   The *Bacillus thuringiensis (Bt)* endotoxin is a crystalline protein. It was first identified in 1911 when it was observed to kill the larvae of the flour moth. It was registered as a biopesticide in the United States in 1961. Bt is very selective in action—that is, one strain of the bacterium kills only certain insects. Formulations of whole sporulated bacteria are widely used as biopesticide sprays for biological pest control in organic farming. There are several major varieties of the species that produce spores for certain target pests: *B. thuringiensis* var *kurstaki* (for controlling lepidopteran pests of forests and agriculture), var *brliner* (wax moth), and var. *israelensis* (dipteran vectors of human disease). The most commercially important type of the crystalline proteinaceous inclusion bodies are called **δ-endotoxins**. To become toxic, these endotoxins, which are predominantly protoxins, need to be proteolytically activated in the midgut of the susceptible insect to become toxic to the insect. These endotoxins act by collapsing the cells of the lining of the gut regions.

Endotoxin   In its strictist usage, the term refers to the lipopolysaccharide complex associated with the outer membrane of Gram-negative bacteria.

Bt resistance development has been targeted especially at the European corn borer, which causes significant losses to corn in production. Previous efforts developed resistance in tobacco, cotton, tomato, and other crops. The effort in corn was more challenging because it required the use of synthetic versions of the gene (rather than microbial *Bt* per se) to be created.

Two genes, *cryB1* and *cryB2*, were isolated from *B. thuringiensis* subsp *kurstaki* HD 1. These genes were cloned and sequenced. The genes differed in toxin specificities, *cryB1* gene product being toxic to both dipteran *(Aedes aegypti)* and lepidopteran *(Manduca sexta)* larvae, whereas *cryB2* affects only the latter. The Bt toxin is believed to be environmentally safe as an insecticide. In engineering Bt resistance in plants, scientists basically link the toxin to a constitutive (unregulated) promoter that will express the toxin systemically (i.e., in all tissues).

Transgenic plants expressing the δ-endotoxin gene have been developed. The first attempt involved the fusion of the *Bt* endotoxin to a gene for kanamycin resistance to aid in the selection of plants (conducted by a Belgian biotechnology company, Plant Genetic Systems, in 1987). Later, Monsanto Company researchers expressed a truncated *Bt* gene in tomato directly by using the CaMV 35S promoter. Agracetus Company followed with the expression of the *Bt* endotoxin in tobacco with the CaMV 35S promoter linked to an alfalfa mosaic virus (AMV) leader sequence. Since these initial attempts, modifications to the protocols have increased expression of the toxin in transgenic plants. Transformation for expressing the chimeric *Bt* genes was *Agrobacterium*-mediated, using the TR2′ promoter. This promoter directs the expression of manopine synthase in plant cells transformed with the TR DNA of plasmid pTiA6.

The original *Bt* coding sequence has since been modified to achieve insecticidal efficacy. The complete genes failed to be fully expressed. Consequently, truncated (comprising the toxic parts) genes of *Bt* var *kurstaki* HD-1 *(cry1A*[b]) and HD-73 *(cry1A*[c]) were expressed in cotton against lepidopteran pests. In truncating the gene, the *N*-terminal half of the protein was kept intact. For improved expression, various promoters, fusion proteins, and leader sequences have been used. The toxin protein usually accounts for about 0.1% of the total protein of any tissue, but this concentration is all that is needed to confer resistance against the insect pest.

Genetically engineered *Bt* resistance for field application is variable. For example, Ciba Seeds has developed three versions of synthetic *Bt* genes capable of selective expression in plants. One is expressed only in pollen, another in green tissue, and the third in other parts of the plant. This selectivity is desirable for several reasons. The European corn borer infestation is unpredictable from year to year. The lifecycle of the insect impacts the specific control tactic used. The insect attack occurs in broods or generations. The *Bt* genes with specific switches (pollen and green tissue) produce the *Bt* endotoxin in the parts of the plants that are targets of attack at specific times (i.e., first and second broods). This way, the expression of the endotoxin in seed and other parts of the plant where protection is not critical is minimized. Monsanto's YieldGard™ corn produces *Bt* endotoxin throughout the plant and protects against both first and second broods of the pest. The commercially available *Bt* corn cultivars were developed by different transformation events, each with a different promoter.

Bt cotton is another widely grown bioengineered crop. The pest resistance conferred by the *Bt* gene has led to a dramatic reduction in pesticide use and consequently has reduced the adverse impact on the environment from agropesticides. As indicated previously, *Bt* sprays are widely used in organic farming for pest control. However, such application is ineffective if the insect bores into the plant. Further, *Bt* sprays have short-duration activity.

Engineering Viral Resistance Even though viruses may utilize DNA or RNA as hereditary material, most of the viruses that infect plants are RNA viruses. One of the most important plant viruses in biotechnology is the Cauliflower Mosaic Virus (CaMV), from which the widely used 35S promoter was derived (CaMV 35S promoter). As previously described, a virus is essentially nucleic acid encased in a protein coat. The primary method of control of viral infections is through the breeding of resistance cultivars. Also, plants can be protected against viral infection by a strategy that works like inoculation in animals. Plants may be protected against certain viral infections upon being infected with a mild strain of that virus. This strategy, called **cross-protection**, protects the plant from future, more severe infections.

Engineering transgenic plants with resistance to viral pathogens is accomplished by the method called **coat protein-mediated resistance**. First, the viral gene is reverse transcribed (being RNA) into DNA, from which a double-stranded DNA is then produced. The product is cloned into a plasmid and sequenced to identify the genes in the viral genome. A chimeric gene is constructed to consist of the open reading frame for the coat protein to which a strong promoter is attached for a high level of expression in the host. This gene construct is transferred into plants to produce transgenic plants.

Successes with this strategy have been reported in summer squash (the first product developed by this approach) and for resistance to papaya ringspot virus (a lethal disease of papaya), among others.

Adoption

The primary insect pest targeted by *Bt* toxins in corn production is the European corn borer *(Ostrinia nubilalis)*. Other insect pests are the corn earworm *(Helicoverpa zea)*, southwestern corn borer *(Diatraea grandiosella)*, and lesser cornstalk borer *(Elasmopalpus lignosellus)*. The first commercial *Bt* corn hybrid was first marketed in 1996 by Mycogen Corporation, in conjunction with Ciba Seeds. Currently, genes for three *Bt* toxins have been incorporated into corn varieties by various companies (Table 13–1). The toxin is expressed in different tissues of corn, and varies in levels among tissues and growth stages. The *cry1Ab* gene is present in transgenic varieties marketed by CIBA Seeds (Norvartis) and Mycogen as KnockOut and NatureGuard. Monsanto's and Northrup King's (Norvatis) YieldGard contains the *cry1Ac* gene.

Corn rootworm *(Diabrotica* spp) can cause severe injury to roots. Rootworm-resistant corn was approved in 2003. It was developed by Monsanto Corporation to express the *Bt* proteins in corn roots in levels that are high enough to kill corn rootworm larvae. The damage by the larvae is high in continuous culture systems.

Monsanto Corporation released the *Bt* cotton Bollgard in 1996. This variety provides resistance to tobacco budworms and pink and cotton bollworms. Several companies use the Bollgard technology under license. Cotton is a relatively difficult crop to

Table 13–1 Types of Commercially Introduced *Bt* Corn. Some Companies have Licensed Their *Bt* Inventions to Other Companies for Development of Various *Bt* Products

Trade Name	Bt Gene	Transformation Event	Company
Knockout	*cry1A (b)*	1st/2nd	Ciba (Novartis)
NatureGard		Generation 176	Mycogen
YieldGard	*cry1A (b)*	*Bt* 11	Northrup King (Novartis)
YieldGard	*cry1A (b)*	MON 810	Monsanto
Bt-Xtra	*cry1A (b)*	DBT 418	Dekalb

Note: The *Bt* product produced by Aventis (called StarLink™) used the *cry 9 (c)* gene, and was approved for feed only. KnockOut and NatureGard have been discontinued as commercial products.

produce; consequently, growers appreciate any efforts to make their tasks easier and are likely to be willing to pay the high premiums for GM cotton varieties. It is predicted that GM cotton will dominate most of the cotton acreage in the next decade.

Management of Bt Resistance in the Field

The introduction of Bollgard cotton faced an unexpectedly jerky start in 1996. An abnormally high incidence of bollworm infestation in the Cotton Belt made it necessary for producers, especially in Texas, to spray in midsummer to keep the pest under control. Following this incidence, Monsanto decided to issue a spray advisory for Texas, Arkansas, Oklahoma, Louisiana, and Mississippi.

Insects are known to have relatively short lifecycles and hence are susceptible to mutations. There are concerns about development of resistance to *Bt*. There are concerns that a widespread use of *Bt* varieties would accelerate the development of resistance in the target pests. Consequently, the EPA mandates the use of a refuge, an area planted to a non-*Bt* variety in physical proximity to a field planted to a *Bt* variety, in the management of *Bt* pests with transgenics to slow the adaptation of insects to *Bt* cotton.

In corn, a refuge should comprise at least 20% of the total acreage of the crop to a non-*Bt* variety. In cotton, at least 4% of the crop should be non-transgenic. The refuge is not treated to any insecticide. The rationale of a refuge is that the few surviving insects (resistant) would be swamped by the numerous insects in the susceptible refuge, thereby retarding the evolution of insects resistant to the *Bt* gene. The potential problems of refuges include the possibility that the frequency of *Bt*-resistant alleles in the insect populations may be greater than assumed in refuge models and that the resistance to *Bt* in the European corn borer may be semidominant rather than recessive. There is also the possibility that the insects surviving in the *Bt* field will mature several days later than susceptible insects in the refuge, thereby preventing any mating from occurring.

The Issue of Technology Fees

Growers of transgenic crop varieties are required to sign a license agreement to honor the restrictions on the use of proprietary materials and to implement governmental regulations prescribed for the safe use of GM materials. A contract normally binds a grower to growing the purchased seed for only one season (i.e., the farmer cannot save seed from the current season for planting the next season's crop), to growing and maintaining a refuge (as prescribed by the EPA for the species), and to paying a royalty fee to the company (the so-called technology fee).

The technology fee depends on the specific technology in the cultivar (e.g., herbicide resistance, insect resistance, or multiple traits), the species, the type of product, and the production region. Prior to 1998, the technology fees were a flat fee of $8/acre for picker cotton, $32/acre for Bollgard, and $40/acre for Roundup Ready®/Bollgard® *Bt*. A new method of calculating the technology fees uses the Seed Drop Rate (SDR) factor developed by Monsanto. The SDR is the number of seeds dropped from the planter to obtain a final stand count as plant population. Cotton varieties containing the Bollgard/Roundup® genes have been placed into groups containing 4,200, 4,700, or 5,400 seeds per pound, according to seed size. The technology fee is calculated as follows:

Number of seeds/lb × number of 50 lb bags purchased/SDR =
number of acres on which the technology fee will be based

For example, given an SDR of 52,000 seeds/acre and cotton variety of group 4,700 seeds/lb, the acreage subject to a technology feed is

4,700 seeds/lb × 50 lb (one bag) = (235,000 seeds/bag)/52,000 SDR = 4.5 acres/bag

That is, the grower will pay a technology fee on 4.5 acres/bag.

The SDR differs for the production regions. For example, it is 67,500 seeds/acre for Georgia, Florida, and southern Alabama; 60,000 seeds/acre for Arizona and California; and 67,000 seeds/acre for east Texas. The method of SDR allows the grower to know upfront what fee to pay rather than waiting to reconcile crop acreage after the planting season. It allows growers to be charged a fee based on what is agronomically best for the region of production and priced on a per-acre basis. For the 12 to 14 million hectares of glyphosate-resistant soybeans planted in 1999, an estimated $240 to $280 million in technology fees accrued to the seed companies.

Other Transgenic Technology

Multiple traits (gene stacking)   The crop biotechnology industry is narrowly dependent on two genetically engineered traits (herbicide and insect resistance) concentrated primarily in two crops (soybean and corn). These traits benefit the grower. Quality traits (e.g., enhanced nutritional traits, such as protein content) are being developed (e.g., Golden rice). Transgenics have been developed in other crops (e.g., cotton, canola, potato, tomato). These crops are currently grown on a small percentage of the acreage devoted to transgenic crops. A practice that is gaining attention is the incorporation of multiple transgenes into one cultivar, a practice called gene stacking. Some of the multiple-trait cultivars feature genes for both insect resistance *(Bt)* and herbicide resistance, as in cotton. In 1998, about 9% of the global acreage for transgenics had this combination of genes. Other stacks available or being developed include BXN/Bollgard®, Roundup Ready®/Bollgard®, both in cotton, and the first dual herbicide-resistant hybrid in corn—resistance to both Liberty® and IMI herbicides—is being developed by ICI/Garst.

Engineering fiber color   Regarding cotton, Calgene has received the first U.S. patent for transgenic color alteration without the use of dyes.

Engineering nutritional quality   Molecular techniques used in enhancing the nutritional quality of plant products may be categorized as follows:

1. Altering the amino acid profile of the seed
2. Selectively enhancing the expression of existing genes
3. Designing and producing biomolecules for nutritional quality

One of the successes in this area is the creation of Golden rice. Golden rice is so-called because it has been genetically engineered to produce β-carotene (responsible for the yellow color in certain plant parts, such as carrot roots) in its endosperm. This rice produces β-carotene or pro-vitamin A, the precursor of vitamin A, which does not occur in the endosperm of rice. The scientific feat accomplished in engineering β-carotene into rice is that it marks the first time a metabolic pathway has been engineered into an organism. Rice lacks the metabolic pathway to make β-carotene in its endosperm. Potrykus and Beyer had to engineer a metabolic pathway consisting of four enzymes into rice. Immature rice endosperm produces geranylgeranyl-diphosphate (GGPP), an early precursor of β-carotene. The first enzyme engineered was phytoene synthase, which converts GGPP to phytoene (a colorless product). Enzyme number 2, called phytoene desaturase, and enzyme number 3, called ζ-carotene desaturase, each catalyzes the introduction of two double bonds into the phytoene molecule to make lycopene (has red color). Enzyme number 4, called lycopene β-cyclase, converts lycopene into β-carotene. A unit of transgenic construct (called an expression cassette) was designed for each gene for each enzyme. These expression cassettes were linked in series, or stacked in the final construct.

Engineering fruit ripening   Certain fruits exhibit elevated respiration during ripening, with concomitant evolution of high levels of ethylene. Called **climacteric fruits** (e.g., apples, bananas, tomatoes), these fruits' ripening process involves a series of biochemical changes leading to fruit softening. Chlorophyll, starch, and the cell walls are degraded. There is an accumulation of lycopene (red pigment in tomato), sugars, and various organic acids. Ripening is a complex process that includes fruit color change and softening. Ripening in tomato has received great attention because it is one of the most widely grown and eaten fruits in the world. Ethylene plays a key role in tomato ripening. When the biosynthesis of ethylene is inhibited, fruits fail to ripen, indicating that ethylene regulates fruit ripening in tomato. The biosynthesis of ethylene is a two-step process, in which S-adenosyl methionine is metabolized into aminocyclopropane-1 carboxylic acid, which in turn is converted to ethylene. Knowing the pathway of ethylene biosynthesis, scientists may manipulate the ripening process by either reducing the synthesis of ethylene or reducing the effects of ethylene (i.e., plant response).

In reducing ethylene biosynthesis, one successful strategy has been the cloning of a gene that hydrolyzes s-adenosyl methionine (SAM), called SAM hydrolase, from a bacterial virus, by Agritope Company of Oregon. After bioengineering the gene to include, among other factors, a promoter that initiates expression of the gene in mature green fruits, *Agrobacterium*-mediated transformation was used to produce transgenic plants. The effect of the chimeric gene was to remove (divert) SAM from the metabolic pathway of ethylene biosynthesis. The approach adopted by researchers was to prevent the aminocyclopropane-1-carboxylic acid (ACC) from being converted to ethylene. A gene for ACC synthase was isolated from a bacterium and used to create a chimeric gene, as in the Agritope case.

The technology of antisense RNA (a complementary RNA sequence that binds to a naturally occurring mRNA molecule, blocking its translation) has been successfully used to develop a commercial tomato that expresses the antisense RNA for ACC synthase and ACC oxidase. USDA scientists pioneered the ACC synthase work, whereas scientists from England in collaboration with Zeneca Company pioneered the ACC oxidase work. Because the transgenic tomato with an incapacitated ethylene biosynthetic pathway produced no ethylene, it failed to ripen on its own, unless exposed to artificial ethylene sources in ripening chambers. The technology needs to be

Climacteric fruits   The period in the development of some plant parts involving a series of biochemical changes associated with the natural respiratory rise and autocatalytic ethylene production.

perfected, so that fruits can produce a minimum amount of ethylene for autocatalytic production for ripening over a protracted period.

The Flavr Savr Tomato   Another application of antisense technology is in preventing an associated event in the ripening process, fruit softening, from occurring rapidly. Vine-ripened fruits are tastier than green-harvested and forced-ripened fruits. However, when fruits vine ripen before harvesting, they are prone to rotting during shipping or have a short shelf life in the store. It is desirable to have fruits ripen slowly. In this regard, the target for genetic engineering is the enzyme **polygalacturonase (PG)**. This enzyme accumulates as the fruit softens, along with cellulases that break cell wall cellulose, and pectin methylesterase which, together with PG, break the pectic cross-linking molecules in the cell wall. Two pleiotropic mutants of tomato were isolated and studied. One mutant, never ripe *(Nr)*, was observed to soften slowly and had reduced accumulation of PG, whereas the second mutant, ripening inhibitor *(rin)*, had very little accumulation of PG throughout the ripening process. This and other research evidence strongly suggest a strong association between PG and fruit ripening. PG is biosynthesized in the plant and has three isoenzymes (PG1, PG2, PG3).

Polygalacturonase   In tomato, this is a cell wall enzyme that is secreted in large amounts during fruit ripening.

This technology was first successfully used by Calgene to produce the **Flavr Savr tomato**, the first bioengineered food crop, in 1985. As previously noted, this pioneering effort by Calgene flopped because of a poor decision to market a product intended for tomato processing as a fresh market variety.

Benefits of Transgenics

The general goal of transgenics in crops is to increase crop productivity primarily by reducing inputs and hence production costs. In one study, growers using glyphosate-resistant soybean for improved weed control experienced a 30% lower herbicide cost than for non-GM soybean. This technology also allows growers flexibility in weed management in terms of timing of application. Resistant weeds can be managed in transgenic cropping systems. Transgenics with enhanced compositional traits (e.g., high nutrition) are advantageous in resolving the malnutrition problems in the world.

Risks Associated with Transgenic Crop Production

The following are the risks associated with transgenic crop production:

1. A transgenic crop could become a volunteer weed.
2. Gene transfer to non-GM crops is possible.
3. Herbicide drift of non-selective herbicides to neighboring crop occurs.
4. Yield potential of GM and non-GM crops may not be the same. A 4% yield decline is generally associated with the glyphosate resistance traits.
5. Economic returns from using GM often do not exceed those of non-GM crops.
6. Because of unfavorable export markets there is a loss of marketability for GM grain (e.g., European Union non-acceptance of GM crops).

Discontinued Trangsgenic Products

Some of the major discontinued GM crops and their characteristics are summarized in Table 13–2. Some of these products made headlines when they were introduced on the market (e.g., Flavr Savr tomato, StarLink corn, and the NewLeaf potato).

Table 13–2 Some Discontinued Transgenic Products

Product	Company/Organization	Reason(s)
FlavrSavr tomatoes	Calgene	Poor genetic background
Tomato paste	Zeneca	European negative opinion
NewLeaf potatoes	Monsanto	Decline by fast-food chains
Triffid flax	University of Saskatchewan	EU market decline
StarLink corn	Aventis	Protein was allergenic
Bt 176 corn—Knockout	Norvatis	Pollen toxic to Monarch butterfly
—NatureGard	Mycogen	

Summary

1. Adoption of genetically modified (GM) crops is on the rise.
2. The major growers of GM crops are the United States, Argentina, Canada, South Africa, and Australia.
3. The most important transgenic crop species are corn, cotton, canola, potato, squash, and papaya.
4. About 25% of corn and 60–70% of soybean in the United States are GM.
5. Three traits dominate the current GM market: herbicide tolerance, *Bt* insect resistance, and virus resistance.
6. Resistance to two major pesticides, Roundup® and Liberty®, dominate the market.
7. *Bt* products offer resistance against the European corn borer, corn ear worm, and southwestern corn borer.
8. Bollgard® is a GM cotton that is widely grown by producers.
9. Producers using GM products have to pay technology fees.
10. Some GM products have been discontinued.

References and Suggested Reading

Acquaah, G. 2004. *Understanding biotechnology: An integrated and cyber-based approach.* Upper Saddle River, NJ: Prentice Hall.

Tauer, L., and J. Love. 1989. The potential economic impact of herbicide-resistant corn in the U.S.A. *J. Prod. Ag* 2, 202.

Turner, M. 1999. How will GMO restrictions affect corn and soybean exports? *Crop Watch* 99–13, 120.

Ye, X., S. Al-Babili, A. Kloti, J. Zhang, P. Lucca, P. Beyer, and I. Potrykus. 2000. Engineering the provitamin A biosynthetic pathway into rice endosperm. *Science* 287:3003–3005.

Selected Internet Sites for Further Review

http://www.whybiotech.com/index.asp?404;http://www.whybiotech.com/en/default.asp

Site of the Council on Biotechnology

http://www.sciencemag.org/cgi/content/abstract/287/5451/303?ijkey5/wflieVWVyTQA

The science behind the Golden Rice

http://www.nal.usda.gov/bic/

Links to biotechnology information

http://www.agbioworld.org/biotech_info/topics/agbiotech/agbiotech.html

Agbiotech issues in developing countries

OUTCOMES ASSESSMENT

PART A

Answer these questions true or false.

1. T F   Roundup Ready® GM crop has protection against the corn ear worm.
2. T F   The United States is the leading adopter of GM crops.
3. T F   Herbicide tolerance is the leading trait incorporated into GM crops.
4. T F   Some GM crops have multiple transgenic traits.
5. T F   The flavrsavr is a GM potato.

PART B

Answer the following questions.

1. The acronym GM stands for _____

   _____

2. Give the top three major GM adopting countries in the world. _____

   _____

3. Give the three most common traits incorporated into GM crops. _____

   _____

4. Define biotechnology. _____

   _____

PART C

Write a brief essay on each of the following topics.

1. Discuss the management of *Bt*-resistant crops in the field
2. Discuss the discontinuation of GM crops.

3. Discuss the benefits of transgenic crops.

4. Discuss the risks associated with transgenic crop production.

5. Discuss the concept of gene stacking in the development of GM crops.

PART D

Discuss or explain the following topics in detail.

1. Discuss the development of *Bt*-resistant crops.

2. Discuss the adoption of GM crops in the world.

3. Discuss the issue of technology fees in the adoption of GM crops.

14

Rangeland and Pastures and Their Management

PURPOSE

The purpose of this chapter is to discuss the establishment, management of pastures, and the management of rangelands for the production of livestock.

EXPECTED OUTCOMES

After studying this chapter, the student should be able to:

1. Define and distinguish between the terms grazingland, rangeland, and pasture.
2. Give examples of specific rangelands in the United States.
3. Distinguish between native and tame pasture.
4. Discuss the establishment of a permanent pasture.
5. Distinguish between continuous and rotational stocks and their management.
6. Discuss poisonous plants in the rangeland.

KEY TERMS

Browsing	Native pasture	Rangeland
Continuous stocking	Pasture	Rotational stocking
Decreasers	Permanent pasture	Sprigging
Grazingland	Range condition	Tame pasture
Increasers	Range site	

TO THE STUDENT

Rangeland and pastures are distributed widely throughout the United States. They are important in the nation's agriculture because they support livestock and wildlife. Rangelands consist of a wide variety of species of plants and are managed as natural ecosystems. When there is an imbalance in the system as a result of improper stocking rate, time of grazing, livestock preference, certain species become overgrazed, changing the range condition. Sometimes, livestock producers deliberately establish pastures using carefully selected species. Called tame pastures, these areas are managed as monocultures or mixtures of a few species. Livestock use of pastures may be controlled by improving a restriction on the movement of livestock through fencing and other approaches. This chapter is different from the others in the sense that it includes a discussion of animals in association with plants, as opposed to a discussion of only plants.

14.1: KINDS OF GRAZINGLANDS

Grazingland Any land that is grazed or has the potential to be grazed by animals.

Browsing The act of feeding by nibbling leaves and succulent material from trees and shrubs.

A **grazingland** is any land that is grazed or has the potential to be grazed by animals. Grazing is the term for the consumption of standing forage (or edible grasses and forbs) by animals (livestock and wildlife). **Browsing**, on the other hand, is the consumption of edible leaves and twigs from woody species (trees and shrubs) by animals.

Grazinglands differ by several factors, especially the kinds of vegetation (e.g., woody perennials, grasses, forbs, shrubs, agricultural crops) and origin (indigenous or introduced plants). The kinds of grazinglands and their major characteristics are summarized in Table 14–1. They include agroforestry, agro-silvo-pastoral, forestland, pastureland, rangeland, range, and cropland. In this book, detailed attention has been given to a discussion of agroforestry practices in Chapter 11. The discussion in this chapter is devoted primarily to rangeland and pastures and their management.

14.2: RANGELANDS

Rangeland A land on which indigenous vegetation consists primarily of grasses, forbs or shrubs.

A **rangeland** may be defined as an uncultivated land that provides a habitat for animals (Figure 14–1). It consists of indigenous vegetation that includes trees, shrubs, forbs, and grasses. The specific types (15 types identified in the United States) of range vegetation includes tallgrass prairie, mixed grass prairie, shortgrass prairie, post oak–blackjack oak (Cross Timbers) savannah, sand sagebrush grasslands and mesquite grasslands (Table 14–2). Rangelands are the dominant land type in the world, comprising about 70% of the total land area of the world. In the United States, rangelands supply about 50 to 65% of the total needs of domestic ruminants and about 95% of wild ruminant feed.

Rangeland communities are often diverse, consisting of many species of plants and animals. In some cases, some of the native grazing species (e.g., bison and elk) have been replaced by domestic grazing species (e.g., cattle). Furthermore, some rangeland is plowed up for use as cropland. In this regard, the amount of rangeland varies from year to year, according to prevailing socioeconomic factors. Over the past century, the Central Great Plains region of the United States has experienced shifts between rangeland

Table 14-1   Kinds of Grazinglands and Their Characteristics

Agroforestry
A land use system in which woody perennials are grown for wood production with agricultural crops, with or without animal production

Agro-Silvo-Pastoral
A land use system in which perennials are grown along with agricultural crops, forage crops, and livestock production

Forestland
Land dominated by forestland or woodland and used for both wood production and animal production; the animal may graze indigenous species forage or the vegetation can be managed as an indigenous forage

Pastureland
Land that is devoted to the production of indigenous or introduced forage, mainly to be grazed by animals

Rangeland
Land on which the indigenous vegetation consists primarily of grasses, forbs, or shrubs; it is managed as a natural ecosystem

Range
Land that supports indigenous vegetation and is grazed or has the potential to be grazed by animals; it is managed by humans as a natural ecosystem; this grazing land includes grazable forestland and rangeland

Cropland
Land that is used primarily for production of cultivated crops but may be used to produce forage crops

FIGURE 14-1   Cattle grazing on a rangeland in California. (Source: USDA)

Tallgrass Prairie
- Occurs in the central United States
- Currently occurs mainly in the Osage Hills of Oklahoma, the Flint Hills of Kansas, the Nebraska Sandhills, and Texas Coastal Prairie
- The climate is subhumid and temperate.
- Most of the rainfall comes in the summer growing season.
- The soils are mainly Mollisols.
- The major species of grasses are little bluestem *(Schizachyrium scoparium)*, big bluestem *(Andropogon gerardii)*, yellow indiangrass *(Sorghastrum nutans)*, and switchgrass *(Panicum virgatum)*.
- The tall grasses are coarse, unpalatable, and low in nutritive value in winter.
- Favor the production of grass crops—corn, wheat
- Rotation grazing schemes are more successful on these lands.

Southern Mixed Prairie
- Most important of the western range types for livestock production
- Occurs from eastern New Mexico to eastern Texas and from southern Oklahoma to northern Mexico
- Soils are mainly Mollisols, Entisols, and Aridisols.
- Precipitation ranges between 300 and 700 mm per year.
- The important species of plants include blue grama, buffalograss *(Buchloe dactyloides)*, little bluestem, three-awned grasses *(Aristida* sp.), and silver bluestem *(Bothriochloa saccharoides)*.
- Mesquite *(Prosopis* sp.) invasion is heavy in the Texas region. It reduces forage production but is valuable for wild animal use.

Northern Mixed Prairie
- Occurs in the Great Plains extending northward from the Nebraska-South Dakota state line
- Precipitation ranges between 300 and 650 mm per year.
- Soils are mainly Mollisols but Entisols and Vertisols occur in some parts.
- Mostly used as rangeland, since periodic drought and low precipitation makes it unsuitable for crop production
- It supports short-, mid-, and tall grasses as well as cool and warm season grasses, making it the type of rangeland that supports the highest diversity of grasses in the western range types.
- Suitable for domestic and wild animal production from nutritional perspective
- Major species include bluebunch wheatgrass *(Agropyron spicatum)*, various bluegrasses *(Poa* sp.), silver sagebrush *(Artemisia cana)*, skunkbrush sumac *(Rhus trilobata)*, ponderosa pine *(Pinus ponderosa)*, crested wheatgrass *(Agropyron critatum)*, prickly pear cactus *(Opuntia polyacantha)*, and many others.
- Common use of cattle and sheep is favored.

Shortgrass Prairie
- Extends from northern New Mexico into Wyoming
- Precipitation is low (300–500 mm per year) and hence much of the area is still rangeland.
- Soils are mostly Mollisols.
- Dominant species are blue grama, buffalograss, and western wheatgrass.
- One important shrub is winterfat *(Ceratoides lanata)*.
- Generally flat terrain, good water distribution, long growing season, and high nutritional quality of species makes this rangeland type suited for grazing by both cattle and sheep.

California Annual Grassland
- Occurs mainly west of the Sierra Nevada Mountains
- Climate is Mediterranean, with mild, wet winters and long, hot, dry summers.
- Rainfall ranges between 200 and 1,000 mm per year.
- Soil types are mainly Mollisols, Aridisols, and some Inceptisols.
- One of the oldest livestock grazing lands in the United States

- Major species are annuals—including slender oat *(Avena barbata)*, wild oat *(Avena fatua)*, ripgut brome *(Bromus rigidus)*, foxtail brome *(Bromus rubens)*, and little barley *(Hordeum pusillum)*; also pointleaf manzanita *(Arctostaphylos pungens)*, blue oak *(Quercus douglasii)*, wedgeleaf ceanothus *(Ceanothus cuneatus)*.
- Being predominantly annual species, they are not very responsive to grazing intensity.
- Forage quantity is inadequate in fall, winter, and summer. Supplementation with protein and phosphorus is helpful.
- Because forage quantity fluctuates, the operator must adjust stocking rates rapidly or have a good reserve of harvested forage.

Palouse Prairie
- Also called the northwest bunchgrass prairie
- Used mainly for wheat production
- Extends from eastern Washington, north-central and northeastern Oregon, and west Idaho
- Rainfall ranges from 300 to 640 mm per year, occurring mainly in winter.
- Soils are mainly Mollisols.
- One of the most productive and beautiful grasslands of the world
- Species include bluebunch wheatgrass, Sandberg bluegrass *(Pos sandbergii)*; forbs include arrowleaf balsamroot *(Balsamorhiza sagitata)*, western yarrow *(Achillea millefolium)*.

Hot Desert
- One of the largest of the western range types
- Occurs in southern California, Arizona, New Mexico, southern Texas, and Nevada
- Precipitation ranges between 130 and 500 mm per year, increasing with elevation above sea level.
- Key deserts are Mojave, Sonoran, and Chihuahuan.
- Soil are mixed but mainly Aridisol.
- Species include mesquite *(Prosopis* sp.), catclaw *(Acacia* sp.), creosotebush *(Larrea tridentate)*, black grama *(Bouteloua eriopoda)*, *Aristida, Hilaria.*

Cold Desert
- Also called Great Basin
- Consists of two types—sagebrush grassland (e.g., Oregon, Nevada, Idaho, Montana, Utah, Wyoming) and salt desert shrubland (e.g., Colorado, New Mexico, Oregon, Idaho, Montana, Wyoming)
- Precipitation ranges between 200 and 500 mm per year in sagebrush and 80 and 250 mm per year in salt desert.
- Species in sagebrush grassland include big sagebrush, bluebunch wheatgrass, and bottle brush squaretail *(Sitanion hystrix)*.
- Species in salt desert include those belonging to Chenopodiaceae and Asteraceae.
- Salt desert is one of least productive rangetypes in the United States; used as winter grazing for sheep.

Pinon-Juniper Woodland
- One of most widely distributed range types in the western United States
- Mostly occurs in Utah, Colorado, New Mexico, and Arizona
- Trees are small.
- Very depleted; few areas have good grass understory.
- Species include pinon pine *(Pinus* sp.), juniper *(Juniperus* sp.), and Gambel oak *(Quercus gambelii)*.
- Used for fenceposts, firewood, pulpwood, and Christmas trees

Mountain Browse
- Occurs in Colorado, Utah, Oregon, and Idaho as narrow, intermittent strips between the uppermost part of the grasslands and the coniferous forest types
- Precipitation ranges between 450 and 500 mm per year.
- Soils are mainly Entisols and Inceptisols.
- Species include chockecherry *(Prunus virginiana)*, buckbrush *(Ceanothus)*, and Gambel oak.
- Used for big-game animals

continued

Western Coniferous Forest
- Widely distributed all over the interior of the western United States
- Important species include ponderosa pine, Douglas fir *(Pseudotsuga menziesii)*, and pinon-juniper.
- Precipitation ranges from 450–650 mm per year.
- Soils are mainly Entisols, Inceptisols, and Mollisols on gentle topography.

Southern Pine Forest
- One of the largest and most important range types in the United States
- Species include oak-hickory *(Carya* sp.), pine *(Pinus palustris)*, shortleaf pine *(Pinus echinata)*, loblolly pine *(Pinus taeda)*, *Andropogon, Panicum, Aristida, Cynodon,* and *Paspalum.*
- Cattle are most important on this rangeland.
- Wildlife (white-tailed deer, turkey, bobwhite quail) also important

Eastern Deciduous Forest
- Occurs mainly in Missouri, Indiana, Ohio, Kentucky, Virginia, and Wisconsin
- Precipitation ranges between 800 and 2,000 mm per year
- Soils are mainly Alfisols.
- Key species include maples *(Acer* sp.), birches *(Betula* sp.), oak *(Quercus* sp.), hickories *(Carya* sp.), beeches *(Fagus* sp.), and *basswood (Tilia* sp.); also bluestem *(Andropogon),* fescue *(Festuca),* bluegrass *(Poa),* ryegrass *(Lolium),* and brome *(Bromus).*

Oak Woodland
- Main types are shinnery oak (occurring in southeastern New Mexico, Texas, Chihuahua, Mexico), Gambel oak (central and southern Rockies), and open savannah (in California, Oregon, southern Arizona, central Texas).
- Rainfall ranges from 650–1,000 mm per year.
- Important to wildlife

Alpine Tundra
- Occurs at the highest range altitude
- Occurs in Alaska, Colorado, Washington, Montana, and California
- Precipitation is between 1,000 and 1,500 mm per year
- Temperatures are very cold.
- Soils are shallow, rocky, mainly Entisols and Histosols.
- Species include bluegrasses, sedges, buckwheat, saxifrage, rose, and mustard families.
- Grazed mainly by sheep

and cropland, depending on economic conditions—specifically, the price of beef versus the price of grain. Cultivable rangelands generally are marginal lands (e.g., thin soils, low precipitation, rugged topography) and hence unable to sustain cultivation.

Rangelands are managed as natural ecosystems. In the past two decades, they have been managed more intensively, with a heavy involvement of fencing. Furthermore, introduced species, such as crested wheatgrass *(Agropygron cristatum),* buffelgrass *(Cenchrus ciliaris),* and Lehman lovegrass *(Eragrostis lehmanina),* persist in many parts of the western United States without any periodic cultivation or other agronomic inputs and are considered rangelands. A significant difference between rangeland management and pasture management is that rangeland management tends to focus on maintaining multiple species stand (comprising grasses, forbs, shrubs, trees), whereas pasture management tends to focus on monoculture of forage for high quality and quantity.

14.3: RANGE CONDITION CLASSES

A **range site** is defined by the Society for Range Management as a distinctive kind of rangeland which, in the absence of abnormal disturbance and physical site deterioration, has the potential to support a native plant community typified by an association of species different from that of other sites. With time, the species composition of a site can change. **Range condition** is defined by a departure from a certain conceived potential for a specific range site (i.e., a departure from what it is naturally capable of producing). It describes the state of health of the rangeland by comparing the kind of native vegetation currently on the site with the kind of vegetation originally on that site. Changes in range condition are influenced especially by intensity and time of grazing.

Plant species at a range site may be classified as decreasers, increasers, and invaders. **Decreasers** are the most desired plants at a range site. They tend to decrease in amount from the original vegetation when animals continuously overuse the vegetation. **Increasers** are less desirable plant species and consequently tend to increase as the decreasers are preferentially grazed. **Invaders** are opportunistic species that invade the range site when both decreasers and increasers are continually overgrazed. These species are weeds and poor-quality grasses, such as broomweed, ragweed, and threeawn.

Based on the proportions of decreasers and increasers relative to the original vegetation, range sites may be classified as *excellent* (76 to 100% of climax or original vegetation present), *good* (50 to 75% of original decreasers and increasers present), *fair* (26 to 50% original vegetation), and *poor* (0 to 25% original vegetation).

14.4: PASTURES

Pasture is a type of grazing management unit that is enclosed and separated from other areas by fencing or other appropriate barriers and devoted to the production of forage to be harvested mainly by grazing animals. Over 50% of U.S. land is in permanent pastures, with another 30% in forests and woodland (Figure 14–2).

Pasture A grazing management unit devoted to the production of forage to be harvested by domestic grazing animals.

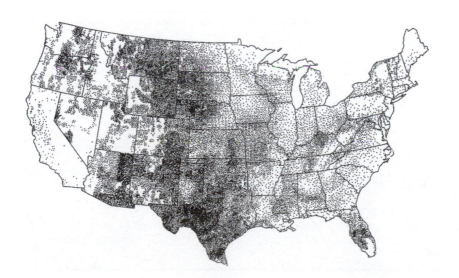

FIGURE 14–2 Figure showing the total land pastured in the U.S., including cropland used only for pasture, woodland pasture, and other pasture (Source: USDA)

Unlike rangeland, pastures are often managed as monocultures (or mixtures of few species) for high quantity and quality, as previously stated. The species are introduced and intensively managed with various agronomic inputs for high productivity. These introduced (or tame) pastures are usually introduced not as a replacement of native pastures but as a complement to them in a rotational grazing system.

14.5: KINDS OF PASTURE

Tame pasture A pasture established by seeding to domesticated pasture species.

There are two basic kinds of pasture—**tame** and **native**.

1. *Tame pastures.* Tame pastures are lands that have been cultivated and seeded to domesticated pasture species for the specific purpose of being grazed by livestock. There are different kinds of tame pasture—**permanent**, **supplemental**, **rotational**, **annual**, and **renovated**. The characteristics of these pastures are summarized in Table 14–3. Permanent pastures are usually composed of perennial species planted on marginal cropland and will persist on the site indefinitely. They cost less to establish and maintain. Because the ground is covered all the time, permanent pastures reduce soil erosion, add organic matter to the soil, and improve soil structure. However, the productivity of perennial pastures tends to decrease after the second or third year as a result of the plants being weakened by grazing pressure, diseases, weeds, and insect pests. The Corn Belt has the most productive grazinglands. The Great Plains also have good grazinglands, though less productive than those of the Corn Belt because of low rainfall.

Table 14–3 Tame Pastures and Their Characteristics

Permanent Pastures
- These are fields of grazinglands that are occupied by perennial pasture plants or by self-seeding annuals that remain unplowed for extensive periods of time (5 or more years).

Supplemental Pastures
- These pastures are established for use in grazing when the permanent or rotation pastures are unproductive or do not yield adequate feed for the livestock needs.

Rotational Pastures
- These are fields used for grazing that were seeded to perennials or self-seeding annuals.
- These species form units in the crop rotation plan and are plowed within a 5-year or shorter interval.

Annual Pastures
- These pastures are seeded annually to take the place of permanent pastures, completely or partially.
- The species used include annual crops, such as rye, oats, barley, soybeans, vetch, and rape.

Renovated Pastures
- These are old pastures that are restored to their former levels of productivity through implementing various agronomic practices, such as tillage, mowing, reseeding, and fertilizing.

2. *Native.* Also called **natural pastures, native pastures** are uncultivated lands that carry native, or indigenous, species that are grazed or have the potential to be grazed by animals. The major types include ranges, brush pastures, woodland pastures, and stump pastures. Figure 14–2 shows the general areas where woodland pastures occur.

14.6: PASTURE SPECIES AND MIXTURES

Tame pastures may be established as monocultures or mixtures (Fig. 14–3). Temporary or annual pastures are usually established as pure stands. However, when establishing a permanent pasture, it is usually best to seed the land to multiple species (grasses and legumes) than a single species for the following reasons:

1. Including legumes is beneficial because of their capacity for biological nitrogen fixation to enrich the soil.
2. Mixtures provide the species diversity needed to exploit the diversity of soil conditions for high productivity and uniform stand.
3. Species differ in their growth and dormancy characteristics; hence, mixtures ensure that herbage is available uniformly over the season.
4. Mixtures of species (grasses and legumes) provide more balanced nutrition to grazing animals than do monocultures.

14.7: SELECTING PASTURE SPECIES

The producer should determine which species best fit his or her forage program. The selected species should be adapted to the production region (climate, soil) and have high yield potential, so that it can produce optimally in response to the pasture management to be imposed. The species should have desirable longevity for sustained high productivity over many seasons if it has to be successful for establishing a permanent pasture.

FIGURE 14–3 Cattle grazing on tame pasture. (Source: USDA)

The species chosen should not only be palatable to the livestock but also be of high nutritional quality. They should be tolerant to the grazing management or system to be used, so that the balance in the mixture is maintained for a long time. To this end, the species in a mixture should be compatible, so that none of the species are at a competitive disadvantage. It is important that the species in a pasture be resistant to the major diseases and insect pests in the production area to reduce or eliminate the need for the use of pesticides that could interfere with grazing cycles or even be toxic to livestock.

Although mixtures are desirable for the seasons presented in this section, there are some disadvantages to having legume-grass mixtures. First, many legumes cause bloat in ruminants and hence should not constitute more than 50% of the mixture. Second, complex mixtures present challenges to pasture management, because no two species have identical requirements for management inputs (fertilizer, water, grazing scheme). Third, more palatable species are predisposed to overgrazing and subsequent elimination from the stand. Two-species mixtures are most manageable. The recommended pasture mixtures for various regions in the United States are summarized in Table 14–4. The pasture composition should be adjusted to fit a particular mix of climate, soil type, management system, and kind of livestock to be grazed.

Table 14–4 Adapted Pasture Plants and Recommended Pasture Mixtures for Production Regions of the United States

Northeast
- Grasses: orchardgrass, bromegrass, timothy, tall fescue, Kentucky bluegrass, perennial ryegrass
- Legumes: Ladino clover, alfalfa, red clover, birdsfoot trefoil, white clover, Korea lespedeza
- Seeding rate: 20–25 lb/acre

Northern Great Plains (Eastern)
- Grasses: crested wheatgrass, bromegrass, wild rye
- Legumes: alfalfa
- Seeding rate: 12 lb/acre

Northern Great Plains (Western)
- Grasses: crested wheatgrass, blue grama, buffalograss, sideoat gramma, western wheatgrass
- Legume: sweetclover
- Seeding rate: 10 lb/acre

Northern Intermountain (Irrigated)
- Grasses: bromegrass, orchardgrass
- Legumes: alsike clover, Ladino clover, alfalfa
- Seeding rate: 16–25 lb/acre

Northwestern (Irrigated)
- Grasses: bromegrass, tall fescue, orchardgrass, tall oatgrass
- Legumes: alfalfa, birdsfoot trefoil, alsike
- Seeding rate: 16–25 lb/acre

Southwestern (Irrigated)
- Grasses: orchardgrass, perennial ryegrass, tall fescue
- Legumes: Ladino clover, alfalfa, burclover, sweetclover
- Seeding rate: 13–30 lb/acre

Western Rangelands
- Grasses: crested wheatgrass, bluestem wheatgrass, bromegrass, tall oatgrass, slender wheatgrass, wild rye
- Seeding rate: 5–13 lb/acre

Southwestern Rangelands
- Grasses: crested wheatgrass, western wheatgrass, bromegrass, ryegrass, galletagrass, weeping lovegrass
- Legumes: yellow sweetclover, subterranean clover, alfilaria
- Seeding rate: 8–10 lb/acre

Pacific Northwest
- Coastal wetlands
  - Grasses: tall fescue, reed canarygrass
  - Legumes: birdsfoot trefoil, big trefoil
  - Seeding rate: 12–20 lb/acre
- Humid uplands
  - Grasses: perennial ryegrass, orchardgrass, tall oatgrass, tall fescue
  - Legumes: big trefoil, birdsfoot trefoil, subterranean clover, New Zealand clover
  - Seeding rate: 14–20 lb/acre
- Great Basin semiarid
  - Grasses: crested wheatgrass, slender wheatgrass, bromegrass, big bluegrass
  - Seeding rate: 6–12 lb/acre

Southern Great Plains
- Grasses: blue grama, sideoat grama, sand lovegrass, sand bluestem, switchgrass, indiangrass, little bluestem, weeping lovegrass
- Seeding rate: 9–19 lb/acre

Southeast
- Grasses: coastal bermudagrass, tall fescue, dallisgrass, bahiagrass, St. Augustinegrass
- Legumes: lespedza, white clover, hop clover, crimson clover, burclover, kudzu
- Seeding rate: 9–32 lb/acre

Source: USDA.

14.8: ESTABLISHING A PERMANENT PASTURE

14.8.1 LAND PREPARATION

Land preparation is necessary if the pasture is to be irrigated, in which case the land is prepared according to the types of irrigation system to be used.

14.8.2 SEEDBED PREPARATION

Seedbed preparation is important for seeding all crops but is even more so for small-seeded species, as are most pasture species. The bed should be of fine tilth to provide effective seed-soil contact for rapid germination and stand establishment. Seeding into a moist soil is desirable for a more uniform stand. Irrigating a field that has been seeded dry causes seed movement and crusting, which might impede the germination of tiny seeds.

14.8.3 TIME OF SEEDING

The soil temperature should be appropriate. Cool season grasses are best sown in early fall. They germinate well when soil temperature reaches 45°F. Warm season grasses germinate when soil temperature is 55°F or higher. Hence, they should be sown in spring and up to midsummer. This will allow an establishment of a good root system and upper vegetative material before the freezing temperature arrives. Weed competition is usually less in fall than in spring. Similarly, plant water needs are less in fall (reduced evaporation and transpiration) than in spring. However, fall planting is more prone to winter freeze damage because the seeding date is close to the first killing frost.

14.8.4 METHODS OF SEEDING

Sprigging Establishing a grass stand using live vegetative plant parts derived from rhizomes, stolons, or roots.

Most pastures are established by seed. However, some species, such as bermudagrass, are best established by vegetative methods (stem or root cutting or sod pieces). Seeding may be by broadcasting or drilling, the latter giving more uniform seed distribution and coverage. Vegetative propagation may be done by using sprigs (called **sprigging**). The material is scattered by a sprigging machine and then lightly disked or rototilled into the soil. The method of seeding notwithstanding, it is best to pre-irrigate the soil, so that seeding can be done into moist soil to avoid seed displacement from post-planting irrigation. Drilling or broadcast operations may be conducted simultaneously with fertilization. Sometimes, producers seed the pasture species together with a nurse crop (e.g., a small grain, such as wheat, oat, or barley) to suppress weed growth. This practice is not recommended for fall seeding. The field should be watered frequently during the first 7–15 days for quick germination and stand establishment.

14.8.5 SEEDING DEPTH

Being mostly small-seeded, pasture legumes and grasses are seeded at a shallow depth of between 1/4 inch and 3/4 inches.

14.8.6 FERTILIZATION

Pasture yield, palatability, and nutritive role can be increased by proper fertilization. Furthermore, proper fertilization improves stand life, weed control, disease and insect pest control, and water use efficiency. Most pastures respond to nitrogen, phosphorus, and potassium fertilization. A ton of dry matter in legume hay contains about 50–70 lb of nitrogen, 5–20 lb of P_2O_5, and 20–25 lb each of K_2O and CaO. Nitrogen applications hasten vegetative growth so that pastures are grazable earlier than normal. It also increases the protein content. Fertilization can also be used to alter the botanical composition of grass-legume mixed pastures. Low nitrogen application and high phosphorus application tends to promote the dominance of legumes over grasses. Pure stands of grasses can respond to very high amounts (300–400 lb/acre) of nitrogen and high amounts (60–90 lb/acre) of phosphorus fertilization. Tall fescue and other cool season grasses require 200–250 lb/acre of nitrogen and 60–80 lb/acre of P_2O_5.

14.8.7 IRRIGATION

About 40–60 acre-inches of water are required annually to produce a good crop of irrigated pasture. The amount is dependent on prevailing temperature, humidity, wind velocity, soil types, and plant species. The consumptive use of most plants is between 0.1 and 0.4 inch/day. Pasture irrigation is done commonly by flood (border) or sprinkler methods. The frequency of irrigation depends on soil texture, being more frequent (7–10 days at 1–3 acre-inches) on sandy soils and less frequent (10–15 days) on medium-textured soils.

14.8.8 WEEDS

The key to keeping weeds out of pastures is good management that includes proper choice of species, the proper planting density, good fertilizer and water management, and the proper grazing system. The pasture should be well established and well maintained such that the ground is well covered with vigorous plants. It should not be overgrazed. Sometimes, it is necessary to use herbicides to control weeds, especially perennial weeds.

14.9: RANGELAND AND PASTURE MANAGEMENT CONCEPTS

As previously discussed, rangelands consist of mature plants, whereas tame pastures consist of introduced species. Introduced pastures are generally managed intensively (i.e., high cost/acre of inputs), whereas rangelands are generally managed extensively (at low cost/acre of inputs). Consequently, the management approach to introduced pastures is different from the approach to rangelands. However, certain fundamental physiological principles pertaining to plant behavior apply equally to both types of grazinglands:

1. Forage has its highest nutritional quality (especially highest protein content) at the youngest stage of growth. Consequently, forage producers must compromise between achieving high forage yield and achieving high forage quality.
2. Excessive grazing adversely impacts leaf and root development, reducing the two organs proportionally as the root carbohydrate reserves are depleted.
3. The ability of plants to withstand defoliation is in proportion to their ability to develop side shoots in spite of defoliation.
4. The adverse impact of excessive defoliation is most severe when it occurs when active growth is starting rather than at other times during plant growth. Consequently, plants that emerge very early in the spring are most prone to being defoliated in early active growth.
5. Plants that have a high yielding ability tend to have a high tolerance for abuse.
6. Generally, plants that are erect and more accessible to animals are more susceptible to grazing.

14.10: GRAZING SYSTEMS AND THEIR MANAGEMENT

A grazing system may be defined as the combination of pastures, livestock, fences, and management used to control forage production and harvest (Fig. 14–4). There are two basic types of grazing systems—**continuous stocking** and **rotational stocking**.

14.10.1 CONTINUOUS STOCKING

Continuous grazing (or, more precisely, continuous stocking, since animals do not graze continuously) entails grazing a pasture for a very long time without restriction. The advantages of this system include low fencing costs and minimum daily management. Also, studies show that, at moderate livestock stocking rates, the actual grazing pressure during the critical growing season is light. Furthermore, research also shows that, contrary to belief, continuous grazing does not place grasses in jeopardy of overgrazing. This is because animals tend to select a variety of plants during the critical growing season. At the proper stocking rate, continuous stocking results in good animal gains per head.

Some disadvantages of continuous stocking include a lack of control over the timing and intensity of grazing. Livestock have preferred areas of grazing in the pasture. These

Continuous stocking A grazing management system in which animals are grazed for a very long time on the same grazing management unit without restriction.

Rotational stocking A grazing management system in which two or more pastures are grazed in a predetermined sequence.

FIGURE 14–4 Fencing is a key activity in rotational grazing systems. Fences may be temporary or permanent, and constructed of different kinds of fencing materials. (a) Electric fences may be used for temporarily dividing pasture; (b) Barbed wire fencing is used by some producers. The field on the left is grazed whereas the field on the right is not. (c) A water tank my be located such that four pastures in rotation can receive watering at the appropriate time. (Source: USDA)

(a)

(b)

(c)

areas occur where watering points, forage, and cover are in close proximity. However, where the terrain is relatively flat and the watering points are in close proximity, there is less tendency for livestock to congregate and linger in the most convenient parts of the pasture. Continuous stocking tends to produce poor forage utilization in the spring when plant regrowth is rapid. This tendency may be reduced by adopting a variable stocking rate during the period, using a higher rate in spring and a lower rate in summer. Overstocking under continuous stocking leads to open swards in the field that are prone to erosion and weed infestation.

Certain species are not suitable for continuous stocking. These include alfalfa, timothygrass, and bromegrass pastures. Species that are tolerant of continuous stocking include bluegrass, orchardgrass, tall fescue, white clover, and some birdsfoot trefoil. Continuous stocking is advantageous when the farmer has plentiful pasture and does not intend to increase the herd of livestock. The types of livestock enterprises that can benefit from this system of grazing include dry cows, sheep, growing heifers, and beef cows (especially those with low to moderate milking ability). The stocking rate under continuous stocking should allow adequate forage in July and August.

Individual livestock weight gains on continuous stocking grazing systems in the Great Plains prairie and California annual grasslands have been found to be superior to rotational grazing systems. The reason for this is the fact that continuous stocking allows livestock to maximize forage selectivity, while minimizing livestock disturbance due to behaviors such as gathering and trailing, as well as rapid change in forage quality. Livestock production per unit area is generally higher under continuous stocking than rotational stocking, except under conditions such as humid grassland range types where average annual rainfall is above 500 mm.

14.10.2 ROTATIONAL STOCKING

By definition, a rotational grazing, or stocking, system of grazing pastures is one in which two or more pastures are grazed in a predetermined sequence, with planned rest periods between grazing. Livestock are moved from one pasture to another during a single grazing season. The new period depends on the pasture species and growing conditions.

The advantages of continuous stocking include the control the farmer has over the timing and intensity of grazing. Such a control can enhance forage regrowth and utilization during the grazing period. Plants maintain a healthy growth with better-developed roots, that help them better survive during the summer dry period. The pasture species are better able to compete with weeds. Rotational stocking is a more flexible system, allowing a producer to use livestock to manage pasture species and weeds.

The disadvantages of rotational stocking include the expenses of fence construction and maintenance and the time spent in moving livestock from one paddock to another. This approach to livestock grazing is suited to a producer who desires to increase animal production per unit area. The operator who also wants to reduce production cost by using animals to harvest forage directly, rather than harvesting by machinery, can benefit from this system. Specific livestock enterprises that are suited to this pasture management system include dairy cattle and high-producing beef cattle; sheep can also be managed intensively using this grazing system. Rotational stocking varies in intensity regarding the number of paddocks involved.

14.10.3 SPECIALIZED GRAZING SYSTEMS

Various specialized grazing systems are in use, including deferred rotation, the Merrill three-herd/four-pasture system, seasonal suitability grazing, high-intensity/low-frequency grazing, and short-duration grazing. Rotation is the critical feature of all specialized grazing systems. These systems are useful under conditions such as rugged terrain, poor water distribution, vegetation with low grazing resistance, wildlife needs, and poor precipitation.

1. *Deferred rotation.* In this method of growing, the range is divided into two pastures. Each pasture receives deferred grazing every other year. Every 2–4 years, each pasture receives a deferment. Deferred rotation allows plants and areas to gain and maintain vigor more effectively than continuous stocking systems.
2. *Merrill three-herd/four-pasture system.* Developed in south-central Texas, this grazing system involves three herd and four pastures. At any given time, three pastures are being grazed in a continuous fashion, while the fourth is allowed to recover. Each pasture is allowed a 4-month deferred period, which is rotational among the four pastures. The non-use period allows the pasture to recover. It is suitable for areas where precipitation and effective plant growth can occur at anytime during the year. It allows sustained livestock, forage, and wildlife production.
3. *Seasonal suitability grazing.* In this system, the range is divided into pastures according to vegetation types. These vegetation types are fenced off and integrated into a grazing system based on vegetation (forest, grassland, meadow) and livestock requirements. These vegetation types are used at the times of the year when they are most suitable.

4. *High-intensity/low-frequency grazing.* This system was designed to force livestock to use less palatable or less preferred forage species as well as the palatable ones by stocking at a high density. To accomplish this, three or more pastures are used for grazing one combined herd of animals. The grazing periods are more than 2 weeks, with more than 60 days of non-use. The goal of this system is to improve the pasture, rather than to obtain high animal performance, since the animals are forced to utilize old, coarse, and unpalatable forage.
5. *Short-duration grazing.* This system is also called a rapid-rotation, time-control, or cell grazing system. It works best in areas where there is at least 3 months of good weather for plant growth and more than 25 inches of precipitation. The pasture is intensively grazed for a short period. Such a system should involve no fewer than eight paddocks. A wagon-wheel arrangement facilitates the rotation of grazing.

14.11: POISONOUS PLANTS IN THE RANGELAND

Poisonous plants occur in rangelands. These species contain various toxic principles, including cynogenic glucosides (e.g., in arrowgrass); alkaloids, such as zygadenine (e.g., in deathcamas); delphinine (in larkspur); lupinine (in lupines); equisetin (in horsetail); and many other toxic compounds. Animals turn to poisonous species when the safe, nutritious, and palatable species are scarce in the range. Some of the common toxic plants in the range in the United States are shown in Table 14–5.

14.12: BLOATING

Bloating, a distention of the stomach, is associated with the consumption of certain legumes (e.g., alfalfa, white clover, sweet clover). Grasses and some legumes, such as sainfoin and birdsfoot trefoil, do not promote bloating. Pastures that contain a high proportion of legumes are more likely to cause bloating in livestock. Grazing pure stands of these legumes should be avoided. A good pasture mixture that reduces the incidence of bloating may be 60:40 of

Table 14–5 Poisonous Plants in the U.S. Rangeland

Common Name	Scientific Name
Arrograss	*Triglochin maritime*
Larkspur	*Zygadenus* sp.
Loco	*Oxytropis* sp.
Lupine	*Lupinus* sp.
Milkweed	*Asclepia galioides, A. mexicana*
Poison vetch	*Astragalus* sp.
Water hemlock	*Cicuta* sp.
Snakeroot	*Eupatorium urticaefolium*
Horsetail	*Equisetum* sp.
Sneezeweed	*Helenium* sp.
Crazyweed	*Oxytropis* sp.
Halogeton	*Halogenton glomeratus*

grass:legume. Even though the exact cause of bloating is unknown, the condition occurs when foam forms and accumulates in the rumen of the animal, preventing the release of the gas through belching. Unrelieved gas pressure can suffocate an animal. Animals vary in susceptibility to bloating; the rumen microbiology is variable among animals. Bloating is more common as temperature drops during the year or when the plants grazed are younger. Farmers may use bloat preventatives (e.g, poloxalene) for cattle and nutrient blocks to reduce the incidence of bloat. Also, livestock should be fed dry roughage before they graze a pasture.

14.13: Pasturing Winter Wheat

Livestock may graze winter wheat fields safely until a critical stage; then the animals are removed to allow the crop to flower to produce grain. This cultural practice is common in the Central and Southern Plains. Most of the wheat acreage in Kansas is pastured. Winter grazing occurs between mid-October and mid-December. Moderate grazing does not adversely impact yield. Wheat may be grazed in spring from about March to May, but before the jointing stage starts. Once the growing point is above the ground surface, grazing should be discontinued. Damaged culm tips do not recover. In the Oklahoma and Texas panhandles, wheat pasture poisoning ("grass tetany," as it is called) may occur when wheat is grazed, especially when wheat growth is luxuriant. Cows with calves by their side or in late pregnancy are most susceptible. Injecting animals with calcium gluconate may treat the afflicted cows. To reduce the incidence of poisoning, the cows should be removed from the pasture after a period of grazing.

14.14: Pasture Renovation

Periodic renovation of pastures improves both forage yield and livestock performance. Pasture renovation is done by introducing desirable forage species into an existing pasture (Fig. 14–5) In addition to seeding new materials (e.g., legume, legume-grass mixture),

FIGURE 14–5 Interseeding native grasses into cool season pasture (Source: USDA)

weed control is part of the renewal process of a pasture. Pasture renewal should be a part of pasture management because legumes are relatively short-lived, compared with grasses in a pasture. Consequently, with time, the legume species decline or are eliminated. Nutritional deficiency affects legumes more than grasses.

SUMMARY

1. A grazingland is any land that is grazed or has the potential to be grazed by animals.
2. A rangeland is an uncultivated land that provides a habitat for animals.
3. Thirteen specific types of range vegetation are identified in the United States.
4. Pastures may be native or tame.
5. Pastures may be established as monocultures or pure stands.
6. Rangelands are managed as natural ecosystems.
7. Stocking systems may be continuous or rotational.
8. Pastures can be renovated after a period of time.
9. Using certain legumes in the pasture may cause bloating in animals

REFERENCES AND SUGGESTED FURTHER READING

Holecheck, J. R. 1984. *Comparative contribution of grasses, forbs, and shrubs to the nutrition of range ungulates.* Rangelands 6:245–248.

Holecheck, J. R., R. D. Pieper, and C. H. Herbel. 5th ed. 2004. *Range management: principles and practices.* Upper Saddle River: Prentice Hall.

Lorenz, R. J. and G. A. Rogler. 1962. *A comparison of methods of renovating old stands of crested wheatgrass.* J. Range Management. 15:215–219.

SELECTED INTERNET SITES FOR FURTHER REVIEW

http://www.rangelands.org/srm.shtml

Site of the Society of Range Management.

http://www.ext.colostate.edu/pubs/natres/06105.html

Glossary of range management terms.

http://rangelandswest.org/

Detailed site on various rangeland topics.

OUTCOMES ASSESSMENT

PART A

Answer the following questions true or false.

1. T F   Consumption of standing forage by animals is called grazing.
2. T F   Rangelands are managed as natural ecosystems.
3. T F   Increasers are the most desired plants at a range site.

4. T F   A range condition classified as excellent would have up to 25% of the original vegetation.
5. T F   Using certain legumes in the pasture may cause bloating of livestock.
6. T F   Fencing costs are less in rotational grazing than in continuous grazing systems.
7. T F   Tame pastures may be established as mixtures or monocultures.

PART B

Answer the following questions.

1. Please define the rangeland. _____

2. Distinguish between grazingland and pasture. _____

3. A pasture can be native or _____ .

4. There are two basic kinds of stocking in a grazing system _____

   or _____

5. Give the advantages of rotational stocking. _____

6. Discuss poisonous plants in the rangeland. _____

7. Define the terms range site and range condition. _____

8. Define a grazing system. _____

PART C

Write a brief essay on each of the following topics.

1. Discuss continuous stocking in a grazing system.

2. Discuss the issue of bloating of livestock in a grazing system.

3. Discuss the classes of plants at a range site.

4. Discuss the practice of rotational stocking in a grazing system.

5. Discuss the selection of pasture species for establishment of permanent pasture.

6. Discuss the practice of pasture renovation.

PART D

Discuss or explain the following topics in detail.

1. Discuss the management of rangelands as natural ecosystems.

2. Give an overview of the diversity of range vegetation in U. S. rangelands.

3. Discuss the establishment of a permanent pasture.

4. Briefly discuss the major specialized grazing systems used in the United States.

15

Tillage Systems and Farm Energy

PURPOSE

Land preparation precedes seeding of a crop. In this chapter, tillage and tillage systems are described. Farm energy and machinery are major inputs of modern crop production. These factors are also discussed.

EXPECTED OUTCOMES

After studying this chapter, the student should be able to:

1. Discuss the purposes of tillage.
2. Discuss tillage systems—conventional and conservation tillage.
3. Describe primary tillage and the equipment used.
4. Describe secondary tillage and the equipment used.
5. Discuss the considerations in choosing a tillage system.
6. Discuss energy use in crop production.
7. Discuss farm machinery and equipment use in U.S. crop production.

KEY TERMS

Conventional tillage	Primary tillage	Tillage
Deep tillage	Secondary tillage	Tillage systems
Fossil fuels	Subsoiling	Tilth
Leveling	Terracing	Zero till
Minimum till		

TO THE STUDENT

Before seed is planted, the land must be prepared in some fashion to provide an adequate seed bed. Land preparation entails removing vegetation and loosening the soil to facilitate seed germination and root penetration. The extent of soil disturbance depends on the type of technology, socioeconomics, knowledge of the producer, and the environment. Plowing is an age-old method of land preparation for seeding. Tillage systems have changed over the years. Some of the old practices have been revisited as more knowledge has become available. For example, crop residue was believed to be harmful to crops and was thus plowed under, leaving a clean field. Currently, various systems of tillage have been developed to encourage leaving stubble or plant debris on the soil surface.

Crop production, in a way, is about energy. The ultimate source of energy in the agroecosystem is the sun. The goal of the crop producer is to harness this energy in the most efficient and effective manner. This is accomplished through implementing appropriate cultural practices. In order to harness solar energy (through photosynthesis), the crop producer expends energy from other sources. As society becomes more technologically advanced, the sources of energy similarly evolve. Energy cost is a major production cost. In this chapter, you will learn how agricultural energy sources have changed over the years. Crop production is facilitated by farm machinery. Not only does it reduce the tedium in field production, but machinery and equipment also allow the cultivation of large acreage and increase the efficiency with which producers conduct their operations. Mechanical technology is expensive to acquire but certainly advantageous. As you study this chapter, note the rationale behind the methods of tillage and the role of technology in the adoption and implementation of a tillage practice.

15.1: LAND PREPARATION

Some forms of land preparation are undertaken before field seeding. A critical aspect of crop production is seedbed preparation. Crop productivity depends upon how the soil is managed. Soil management for sustainable crop production entails six essential practices:

1. Adoption of an appropriate tillage system
2. Maintenance of an appropriate supply of soil organic matter
3. Maintenance of a proper supply and balance of nutrients
4. Control of soil pollution
5. Maintenance of proper soil reaction of pH
6. Control of soil degradation

The first step in preparing a virgin land for crop production is clearing it to remove existing vegetation, either completely or partially. Removing the natural (climax) vegetation should be done properly to avoid soil degradation from soil erosion. The amount of vegetation to be removed depends on the production region (grassland, shrubland, forest) and the production system to be adopted (mechanized or non-mechanized). It also depends on the crop to be produced. In developing countries, the bush is usually cleared and burned. Some trees may be left standing intact or cut down to the trunk. In more advanced economies, where production is highly mechanized, not only are trees cut down, but also the stumps are removed. All obstructions in the soils (e.g., roots and rocks) are removed and burned or carted away. Obstructions in the soil interfere with the operation of tillage implements. Once the original vegetation is cleared to the desired extent, the land is ready to be prepared for seeding.

One of the land preparation activities that is undertaken on a virgin land in some cases is *terrain modification*. Terrain modification may be undertaken even on cultivated land. This is undertaken for several purposes, including the following:

Leveling. A terrain modification procedure for rendering the land surface even and level to reduce surface runoff.

Terracing. Land preparation involving terrain modification through reduction in slope, designed to prevent accelerated erosion and to facilitate the use of the land for cropping.

Tillage. The mechanical manipulation of soil for any purpose.

Tilth. The physical condition of a soil as related to its ease of tillage, fitness as a seedbed, and impedance to seedling emergence and root penetration.

1. **Leveling.** The land may require leveling to improve surface drainage, to install irrigation equipment, or to facilitate the use of farm machines and equipment. Leveling should be done professionally to conserve the fertile topsoil.
2. **Terracing.** Cropping steep slopes pre-disposes the soil to rapid deterioration from soil erosion. In mountainous production areas and places where undulating relief exists, the land may be terraced. Terracing makes it easier for farmers to produce crops on steep slopes.
3. Poor drainage may be corrected before crop production is commenced. This may be accomplished by surface drainage methods or underground methods.

Before seeding, the land is prepared by the process of tillage. **Tillage** is the manual or mechanized manipulation of the soil to provide a medium for proper crop establishment and growth. There are several purposes of tillage, the major ones being

1. *Land leveling.* Land is leveled for several purposes, as previously discussed.
2. *Seedbed preparation.* In order for proper germination to occur, the seed must make good contact with the soil to be able to imbibe moisture. Seedbed preparation is done according to the seed characteristics, especially size. The fineness of the soil after tillage is called its **tilth**. A fine tilth is required for seeding small seeds.
3. *Incorporation of organic matter and soil amendments.* Stubble left after crop harvesting can be mixed in the soil to improve its physical characteristics. Fertilizers, organic and inorganic, and soil amendments, such as lime, may be added to the soil during preparation prior to seeding.
4. *Weed control.* Weeds compete with crop plants for growth factors and may harbor diseases and insect pests. Weeds are controlled at various stages in crop production.
5. *Improved soil physical conditions.* Soil texture and structure are important in crop production. Soil structure can be destroyed with time for a variety of reasons. Heavy traffic (by farm vehicles, farm animals, and humans) can compact the soil and create an impervious soil barrier called a pan. Tillage can be used to break up the hard pan for crop root growth and development.
6. *Erosion control.* Tillage may be conducted in a certain way to provide a rough soil surface to impede the actions of the agents of soil erosion (conservation tillage).
7. *Shaping of soil.* Tillage is used to create raised beds for planting or to create furrows for irrigation.

15.2: TILLAGE SYSTEMS

Tillage systems. The nature and sequence of tillage operations used in preparing land for planting a crop.

Is it always necessary to plow the land and make seedbeds before planting a crop? **Tillage systems** describe the nature and sequence of tillage operations used in preparing a seedbed for planting. Tillage systems differ in the degree of soil stirring and nature of the finished product. They may be classified into two basic groups, according to the degree of soil stirring—conventional tillage and conservation tillage.

Earlier in this chapter, two models of agriculture (conventional or industrial versus sustainable or conservation) were discussed. One of the key ways in which the two models differ is in the preparation of the land for planting a crop. Tillage is accompanied by using implements, a generic term for basically any tool used in agricultural production but often implied to be equipment attached to power units, such as tractors.

15.2.1 CONVENTIONAL TILLAGE

In **conventional tillage**, the entire field is stirred up to a certain depth (called the plow depth). This requires using various kinds of implements. The final condition depends on the purpose of tillage and the crop to be produced. Conventional tillage incorporates two basic methods: clean till (in which no debris or plant remains are left on the soil surface) or mulch till (some debris left on the soil surface). Mulch till is also a conservation tillage method.

Conventional tillage. The tillage system that leaves less than 15% of the soil covered with plant residue.

Conventional tillage consists of three general steps.

1. The land is first cleared to remove large pieces of debris and trees and shrubs. This is necessary to facilitate the use of tillage implements that are prone to damage from these obstacles or are impeded in efficiency by such materials. Low-growing grasses can be readily plowed under without the need for a pre-plowing clearing operation.
2. Primary tillage implements are then used to bury the remaining plant materials on the soil surface.
3. The conventional tillage operations are completed by conducting a secondary tillage of the soil, using lighter implements to pulverize the clods left over by the primary tillage, to obtain a fine tilth for a seedbed.

During tillage, the soil can be stirred to varying depths. In this regard, there are two general classes of tillage operations—primary tillage and secondary tillage.

Primary Tillage

In **primary tillage**, the topsoil is plowed to a depth of 6 to 14 inches (15 to 36 centimeters) and inverted, burying the vegetation and debris on the soil surface. The soil surface after the operation is rough with clods and unsuited for a seedbed for most seeding operations (Figure 15–1). The time of primary tillage depends on the soil type, soil moisture, climatic conditions, and the time of seeding the crop. Heavier (clay) soils are best tilled in the fall to allow a period of time of exposure of the inverted soil to weather from the freezing and thawing that occur in the season. This improves soil structure for seeding in spring. The depth of tillage also depends on the amount and nature of the plant residue on the soil, the soil type, and the producer's preference. The deeper the depth of tillage, the more the cost of the operation, since more fuel and a higher tractor power will be needed to operate the heavy implements used. As previously described, plowing is the primary tillage operation used to invert the top soil to cover stubble and weeds.

Primary tillage. The mechanical manipulation of the soil that produces a rough finish unsuitable for seeding; usually precedes secondary tillage.

Secondary Tillage

Secondary tillage generally follows primary tillage (plowing) to produce a finer tilth for seeding (Figure 15–2). It is done to a shallower depth (from 2 to 6 inches, or 5 to 15 centimeters). The implement pulverizes the clods left by primary tillage. The producer may also use this tillage for leveling and firming the soil.

Three tillage operations are often conducted in a certain sequence: *plowing, disking,* and *harrowing,* the last two being secondary tillage operations. The sequence does

Secondary tillage. The mechanical manipulation of soil to produce a finer tilth for preparing a seedbed; usually follows primary tillage.

FIGURE 15–1 Primary tillage of the land often precedes secondary tillage. It produces a rough finish that is often unsuitable for direct seeding.

FIGURE 15–2 Secondary tillage produces a finer tilth for direct seeding.

not have a specified time between tillage operations. There may be several weeks between consecutive operations in the sequence. Disking pulverizes the clods that remain after plowing. In certain situations, the producer may skip plowing and prepare the soil by disking alone for seeding. When necessary, harrowing follows disking, an operation that further pulverizes the topsoil for the finest tilth for seeding.

Disking and harrowing may be done in one operation by hitching the appropriate implements together. In certain cultural systems, crops are planted in rows and on beds or ridges. Where raised beds are needed, *listing* is an operation that may follow primary tillage to create these raised beds. In practice, a plow equipped with two moldboards may be used in one operation to create the raised beds.

Conventional tillage has advantages and disadvantages, including the following:

Advantages

1. Even though tillage may cause compaction, it is the most convenient method of managing soil compaction when it occurs.

2. It is easier to apply fertilizers and perform other agronomic operations when the land is clean.
3. The lack of crop residue on the soil surface reduces the opportunity for overwintering/oversummering of pests.

Disadvantages

1. *Erosion.* The soil is left clean and exposed to agents of soil erosion.
2. *Compaction.* Several trips are required by the machinery over the land during the operations, thus predisposing the soil to compaction. Excessive and repeated use of primary tillage implements at the same depth places pressure on the soil in that region, resulting in compaction, creating a plow pan. Pan formation means the producer may have to periodically conduct a deep plow tillage operation to dislodge and break up these obstructions in the soil.
3. *Cost.* Conventional tillage is expensive, requiring different implements and several energy-consuming passes over the field to complete the job.
4. *Soil organic matter loss.* Soil organic matter decreases over time.

15.2.2   CONSERVATION TILLAGE

The other basic tillage system is called *conservation tillage;* it entails practices in which some crop residue remains on the soil surface after the operation. The chief goals of this tillage practice are to reduce soil erosion and conserve moisture. There are various types of conservation tillage practices that vary in the degree of soil disturbance and the amount of crop residue on the soil surface. They are sometimes called crop *residue management* systems and are widely used in U.S. crop production (Figure 15–3). These types of tillage systems are not always clearly distinct from each other. They may overlap in certain ways. The methods of managing crop residue are variable among the types of conservation tillage.

The common types of conservation tillage include zero tillage, mulch tillage, strip tillage, ridge tillage, and minimum tillage.

Zero Tillage

Zero tillage (also called *no-till, direct drilling,* and *direct seeding*) describes a practice in which soil disturbance is limited only to the spot where the seed would be placed. Weeds are controlled by herbicides prior to seeding. For example, corn may be seeded into a wheat cover crop (grown to protect the soil and to retain leachable nutrients), or soybeans may be seeded into grain stubble (Figure 15–4).

Zero tillage.   A system of cropping whereby a crop is seeded directly into a seedbed not tilled since the harvest of the previous crop.

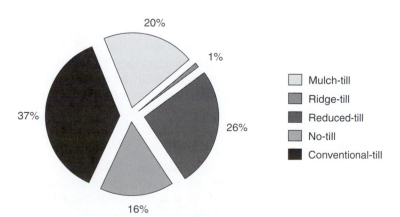

FIGURE 15–3 Conservation tillage, or crop residue management practices, accounted for about 84% of all practices used for preplant land preparation in the United States in 1997. Reduced-till was the most dominant method of land preparation for crop.

Mulch-till
Ridge-till
Reduced-till
No-till
Conventional-till

20%
1%
26%
16%
37%

FIGURE 15–4 No-till systems involve sowing into fields without plowing. This may involve planting into fields with decaying plant material or into recently harvested field, such as seeding soybean into recently harvested small grain field. (Source: USDA)

Mulch Tillage

Mulch tillage systems vary, but the common objective is to leave crop residue to serve as mulch. A variety of implements are used to incorporate a part of the crop residue into the soil, the remainder left on top. One type of mulch till is stubble-mulch till, in which the goals are to conserve moisture and to protect the soil from wind and water erosion by leaving crop residue on the soil surface. Implements such as sweep plows, chisel plows, stubble mulch plows, and rod weeders are used in various mulch-tillage systems. This tillage is commonly used in dryland wheat production areas of the Great Plains and the Pacific Northwest states. These areas experience low rainfall and thus producers frequently include a fallow in the cropping system. The stubble reduces erosion from wind and water during the period in which the land is not cropped and weed-free.

Strip Tillage

Strip tillage (also called *strip-till* or *zone till*) entails the disturbance of narrow strips in the soil where seeding is done. The interrow zone remains undisturbed and covered with crop residue. It is the soil preparation stage in strip cropping, a practice that was previously discussed. This tillage system promotes drying and warming of the soil and hence is useful where the soil is poorly drained.

Ridge Tillage

In ridge tillage, a small band of soil on the ridge is tilled (Figure 15–5). The soil from the top of the ridge is mixed with crop residue between ridges. The debris reduces soil erosion and increases water retention. The soil is left undisturbed from harvest to planting, except fertilizer application. Crops are planted on ridges formed in the previous growing season. The system is also called ridge-plant or till-plant.

Minimum tillage. The land preparation procedure whereby the soil is disturbed only to the minimum extent suitable for planting a crop under existing conditions.

Minimum Tillage

Minimum tillage involves considerable soil disturbance, though to a much lesser extent than that associated with conventional tillage. Some crop residue is left on the soil surface. Minimum till is also called *reduced till*. It is non-specific in terms of soil disturbance and hence not a very meaningful term.

FIGURE 15–5 Ridge-till entails tilling top of a ridge for seeding. (Source: USDA)

Advantages of Conservation Tillage

Conservation tillage has advantages, including the following:

1. *Reduced soil erosion from wind and water.* Crop residue impedes soil surface movements, trapping loose soil and slowing down runoff.
2. *Reduced soil compaction.* Reduced use of tillage machinery reduces soil compaction. Further, the cover provided by crop residue reduces compaction from rain and sprinkler irrigation.
3. *Applicability to steep slopes.* Because the soil is least disturbed and the surface protected, steep slopes can be safely tilled with minimal consequence.
4. *High soil infiltration and moisture conservation because of a large amount of crop residue.* Surface residue reduces evaporation and increases infiltration, thus maintaining higher soil moisture content.
5. *Reduced cost of tillage.* Conventional tillage involves several passes with farm machinery and thus is more costly in terms of fuel use.
6. *Soil temperature moderation.* Crop residue has an insulating effect on the soil. It shades the top soil and reduces temperature fluctuation in this layer. Temperature reduction is advantageous in summer when sunlight is intense. However, in spring, the temperature reducing effect makes soils under conservation tillage cooler, thus slowing down seed germination.
7. *Increased soil organic matter over prolonged periods of no tillage.*

Disadvantages of Conservation Tillage

The disadvantages of conservation tillage include the following:

1. *Dependence on chemicals.* Drastically reduced soil stirring means chemicals are depended upon in no-till operations for weed control. However, much of the herbicide used remains on the crop residue, thus reducing runoff and contamination problems.
2. *Cost.* The conventional equipment used for seeding under conventional tillage systems cannot be used in untilled soil. Special planters are needed for no-till seeding
3. *Higher risk of insect pests and pathogens in early crop establishment because of soil-borne pathogens and soil surface insects.* The high moisture stability favors

the survival of soil pathogens, such as *Rhizoctonia* and *Phythium*. The crop residue provides an opportunity for the overwintering and oversummering of pathogens. Further, insects, rodents, grubs, and other pests thrive on the crop residue.

4. *Susceptibility to leaching.* The higher soil moisture increases the chance of leaching of water-soluble bases (such as NO_3^-). These bases are replaced by H^+ ions. Soils thus tend toward acidity over time.

5. *Obstruction to farm operating.* Crop residue impedes the application of fertilizers.

6. *High risk of herbicide resistance.* High levels of herbicide use increase the opportunities for the development of herbicide resistance. Also, new weed problems may emerge under conservation tillage.

15.2.3   RESIDUE COVER

The amount of crop residue left after harvesting a crop depends on the crop itself, tillage system used in production, method of harvesting, crop yield, and fertilizer application methods. Crop residue may be classified as *fragile* (easily decomposed) such as green peas, potatoes, and vegetables) or *non-fragile* (i.e., difficult to decompose or bury by tillage operation—durable), such as corn, sorghum, barley, and alfalfa.

Crops with high grain yield usually produce high residue on harvesting. Some of the crops with high residue are corn, sorghum, and wheat, in the decreasing order of amounts of residue left after harvesting. Soybean yields relatively less plant residue after harvesting. The harvesting method determines the amount of plant residue after harvesting. For example, corn harvested for silage leaves little residue, whereas corn for grain leaves a high amount of residue. The growing season impacts residue quantity in two ways. It can increase or decrease yield and hence plant residue accordingly, depending on whether it is favorable or not. If the weather is cool, the residue decomposition rate will be reduced; however, warm, wet weather accelerates plant residue decomposition, leaving less plant residue on the soil surface. The amount of cover depends on the number of tillage operations (passes), the implement used (e.g., harrow, field cultivator, chisel plow), and the crop species.

The amount of residue left on the soil surface is determined by the line-transect method. A rope (or tape) with 100 equally spaced knots or marks is stretched diagonally across the crop rows. The percentage residue cover is determined by counting the number of knots or marks that are directly over pieces of crop residue. The procedure is repeated at least three times for accurate estimation of crop cover (Figure 15–6).

15.3: TILLAGE IMPLEMENTS

A wide variety of implements are used in land preparation. Tillage implements differ in size and complexity. Some are as simple as hand-held hoes for small-scale work, whereas others are as complex as some of the tractor-drawn implements used in very large production operations. Conventional tillage implements may be placed into two categories, based on the type of tillage operation. These are *primary tillage implements* and *secondary tillage implements.* Generally, primary tillage implements are used before secondary tillage implements. Further, the use of some of these implements may be switched—that is, primary tillage implements used for secondary tillage.

(a)

(b)

FIGURE 15–6 Residue cover management is a key activity in conservation tillage. (a) The amount of plant remains on the soil surface may be light or heavy. (b) The amount of plant debris is estimated by the line method. A string with 10 knots or markings spaced 12 inches apart is stretched diagonally across the field. The number of knots that fall on plant debris are counted (e.g., 5 = 50 percent). The activity is repeated a number of times over several parts of the field, and the numbers summed and averaged to obtain the percent cover for the field. (Source: USDA)

15.3.1 PRIMARY TILLAGE IMPLEMENTS

Primary tillage implements are usually heavy implements that are used to stir the soil at low depths. Common examples are the moldboard plow, disk plow, chisel plow, powered rotary tiller, sweep plow, and lister/bedder (Figure 15–7).

Moldboard Plow

This plow is designed to cut slices of soil to varying depths and to invert them so they completely or partially bury crop residue on the soil surface. The plow may consist of one moldboard unit or several units rigged together. For best results, the soil should be moist when using a moldboard plow.

Disk Plow

A disk plow differs from a moldboard plow in that a set of heavy concave disk blades are used in place of moldboards. Further, this plow is capable of turning soil only partially. It is often used where a moldboard plow is not effective, such as on virgin soil with stumps, roots, and other obstructions; heavy soil; dry soil; or soil with pan. It is a multipurpose implement that may be used on a soil that already has a fairly good tilth, requiring minimal additional preparation for seeding.

Chisel Plow

A chisel plow is especially suitable for breaking up hardpan. It consists of curved chisels (that differ in size and shape) attached to curved shanks. This implement is unable to cover up residue. Disking may precede its use.

(a)

(b)

(c)

(d)

(e)

FIGURE 15–7 Selected primary tillage equipment: (a) moldboard plow, (b) notched disk plow, (c) offset disk plow, (d) rotary hoe, (e) chisel plow.

Powered Rotary Tiller

The rotary tiller has rotating blades that cut, lift, and mix the soil into a fine tilth for a seedbed. This implement may be used for secondary tillage when operated at a shallower depth.

Sweep Plow

Also called a stubble mulch plow, the sweep plow operates by undercutting the soil at a shallow depth (about 3 to 4 inches, or 8 to 10 centimeters) such that the roots of the plants and the stubble from the previous planting are left in place on the soil surface. This tillage operation is usually used in areas prone to drought. The stubble allows rapid water infiltration and impedes soil erosion.

Lister/Bedder

This is an implement used to create ridges or beds for seeding.

15.3.2 SECONDARY TILLAGE IMPLEMENTS

Secondary tillage implements are used after primary tillage implements (Figure 15–8). They are intended to be used to prepare the soil to a finer tilth for seeding. These implements are operated at shallow depths (4 to 6 inches or 10.2 to 15.2 centimeters). Common secondary tillage implements include the disk harrow, harrows, and the field cultivator.

Disk Harrow

The disk harrow consists of rows of disks dragged over the soil to pulverize clods left after primary tillage to produce a fine tilth.

(a)

(b)

(c)

(d)

FIGURE 15–8 Selected secondary tillage equipment: (a) spike harrow, (b) harrow, (c) field cultivator, (d) chisel plow in operation.

Harrows

There are several designs of harrows that vary especially in the shape and design of the teeth. The teeth may be in the form of tines, spikes, or springs (as in spring-tooth or spike-tooth harrow).

Field Cultivator

The field cultivator is used widely for weed control and seedbed preparation. It is similar in design to chisel plows but much lighter.

Other Implements

Other tillage implements include the rod weeder (undercuts weeds) and cultipacker (firms seedbeds for grasses and small legume seeds).

15.3.3 SPECIAL TILLAGE OPERATIONS

Under certain conditions of crop production, a combination of climate, weather, and cultural practices may create adverse soil physical conditions (namely, compaction) that impede crop growth. A compacted soil drains poorly. To remedy this, some producers use special tillage techniques such as deep tillage and subsoiling.

Deep Tillage

Deep tillage. The mechanical manipulation of the soil at depths below those attainable with ordinary plows, for the purpose of breaking down impervious layers.

Deep tillage is a primary tillage operation in which special and heavy implements are used to stir the soil at depths of 4 to 6 feet (1.2 to 1.8 meters). It is used to break up the plow pan where it occurs. Sometimes, serious floods deposit debris over topsoil. Deep tillage is a technique that may be used to bring the buried topsoil back to the top. In regions where surface salts are a problem, deep tillage can be used to bury the surface salt deposits.

This technique of deep tillage is very expensive. The implements used include larger models of conventional implements such as moldboard plows and deep plowing chisels (such as a *ripper* or a *slip plow* [a V-shaped blade that is dragged horizontally underground]) (Figure 15–9).

FIGURE 15–9 A ripper, or subsoiler, is used to break up impervious layers by stirring the soil at a low depth.

Is it always worth investing in some of these special tillage operations? Deep tillage is not always cost-effective in terms of increasing crop productivity. Further, the soil so tilled frequently returns to its compacted state in about 3 to 5 years.

Subsoiling

Subsoiling is done to a shallower depth than deep plowing. It is done to break up impervious layers located not too deeply in the soil (e.g., plow pan). A chisel-type implement is used for this soil stirring. It improves soil drainage but can revert to its compacted status if tillage implements continue to be used on the soil. Its cost-effectiveness is also debatable.

Subsoiling. The use of machinery to mechanically manipulate the soil at lower depths than deep tillage to break impervious layers.

15.3.4 SELECTED CONSERVATION TILLAGE EQUIPMENT

Conservation tillage equipment is designed to turn the soil at a shallow depth and only to invert it partially. This strategy allows plant remains to be only partially covered (Figure 15–10).

15.4: CHOOSING A TILLAGE SYSTEM

A tillage system should be selected after careful consideration. What should an operator take into consideration in selecting a tillage system to adopt? In selecting a tillage system, the producer has to consider agronomic and economic factors. The conventional tillage system is the standard against which others are compared. The comparison is based on several criteria, including the following: proper seed placement, water infiltration, pest control, soil deterioration, cost of machinery and equipment, and crop yields. Systems that leave large amounts of crop residue on the soil surface tend to interfere with proper seed placement and covering. Similarly, conservation tillage systems interfere with fertilizer placement, while encouraging the growth of perennial weeds. More efficient implements have been developed to overcome some of these problems.

FIGURE 15–10 Sweep plow for conservation tillage

From the economic standpoint, the producer has to consider crop yield, cost of machinery and equipment, and soil conservation. On the basis of yield, conservation tillage practices and other special practices are known to support yield to extents that are comparable to, or slightly lower than, those obtained under the conventional system. However, conservation tillage systems have the added advantage of protecting the soil. Changing from conventional to conservation tillage may require new implements to be purchased at additional production cost.

15.5: ENERGY IN CROP PRODUCTION

15.5.1 ENERGY IN AGRICULTURE MAY BE CLASSIFIED AS DIRECT OR CULTURAL

Solar energy is the most important source of energy in crop production. It is converted into chemical energy by the process of photosynthesis. Photosynthesis is the single most important biochemical reaction in crop productivity. This source of energy is inexhaustible and free to all crop producers. However, since there are factors that prevent the full potential of this energy from being realized, the crop producer has to adopt farming methods for efficient and effective harnessing of solar energy when it is most available. The crop should be planted at the right time during the growing season. The crop cultivar used should have the proper architecture and be properly distributed (spaced) and arranged in the field to intercept most of the sunlight. The cropping environment should enhance the factors that promote photosynthesis (water, carbon dioxide, nutrients).

In addition to solar energy that only has to be harnessed, the producer has to provide *cultural energy* for a variety of production activities. These include tillage, pest control, fertilization, irrigation, transportation, and storage. The energy used this way is described as *direct energy*. On the other hand, large amounts of energy are used in manufacturing or processing food and manufacturing production inputs (agrochemicals, machinery). Energy used in the latter way is called *indirect energy*. The types of direct energy include natural gas, oil, electricity, and coal. Refined petroleum products or **fossil fuels** are used as sources of energy for farm activities such as tillage, planting, fertilizer application, pesticide application, harvesting, and transportation. Electrical energy is used for activities such as irrigation and refrigeration of storage houses.

Fossil fuels. Chemical energy derived from plant remains that have been preserved in the earth's crust.

During the beginnings of agriculture, the primary cultural energy source was human labor. This was available in the form of hired farm laborers, forced labor, and household labor. With time, this source of energy was integrated with animal power (draft) in the form of oxen, mules, horses, and other beasts of burden. With the coming of the Industrial Revolution, farm energy was dramatically transformed from human and animal power to mechanical power. Farm machines depend on fossil fuels to operate. Mechanized production meant larger acreages could be farmed. The average farm size in the United States increased while the number of farms declined as the smaller operators, unfortunately, became increasingly less competitive.

Food production soared to new heights. However, the dependence on petroleum energy for crop production meant that the cost of production was dependent on the dynamics of the petroleum supply. In the early 1970s, there was a period of shortages and threats of shortages of petroleum products that translated to higher fuel prices and a rising cost of crop production. Therefore, there was a move to abandon the trend of intensive cultural energy inputs. There was an urgent need to discover more efficient ways of using farm energy in order to make farming more profitable.

15.5.2   U.S. CROP PRODUCTION USES A LARGE AMOUNT OF ENERGY

By 1978, U.S. agricultural energy use had peaked at 2.2 quadrillion Btu. The oil price hikes and attendant problems that lasted from the 1970s through the early 1980s compelled farmers to adopt energy-saving production practices. They switched from gasoline-powered equipment to more fuel-efficient diesel-powered engines. Multi-function machines (combines, planter/fertilizer combos, etc.) were adopted, while conservation tillage operations such as zero or minimum tillage were also considered. Energy-saving methods of irrigation, drying, and storing of products were adopted. This resulted in a decline of 25% in the use of energy by agriculture between 1978 and 1993. There was also a significant change in the types of energy used during this period. Gasoline use declined from 42% in 1965 to 11% in 1993, while diesel use increased from 13 to 29% (Figure 15–11). However, this decline in energy did not adversely affect agricultural output. Between 1978 and 1993, agricultural output increased by 47%.

Historically, it is estimated that for each 1% rise in U.S. imported crude oil, there is a corresponding 0.7% rise in farm price of fuel (gasoline and diesel). World oil price dynamics in the past three decades have been impacted by world crises. Hikes occurred during the Arab oil embargo (1973 to 1974) and the Iranian crisis (1979). An oil glut caused a sharp drop in prices in 1986. This was followed by a rise in prices after the Persian Gulf War (1990 to 1991). Trends in fuel use on the farm are presented in Figure 15–12. In 1994, the farming regions that used the most fuel were the Corn Belt, the Northern Plains, and the Great Lakes states, in that order.

Fossil fuels sometimes not only are expensive but also produce pollutants that are harmful to the environment. Exhaust fumes from automobiles are implicated in global warming and the depletion of the ozone layer. Smog that hangs over certain cities is largely due to automobile emissions. Increasing efforts are being made to decrease pollution by building automobile engines and accessories that produce less pollution, as well as fuels that burn much cleaner.

Agricultural biomass consisting of plant materials and waste products from livestock is being used as a source of energy for various activities. In some underdeveloped

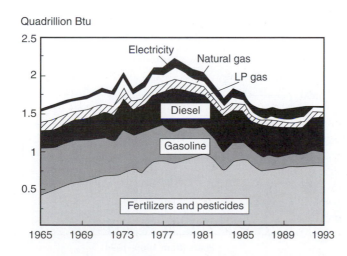

FIGURE 15–11 Composition of energy use in U.S. agriculture showed a change from gasoline to diesel in the late 1970s. Agrochemicals (fertilizers and pesticides) continue to be the dominant component of the energy budget. (Source: USDA)

economies, livestock manure (cow dung) is dried and burned like wood in the fireplace for cooking. Wood is more widely used as a source of fuel in many countries for domestic purposes (cooking, heating) and for industrial use. Wood is partially combusted to produce charcoal for domestic use.

In 1993, U.S. energy from wood accounted for 87% of the total biomass energy consumption, while corn-ethanol accounted for 3%. Also, only about 3.7% of the total energy consumed in the United States came from biomass. The use of biomass to generate electricity received attention in the 1980s but has since experienced slow growth caused by several factors, including low fossil fuel prices and improved technology for more efficient use of conventional fuels with less pollution.

Corn is currently the primary crop in the grain-based *fuel ethanol* industry. Ethanol is obtained by two standard production processes—*dry milling* and *wet milling*. About 71% of the production capacity of ethanol fuel is located in the Corn Belt. Feedstock prices greatly influence the cost of production of ethanol. Corn prices over recent years have ranged from approximately $1.50 to $3.20 per bushel, reaching even as high as approximately $5.00 per bushel in 1996. Ethanol producers derive income from the byproducts of crop production (e.g., oil, corn meal). Production of ethanol fuel is on the rise (Figure 15–13). This situation was caused by several factors. The world energy

FIGURE 15–12 Most of the fuel used in agriculture is accounted for by production in the Corn Belt and the Northern Plains. The Corn Belt spent a total of $1.02 billion on farm fuel in 1994. The Delta states had the lowest farm fuel bill of $281,000. (Source: USDA)

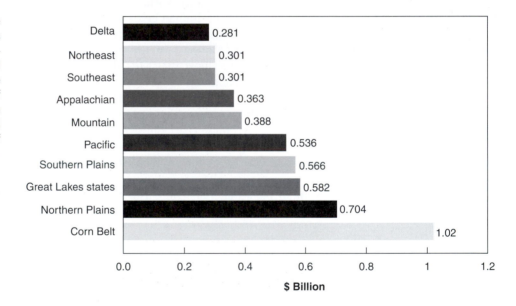

FIGURE 15–13 Fuel ethanol production continues to increase from just a few thousand barrels in 1980 to over 1 billion barrels in the 1990s. Most of the production capacity for fuel ethanol is in the Corn Belt and the Northern Plains. (Source: USDA)

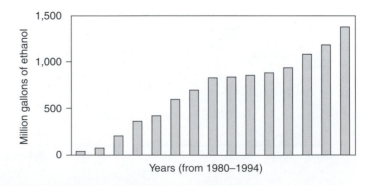

crises and activism against environmental pollution compelled the government to pass legislation that favors the development of alternative energy sources. The Environmental Protection Agency (EPA) in 1990 began a phase out of lead in gasoline. This led to the consideration of ethanol as a gasoline extender. Also, the EPA's passage of the Clean Air Act Amendment of 1990 established the Oxygenated Fuels Program and the Reformulated Gasoline Program, which benefited the development of alternative fuels. These programs were designed to control carbon monoxide pollution and to curtail ground-level ozone problems. Ethanol and its derivative ethyl tertiary butyl ether are among the top three fuel enhancers (the third being methyl tertiary butyl ether, which is derived from methanol from natural gas). The chemicals raise the oxygen levels of the fuel to EPA standards of 2.7% (by weight) for gasoline and 2.0% for reformulated gasoline.

Energy efficiency in agriculture may be improved in several ways, including the following:

1. Use of energy efficient farm machines (diesel instead of gasoline)
2. Development and use of efficient cultural practices (e.g., zero tillage)
3. Crops that require minimum processing before sale or use
4. Plants that have high photosynthetic efficiency

15.6: FARM MACHINERY AND EQUIPMENT

Farm machinery and equipment make large-scale production feasible by enabling the cultivation of large acreages and reducing the tedium in field production. They have also increased the efficiency of farmers in conducting various cropping operations. As technology advances, farm machinery and equipment also become increasingly more complex. Productivity gains in U.S. agriculture are attributable in part to the increasing complexity in farm machinery and equipment. As expected, the prices of farm machinery and equipment rise with their complexity. Frequently, complex machines also are bigger in size.

The evolution of mechanical technology is necessitated and prompted by advances in agronomic technologies. Both technologies are fueled by research. For example, the increasing emphasis on crop residue management in crop production has led to a surge in interest in associated technologies such as conservation tillage and no-till. These agronomic practices have led to the development of new mechanical technology such as *air drill* and *coulter chisel plow.* The cutting-edge technology of GIS and GPS has been incorporated into the design and manufacture of machinery and equipment to implement the agronomic technology called variable rate technology (as embodied in the technology of precision farming).

The major farm machinery and equipment associated with crop production are listed in Table 15–1. There are various types of equipment for various operations at all stages of field crop production, from land preparation to harvesting (Figure 15–14). Of all of these, the tractor and combine sales are indicators of the general farm machinery economy. The tractor is perhaps the single most important piece of farm machinery. It is available in various horsepower classes (Table 15–2). Land preparation, planting, fertilizing, and pest control operations are performed with implements attached to the tractor. Trends in tractor and combine sales are shown in Figure 15–15.

The demand for machinery is affected by such factors as machinery prices, interest rates, farm equity, and farm income. Equity (assets minus debt) determines the collateral a farmer has to back a loan taken out toward machinery purchase. Farm income (cash receipts minus production expenses) indicates the cash flow available for use by the farmer

Table 15–1	Major Farm Machinery and Equipment
Tractors	Tillage equipment
Wheel tractors	Moldboard plows
Crawlers	Disc and other plows
Combines	Harrows
Self-propelled combines	Cultivators
Others	Seeders and planters
Trucks, trailers, wagons	Fertilizing equipment
Haying equipment	Spraying equipment
Balers	Irrigation equipment
Mowers	Cleaning and grading equipment
Others	Agricultural engines

(a)

(b)

(c)

FIGURE 15–14 Tractors are the backbone of mechanized crop production. They come in various sizes and complexity, with four wheels (a), six wheels (b), eight wheels (c), and even as crawlers.

Table 15–2 U.S. Farm Machinery Unit Sales in the 1990s

	Units Sold						
Machinery Category	1990	1991	1992	1993	1994	1995	1996
Tractors							
Two-wheel drive							
40–99 hp	38,400	33,900	34,500	35,500	39,100	39,700	41,200
100 hp+	22,800	20,100	15,600	19,000	20,400	20,500	21,400
Four-wheel drive	5,100	4,100	2,700	3,300	3,300	4,400	4,400
Total tractors	66,300	58,100	52,800	57,800	63,200	64,600	67,000
Self-propelled combines	10,400	9,700	7,700	7,850	8,500	9,200	9,000

Source: Extracted from USDA, ERS data.

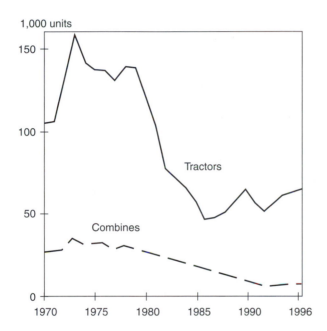

FIGURE 15–15 Tractor sales plummeted to a low in the mid-1980s. Since then, sales have been rising, though not steadily. Similarly, sales of combines declined to a low in the mid-1990s, followed by a slow and erratic rise. (Source: USDA)

to purchase machinery and equipment. Farm machinery nominal and real interest rates lag behind the prime rate. In 1996, the nominal and real interest rates fell to 9.7% and 7.6%, respectively. Farm equity also rose from $705 billion in 1991 to $880 billion in 1996, primarily because of an increase in real estate assets. Farm business debt rose 3% ($4.6 billion) in 1996 but debt to asset ratio decreased 15% from 1995 to 1996, making it easier for farmers to borrow. Net farm income rose 7% in 1996 to $51.7 billion.

Agriculture is a hazardous occupation in the United States. The U.S. Department of Human and Health Services (DHHS) estimates that 26 to 50 workers per 100,000 die each year from farm accidents. This rate far exceeds the average for all industries combined, which stands at only 11 deaths per 100,000. According to the U.S. Department of Census, there were 64,813 farm-related injuries and 673 farm-related deaths in 1992. Machinery and equipment manufacturers continue to incorporate safety measures into their designs. Some of these improvements include rollover protection for tractors, fully enclosed cabs for tractors and combines, and power-take-off (PTO) shields for tractors.

SUMMARY

1. Land preparation for planting may involve terrain modification.
2. Before seeding the land is prepared by tillage.
3. There are two general categories of tillage systems—conventional tillage and conservation or zero tillage—the former involving more stirring of the soil.
4. There are two basic types of conventional tillage—primary and secondary.
5. Different types of implements are involved in the two types of tillage, the product of secondary tillage being finer than primary tillage.
6. Seeding is usually conducted after secondary tillage.
7. Crop production is powered by solar energy and cultural energy.
8. Solar energy is the most important source of energy in crop production. It is free and inexhaustible but controlled by nature. The producer has to devise effective strategies to harness it.
9. Solar energy in crop production is used primarily by direct interception by plants and conversion of the radiant energy into chemical energy by photosynthesis.
10. Cultural energy is provided by the crop producer for operations such as tillage, planting, pest control, fertilization, harvesting, transporting, and storing.
11. Direct energy (consumed on the farm) is provided by natural gas, oil, electricity, and coal.
12. Direct energy has evolved from laborious human labor to animal power to its modern mechanical form.
13. Petroleum is the primary source of farm energy. Over the years, fuel-efficient diesel has replaced gasoline engines in farm machinery. Multi-function machinery (e.g., combines) have reduced energy use in crop production.
14. Energy use in crop production contributes to air pollution.
15. Agricultural biomass in the form of wood, charcoal, and biofuels, such as ethanol from corn, are significant sources of energy for industrial and domestic purposes.
16. Tractors and combines are the heart of mechanized crop production.
17. As agronomic technologies continue to be improved, farm machinery and equipment are developed to facilitate the implementation of the new technologies.
18. Modern agricultural machinery is complex and big.

REFERENCES AND SUGGESTED READING

Brady, N. C., and R. R. Weil. 1999. *The nature and properties of soils.* 12th ed. Upper Saddle River, NJ: Prentice Hall.

Economic Research Service/USDA. 1997. *Agricultural resources and environmental indicators, 1996–97.* Agricultural Handbook No. 712.

Green, D. E., D. G. Wooley, and R. E. Mullen. 1981. *Agronomy: Principles and practices.* Edina, MN: Burgess.

McCarthy, D. F. 1993. *Essentials of soil mechanics and foundations.* 4th ed. Englewood, Cliffs, NJ: Prentice Hall.

Metcalfe, D. S., and D. M. Elkins. 1980. *Crop production: Principles and practices*. 4th ed. New York: Macmillan.

Unger, P. W., and T. C. Kaspar. 1994. Soil compaction and root growth: A review. *Agron. J.* 86:759–766.

SELECTED INTERNET SITES FOR FURTHER REVIEW

http://www.greatachievements.org/greatachievements/ga_7_3.html

Agricultural mechanization—history/timeline.

http://www.farmphoto.com/album2/html/noframe/PO1389.asp

Photos of tillage implements.

http://www.farmequip.com/cihfield.jpg

Photos of various equipment.

OUTCOMES ASSESSMENT

PART A

Answer the following questions true or false.

1. T F   Seeding is usually conducted after secondary tillage.
2. T F   Primary tillage is performed to a lower depth than secondary tillage.
3. T F   A moldboard plow is a secondary tillage implement.
4. T F   Harrows are primary tillage implements.
5. T F   Herbicide use is a major part of conservation tillage.
6. T F   Ethanol is a fossil fuel.
7. T F   Solar energy is converted into chemical energy by plants in crop production.
8. T F   Commercial production of fuel ethanol depends mainly on corn.
9. T F   Most of ethanol production occurs in the Cotton Belt.
10. T F   Historically, for each 1.1% rise in U.S. imported crude oil, there is a corresponding 2% rise in the price of fuel.
11. T F   Biofuels burn cleaner than fossil fuels.
12. T F   Tractors and combines are the most purchased crop production machinery.

PART B

Answer the following questions.

1. Conventional tillage involves two basic methods. Name them.

2. Minimum tillage is also called _____ .

3. What is tilth?

4. Give three specific examples of secondary tillage implements.

5. Give three specific examples of primary tillage implements.

6. The two general categories of energy used in crop production are
_____ and _____ .

7. Give five specific ways in which cultural energy is used in crop production.

8. Give three examples of direct energy in agricultural production.

9. Give the three agricultural regions in the United States that use the most fuel.

10. Give the EPA approved oxygen levels for gasoline and reformulated fuels.

11. _____ and _____ sales are indicators of the farm machinery economy.

PART C

Write a brief essay on each of the following topics.

1. Discuss primary tillage.

2. Discuss the factors affecting the choice of a tillage system.

3. Discuss terracing of land for cropping.

4. Discuss the advantages of conservation tillage.

5. Describe the three general steps in conventional tillage.

6. Describe how a virgin land may be prepared for seeding.

7. Briefly discuss the role of solar energy in crop production.

8. What are biofuels?

9. Discuss the impact of international affairs on fuel use in agriculture.

10. Discuss the environmental impact of agricultural fuel use.

11. Discuss the impact of the Industrial Revolution on agricultural production.

12. In what ways can the efficiency of cultural energy in crop production be increased?

13. Discuss the evolution of sources of energy for agricultural production.

14. Describe how fuel ethanol is produced.

PART D

Discuss or explain the following topics in detail.

1. Can U.S. agriculture be held hostage by world oil crises?

2. How important are biofuels in U.S. agriculture?

3. Is agricultural machinery becoming too complex?

16

Seeds, Seedlings, and Seeding

PURPOSE

The purpose of this chapter is to describe the characteristics of the seed that impact germination and crop establishment and to discuss the factors affecting crop establishment. Methods of seeding are also discussed.

EXPECTED OUTCOMES

After studying this chapter, the student should be able to:

1. Describe the purpose of seed analysis.
2. List and describe the various tests involved in seed analysis or testing.
3. Describe the factors affecting seed longevity.
4. Discuss the factors affecting seed germination.
5. Discuss the two modes of germination.
6. Discuss the factors affecting establishing a crop stand.
7. Describe the methods of seeding field crops.
8. Discuss the factors affecting seeding rate and how it is calculated.
9. Discuss the methods of plant distribution.

KEY TERMS

Broadcasting
Chemical dormancy
Chemical scarification
Direct-seeded
Epigeal (epigeous) emergence
Hypogeal (hypogeous) emergence

Mechanical dormancy
Mechanical scarification
Physical dormancy
Precocious germination
Seed dormancy
Seed drilling

Seed purity
Seed scarification
Seed viability
Seed vigor
Seeding rate

To the Student

Most field crops are propagated by seed. Agronomy of field crop production is often called "seed agriculture." Seed quality is critical to crop establishment. The producer should always use high-quality seed (excellent germination and vigor, etc.) for quick establishment in the field. Poor establishment (sporadic stands with empty spots), caused by poor germination, will reduce crop yield. Seedlings of low vigor grow and develop slowly. This may allow weeds to overtake the crop plants. Poor seed may also cause uneven germination and crop establishment. Hence, the crop will mature unevenly. Poor seed may also harbor diseases and include weed seeds. Simply put, garbage in, garbage out. You literally reap what you sow. If the seed is poor, no amount of management can turn it around for high productivity. On the contrary, by selecting seed of the proper cultivar that is adapted to the production area and has excellent germination and vigor, yield can be greatly enhanced through management of the production environment. In crop production, and from the perspective of the plant, the seed comes first. Crop productivity begins with good crop establishment, which is a factor of seed quality and favorable environment. In addition to seed quality, seeding should be conducted such that land use is optimized. This is accomplished by adopting appropriate plant distribution or spacing in the field. Appropriate seeding enables plants to develop optimally without adverse competition for high productivity. As you study this chapter, pay attention to the importance of plant anatomy, adult plant size, soil properties, technology, and other factors in field seeding and plant distribution. Take note of how seed quality and the germination environment are critical to crop establishment.

16.1: Seed Analysis

A good crop stand starts with good seed. In less developed economies, crop producers often save seed from the current season's crop for planting in the next season. The farmer usually saves seed from plant(s) with the most desirable features (e.g., large cobs, large fruits, large number of fruits). In the case of a cross-pollinated plant (receives pollen from other plants in the vicinity) such as corn, the farmer will have a constantly changing cultivar (genotype) year after year.

In more technologically advanced countries, farmers usually purchase fresh seed for each cropping season. Varietal purity is the same from season to season. In such a case, the producer should purchase seed from a reputable source. In the United States, there are commercial seed breeders and producers for numerous crops. These companies employ researchers who develop cultivars for use in various growing areas.

A good crop stand and establishment depend on the quality of seed planted and the conditions under which the seed was planted. How can one tell if the seed purchased is of high quality? *Seed testing* is the procedure for gathering pertinent information about seed, as pertaining to its capacity for establishing a stand of seedlings. Several procedures exist for seed testing. In the United States, the Federal Seed Act of 1939 requires commercially marketed seeds to display certain basic information on the label that accompanies the merchandise (Table 16–1). This includes cultivar name, germination percentage, percent pure seed, and presence of weed seed (Figure 16–1). Most seed testing labs will provide information on five major aspects of *seed quality*—*viability, purity, vigor, seed health,* and *presence of noxious weed seed.* Seed quality has physiological, genetic, physical, pathological, and mechanical aspects.

Table 16–1   Understanding the Commercial Seed Tag

Item	What It Means
Kind and variety	States variety name; term *mixture* is used where seed has more than one component
Lot identification	Identifies a specific amount of seed of uniform quality associated with a specific seed test
Pure seed	Percent of purity related to the kind and variety of crop indicated
Other crop seed	Percent by weight of other crop seeds as contaminants
Weed seed	Percent by weight of weed seeds present
Inert material	Percent by weight of foreign material such as chaff, stones, cracked seed, and soil
Prohibited noxious weeds	Presence of weed seeds prohibited in variety to be sold (e.g., field bindweed, Canada thistle)
Restricted noxious weeds	Presence of weed seeds that may be present up to an allowable limit (90 seeds per pound of pure seed)—e.g., quackgrass, dodder, and hedge bindweed
Germination	Percent of seed that germinates to produce normal seedlings during a standard seed analysis
Date of test	Year and month in which seed test was conducted
Origin	Source of seed—where seed was produced
Name and address	Name of seed company or seller of the seed

16.1.1   SEED VIABILITY TESTING

Seed viability is the capacity of seed to germinate under favorable conditions. A viable seed is one capable of germinating to produce a healthy, normal seedling. It is not enough for tissues in the seed to be viable; the seed must be able to grow to produce a seedling. There are several viability-testing techniques in use.

Seed viability.   The capacity of non-dormant seed to germinate under favorable conditions.

Standard Germination Test

The *standard germination test* is required before seedmen (commercial seed producers) sell seed to farmers. There are several techniques for conducting this test, including the *petri dish test* and the *rolled-towel test*. Seeds are placed on absorbent material in the dish. Small seeds may be sandwiched between two layers of absorbent material. In the rolled-towel test, seeds are arranged in rows and then rolled up. Other kinds of absorbent material may be used, such as sand and absorbent cotton. All materials and water used to wet the medium should be sterilized in order to reduce the incidence of microbial growth. After placing the seeds in the appropriate medium, the rolled material is placed in a germinator at a relative humidity of 90% or greater and a temperature of 20°C for 16 hours, followed by further exposure for 8 hours at 30°C for one to several weeks.

Seed absorbs water by the process of imbibition (the absorption of liquids or vapors into the ultra microscopic spaces, such as cellulose). Scoring is done by grouping seedlings into different categories as normal, hard seed (no imbibition), dormant seed (imbibition without proper sprouting), abnormal seed (malformed seedling), and dead or decaying seed.

(front)

CAUTION: MAXIM™* (FLUDIOXONIL) & APRON®* (METALAXYL) TREATED
Do not use for food, feed or oil purposes. TREATED AT THE MANUFACTURERS'
RECOMMENDED RATE. Keep out of reach of children.
DO NOT GRAZE UNTIL 30 DAYS AFTER PLANTING
*Trademarks of Ciba-Geigy Corporation.

HAZARD COMMUNICATION DATA
TREATED SEED CORN – CAUTION

Causes eye irritation. May cause skin irritation. Avoid contact with eyes, skin, or clothing. Harmful if inhaled or absorbed through the skin. Wash thoroughly with soap and water after handling. Remove contaminated clothing and wash before reuse. If contact occurs, flush eyes with plenty of water for at least 15 minutes. Seek medical attention if eye or skin irritation persists. Seek prompt medical attention if swallowed. Note to physician: if ingested, induce emesis or lavage stomach. Treat symptomatically. For more information call Pioneer at 1-800-247-6803, extension 4166.

THE FOLLOWING PROVISIONS ARE PART OF THE TERMS OF
PURCHASE OF THIS PRODUCT

Pioneer Hi-Bred International, Inc., (Pioneer) warrants that the seed purchased from it conforms to the descriptions on the label within tolerances, if any, established by law. THIS EXPRESS WARRANTY EXCLUDES ALL OTHER WARRANTIES, EXPRESS OR IMPLIED, INCLUDING ANY WARRANTY OF MERCHANTABILITY AND OF FITNESS FOR A PARTICULAR PURPOSE. Buyer's sole remedy shall be limited to the purchase price of the seed. Buyer agrees to these terms (and the more fully detailed limitation of warranty and use provisions on the bag) as being part of the purchase terms.

Pioneer sales representatives (agents) have no ownership interest in the PIONEER® brand products which they distribute. Consequently, a sales representative has no ability to grant a security interest in those products to his or her creditors.

UNAUTHORIZED EXPORT IS PROHIBITED.

This hybrid is produced and licensed under one or more of the following issued U.S. patents:

4731499	5097093	5220114	5349119	5435564	5491295	5527986
4806652	5097094	5245125	5354941	5436390	5491387	5530180
4806669	5097095	5276265	5354942	5444178	5495065	5530184
4812599	5097096	5285004	5365014	5463173	5495066	5602317
4812600	5157206	5304719	5367109	5476999	5495069	5608138
5082991	5157208	5304720	5387754	5491286	5502272	5608140
5082992	5159132	5347079	5387755	5491289	5506367	5618987
5095174	5159133	5347080	5416254	5491290	5506368	5639946
5097092	5159134	5347081	5434346	and/or other patents pending.

License is granted solely to produce grain and/or forage. For other licenses, contact Pioneer Hi-Bred International, Inc.

PIONEER® brand products are provided subject to the terms and conditions of purchase which are part of the labeling and purchase documents. Pioneer is a brand name; numbers identify varieties and products.
® Registered trademark of Pioneer Hi-Bred International, Inc., Des Moines, Iowa, U.S.A.
©1997 Pioneer Hi-Bred International, Inc

PIONEER HI-BRED INTERNATIONAL, INC.
SUPPLY MANAGEMENT • 6900 N.W 62ND AVE. • P.O. Box 256 • JOHNSTON, IOWA 50131-0256

(back)

Tetrazolium Test

The *tetrazolium test* is a colorimetric test in which a biochemical reaction causes the test solution to change color under certain conditions. Certain enzymes (including dehydrogenases) become active when viable seed (living tissue) imbibe water and start respiration. Hydrogen ions are released when the dehydrogenases act on substrates (seed tissue such as cotyledons). Tetrazolium (2,3,5-triphenyltetrazolium chloride) solution is colorless but changes into a red, insoluble compound called a formazan upon being reduced by hydrogen ions. Respiring and viable seeds will change color to red; dead or nonrespiring seeds remain colorless.

Scoring the results of this test is based on the pattern of staining of the tissue (called topographical staining). Complete staining of cotyledons in dicots, or embryo in monocots, indicates that the seed has germinability. Seeds are still deemed germinable if the extremities of the radicle or scutellum are unstained. Completely unstained embryo or cotyledon is the worst indication of non-germinability.

The tetrazolium test is quick and reliable. Seeds are first soaked in water for 24 hours and dissected longitudinally through the embryo; then the halves are soaked in 0.1% tetrazolium chloride for 1 hour. However, the test is difficult to score and interpret correctly without much experience.

16.1.2   SEED PURITY TEST

Seed purity is the percentage of pure seed (only the seed of the desired kind without contaminants) in the sample tested. Contaminants include the following:

1. *Seed of other crops.* For example, soybean seeds may be found in corn seed if the combine harvester was not properly cleaned after harvesting soybean before corn.
2. *Weed seed.* Especially, the presence of noxious weed seed should be noted. *Noxious weeds* are the most undesirable and difficult to control. They may be placed into two categories: *primary noxious weeds* (also called *prohibited weeds*) and *secondary noxious weeds* (also called *restricted weeds*). Prohibited weeds are intolerable and will disqualify a seed lot from being offered for sale.
3. *Inert matter.* Inert matter includes materials (collectively called *foreign matter*) such as small stones, pieces of wood, and other plant materials.

Seed purity. The percentage of pure seed of a specified kind in the sample tested.

16.1.3   PURE LIVE SEED

Pure live seed (PLS) is the percent of the seed of the desired cultivar that will germinate. It is a function of both percent purity and percent germination. It is calculated as

$$\% \text{ PLS} = [\% \text{ germination} \times \% \text{ purity}] / [100]$$

16.1.4   SEED VIGOR TEST

An operational definition of **seed vigor** may be the properties of the seed that determine its potential for rapid, uniform emergence and the development of normal seedlings under a wide range of field conditions.

Seed vigor. The vitality or strength of germination.

The vigor of seeds is influenced by several conditions, including genetic factors (e.g., hard-seededness and hybrid vigor) and external environmental conditions during seed development and maturity, at harvest and in storage. An environment of high temperature and humidity adversely affects seed vigor.

There are several tests for seed vigor. Examples are the *cold test* and the *accelerated aging test:*

1. *Cold test.* This test places the seed samples on an appropriate medium and then places the setup in a cold environment at 10°C (50°F) for 7 days. After that, it is returned to an environment maintained at 25°C (77°F) for 4 days. The seeds that emerge are counted.
2. *Accelerated aging test.* In this test, imbibed seeds are kept at high temperature (45°C, or 113°F) and high relative humidity (100%) for about 3 to 4 days. The seeds are then placed under optimal germination conditions for germination to occur. Vigorous seeds survive this harsh treatment.

16.1.5   SEED HEALTH

Seed health is an evaluation of the presence of pathogens and insect pests on the seed. Seed health may be evaluated visually (e.g., for changes in testa color, presence of

There are four classes of commercial seed in seed certification. They differ primarily in the quantity of material (seed), and the person(s) responsible for maintaining genetic purity and other quality factors.

1. *Breeder seed.* This is seed in the custody of the plant breeder, or originating institution. This is usually a small quantity of a newly released cultivar. The breeder is responsible for maintaining this material. Depending on the crop, the breeder may have to increase the seed over several years before having enough to pass on to the next step in the certification process (the *foundation seed*).
2. *Foundation seed.* When the breeder seed is received, it undergoes further increase, usually by an agricultural experimental station, as in the case of breeders associated with academic institutions. Commercial seed companies will also conduct this increase in order to ensure genetic purity and identity.
3. *Registered seed.* The foundation seed is usually contracted out to selected farmers called *registered seed growers.* Hundreds to several thousand hectares of the foundation are grown.
4. *Certified seed.* Tens of thousands of hectares of registered seed are increased in this final stage in the seed certification and release process. During the production of this class of seed, certifying agents from, for example, the State Crop

Improvement Association, inspect the crop in the field for diseases, weeds, and off-types or atypical plants. In addition to these, the seed is further subjected to rigorous quality tests to determine the type and number of seeds of other crops, amount of inert matter (e.g., chaff), total weed seed, amount of noxious weeds, and germination percentage. The standards for the certification process differ from one state to another. When the seed lot passes the tests, it receives the stamp of approval and becomes *certified seed.* This seed is what is legal to sell to crop producers.

Protection of plant variety

There are laws in place for scientists to obtain legal property rights (patent) to their creations, in order to protect the investments of commercial seed companies and plant breeders. The *Plant Variety Protection Act of 1970* protects crop varieties from unauthorized use or sale. Biotechnology is enabling scientists to develop cultivars with alien genes in their genetic makeup for various purposes (e.g., disease resistance, seed quality). Often, the development of such genotypes is very expensive. Some companies incorporate unique genetic markers to deter unauthorized use of their genotypes by competitors. Sophisticated genetic tests are available to catch those who illegally use proprietary plant materials.

Seed produced by plant breeders move down a chain of growers who increase the number of seed each subsequent step. The farmer has access to certified seed for planting a crop.

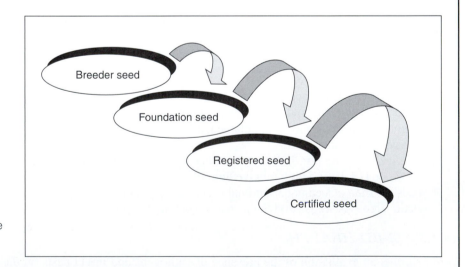

spores, etc.) after incubating on an appropriate medium for disease development. It may be determined by a biochemical test such as Enzyme-Linked Immunosorbent Assay (ELISA).

16.1.6 MECHANICAL SEED DAMAGE

Seed quality is adversely affected by mechanical or physical damage inflicted during harvesting and handling of seed. Damage may include readily visible splits or cracks in the testa or chips of the cotyledon. Sometimes, the embryo is shaken loose from its attachment to the cotyledons. Physically damaged seeds are prone to rotting when planted in the soil, resulting in poor crop stand and establishment. Broken seeds can be sorted out readily and discarded. Subtle damages are more difficult to detect.

Mechanical damage to seed may be evaluated in the laboratory by soaking a sample of seed in 0.1% household bleach (sodium hypochlorite) for 15 minutes. Seeds with cracks in the testa will imbibe the solution, and the testa will separate from the cotyledon.

16.2: FACTORS AFFECTING SEED QUALITY

How long should seed be kept before it is too old to plant? Crop production does not end with harvesting. The quality of seed declines soon after harvesting, some species more rapidly than others. Significant losses can occur at harvest time and postharvest (in storage). For certain uses, it does not matter if the seed is viable or not. However, for planting, seed viability is imperative.

Seed deterioration, just like aging, is not a reversible process. However, it can be slowed down. The whole seed does not deteriorate uniformly. Deterioration starts at the physiological level and manifests itself at the whole-seed morphological level, and eventually in seed germination and seedling growth.

Readily visible signs of deterioration are observable in both the physical and chemical changes in the seed, as well as in performance when tested for germinability. Physically, the seed coat of legumes changes color. It is believed that a combination of high temperature and humidity causes certain oxidative reactions to occur in the testa, resulting in the darkening of the testa. The cell wall weakens and leaks cellular fluids. Deteriorating seed also loses enzyme activity and experiences a reduction in respiration. When sown, such seed emerges slowly and has decreased vigor. Numerous factors have been proposed to be possible causes of seed deterioration, including lipid peroxidation (which involves the release of a free radical), genetic degradation, depletion of food reserves, and accumulation of toxic compounds.

Some storage (however brief) is usually required for harvested seed, whether it is to be used for planting or processing. Decline in certain aspects of seed quality starts soon after harvesting. The rate of decline depends on various factors.

16.2.1 SEED FACTORS

Major factors of decline in seed quality that are caused by seed characteristics are:

1. *Physical condition and mechanical damage.* Seeds that are damaged (e.g., bruised, broken, or cracked) from harvesting and handling operations are more prone to deterioration. Hard-seededness, on the other hand, promotes seed longevity.
2. *Genetic factors.* Characteristics such as hard-seededness and impermeable seed coats are more prevalent in certain species of *Lupinus* and *Lotus*. These species

are known to be long-lived, remaining viable for hundreds of years. On the other hand, species of onion, lettuce, and other agronomic crops are short-lived. Seeds with high oil content are generally known to be short-lived. Cultivars within one species vary in longevity.

3. *Seed maturity.* Physiologically immature seeds generally store poorly.
4. *Nutritional status.* Seed that is deficient in phosphorus, potassium, and calcium stores poorly.
5. *Seed moisture.* Seed moisture content during storage is known to contribute to storability or longevity. Moisture content below 5% and above 14% favor deterioration of seed through respiration and pathogenic attack.

16.2.2 ENVIRONMENTAL FACTORS

Seed longevity is affected by both abiotic and biotic factors.

1. *Abiotic factors.* Relative humidity and temperature are the two most important environmental factors that influence seed longevity. These two factors work together. Temperatures of about 25° to 30°C coupled with high relative humidity of 80% or more cause rapid loss of seed viability in most crop plants. Most crop plant seeds remain viable for over 10 years, when stored at 5°C (41°F) and 45 to 50% relative humidity.
2. *Biotic factors.* Fungi thrive at high relative humidity (90% or higher) and high seed moisture content (30% or higher). In storage, species of *Penicillium* and *Aspergillus* succeed in invading seed embryos and other tissue, causing deterioration.

16.3: SEED GERMINATION AND EMERGENCE

Good seed germination is critical to the establishment of a good crop stand. A seed contains an embryo that is equivalent to a miniature plant. The Association of Official Seed Analysts (AOSA) defines *seed germination* as the emergence and development from the seed embryo of those essential structures that, for the kind of seed in question, are indicative of the ability to produce a normal plant under favorable conditions. The first physical sign of seed germination is the exit of the radicle through the seed coat.

16.3.1 THE PROCESS OF GERMINATION

Seeds will germinate only if three conditions are met:

1. *Viability.* The seed embryo must be capable of germinating.
2. *Appropriate environment.* The environmental conditions required for germination (water, proper temperature, oxygen, and sometimes light) must be adequate.
3. *No primary dormancy.* Seed dormancy should be overcome.

16.3.2 PHASES OF SEED GERMINATION

There are certain general phases of germination:

Phase I—Water Uptake by Imbibition

Most seeds are dry (less than 15% moisture) at the time of planting. Before a seed germinates, it must first imbibe water to about 10 to 100% of its weight, depending on the species. This physical process occurs through the water-permeable seed coat. The initial stages of imbibition are very rapid, occurring over 10 to 30 minutes. This is followed by a slower wetting stage that is linear and occurs over a few to several hours, depending on the seed size (up to an hour for small seeds and 5 to 10 hours for large seeds). During the imbibition process, several compounds are leaked out of the seed (ions, sugars, amino acids, etc.). This is the basis of the electrolyte leakage test of seed vigor.

Phase II—Lag Phase

During this phase, mitochondria rehydrate and become enzymatically active. Protein synthesis occurs while storage reserves are metabolized (by respiration). Cell wall–loosening enzymes are produced to loosen the seed coat to prepare for the exit of the radicle.

Phase III—Radicle Emergence

Radicle (embryonic root) emergence is the first evidence of germination. It is initiated by cell enlargement followed by radicle elongation. Radicle emergence may occur within 1–2 days (in species that germinate quickly). It is followed by the plumule (young shoot).

16.3.3   SEED EMERGENCE

The term *seed emergence* is used to describe the appearance above the soil surface of some of the essential structures of the embryo. There are two basic modes of seed emergence:

1. *Epigeal.* In **epigeal** (or **epigeous**) **emergence**, the cotyledons (seed leaves) emerge above the soil, preceded by a characteristic arching of the hypocotyl (Figure 16–2). The cotyledons gradually change color from creamish to green as they assume photosynthetic functions. After a period, they shrivel and fall off the plant when true leaves appear.
2. *Hypogeal.* **Hypogeal** (or **hypogeous**) **emergence** is characteristic of many grass species. The coleoptile emerges above the ground, leaving the cotyledons or appropriate storage organs underground (Figure 16–3). Can you think of any cultural considerations the producer may make to take into account the type of seed emergence for good crop establishment?

Epigeal (epigeous) emergence. The type of seedling emergence associated with dicots in which the cotyledons rise above the soil surface.

Hypogeal (hypogeous) emergence. The type of seedling emergence in which the cotyledons remain below the soil surface.

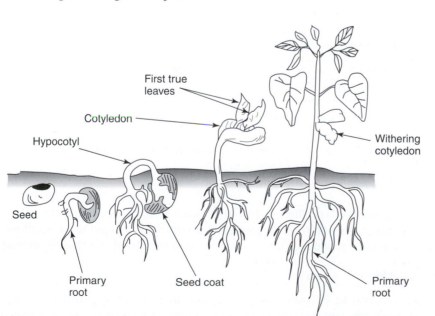

First true leaves

Cotyledon

Hypocotyl

Seed

Primary root

Seed coat

Withering cotyledon

Primary root

FIGURE 16–2 Epigeal, or epigeous, seed emergence is common in dicots. Exceptions include the pea plant. The arching of the hypocotyl characterizes this type of seed emergence.

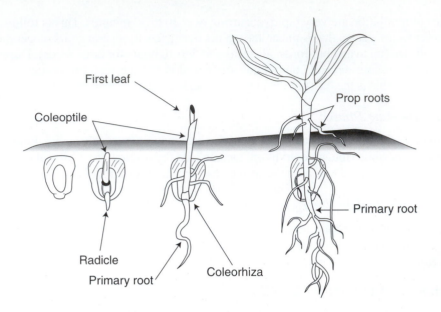

First leaf

Coleoptile

Prop roots

Primary root

Radicle

Primary root

Coleorhiza

16.3.4 SEED GERMINATION IS AFFECTED BY INTERNAL AND EXTERNAL FACTORS

Desirable field germination should produce a stand of at least 90%. Unfortunately, in certain species such as cotton and sorghum, this level of germination is difficult to attain, due to the fact that seeds are prone to fungal diseases. The factors that affect seed germination can be divided into three categories: *seed internal conditions, external abiotic conditions,* and *external biotic conditions.*

Internal Factors

Maturity   The main internal factor that affects seed germination is seed maturity. How mature should a seed be before it can germinate? It is not necessary for most seeds to attain physiological maturity (dry matter no longer added) before they can germinate. However, for field planting, physiologically mature seeds are preferred for a number of reasons:

1. In certain plant species such as soybean, certain critical enzymes for germination (e.g., malate synthase and isocitric lyase) are produced only as the seed slowly matures and dries down.
2. Immature seeds lack the protection provided by the seed coat or testa against pathogens.
3. If immature seeds must be used, they require special storage conditions for protection against environmental damage.
4. Immature seeds are difficult to handle in field planting.

Seed Size   Large seeds produce more vigorous seedlings that help tolerate adverse conditions. In wheat, shrunken seeds germinate poorly and produce weak seedlings such that it is not recommended to sow seeds with a test weight of less than 50 lb/bushel.

External Abiotic Factors

The main external or environmental conditions that influence germination are *water (moisture), air,* and *temperature. Light* is not a universal requirement for all seeds, but certain aspects of this factor are critical to germination of certain species. These environmental factors may interact to produce an effect on germination. For example, the interaction of moisture and temperature produces a very significant effect on germination.

Water (Moisture)   Prior to seed germination, all seeds must imbibe some water. Water is imbibed through the micropyl, through the hilum, or directly through the testa. In sweet clover, the caruncle or strophiole is the entry point for moisture. Water is needed to activate enzymes in the seed to initiate the breakdown and translocation of stored food in the cotyledons and other storage materials of the seed. Depending upon the species, the seed may have to imbibe between 25 and 75% of its dry weight. Sorghum requires about 25% moisture imbibition, while soybean imbibes about 75% of its weight. Even though a certain minimum amount of moisture is needed for germination by all seeds, abundant moisture accelerates the process. By the same token, excessive drought or excessive moisture conditions are detrimental to seed germination.

Seeds may sprout or partially germinate when the environment is highly humid. Under such conditions, certain species, such as soft white winter wheat, experience losses from the phenomenon of **precocious germination**, the sprouting of seeds while still in the head, pod, or cob. When water supply after planting is not adequate or is erratic, seeds may undergo a cycle of wetting and drying. Under the wet cycle, the seed imbibes moisture and begins the germination process. However, the moisture is inadequate to sustain the process. Species such as corn, oats, and wheat can tolerate several sprout-dry-resprout cycles. Unfortunately, the germination percentage is decreased after each episode of germination interruption.

Precocious germination. The premature germination of seed prior to harvesting while still in the fruit and attached to the plant.

Temperature

Temperature Range   Temperature is an important factor whenever biochemical reactions occur. It is a basis for adaptation of plants (cool and warm season plants).

Most seeds will germinate at 15° to 30°C (60° to 86°F), with the optimum being 25° to 30°C (77° to 86°F). The maximum temperature for most seeds is 30° to 40°C (86° to 105.2°F). Generally, cool season species require a lower temperature for seed germination than warm season ones. These requirements become more exacting when seed quality is poor.

Temperature Adaptation   Seeds, just like plants, may be divided into two broad adaptive groups:

1. *Cool-temperature-requiring seeds.* These seeds require temperatures lower than 25°C (77°F) in order to germinate. Examples of crops with this requirement include onion and lettuce. These seeds, however, are not tolerant of cold temperatures as much as species such as carrot, cabbage, and broccoli.
2. *Warm-temperature-requiring seeds.* Such seeds will germinate only when the soil temperature is at least 10°C (50°F). Examples of crops with this requirement include sweet corn and tomato. In some species, including soybean, cotton, and sorghum, a high soil temperature of about 15°C (60°F) is required for proper germination. Warm-temperature-requiring seeds are frost sensitive and may be injured at temperatures below 10°C (50°F) during imbibition.

Temperature Periodicity   Most seeds will germinate normally at constant temperature. However, exposing seeds to alternating cold and warm temperatures during germination is known to enhance both germination and seedling growth. Seeds with dormancy (failure of viable seed to germinate under favorable conditions) factors benefit from this treatment.

Early-maturing cultivars are more susceptible to injury from imbibitional chilling than late-maturing ones. It is suggested that pre-soaking seeds prior to planting may allow growers to plant seeds early in cold soil. In warm-temperature-requiring seeds, early planting may be accomplished by draining the soil to increase aeration for warming. This may be accomplished by planting on raised beds or under mulch.

Moisture and temperature interact in affecting the process of germination and seedling emergence. For best results, seeds must imbibe and hold moisture for a certain period above a critical temperature value, called the *thermal value,* prior to germination. Plant species differ in the duration of thermal time needed for germination to occur.

Air Seed germination is generally an aerobic (requires oxygen) process. A good exchange between the soil air and atmospheric air is necessary to keep soil air composition in proper balance. There should be adequate oxygen in the germination environment in order for physiological processes associated with germination to proceed. The embryo depends on respiratory processes in order to develop and emerge from the seed. The amount of oxygen required depends on the species. A rice species has been reported to germinate under extremely low oxygen (near zero oxygen) content of the atmosphere. On the other hand, carbon dioxide is inhibitory to germination, especially when it occurs in a concentration above 0.03% of the air. Carbon dioxide is a product of respiration. Thus, improper gaseous exchange between the soil and the atmosphere leads to carbon dioxide accumulation for respiration of soil organisms. Such conditions occur under waterlogged conditions when poor soil drainage occurs. The producer may not be able to prevent a temporary anaerobic (lack of oxygen) situation due to excessive moisture from a torrential rain. However, when supplemental moisture is applied through irrigation, the grower should apply the moisture correctly (right rate and amount) to avoid pooling of water in the field. In indoor production under hydroponic or water culture, the nutrient medium needs to be aerated frequently to prevent anaerobic conditions from developing.

Light Light is not a universal requirement for germination as are temperature and moisture. In species that need it, the effect of light on seed germination is manifested through its intensity, wavelength (quality), and duration (photoperiod). Of these three, intensity is least important to germination. Even on cloudy and overcast days, crops in the field receive over 10,000 lux of light. High intensity is desired after seedling emergence for proper growth and development to avoid etiolation or spindly growth.

In terms of wavelength, germination is most stimulated by red light (optimum at 600 nanometers) but inhibited by far red light (at about 700 nanometers). Germination is also inhibited at wavelengths of less than 290 nanometers. *Phytochrome* (a light-sensitive plant pigment) effect on germination was first documented in lettuce seed. When imbibed seeds were alternatively exposed to red and far-red light, germination occurred only when the last treatment the seed received in the sequence was far-red (Figure 16–4). In other words, lettuce seed germination is photoreversible. The effect of photoperiod is associated with phytochrome activation. In carrot, seed germination is inhibited by continuous light.

External Biotic Factors

Seed germination is affected by disease, some of which is encouraged by inappropriate abiotic environmental factors such as high humidity or excessive moisture, coupled with

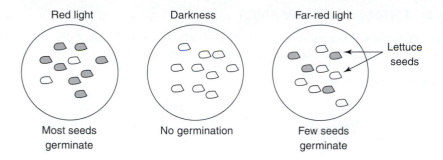

FIGURE 16–4 Phytochrome influences seed germination. Red light stimulates germination, while darkness inhibits it. Far-red light is only mildly stimulatory of loose germination of lettuce seed.

high temperature. Some of the pathogens are seedborne; others occur in the soil or general environment.

The most common disease of germinating seedlings is *damping off*. This fungal disease results in seedling death and is caused by one of several fungi, the most important being *Pythium ultimum* and *Rhizoctonia solans*. Others are *Botrytis cinerea* and *Phytophthora* spp. The last two fungi produce spores that are disseminated by water. Others have mycelia that live in the soil or on seed and other plant materials.

Damping off occurs under warm and humid conditions. Thus, warm-temperature-requiring seeds are the most prone to the disease at the time of germination. A practical way of reducing the incidence of the disease in such crops is to plant when the soil is warm and to eliminate the pathogen during propagation. The seed may be treated with a fungicide. The soil may also be treated directly with a fungicide (e.g., methyl bromide).

16.4: SEED DORMANCY

Apart from appropriate environmental conditions, what other factors influence the germination of viable seeds? The factors affecting seed germination that have been discussed previously assume that the seed is quiescent and will germinate, provided that factors are present in the appropriate amounts or degrees. Sometimes, seed fails to germinate even when provided ideal conditions.

When all the requisite conditions for germination are provided and a still-viable seed fails to germinate, the seed is said to be in a state of **seed dormancy**. This genetic condition is intensified by the effect of the environment and is often associated with wild plants. Thus, it occurs with less frequency in domesticated crop plant seeds. Seed dormancy is a survival mechanism whereby seed-propagated plants in the wild obtain some insurance against extinction from exposure to unfavorable environmental conditions. Seeds fail to germinate unless environmental conditions can support them beyond sprouting and through seedling establishment.

Seed dormancy. The lack of germination of a viable seed because of unsatisfactory environmental conditions or certain factors inherent in the seed.

In the production of domesticated crop plants, this germination-delaying mechanism is generally undesirable, unless in specific instances such as in winter cereals, which are prone to quality-reducing precocious germination. In modern crop production, seeds should germinate on schedule and have uniformity of stand. An advantage in crop cultivation occurs during storage, handling, and transportation.

There are different kinds of seed dormancy mechanisms, based on physiological and physical mechanisms. The mechanisms may be grouped into two categories: *primary dormancy* and *secondary dormancy*.

16.4.1 PRIMARY DORMANCY

Primary dormancy is the most dominant and widely occurring kind of dormancy. There are two forms, exogenous dormancy and endogenous dormancy.

Exogenous Dormancy

The factors involved in exogenous dormancy operate from outside the embryo and manifest themselves primarily through imposing physical barriers between the embryo and the factors for germination (water, temperature, air, and light).

Physical dormancy.
Dormancy imposed by an impervious seed coat that prevents access to the external environmental factors required for germination.

Seed Coat (Physical) Dormancy   Sometimes called **physical dormancy**, *seed coat dormancy* is a form of exogenous dormancy that occurs when the seed covering is significantly modified to render it impervious to external environmental factors. The seed in this form of dormancy is said to be *hard (hard seed)*. This seed hardening occurs by different mechanisms, depending on the species. For example, in leguminous species, suberization (deposition of suberin in the cell wall) of the testa renders it impervious to water and gases. Seed hardness occurs in cultivated herbaceous legumes such as alfalfa and clover. Though genetic, the condition is intensified by environmental factors such as drying at high temperatures during seed maturation.

Mechanical dormancy.
Dormancy imposed by mechanical barriers in the seed to the emergence of the embryo.

Mechanical Dormancy   Another form of exogenous dormancy is **mechanical dormancy**, in which dormancy is imposed through the mechanical restriction of the quiescent embryo from expanding during germination. Due to the strength of mechanical structures that encapsulate the seed (e.g., in pits of stone fruits), the embryo cannot rupture the barrier to emerge.

Chemical dormancy.
Seed dormancy imposed by the presence of certain chemicals in the seed that inhibit germination.

Chemical Dormancy   The mechanical restriction of the seed coverings is the cause of yet another form of primary dormancy, called **chemical dormancy**. Certain chemicals accumulate in the seed during its development. These chemicals are prevented from being leached out of the seed by the retaining wall. These chemicals eventually act as germination inhibitors in species including flax.

Overcoming Exogenous Dormancy   Can dormant seed be made to germinate? Seeds with exogenous dormancy may successfully germinate when they are sown in the field and acted upon by several natural factors. Soil microorganisms are capable of acting on the thick seed covering to loosen it. Seeds that are ingested by animals are exposed to harsh digestive chemicals, which can loosen the coat of a hardened seed. Forest fires as well as the natural acidity of the soil are means by which hard seeds are aided to overcome exogenous dormancy.

Seed scarification.
Mechanical or chemical techniques employed to loosen a hard seed coat to permit access to environmental factors required for seed germination.

Seed scarification are methods used to loosen a hard seed coat. Crop producers may artificially subject hard seed to mechanical abrasion through impacting (shaking) or churning in a drum containing coarse abrasive materials such as sand. These methods are called **mechanical scarification** and are distinguished from **chemical scarification** that involves the use of harsh chemicals such as hydrogen peroxide and sodium hypochlorite. Scarification can damage seeds irreversibly and should be done with caution. Chemical scarification is not commonly used in crop production.

Endogenous Dormancy

Endogenous dormancy is the most common form of seed dormancy. The factors that impose this condition originate in the embryo. Any one or a combination of several factors causes it. In rudimentary embryo dormancy, the underdeveloped embryo requires a protracted period of germination. In species such as carrot, in which the inflorescence is an umbel, seeds develop in a sequential order, those in the older primary umbel being more developed than the latter ones in the secondary and tertiary umbels. The position of the seed on the parent plant thus affects dormancy. Other endogenous factors include age of the plant at the time of flowering (i.e., dormancy increases with age) and moisture content of the mother plant (i.e., water deficiency during maturation reduces the level of dormancy).

Physiological Dormancy

Physiological dormancy (a form of endogenous dormancy) is manifested in several forms. In most cultivated cereals in the temperate region, germination will occur only after a certain period (about 1 to 6 months) of overwintering. This is the process by which seeds with this condition lose dormancy. This may also occur during dry seed storage at about 15° to 20°C (59° to 68°F). Seeds contain certain phenolic chemicals such as coumarin. These chemicals that induce dormancy inhibit germination. Abscisic acid induces bud dormancy in trees and seed. To overcome these dormancy mechanisms, the inhibitors may be leached. In certain forms of physiological dormancy, seeds need moist chilling, a practice whereby seeds are placed between layers of aerated and moist sand and exposed to chilling temperatures. The process is called *stratification*.

16.4.2   SECONDARY DORMANCY

Secondary dormancy is a condition in which non-dormant seeds can become dormant after exposure to certain environmental conditions. For example, storing winter barley at high moisture content at 20°C (68°F) for 7 days imposes some form of dormancy on seeds. Temperature *(thermodormancy)*, light *(photodormancy)*, darkness *(skotodormancy)*, and others can impose secondary dormancy.

Germination is affected by other factors such as mechanical damage. Mechanically harvested and processed seed (e.g., by threshing) is prone to physical or mechanical injury of varying degrees that influence germination. This damage can range from microscopic cracks to large cracks or splits in the seed. Some seeds may appear whole and intact on the surface but violent shaking may have damaged the internal structures, particularly the embryo. Mechanically damaged seeds are susceptible to pathogenic attack. Large seeds and very dry seeds (less than 15% moisture) are more prone to mechanical damage than small and less dry seeds.

16.5: FACTORS AFFECTING CROP STAND

How much of the successful establishment of a crop stand depends on the crop producer? A good stand establishment depends on crop genetic and environmental factors, coupled with good management. In order for a good crop stand to occur, the following factors must be taken into account.

16.5.1   PLANT FACTORS

The seed should be of the proper genotype, a cultivar that is well adapted to the region of production. It should have high genetic purity and be free from contaminants (admixture) and weeds. It is imperative that the crop producer always obtain seed from a reputable supplier or source. A weed-infested crop stand is doomed to fail, since the weeds sooner or later will aggressively outcompete the crop, or additional production cost will be incurred to control the weeds. The seed should have good quality. It should have high germination percentage and be viable and disease-free. It should germinate rapidly and establish vigorously. These quality factors were discussed in detail previously in this chapter.

For a good and uniform stand, the seed should germinate uniformly. It should be free of germination inhibitors or regulators such as dormancy. This may require pre-germination treatment of seed or planting at certain times that are more favorable.

16.5.2 ENVIRONMENTAL FACTORS

There are certain critical environmental factors for seed germination—adequate water, proper temperature, adequate oxygen, and proper light or dark conditions. The moisture should be adequate to sustain the seed through emergence to seedling establishment. Rodents may remove seeds, while soil-borne diseases may cause rot or seedling death.

16.5.3 MANAGEMENT FACTORS

The choices of cultivar, quality of seed, and germination conditions all have management implications. In addition, the producer should sow the seeds at the proper time and proper depth and protect them from diseases and pests, including weeds. The seedbed should be well prepared to be warm, of good tilth, and well-draining. The grower should firm the soil around the seed adequately for effective contact with soil in order for imbibition of water to occur.

16.5.4 IMPORTANCE OF A GOOD SEEDBED

A seedbed is the key cultural practice in achieving a good crop stand establishment. A good seedbed provides the environmental factors needed by the seed. It should have the following qualities associated with good soil texture and structure:

1. *Well drained*—to avoid waterlogging that causes rot
2. *Well aerated*—to provide adequate oxygen for respiration needed for germination
3. *Good water-holding capacity*—to provide adequate moisture for seed imbibition
4. *Good tilth*—for effective contact of soil with seed for imbibition
5. *No crusting*—crusting impedes seedling emergence

Other factors (non-soil) that impact seedbed quality are absence of weeds and excessive litter, as well as availability of adequate moisture.

16.6: FIELD SEEDING

Direct-seeded. Planted in a spot in the field where it will grow to maturity.

Field or agronomic crops such as cereal grains, grain legumes, forage crops, oil crops, and fiber crops are generally **direct-seeded**—that is, the seeds are placed where they will grow and develop to maturity. This is opposed to *transplanting,* in which seedlings are raised in nurseries and then transferred to permanent locations in the field. Tree crops and many vegetables are transplanted. Some vegetables are also direct-seeded. After selecting the appropriate cultivars and preparing the seedbed, the producer has five major decisions to make toward crop establishment. These are depth of seed placement, plant density, plant arrangement, time of planting, and method of planting.

16.6.1 DEPTH OF SEED PLACEMENT

If seeds are sown at different depths, germination will be uneven, resulting in an uneven crop stand which will, in turn, affect later crop production activities, such as harvesting.

Table 16–2 Seeding Depths for Selected Field Crops

Seeding Depth (Inches)	Seed Size (Number of Seeds per Pound)	Examples of Crops
0.25–0.50	300,000–5 million	Tobacco, white clover, bluegrass, and timothy
0.50–0.75	150,000–300,000	Alfalfa, lespedeza, red clover, and ryegrass
0.75–1.50	50,000–150,000	Flax, sudangrass, and beet
1.50–2.00	10,000–50,000	Wheat, oats, rice, sorghum, and barley
2.00–3.00	400–10,000	Corn, cotton, and pea
4.00–5.00	4–20 (tubers or pieces)	Potato

Source: Extracted from Principles of Field Crop Production. MacMillan. 2ed. NY.

The crop will mature unevenly and hence present a problem for mechanized harvesting. Depth of seed placement is influenced by several factors, including *seed size, type of seedling emergence, soil type,* and *soil moisture.*

1. *Seed size.* Large seeds have more food reserves and can emerge from lower depths in the soil while depending on stored energy. Small seeds have limited food reserves and are planted at shallow depths (Table 16–2).
2. *Type of seedling emergence.* Species with epigeal germination need to emerge above the soil to commence seedling establishment. If seeds are planted too deeply in the soil, emergence may be greatly delayed, and seeds may rot in the process.
3. *Soil type.* A heavy soil (clay) is cold, poorly drained and aerated, and prone to crusting. Seed placement in such a soil type should be shallow in order to provide a more favorable environment for germination. Light soils (sandy), on the other hand, drain freely and are prone to drying, especially in the soil surface. They also provide less impedance to emergence. Seed may be planted deeply in such soils. Shallow planting may cause seeds to dry after imbibing water.
4. *Depth of soil moisture available.* This feature is associated with soil type. Clay soils have a higher water-holding capacity than sandy soils. The top of sandy soils is prone to drying. Where supplemental moisture will not be applied during production, the depth of soil moisture will be low in sandy soils, necessitating a much deeper seed placement. Seeds that may be deeply planted (about 2 inches [5.1 centimeters] deep) include corn and peas. Wheat is planted at a medium depth of about 1.5 inches (3.8 centimeters), while alfalfa and red clover are planted at about 1 inch deep. As a rule of thumb, seeds are planted to a depth of about three to four times their size.

16.6.2 PLANT DENSITY (SEEDING RATE)

Plant density is determined by the **seeding rate** of a crop, or the number of established plants per unit land area. The seeding rate should be estimated as closely as possible for optimal crop stand establishment. Overseeding (which causes intense competition among plants) or underseeding (which results in underutilization of resources) can reduce crop productivity. The seed needed to plant an area is estimated as a weight, not a count.

Seeding rate. The number of seeds per unit area adopted in a seeding operation.

Weight of seed/unit area = [(Plants per unit area desired) / (number of seeds per unit weight × % germination × % purity)]

Seeding rate for a specified percent live seed (PLS) per hectare is calculated as

$$PLS/ha = [desired \ kilograms \ of \ PLS/ha]/[\% \ PLS/100]$$

$$where \ \%PLS = [\% \ germination \times \% \ purity/100]$$

Since this is only an estimate, it is customary for producers to adjust upward to accommodate reasonably expected losses from their experience.

In order to be most effective, an estimate of seeding rate should consider certain factors:

1. *Pure live seed ratio.* Pure live seed ratio is a combination of two factors—purity and percent germination. This is a realistic indication of seeds that actually have a chance of contributing to the crop stand.

2. *Plant's capacity for competition.* Crop production involves the cultivation of plants in a population as a pure stand (same crop species) or in a mixture (different crop species). The plant's competitive ability depends on certain plant characteristics, including plant architecture. Large plants with large canopies are spaced wider in the field than plants with smaller canopies, because they need and use more space. Closer spacing reduces branching and tillering. Indeterminate cultivars require more space than determinate, or "bush," cultivars. When crop plant species are grown in mixtures, as occurs often in forage production, plant competitive ability becomes more critical than in pure stands. Weak competitors such as birdsfoot trefoil are prone to smothering by more aggressive species.

3. *Cultural conditions.* Plant density adopted by the producer is influenced by the kind of production system regarding the level of inputs. Under high fertility and irrigation, the producer can afford to use higher plant densities. The native nutrition will be supplemented, thereby reducing competition for resources. Lower plant densities are recommended under rainfed conditions. Generally, the amount of seed per acre doubles if crops are irrigated. For example, 10 pounds of seed per acre of corn may be used under rainfed conditions, while 20 pounds of seed per acre may be planted under irrigated conditions. These seeding rates produce about 13,000 plants and 26,000 plants per acre, respectively (Table 16–3).

   Temperature and moisture are critical to successful crop production. The season and the time of the season influence the plant density used. Moisture stress is higher in summer production; hence, lower density of plants is more desirable.

Table 16–3   Common Seeding Rates of Selected Crops

Crops	Pounds/Acre of Seed		Plant Density (Plants/Acre)	
	Dryland	Irrigated	Dryland	Irrigated
Alfalfa	5	20	300,000	1,000,000
Corn	10	20	13,000	26,000
Cotton	15	30	50,000	100,000
Peanut	20	40	20,000	40,000
Sorghum	2	6	40,000	120,000
Soybean	15	30	60,000	120,000
Wheat	30	120	500,000	2,000,000

Figures represent a range, the irrigated values indicating the higher end, with various mid-values between the extremes according to specific cultural practices.

16.6.3 PLANT ARRANGEMENT

Plant arrangement in the field is influenced by some of the factors that influence plant density. Seeds may be distributed in the field according to a predetermined, constant, and uniform spacing pattern or randomly distributed. The distribution pattern depends on factors including seed size and production system. The common types of distribution of plants in the field are *unstructured distribution* and *structured (patterned) distribution*.

Unstructured Distribution

Broadcasting. The seed distribution process employed in direct seeding whereby seeds are distributed randomly without any particular pattern.

Unstructured seed distribution is called **broadcasting**. Small seeds are difficult to plant individually. They are thus distributed without any predetermined interplant spacing or pattern. Field crops seeded this way include wheat, oats, rice, forage grasses, and forage legumes. Crops that are broadcast-planted are planted at a high density. Due to lack of a pattern, such operations are not readily amenable to some post-planting activities such as cultivation and fertilizing without endangering the plants by trampling.

Advantages The advantages of broadcast seeding include the following:

1. A small grower may manually broadcast seed.
2. It is suitable for small-seeded crops that are difficult to plant separately.

Disadvantages Broadcast seeding leads to uneven seed establishment for several reasons. If a mechanized operation is used,

1. Seed metering is not at a uniform rate.
2. Poor distribution of seed occurs, especially under windy conditions and over uneven terrain.
3. Uneven depth of seed placement results from non-uniform seed covering from the harrowing operation that is required after seed distribution.
4. Some seeds may not be covered and hence may not germinate.
5. It requires a higher seeding rate to make up for potential losses.

Structured (Patterned) Distribution

Field seeding may be done in a predetermined fashion to accommodate postemergence operations and to optimize the land resources, while minimizing undesirable interplant competition. Several patterned distributions are in use.

Seed drilling. The planting of seed with a mechanical seed drill.

Drill Pattern **Seed drilling** is characterized by very high plant population rates and very close spacing, within and between rows, called solid drilling (Figure 16–5). Larger seeds (e.g., soybean) may be drilled as well. Row spacing in drilling may be

FIGURE 16–5 Drilled and narrow-row spacing in soybean has doubled since 1990, with a corresponding decrease in the area planted in 24-inch or wider rows. (Source: USDA)

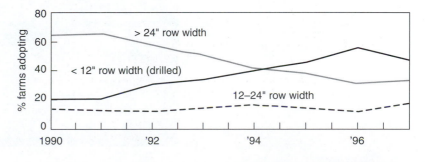

5 to 20 inches apart. Narrower spacing prohibits postemergence operations. Drilling is done with a mechanical planter and is widely used to plant small seeds such as those that may be broadcast planted (e.g., barley, oats, rice, wheat, and forage crops). This distribution pattern allows seeds to be planted at a uniform and predetermined depth. Where interrow spacing is wider, drilling may allow harvesting and other postemergence cultural operations to be performed without difficulty. The spacing between rows is set to accommodate the equipment to be used during the postemergence productions (cultivation, pesticide application, fertilization, and harvesting).

Row Planting Row planting entails more accurate spacing between seeds in a row and between rows. Rows are adequately spaced for postemergence activities. Seed may be placed in rows in one of three general patterns and at predetermined intervals (Figure 16–6). The patterns differ according to the number of seeds placed at one spot.

1. *Drill planting.* Drill planting is different from drilling crops, which was discussed previously. Seeds are placed individually at predetermined spacing between

FIGURE 16–6 (a) Most field crops are planted by drilling in which single seeds are placed in the furrow at predetermined spacing. A less used variation is the hill drop, in which several seeds are placed at a spot. (b) The intra- and interrow spacing varies according to the crop and purpose of production. This figure shows close spacing wheat planting. (Source: (b) USDA)

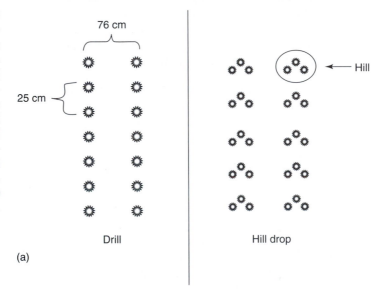

(a)

(b)

(a)   (b)

adjacent seeds in a row. This is the most commonly used method of planting most row crops. Seeds may be drilled in single or double rows (Figure 16–7).

2. *Hill drop.* Two or more seeds may be placed in a group in one planting hole or spot, called a hill. The advantage of this pattern of distribution is that at least one seedling is likely to emerge and become established, thus ensuring a high likelihood of complete crop stand. However, should all seeds germinate, the crop stand faces competition among plants in the hills. This may adversely affect crop productivity or necessitate thinning. Hill planting hence results in equidistant spacing in the field of crops planted by this method.

16.6.4   TIME OF PLANTING

Field seeding should be done at the appropriate time. Field cropping is a seasonal operation. Weather conditions are not conducive for cropping all year round. There is a window of opportunity that must be exploited for best results. Many crop production regions have a longer major cropping season and a shorter minor cropping season. Time of seeding is especially critical if production is to be rainfed. The optimal time of seeding is chosen for several reasons, including the following:

1. Optimal soil conditions for germination occur at certain times in the growing season.
2. Diseases and pests that destroy seeds and seedlings are more prevalent at certain times in the growing season.
3. There are best times for seeding in the field to make the most of the growing season (i.e., to avoid frost and have the maximum period for photosynthesis) and avoid weather and pest problems at harvest.

Producers may sometimes be able to manipulate the soil environment for early planting (e.g., by raising the seedbed or mulching to warm the soil). Early planting may jeopardize crop establishment if unexpected adverse weather (e.g., frost) occurs. Late planting, similarly, may produce low yields because of the loss of part of the growing season. Cool season field crops such as spring grains are seeded early, since they can tolerate the cooler environment during the early part of spring.

16.6.5 METHOD OF PLANTING

Seeds may be planted manually or by using mechanized systems.

Manual Seeding

Perhaps the oldest method of manual seeding is *broadcast seeding,* whereby seed is held in a sack over the sower's shoulder. This was the basis of the biblical account of Jesus' "Parable of the sower" in Matthew 13:3–8. Broadcast seeding may be accomplished by hand or by using a simple mechanical seeder such as a cyclone seeder. Manual seeding may also be accomplished in a pattern or no pattern using tools that may be as simple as a stick that is used to dig a shallow hole in which seeds are placed. The seeds are covered with soil and lightly firmed by, for example, applying pressure with the foot. There is also the manual jab planter, in which a hand-held tool is used to dig the hole, instead of a stick. Certain planting operations may not be completely manual. Machines may be used to perform certain portions of the operation (e.g., digging holes for seeds, seedlings, or cuttings). Manual planting is practical for seeding small fields.

Advantages   Advantages of manual seeding include the following:

1. No major equipment is needed for manual planting.
2. This planting method is inexpensive and adaptable to small plots of land (e.g., home garden).

Disadvantages   Disadvantages of manual seeding are as follows:

1. Human judgment is used in manual planting, leading to errors, especially when the operator becomes exhausted.
2. Manual planting is a tedious operation and not readily applicable to large acreages, unless in areas where labor is inexpensive enough to allow economic hiring for profitability.
3. The operation is slow.
4. It results in uneven depth of seed placement, leading to uneven germination and hence uneven crop establishment.
5. Uneven spacing and plant distribution are also weaknesses. Further, it is not convenient to seed and fertilize simultaneously with this method.
6. It is an indiscriminate method with much seed wasted when it falls on rocks, fence rows, or other unsuitable areas.

Mechanized Seeding

Mechanized seeding is accomplished with three categories of implements: *broadcast seeders, grain drills,* and *row crop planters* (Figure 16–8). Whereas equipment from any of the three categories described may be calibrated for planting different kinds of seed, a fourth category of equipment, called *specialized planters,* is designed to plant specific crops. Examples of this type of equipment are potato planters and tobacco planters. Seeders may be operated simultaneously with other crop production implements (such as fertilizer applicators), in order to combine different jobs in one operation. This is time-saving and reduces the number of passes over the soil with implements (a cause of compaction of soil).

A mechanized seeder has three main parts: a *hopper* for carrying the seed and a *metering device* to deliver seed to a *drill* that opens for seed placement. The seed meter-

(a)

(b)

(c)

(d)

FIGURE 16–8 Direct-seeding in the field may be accomplished with a wide variety of planting equipment. Some of these are designed to perform both seeding and chemical application simultaneously. A regular drill (a) is used for seeding under conventional tillage, while no-till drills (b, c, d) are used under conservation tillage.

ing devices operate in different ways. Some have a fluted wheel, others have wheels with cells and brush cut-outs, and others have agitated free-flow of seeds through an opening in the bottom of the hopper. In addition, these implements are equipped with *furrow openers (double-disk openers,* or *runner openers).* Double-disk openers are more capable of cutting through crop residue and hence recommended for planting fields that have been conservation tilled (e.g., in mulch tillage systems). Runners are best used in conventional tillage systems. Further, drag chains for covering the seed with soil and *press wheels* or *rollers* for firming the soil after planting are used.

Mechanical seeders vary in efficiency. *Precision seeders* are capable of metering seed to a high degree of accuracy for equidistant distribution. *Random seeders* are less precise in seed distribution and are usually designed to use gravity to distribute seed.

Broadcast Seeders   As previously discussed, broadcasting can be done by hand. It can also be done from the air by an airplane or on the ground by a tractor-drawn implement or shoulder-supported cyclone seeder®. Seeds are distributed in a random but uniform manner. The seedbed must be well prepared for this operation to be successful. Broadcast seeders have different designs, such as *endgate seeder* or *centrifugal throwers*. Since broadcast seeders have no devices for opening the soil, broadcasting is usually followed by harrowing to cover the exposed seed with soil.

Grain Drills   Grain drills are used when a high seeding rate is required, as in the seeding of small grains. In this seeding operation, the rows are too close together (solid seeding) to permit any after-establishment cultural practices such as mechanized cultivation. In some cases, seeds are drilled at extremely close spacing, while in larger grains such as soybean, seeds are placed at close but even and predetermined spacing. Grain drills can be fitted with accessories such as a fertilizer hopper to dispense fertilizer and seed simultaneously. A drag chain and pressure wheels may also be attached for covering the seed and firming the soil after placement.

Row-Crop Planters   These implements are used to plant seed such that the crop can receive post-establishment husbandry by mechanized methods. These implements can be adjusted for planting seed in a drill pattern (i.e., close in the rows). Just like grain drills, row crop planters may be fitted with accessories for distributing fertilizer or herbicide simultaneously with seeding. They may be adapted for planting in reduced tillage systems by being fitted with coulter blades for cutting trash and down-pressure springs for soil penetration.

16.6.6   CALIBRATION OF SEEDING EQUIPMENT

Manufacturers provide seeding rate guides for the metering devices attached to seeding equipment. These devices periodically need to be recalibrated for accuracy. Further, even though seeding equipment is set to seed soybean, for example, there are many cultivars of soybean that differ in size and other physical characteristics that may have implications in mechanical seeding. It is best to calibrate the seeding equipment with the cultivar to be planted. Equipment calibration is done off the field.

Row Planter Calibration

To calibrate a row planter, drive the planter over a distance (e.g., 100 ft) and collect the total number of seeds delivered from one row. Note the distance traveled between adjacent units on the planter (e.g., 30 inches apart). The area available, then, to the seeds is 250 ft^2 (i.e., 100 ft $\times$ 30 inches, or 2.5 ft). Suppose 150 seeds are collected. The seeding rate is calculated as follows:

$$[150 \text{ seeds}/250 \text{ ft}^2] \times [43,560 \text{ ft}^2/\text{acre}] = 26,136 \text{ seeds/acre}$$

Grain Drill Calibration

Since grain drills seed at high densities, seed collected is not counted but weighed. Using a 16 ft grain drill, for example, the diameter of the wheel is determined first (e.g., 2 ft). It is turned a predetermined number of times (e.g., 20 times). Seed is collected from the hoppers on one-half of the equipment (e.g., 8 ft length). The seed is discharged from the hoppers onto a plastic sheet (or other material) and gathered together (e.g., 2 lb). The length traveled by 20 rotations of the wheel is calculated as

$$\text{Number of rotations} \times \text{diameter} \times 3.14$$

$$= 20 \times 2 \times 3.14$$

The area covered is obtained as

$$125.6 \times \text{width of equipment}$$
$$= 125.6 \times 8 \text{ ft}$$
$$= 1004.8 \text{ ft}^2$$

The seeding rate is determined as

$$[2 \text{ lb seed}/1004.8] \times [43,560 \text{ ft}^2/\text{acre}]$$

If the rate of seeding is not appropriate, the necessary adjustment should be made and the process repeated.

16.7: SEEDING-RELATED ACTIVITIES

Certain operations are done in conjunction with seeding to save time and reduce soil compaction from multiple passes:

1. One common seeding-related activity is row fertilization. Some grain drills and row planters have double hoppers, one for seed and the other for fertilizer.
2. Herbicides may also be applied in a similar fashion as fertilizers by attaching appropriate hoppers.
3. To ensure proper germination, supplemental irrigation may be necessary in regions with erratic rainfall.
4. Thinning is also an activity associated with seeding and crop stand establishment. Under certain conditions, the field is deliberately overseeded and then thinned to the required plant population density. This is expensive and time-consuming. To avoid this step, the producer should start with high-quality seed and use a precision seeder for best seed distribution and proper seeding rate. Certain operational terms are in use for describing thinning in certain crops. These include *hoeing corn, chopping cotton,* and *walking beans.*
5. For a good crop stand, all competitive factors should be removed or minimized. The producer should control weeds and other pests. Seeds may be treated with fungicides (seed dressing) prior to planting, especially in soils infested with fungi.
6. Seeding of legumes, especially soybean, may involve the treatment of the seed with *Rhizobium japonicum* soybean inoculant (called inoculation) to enhance nodulation for symbiotic nitrogen fixation.

SUMMARY

1. Seed produced by plant breeders reaches the grower only after going through a seed-certification process that is conducted according to certain standard procedures determined by state and federal guidelines.
2. There are four classes of commercial seed: breeder seed, foundation seed, registered seed, and certified seed.
3. Farmers have access only to certified seed for crop establishment.
4. Plant breeders receive protection of their creation (cultivars) under the federal Plant Variety Protection Act of 1970.

5. Prior to sale, commercial seed producers have to analyze their seed and provide certain standard pieces of information pertaining to quality, including percent germination, seed purity, seed vigor, and mechanical damage.
6. Seed of high purity should be free of contaminants, including other crop seed, weed seed, and inert matter.
7. Seed longevity is affected by seed genetic factors, maturity, nutritional status, moisture content, physical conditions, and environmental factors.
8. Seed germination is affected by environmental factors, including temperature, air, and moisture. Light is not a universal requirement for germination.
9. Seeds that are healthy may fail to germinate because of dormancy problems. Endogenous dormancy involving an undeveloped embryo is most common. Others are physical (hard seed), mechanical, chemical, and physiological dormancy.
10. A good stand establishment depends on crop genetic end environmental factors, coupled with good management.
11. A good seedbed is critical to a good crop stand establishment.
12. Many field crops are propagated by seed rather than vegetatively or by transplanting.
13. Depth of seed placement depends on seed size, type of seed emergence, soil type, and soil moisture available.
14. Seeding rate depends on pure live seed, a plant's capacity for competition, and cultural conditions
15. Plants may be distributed randomly or according to a predetermined pattern (e.g., spaced and in rows).
16. Time of seeding depends on soil conditions, the diseases and pests endemic in the region, and the best time for seeding for the species during the cropping season.
17. Seeding may be done manually or by mechanized methods.
18. Seeding equipment needs to be calibrated for the proper seeding rate.
19. Fertilization, irrigation, and other activities may be conducted during seeding to improve crop establishment.

REFERENCES AND SUGGESTED READING

Association of Official Seed Analysts. 1993. Rules for testing seeds. *Jour. Seed. Tech.* 16:1–113.

Basra, A. S. 1995. *Seed quality: Basic mechanisms and agricultural implications.* New York: Food Products Press.

Copeland, L. O., and M. B. McDonald. 1995. *Principles of seed science and technology.* 3d ed. New York: Chapman and Hall.

Green, D. E., D. G. Wooley, and R. E. Mullen. 1981. *Agronomy: Principles and practices.* Minneapolis, MN: Burgess.

Hartmann, H. T., D. E. Kester, F. T. Davies, Jr., and R. L. Geneve. 1997. *Plant propagation: Principles and practices.* 6th ed. Upper Saddle River, NJ: Prentice Hall.

Mayer, A. M., and A. Poliakoff-Mayber. 1975. *The germination of seeds.* 2d ed. Oxford: Pergamon Press.

Stelk, B. 1993. Seed sowing starts with the right equipment. *Grower's Talks* 67:33–37.

SELECTED INTERNET SITES FOR FURTHER REVIEW

http://www.farmphoto.com/album2/html/noframe/fld00154.asp

Photos of planters.

http://www.farmphoto.com/album2/html/noframe/fld00118.asp

Photos of planters.

OUTCOMES ASSESSMENT

PART A

Answer the following questions true or false.

1. T F   A viable seed is one that is capable of germinating.
2. T F   The tetrazolium test is designed for testing seed viability.
3. T F   Farmers use "registered seed" in establishing a crop.
4. T F   Light is required by all seeds for germination.
5. T F   In hypogeal germination, the cotyledon or appropriate storage organ remains underground.
6. T F   The larger the seeds, the deeper the depth of placement.
7. T F   Broadcast planting is suited to large-seeded crops.
8. T F   Placing two or more seeds in a group in one planting hole is called drill planting.
9. T F   Small grains are commonly drill planted.
10. T F   Seeds with epigeal emergence are generally not planted too deeply in the soil.
11. T F   Airplanes can be used for the drilling of small grains.

PART B

Answer the following questions.

1. Primary noxious weeds are also called prohibited weeds; secondary noxious weeds are called _____ .

2. The percentage of seed in a sample that will germinate is called the _____ .

3. Give three tests conducted in a seed analysis.

4. Give the different kinds of seed dormancy.

5. Give the four commercial seed classes involved in the seed certification process.

6. The method of overcoming seed dormancy by scratching the seed is called _____ .

7. Give three specific examples of crops that are direct-seeded.

8. How is seeding rate calculated?

9. Give the three categories of seeding equipment.

10. Give the three main parts of a mechanized seeder.

PART C

Write a brief essay on each of the following questions.

1. Discuss the factors affecting seed longevity.

2. Discuss the factors affecting seed germination.

3. Discuss the role of temperature in seed germination.

4. Discuss dormancy in seeds.

5. Describe the process of seed certification.

6. Discuss the importance of a good seedbed in crop stand establishment.

7. Discuss factors affecting time of seeding of field crops.

8. Discuss the advantages of mechanized seeding.

9. Discuss the disadvantages of manual seeding.

10. Describe the plant factors that are critical to a good crop establishment.

11. How does type of seedling emergence affect the depth of seed placement?

12. What are the disadvantages of seed broadcasting?

13. Discuss the management factors critical to good crop establishment.

14. Describe how seed metering devices operate.

15. Describe how a row planter is calibrated.

SECTION D

Discuss or explain the following topics in detail.

1. Should seed companies be allowed to patent germplasm?

2. Discuss the importance of biotech seeds in crop production.

Harvesting and Storage of Crops

17

PURPOSE

This chapter is devoted to the discussion of the importance of harvesting a crop (grain or forage) at the appropriate time, and presenting the produce in the best quality to the consumer. The importance of storage of produce is also discussed.

EXPECTED OUTCOMES

After studying this chapter, the student should be able to:

1. Distinguish between physiological maturity and harvest maturity.
2. Describe the methods of harvesting field crops.
3. Discuss the nature of spoilage and the role of microbes in spoilage.
4. Describe the types of grain storage.
5. Describe the methods of drying grain in storage.
6. Describe the methods of silage preparation.

KEY TERMS

Bin burn	Curing	Silage (ensilage)
Combines (combine harvesters)	Harvest maturity	Straw
	Physiological maturity	Windrowing (raking)

TO THE STUDENT

The time of harvesting a crop product has implications in its yield, use, quality, and storage. Delaying harvesting or harvesting prematurely often adversely affects product quality. Premature harvesting is desirable for certain products for certain markets. As plant cells mature, the cell wall often becomes lignified. As a result, products that are preferred juicy and succulent may become fibrous and tough. Sugar may be converted to starch, reducing the sweetness of some products. The processing (e.g., milling) of cereal grains is impeded if the crop is not at the proper stage of maturity and moisture content. If the crop is left in the field for too long after maturity, it becomes predisposed to pest attack and deterioration from the vagaries of the weather. Premature harvesting may mean improper filling of the seed or grain, leading to reduced yield and quality of product. If poor-quality produce is placed into storage, poorer-quality produce will come out of storage. It should be clear by now that timeliness of crop harvesting is of the essence. As you study this chapter, pay attention to the factors affecting crop harvesting and how producers manage them for better product quality and yield.

17.1: Time to Harvest

How important is the harvesting operation in the profitability of a crop production enterprise? Harvesting is literally "reaping what you sow." The potential yield expectation from a production operation depends on several factors. For a set of production conditions (cultivar, production inputs, weather factors, husbandry practices, etc.), a certain potential yield is expected of a crop. In order to obtain this potential yield or harvest, the crop must be harvested at the right time (when the economic product is at its optimal quantity and quality), and by the right method (to minimize harvest losses), and stored in the right way (to minimize post harvest losses).

What is the best time to harvest a crop? What factors influence the decision to harvest a crop? Generally, crops are harvested when they are said to be "mature." The concept of crop maturity is discussed in this section. The best time to harvest a crop depends on a number of factors, including the economic part (product), utilization of the product, and postharvest storage.

1. *The economic part.* The economic part or product of the crop plant can be the root, leaves, stem, grain, or other parts. These plant parts have different times when it is best to harvest them. At certain times in the plant lifecycle, there is partitioning of dry matter to various parts of the plant. Repartitioning of assimilates occurs under certain conditions. Translocation of the stored food from the plant part of economic importance to parts of no economic use will reduce the yield of the desired products. Crops should be harvested when the desired product is at peak quality and quantity.

2. *Utilization.* The economic product may come from the same part of the plant but, on one occasion, it may be desirable to harvest it fresh, while on another it may be best to harvest it dry. For example, corn may be harvested fresh or dry, depending on the intended use. The purpose of growing the crop determines when it is best to harvest it in order to have the highest quality and quantity of the desired plant product. For example, alfalfa may be grown for forage or for seed, while corn may be grown for silage or grain.

3. *Storage method.* Harvested products often require some form of storage at the site of production while awaiting shipment to the market. The product may deteriorate in storage if harvested at an improper moisture content. Cold storage is required for grain with high moisture content. Many grain producers have supplementary drying facilities for drying the harvested product to a "safe" moisture content for storage.

17.1.1 CROP MATURITY AND THE TIME TO HARVEST

What is crop maturity? There are several operational and technical categories of crop maturity used by scientists and crop producers. The common ones are physiological maturity, harvest maturity, and storage maturity.

Physiological Maturity

Scientists define **physiological maturity** as the stage of development in the lifecycle of the plant when the plant reaches maximum dry weight. At this stage, increasing production inputs does not produce any gains in yield. In grain crops, there is cessation in growth and grain filling at this stage. The grains at this stage have about 40% moisture content and are of *hard dough* consistency. Since grain ripening within a head and among different heads on the plant does not occur at a uniform rate, growth may continue in certain species until the moisture content is less than the average of 40%. If grain is harvested prior to its physiological maturity, it has low dry matter and lower quality (low starch content), and it shrivels upon drying. Producers of various crops use specific indicators of maturity. In corn, for example, the development of a *black layer* at the base of the kernel signifies the onset of physiological maturity.

Physiological maturity. The stage of development of the product at which maximum dry weight has been attained by the plant and consequently no gains in product yield can occur with increased production inputs.

Harvest Maturity

Harvest maturity for a crop is when the product of interest is at peak quality and quantity (i.e., maximum yield). Crop producers normally determine when harvesting a crop will produce the highest yield of the product of interest by using certain indicators acquired through experience or knowledge of the crop. In certain grains, harvesting is done at 25 to 35% moisture. Mechanized harvesting may result in considerable losses unless the crop is at the appropriate maturity. In some cases, this means the grain should ripen to moisture content of about 15 to 18%. Corn can be shelled at 27% moisture content but then must be dried to 13 to 14% moisture before safe storage. Soybean, on the other hand, must dry in the field to at least 16 to 18% moisture content before it can be successfully harvested.

Harvest maturity. The stage of harvesting a product to obtain peak quality and quantity, as determined by the producer.

When crop plants are grown for forage and pasture, the best time to harvest them is when the crop has attained the maximum vegetative yield, coupled with high quality (high protein and digestibility). In cases where multiple harvests will be made during the year, another factor to consider is a healthy and well-maintained stand after each harvest. The stage of maturity for cutting grain crops for silage is variable among species. It is best to cut barley or wheat in the milk stage, whereas sorghum is cut in the medium to hard dough stage.

Storage Maturity

In production systems where postharvest drying is not available, the crop is harvested at a stage when it can be directly hauled into storage. For grains, this means allowing ripening (desiccation) to occur to less than 14% moisture content.

What may happen if a crop is not harvested on time? Sometimes, inclement weather, equipment failure, or other eventualities delay harvesting of field crops. Crops may be harvested prematurely or when overmatured, each with consequences:

1. *Crop yield reduction.* Harvesting early means seed development and filling is curtailed prematurely. The attainable dry matter will not be realized, leading to lower yield. Delayed harvesting may cause lodging or crinkling of the stem and shattering of seeds. These events lead to increased harvest losses from mechanization.

2. *Grain quality reduction.* Prematurely harvested grains are shriveled and have low starch content. Delayed harvesting in the field causes field weathering of grains, leading to reduced germinability and storability. Delayed harvesting may cause certain products to be fibrous and undesirable.

3. *Loss of value.* Weathered grain (sun-bleached) attracts low prices. It is generally rated low on the quality grading scale. In certain species, precocious germination may occur, leading to loss in yield and quality and hence market value.

17.2: Methods of Harvesting

There are two basic ways in which field crops are harvested: by grazing animals and by humans.

1. *Grazing animals.* Pasture is field that is seeded to forage species for livestock to graze or browse. Animals can also graze or browse plants on the range. Producers of cereal crops, such as wheat, may allow livestock to graze the crop for a period during the growing season.

2. *Humans.* There are two ways in which humans harvest crop products—through either manual harvesting or mechanized harvesting.

17.2.1 MANUAL HARVESTING

Manual harvesting is routine in most developing countries and in small operations. In the United States, certain crops are harvested manually for highest quality and premium prices (e.g., delicate vegetables fruits). Certain crops are handpicked using no tools. For small grains, hand-harvesting tools include the scythe, sickle, and cutlass. Manual harvesting is tedious and labor intensive.

17.2.2 MECHANIZED HARVESTING

Curing. The preparation of harvested products for handling and storage that involves drying.

Machines used for mechanized harvesting may be placed into several operational categories according to how they are used. How crop plants are harvested depends on several factors, including the economic part of the plant, the harvest maturity, and the field **curing** required. Certain field crops require a period in the field, after harvesting, in order for the product to cure to attain a certain quality.

In one category of harvesting, the whole plant is cut at or near the ground level, before the economic part is retrieved immediately or at a later date (e.g., in soybean harvesting). In another situation, the plant is left standing, while only the economic part is removed (as in picking cotton or corn).

In another categorization, mechanized harvesting may be grouped into *one-step* and *multi-step.* In the one-step operation, harvesting the economic product is obtained in a state that can be placed directly into storage (e.g., in harvesting grain crops such as corn and wheat). Several operations are conducted in one pass of the machine (combination harvesting or combine harvesting). In multi-step harvesting, the economic part may be harvested along with other plant parts and left in the field for a period of curing or drying (e.g., alfalfa). The material is later collected and further processed for storage.

FIGURE 17–1 Combines facilitate grain harvesting. They vary in size and design, and some can be adapted for harvesting more than one kind of crop.

Combine Harvesting

Combines (or **combine harvesters**) are routinely used to harvest field crops (e.g., soybeans, small grains, beans, dry peas). The combine is a complex machine that is capable of performing multiple harvesting operations (Figure 17–1). As the name implies, this machine combines several operations (harvesting and threshing) in one pass over the field. Originally designed for small grains, the combine is now widely used for other crops, including corn and soybean.

Combines (combine harvesters). Machines that are capable of multiple operations, usually harvesting, shelling, and cleaning.

Cutting and Gathering   The front end of a combine is fitted with *cutter bars* or sickles that can be set to different heights. Once cut, the plant material is gathered and conveyed into the *threshing chamber.* Grain loss can occur with improper height of cutters. If cutters are set too high, some low-located pods (e.g., in beans) may be left unharvested. Similarly, if set too low, the cutter may slice some soil, thereby introducing foreign matter into the harvested crop material. Grain loss can also result from using dull sickles that may not cut certain plants or may violently agitate the plant to cause shattering of grain. Harvest losses also depend on the speed of the combine. The operator should be sure to operate the combine at appropriate settings and speed. Front-end harvest losses are due primarily to these factors, as well as improper ground and reel speeds.

Threshing   In the *threshing chamber,* a cylinder rotates to thresh the crop. For best results, the cylinder should be operated at an appropriate speed and spacing (clearance). If the rotating cylinder space is set too wide, incomplete threshing will occur. On the other hand, if the space is too narrow, the grain may be physically damaged (chipped or cracked). Similar damage and losses are also attributable to combine cylinder speed—a slow speed causing incomplete threshing, a fast speed causing physical damage.

Cleaning   The grain falls through *sieves* of appropriate sizes to clean out foreign material such as gravel. Losses occur also at the threshing and cleaning stages, caused by improper fan speeds and sieve settings.

Separating   The chaff (glumes, lemma, palea) and straw (stem and leaves) are shaken out of the combine onto the ground.

Handling   The clean grain is transported into the holding tank by means of augers and elevators. Grain is stored temporarily in this tank until emptied into a truck for hauling to the grain storage barn.

Windrowing

Windrowing (raking). A method of multi-stage harvesting in which the crop is cut, gathered, and left in the field for a period to undergo additional ripening, before being picked up for threshing.

Windrowing (or **raking**) is the method of harvesting whereby the crop is cut and gathered by an implement called a *windrower* (Figure 17–2). The material gathered is left in piles in rows in the field. The harvest is left for a period before it is gathered for further processing. Windrowing may be appropriate when producing crops that ripen unevenly. Leaving harvested materials in the field for a period brings about the desired uniformity in ripening of the product. The rows of harvested material are picked up for threshing. Sometimes, when early harvesting is desired, the crop is windrowed. Windrowing is also

(a)

(b)

(c)

(d)

FIGURE 17–2 Haymaking involves the slashing and raking of the material after it has dried in the field. Windrowing equipment is used for gathering the hay for baling. Haymaking equipment includes (a) disk mower/condition, (b) swather, (c) rake, and (d) round baler.

used when harvesting occurs under heavy weed infestation. Windrowing is applicable to forage species such as alfalfa, small grains (e.g., barley and oats), legumes (e.g., peas and beans), and commercial crops (e.g., flax). It is used to sun-cure hay to facilitate baling.

Picking Machines

Pickers are used to remove only the economic part of the plant that is located on an aerial part. There are pickers for crops such as corn and cotton.

17.2.3   HARVESTING IS A "THRASHY" OPERATION

After the economic part of the crop is harvested, a variety of vegetative material is left over in the field. The material depends on the type of crop. Three types of remnants are straw, stover, and stubble.

Straw

Whether harvesting is done by completely cutting the plant and separating the economic part, or by picking only the economic part from the plant, some plant material remains on the field after the operation. **Straw** is the term used for all the dried fine stems and other plant parts left after the seed has been threshed at harvest maturity. The remnant plant material may be baled for later use. Straw may be used as mulch. It can also be plowed under to improve the soil's physical qualities. Straw from leguminous species (e.g., cow pea and soybean) is of high nutritional and soil amendment value. The carbon:nitrogen (C:N) ratio of straw is very high. Due to this characteristic, the use of straw as a soil amendment should be accompanied by nitrogen fertilization in order to avoid yield depression. When bacteria are decomposing straw, they mine the native nitrogen in the soil during the process. The nitrogen level in the soil returns to normal when mineralization occurs. Straw is also useful as bedding material in animal barns.

Straw. The remnant material after the economic part has been removed, consisting of finer stemmed grasses and legumes.

Stover (Stalks)

Stover is the stalks of corn or sorghum plants that remain after the ear or head is picked or removed.

Stubble

Stubble refers to both straw and stubble, especially those that remain rooted in the soil after harvesting. Stubble may be used as mulch (stubble mulch).

17.3: Hastening Harvest Maturity and Facilitating Harvesting

Sometimes, it is desirable to hasten the time of crop harvesting. After a grain crop reaches physiological maturity, any delay in harvesting may cause deterioration in quality and yield reduction. A number of direct and indirect methods may be adopted by crop producers to hasten the time of crop harvesting.

17.3.1 DIRECT METHODS

Defoliation and Desiccation

Once a plant has attained physiological maturity, it cannot accumulate any more dry matter. Leaving it in the field longer predisposes it to weathering. Sometimes, inclement weather threatens the quality of the harvest, thereby compelling the producer to make adjustments in the harvesting schedule. Such adjustments include "inducing" early harvest maturity by applying a *defoliant* or *desiccant*. A defoliant causes leaf-drop, while a desiccant causes the plant to dry out in the field and die. In cotton production, green leaves tend to stain the fibers. Defoliation of cotton prior to mechanized harvesting reduces not only the amount of undesirable plant debris in the harvested fiber but also the chance of tainting it with plant pigments. In haymaking, a chemical treatment (e.g., with Endothal) hastens drying of the plant material in the field.

Topping

Topping is the preharvest reduction of vegetative material (mainly leaves). This slows photosynthetic activity and hastens drying of the economic part (e.g., pods).

17.3.2 INDIRECT METHODS

Generally, crops are harvested when the conditions of the economic product are such that the product can be stored for a reasonable period of time without deterioration. However, if the producer has a facility for drying, the crop may be harvested sooner than normal harvest maturity, then dried to storable moisture content at a later date.

17.4: STORING GRAIN

Some type of storage is usually an integral part of a crop production enterprise. Harvested grain needs some type of on-farm storage after harvesting, unless the product is to be shipped immediately to the user or market. The duration of storage may be a few days, weeks, or even years. Some producers deliberately hold their grain in hope of higher prices in the future.

There are four general ways in which seed can be stored for varying lengths of time. The method chosen depends on the duration over which it is desired for seed to maintain its quality, among other factors. Seed quality cannot be improved during storage, since quality declines with time.

17.4.1 CONDITIONED STORAGE

Seeds are maintained in a dry and cool environment. For most grain crops such as corn, wheat, and barley, seed moisture at storage should be about 12 to 13% and the temperature 20°C (68°F) or less. These conditions can hold seed quality for about 1 year. Many commercial seed companies operate such a facility.

17.4.2 CRYOGENIC STORAGE

Cryogenic storage of seed is used when seed needs to be stored for a very long period. Such seeds are held in liquid nitrogen at −196°C (−295°F). The practical use of this method of storage is limited by the small size of the cryogenic tank. It is widely used by germplasm banks for long-term storage.

17.4.3 HERMETIC STORAGE

Seeds under this type of storage are sealed in moisture-resistant containers. Metal containers are used when very long storage periods are desired. Before the container is sealed, the ambient air inside may be replaced with an inert gas (e.g., argon or nitrogen) for best results.

17.4.4 CONTAINERIZED SEED STORAGE

In containerized seed storage, seeds (usually high-value germplasm) are maintained in specially constructed rooms, equipped with dehumidifiers and other environmental control systems. Sometimes, a desiccant is used to control the level of humidity of the environment.

Grain may be stored in large amounts, unpacked, or in *grain bins* or *grain elevators*.

Grain Bin Storage

On a large commercial scale, grain is usually stored in bulk in bins of different sizes and materials (Figure 17–3). The bin may be made of concrete, steel, or wood. It may be airtight or ventilated. Airtight storage reduces respiration and subsequent spoilage resulting from the absence of oxygen. However, cold air from the atmosphere can cause moisture accumulation in the top layer of the grain in airtight storage. This moisture results from condensation of warm air rising from the bottom of the bin. The use of ventilation (e.g., using a fan) circulates air through the grain and prevents temperature rise and heat buildup.

Grain Elevators

In large grain-producing areas, a number of large silo-type bins are constructed together to create what is called a *grain elevator* (Figure 17–4). These structures are usually constructed at railroad sidings or waterfronts in order to facilitate grain transportation. The storage bins are loaded in various ways, including the use of elevators, pneumatic systems, and conveyor belts.

FIGURE 17–3 Storage may be provided on the farm or away from the farm. Storage bins differ in structure and material, common materials being metal and concrete.

FIGURE 17–4 Grain elevators are large storage facilities that are often constructed near major transportation routes. (Source: USDA)

17.5: THE GOALS OF GRAIN STORAGE

Grain products coming out of storage should be of good quality. The goal of storage is to preserve product quality and quantity (prevent losses) to meet the demands of the end users. The end results depend significantly on the condition of the product prior to storage. A high-quality product from storage should have certain characteristics:

1. *The product should have high purity.* Before storage, the seed should be cleaned to remove contaminating seed (admixture) such as weeds, other crop seeds, and chaff. In practice, mechanical harvesting unavoidably includes some impurity in the form of weed seeds, other crop seeds, plant debris, and even soil. The quantity and the type or source affect the usefulness of the harvest. The presence of toxic crop seeds such as castorbean (*Ricinus communis* L.) and weed seeds such as crotalaria (*Crotalaria spectabilis* Roth) and jimsonweed (*Datura stramonium* L.) makes the seed unsafe for food or feed.
2. *It should be in good physical condition (shape, size, color).* There should be no shrinkage, distortion, discoloration, or heat damage. Improper temperature can cause shrinkage, while mold infestation can produce discoloration of grain. Discoloration or dullness in color and shriveling reduce the eye-appeal. For certain crops, artificial drying produces less attractive product than natural drying.
3. *There should be minimal mechanical damage (breakage, cracks, and splits).* Mechanical harvesting may cause subtle internal injuries to the seed, thus accelerating its deterioration in storage. More visible damage includes split seeds and chipped, cracked, or broken seed. Soundness (lack of mechanical damage) of seed is a quality index of paramount importance in seed evaluation.
4. *The seed should have high viability.* This is critical when stored grain is intended for use as planting seed. For most uses, stored grain does not need to remain viable. However, low viability indicates problems in storage (e.g., improper temperature or moisture content) that may lead to spoilage caused by microbial invasion.
5. *There should be no damage from insects (holes, devoured contents).* Insect damage is common when grain is stored in large quantities. Storehouse insect

damage includes loss of harvested produce weight caused by feeding on cotyledons or endosperm, excreta, dead insect parts, loss of nutrients, and others. Insect damage can reduce grain quality and income to the producer or processor.

6. *There should be no molds or disease infection.* Drying to safe moisture content prior to storage reduces the incidence of molds and the growth of other pathogens. Whereas mold damage in storage is practically impossible to eliminate, it can be prevented from increasing. Mold infestation reduces seed viability while discoloring seed and creating a foul odor. Molds may also cause toxic substances to develop in the product, making it unsafe for food or feed.

7. *There should be no contamination from rodent droppings.* Storage facilities attract rodents. Mechanical traps and baits should be used to trap and destroy these pests.

8. *There should be no pesticide residue.* Sometimes, it is necessary to manage storage pests by using chemicals. The choice of pesticides should be made carefully to avoid tainting the stored product.

9. *There should be no toxic microbial metabolites.* Toxic metabolites are released as a result of mold infestation and the presence of insects under high moisture conditions in storage.

10. *There should be no loss of flavor.* Off flavor may be caused by bin burn (and charring from excessive heat buildup in storage), fermentation, and other factors. Stored bulk grain develops considerable heat, resulting from the use of hexose sugars by microbes according to the following reaction: $C_6H_{12}O_6 + O_2 \rightarrow 6CO_2 + 6H_2O$ + heat. Improper drying (high moisture content) predisposes stored grain to heat damage. The affected grain has darkened pericarp or seed coat. Wheat is very heat-sensitive and can be damaged in storage to the extent of not being fit for flour production. Heat damage also makes grain brittle and more predisposed to further mechanical damage during handling. Grain also loses viability. In order to be usable for certain products, the grain chemical characteristics are critical. Sometimes, protein and starch deteriorate in storage.

11. *There should be no foul odor.* Spoilage caused by high moisture and disease infection produces odor in the grain.

12. *There should be adequate moisture in the product.* Depending on the storage environment, stored grain may lose or gain moisture. It is important to store produce at a moisture content that will prevent fungal growth and molding. Grains from different parts of the plant may have different moisture percentages at harvest. When grains are dried artificially, it is often difficult if not impossible to have uniform drying. Sometimes, certain producers deliberately mix different lots of grain in an attempt to achieve certain desired average moisture content. This method is not recommended from the standpoint of moisture content. The maximum moisture content allowed is set by industry in order to have a standard for fair trade. For soybean, for example, the maximum percentage of moisture allowed for U.S. grades 1, 2, and 3 are 13, 14, and 16%, respectively.

17.6: DRYING GRAIN FOR STORAGE

Grain must be dried to a safe moisture content before storage. Drying is necessary after harvesting because certain grains, such as corn, are difficult to dry in the field to desired moisture before harvesting. In rice, the grain can be safely combine-harvested if moisture is high; otherwise, checked kernels will occur during milling. Grain can be dried in three ways: natural air-drying; drying with forced, unheated air; and drying with heated air.

17.6.1 NATURAL AIR-DRYING

Natural air-drying is accomplished by exposing the grain to the heat of the sun and gentle breeze. It is an inexpensive way of drying grain, especially in arid and semiarid regions where there is abundant sunshine and dry winds. Grain harvested in humid seasons usually has excess moisture and requires drying prior to storage.

Cereal crops may be harvested and bundled into shocks and left to stand in the field to dry. In certain cultures, the bundled material is hung to dry on fences or a line. Damp grain may be spread on a mat, concrete floor, or tray to a shallow depth for drying in the sun. During the night, the grain is covered to avoid reabsorption of moisture in the cool of the night. The grain or produce is stirred often by raking and mixing to bring the lower layer up for more exposure to the sun. Grain at about 18% moisture content can be air-dried to about 14% in about 3 weeks.

Grain in storage can be dried by ventilation through perforations in the walls or floor of the bin. Some bins have perforated tubes or screen flues inserted through holes in the sides and across the bin at certain intervals. Drying grain this way is effective when the moisture at harvest is not very different (less than 2%) from the ideal storage moisture. The disadvantage of ventilated bins with cowls (either wind pressure or exhaust wind types) is that the stored grain absorbs moisture during damp periods, unless there is a way to close these vents.

17.6.2 DRYING WITH FORCED, UNHEATED AIR

Binned grain may be mechanically dried by forcing unheated air through the grain. The bin may have perforated ducts or false floors. A fan is used to draw air (about 1 to 6 cubic feet per minute per bushel of grain). The humidity of the air should be below 70%. This level of humidity usually occurs from the late morning (10 A.M.) to late afternoon (about 6 P.M.). If drier grain (less than 13% moisture) is needed, the air humidity should be lower than 60%. To be efficient, the grain to be dried this way should be clean and free from debris that impedes air flow through the mass of grain.

17.6.3 DRYING WITH HEATED AIR

Drying grain in storage with heated air is more rapid and dependable than the other methods but is also more expensive. Heaters that burn natural gas or petroleum fuels are frequently used for drying grain. They are more efficient in drying grain during the summer months than in winter. Since these heaters are capable of delivering heat at high temperatures, it is very important for the producer to know the end use of the product in storage in order to dry at the appropriate temperature. Temperatures in excess of 130°F (54.4°C) destroy the grain quality of wheat, corn, and sorghum. The starch is damaged to the extent that starch separation is low. Grain for feed can tolerate higher temperatures, but the protein digestibility can be severely impaired.

Construction designs differ, but commercial dryer design is usually the column design, in which grain is moved downward slowly through perforated columns. Hot-air fans force heated air into chambers between the columns as the grain slowly streams down the columns. Column-type dryers are also used on farms.

17.7: Temperature and Moisture Effects on Stored Grain

Notwithstanding the storage facility, two factors are critical to the retention of grain quality for a long period of time: the temperature and moisture content of grain. Seeds in a single head do not ripen uniformly but over a period of about 3 to 10 days. At this stage, the average moisture content of the seed is about 25 to 30%. This moisture is too high for bin storage. For storage in winter, the grain moisture should be more than 14% in most cases. In the Coastal Plains and Gulf Coast, a lower moisture content of 11 to 12% is recommended. Lower percent of moisture (1% lower on the average) is required in the summer months. If storage will be in cribs, where natural drying can occur in fall and spring, ear corn can be stored at 20 to 21% or less moisture content in winter. The appropriate crib size should be used. Larger cribs (10 feet) are recommended for drier regions, while smaller cribs (3 to 6 feet) are used in cool and humid areas.

A combination of high moisture (13% or greater) and high temperature (70°F, or 21.1°C) promotes infestation by microorganisms and insects that leads to spoilage. Under such conditions, the grain respires, producing heat, carbon dioxide, and water. This additional water further increases grain moisture content. The organisms that are introduced into the storage facility from the field or the air include bacteria and fungi, especially *Aspergillus* spp. (e.g, *A. glaucus, A. flavus,* and *A. candidus*) and *Penicillium, Helminthosporium,* and *Fusarium.* These organisms create moldiness in stored grain. Some, like the *A. flavus,* produce deadly toxins (e.g., the mycotoxin *aflatoxin* in grains such as corn and crops such as peanuts).

The respiratory activities of the organism raise the heat in the grain bin, sometimes to 90° to 160°F. The intense heat is responsible for the browning of grain (called **bin burn**) and even a charred appearance and off-taste. At temperatures above 130°F (54.4°C), heat sterilization occurs in the bin, killing insects and inactivating microbes.

Bin burn. The brown discoloration of grain in storage caused by intense heat generated from the respiration of organisms in the enclosed storage bin.

Under conditions of high moisture (greater than 14%) and low oxygen, fermentation occurs in the bin. This is more of a problem when deep bins (such as those used in terminal elevators) are used for grain storage. It is important to aerate these bins to avoid condensation of moisture. This happens by convectional current (warm air at the bottom of the bin rising through the grain and being replaced by cold air). The rising warm air may condense near the top of the bin where the cool grain is. To avoid this condition, an exhaust fan is installed to draw up the warm air at the bottom of the bin through a metal pipe. This pipe has perforations near the bottom and is installed in the center of the bin to draw up the warm air at the bottom of the bin.

17.8: Storage Pests

How does stored grain lose quality? A variety of pests cause grain spoilage in storage. Spoilage of grain takes several forms: mold growth, decrease or loss of viability, increase in moisture content of grain, change in color or discoloration of grain (bin burn), change in chemical content (carbohydrates, protein, fats), decay, and direct insect damage. If the bin is not protected adequately, rodent attack can become a big problem. The taste of grain may change to become sour and the grain may develop an odor as a result of fermentation. The grain can also be lost through eating by rodents and insects (Figure 17–5).

FIGURE 17–5 Storage pest: grain weevil. (Source: USDA)

In certain grains, such as wheat that is cultivated principally for flour, the characteristic elastic properties caused by the protein gluten is lost, reduced, or completely destroyed by pests eating it. Such grain is no longer fit for making bread flour.

The most common storage pests that damage stored grain include the rice weevil *(Sitophilus oryzae),* granary weevil *(Sitophilus granarius),* Angoumois grain moth *(Sitotroga cerealella),* Australian wheat weevil *(Rhizopertha dominica),* and the lesser grain borer found in elevators. Others are *Ploidia interpunctella,* which damages corn in particular; cadelle *(Tenebroides mauritanicus);* khapra beetle *(Trogoderma granarium);* saw-tooth grain beetle *(Oryzaephilus surinamensis);* and confused flour beetle *(Tribolium confusum).*

Saprophytic fungi (feed on dead tissue) are the most important microbes in quality deterioration of stored grain. Molds are more important than yeast, the latter requiring high humidity of at least 88% (not usually present in storehouses) to develop.

Fungi may infect grain in the field, the major species being *Alternaria, Cladosporium, Helminthosporium,* and *Fusarium.* Their role in quality deterioration in storage is minimal. Storage fungi of importance are *Aspergillus, Penicillium,* and to a lesser extent *Sporendonema.* Fungal growth in storage is influenced by seed moisture content, relative humidity of the storage environment, temperature, air quality (O_2/CO_2), and duration of storage. These factors interact. Moisture, especially interseed moisture, is the most critical in seed quality deterioration during storage. Equilibrium relative humidity of 65% or less is considered safe for long-term (6 months) storage. The seed moisture content and storage area environment should equilibrate at about 65%. Longer storage should be at a lower temperature and seed moisture content.

17.9: STORING FRESH OR UNPROCESSED PRODUCE

Dry products, such as nuts and grains, store for long periods of time in dry environments. Fruits and vegetables have a much shorter shelf life, unless preservation measures are taken to prolong it. The storage conditions (especially temperature, humidity, and light) and the kind of crop (regarding quality characteristics and condition at time of storage) affect the duration of storage the crop can withstand before deteriorating. Even under the best conditions, poor quality of produce at the time of storing will cause rapid deterioration. The general goals of storage are to slow the rate of respiration occurring in living tissues (which also slows the rate of microbial activity) and to conserve moisture in the tissues (to prevent excessive dehydration). These goals are accomplished by providing the appropriate temperature (usually cool to cold), maintaining good levels of oxygen and carbon dioxide, and controlling humidity. Bruised products respire at a higher rate than intact ones. Further, the areas around the wounds become discolored. Certain crops have inherent genetic capacity for prolonged storage. Those with dormancy mechanisms have a reduced respiration rate in storage.

As a general rule, cool season crops are stored at low temperatures (32° to 50°F), whereas warm season crops are stored at warmer temperatures (50° to 54°F). However, sweet corn, a warm season crop, should not be stored at warm temperatures that cause the conversion of sugar into starch, an event that reduces the sweetness of the corn. Instead, sweet corn that is not going to be used soon after harvesting should be placed in cool storage. Fresh fruits and vegetables should be stored at high relative humidity to keep their succulence and general quality. Crops such as lettuce and spinach require higher relative humidity (90 to 95%) than crops such as garlic and dry onion (70 to 75%). Storage should take place in darkness or in subdued light. Light may cause produce such as potato tubers to green (from the development of chlorophyll). There are two general methods for storing unprocessed produce: the low-temperature method and the low-moisture method.

17.9.1 LOW-TEMPERATURE METHOD

Fresh products retain the capacity for certain physiological activities, such as respiration. Because respiration is accompanied by the evolution of heat, ventilation in storage is critical for fresh produce to avoid excessive heat buildup to cause rotting. The respiration rate of crops such as spinach is very high. At a given temperature, strawberries can respire about six times as much as lemons. Temperature is known to affect the rate of respiration, lower temperatures slowing down all biochemical and enzymatic reactions. Temperate or cool season crops generally tolerate lower temperatures than do tropical crops that are readily injured by cold.

Whether at home or in a commercial setting, the *mechanical refrigerator* is the workhorse for cooling. Refrigerated trucks and containers are used to transport fresh horticultural produce over long distances. For produce and products, such as cut flowers, strawberries, and lettuce that should be stored dry (no contact with moisture), a *forced-air cooler* system is used to pass cooled air through a stack of the produce in a cold room. Some commercial growers use *vacuum cooling* for the direct field packing of leafy crops. *Package icing* involves the use of slush ice. In the fall, vegetable produce may be stored outside in earthen mounds or trenches.

The rate of respiration is affected by the concentration of carbon dioxide and oxygen in the environment. Where carbon dioxide levels are very high (low oxygen), respiration slows down. The normal levels of oxygen and carbon dioxide in the air are 21% and 0.03%, respectively, nitrogen being 78%. In an airtight room full of fresh fruits, such as apples, the oxygen soon gets used up in respiration and is replaced by the byproduct of respiration (carbon dioxide). After a period of time in storage, the carbon dioxide level reaches about 21% (the previous level of oxygen). At this stage, the fruits respire anaerobically (fermentation), a process that produces alcohol. This is undesirable; hence, growers should ventilate such a storage room before anaerobic respiration sets in. It has been determined that, when the carbon dioxide level is raised, fruits can be stored at high temperatures of 37° to 45°F instead of the low 30° to 32°F.

The gaseous environment during storage can also be enriched with a variety of volatile organic compounds in order to influence ripening. One of the most common is ethylene, which is used to commercially ripen bananas. Certain fruits produce ethylene naturally as they ripen. Since the gas is harmful to cut flowers, fruits (e.g., apples) should not be stored in the same room with cut flowers. The storage environment should be maintained at an appropriate humidity level in order to avoid excessive moisture loss from fresh produce. High humidity predisposes stored produce to decay. A relative humidity of 90 to 95% is appropriate for most fruits, including apples, bananas, pears, and pineapples. In the case of leafy vegetables that are prone to wilting, high relative humidity (RH) of 95 to 100% is recommended. Examples of such crops are lettuce, broccoli, celery, and root crops such as carrots and turnips. On the other hand, vegetables such as garlic, dry onions, and pumpkins store better at 75 to 85% RH. It is important to maintain good ventilation when manipulating RH to avoid condensation and the accumulation of undesirable gases.

17.9.2 LOW MOISTURE METHOD

Many crops, including grapes, plums, dates, figs, and apples, may be preserved for long periods of time by drying. *Solar dehydration* is a relatively inexpensive method for drying in areas where a long dry and reliable sunny period occurs. The produce is spread in appropriate containers and exposed to open, dry, warm air. For more rapid dehydration of large quantities of produce, the *forced hot air* method, which involves air heated to 140° to 158°F is used. This method removes water by dehydration. However, water may be removed by *sublimation* or ice at temperatures below freezing point by using the *freeze-drying* method. This method is expensive, but the product quality is restored by rehydration to the level of the quality of products stored by freezing. As occurs in cold storage, the oxygen concentration of the storehouse may be reduced by increasing the carbon dioxide concentration. Fresh produce stored in reduced oxygen environments respire slowly and thus deteriorate slowly.

The succulence of fresh produce depends on how well it retains its moisture content. A high relative humidity reduces the rate of water loss from plant tissue; however, in the presence of high temperature, the combination might encourage the growth and spread of disease-causing organisms. The provision of ventilation in a storage room ensures that condensation of moisture does not occur, while harmful gases do not build up in the room.

Rodents are usually controlled by using mechanical traps and baits. Insect pests are commonly managed by the use of fumigants. To reduce pests and diseases, the grain should be dried to the appropriate moisture content prior to storing. The storehouse should be adequately ventilated. The method of control of insects and other pests in stored grain is discussed further in Chapter 10. Storage insects are more prevalent in certain parts of the United States than others. Storage insects are more difficult to control in the southern states.

17.10: MECHANICALLY HARVESTED FORAGE

Livestock producers grow forage crops in the field to be grazed by livestock. Sometimes, they graze animals on native (or wild) pasture. Green pastures are not available year-round to livestock producers because of seasonal factors and the vagaries of the weather. Forage species grow slowly or become dormant in winter, or they may not be accessible to livestock (e.g., because of rainfall). Most forage species are productive for 7 months. Hence, additional feed is needed for 3 to 4 months. Livestock producers must have feed year-round for their animals. Some forage crop is hence mechanically harvested and stored in various forms for use during the off-season.

Mechanically harvested forage costs about twice as much as grazed forage. It is costly in terms of labor, machinery, and time. Consequently, the first choice of forage managers is to extend the growing season—for example, by planting mixtures of forage species or by overseeding a winter annual on fields that have dormant warm season species.

There are two main products of mechanically harvested forage—hay and **silage** (or **ensilage**). A third product, *haylage,* is considered to be a cross between hay and silage. There is also *green chop,* or *soilage,* fresh forage harvested each day for livestock feeding.

17.10.1   MAKING HAY

Hay is herbage of grasses or other fine-stemmed plants that are harvested and cured for forage. It is the most durable of the harvested forages, keeping its quality for a long time in storage. Hay is made literally "while the sun shines" and under aerobic conditions.

Wild hay is harvested from mature species that grow in the wild. Crop species can be deliberately planted in pure stands or mixtures and harvested as hay. The most widely used hay crops are alfalfa (pure and mixtures), clover and timothy grass (or other grass) mixtures, wild hay, grain hay, and lezpedeza. Others are tall fescue, bromegrass, sudangrass, sweet clover, and vetch. Just like grain crops, the quality of hay depends on the stage of maturity of the plant material.

Hay production in 2001 totaled 157 million tons harvested from 63.5 million acres, with an average yield of 2.47 tons/acre. Texas led the United States in total production of all hay (accounting for 10.8 million tons), followed by South Dakota, California, and Kansas, in order of decreasing production. Alfalfa hay production in 2001 totaled 80.3 million tons, harvested from 23.8 million acres and averaging 3.37 tons/acre. The production of all other hay totaled 76.4 million tons.

Stage of Development of Plant Material

Young, immature plant material may yield low dry matter but is high in nutritional quality and palatability. Crops for hay are cut at certain intervals. This can be labor-intensive and time-consuming. However, delay in cutting plants can result in an increase in plant fiber content (lignin increases), resulting in decreased nutritional quality. Grasses for hay are best cut between early bloom and full bloom, while sweet clover is best cut when in the bud stage. Cereal crops may be grown for hay. They are most desirable when cut at the soft- to medium-dough stage of grain development. Since hay crops are cut several times during the growing season, it is critical that each cut does not endanger regrowth for the next round of cutting. Improper cutting can reduce plant vigor, reducing the yield of subsequent cuts. For example, alfalfa is more nutritious when cut prior to bloom. However, cutting in the bud stage reduces plant vigor more than when cutting is delayed until full bloom. The nutritional quality of hay changes as the plant matures (Table 17–1). In alfalfa, for example, prebloom crude protein is about 22%, while seed storage protein is about 14%. Conversely, prebloom crude fiber is about 25%, while seed-stage level is about 37%. The trend is the same in grasses.

Curing

Hay is commonly cut by a tractor mower, windrower, or mower-crusher-windrower and left in the field to cure in the sun (sun-cured hay). To hasten the curing process and minimize losses, hay crushers or conditioners may be used to crush the herbage between rolls. The goal of curing is to dry the crop to 25% moisture or less. The drying process should occur such that the herbage retains its nutrients and good green color. In the event of rain during the curing process, the hay may be stirred using a machine called a *tedder.* Rapid drying reduces field-curing losses such as leaf loss, dry matter loss, loss of nutrients, spoilage, reduced palatability, and loss of color. Leaf loss is very serious, since leaves contain most of the nutrients (proteins, vitamins, calcium, and others).

Table 17–1 The Effect of Stage of Maturity on Chemical Composition of Hay

Hay Crop and Selected Stages of Maturity	Chemical Composition (%)				
	Ash	Crude Protein	Crude Fiber	Nitrogen Free Extract	Ether Extract (Fat)
Legume (Alfalfa)					
Prebloom	11.24	21.98	25.13	38.72	2.93
Half bloom	10.69	18.84	28.12	39.45	2.90
Full bloom	9.36	18.13	30.82	38.70	2.99
Seed stage	7.33	14.06	36.61	39.61	2.39
Grass (Timothy)					
No head showing	8.41	10.18	26.31	50.49	4.61
Beginning to head	7.61	8.02	31.15	49.14	4.07
Full bloom	6.10	5.90	33.74	51.89	2.38
Seed (dough stage)	5.38	5.06	30.21	56.48	2.87
Seed (mature)	5.23	5.12	31.07	55.87	2.72

Source: Extracted from *Principles of field crop production*. 2nd ed. NY. Macmillan.

Hay may be artificially cured. This method has reduced losses, compared with sun-curing. The hay is first left to dry to about 35 to 40% moisture, then dried artificially in the storage structure. Unheated air is forced through the hay using blowers. Drying may also be accomplished by using heated air.

Certain losses accompany field-curing of hay. These losses can be minimized with rapid curing. There are nutrient losses from loss of leaves (they contain most of the plant's protein, vitamins, phosphorus, and calcium). In alfalfa, leaf loss may be about 6 to 9% of total weight. There is also 10 to 25% loss of dry matter. Oxidation causes carotene to decompose immediately.

Dehydrated Hay

Hay dehydration is an expensive process. Chopped hay is placed into a rotary drum drier at 760° to 815°C (1,400° to 1,500°F). The product from this process is of high quality.

Raking (Windrowing)

The third activity in hay harvesting is raking. When hay is mowed in strips (called *swaths*), a side delivery rake or a buncher attached to the mower may be used to windrow the herbage.

Processing (Packaging)

After curing, the hay is packaged for storage. Hay may be processed in one of several ways:

1. *Baled hay.* Baling is the most common packaging for cured hay. It is done by using a *field baler.* This implement gathers the windrowed hay, compresses it, and rolls it into a cylinder (called *rolled bale* or *round bale*). The bale is tied with twine or wrapped with plastic (Figure 17–6). Some round bales may weigh up to 2,500 lb (1,135 kg). Baling equipment may also be designed to produce square-shaped bales *(square bales)* (Figure 17–7). Square bales weigh up to 80 lb (36 kg). Some of the modern square bales weigh 400 kg or more.

FIGURE 17–7 Square bales vary in size, but they are much easier to handle by one person.

2. *Stacked hay.* Before baling was introduced, hay was simply stacked in piles, using a hay stacker. A stack may be as heavy as 6 tons. Instead of baling, hay may be compressed and stacked. Hay may also be stored loose in the loft or mow of the barn.

3. *Chopped hay.* Hay may be chopped into small pieces by a field forage harvester and blown into a trailer. This may be dehydrated and then pelleted or pressed into wafers of 12 to 16% moisture content. Chopped-hay production can be completely mechanized.

17.10.2 MAKING SILAGE

As previously mentioned, silage is a forage crop preserved in succulent condition by the process of fermentation (i.e., under anaerobic conditions). In what fundamental ways is haying different from ensiling? Silage is made under anaerobic conditions. To prepare silage, green forage is chopped and stored in a silo. If silage is properly prepared and stored, nutritive losses are minimal.

Storage Containers

Bunker silos are generally less expensive to construct. Trench silos are suited to semiarid and arid regions because of both low cost and less interference from rain during filling or emptying of the containers. Trench and bunker silos are amenable to automatic feeding by means of movable feeding racks at the end. The animals reach through the rack for the feed and move it forward as the feed is consumed. Automatic unloaders may be used to unload tower silos.

Silage may be stored in one of six containers:

1. *Bunker silos.* Bunker silos are erected above the ground with concrete or wood.
2. *Trench silos.* Trench silos are constructed such that the container top is level with the ground level.
3. *Stack silos.* Stack silos entail piling up the silage in stacks on the ground and covering the stacks with sheets of plastic.
4. *Tower silos.* Tower silos are circular structures with vertical walls reaching about 60 feet (18.2 meters) high and 30 feet (9.1 meters) in diameter.
5. *Plastic sacks.* Silage may be stored in commercially prepared sacks such as AGBAG®.
6. *Pit silos.* These are like giant cisterns located in the ground.

Crops Used for Silage

Corn silage production in 2001 was estimated at 102.4 million tons, harvested from 6.2 million acres, with an average yield of 10.6 tons/acres. Sorghum silage production was estimated at 3.7 million tons.

Corn and sorghum are the two principal silage crops. Alfalfa is used significantly in certain regions (e.g., Midwest). However, any carbohydrate-rich crop (i.e., more than two parts of carbohydrate to protein) may be used. Corn is the crop of choice for silage. It yields the best product when ensiled when the grain is in the dough stage. There is a loss of dry matter when more mature corn is used. Under drought conditions, sorghum is preferred over corn as a silage crop. For sorghum silage, it is best to harvest at a later maturity (stiff dough stage) than corn. The product at this stage is less palatable to animals.

Other silage crops include forage grasses (e.g., timothy) and legumes (e.g., alfalfa, clover, vetch, and soybean). Alfalfa is comparable to corn in palatability when used for silage. Sugar beet and certain wild plants are also usable. Silage prepared from coarse materials such as stems of corn and sorghum after harvesting is called *stalkage*.

Advantages of Silage

Crops used for silage can be harvested even under wet conditions, since no drying is required. Silage making is a means of preserving succulent roughage for winter feedings as well as saving forages that otherwise would be wasted, damaged, or completely lost. Whereas ensiling is relatively more expensive than haying, silage has the advantage of preserving a greater proportion of the nutritive value of the green plant.

17.10.3   MAKING SILAGE BY FERMENTATION

Silage is formed under anaerobic conditions. The silage crop is chopped and stored in a tight silo at 60 to 70% moisture content. The plant material is packed into silos. It may be necessary and desirable to trample by feet or with a tractor. The trench silo is more durable in dry regions.

After closing the silo, the microorganisms respire aerobically during the first few hours (3 to 5 hours). This activity increases the carbon dioxide concentration in the chamber to about 70% (the remainder is mainly nitrogen). This condition favors the rapid growth and multiplication of lactic acid-forming bacteria, reaching millions per gram of silage.

The bacteria convert glucose in the plant tissue to alcohol and then to mostly lactic acid (and others such as acetic and succinic acids, in minute quantities, temperature reaching 60°C, or 140°F). This occurs in about 3 to 4 days. Corn and sorghum content are easily fermented, producing acidity in the process. Due to low content of basic elements, the pH of the environment reaches about 4 or lower. Legume species produce high pH due to their low soluble carbohydrate content and high calcium content. The most desirable pH for silage is 4.6. Poor-quality silage contains significant amounts of other acids (e.g., butyric acid, acetic acid, succinic acid, and ammonia). The odor of poor-quality silage comes from these additional chemicals, especially butyric acid and ammonia. Further, proteins and amino nitrogen are broken down to ammonia forms. These forms are less digestible and, coupled with the odor, make poor-quality silage less palatable.

Leaf color changes slightly during fermentation, even though the carotene content is relatively unchanged. Ensiling is completed in about 12 days. The finished product, if properly done, has a pleasant odor and a sour taste.

Certain energy and material losses occur during the fermentation process. There is a loss of energy as glucose is changed to lactic acid and then to butyric acid. There is also a loss of dry matter (5 to 20%) through gases and seepage. Exposure to air (through cracks in the container) can result in spoilage.

17.10.4   CAUSES OF POOR-QUALITY SILAGE

The principal causes of poor-quality silage are improper anaerobic conditions, poor-quality crop material, and improper moisture content.

Improper Anaerobic Conditions

Good-quality silage has an alcoholic odor and sour taste. Poor-quality silage, on the other hand, has a strong odor caused by high concentration of ammonia, hydrogen sulfide, and butyric acid. The two key organic acids associated with silage production are butyric acid and lactic acid. Glucose is converted to lactic acid, which in turn is converted to butyric acid. The amount of lactic acid increases rapidly as fermentation progresses, producing good-quality silage. However, under unfavorable conditions, lactic acid is rapidly converted to butyric acid. The high amount of butyric acid (and the low amount of lactic acid) is the key factor in the poor quality of silage.

Poor-Quality Crop Material

Garbage in, garbage out! Ensiling, contrary to popular misconception, does not improve the nutritional quality of the product. If the plant material is not at the proper stage of maturity, the product will be of low quality. From a nutritional standpoint, it is critical that the total digestible nutrients (i.e., protein content, fiber content, and dry matter content) be optimal in the plant material at harvest. Poor-quality material may also be due to the presence of foreign matter.

Fermentation is the key biochemical process in ensiling. The task of the silage producer is to manage fermentation by providing optimal conditions for this anaerobic process to occur. The ensiling process is not identical for all crop species, even though the principles are the same. For best results, the producer should be familiar with the fine details regarding optimum conditions for the crop species of interest. Changes that occur in conditions during the ensiling process involve moisture, temperature, and air composition. There are five basic phases of ensiling:

Phase 1: Loading the silo (aerobic phase). The plant material is chopped and placed in the silo. It is important to load quickly, distribute evenly, and pack tightly to reduce air damage. Plant cells continue to respire while bacteria use up the oxygen to digest the carbohydrates in the plant material. With time, the oxygen concentration drops while carbon dioxide concentration increases. There is a dramatic rise in temperature in the silo. Too high a rise in temperature (above 100°F) may cause decomposition of silage and a poor-quality product.

Phase 2: Acetic acid production (anaerobic phase). All the oxygen is used and plant cells die. Digestion of fermentable carbohydrates continues. Organic acids, primarily acetic acid, are produced, resulting in the lowering of pH from 6.0 to 4.2.

Phase 3: Lactic acid formation. This phase starts on the third day. Acetic acid production declines while lactobacillus bacteria multiply rapidly, causing the production of lactic acid.

Phase 4: Lactic acid formation continues. Lactic acid production continues for about 2 more weeks. The temperature gradually drops to the 80s; pH drops to 3.8 as a result of organic acid produced. At this acidity, bacterial digestion of fermentable carbohydrates ceases.

Phase 5: Steady phase. The pH holds steady and the acids in the system prevent further breakdown of carbohydrates in the silage. At proper storage conditions, the silage will be preserved by the organic acids generated for a long period.

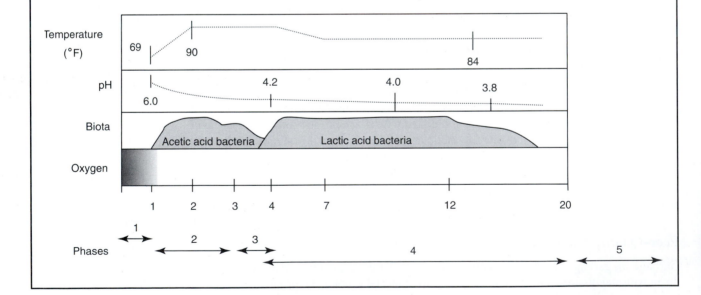

Improper Moisture Content

The moisture content of beginning plant materials should be 60 to 70%. Materials with high moisture (above 70%) have low carbohydrate content and may not attain the desired anaerobic environment needed for fermentation. Undesirable organic acids may form under such conditions. Further, the desirable low pH (3.6 to 4.6) may not occur. However, when the moisture content of the material is low (below 60%), anaerobic conditions may not develop at all (unless the material is stored in an airtight container). This situation may cause molds to grow, leading to poor-quality silage.

17.10.5   ENHANCING SILAGE QUALITY

The quality of silage produced may be enhanced by several means:

1. Use finely chopped (0.25 to 0.50 inches, or 0.63 to 1.27 centimeters) plant material and compact it very well in the silo to eliminate air for high anaerobic conditions to occur.
2. Include silage additives such as phosphoric or hydrochloric acids to lower the pH of the environment to the optimal level.
3. Increase the carbohydrate content of the silage material by adding ground grains and molasses.
4. Decrease the moisture in the silage material, if necessary, by using partially withered (60 to 70% moisture content) grasses and legumes instead of using all fresh materials.

17.10.6   HAYLAGE

Hay is dried to a low moisture content (about 10%), whereas silage retains a high moisture content (70 to 80%). Haylage is forage that has been cut and rapidly dried to about 40 to 60% moisture content. It is sometimes described as specialized forage for horses. Because hay has a low moisture content, mold spores easily become air-borne and may cause an allergic respiratory disease, called chronic obstructive pulmonary disease (commonly called broken wind, dust cough, or dust allergy), that is prevalent in the horse industry. Once affected, a horse never recovers and requires special care to manage the problem. Haylage reduces this disease incidence.

Haylage is semiwilted forage that has been allowed to dry for 1 or 2 days, then compressed and packed into heat-sealed bags (layers of stretch film). Hence, it is sometimes called *round bale silage*. The forage then undergoes rapid fermentation, producing a final pH of 5–5.5. The curing period is less than that of silage production. The moisture at baling is critical (50 to 60%); if it is too high, the feed quality will be reduced. However, lower moisture reduces fermentation and increases mold production. An inoculant (e.g., *Lactobacillus plantarum* MTDT) may be added to control yeast and molds and to improve the efficiency of fermentation to lactic acid and dry matter recovery.

17.10.7   SOILAGE (GREEN CHOP)

Sometimes, livestock cannot get to the pasture to graze the forage. The forage may then be harvested fresh each day and taken to the animals. There is no curing or wet preservation involved. Feeding by green chop is a laborious practice. Drought-damaged corn may be used for animal feed. However, because corn is heavily fertilized with nitrogen, the shoot often contains high quantities of nitrates. When such material is used for silage, some of the nitrates are neutralized. However, using drought-stricken corn for green chop can be very hazardous to livestock, even more so when the chopped material sits in the feed bunks or feed wagons in the heat for a long time. This condition causes nitrates to be converted to nitrites, which are about 10 times more toxic to animals than nitrates are.

SUMMARY

1. Potential crop yield depends on the cultivar, production inputs and practices, weather factors, and the methods of harvesting and storage.
2. The best time of harvesting depends on the economic part of the plant, utilization, and the method of storage.
3. Improper timing of harvesting causes yield reduction, reduction in product quality, and loss of economic value.
4. Field crops can be harvested by humans using manual or mechanical methods or by animals (grazing or browsing).
5. Harvest time may be hastened by methods such as defoliation, dessication, and topping.
6. To prevent spoilage in storage, the grain should be in proper condition prior to being placed in storage (e.g., right moisture content of material). The storage conditions should be appropriate (temperature and moisture).
7. Saprophytic fungi are the most important microbes in quality deterioration of stored grain.
8. The quality of hay depends on the stage of maturity of the plant material and the curing process.
9. Corn and sorghum are the two most common silage crops.
10. Some crops are difficult to dry to desirable moisture content in the field and thus require additional drying in storage.
11. Silage making is an anaerobic process. Too high or too low moisture content of materials causes improper anaerobic environment, leading to poor-quality silage.

REFERENCES AND SUGGESTED READING

Anderson, K. B. *Grain producer marketing alternatives.* Stillwater, Oklahoma State University, Current Report, CR-480.

Burton, W. G. 1982. *Postharvest physiology of food crops.* London: Longman.

Hall, C. W., ed. 1980. *Drying and storage of agricultural crops.* Westport, CT: AVI.

Horrocks, R. D., and J. F. Vallentine. 1999. *Harvested forages.* Academic Press.

Richardson, C. W., and L. Rommann. *Harvesting and ensiling silage crops.* Oklahoma State University Stillwater, Extension Facts, No. 2039.

Tilley, D. S., and K. B. Anderson. *Wheat producer's marketing objectives and plans.* Oklahoma State University Stillwater, Extension Facts, No. 472.

SELECTED INTERNET SITES FOR FURTHER REVIEW

http://www.farmphoto.com/album2/html/noframe/fld00102.asp

Silos and storage bins.

http://www.farmphoto.com/album2/html/noframe/fld00099.asp

Hay balers.

http://www.farmphoto.com/album2/html/noframe/fld00114.asp

Forage; baling of hay.

http://www.encarta.msn.com/find/MediaMax.asp?pg=3&ti=761558496&idx=461516721

Harvesting wheat.

http://www.encarta.msn.com/find/MediaMax.asp?pg=3&ti=761558496&idx=461568310

Combine harvesting of corn.

http://www.encarta.msn.com/find/MediaMax.asp?pg=3&ti=761558496&idx=46156518

Baling hay (round bales).

http://www.das.psu.edu/dcn/catforg/396/phases.html

Discussion of silage making.

OUTCOMES ASSESSMENT

PART A

Answer the following questions true or false.

1. T F   The stage of development in the lifecycle of a plant when the plant reaches maximum dry weight is called physiological maturity.
2. T F   The remains of corn or sorghum plants in the field after the ear or head is removed are called the stover.
3. T F   Curing hay increases its succulence.
4. T F   Excessive drying of grain in the field causes bin burn.
5. T F   Seeds in a single head of grain ripen uniformly.

PART B

Answer the following questions.

1. Define physiological maturity.

2. The best time to harvest a crop depends on three main factors. Name them.

3. What is stover?

4. Give four of the major microbes involved in deterioration of grain in storage.

5. Give four of the major storage insect pests.

6. What is bin burn?

7. The method of harvesting whereby the crop is cut, gathered, and left in piles in rows in the field is called _____.

PART C

Write a brief essay on the following topics.

1. Discuss the role of weather factors in grain harvesting.
2. Discuss the methods of mechanical harvesting of field crops.
3. Discuss the consequences of improper timing of crop harvesting.
4. Discuss harvest maturity of field crops.
5. Discuss the desired quality of a grain product.
6. Describe how microbes cause deterioration of grain in storage.
7. Describe the process of curing of hay.
8. Describe the making of silage by fermentation.
9. Discuss the factors that affect longevity of stored grain.

PART D

Discuss or explain the following topics in detail:

1. How important is mechanized harvesting to modern crop production?
2. A high-quality grain product from storage should have certain characteristics. Discuss these characteristics.
3. Distinguish between the methods of storing dry grain and storing fresh produce.

18

Marketing and Handling Grain Crops

PURPOSE

The purpose of this chapter is to discuss the importance of marketing in crop production and how crop produce (emphasis on grain crops) is handled prior to sale. The role of risk management in crop productivity, especially as it pertains to marketing, is also discussed.

EXPECTED OUTCOMES

After studying this chapter, the student should be able to:

1. Define and explain the concept of marketing.
2. Discuss the elements of marketing.
3. Discuss the role of the middleman in marketing.
4. Describe how produce is graded for marketing.
5. Discuss grain producer marketing alternatives.
6. Define risk and discuss the types of risk.
7. Discuss how risk is managed in crop production.

KEY TERMS

Dockage	Liquidity	Middleman
Forward (futures) contract	Market class	Test weight
Grades	Marketing contracts	

TO THE STUDENT

Market is arguably the most important consideration in crop production, if profit is the goal of the enterprise. After all, why produce what you cannot sell? Market determines what a producer grows, how much is grown, and the quality of product produced. In other words, the producer supplies products according to demand. Further, some markets are more discriminating than others, in terms of quality of product. Thus, the producer aims at satisfying the consumer by producing a product for which there is demand. A system of assigning quality to products is called grading. Premium grades fetch higher prices on the market. In this chapter, you will learn how field crops are handled, as well as the marketing strategies adopted to ensure profitability of crop production. Also, you will learn risk management strategies that crop producers may employ to ensure that deviations from expected business returns are minimized.

18.1: MARKETING CROP PRODUCE

The three factors to consider in a commercial crop production enterprise are market, market, and market. Backyard or small-scale crop production is undertaken primarily to produce crops for the table or for roadside sale of surplus produce. On the other hand, large-scale production is geared primarily for the market. Some farmers use some of what they produce on the farm for feeding animals. However, over 90% of wheat and 30% of corn produced in the United States are sold off the farm on which it was produced. It does not make sense to produce large quantities of an agricultural produce, which frequently has a short shelf life, and not be able to sell it. *Marketing* is the system by which a producer supplies satisfactory products to a consumer at a price that is acceptable to both parties.

18.1.1   CONCEPT OF A MARKET

Marketing occurs in a market, which is defined as all the possible buyers and sellers of a product or commodity. The marketing process is full of decisions. *Sellers* decide what to produce; when, where, and how to sell; and at what price to sell the produce. *Buyers* make similar decisions. These decision makers are organized into individual businesses, called *firms,* that may involve only one family, as often is the case in farms, or hundreds to thousands of people, as in large corporations. The market participants who buy to consume are called *consumers.*

Cash Crop

Field crop producers may grow food, feed, fiber, or a combination of these crops on varying scales. Their produce may be used on the farm, used in the home, sold, or a combination of these uses. A cash crop is one that is grown mainly or solely for sale off the farm. In developing countries, cash crops are usually grown for export markets. Because they are grown for sale, cash crops generate higher incomes for the producer than food crops. They also generate employment where they can be processed and generate economic diversification. Frequently, cash crops are mono-cropped.

Commodities and Differentiated Products

Most farmers produce commodities. A *commodity* is any economic good that may be marketed by almost anyone. Crop commodities may be placed into two broad categories: large-acreage-volume crops (e.g., wheat, corn, soybean) and small-acreage specialty crops (e.g., tomato, potato, onions). Because a commodity is not proprietary material, anyone with the resources can begin at any time to produce a commodity to take advantage of profitable prices as they occur. Unfortunately, such a rush to capitalize on the market sooner than later drives prices down because of oversupply.

A *differentiated product,* on the other hand, is an economic good that is owned by a single seller. Frequently, such a product is covered by a patent, copyright, or trademark that belongs exclusively to the seller. Product differentiation promotes monopolistic differentiation (i.e., each seller is the only seller of a particular product). The seller controls the supply of the product and can price it to reap the benefits of promotion. On the contrary, producers of commodities typically cannot set prices; nor can they benefit from individual promotion, because there are other sellers of the same commodity. Also, a commodity producer, as a price taker, can usually sell any quantity of a commodity without depressing the price. However, a seller of a differentiated product usually would like to sell more units than the buyer can purchase.

Branding

Branding is the identification by brand names, trademarks, or other such names of a product or service by a seller. A trademark includes brand names, symbols, and other marks that have been registered and accorded legal recognition as branding devices. Such items are identified by symbols such as ® and ™ (e.g., Roundup®). Branding is often associated with the manufacturer (manufacturers' brands) but retailers may market products under their own brand names (private labels). A successful brand must be meaningful and contain a homogeneous product quality over a period of time, so that buyers can identify and rely on it. It should also be easy to identify and be backed by aggressive promotional efforts. Brands are a communication device.

The Role of Agribusiness in Agricultural Marketing

Most farm commodities are processed before being offered for sale to the consumer, because many farm commodities tend to be produced by agribusinesses instead of farmers. Many products, such as frozen meals, consist of several farm commodities. Agribusinesses provide most of the transportation, storage, and processing that add place, time, and form utilities (i.e., convert commodities to forms more suitable for consumption) as farm commodities are transformed into consumer products and transferred to the final buyer. The most visible and generally most costly part of agricultural marketing are physical functions—transportation, storage, and processing. Storage adds time utility to the product (i.e., extends or spreads out the consumption over a year of an agricultural product that is harvested once a year). *Place utility* refers to a situation where a product is at point A but has more value at point B. In such a case, it is worth transporting the product to point B. The various businesses involved in agricultural marketing are called *agribusinesses.* Agribusiness firms vary in size from one-person operations to large corporate giants.

From the macro (social) perspective, *agricultural marketing* is the performance of all business activities involved in the forward flow of food and fiber from producers to consumers. From the micro (individual firm) perspective, agricultural marketing is the performance of business activities directing the forward flow of food and services to customers and accomplishing the objectives of the firm (i.e., farmer or agribusiness).

18.1.2   MARKETING IS MORE THAN SELLING; IT MUST BE STRATEGICALLY PLANNED AND EXECUTED TO BE SUCCESSFUL

Crop producers develop *production plans,* which may be formal or informal. In a formal plan, the producer prepares a budget containing estimated costs and returns. In this activity, he or she takes into account the planting of seed (cultivar to use and amount), fertilizers (kinds and amounts), other agrochemicals (herbicides and rates), land preparation costs, harvesting methods, storage, and so on.

Similarly, a crop producer should have a *marketing plan.* Market planning entails developing a strategy to deal with alternative situations that may arise during a short-term period of about 12 to 24 months, during which the crop would be produced and marketed. Marketing planning should be part of production planning. As indicated previously, there is no need to produce what you cannot sell. The plan will include objectives to adopt in relation to specific pricing and risk goals. Situations may change during the crop production period and soon thereafter that may warrant modifications of the plan. A sound plan does not guarantee the highest return on risk but increases the chances of higher returns and reduces the risk.

In developing a marketing plan, four key steps are critical:

1. Evaluating the plan and outcomes from the previous production year
2. Identifying and evaluating the production and marketing environment
3. Setting price and risk objectives
4. Developing a market plan

Before developing a new plan, it is wise to evaluate the previous year's plan to determine strengths and weakness. This is possible only if there was a formal plan in which activities were documented and good records kept. An evaluation will enable the producer to determine what strategies need to change and what should be left alone.

Identification and evaluation of the production and marketing environment require accurate forecasting and thus depend on the availability of pertinent information. The producer should know about marketing outlets and options, price expectations, expected government programs, and production potential. On the personal front, the producer needs to know his or her cash flow needs, financial position, and personal preferences. Crop production is subject to weather factors. The producer should have information about potential production and yield variability in the area of operation. This information would enable him or her to effectively use risk management strategies such as forward contracting and futures contracts. It is important for the producer to know who the potential customers are and where they are located. Information about the price outlook for the proposed enterprise will help in selecting the most appropriate marketing alternatives. The producer should know his or her financial situation and obligations (e.g., debt payment). The cash flow needs will affect timing of sales. If a producer is not in a sound financial situation, he or she may be more conservative in terms of price risk taking.

A producer undertakes a production enterprise with certain objectives in mind. These include profit objectives (e.g., maximizing price received), cash flow objectives (e.g., meeting cash flow needs), and risk reduction objectives (minimizing risk of losing money this year). These objectives are set according to the producer's personal preference and the nature of the enterprise.

Once the first three steps have been taken, the producer should proceed to develop schedules of activities (marketing plans) that are designed to increase the probability of achieving the marketing objectives. The specific activities depend on the type of enterprise, the stage in the production, and the marketing season. In making planting decisions, the producer should take into account the variable costs (e.g., seed, planting, care of growing plants, harvesting, and interest on operating capital). It is important to know the total production cost in order to establish breakeven prices. After establishing breakeven prices, the producer can then develop marketing strategies. Some of the product may be forward-priced. The producer must then decide how to handle the remainder—sell at harvest or store.

In deciding to store grain, the producer should consider three key factors:

1. The opportunity cost of stored grain (money that could have been received)
2. The actual cost of storage
3. The risk of quality loss if grain is stored on-farm

18.1.3   VALUE-ADDED AGRICULTURE

Crop produce may be marketed as harvested (i.e, as raw materials). Agribusinesses are engaged in two basic types of marketing—commodity handling and product marketing. In *commodity handling,* the farmers sell commodities to businesses (e.g., at the farm gate) and sell to consumers in a commodity firm after some degree of processing. Usually, the commodity reaches the consumer in unbranded and undifferentiated form. Product marketing agribusiness transforms the raw commodities into other forms (e.g., bread, frozen entrees, mashed potatoes). *Value-added agriculture* converts agricultural outputs into products of greater value. It increases the economic value of an agricultural commodity through changes in genetics, processing, or diversification (e.g., wheat → flour → bread). Value added increases the consumer appeal of an agricultural commodity. In short, value added is the enhancement added to a product or service by a firm or company before it is offered to the consumer.

18.2: MARKETING ALTERNATIVES FOR GRAIN

What are some of the major marketing alternatives available to a U.S. crop producer? The producer must know what marketing alternatives are available, what they offer, and how the alternatives meet his or her needs. Marketing is a systematic approach toward achieving a reasonable return on investment in terms of capital land management and labor. As previously indicated, marketing requires planning based on estimated production costs, cash flow needs, and the market price outlook. It is critical to have a target price. There are a number of marketing alternatives used by crop producers, each with advantages and disadvantages.

18.2.1   SELL AT HARVEST FOR CASH

Selling at harvest for cash, though once most popular for many crops (cash crops such as wheat) but not all, is losing popularity because of price inflexibility and seasonally low harvest prices. This method of marketing is easy to conduct. It eliminates the need for storage. Further, the producer sets the price immediately and receives the money promptly. However, this method has no pricing flexibility, and the producer has to accept prices that are usually lower than those later in the year. In addition, local elevators are usually crowded at harvest time, necessitating long waits to be served.

18.2.2   HOLD GRAIN IN ON–FARM STORAGE AND SELL LATER

Some producers have on-farm storage facilities that give them the option to store grain temporarily and sell later. This system of marketing has pricing flexibility that allows the producer room to decide on price as well as customers. However, stored grain may deteriorate in storage and lose quality and hence fetch a lower price. Further, storage is costly and adds to the production cost of the product. The producer continues to speculate on the cash market after harvest, instead of knowing the price immediately.

18.2.3 STORE GRAIN IN A COMMERCIAL ELEVATOR AND SELL LATER

Commercial elevators provide long-term quality storage. The producer who elects to pursue this marketing alternative does not have to worry about grain deterioration or physical management of the product. The producer has pricing flexibility, allowing him or her to sell at any time prices are high. However, commercial storage is expensive, and the producer's income is tied up for a period of time after harvest.

18.2.4 CASH FORWARD CONTRACTS

Forward, or futures, contracts are agreements to deliver a specified amount of grain (or turn over ownership of grain in commercial storage) at a predetermined price to a specified location. The producer, in doing this, locks in a price that is not affected by market fluctuations, especially low prices. This method also eliminates price risk. It is widely used and legally binding for as long as the parties involved duly sign the contract. However, once the contract is signed, it must be honored. Any attempts to change it will usually attract a penalty. Further, if prices should rise, the benefit goes to the elevator or customer. The producer is also obligated to produce and deliver the contractual amount and, if the production is not in progress at time of signing the contract, the producer may be in jeopardy of not meeting the stipulated amount of product in the contract.

18.2.5 DELAYED PRICING

Delayed pricing is the marketing method whereby the producer delivers the grain to the elevator and signs a contract agreeing to price it before a specified date. Prior to pricing, the elevator takes ownership of the grain and usually pays a percentage of the current price when the contract is signed. If the producer fails to price the grain on time, the default price is the price posted by the elevator for the commodity at closing on the contract maturity date. The disadvantages of this strategy include the fact that the producer pays interest cost on the grain in storage and stands to lose, should the elevator go bankrupt before the pricing. By selecting this marketing alternative, the producer, in effect, gives an unsecured loan to the elevator operator. Should the elevator go bankrupt, the producer would have difficulty receiving payment.

18.2.6 BASIS CONTRACT

Basis is the difference between current cash price and the nearby futures contract price. This marketing alternative is similar to delayed pricing. The difference is that the price is not determined by the elevator's posted price but, rather, is based on prices posted by futures markets such as the Kansas City or Chicago board. To use this to advantage, the producer must have access to historical basis information. It is advantageous to use this alternative when the basis is unusually high. It eliminates the basis risk but the producer still has price risk and will pay interest costs. If the basis gains, the advantage goes to the elevator that takes title to the grain when the agreement is signed.

18.2.7 HEDGING IN THE FUTURES MARKET

The futures market is discussed later in this chapter as one of the tools of managing risk in crop production. The marketing alternatives that have been previously discussed are *cash marketing* strategies that allow the producer to deal directly with the grain industry. The elevators assume some risk on behalf of the producers, for which they charge a fee.

Forward (futures) contract. An agreement to deliver a specified amount of grain or turn over ownership of grain in commercial storage at a predetermined price to a specified location.

To avoid any price risk, the elevators normally would sell the grain in what is called *back-to-back cash selling,* or cover *(hedge)* the cash position in the futures market. In hedging, the producer trades price risk for basis risk. Basis is more predictable than cash price. The producer can also sell futures contracts and buy them back if market situations change, giving him or her market flexibility. However, the producer has to sell the grain in certain increments (e.g., 1,000- or 5,000-bushel increments) and accept basis risk. Further, the producer is required to maintain a margin account at a specified level.

18.2.8 OPTION CONTRACTS

This is a strategy producers may use to protect themselves from declining prices without giving up the opportunity to benefit from rising prices. In the event of price decline, producers may exercise their option by accepting a *sell* position in the futures market. On the other hand, should prices rise, they let the option *expire,* thereby losing only the amount of the option premium. The disadvantages of this strategy include the need for the producer to be knowledgeable in basis patterns, prohibitive option premiums for acceptable prices, and the requirement to purchase and sell options in specified increments (e.g., 5,000 bushels).

18.3: MARKETING ALTERNATIVES
FOR FRESH PRODUCE

Two basic marketing alternatives are available to the producer—*direct marketing* and *non-direct marketing.* Both have advantages and disadvantages.

18.3.1 DIRECT MARKETING

The producer may market produce directly to consumers via one of several mechanisms, the common ones being pick-your-own, roadside stands, and farmers' markets. The producer interacts with customers on a one-on-one basis and should be ready to provide honest and convincing information about the operation.

Pick-Your-Own

In pick-your-own (u-pick) operations, consumers go to the farm, where they harvest their own produce (usually fruits and vegetables) and present them to the farmer for pricing. Pricing may be on weight, volume, or count basis, depending on the produce. To be successful, the crop should mature uniformly, be of very high quality, and the farm should be located near potential customers. It is advantageous if the farmer produces different varieties of the crop or different crops, so that customers have choices. Consumers patronize this system of marketing for the high quality of fresher produce they expect.

The layout of the farm should facilitate easy and safe movement of customers through the farm. Some fruits may require some supervision for proper and safe harvesting by customers. The sale should be advertised to the public through local newspapers, fliers, radio, and so on. The most effective advertisement is probably by word of mouth from satisfied customers. Signs should be placed near roadsides and sometimes throughout the farm to direct customers. Furthermore, adequate parking should be available.

Proper pricing is critical to the success of such an operation. Customers expect to pay less than they would pay in grocery stores because they are aware that overhead cost is low (no transportation, no labor for harvesting, no packaging, no storage, etc.). However, if the quality of the produce is excellent, the farmer can obtain a high price for it. Some farmers allow some haggling, whereas others are firm with pricing. When selling by volume, it is advantageous to provide containers to customers, even though some will find a way to overfill a container.

The checkout station should be readily visible to customers and provide some storage space, since some customers may make several trips to the field. It may help to have chairs and tables for weary customers to rest. For a large operation, farm assistants should be available to assist customers. The disadvantages of pick-your-own include liability for accidents assumed by the producer and the need to work long hours during harvesting season. Also, inclement weather would prevent an open field sale from taking place.

Roadside Stands

Heavy traffic may create congestion and discourage some potential customers from stopping to buy items from roadside stands. Weekends (when people are not hurrying to or from the workplace) may attract more customers. Furthermore, locating the operation along roads with low or moderate speed limits (35–50 mph) is safer for drivers to pull over and attracts more customers. The produce should be attractively displayed to draw customers. Sales increase when a diversity of products are offered for sale. Pre-packaging of produce diminishes from the fresh-farm image of roadside marketing.

Farmers' Markets

Farmers' markets are owned by a grower organization, community development groups, or a state or local government. The partners in the operation share the operating costs (insurance, advertising, facilities, etc.). Consumers associate farmers' markets with lower prices and fresh, high-quality produce. Unlike pick-your-own, the business hours of farmers' markets are set by the coordinator or organizer. The produce should be competitively priced, since there may be other producers selling the same produce.

18.3.2 NON-DIRECT MARKETING

In non-direct marketing, the producer reaches the consumer through a third party, which may include terminal market firms, brokers, processors, cooperatives, and buyers for retail outlets.

Terminal Market Firms

Terminal market firms purchase produce for chain stores or large wholesalers. They are usually interested in doing business with producers who can supply large volumes of consistently high-quality produce. However, they have specific stipulations concerning product quality, quantity, and packing. Furthermore, because prices are based on current retail prices, the prices they pay for produce are variable. Producers are also responsible for delivering produce to the designated terminal market.

Brokers

Brokers are agents, individuals, or firms that negotiate sales contracts between producers (sellers) and buyers, without taking physical custody of the produce. The producer retains responsibility for most of the marketing functions (e.g., grading, handling, packing). The broker (selling or buying) locates the produce of desired quality and quantity at a price acceptable to both seller and buyer and arranges the appropriate contract.

Though brokers may be involved in invoicing and collecting payments, they are not responsible for payment if the buyer defaults on the contract.

Processors

Processors add value to fresh produce, which they often purchase in large amounts. They contract some of their needs to growers. The producer has the advantage of an assured sale of large quantities of produce. Often, the processor provides some technical expertise to the producer to ensure that the desired produce quality is attained. On the other hand, the producer is expected to supply produce at a specified time and quality, demands that may be hard to satisfy all the time.

Cooperatives

Producers of specific produce may join together to form cooperatives for collective bargaining for high prices and guarantee of markets. Members are able to benefit from the use of communal equipment and facilities (e.g., for harvesting, grading, storage, and transportation), some of which they could not afford on their own. By coming together, small producers are able to pool their outputs to satisfy the needs of processors and other purchasers of large quantities.

Whereas cooperatives provide bargaining power for producers, the producers lose their independence to sell for better prices when consumers demand changes. Furthermore, the more efficient producers subsidize the less experienced producers, not being able to receive optimal returns for their produce.

Retail Outlets

Retailers usually service relatively smaller markets, such as schools, hotels, and prisons. The produce may not require special packaging prior to delivery. Some of these customers may be niche markets that are courted and developed over time.

18.4: ELEMENTS OF MARKETING

Marketing is a complex activity. What are the key processes involved in a marketing operation? Five basic elements characterize produce marketing:

1. *Packaging.* The produce has to be appropriately packaged (in containers or wrapping) for ease of sale, protection from damage or contamination, case of transportation, and other factors. Some grain producers transport their produce directly to the elevator or consumer in trucks without any packaging.
2. *Storage.* Storage and transportation are the two critical components of a market-oriented crop production enterprise. First, unless there is a ready market for the produce obtained directly from the field, storage is needed until sale or distribution. Some vegetables may be harvested and shipped directly from the field to consumers. This cuts down on production costs by eliminating expensive storage. However, cereal grains need to be stored for varying durations of time prior to sale. The storage facility is useful in another aspect of crop marketing. The producer may deliberately hold back the sale of produce until he or she obtains a high price. It should be borne in mind that crop products are perishable without appropriate storage.

3. *Transportation.* Whether the farm is *metro* or *non-metro* in location, the producer needs to transport the products from the farm to the consumer. Sometimes, specially equipped trucks (e.g., refrigerated trucks) are needed to protect the produce from damage in transit. Transportation can be by rail, land, water, or air. Grain elevators are frequently established near railroads to facilitate transportation.

4. *Distribution.* Marketing depends on the principle of supply and demand of goods and services. Once demand has been established, the produce should be moved in a timely fashion to the customer. An effective and efficient distribution network is key to successful marketing of agricultural produce.

5. *Financing.* Financing is needed to provide a means of storage and transportation of produce.

18.4.1   THE MIDDLEMAN IN MARKETING

Middleman. An intermediary, acting as a broker between the producer and the consumer, who may resell to consumers as purchased or value added.

Some producers have direct contact with the consumer. A producer may transport produce directly to a factory or processor. Sometimes, the produce is sold through a broker, or **middleman**, an intermediary who may resell to consumers as purchased or value-added (processed). The middleman usually has a warehouse, a means of transportation, a packaging system, and other facilities for handling the produce.

Grain producers often sell their produce to grain elevators, from where it is distributed to various consumers. Farmers may also form cooperatives for marketing their produce for good prices. Agricultural produce is also marketed through an exchange.

18.5: GRADING CROP PRODUCE

Quality may be defined as the sum of the attributes of a commodity that influences its acceptability to many buyers and hence the price they are willing to pay for it. Grading may be defined as the sorting of a commodity into quality classifications according to some standards (federal standards). These standards are called grade standards and the resulting classes are called **grades**.

Grades. Designations assigned to reflect the quality of a commodity.

Who determines the grading system for a commodity, and why is such a system necessary in the first place? Uniform grading is necessary for fair trading of crop produce. A producer can obtain a higher price for a commodity if it is of superior quality. Factors upon which grades are based determine the quality or market value for the purpose for which the particular grain is generally used.

18.5.1   GRADING AND QUALITY

Grading is primarily for trading purposes. It is not only for the benefit of the consumer but also for agribusiness firms and farmers. Grades are a means of communication in the marketplace that is ordinarily used voluntarily. Grades are useless unless they have meaning to both the buyer and seller. Because of produce quality deterioration, a grade A product today may be downgraded to grade C after transportation or storage over a period of time. Grading helps both operational and pricing efficiency. In some instances, the use of grades is mandatory. For example, any grain sold for export must be graded according to federal standards. Tobacco has specific grading requirements, as does cotton. The U.S. Grain Standard Act was passed in 1916. The USDA in cooperation with the Trade Department developed standards for specific grains in the 1920s.

18.5.2 THE GRADING PROCESS MAY DIFFER FROM ONE PRODUCT TO ANOTHER

There are four general steps for inspecting and grading cereal grain:

1. *Sampling.* For effectiveness, it is critical that the sample be representative of the seed lot. *Grain probes* and *spout samplers* are commonly used to obtain grain from the lot. The grain probe is capable of sampling from various depths of the bin. The spout sampler (or *pelican*) is used to retrieve some grain at certain intervals as it is being discharged from the elevator spout. Seeds in bags may be sampled with a *trier.*
2. *Preliminary examination.* This visual examination is designed to identify surface qualities of the seed regarding uniformity of grain, odor, bin burn, and presence of insect pests. The material is also classified according to crop species and **market class** (e.g., corn, wheat, barley).
3. *Preparation of sample portions.* The sample obtained from the lot is called the submitted sample (what the producer or customer presents to the crop laboratory). This original sample is subdivided for various analytical determinations (called working samples), making sure to leave a portion (file sample) for further reference (confirmatory test, etc.). Subsampling should be done without bias. For that matter, there are several mechanical halving devices that are used. The common ones are the Gamet Precision Divider and the Boerner Divider, the latter being more commonly used.
4. *Analysis on the working sample (subdividing as desired for various determinations)*

Market class. A grade of grain marketed on the basis of its suitability or utility for a specific purpose.

18.5.3 LABORATORY-DETERMINED NATIONAL STANDARDS AND GRADES FOR VARIOUS CROPS

Market Class

Major field crops are classified into commercial classes based on physical and chemical properties that affect their use. The USDA *Federal Grain Inspection Service* publishes current standards and grades. Wheat will be used to demonstrate the grading system for grain.

Grades and market classes for various crops are discussed in Part II of this book.

Definition of Wheat   Wheat is grain that, before the removal of dockage, consist of 50% or more wheat (*Triticum aestivum* subsp. *Aestivum* L.), club wheat (*T. aestivum* subsp. *Compactum* Host.), and durum wheat (*T. turgidum* subsp. *durum* Desf.) and not more than 10% of other grains for which standards have been established under the U.S. Grain Standards Act and that, after the removal of the dockage, contains 50% or more of whole kernels of more of these wheats.

Classes   There are seven classes of wheat: durum wheat, hard red spring wheat, hard red winter wheat, soft red winter wheat, white wheat, unclassed wheat, and mixed wheat.

Subclasses

1. *Durum wheat* (all varieties of white [amber] durum wheat)
   a. *Hard amber durum wheat* (durum wheat with 75% or more of hard and vitreous kernels of amber color)

     b. *Amber durum wheat* (durum wheat with 60% or more but less than 75% of hard vitreous kernels of amber color)

     c. *Durum wheat* (durum wheat with less than 60% of hard and vitreous kernels of amber color)

  2. *Hard red spring wheat* (all varieties of hard red spring wheat)

     a. *Dark northern spring wheat* (hard red spring wheat with 75% or more dark, hard, and vitreous kernels)

     b. *Northern spring wheat* (hard red spring wheat with 25% or more but less than 75% dark, hard, and vitreous kernels)

     c. *Red spring wheat* (hard red spring wheat with less than 25% of dark, hard, and vitreous kernels)

  3. *Hard red winter wheat* (all varieties of hard red winter wheat); no subclasses

  4. *Soft red winter wheat* (all varieties of soft red winter wheat); no subclasses

  5. *White wheat* (all varieties of white wheat)

     a. *Hard white wheat* (white wheat with 75% or more hard kernels; may contain no more than 10% of white club wheat)

     b. *Soft white wheat* (white wheat with less than 75% of hard kernels; may contain no more than 10% of white club wheat)

     c. *White club wheat* (white club wheat containing no more than 10% of other white wheat)

     d. *Western white wheat* (white wheat containing more than 10% of white club wheat and more than 10% of other white wheat)

  6. *Unclassed wheat* (any variety of wheat that is not classifiable under other criteria provided in the wheat standards; includes red durum wheat and any wheat that is other than red or white in color)

  7. *Mixed wheat* (any mixture of wheat that consists of less than 90% of one class and more than 10% of one other class, or a combination of classes that meet the definition of wheat)

Test Weight

Test weight. A measure of the plumpness and flour or starch yield of the grain.

Test weight per bushel of wheat, for example, depends on its plumpness, shape, density, and moisture content. For a given seed lot, the test weight varies inversely with seed moisture content. It increases by cleaning (breaks away awns of small grains, for example, allowing for closer packing in test kettle). Test weight indicates kernel plumpness and flour or starch yield. Its units are kilograms per hectoliter. A high test weight indicates that the grain has a low percentage of hull, bran, or both (i.e., low percentage of crude fiber) and thus is of higher feed value.

Dockage

Dockage. A measure of the presence of foreign matter in a grain sample.

Dockage measures the presence of foreign matter in the sample. Corn and soybean need dockage, but this determination may not be needed for all grains.

Moisture

Moisture content of the grain can be determined almost instantaneously by using a moisture meter. It may also be determined in the laboratory by drying the seed and noting moisture content before and after drying.

% moisture = [(weight before drying − weight after drying)/weight before drying] × 100

Special Grades Assignments

A numerical grading system is used, whereby grade 1 is superior to grade 5, for example. In addition to these designations, the seed analysis may include additional descriptions of grain quality, such as sour, musty, heated, moldy, disagreeable odor, bin burnt, and infected. In wheat, special grades are

1. *Ergoty wheat*—contains more than 0.30% of ergot
2. *Garlicky wheat*—contains more than two green garlic bulblets per 1,000 g portion
3. *Light smutty wheat*—has the unmistakable odor of smut or contains more than 14 smut balls but less than 30 smut balls per 250 g portion
4. *Smutty wheat*—contains more than 30 smut balls per 250 g portion
5. *Treated wheat*—wheat that has been scoured, limed, washed, sulfured, or treated such that the quality is not reflected by either the numerical grade or the U.S. sample grade designation alone

Analytical Determinations

Analytical determinations differ from one crop to another. They include broken kernels, foreign matter, splits, seed colors, and heat damage.

18.5.4 ASSIGNING THE FINAL GRADE

The final composite grade is determined from a summary of individual factors calculated according to the grade requirements established by the official U.S. standards for grain. For soybean, the grade ranges numerically from U.S. No. 1 (premium, best quality) to U.S. No. 5 (the poorest quality). Grain is assigned the U.S. sample grade when it does not meet the requirements for any of the four grades. Some of the factors are graded based on the minimum allowable for the grade (e.g., test weight), the minimum limit for U.S. No. 1 being 56 lb, while others are based on the maximum observable in the seed lot (e.g., splits, total damage, foreign matter). The final grade is determined by that individual factor (or two) with the lowest (worst) grade. That is, if out of six factors used in the grading process, five are graded No. 2 but one is graded No. 4, the overall grade of the product is U.S. No. 4.

The customary way of writing the final grade is

1. "U.S. No. X," where X is the numeric grade (1–4) or "sample grade"
2. Class/subclass (where applicable)
3. Special grade designation (e.g., infested)
4. Factors that determine the grade

A grade for corn may be *U.S. No. 1 Yellow corn.*

18.6: MANAGING RISK IN CROP PRODUCTION

Crop production is risky business. Crop production depends on the weather, thereby making timeliness of operations critical to success. Yield can be drastically reduced if windows of opportunity are missed (regarding implementation of certain production activities). Unlike manufacturing where, for example, a misassembled car can be disassembled and reassembled correctly, once a crop is planted, certain problems that arise are impossible to correct.

The producer operates in a complex and dynamic environment in which variables include weather, world trade, and government regulations. Agricultural products are generally perishable to some degree. Further, it is often impossible or difficult to halt or change the nature of the enterprise midstream, once the production starts. The crop producer makes decisions whose outcomes or impacts are frequently difficult to predict with certainty. The consequence of this is that the realization from the enterprise may be better or worse than predicted or expected. A glut on the world market immediately translates into lower prices of produce. On the other hand, a shortage due to, for example, increased demand or adverse weather in certain competitive production areas will bring higher returns to producers in certain areas.

Risk is an uncertainty that affects an individual's well-being. It is the possibility of adversity or loss (e.g., of money, endangerment to human health) that, when it occurs, is of consequence to an individual. In other words, risk is uncertainty that matters to the risk taker. Even though uncertainty is a prerequisite for risk to occur, it does not of necessity produce a risky situation.

18.6.1 THERE ARE MANY SOURCES OF RISK IN AGRICULTURAL PRODUCTION

How risky a business is crop production? The sources of concern depend upon the enterprise, the geographic area, government policies, and the individual circumstances, among others. The major categories of sources of risk in crop production include *production or yield risks, price or market risks, institutional risks, personal or human risks,* and *financial risks.*

Production or Yield Risks

Production or yield risks are of primary concern to field crop producers, who are at the mercy of the weather. Unpredicted fluctuations in weather (e.g., excessive rainfall, drought, hail, frost) can wreak untold havoc on crops in the field. Floods wash soil away (erosion) and reduce soil fertility, while drought impedes crop growth and development. Hail and frost cause damage to plant vegetative and reproductive parts, resulting in yield losses. Apart from the physical aspect of the environment, biological factors (e.g., pathogens, weeds, and insect pests) are sources of yield risk in production. If not properly managed, an otherwise healthy crop stand can be devastated by biopests. With advances in technology (e.g., mechanization), farmers face yet another risk whereby an adoption of new technology may not produce expected results. If even successful, this may be for a limited period of time. The producer may be unable to upgrade his or her technology in terms of acquisitions of new and modern machinery. One of the risks of concern to major field crop producers is the possibility of lower than expected yield at harvest time. Yield variability for a crop differs from one geographic region to another. Yield risks are less important to producers who depend on irrigation, since they are able to control this key production factor. In the Corn Belt, yield variability for corn is low in Nebraska, Iowa, and Illinois (i.e., the central states in the Corn Belt), where the soils and climatic conditions are ideal and irrigation is available for production.

Price or Market Risks

The crop producer faces risks associated with unexpected changes in certain market forces that may cause prices to change during the production cycle. Sometimes, certain production inputs experience price increases, while prices of outputs or products may drop. Since these are not anticipated, the profitability of the enterprise is adversely affected. If a crop production operation is geared toward a competitive market, the profitability will be affected by events at local, regional, national, and international levels that have a bearing on supply and demand. Even a trade embargo can cripple a production enterprise.

Like yield variability, price variability differs among commodities. Between 1987 and 1996, dry edible beans, rice, sorghum, and potato experienced price variability of 20% above or below the mean price. Year-to-year changes in price variability are caused by various factors, including crop prospects, changes in government program provisions, and shifts in world supply and demand. Corn price variability was high in the 1920s and 1930s, and again in the 1970s and 1980s. The collapse of grain prices after World War I, coupled with low yields in 1934 and 1936, were primary causes of variability in the period. Government support, stable yield, and consistent demand stabilized corn prices in the 1950s and 1960s. Low yields in 1983 and 1988 were major factors in corn price variability in the 1980s, as was the ill-fated, ineffective grain and oil seed embargo, put into effect by the Carter administration as a punishment of the Soviet Union for its invasion of Afghanistan.

Institutional Risks

Government intervention in agricultural production is a major source of risk for producers. Changes in agricultural policies and regulations can occur at any time during a production cycle. For example, the 1996 Farm Act reduced government intervention in markets for *program crops* such as wheat, corn, and cotton. Such changes may affect producers of specialty crops in very dramatic ways. New laws may be instituted to restrict the use of certain production inputs for certain environmental reasons, new tax policies may be introduced, growers may be compelled to implement certain conservation practices, and new capital lending policies may be introduced that increase production costs.

A survey conducted by the USDA in 1997 (following the Farm Act of 1996) indicated that the top three risk factors of concern to producers were price risk, yield risk (decline in yield), and institutional risk (changes in government law and regulations).

Personal or Human Risks

Agricultural production is subject to human risks that are common to other operations. Sickness, death, personal injury, or problems associated with personal family life are some unexpected events that can destabilize an enterprise, causing loss in profitability. Acts of God and accidents such as fire or theft are sources of asset risk. Critical stages in crop production may be jeopardized by such events.

Financial Risks

Farmers acquire capital from commercial banks for their operation. Interest rates vary from time to time. A higher interest rate raises the input costs and may cause a producer to change production plans. Unexpected expenses may cause the producer to experience cash flow problems and be in danger of defaulting on a loan.

18.6.2 THE GOAL OF RISK MANAGEMENT IS REDUCING THE EFFECT OF RISK BY CHOOSING AMONG AVAILABLE ALTERNATIVES FOR REDUCING THE EFFECT

How can a producer reduce risk in crop production? To manage risk effectively, the producer usually has to evaluate the tradeoff between factors such as changes in risk, expected returns, and entrepreneurial freedom. These alternatives consist of various combinations of production activities with uncertain outcomes and varying levels of expected return. Using risk management does not in and of itself avoid risk. Rather, it balances risk and return according to a producer's capacity to withstand a range of outcomes from a production enterprise. There are several options available to a farmer in risk management. The producer's attitude toward risk is important in selecting an option. Some producers are less risk-averse than others and, further, producers face different degrees of variability in price and yield. Thus, there is no one best approach to risk management for all farmers.

Enterprise Diversification

Enterprise diversification is a risk management approach whereby the producer adjusts the enterprise mix so as to participate in more than one activity simultaneously. The rationale for this approach is that, should one of the component activities fail, the other(s) might succeed to avert a total loss. In effect, enterprise diversification reduces risk within the farming operation. To this end, a crop producer may grow more than one crop—for example, corn and soybean or wheat and barley. In some cases, a crop producer may include livestock. The degree of diversification varies from one farming region to another. In the United States, the Northern Plains and Corn Belt are among the most diversified of the 10 farming regions. Cotton farmers are also very diversified.

Farmers are more likely to diversify if they are faced with high risk in their operations. Sometimes, they are motivated by anticipated profits to diversify. Diversification does not automatically mean profitability. Sometimes, returns decline with diversification. This is because diversification comes with new management problems. The producer may not have adequate skills to manage the new problems. Further, there is often the need to acquire additional equipment and labor and to locate new markets for the new products. Production specialization has its advantages. The producer acquires more skills in conducting the enterprise and can operate more efficiently. Farmers with high net worth tend to be more specialized in their operations than those with lower net worth. Beginning farmers also tend to specialize and may diversify when they expand their operation.

Climatic conditions and production resources limit enterprise diversification. Gross income among crops in a diversified enterprise is highly correlated. Diversification beyond two to three crops may not be profitable. Another factor in enterprise diversification is geographic diversification (i.e., conducting the operation in different regions simultaneously), which tends to reduce the climatic effect. However, such an approach entails additional transportation and managerial costs, among other factors.

Vertical Integration

Vertical integration is a production system in which the producer retains ownership control of a commodity across two or more levels of activity. This strategy also reduces risk within the farming operation. In animal production, this means that a farmer may raise the corn and hay needed to feed the livestock, instead of renting pasture or purchasing feed. In poultry production, the farmer may have a hatchery to produce chicks, a feed mill, and even a processing plant. In crop production, wheat farmers may form a cooperative to develop a flourmill to process their grain for value added to increase profits. Some cooperatives may even go a step further to include a bakery or some other product such as pasta. This vertical integration reduces risk and enhances profitability of the enterprise.

18.7: PRODUCTION CONTRACTS

A *production contract* is a system of production by contractual agreement in which a contractor or the buyer of the commodity retains control of the product during production and provides specific guidelines for production for a certain desired quality. The end product is designed for the special need of the contractor. This strategy transfers a share of the risk outside the farm. Frequently, specialized equipment, inputs, and management are needed to produce the product in a timely fashion. The advantages of this strategy are lower start-up costs, lower income risk, and guaranteed market access. However, production contracting can limit entrepreneurial capacity of crop producers and, more important, these contracts can be terminated on short notice. In 1997, 31.2% of total U.S. production was covered by contracts.

The types of production contracting differ in the roles of the contractor and producer in terms of control and the risk and uncertainty both parties assume.

18.7.1 PRODUCTION MANAGEMENT CONTRACTS

In the production management contract system, the producer and the contractor agree upon a certain price at the time of signing the contract. In return, the contractor has control over production activities such as time of planting and the cultivars to plant. By this arrangement, the contractor ensures that the end products will meet all of his or her special needs. If the product quality is as desired, the producer receives the agreed payment. However, if quality drops, the price may be discounted. Further, in times of production shortfalls due to crop failure, the producer receives little or no income. Vegetable producers often enter into this type of agreement.

18.7.2 RESOURCES-PROVIDING CONTRACTS

In *resources-providing contracts,* the producer provides the infrastructure (e.g., housing and land) and is responsible for maintenance and repairs, labor, and utilities. The contractor provides other inputs and management. In the poultry industry, for example, inputs include chicks, feed, veterinary care, transportation, and management. The producer, in actuality, acts as a custodian of the operation for the contractor.

The benefits of this arrangement for the producer include guaranteed market access and improved efficiency, access to technical advice, managerial expertise, and quality genotypes. On the other hand, the entrepreneurial capacity of the producer is diminished. Further, contracts can be terminated on short notice. The contractor may demand renovations and upgrading of the production facilities in the future. When the contractor manages the production, the producer may not be entirely satisfied with accounting procedures and other production decisions.

18.7.3 MARKETING CONTRACTS

Marketing contracts are pricing arrangements made (written or verbal) between a contractor and a producer before harvesting a crop or marketing the produce. In 1997, 21.7% of commodities in the United States were covered by marketing contracts. The arrangement is more commonly used for pricing field crops and also specialty crops. The contractee has ownership of the commodity during production, and is responsible for most or all of the production decisions (e.g., crop cultivars to plant, supplies, finance, and other inputs). The contractor purchases a quantity and quality of product at a prenegotiated price. Marketing contracts take various forms, one of the most common being a *forwarding contract.*

Marketing contracts. Agreements by which a producer offers commodities for sale.

Price risk is completely eliminated, however, only when an exact price to be paid to the producer is established prior to delivery. The producer assumes all production risks. However, a pricing arrangement reduces price risk. In addition to price risk, forward contracting also has yield risk. Farmers should not forward contract 100% of their produce. This is so advised because adverse weather may cause yield shortfalls. When this happens, the producer will be compelled to purchase the deficit from other sources to satisfy contractual obligations. A type of market pricing called *deferred (delayed) price contract* entails negotiating a price when grain is delivered. Yield risk is hence eliminated by this pricing arrangement.

18.7.4   HEDGING IN FUTURES

As mentioned previously, a forward, or futures, contract is a risk-reducing strategy whereby risk is shifted from a party that desires less risk (called the *hedger*) to one who is willing to accept the risk in exchange for an expected profit (the *speculator*). That is, a futures contract is a commitment to trade in the future. It is an agreement between both parties that is priced and entered on an *exchange*. There are a variety of exchange institutions in the United States that trade in agricultural futures contracts. These include the Chicago Board of Trade, the Chicago Mercantile Exchange, and the Minneapolis Grain Exchange. One of the roles of these exchanges is to enforce contract terms and guarantee the contracts. Even though actual delivery and payment are delayed until the contract matures, the parties to the contract are required to make marginal deposits with their brokers to guarantee their respective contractual obligations. Hedging is usually ineffective as a risk management strategy as yield variability increases and the relationship between prices and yields becomes more negative.

18.7.5   FUTURES OPTIONS CONTRACTS

A *futures options contract* is a strategy similar to that used by insurance agencies. The holder of a commodity option (the hedger) pays a certain premium that, like insurance, protects the hedger against unfavorable events. Futures prices are dynamic and may decrease or increase. If the price change is favorable, the hedger may exercise the option or sell at a profit. If the price is low, the hedger may allow the option to expire. In other words, a futures options contract gives the hedger the right but not the obligation to take a futures position at a specified price before a specified date.

18.7.6   MAINTAINING FINANCIAL RESERVES AND LEVERAGING

The crop producer may use debt to finance an operation. This strategy is called *leveraging*. A high degree of leveraging increases the potential for bankruptcy, especially in a season of low farm returns. The producer increases the farm leverage through borrowing of capital for business expansion and sustenance. However, borrowing increases indebtedness to the financial institution and increases the risk of loan default, due to the fact that field crop production is a highly risky operation. Producers who are highly leveraged are more likely to adopt the strategy of hedging in risk management.

Liquidity. A measure of one's ability to generate cash quickly and efficiently, through asset disposal, to meet financial obligations.

18.7.7   LIQUIDITY

Liquidity is the ability of the producer to generate cash quickly and efficiently to meet financial obligations. This requires certain assets to be converted to cash through sale. Assets such as grain in storage are said to be *liquid* because, when sold, they generate a cash amount that is equal to or greater than the reduction in the value of the farm due to the sale. On the other hand, fixed assets such as land and machinery are said to be *illiquid* because they generally cannot be quickly sold without the owner accepting or discounting price. It is important that, in case of adversity, the producer have assets that can be quickly converted to cash to be used to meet financial obligations.

18.7.8   LEASING INPUTS AND HIRING CUSTOM WORK

Instead of purchasing and owning all production inputs, a crop producer may manage production risk by leasing, for example, land, equipment, or machinery, or by hiring labor to assist during periods of specific activities such as harvesting. In leasing, the leasee has control over someone else's asset during the contractual period. The leasee is also obligated to make regular payments to the owner of the asset. However, payments do not have to be made over the long term, only during the contractual period. Leasing limits fixed cost and enables a producer to operate at reduced capital investment, thus reducing financial risk.

18.7.9   INSURING CROP YIELDS AND CROP REVENUES

The concept of *insurance* is that the insured agrees to exchange a fixed, small, regular payment, called the *premium,* for protection in the future from uncertain but potentially larger losses. There are different types of policies. This strategy works through risk pooling, in which a large number of participants in the program contribute, through premium payment, to the fund from which benefits are paid to a participant during a loss.

Through the Multiperil Crop Insurance program, the government is able to compensate producers in time of adversity. Through the Federal Crop Insurance Reform Act of 1994, the government increased the premium subsidy given to producers. Grower participation also increased because of such subsidies. Crop revenue insurance pays indemnities to crop producers based on revenue shortfalls instead of yield or price shortfall. There are a variety of such revenue insurance programs, including Crop Revenue Coverage, Income Protection, Revenue Assurance, Group Risk Income Protection, and Adjusted Gross Revenue.

18.7.10   OFF–FARM EMPLOYMENT AND OFF–FARM INCOME

Crop producers may supplement their farm income by engaging in off-farm activities. Off-farm income is a form of diversification (of income sources). In Chapter 1, it was indicated that the average farm operator supplemented his or her income by 13% from off-farm business income in 1995. In 1996, 82% of all farm households reported off-farm income exceeding farm income.

18.7.11   CULTURAL PRACTICES

The use of early-maturing cultivars allows the producer to reduce yield risk by producing within the season to avoid the effects of abnormal weather. Where production occurs under drought-prone conditions, supplemental information can be used to protect against yield loss.

18.8: U.S. AGRICULTURAL PRODUCT MARKETING

The U.S. has one of the most successful agricultural production systems in the world. Whereas much of the production is geared toward local markets, the U.S. is engaged in extensive agricultural export trade with countries all over the world.

Table 18–1   U.S. Agricultural Exports: Forecast and Recent Performance

Commodity	1997	1998	1999	2000	2001	2002	2003
	Billion Dollars						
Grains and feeds	16.5	14.1	14.4	13.9	13.9	14.1	16.5
Oilseeds and products	11.5	11.1	8.7	8.6	8.8	9.6	9.8
Livestock products	7.6	7.5	7.1	8.5	8.8	8.7	8.9
Poultry and products	2.9	2.7	2.1	2.2	2.5	2.4	2.6
Dairy products	0.8	0.9	0.9	1.0	1.1	1.0	1.1
Tobacco, unmanufactured	1.6	1.4	1.4	1.4	1.2	1.2	1.3
Cotton and linters	2.7	2.5	1.3	1.8	2.1	2.3	2.7
Seeds	0.8	0.8	0.8	0.8	0.7	0.8	0.8
Horticultural products	10.6	10.3	10.3	10.5	11.1	11.2	11.5
Sugar and tropical products	2.1	2.1	2.1	2.3	2.6	2.3	2.4
Total value	57.3	53.6	49.1	50.7	52.7	53.5	57.5

Source: Economic Research Service, USDA.

18.8.1   EXPORT AND IMPORT OF AGRICULTURAL PRODUCTS

Agriculture is a major contributor to the U.S. economy. The projected value of U.S. exports in fiscal 2003 was $57.5 billion, representing a gain of 7.5 percent above the 2002 export. The increase is attributed primarily to gains in bulk commodity exports as well as high-value products (Table 18–1). Similarly, imports of agricultural products for 2003 was projected to increase from $42 billion in 2002 to $43.5 billion, representing a 4 percent increase. Most of the projected increase in imports consists of horticultural product (fruits, juices, wines, malt-beverages), and comes from Canada and Mexico.

Bulk product exports from the U.S. include wheat, rice, coarse grains (corn, sorghum, barley), soybeans, cotton, and tobacco. The projected increase in agricultural exports is attributed primarily to gains from bulk commodities that are estimated at $21 billion, an increase of 14 percent over the 2002 values.

18.8.2   U.S. AGRICULTURAL EXPORT MARKETS

Canada is the single most important market for U.S. agricultural products. The largest European markets for U.S. agricultural products are the European Union (especially Netherlands, United Kingdom, Germany), and the Former Soviet Union (Table 18–2). In Asia, the key markets for U.S. agricultural products are Japan, the Republic of Korea, China, Taiwan, Hong Kong, Philippines, and Indonesia. North Africa, Egypt, Nigeria, and South Africa are the largest markets for U.S. agricultural products in Africa, whereas Mexico, Central America, Colombia, Venezuela, and Brazil are the most important markets in Latin America. The United States does a considerable amount of agricultural trades with the Caribbean Islands as well.

18.8.3   VALUE ADDED TO THE U.S. ECONOMY

In terms of field crops, feed crops and oil crops are the most important contributors to the U.S. economy, accounting for $23.2 billion and $14.3 billion in 2001, respectively. Food grains are also important, accounting for $6.6 billion to the U.S. economy in the same year. Fruits and tree nuts, and vegetables are also important crop products, contributing $11.7 billion and $5.5 billion, respectively, in 2001 (Table 18–3).

Table 18–2 U.S. Agricultural Exports by Region

Region and Country	2000	2001	2002F
		$ Million	
Western Europe	6,532	6,761	7,300
European Union	6,193	6,249	6,500
Belgium-Luxembourg	514	625	-
France	348	352	-
Germany	910	907	-
Italy	559	509	-
Netherlands	1,388	1,398	-
United Kingdom	1,028	1,048	-
Spain (plus Canary Islands)	641	590	-
Other Western Europe	340	512	800
Switzerland	250	422	-
Eastern Europe	168	201	200
Poland	47	83	-
Former Yugoslavia	67	44	-
Romania	12	24	-
Former Soviet Union	921	1,029	900
Russia	659	823	700
Asia	21,917	22,271	21,900
West Asia (Mideast)	2,364	2,190	2,600
Turkey	701	564	800
Iraq	8	8	-
Israel (plus Gaza, W. Bank)	459	435	-
Saudi Arabia	481	470	400
South Asia	415	570	900
Bangladesh	82	104	-
India	185	294	-
Pakistan	93	97	-
China	1,465	1,875	1,700
Japan	9,301	8,942	8,100
Southeast Asia	2,580	2,907	2,800
Indonesia	675	877	800
Philippines	866	836	800
Other East Asia	5,791	5,786	5,700
Korea (Republic)	2,531	2,541	2,800
Hong Kong	1,249	1,252	1,100
Taiwan	2,002	1,986	1,900
Africa	2,236	2,126	2,500
North Africa	1,522	1,464	1,700
Morocco	139	120	-
Algeria	254	211	-
Egypt	1,056	1,004	1,100
Sub-Sahara	715	662	800
Nigeria	160	233	-
South Africa	165	108	-

continued

Table 18–2   U.S. Agricultural Exports by Region—*continued*

Region and Country	2000	2001	2002F
	$ Million		
Latin America and Caribbean	10,614	11,561	11,700
Brazil	253	219	300
Caribbean Islands	1, 463	1,398	1,500
Central America	1,132	1,191	1,200
Colombia	427	442	500
Mexico	6,307	7,277	7,100
Peru	200	182	-
Venezuela	405	416	300
Canada	7,512	7,994	8,500
Oceania	487	472	500
Total	**50,744**	**52,699**	**53,500**

Source: Economic Research Service, USDA.

Table 18–3   Valued Added to the U.S. Economy by the Agricultural Sector

	1998	1999	2000	2001	1992-2001 (average)
	$ Billions				
Final crop output	101.7	92.4	95.0	93.9	97.8
Food grains	8.8	7.0	6.8	6.6	8.7
Feed crops	22.6	19.6	20.8	23.2	22.6
Cotton	6.1	4.6	3.8	5.0	5.7
Oil crops	17.4	13.4	13.8	14.3	15.2
Tobacco	2.8	2.3	2.3	1.9	2.6
Fruits and tree nuts	11.8	12.0	12.6	11.7	11.5
Vegetables	15.2	15.1	15.6	15.5	14.5
All other crops	17.2	18.0	18.4	18.2	16.2
Home consumption	0.1	0.1	0.1	0.1	0.1
Value of inventory adjustment	-0.3	0.4	0.8	-2.7	0.8
Final animal output	94.2	95.3	99.3	106.3	94.1
Meat animals	43.3	45.6	53.0	53.3	47.9
Dairy products	24.1	23.2	20.6	24.7	21.5
Poultry and eggs	22.9	22.9	21.8	24.6	20.7
Miscellaneous livestock	3.7	3.9	4.2	3.9	3.5
Home consumption	0.3	0.4	0.4	0.4	0.4
Value of inventory adjustment	-0.3	-0.6	-0.6	-0.5	0.0
Service and forestry	23.8	25.2	24.4	25.5	21.2
Machine hire and customwork	2.2	2.0	2.2	2.0	2.1
Forest products sold	3.1	2.8	2.9	2.8	2.7
Other farm income	8.7	10.2	8.7	10.1	7.0
Gross imputed rental value of farm dwellings	9.9	10.2	10.7	10.6	9.4

Source: Economic Research Service, USDA.

18.8.4 THE IMPACT OF NAFTA

Implemented on January 1, 1994, the North American Free Trade Agreement (NAFTA) provides for the progressive dismantling of most of the barriers to trades and investment among Canada, Mexico, and the United States, over a 14-year period, ending on January 1, 2008. The agreement incorporates the Canada-U.S. Free Trade Agreement (CFTA), whose implementation was completed on January 1, 1998. The implementation of NAFTA has generally been beneficial to the U.S. agriculture and related industries, being credited at least in part for the doubling of U.S. trade with Canada and Mexico in the 1990s (Figure 18–1).

NAFTA's impact on trade is substantial (about 15 percent or more) for certain commodities, but negligible for others (Table 18–4). The products whose trade was severely restricted before the implementation of CFTA and NAFTA are the ones that have experienced significant boost in trade. The high decrease in sorghum import by Mexico is attributed to the high increase in corn import that has caused some livestock producers to switch from using sorghum as feed to corn, because of the increased availability of the latter. Also, even though U.S. wheat imports from Canada have significantly increased, exports of the crop to Canada is negligible.

Generally, the United States is the principal foreign supplier of rice to Mexico. This is because the United States meets the strict phytosanitary standards demanded by Mexico, conditions that other exporters from Asia are not able to satisfy. Mexico imports mainly rough rice (unmilled), a product that Asian producers do not currently provide because the

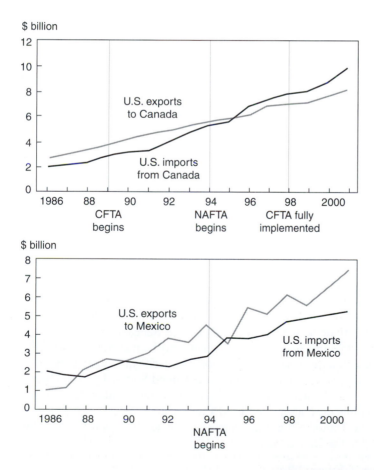

FIGURE 18–1 The impact of NAFTA on trade United States and partners—Canada and Mexico. (Source: USDA)

Table 18-4 Changes in Trade Attributed to the Implementation of NAFTA

	Annual Average of Actual Trade					
	Value (U.S. $ Million)		Volume (1,000 Units)			Estimated Effect of NAFTA
Selected Commodities	1990-93	1994-00	1990-93	1994-00	Units	
U.S. exports to Canada						
Beef and veal	349	317	82	92	Mt	up, high
Wheat products	22	48	27	66	Mt	up, high
Cotton (plus linters)	62	91	42	60	Mt	up, medium
Processed tomatoes	71	109	NA	NA	NA	up, medium
U.S. exports to Mexico						
Rice	41	87	161	386	Mt	up, high
Dairy products	151	160	NA	NA	Mt	up, high
Cotton (plus linters)	102	314	80	234	Mt	up, high
Processed potatoes	10	37	12	40	Mt	up, high
Fresh apples	28	61	54	112	Mt	up, high
Fresh pears	16	26	31	51	Mt	up, high
Corn	178	521	1,557	4,322	Mt	up, medium
Oilseeds	401	739	1,662	2,953	Mt	up, medium
Beef and veal	149	306	50	106	Mt	up, medium
Sorghum	402	336	3,687	3,073	Mt	down, high
U.S. imports from Canada						
Wheat (minus seed)	136	268	1,109	1,920	Mt	up, high
Wheat products	38	98	72	185	Mt	up, high
Beef and veal	260	638	111	264	Mt	up, high
Corn	21	31	218	268	Mt	up, medium
Fresh/seed potatoes	51	77	274	380	Mt	up, medium
Processed potatoes	51	209	92	322	Mt	up, medium
Cattle and calves	741	857	1,063	1,185	Hd	down, high
U.S. imports from Mexico						
Wheat products	4	14	6	22	Mt	up, high
Cattle and calves	388	300	1,144	965	Hd	up, high
Peanuts (all)	*	3	*	4	Mt	up, high
Sugar (cane and beet)	1	17	2	49	Mt	up, high
Fresh tomatoes	264	470	322	608	Mt	up, medium
Processed tomatoes	15	16	NA	NA	NA	up, medium
Cantaloupe	40	47	120	136	Mt	up, medium

Up = increase; Mt = metric tons; NA = not available; Hd = heads.
Source: Economic Research Service, USDA.

processing of rough rice provides the much needed jobs for the countries. NAFTA has boosted U.S. soybean exports to Mexico in response to demand by the livestock industry. The increase in trade between Canada and United States has been two-way, and primarily involves vegetable oil. The exports of peanuts to United States from Mexico have substantially increased because of the creation of a tariff rate quota (TRQ) for raw peanuts. Mexico exports peanut butter and past to the United States. The TRQ barrier currently remains between the United States and Canada for peanut butter. United States has experienced a doubling in cotton exports to Canada and Mexico, while Mexico's export of apparels to the United States has increased. Canada enjoys a boost in fresh and frozen potato export to the United States.

Table 18-5 Prices of Principal U.S. Agricultural Trade Products

	1999	2000	2001
Export Commodities			
Wheat, f.o.b. vessel, Gulf ports ($/bu.)	3.04	3.17	3.50
Corn, f.o.b. vessel Gulf ports ($/bu)	2.29	2.24	2.28
Grain sorghum, f.o.b. vessel, Gulf ports ($/bu)	2.14	2.23	2.42
Soybeans, f.o.b. vessel, Gulf ports ($/bu)	5.02	5.26	4.93
Soybean oil, Decatur ($/lb)	17.51	15.01	14.49
Soybean meal, Decatur ($/ton)	141.52	174.69	168.49
Cotton, 7-market avg. spot (c/lb)	52.30	57.47	39.68
Tobacco, avg. price at auction (c/lb)	177.82	182.73	186.21
Rice, f.o.b. mill, Houston ($/cwt)	16.99	14.83	14.55
Inedible tallow, Chicago (c/lb)	12.99	9.92	12.50
Import Commodities			
Coffee, NY, spot ($/lb)	1.05	0.92	0.55
Rubber, NY, spot ($/lb)	36.66	37.72	33.88
Cocoa beans, NY ($/lb)	0.47	0.36	0.47

Source: Economic Research Service, USDA

18.8.5 PRICES OF PRINCIPAL US AGRICULTURAL PRODUCTS

The prices of selected products involved in agricultural trade between the United States and other nations are summarized in Table 18–5. The prices of rice and cotton have significantly declined over the past several years. The port prices for wheat and tobacco have also increased over the past several years.

SUMMARY

1. Marketing is critical to the success of a commercial crop production enterprise.
2. Storage and transportation are critical components of a marketing strategy. This is largely because agricultural produce is generally perishable, and products need to be transferred from the farm to customers some distance away.
3. For fair pricing and standardization, crop produce is graded or placed into quality classes.
4. The grading process involves various laboratory analyses for characteristics including dockage, moisture content, market class, and test weight.
5. Frequently, a middleman who provides a variety of services, including processing, packing, and warehousing, separates the producer and customer.
6. Farm management involves decision making, problem solving, and resource allocation, for productivity and profitability.
7. The crop producer operates as a farm manager.
8. Some production factors are within the control of the producer, others are outside his or her control, while some can be manipulated within limits.
9. The basic farm management problems are what to produce, how much to produce, and how to produce the products.
10. An economic problem is characterized by three basic factors: production goals or objectives, limited resources, and alternative uses.

11. A good decision process involves certain steps: identification of a problem, data and information collection, identification and analysis of alternatives, decision making, implementation, and evaluation.
12. There are two categories of decisions—organizational (that are long-lasting) and operational (that are short-term).
13. The functions of a manager are the planning, implementation, and control of production processes.
14. Crop production depends on the weather for success. Risk management is a key activity in production management. There are several types of risk, including production, market, institutional, human, and financial risks.
15. In managing risks, a producer may adopt enterprise diversification or vertical integration.
16. Producers may also manage risk through marketing contracts, futures contracts, liquidity, and crop insurance.

REFERENCES AND SUGGESTED READING

Agricultural Outlook. 2000. *Managing farm risk: Issues and strategies.* Washington, DC: Economic Research Service/USDA.

Anderson, K. B. *Grain producer marketing alternatives.* Oklahoma State University, Current Report, CR-480.

Harwood, J., R. Heifner, K. Coble, J. Perry, and A. Somwaru. 1999. *Managing risk in farming: Concepts, research, and analysis.* Washington, DC: Economic Research Service/USDA Report 774.

Rhodes, V. J., and J. L. Dauve. *The agricultural marketing system.* Scottsdale, AZ: Holcomb Hathaway.

Tilley, D. S., and K. B. Anderson. *Wheat producer's marketing objectives and plans.* Oklahoma State University, Extension Facts, No. 472.

SELECTED INTERNET SITES FOR FURTHER REVIEW

http://www.extension.umn.edu/distribution/horticulture/DG7618.html

marketing fresh produce

http://www.agednet.com/lesmk.htm

presentations on marketing of crops

http://www.cbot.com/cbot/www/main/0,1394,00.html

Chicago Mercantile Exchange

http://www.kcbt.com/

Kansas City Board

http://www.mgex.com/

Minneapolis Grain Exchange

http://www.nybot.com/

New York Board of Trade

OUTCOMES ASSESSMENT

PART A

Answer the following questions true or false.

1. T F   Soft red winter is a valid market class of wheat.
2. T F   In grain grading, the higher the dockage, the higher the grade.
3. T F   The final grade of a product depends on the individual factor in the composite rating with the lowest score.
4. T F   A grade of 1 indicates a product superior to grade 2.
5. T F   Grain in storage is considered a liquid asset.
6. T F   Futures contracts are entered on an exchange.
7. T F   Enterprise diversification automatically means profitability.
8. T F   The crop producer is a manager of production resources.

PART B

Answer the following questions.

1. What is dockage?

2. Give the market classes of wheat.

3. Give the steps in grain grading process.

4. What is grain test weight?

5. How is seed moisture determined in the laboratory?

6. Give the three categories of management.

7. Give the three basic functions of a manager.

8. What is risk?

9. Define liquidity.

10. The two parties in a futures contract are (a) _____ and (b) _____ .

11. Give one specific example of each of the following crop production factors:
    a. Factors within total control of the producer
    b. Factors that can be manipulated by the producer
    c. Factors that are outside the control of the producer

PART C

Write a brief essay on each of the following topics.

1. Discuss the role of the middleman in crop production.

2. Discuss the role of transportation in the marketing of produce.

3. Give the importance of sampling in the grain grading process.

4. Discuss, using wheat as an example, market class of a product.

5. Describe how grain is sampled for analysis for grading.

6. What is farm management?

7. Briefly discuss the three basic types of management problems (what to produce, how much to produce, and how to produce).

8. Give the steps involved in a management decision-making process.

9. Distinguish between organizational and operational decisions in management.

10. What are marketing contracts?

11. Give the types of risk.

12. Discuss production risks.

PART D

Discuss or explain the following topics in detail.

1. Is crop production worth the risk involved?

2. What preparation or experience would make a crop producer an effective risk manager?

PART TWO
COMMERCIAL PRODUCTION OF SELECTED FIELD CROPS

19
Wheat (Common)

TAXONOMY

Kingdom	Plantae	Subclass	Commelinidae
Subkingdom	Tracheobionta	Order	Cyperales
Superdivision	Spermatophyta	Family	Poaceae
Division	Magnoliophyta	Genus	*Triticum* L.
Class	Lilliopsida	Species	*Triticum aestivum* L.

KEY TERMS

Custom harvesting	Awned floret	Milk stage
Winter wheat	Awness floret	Hard Hough stage
Spring wheat	Boot	Patent flour
Gluten	Flag leaf	

19.1: ECONOMIC IMPORTANCE

Wheat is the most important cereal grain crop in the world. It is the principal cereal grain crop used for food consumption in the United States and most other parts of the world. In the United States, it usually ranks fourth after corn, hay, and soybeans in importance (Figure 19–1). Wheat is grown commercially in nearly every state in the United States, with a concentration of production in the Great Plains, an area spanning

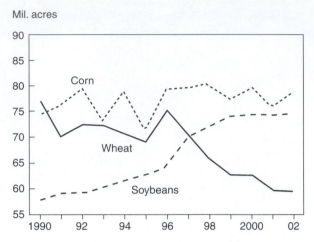

Mil. acres

FIGURE 19–1 Acreage trends of selected major field crops in the United States show a general decline for wheat after 1996, whereas soybean acreage has been on the rise. Corn acreage has been erratic. (Source: USDA)

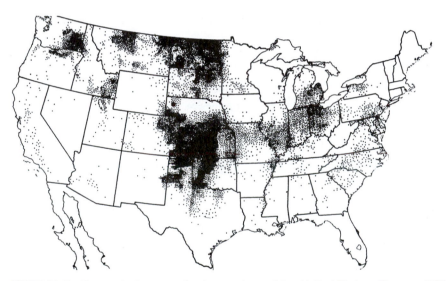

FIGURE 19–2 General wheat production regions of the United States. (Source: USDA)

from Texas to Montana. USDA production trends indicate that, in 1866, wheat was harvested from an area of 15.4 million acres, yielding an average of 11 bushels per acre. By 1950, production had occurred on 61.6 million acres, with an average yield of 16.5 bushels per acre. In 1990, the acreage was 69.2 million acres, yield average being 39.5 bushels per acre. The Central and Southern Plains (Texas, Oklahoma, Kansas, among others) produces more than the Northern Plains (e.g., Montana, North Dakota), the two regions accounting for two-thirds of U.S. wheat production and about 80% of wheat acreage (Figure 19–2). Kansas leads all states in wheat production.

World production of wheat in 2001 was 583.9 million metric tons, occurring on 219.5 million acres. World wheat consumption in that period was 590.6 million tons. Developing countries (excluding those in Eastern Europe and Russia) account for nearly

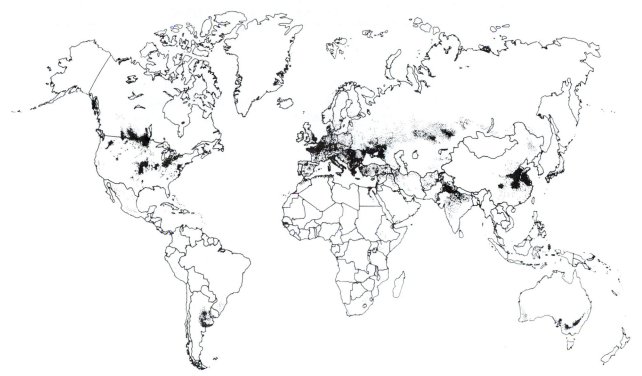

FIGURE 19–3 General wheat production regions of the world. (Source: USDA)

50% of the world's wheat production, the leading producers being China, India, Turkey, Pakistan, and Argentina (Figure 19–3). The success of wheat production in these countries is credited to the impact of the Green Revolution, which occurred in the 1960s and 1970s. In 2000, China produced 111.9 million metric tons, whereas India produced 26.5 metric tons. Latin America and Asia (excluding China and India) each produce about 20 million metric tons a year. Wheat is produced in Europe, including in the United Kingdom, Denmark, the Netherlands, Belgium, Switzerland, and Germany.

19.2: ORIGIN AND HISTORY

Wheat is believed to have originated in southwestern Asia. A cross between wild Emmer wheat *(Triticum dicoccoides)* and *Aegilops squarrosa,* a grass, produced a speltlike plant, suggesting that spelt *(T. aestivum* subspp *vulgare)* may have originated as a natural cross between these two species. Further, this suggests that the common, or bread, wheat *(T. aestivum)* is descended from a cross between spelt and the progenitor of Persian wheat *(T. persicum)*. Persian wheat grows in the wild in the Russian Caucasus. The Persian wheat probably is descended from the wheat of the Neolithic Swiss lake dwellers, which in turn might have originated from a cross between einkorn and a grass, *Agropyron triticum*. Archaeological findings indicate that Emmer wheat was cultivated before 7000 B.C. Similarly, wheat was cultivated in Europe in prehistoric times. In the United States, wheat was first cultivated along the Atlantic Coast in the early seventeenth century, moving westward as the country was settled.

19.3: ADAPTATION

Wheat is best adapted to cool, temperate climates where rainfall is not excessive (15–24 inches per annum). High rainfall coupled with high temperature predisposes the crop to diseases (e.g., *Septoria*), lodging, harvestime problems, and postharvest deterioration. However, insufficient rainfall is the main limiting factor in wheat production in the western United States. Wheat produced in the United States may be categorized according to the season in which it is sown, as *winter wheat* or *spring wheat*.

Winter wheat is concentrated in the Great Plains (Oklahoma, Texas, Kansas) and the North Central (Michigan and Ohio) and Pacific regions. It is sown in the fall, so that it can have some growth before the onset of cold weather in winter. Growth ceases and the plants remain dormant through winter, resuming growth in the spring for harvesting in summer. About two-thirds of U.S. wheat is winter wheat. Spring wheat is planted in early spring and harvested in July–August. Winter wheat can survive cold temperatures as low as −40°F if protected by snow.

Spring wheat is produced in the Northern Plains (Montana, North Dakota, South Dakota, and Minnesota). It is less tolerant of low temperatures and is damaged by even a light frost of 28° to 30°F.

Wheat is a longday plant. Short days of high temperatures stimulate tillering and leaf formation but delay the flowering of wheat plants. Early-maturing varieties are available for production under any photoperiod conditions. However, the quality (for nutritional uses, such as baking) of wheat is influenced by the production environment. For example, growing hard wheats in soft wheat regions results in grains that are starchy or "yellow berry" (soft and starchy).

The ratio of protein to starch in the wheat kernel is a function of the moisture, temperature, and available soil nitrates at the time of blooming as well as after flowering (fruiting). When the weather is cool with high rainfall and high relative humidity, the fruiting period is extended. The kernels produced tend to be plump, starchy, and low in protein. When the soil nitrates are sufficient but the fruiting period is not extended, the kernels are plump and fairly high in gluten. Hot, dry weather results in dark, hard, vitreous bread wheat kernels that are high in protein content. Generally, high protein wheat of kernels occurs in soils of high nitrogen content and low moisture at the time of maturity. Soils that are fertile and well drained, silt, and clay loams produce the best wheat crop. Wheat produces poorly on poorly drained or very sandy soil.

19.4: BOTANY

Wheat (*Triticum* spp) is an annual plant (Figure 19–4). Its spikelet inflorescence consists of a sessile spikelet at each of the notches of a zigzag rachis. A spikelet is made up of two broad glumes and has several florets. A floret is composed of a lemma, a palea, and a caryopsis, or grain, with a deep furrow and a hairy tip or brush. The floret may be *awned* or *awnless*. Awned varieties are common in regions of low rainfall and warm temperatures. The presence of awns also tends to influence transpiration rate, accelerating the drying of ripe grain. Consequently, the tips of awnless spikes tend to be blasted in hot, dry weather. The grain may also be amber, red, purple, or creamy white in color.

Under normal high-density production conditions, a wheat plant may produce 2–3 tillers. However, when amply spaced on fertile soils, a plant may produce 30–100 tillers. The spike (head) of a plant may contain 14–17 spikelets, each spike containing 25–30

FIGURE 19–4 A wheat field (left). A head of mature wheat (right). (Source: USDA)

grains. Large spikes may contain 50–75 grains. The grain size varies within the spikelet, the largest being the second grain from the bottom and decreasing in size progressively toward the tip of the spike.

Wheat is predominantly self-pollinated. Anthers assume a pendant position soon after the flower opens. Blooming occurs at temperatures between 56° and 78°F, starting with the spikelet around the middle of the spike and proceeding upward and downward. The wheat kernel or berry is a caryopsis varying between 3 and 10 mm long and 3–5 mm wide. It has a multi-layered pericarp that is removed along with the testa, nucellus, and aleurone layers during milling. The endosperm makes up about 85% of a well-developed kernel. Below the aleurone layer is a complex protein, called *gluten,* that has cohesive properties. It is responsible for the ability of wheat flour to hold together, stretch, and retain gas as fermented dough rises. This property is available to the flour of only one other species, rye flour.

Wheat is classified based on three primary characteristics—agronomic production needs, kernel color, and endosperm quality (kernel hardness). In terms of agronomic needs, wheat may be classified as either winter wheat or spring wheat, depending on temperature adaptations. There are two seed coat colors—*red* and *white.* Red is conditioned by three dominant genes, the true whites comprising recessive alleles of all three genes. Most wheat varieties in the United States are red. Kernel hardness is classified into two—*hard* or *soft.* Hard wheats have a proteinacious material on the surface of the cell walls. Upon milling, hard wheat yields coarse flour. White wheats, lacking in this starch-protein complex, produce a higher yield of fine flour upon milling. Hard wheat is used for bread making because its gluten protein is cohesive and elastic.

19.5: Species of Wheat

A wide variety of species of wheat of various ploidy levels are known (Table 19–1). The common wheat, *Triticum aestivum,* is a hexaploid. It is usually dorsally compressed. Even though it consists of spikelets that may have two to five flowers, a mature spikelet

Table 19-1 Species of Wheat and Their Ploidy Levels

1. Einkorn Group (number of chromosomes = 7 pairs): diploid wheats
   Wild form with fragile rachis and kernel in hull
       Triticum boeoticum
   Cultivated form with fragile rachis and kernel in hull
       T. monococcum Einkorn
2. Emmer Group (number of chromosomes = 14 pairs): tetraploid wheats
   Wild form with fragile rachis and kernel in hull
       T. dicoccoides
   Cultivated form with partly fragile rachis and kernel in hull
       T. dicoccum Emmer
       T. timopheevi Timopheevi wheat
   Cultivated form with tough rachis and free kernel
       T. durum (durum wheat)
       T. turgidum (Poulard wheat)
       T. polonicum (Polish wheat)
       T. carthlicum or *persicum* (Persian wheat)
       T. turanicum or *orientale*
3. Vulgare Group (number of chromosomes = 21): hexaploid wheats
   Wild form non-existent
   Cultivated form with partly fragile rachis and kernel in hull
       T. aestivum subspp. *spelta* (spelt)
       T. aestivum subspp. *vavilovi*
       T. aestivum subspp. *macha*
   Cultivated form with tough rachis and free kernel
       T. aestivum subspp. *vulgare* (common wheat)
       T. aestivum subspp. *compactum* (club wheat)
       T. aestivum subspp. *sphaerococcum* (shot wheat)

often contains two or three kernels. The spikes of durum wheat are compact and laterally compressed. The varieties are adapted to spring cultivation and are tall. It is often confused with barley because of long, stiff awns; thick, compact spikes; and long glumes. Club wheat may be grown in winter or spring. The spikes are short, compact, and laterally compressed. U.S. producers grow only the common, durum, and club wheat species. Spelt, Emmer, and einkorn wheats are normally not commercially important in the United States. Most of their kernels remain enclosed in the glumes after threshing.

19.6: COMMERCIAL WHEAT CLASSES

The varieties of wheat grown in the United States may be grouped into seven classes based on the time of year they are planted and their kernel characteristics (hardness, color, shape). The classes and characteristics of wheat types are described in Table 19–2. However, for commercial production, the varieties may be narrowed down to six basic classes: hard red winter, hard red spring, soft red winter, hard white, soft white, and durum wheat. The hard red winter wheat accounts for about 40% of total U.S. wheat production and is the dominant class in U.S. wheat export. The distribution of the major classes are shown in Figure 19–5.

Table 19–2   Commercial Wheat Classes in the United States

Hard Red Winter Wheat
– Fall-seeded
– Accounts for more than 40% of total U.S. crop
– Wide range of protein content (10–13.5%)
– Used to produce bread, rolls, and all-purpose flour

Hard Red Spring Wheat
– The highest protein content (13–16.5%)
– Excellent bread qualities with superior milling and baking qualities
– Spring-seeded

Soft Red Winter Wheat
– Low to medium protein content
– Used for making pastries, cakes, flat breads, and crackers

Hard White Wheat
– Newest class of wheat to be grown in the United States
– Similar to hard red, except grain color and flavor
– Used in yeast breads, hard rolls, tortillas, and oriental noodles

Soft White Wheat
– Similar uses as soft red winter wheat
– Low protein and high yields

Durum Wheat
– Hardest of all U.S. wheats
– Spring-seeded
– High protein content (12–16%)
– Used for making pasta products

1. *Hard red winter wheat.* It is grown mainly in the Great Plains (Kansas, Oklahoma, Nebraska, Texas, Colorado). It is also grown in Russia, Argentina, and the Danube Valley of Europe. It is used for bread flour.
2. *Hard red spring wheat.* This class of wheat is grown in regions with severe winters in the North Central states (North Dakota, Montana, South Dakota, Minnesota). It is also produced in Canada, Russia, and Poland. It is the standard wheat for bread flour.
3. *Soft red winter wheat.* This class of wheat is grown predominantly in the eastern and midwestern states of the United States (Ohio, Missouri, Indiana, Illinois, Pennsylvania). It is also grown in western Europe. Soft red winter wheat is used mainly for pastry, cake, biscuit, and household flour. For bread making, it needs to be blended with hard red wheat flour.
4. *Hard or soft white wheat.* White wheat (hard or soft) is produced in the far western states and in the northeastern states (e.g., Washington, Oregon, Idaho, California, New York) and also Michigan. Some of this is club wheat. It is produced also in northern, eastern, and southern Europe, Australia, South Africa, South America, and Asia.
5. *Durum wheat.* Durum wheat is grown mainly in North Dakota, Minnesota, and South Dakota. Smaller-production states include California, Arizona, Oregon, and Texas. Elsewhere, it is grown in North Africa, southern Europe, and Russia.

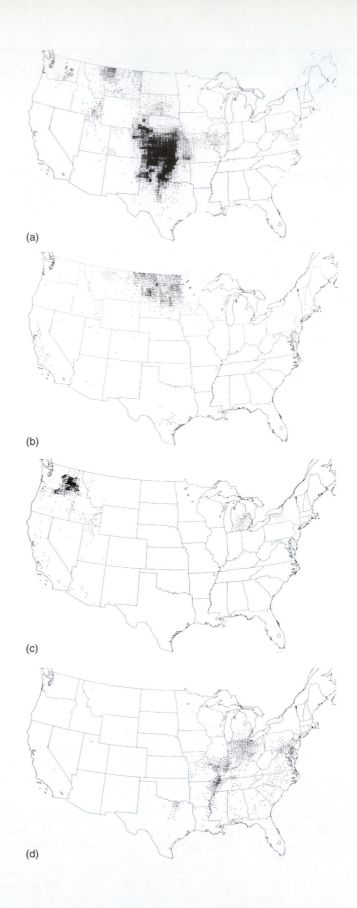

FIGURE 19–5 The general distribution of commercial wheat classes in the United States: (a) hard red winter wheat, (b) hard red spring wheat, (c) white wheat, (d) soft red winter wheat. (Source: USDA)

(a)

(b)

(c)

(d)

Durum wheat is used in making semolina, which is used for producing products such as macaroni and spaghetti.

19.7: WHEAT CULTIVARS

Producers choose wheat cultivars based on various characteristics, including yield potential, lodging resistance, earliness, and disease resistance. Many of the wheat cultivars in the United States have the semidwarf genes, making them more responsive to high nitrogen fertilization. Winter hardiness is important to wheat producers. Winter wheat is sensitive to extreme cold. Growers in the Great Plains and Great Lakes regions need to use winter-hardy cultivars. The protection against extreme cold is at the expense of grain yield. Cultivars that are hardy cultivars yield higher but are more susceptible to severe frost. Many wheat cultivars are photoperiod-sensitive. Information pertaining to this characteristic is available to producers through their local Agricultural Extension Services. Cultivars that are early-maturing tend to have little vernalization requirement and are photoperiod-insensitive.

19.8: CULTURAL PRACTICES

19.8.1 TILLAGE

Tillage in wheat production includes plowing, disking, cultivation, and harrowing, as well as the use of implements such as bedders, shapers, and packers. The tillage system used for seedbed preparation for seeding wheat varies among the regions of the United States and within regions. The system selected is based on the producer's need for moisture conservation, the residual from the previous crop, and the producer's preference for or attitude toward tillage. The four general systems of tillage used by wheat farmers are conventional tillage with moldboard plow, conventional tillage without moldboard plow (primarily using chisel and disk plows), mulch tillage, and no-till. Conventional tillage without moldboard plow is the dominant system used by wheat growers in all regions of the United States, especially in the Central and Southern Plains, where about 80% of the producers use this system. About 28% of all farms surveyed in 1994 used conservation tillage systems (mulch or no-till), the practice being most common in the North Central regions. Tillage operations may be used for incorporating fertilizers and improving the soil's physical properties to control erosion, reduce water loss, and control pests (insects, diseases, and weeds). The goal of tillage is to produce a firm seedbed that provides good seed-to-soil contact at minimum cost. The cost of tillage is related to the number of field passes or *times over* (the acreage covered in the operation divided by the total acreage planted in the crop).

The number of times over was highest in a 1994 survey in the Pacific region because of more harvesting and application of agrochemicals. Another source of variation in the tillage system adopted is the time of planting. Regions that produce spring wheat tend to use fallow, whereas the adoption of no-till and conventional tillage without moldboard plow has gradually increased since 1988 in regions where winter wheat production occurs. It is important that residue be properly managed to avoid diseases such as *Septoria* leaf blotch. An initial one or two diskings or a chisel operation (followed by another disking or the use of a field cultivator near planting time) may be adequate to incorporate most residues.

19.8.2   SEEDING

Method

Wheat in the U.S. production system is primarily drilled by using grain drills with furrow openers in humid regions and hoe drills in drier regions.

Seeding Date

A crop is seeded at a time of favorable soil and climatic conditions. For winter wheat, the crop should be seeded at a time that would allow the seedlings to have well-established crown roots and about three to five tillers before winter dormancy to minimize winter kill damage. Late seeding leads to higher winter injury, reduced tillering, and delayed ripening. However, seeding early predisposes the crop to infestation with the Hessian fly, leaf rust, and wheat streak mosaic virus. However, in production systems where wheat is grazed, seeding 2–3 weeks early is desirable for wheat pasture. The optimal time for seeding in the Great Plains is about September 1 in Montana and progressively later southward to October 15 in northwestern Texas. As a rule of thumb, wheat is seeded 7–10 days earlier than the safe dates where the Hessian fly is absent. Seeding is done in mid-September in the Pacific Northwest, whereas November to early December is desirable for seeding in California and southern Arizona. Where producers double-crop after row crops, wheat may be planted later than optimum. However, delayed planting results in reduced fall growth (reduced root and tiller formation) and increased susceptibility to wind damage and winterkill damage.

Early seeding of spring wheat often produces high yield, because the crop escapes damage from stresses (such as drought, heat, diseases,) that often increase in intensity toward the end of the season. Growers in Nebraska and Colorado seed in March, whereas those in South Dakota and southern Minnesota seed in April; Washington and Oregon growers also seed in March.

Seeding Rate

The seeding rate depends on the variety, the seed quality, the time of seeding, cultural methods, the soil moisture, and the locality. Plant density affects tillering and crop yield. Seeding rates in drier regions are lower than in wetter regions. For example, in western Kansas, the seeding rate is 40–60 lb/acre, giving a plant population of 600,000–900,000/acre (1 pound of seed contains about 15,000 seeds). The seeding rate is 50–60 lb in central Kansas and 60–90 lb/acre under irrigation. The grower may adjust the recommended seeding rate upward for quicker establishment of ground cover, or when delaying planting date, and when using varieties with lager kernels. Also, when double-cropping or planting wheat to graze, the seeding may be increased 50–100%.

Depth

A planting depth of 1–2 inches is optimal for most varieties. Wheat may be seeded deeper (3 inches) where the soil dries quickly. Varieties with a long coleoptile can survive deeper seeding depth; however, deep seeding produces weakened seedlings.

19.8.3   WHEAT GROWTH AND DEVELOPMENT

The growth and development of cereals have been described in stages by researchers. The Feekes Scale of developmental stages of wheat have been summarized in Table 19–3. These stages can be grouped into four main categories—tillering, stem extension, heading, and ripening stages. During the production of wheat, various production practices coincide with the different stages of development.

Table 19–3 Feekes Scale of the Developmental Stages of Wheat

1. Tillering
   - Stage   1   One shoot emerged.
   -            2   Tillering begins.
   -            3   Tillers formed.
   -            4   Leaf-sheaths lengthen.
   -            5   Leaf-sheaths are strongly erected.
2. Stem extension
   - Stage   6   First node of stem is visible.
   -            7   Second node is visible.
   -            8   Last leaf (flag leaf) is just visible.
   -            9   Ligule of the last leaf is just visible.
   -           10   Boot is just swollen.
3. Heading
   - Stage 10.1   First spikelet of inflorescence is visible.
   -         10.5   Emergence of inflorescence is completed.
4. Ripening
   - Stage 11.1   Kernel in early milk stage.
   -         11.2   Kernel in hard dough stage

19.8.4 CROPPING ROTATIONS AND PRODUCTION PRACTICES

Wheat may be continuously cropped but also in rotation with other crops. Continuous cropping is predominantly conducted in the Central and Southern Plains. Double-cropping is predominantly done in the South (with soybean) and North Central region, whereas grazing is predominantly done in the Central and Southern Plains. Crop rotations associated with wheat production include fallow-wheat (in the Northern, Central, and Southern Plains and the Pacific region), corn-soybean (in the North Central region and the Southeast), soybean-corn (predominantly in the North Central and Northern Plains). Wheat often follows soybean, especially in the North Central region, the Southeast, and the Northern Plains. In the Corn Belt, a rotation may be wheat-clover + timothy-corn-oat (or soybean). In the Corn Belt, wheat is followed by cowpeas (or soybean). In the Cotton Belt, wheat is followed by cowpeas (or soybean) and corn. In Kansas, a rotation may be clover-corn-oat-wheat. Spring wheat may be followed by corn, potato, or beet in the Northern plains.

19.8.5 FERTILIZATION

Nitrogen is the nutrient element that is limiting for optimal wheat production. Wheat removes a little more nitrogen per bushel of yield than corn or grain sorghum. This is attributed to the fact that wheat has higher protein content than corn or sorghum. Nitrogen may accumulate in the soil profile under various production practices, such as under summer fallow. In order to determine the proper rate of nitrogen application, the producer should take into account the cropping sequence. Where the previous crop in a rotation is a legume (e.g., alfalfa, clover), wheat can benefit substantially from a nitrogen credit. However, when a wheat crop is preceded by soybean, the nitrogen credit is zero because the nitrogen needs for wheat comes early in the spring before the nitrogen in the soybean residue is mineralized. Wheat producers apply an average of about 57 lb of nitrogen per acre during production. Growers in the Northern Plains apply about 50 lb/acre, whereas growers in the North Central region apply a higher rate of about 85 lb/acre. Furthermore, the protein content of corn grains can be increased with a higher than optimum rate of nitrogen application. However, excessive nitrogen, especially early in the season, predisposes wheat to lodging problems.

Wheat responds also to phosphorus and potassium. When phosphorus and potassium are deficient, wheat does not tiller properly and is more susceptible to winterkill. Potassium is most likely to be deficient on sandy soils. On average, phosphorus is applied at about 22 lb/acre, whereas potassium is applied at a lower average rate of about 8 lb/acre. On the whole, fertilizer application is higher in the Eastern region, where wheat is double-cropped and where straw harvest is substantial. Also, growers in the Eastern region apply more manure than any other region. Heavy fertilization of wheat also occurs in irrigated fields of the Pacific region. Growers in the Central and Southern Plains generally apply lower rates of fertilizer than do growers in other regions.

19.8.6   IRRIGATION

Wheat is adaptable to extreme weather conditions. However, irrigation substantially increases crop yield. Some reports indicate a 50% increase in yield with irrigation. Because irrigation is a high-cost production practice, a survey of farms in 1999 indicated that only 5% of the producers used irrigation. Most irrigation of wheat (25%) occurs in the Pacific region and the Central and Southern Plains, where 5% of the acreage is irrigated. In Kansas, wheat is the second most irrigated crop (after corn). The water use curve of wheat indicates that rapid water use starts in early spring, the critical period of water need for the crop being between the boot and heading stages, and peaking at grain development, when daily needs could be about 0.35 inch per day. For winter wheat, an estimated 20% water use in production occurs between the emergence and the beginning of spring growth, 20% from spring growth to jointing, 10% from jointing to boot, 12% from boot to flower, 15% from flower to milk, 8% from milk to dough stage, and 5% to maturity.

19.8.7   WEED MANAGEMENT

Wheat is considered the least chemical-intensive of the major field crops. Practices such as crop rotation and summer fallow reduce weeds in wheat fields. Winter annual weeds of economic importance include cheatgrass and jointed goatgrass. Summer annuals include wild buckwheat, kochia, Russian thistle, and sunflower. About 40% of wheat producers in the United States use herbicides in production. Weed pressure is highest in areas where spring wheat and durum wheat production occurs. More than 85% of wheat producers in the Northern Plains and Pacific regions use herbicides, whereas only about 20% of producers in the North Central region use herbicides. The most widely used herbicide in wheat production in a 1994 survey was MCPA, followed by 2,4-D and glyphosate. Growers in the Northern Plains and Pacific region use more MCPA, although chlorosulfuron is preferred in the Central and Southern Plains and 2,4-D in the North Central region.

19.8.8   DISEASES

Wheat is attacked by fungal diseases, the chief being stem rust (caused by *Puccinia graminis tritici*), leaf rust (caused by *P. recondite*), and stripe rust (caused by *P. glumarum*). Stem rust is the most destructive of wheat diseases, causing the affected grains to be severely shriveled. Wheat is also attacked by smuts—such as bunt or stinking bunt (caused by *Tilletia caries* and *T. foetida*), especially in the Pacific Northwest, as well as loose smut (caused by *Ustilago tritici*) and Karnal bunt (caused by *Tilletia indica*), which affects flour quality. Wheat scab (caused by *Gibberella scanbinetii*) is a serious disease in the northern and eastern states, particularly in the Corn Belt. The characteristics and economic importance of selected diseases are summarized in Table 19–4.

19.8.9 INSECTS

Insect pests of economic importance include the Hessian fly *(Phytophaga destructor)*, wheat jointworm *(Harmolita tritici)*, and wheat strawworm *(Harmolita grandis)*. The major insect pests are summarized in Table 19–5.

Only about 5% of farms surveyed in 1994 used insecticides or fungicides in wheat production. Producers in the Southeast were the biggest users of insecticides and fungicides, followed by those in the Pacific region and the Central and Southern Plains. The most common insecticides used in wheat fields were methomyl, chloropyrifos, dimethoate, and parathion. The most common fungicide was Tilt.

Table 19–4 Common Diseases of Wheat in Production

1. Stem rust *(Puccinia graminis tritici)*
   - Most destructive wheat disease.
   - Grain severely shriveled.
   - Stems and leaves covered with pustules containing brick-red spores
   - Spores overwinter on straw and stubble and infect barberry bush in spring, infecting wheat plants that grow near these plants sooner than others.
   - Reduce disease by controlling barberry bushes
   - Control by spraying fungicides or using resistant cultivars.
2. Leaf rust *(Puccinia recondite)*
   - Less destructive than stem rust.
   - Reduces yield by reducing number of kernels in the spike without shriveling.
   - Overwinters as pustules or mycelium within leaf tissue.
   - Control by using resistant cultivars or dusting with sulfur.
3. Bunt or stinking bunt *(Tilletia caries* and *T. foetida)*
   - Fungus infects young seedlings, grows within them, and produces smut ball completely filled with black spores instead of kernels in the wheat head.
   - Spores have odor of stale fish.
   - Spores overwinter in soil.
   - Control by seed treatment.
4. Loose smut *(Ustilago tritici)*
   - Most common under humid or irrigated production.
   - Floral bracts are almost completely replaced by black smut masses.
   - Control by using resistant cultivars.
5. Wheat scab *(Gibberella saubinetti)*
   - Pinkish-white fungal growth occurs around dead tissue.
   - Grain is shriveled and scabby in appearance.
   - Most severe when wheat follows corn in a rotation.
6. Take-all root rot *(Gaeumannomyces graminis)*
   - Usually occurs in patches in the field.
   - Discolored stem base, black rotten roots, and stunting of the plants.
   - Continuous wheat, early planting, and liming favor disease.
   - Control by crop rotation, delayed planting, and destruction of wheat crop residue.
7. Mosaics
   - Cause dwarfing or resetting of plants or mottling of leaves.
   - Control with resistant cultivars.
8. Nematodes *(Anguinea tritici)*
   - Nematodes enter roots and grow into young developing heads; ovary does not develop into a grain but into a hard, black gall filled with nematodes.
   - Control by rotation with crops other than rye.

Table 19–5 Common Insect Pests of Wheat in Production

1. Hessian fly *(Phytophaga destructor)*
   - One of the most economically important insect pests of wheat.
   - Damage occurs in both fall and spring.
   - Larvae injure the plant between the sheath and stem, where they extract juices, killing the plants or promoting lodging.
   - Control by cultural methods, such as rotating crops, plowing under infested stubble, or sowing winter wheat late so that the main brood of flies dies before young wheat plants emerge above the ground.
2. Wheat jointworm *(Harmolita tritici)*
   - Second to the Hessian fly in economic importance.
   - Characterized by wartlike swelling on the stem often above the joint of the node.
   - Causes lodging of stalks.
   - Control by plowing under wheat stubble.
3. Wheat strawhorn *(Harmolita grandis)*
   - Two complete generations of insects occur each year, the first generation (spring form) being capable of killing plants outright and hence most injurious to spring wheat.
   - Propagation of the worm is favored by infested stubble on the ground when tillage implements are used.
   - Spring form is wingless and hence can be controlled if wheat is sown 75 yards or more from wheat stubble.
4. Green bug *(Toxoptera graminum)*
   - More damaging to oats and barley than wheat.
   - A sucking insect that reproduces rapidly, by vivipary or eggs, with or without fertilization.
   - Can be controlled by insecticides or by destroying volunteer wheat, oats, and barley in summer and fall in each community.
   - Resistant cultivars are also available.
   - More destructive to barley, corn, and sorghum than wheat.
5. Wheat-stem sawfly *(Cephus cinctus)*
   - Adult female fly splits the wheat stem and deposits eggs inside. Larvae feed down the stem, causing it to lodge.
   - Crop rotation and burying of stubble help control the insects.
   - Resistant cultivars are available.
6. Wireworms *(Otenicera pruinina)*
   - Worms inhabit the soil, surviving there for 3–10 years.
   - They eat the seeded grain from underneath the soil, killing young plants and thinning the stand.
   - Seed treatment may be used, as well as crop rotation, summer fallow, and seeding only when the soil is moist so that germination is rapid.

19.8.10   HARVESTING

Wheat harvesting in the United States is combined. *Custom harvesting* (contracting harvesting operations to an independent company) and hauling are most common in the Central and Southern Plains and the Pacific region (35% of farms), followed by the Southeast (27%). Custom harvesting is used less in the North Central region (13%). World acreage planted to wheat in 2000/01 was 219.5 million hectares, with a total production of 583.9 million metric tons. The average and yield of wheat in the United States have been on the decline in the past five years (Figure 19–6). Similarly, the prices received by farmers have been on the decline (Figure 19–7). However, the price of wheat on the world market has been on the rise (Figure 19–8).

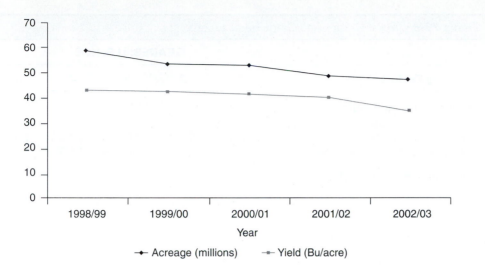

FIGURE 19–6 The acreage and yield of wheat have been on the decline over the past five years. Drawn from USDA data.

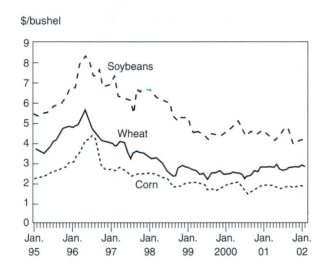

FIGURE 19–7 The prices received by farmers for soybeans, wheat, and corn have generally been on the decline since 1966. (Source: USDA)

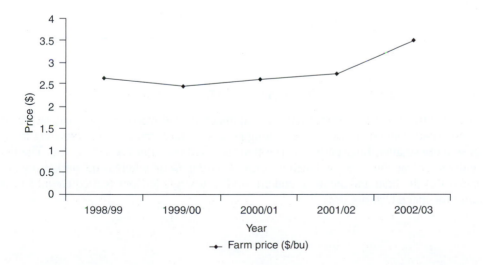

FIGURE 19–8 The price of wheat has been on the rise over the past 5 years. Drawn from USDA data.

Table 19–6 Wheat Grades and Grade Requirements

Grading Factors	GRADES: U.S. NOS.				
	1	**2**	**3**	**4**	**5**
Minimum Pound Limits of					
Test weight (lb/bushel)					
Hard red spring/white club	58.0	57.0	55.0	53.0	50.0
All other classes/subclasses	60.0	58.0	56.0	54.0	51.0
Maximum Percent Limits of					
Damaged kernels					
Heat (part of total)	0.2	0.2	0.5	1.0	3.0
Total	2.0	4.0	7.0	10.0	15.0
Broken kernels and foreign material					
Foreign material (part or total)	0.4	0.7	1.3	3.0	5.0
Shrunken and broken kernels	3.0	5.0	8.0	12.0	20.0
Total	3.0	5.0	8.0	12.0	20.0
Wheat of other classes					
Contrasting classes	1.0	2.0	3.0	10.0	10.0
Total	3.0	5.0	10.0	10.0	10.0
Stones	0.1	0.1	0.1	0.1	0.1
Maximum Count Limits of					
Other material					
Animal filth	1	1	1	1	1
Castor beans	1	1	1	1	1
Crotalaria seeds	2	2	2	2	2
Glass	0	0	0	0	0
Stones	3	3	3	3	3
Unknown foreign substance	3	3	3	3	3
Total	4	4	4	4	4
Insect-damaged kernels in 100 g	31	31	31	31	31

U.S. sample grade
Wheat that
1. Does not meet the requirements for U.S. Nos. 1, 2, 3, 4, 5 or
2. Has a musty, sour, or commercially objectionable foreign odor (except smut or garlic odor) or
3. Is heating or distinctly low-quality.

Source: USDA

19.8.11 MILLING WHEAT AND WHEAT BY-PRODUCTS

Wheat is first cleaned to remove all foreign materials and weed seeds. It may be washed if it is smutty and then dried. The milling process entails cracking and crushing in progressively smaller, finer particles. The flour is sifted out after each reduction. The finer and whiter fractions are combined to form the *patent flour,* whereas the darker fractions consist of the bran and aleurone and are sold as low-grade flour (called *clears* and *red dog*) for special baking uses.

The flour is further bleached to make it whiter and more attractive using chemical agents such as chlorine and hydrogen peroxide. This destroys the yellow pigments, mainly xanthophyl, in the flour. The entire grain may be milled into flour, called *graham* or *whole-wheat flour.* Graham wheat is more easily infested with insects than is white wheat. Most of the wheat sold (97%) in the United States is white wheat.

The by-products of flour milling are often called *offal.* The *middlings* (granular fraction of the endosperm from which the finer particles [flour], bran, and shorts have been separated) are used as feed. The middlings from hard spring wheat may be purified and sold as *cream of wheat,* whereas those from hard winter wheat, called *farina,* are used as breakfast cereal. Purified middlings from durum wheat, called *semolina,* is used to make macaroni, spaghetti, and other foods.

The wheat kernel consists of about 85 percent endosperm. Milling yields about 70 percent straight flour. Yield from light-weight or shrunken wheat is lower. Hard wheat flours are better for bread-making. White flour is lower in protein, fat, ash, and higher in starch than the original wheat or graham flour.

19.8.12 *WHEAT GRADES AND GRADE REQUIREMENTS*

The grades and grade requirements of wheat as prescribed by the USDA Grain Inspection Service are summarized Table 19-6.

REFERENCES AND SUGGESTED READING

Gooding, M. J. and W. P. Davies. 1997. *Wheat production and utilization: systems, quality and the environment.* CABI publishing. CAP International, London: UK.

Wheat production handbook. Manhattan, KS: Kansas State Research and Extension.

SELECTED INTERNET SITES FOR FURTHER REVIEW

http://www.panhandle.unl.edu/personnel/lyon/wheathbk.htm

Wheat production systems by U of Nebraska.

http://www.ext.nodak.edu/extpubs/plantsci/smgrains/eb33w.htm

Wheat production in North Dakota

http://www.agcom.purdue.edu/AgCom/Pubs/AY/AY-244.html

Wheat production in Indiana

http://pearl.agcomm.okstate.edu/plantsoil/crops/f-2080.pdf

Wheat production in Oklahoma

http://wbc.agr.state.mt.us/prodfacts/prodfact.html

Wheat production facts

OUTCOMES ASSESSMENT

PART A

Answer the following questions true or false.

1. T F   Wheat is the most important cereal grain crop in the world.
2. T F   Durum wheat is used for making bread.
3. T F   Hard red winter wheat is produced mainly in the Great Plains.
4. T F   Common wheat is a diploid species.
5. T F   Wheat is a long-day plant.
6. T F   United States No. 1 is the highest wheat grade in the United States.
7. T F   Semolina is made from durum wheat.

PART B

Answer the following questions.

1. Give the scientific name of bread wheat. _____

2. Wheat is classified in the United States according to the growing season as either _____ or _____ .

3. Give four of the top ten wheat producing states in the United States.

4. Give four of the top ten wheat producing nations in the world.

5. _____ is the protein responsible for the cohesion of wheat flour.

6. Give six market classes of wheat in the United States.

7. Give a major insect pest of wheat in crop production. _____

PART C

Write a brief essay on each of the following topics.

1. Discuss the relative importance of wheat as a food crop versus other crops.

2. Briefly discuss the origin of wheat.

3. Describe the general botanical features of the wheat plant.

4. Discuss the time of seeding and seeding rate of wheat.

5. Discuss the use of crop rotation practices in wheat production.

Part D

Discuss or explain the following topics in detail.

1. Briefly discuss the origin of wheat.
2. Discuss the environmental conditions best suited to the production of wheat.
3. Discuss the growth and development of wheat.
4. Discuss the major insect pests in the production of wheat.
5. Discuss the major diseases in wheat production.
6. Discuss the market classes of wheat.
7. Discuss the production of high protein wheat

20

Rice

TAXONOMY

Kingdom	Plantae	Subclass	Commelinidae
Subkingdom	Tracheobionta	Order	Cyperales
Superdivision	Spermatophyta	Family	Poaceae
Division	Magnoliophyta	Genus	*Oryza* L.
Class	Lilliopsida	Species	*Oryza sativa* L.

KEY TERMS

Flood-prone rice	Indica type	Bran
Upland rice	Rough rice	Polished rice
Hull	Milling	Brown rice
Japonica type	Parboiled rice	

20.1: ECONOMIC IMPORTANCE

Rice accounts for about 20% of the world's total grain production, second only to wheat. It is the primary staple for more than 50% of the world's population. An estimated 90% of the production and consumption of rice occurs in Asia, where the per capita consumption is about 100 pounds, compared with a world average of about 60 pounds and a U.S. average of about 20 pounds. World consumption in 1993–1994 was 358.5 million metric tons and has increased steadily to an estimated 408.8 million tons in 2002–2003. Similarly, the acreage devoted to rice increased during the same period from 145.2 hectares to 155.1 hectares in 1990–2000, with a slight drop thereafter. Production trends mirror the area cultivated.

The world's major producers of rice are China, India, Indonesia, and Bangladesh, together accounting for more than 70% of the world's total production (Figure 20–1). China, the world's leading producer, regularly accounts for about 36% of the world's total production. The expansion in rice production is attributed largely to the impact of the Green Revolution, which was implemented in the 1960s and 1970s in Asia and other parts of the tropical world. Other major producers in Asia include Vietnam, Thailand, Japan, and Burma, together accounting for about 12% of the total world production. Outside of Asia, rice is produced in substantial amounts in the United States, Italy, Spain, Egypt, Australia, and countries in the West African region.

Rice production in the United States is dominated by six states—Arkansas, California, Louisiana, Texas, Mississippi, and Missouri, together accounting for about 99% of the total U.S. production. Other minor producers in the United States include Florida, Tennessee, Illinois, South Carolina, and Kentucky (Figure 20–2). Arkansas leads production in the United States, with about 100 cwt of rice produced in 2001 on about 1.61 million acres.

The geographic distribution of rice production in the United States has changed over the years (Figure 20–3). The current production of rice is shown in Figure 20–4.

Thailand leads the world in rice exports and accounts for about 25% of total world exports, representing about 4 million metric tons. The United States is the second largest exporter of rice, exporting about 2.7 million metric tons. Latin America is the largest

FIGURE 20–1 General areas of rice production in the world. (Source: USDA)

FIGURE 20–2 Distribution of rice production in the United States shows three major production regions: California, Arkansas, and Texas-Louisiana. (Source: USDA)

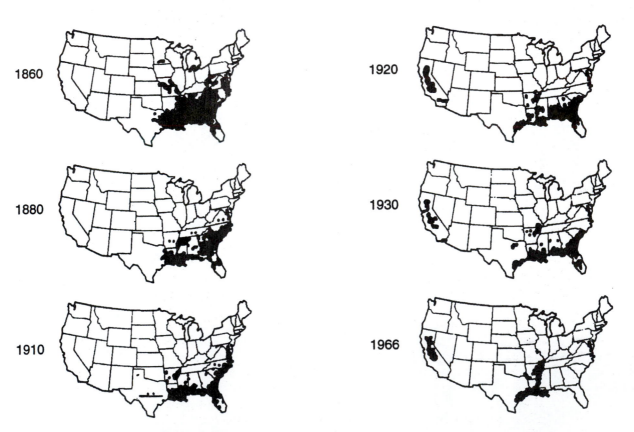

FIGURE 20–3 The geographic production regions of rice in the United States have changed over the years. (Source: USDA)

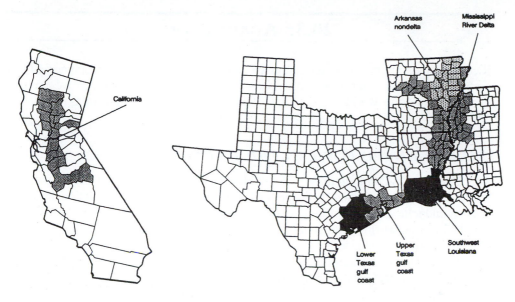

FIGURE 20–4 The current major rice production regions of the United States: (Source: USDA)

market for U.S. rice exports. Other major exporters of rice include Vietnam, Pakistan, China, Australia, Burma, Italy, India, and Uruguay.

20.2: ORIGIN AND HISTORY

The origin of rice is not known. Wild species of rice grow across South and East Asia, including India and south China. Rice was domesticated in the fifth millennium B.C. The cultivated species of rice *Oryza sativa* is believed to have derived from annual progenitors found in a wide area extending from the Gangetic Plains through Burma, northern Thailand, northern Vietnam, and southern China. Rice trade occurred among Egypt, India, and China. The Moors are credited with taking rice to Spain, from which it was introduced to Italy (in the fifteenth century) and subsequently to Central America.

A British sea captain brought rice to the United States from Madagascar. The first commercial planting occurred in South Carolina in about 1685. Early rice production concentrated in the southeastern United States, where South Carolina, Georgia, Louisiana, North Carolina, Mississippi, Alabama, Kentucky, and Tennessee were major producers in the early 1800s. However, in the early nineteenth century, shifts in rice production started in the United States, reaching its highest in the late 1900s (Figure 20–3). Production became more modernized and spread to Arkansas and the Mississippi River Delta, where the flatter lands permitted larger-scale, high-tech production (mechanized, irrigated, etc.). In 1838, South Carolina produced about 75% of the U.S. total rice crop. In 1903, Louisiana and Texas produced 99% of the U.S. crop. However, by 1990, Arkansas, Louisiana, and Mississippi accounted for more than two-thirds of the U.S. rice production, with Texas and California accounting for most of the balance. Rice production in the United States is currently concentrated in the Arkansas Grand Prairie, the Mississippi River Delta, southwestern Louisiana, the coast prairie of Texas, and the Sacramento Valley of California (Figure 20–4).

20.3: ADAPTATION

Rice is adapted to very wide agroecological zones, ranging from dry to submerged root growing conditions. Four general ecosystems can be identified for commercial rice production around the world based on elevation, rainfall pattern, depth of flooding, and drainage.

20.3.1 RAINFED LOWLAND RICE ECOSYSTEM

This ecosystem is common in densely populated rural regions of the world, where producers face severe economic challenges in addition to a burgeoning population. Rice production under these conditions accounts for about 25% of the world's harvested rice area and 17% of the world's production. Producers prepare the land by puddling the soil, bunding, or diking fields to hold water for a variable duration of flooding, according to the rainfall. The soils alternate between flooded and dry conditions during the growing season. Rice is direct-seeded or transplanted into the field.

20.3.2 UPLAND RICE ECOSYSTEM

Upland rice production occurs on well-drained level to steeply sloping farmlands. These soils are frequently moisture-deficient. Upland production occurs in regions of the world where slash-and-burn agriculture is common. The removal of vegetation from these slopes predisposes the soil to physical deterioration and nutrient depletion. Crop yields are generally low. Upland rice production constitutes about 13% of the world's harvested area and only 4% of the total rice produced. Rice is direct-seeded to non-flooded soils.

20.3.3 FLOOD–PRONE RICE ECOSYSTEM

Rice production in certain areas occurs on flooded soils throughout the growing season until harvest time. Rice is direct-seeded or transplanted into flooded fields (50–300 cm deep) during the rainy season. Flooded-rice production occurs widely in South Asia and Southeast Asia, as well as some parts of West Africa and Latin America. Problems of salinity and toxicity from various ions are common in this ecosystem. Crop yields are unpredictable and generally low.

20.3.4 IRRIGATED RICE ECOSYSTEM

The key feature of this system is that moisture is controlled in both dry and wet seasons. Various methods are used to provide and regulate soil moisture. About 55% of the world's harvested area and 75% of total production occurs in an irrigated ecosystem. Production involves the use of modern technology with high production inputs (e.g., fertilizers). Consequently, yields are high, reaching about 5 tons per hectare in the wet season to about 10 tons per hectare in the dry season.

Rice is a warm season species and is successful under a mean temperature of about 70°F or higher. High humidity is undesirable because it encourages diseases. Heavy-textured soil with impervious subsoil for holding moisture is desirable for rice. The tolerable pH ranges between 4.5 and 7.5. Rice is a shortday plant.

FIGURE 20–5 Rice field showing mature heads (left). Harvesting rice (right). (Source: USDA)

20.4: BOTANY

There are two major species of cultivated rice—*Oryza sativa* and *O. glaberrima*—the latter being native to Africa and cultivated in West Africa and Central Africa. An annual grass, rice has erect culms that may reach 6 feet tall in some varieties. It produces about 5 tillers. Rice inflorescence is a loose terminal panicle consisting of 1-flowered spikelets that are self-pollinated (Figure 20–5). A panicle may contain 100–150 seeds. A flower has 6 stamens and 2 long plumose sessile styles. The lemma and palea surround the styles at the base of the flower, where 2 small glumes occur. The rice grain is enclosed by the lemma and palea (constitutes the hull), which may be straw yellow, red, brown, or black. Depending on the variety, the lemmas may be fully awned, partly awned, tip-awned, or awnless. Hulled kernels vary in length from 3.5–8 mm and are 1.7–3 mm in breadth and 1.3–2.3 mm thick. Furthermore, the kernels may be hard, semihard, or soft-textured. The color of the unmilled rice kernel is variable and may be white, brown, amber, red, or purple. However, lighter colors (white, light brown) are preferred in the United States.

20.5: COMMERCIAL CLASSES

Rice varieties are primarily classified according to the length of the grain—short (5.5 mm), medium (6.6 mm), and long (7–8 mm). The shorter-grained varieties are also called **Japonica types** and have short, stiff, lodging-resistant stalks, making them more responsive to heavy fertilization. The longer-grained varieties, called **Indica type**, have taller, weaker stems that lodge under heavy fertilization. The United States produces mainly Indica rice (about 65% of the annual production), primarily in Arkansas, Mississippi, Louisiana, and Texas, but California produces mostly medium- and short-grained rice.

Rice may also be classified in terms of maturity—early-maturing (about 120–129 days), midseason (about 130–139 days), or late-maturing (about 140 days or more). Rice may be scented (aromatic) or unscented, the two most common scented types being basmati and jasmine. Basmati rice has a distinctive odor, doubles in grain size upon cooking, and is non-sticky (its grains remain separate). It is cultivated mainly in the Punjab area of central Pakistan and northern India. Jasmine rice is grown mainly in Thailand and is more preferred by the Asian community in the United States. Cooked jasmine rice is soft, moist, and sticky. The stickiness derives from the types of starch in the grain. The endosperm starch of rice may be glutinous or commonly non-glutinous (non-sticky). The glutinous property is conferred by the amylopectin type of starch.

20.6: CULTURAL PRACTICES

As previously indicated, there are four basic rice ecosystems in the world. The level of technology used in production varies among the regions, ranging from handheld tools and animal-drawn implements in poor regions to highly mechanized production in developed economies. The cultural practices discussed in this section are applicable mainly to mechanized production.

20.6.1 TILLAGE

Tillage of rice fields is estimated to be about 70% of the times over (the acreage covered in the operation divided by the acreage planted to the crop) for all field operations in all production regions in the United States. The farm machinery and implements used in tillage of rice fields include plows, disks, field cultivators, bedders and shapers, and soil packers. The differences in times over among regions are due to the disking, rowing, and packing operations conducted. The tillage systems in use include conventional tillage, water culture, corrugated tillage, and no-till and minimum tillage:

1. *Conventional tillage.* The land is disked and planed with a landfloat. Grading is usually part of the tillage operation. The field is laser-surveyed for levees as 2/10th grade. The levees are constructed with a levee disk.
2. *Water culture.* In water culture, the field is graded to zero slope (or a maximum or 0.01 or 0.02 slope). The field is rolled with a special implement to create a surface texture that will prevent seed drift.
3. *Corrugated tillage.* Corrugated tillage is a modification of the conventional tillage system. The second disking runs in the direction of irrigation flow, making the soil surface very rough or cloddy. Internal levees are not constructed in this system.
4. *No-till and minimum tillage.* No-till cropping systems limit the producer to seeding by the use of seed drills. It is suitable for upland rice production.

20.6.2 SEEDING

Rice seeding is accomplished by using ground equipment (broadcast seeders or grain drills) or airplanes. One of the major considerations in seeding is seed quality—specifically, admixture involving red-rice, a plant of the same species as the proper rice, which can substantially reduce the market quality of milled white rice. Where red rice is a problem, fields are flooded first (to kill the weeds), so that water seeding can be conducted. As previously indicated, a ridged bed surface is needed to reduce seedling drift. Aerial seeding is usually a custom-hired operation. It is the most common method of seeding in California, south-west Louisiana, and the upper coast of Texas. The field is flooded to a depth of about 4–6 inches. The seed is soaked in burlap bags for about 18–24 hours and drained for another 24–48 hours. These pre-soaked seeds are heavier and sink to the bottom to reach the soil. Dry seeds may float or drift and do not give good crop stands. The seeding rate by the aerial method ranges from 99 to 130 lb/acre. The broadcast seeding rate in Arkansas is about 140 lb/acre in the non-delta area; it is 164 lb/acre for water-seeded rice in California.

Seeding by seed drills or broadcast seeders is appropriate where red rice is not a problem. Rice is drilled to about 1–2 inches in a good seedbed. Flooding may follow seed drilling. However, although rice can germinate and emerge through 6 inches or more of water, oxygen becomes deficient when the seed is covered with soil in addition to the water. Consequently, flooded drilled fields are subsequently drained until after the crop is established.

20.6.3   FERTILIZATION

Fertilizers may be applied as aerial applications or ground applications. Flooded fields are usually fertilized by aerial application. Ground equipment may be used to apply fertilizer at the time of seeding. The ammonium, not the nitrate form of nitrogen, is applied, because the nitrate form can be denitrified under anaerobic conditions. Ammonium sulfate is a common source of nitrogen in water culture, as well as urea and anhydrous or liquid ammonia. The rate of nitrogen application varies between 30 and 100 lb/acre. High levels may cause lodging in some varieties. The bulk of the fertilizer is applied as a top dress 30–70 days after seeding. In water culture, algae fix some nitrogen for plant use. Where soils are deficient, an application of moderate amounts of phosphorus and potassium may be beneficial.

20.6.4   CROP ROTATIONS

Continuous cultivation of rice on the same piece of land leads to yield decline and weed problems. Rice producers often adopt crop rotation as part of their cultural practice to reduce pest buildup and rejuvenate the soil. The challenge is finding suitable species to grow, since some rice fields have poor drainage. The specific crop rotation practice differs among the rice ecosystems. After 2–3 years of continuous rice cultivation, farmers may grow a cycle of pasture species—such as grasses, clovers, lespedeza (grown by some Gulf Coast rice producers), and clovers (ladino, burclover, strawberry, grown in California). Field crops such as sorghum, soybean, cotton, and oat may be grown where the land is well drained. Rotations involving crops have a 2–5 cycle, in which rice follows legumes so as to benefit from the nitrogen they fix in the soil. Clean fallow is used to control aquatic weeds in California, whereas some Arkansas growers practice water culture whereby 2–4 feet of water is maintained on the land for about 2 years. The pool of water is used for fish culture during the period.

20.6.5   WEED MANAGEMENT

Some of the weeds commonly found in rice fields are red rice *(Oryza sativa),* watergrass *(Echinochloa crusgalli),* Mexican weed *(Caperonia palustris),* red weed *(Melochia corchorifolia),* coffee weed *(Sesbania* spp.), and sprangletop *(Leptochloa).* Some weeds can be managed with crop rotation. Similarly, flooding the field to a depth of about 6–8 inches can control some weeds. Herbicides may be applied to control weeds in field irrigation canals and on levees. The options are numerous and include Bolero,® Ordram,® and Londax®. Herbicide may be applied by aerial or ground methods.

Table 20-1 Common Diseases of Rice in Production

1. Seedling blight *(Sclerotium rolfsii)*
   - The disease attacks young seedlings in warm weather.
   - Affected seedlings are slightly discolored, with fungal bodies occurring on the lower portions.
   - Early-sown rice is more susceptible than late-sown rice.
   - Disease is checked by submergence or by use of fungicides.
2. Brown leaf spot *(Cochliobus miyabeanus)*
   - Most serious in Louisiana, Texas, and Arkansas.
   - Attacks seedlings, leaves, hulls, and kernels.
   - Brownish discolorations first appear on the sheaths between the germinated seed and soil surface or on roots.
   - Control by using a resistant cultivar or fungicides.
3. Narrow brown leaf spot *(Cercospora oryzae)*
   - Disease often appears late in August and in September.
   - Spots on leaves are long and narrow, reducing leaf area.
   - Control by using resistant cultivars.
4. Blast *(Piricularia oryzae)*
   - Blights rice panicles and rots the stems, causing lodging.
   - Fungus lives on crabgrass and rice straws.
   - Most severe on new rice land.
   - Avoid use of heavy nitrogen fertilization.
   - Use long-stemmed cultivars that can tolerate the injury.
5. Stem rot *(Helminthosporium sigmoideum)*
   - Causes lodging of stem.
   - Control by draining the rice field and keeping the soil saturated but not submerged.

20.6.6 DISEASES

Rice is affected by a number of diseases, the most economic including seedling blight (caused by *Sclerotium rolfsii*), brown leaf spot (caused by *Helminthosporium oryzae*), narrow brown leaf spot (caused by *Cercospora oryzae*), blast (caused by *Piricularia oryzae*) and stem rot (caused by *H. sigmoideum*). Their characteristics and some management practices are summarized in Table 20–1.

20.6.7 INSECTS

Insect pests of economic importance in rice production include the rice stink bug *(Solubea pugnaz)*, rice stalk borer *(Chilo plejadellus)*, sugarcane borer *(Diatraea saccharalis)*, and sugarcane beetle *(Euetheola rugiceps)*. The characteristics of these insect pests and some management practices are summarized in Table 20–2.

20.6.8 HARVESTING

Rice does not mature uniformly in the head. Grain harvest moisture is critical to yield and quality. The recommended grain moisture content is between 23% and 28%. At this grain moisture content, the grains in the top portion of the head are ripe but those in the lower portion are in the hard-dough stage. Harvesting at this stage includes some immature grain, but delaying harvesting increases the chances of shattering and checking the grains in susceptible varieties.

Improper grain moisture content results in poor milling quality. Consequently, some postharvest drying may be necessary. Rice is often dried in stages, allowing grains

Table 20-2 Common Insect Pests of Rice in Production

1. Rice stink bug *(Solubea pugnaz)*
   - Sucks the contents from rice kernels during the milk stage, leaving an empty seed coat or partially empty seed (pecky rice)
   - Control by winter-burning coarse grasses in which insects hibernate, or spray with chemicals.
2. Stalk borer: sugarcane borer *(Diatraea sccharalis)* and rice stalk borer *(Chilo plejadellus)*
   - They tunnel through rice culms, eat inner parts, and cause panicles to turn white without producing grain
   - Grain formation may not be completely inhibited, but panicles may break off.
   - Larger culms are preferred.
   - Insects overwinter in rice stubble.
   - Control by heavy winter grazing of the stubble or flooding of the stubble fields.
3. Sugarcane beetle *(Euetheola rugiceps)*
   - The beetles gnaw off the plants at or just below the soil surface.
   - Attack occurs on young plants before land is submerged, as well as old plants after the field is drained before harvest.
   - Control by avoiding sod land for rice production and treating seed with repellents before planting.

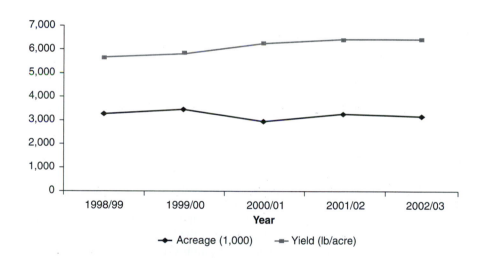

FIGURE 20-6 The acreage of rice has remained stable since 1999. However, rice yield has increased slightly since 1999. Drawn from USDA data.

to equilibrate between stages. Once partially dried, the air temperature may be raised to 130°F to complete the process.

Rice yields are positively correlated with the amount of sunlight received during the growing season. Furthermore, excessive rainfall during the growing period causes the shattering and lodging of plants, leading to reduced yields. In the 1950s, rice yields in the United States averaged 2,800 lb/acre. In the 1980s, the yields averaged about 4,800 lb/acre, with a record yield of 5,749 lb/acre recorded in 1989 (Figure 20-6). High-yielding and disease-resistant varieties as well as early-maturing varieties that avoid the yield losses associated with weather delays at planting and harvesting have been developed. In 1992, California posted a record yield of 8,400 lb/acre. On the whole, rice production in the United States is not subject to as many weather-related variations as impact other field crops. This is largely due to the fact that rice producers irrigate and fertilize their crops. Consequently, between 1983 and 1992, the yield of rice averaged 5,411 lb/acre, with a 5% variation.

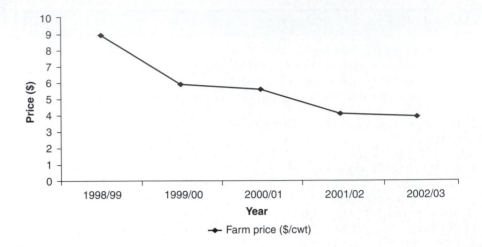

FIGURE 20–7 Farm price for rice has steadily declined over the past five years. Drawn from USDA data.

Farm price for rice has steadily declined over the last five years (Figure 20–7).

20.6.9 MILLING AND TYPES OF RICE

As previously described, the rice kernel is enclosed by the hull, which must be removed before the kernel is used. The process of removing the hull is called **milling** and is preceded by threshing to remove the hulled rice kernels from the panicle. Rice can be categorized according to the stage in postharvest processing and the kind of treatment applied.

1. *Rough rice.* Also called paddy rice, rough rice is whole-kernel rice with the hull intact.
2. *Parboiled rice.* Parboiling involves soaking, steaming, and then drying rough rice before milling. The process partially gelatinizes the endosperm starch. The effect of this treatment is that the milling process produces fewer broken kernels and hence a higher yield of head rice (whole kernels of milled rice). The milled rice cooks up fluffier and less sticky but takes longer to cook than regular rice. Furthermore, the soaking process transfers nutrients (e.g., vitamin B) from the bran and hulls to the endosperm. Parboiled rice keeps better.
3. *Milled rice.* The milling process detaches the hulls, producing various kinds of products and byproducts.
   a. *Brown rice.* This is the whole or broken kernels without the hull but with the **bran** (a layer of light brown covering over the kernel, which contains minerals and vitamins [especially vitamin B complex]). Brown rice may be cooked and eaten and is nutritionally richer than polished rice.
   b. *Head rice.* This rice consists of whole kernels of hulled rice that have a length of at least three-fourths of the length of the whole kernel. Long-grained rice yields about 50% head rice and second head, whereas short-grained rice yields about 60% of these rice categories.
   c. *Second head.* These are the largest broken kernels that are at least half as long but less than three-fourths the length of whole kernels.
   d. *Brewer's rice.* This constitutes the smallest broken rice fragment that will easily pass through a 5.5/64 inch sieve. It is used as a starchy adjunct in the brewing and distilling industries. It is also used in making pet foods and face powders.
   e. *Polished rice.* The process of polishing removes the bran and some of the endosperm. Although polished rice looks attractive on the dinner plate, it has less nutritional value than brown rice.
   f. *Byproducts of milling*

1. *Rice hull.* The outer woody covering of rice is used in making various industrial products, such as fuel, fertilizer additives, and synthetic rubber.
2. *Rice bran.* This consists of the outer cuticular layer and the germ of the rice. It is high in nutrition (vitamin B complex). It is used for feed and industrial products and in foods such as cookies, cereals, and breads.
3. *Rice polish.* Rice polish is produced during the final stages of the milling process and consists of the inner layer of the grain and a small amount of the outer layer. It is high in nutrients and is digestible.
4. *Rice straw.* This is the remaining dry vegetative material after harvesting rice. It is used as bedding material for livestock barns and as roughage in feed.
5. *Rice flour.* This is ground, milled rice. It is used in making baby food, rice cakes, and other snack foods.

20.6.10   RICE GRADES AND GRADE REQUIREMENTS

The grades and grade requirements of rice as prescribed by the USDA Grain Inspection Service are summarized Table 20–3. Grain that does not meet the standards set for any

Table 20–3   Grades and Grade Requirements for the Classes of Milled Rice—Long Grain, Medium Grain, Short Grain, and Mixed Milled Rice

Total		Number in 500 g	U.S. No. 1 2	U.S. No. 2 4	U.S. No. 3 7	U.S. No. 4 20	U.S. No. 5 30	U.S. No. 6 75
Seeds, heat-damaged and paddy kernels	Heat-damaged and objectionable seeds	Number in 500 g	1	2	5	15	25	75
	Red rice and damaged kernels (singly or combined)	Percent	0.5	1.5	2.5	4.0	6.0	15.0
Chalky kernels	In long grain rice	Percent	1.0	2.0	4.0	6.0	10.0	15.0
	In medium or short grain rice	Percent	2.0	4.0	6.0	8.0	10.0	15.0
Broken kernels	Total	Percent	4.0	7.0	15.0	25.0	35.0	50.0
	Removed by a 5 plate	Percent	0.04	0.06	0.1	0.4	0.7	1.0
	Removed by a 6 sieve	Percent	0.1	0.2	0.8	2.0	3.0	4.0
	Removed by a 6 plate	Percent	0.1	0.2	0.5	0.7	1.0	2.0
Other types	Whole kernels	Percent	—	—	—	—	10.0	10.0
	Whole and broken kernels	Percent	1.0	2.0	3.0	5.0	—	—
Color requirements		—	White or creamy	May be slightly gray	May be light gray	May be gray or slightly rosy	May be dark gray or rosy	May be dark gray or rosy
Minimum milling requirements			Well milled	Well milled	Reasonably well milled	Reasonably well milled	Lightly milled	Lightly milled

U.S. sample grade shall be milled rice of any of these classes which (a) does not meet the requirements for any of the grades from U.S. No.1 to U.S. No. 6, inclusive; (b) contains more than 15.0% of moisture; (c) is musty, sour, or heating; (d) has any commercially objectionable foreign odor; (e) contains more than 0.1% of foreign material; (f) contains two or more live or dead weevils or other insects, insect webbing, or insect-refuse; or (g) is otherwise of distinctly low quality.

Source: USDA

FIGURE 20–8 Rice kernel sizes: paddy or rough rice (top row), brown rice (center row), and milled rice (bottom row). (Source: USDA)

Table 20–4 Summary of Official U.S. Rice Grades

Classes	Rough		Brown		Milled		Broken Kernels	
Subclasses	Long		Long		Long		Second heads	
	Medium		Medium		Medium		Screenings	
	Short		Short		Short		Brewers	
	Mixed		Mixed		Mixed			
Grades **Special**	**U.S.** **No.**	**Special**	**U.S.** **No.**	**Special**	**U.S.** **No.**	**Special**	**U.S.** **No.**	
Parboiled	1	Parboiled	1	Parboiled	1		1	
Smutty	2	Smutty	2	Coated	2		2	
Weevily	3		3	Undermilled	3		3	
	4		4	Granulated	4		4	
	5		5	Brewers	5		5	
	6		Sample		Sample		Sample	

Source: USDA.

of the grades (i.e., 1 to 6) is classified as sample grade. Figure 20–8 shows various rice kernel characteristics used in grading. A summary of the official U.S. rice grades for the different rice subclasses is shown in Table 20–4.

REFERENCES AND SUGGESTED READING

Smith, C. W. and R. A. Dilday. (eds). 2002. *Rice: Origin, history, technology, and production*. Wiley and Sons, New York: NY.

Luh, B. S. 1995. *Rice: Production*. 2nd ed. Kluwer Academic Publishers. London: UK.

SELECTED INTERNET SITES FOR FURTHER REVIEW

http://www.cgiar.org/irri/Facts.htm

General information from International arena.

http://www.riceweb.org/countries/usa.htm

Rice production ecosystems

http://agebb.missouri.edu/rice/ricetill.htm

Good information on rice production

OUTCOMES ASSESSMENT

PART A

Answer the following questions true or false.

1. T F Rice is second to wheat in the world's total grain production.
2. T F Indica rice cultivars are long-grained.
3. T F Rice bran is rich in vitamin B complex.
4. T F United States lead the world in rice production.
5. T F Rice is a short-day plant.
6. T F Red rice can be controlled by flooding the field.
7. T F Most of the world's rice is produced under rainfed conditions.

PART B

Answer the following questions.

1. Rice is believed to have originated in _____.

2. Give the scientific name of cultivated rice. _____

3. Rice varieties may be classified according to size as either _____ or _____.

4. The process of removing the hull from rice is called _____.

5. Rice with the hull intact is called _____.

6. The colored layer exposed after removing the hull of the rice kernel is called _____.

7. Soaking, steaming, and drying rough rice before milling is called _____.

8. Give the four ecological zones of adaptation of rice.

_____

_____

_____

_____

PART C

Write a brief essay on each of the following topics.

1. Discuss the relative importance of rice as a food crop versus other crops.
2. Briefly discuss the origin of rice.
3. Describe the general botanical features of the rice plant.
4. Discuss the time of seeding and seeding rate of rice.
5. Discuss the use of crop rotation practices in rice production.

PART D

Discuss or explain the following topics in detail.

1. Briefly discuss the origin of rice.
2. Discuss the environmental conditions best suited to the production of rice.
3. Discuss the growth and development of rice.
4. Discuss the major insect pests in the production of rice.
5. Discuss the major diseases in rice production.
6. Discuss the market classes of rice.
7. Discuss the rice ecosystems of the world.
9. Discuss water seeding of rice.

21 Corn

TAXONOMY

Kingdom	Plantae	Subclass	Commelinidae
Subkingdom	Tracheobionta	Order	Cyperales
Superdivision	Spermatophyta	Family	Poaceae
Division	Magnoliophyta	Genus	*Zea* L.
Class	Lilliopsida	Species	*Zea mays* L.

KEY TERMS

Xenia	Pop corn	Flint corn
Quality protein maize	Flour corn	High lysine corn
Stover	Pod corn	Hybrid corn
Dent corn		

21.1: ECONOMIC IMPORTANCE

Corn, or maize, is the single most important crop in the United States, grown on more than 20% of the cropland. Most of the production occurs in the region of the United States called the Corn Belt, where six states (Iowa, Illinois, Nebraska, Minnesota, Indiana, and Ohio) account for about 80% of the national production (Figure 21–1). Corn is grown in every state except Alaska and on all cropped farms in the United States. Iowa leads the nation, with 22% of the total production. Corn is the fourth most important food crop in the

world, behind wheat, rice, and potatoes in total production. More than 327 million acres of corn are planted each year worldwide. World yields average about 42 bushels per acre. There are six "Corn Belts" in the world—the U.S. Corn Belt, the Danube Basin (southwest Germany), the Po Valley (northern Italy), the plains of northern China, northeastern Argentina, and southeastern Brazil.

In 1866, the USDA reported that 30 million acres of corn was harvested at an average yield of about 24 bushels per acre. In the early 1990s, about 100 million acres were grown, with an average harvested yield of 28 bushels per acre. However, in 1990, the USDA reported 67 million acres of harvested corn, with an average yield of 118 bushels per acre. The dramatic increase in yield over that period is attributed to the adoption of hybrid seed and the use of fertilizers.

About 70 million acres of cropland are devoted to corn, with an average yield of about 92 bushels per acre. A farmer in Illinois is reported to have produced a record yield of 370 bushels per acre on a 20-acre farm in 1985. Some 8.6 million acres are devoted to silage production. Corn has the highest value of production of any crop in the United States, averaging 8 billion bushels worth $20 billion per year.

On the world scene, the United States, China, Brazil, Mexico, France, and Argentina together account for 75% of the world's corn production, the United States accounting for about 40% of this total (Figure 21–2). Other producers include Romania and South Africa.

21.2: ORIGIN AND HISTORY

Corn is arguably the most completely domesticated of field crops. Modern corn is incapable of existing as a wild plant; no wild form is known. Its origin is probably Mexico or Central America. It was produced as early as 6000 B.C. in Tehuacan, Mexico, by the Mayan and Aztec Indian civilizations. It was taken north by the native Americans. Corn was dispersed to the Old World in the sixteenth and seventeenth centuries. The explorers introduced Indian corn to Europe and Africa. Modern varieties have larger cobs and a greater number and weight of kernels per ear than the original Indian corn. The modern cultivated

FIGURE 21–2 General corn production regions of the world. (Source: USDA)

plant is believed to have been obtained through the process of mutation, natural selection, and mass selection by the American Indians. It is proposed that corn's progenitor may be a domesticated version of teosinte, a wild grass that grows in Mexico and Guatemala.

21.3: ADAPTATION

Corn has a wide geographic adaptation. It is grown from as far north as 58°N latitude to 35–45°N latitude. It is grown at below sea level to 13,000 feet. Corn is adapted to warm temperatures. In the United States, the temperatures in the Corn Belt average 70° to 80°F in summer, with a nighttime temperature average of 58°F. Furthermore, this region experiences at least 140 days of frost-free growing season. For optimal yield, corn prefers the June–August temperature average of 68° to 72°F. Higher temperatures, above 80°F, reduce crop yield.

Rainfall in corn production areas range between 10 inches in the semiarid regions (e.g., in Russia) to over 200 inches in the tropics of India. The optimal rainfall amount is 20–40 inches for rainfed production, although corn can grow well in drier conditions where the rainfall is less than 10 inches provided that irrigation is used. Corn is a short-day plant. It produces excessive growth under long days. Varieties should be selected to fit the photoperiod conditions of the production region.

In terms of edaphic factors, corn prefers a soil pH between 5.5 and 8. The soil should be fertile, well drained, and loamy. Sandy soils warm up more quickly in spring and hence are preferred for short-season corn production.

Corn (*Zea mays* L.) is a diploid ($2n = 20$) and a monocot of the family Poaceae (Gramineae), or grass family. The genus has four species: *Zea mays* (cultivated corn and teosinte), *Z. diploperennis* Iltis et al. (diploperennial teosinte), *Z. luxurians,* and *Z. perennis* (perennial teosinte). Of these four species, only *Zea mays* is widely grown commercially in the United States. The closest generic relative of *Zea* is *Tripsacum,* which has seven species, three of which are known to grow in the United States. Teosinte occurs in the wild in Mexico and Guatemala.

Corn is a monoecious annual and one of the largest of the cereals, capable of reaching 15 feet in height (Figure 21–3). The male (staminate) flowers occur in the terminal panicle or tassel at the top of the stalk, whereas the female (pistillate) inflorescence is borne in the axils of the leaves as clusters, called a cob, at a joint of the stalk. Long silks (long styles) hang from the husk of each cob. These pollen tubes are the longest known in the plant kingdom. As pollen receptors, each silk must be individually pollinated in order to produce a fruit or kernel. A fertilized cob (also called an ear) contains eight or more rows of kernels. Furthermore, a stalk may bear one to three cobs.

Corn has a variety of morphological features. Some early-maturing types, maturing in 50 days, may attain a height of 2 feet and produce 8–9 leaves, whereas tall, late-maturing types (330 days) may attain a culm or stalk height of 20 feet and bear 42–44 leaves. The hybrid corn varieties grown in the northern United States attain a height of 3–8 feet, bear 9–18 leaves, and mature within 90–120 days. The central Corn Belt hybrid varieties range between 8 and 10 feet in height, bear 18–21 leaves, and mature in 130–150 days. The varieties used on the Gulf Coast and south Atlantic regions are much taller (10–12 feet), produce more leaves (22–27), and tiller profusely, maturing late (170–190 days).

Corn has both seminal and adventitious roots. The seminal roots number 3–5 and grow downward at the time of seed germination. The crown, or coronal, roots arise from the nodes of the stem, about 1–2 inches below the soil surface and number between 15 and 20 times the number of seminal roots. The aerial roots (buttress, prop, or brace roots) arise at nodes on the stem above the ground.

FIGURE 21–3 Corn plants in the field, with tassels at the top of the plants. (Source: USDA)

Corn is predominantly cross-pollinated. The immediate effect of the pollen parent on the characteristics of the endosperm, embryo, or scutellum is called **xenia**. This effect is manifested when one type of corn is pollinated by another type. For example, when sweet corn is cross-pollinated by flint corn, the resulting kernels are smooth and starchy instead of wrinkled and sugary. Similarly, when yellow corn is pollinated by white corn, the kernels are yellow but are lighter in color and often capped with white.

The number of rows of grain is variable among varieties, ranging between 8 and 28. Each row contains between 20 and 70 kernels. Most of the corn varieties grown in the United States contain 10, 12, 14, or 18 rows of kernels and average about 500 kernels per ear.

21.5: TYPES OF CORN

Corn may be grouped into seven types on the basis of endosperm and glume characteristics—dent, flint, flour, pop-, sweet, waxy, and pod (Figure 21–4). Of these, five are produced commercially—dent, flint, flour, sweet, and waxy corns.

1. *Dent corn (Z. mays indentata).* Dent corn is the most widely cultivated type in the United States. It is characterized by a depression (dent) in the crown, caused by the rapid drying and shrinkage of the soft starch at the crown. Of the multiple colors available, the yellow and the white kernels dominate commercial production.
2. *Flint corn (Z. mays indurate).* Flint corn is predominantly comprised of corneous or hard starch, which encloses the soft starch in the center. The kernels are

FIGURE 21–4 Diversity in corn kernels showing popcorn, sweetcorn, flour corn, flint corn, dent corn, and pod corn. (Source: USDA)

smooth, hard, and usually rounded at the top. This type of corn is grown widely in Europe, Asia, Central America, and South America. It is less widely grown in the United States.

3. *Flour corn (Z. mays amylacea).* As the name implies, flour corn consists almost entirely of soft starch, making the kernels soft. It has the shape of dent corn but shrinks uniformly on drying. It is grown in drier sections of the United States, mainly by American Indians, as well as in the Andean region of Central and South America. Different kernel colors exist, the most common being white, blue, and variegated.

4. *Popcorn (Z. mays everta).* Popcorn is an extreme form of flint corn. It has a very hard, corneous endosperm, with only a small portion of soft starch. The kernels are characteristically small and may either be pointed or have a rounded tip. Different colors exist, most corneous varieties being yellow or white. The kernel pops on heating as a result of the unique quality of the endosperm that makes it resist the steam pressure generated until it reaches explosive proportions.

5. *Sweet corn (Z. mays saccharata).* Sweet corn is characterized by a translucent and wrinkled appearance on drying and a sweet taste when immature. The standard sweet corn is a mutant of dent corn, with a mutation at the *sugary (sy)* locus. This mutation causes the endosperm to accumulate about two times more sugar than field corn. New mutants have been developed—*sugary enhanced (se)* and *shrunken-2 (sh2),* or supersweet corn. Some sweet corn varieties are unable to convert sugar to starch. Sweet corn is grown as a winter crop in the southern United States, especially Florida.

6. *Waxy corn.* Waxy corn has a uniformly dull appearance. Instead of amylose, the starch of waxy corn consists of amylopectin, the result of *waxy (wx)* mutation. Ordinary corn consists of about 78% amylopectin (a high-molecular-weight, branched chain) and 22% amylose (a low-molecular-weight, straight chain).

7. *Pod corn (Z. mays tunicata).* Pod corn has primitive features, each kernel being enclosed in a pod or husk, before the entire ear is enclosed in husks, like other corns. Pod corn versions of the other types of corn (i.e., flint pod corn, dent pod corn, etc.) are available.

21.6: ENHANCED COMPOSITIONAL TRAITS (VALUE-ADDED CORN)

Two of the major areas in which corn has been subjected to intense breeding for enhanced compositional traits are (1) protein content and quality and (2) oil content.

21.6.1 BREEDING HIGH LYSINE CONTENT OF GRAIN

Breeders using conventional methods of ear-to-row selection have been able to increase the total protein content of corn kernels from 10.9% to 26.6%. Unfortunately, because the protein of corn is about 80% zein, and hence nutritionally inadequate, the high increase in total protein was nutritionally unprofitable to non-ruminant animals. The zein fraction of the total protein is deficient in lysine and tryptophan. This deficiency was overcome in 1964, when researchers at Purdue University discovered a mutant, called *opaque-2* or *floury-2,* which increased the lysine content of the kernel. The resulting corn is called **high lysine corn** and has a characteristic soft and starchy endosperm. Consequently, the softer endosperm predisposes high lysine kernels to breakage, cracking, and rot. Generally, high lysine cultivars yield lower than their conventional counterparts. Cross-pollination with normal dent corn reverses the soft endosperm to normal dent endosperm. High lysine corn production must be done in isolated fields. The *opaque-2* recessive gene increased the lysine content of the kernel from about 0.26 to 0.30% to about 0.34 to 0.37%. High lysine has also been transferred into sorghum.

Quality protein maize (QPM) is an extension or improvement of high lysine corn. It is a high lysine product because it uses the *opaque-2* gene. However, it is unlike the traditional high lysine corn because it lacks all the undesirable attributes of high lysine products (i.e., low yields, chalky-looking grain, and susceptibility to diseases and insect pests). It looks like regular corn but has about twice the levels of lysine and tryptophan. QPM was developed by two researchers, K. V. Vasal and E. Villegas, over about three decades. They used conventional breeding methods to incorporate modifier genes to eliminate the undesirable effects of the lysine gene. The two scientists were awarded the World Food Prize in 2001 for their efforts.

QPM has less of the indigestible prolamine-type amino acids that predominate in the protein of normal maize. Instead, QPM cultivars have about 40% of the more digestible glutelins and a balanced leucine-isoleucine ratio for enhanced niacin production on ingestion. Research also indicates that QPM has better food and feed efficiency ratings (grain food intake/grain weight gain) following feeding tests with animals (e.g., pigs and poultry). QPM cultivars have been released for production in more than 20 developing countries since 1997.

21.6.2 HIGH OIL

The major energy-contributing nutrient in corn is starch. The protein fraction makes up only 8.5% and the fat content only 3.7%. Through genetic selection over many years, certain varieties of corn with a higher content of oil have been bred. Cyril G. Hopkins of the University of Illinois initiated the first breeding efforts for corn oil improvement about 100 years ago. Using a classical breeding method, he succeeded in increasing corn oil to 17% by selecting for oil content over 70 years. After 82 generations, the oil content level reached 19%. However, such dramatic increases came at the cost of significant yield reduction (about a 70% yield reduction, compared with commercial hybrids). Hence, these high oil genotypes have not been commercialized. However, breeders have managed to achieve modest increases in oil content (6 to 8%) while maintaining crop yield.

21.7: COMMERCIAL VARIETIES

Corn is naturally open-pollinated. Two types of varieties are grown that differ in the method by which they are produced and maintained: open-pollinated varieties and hybrid corn.

21.7.1 OPEN-POLLINATED VARIETIES

Open-pollinated varieties are heterozygous and heterogeneous and are designed to enable the grower to save seed from the current year's crop for planting the next season. These varieties are widely used in corn production in developing countries.

21.7.2 HYBRID CORN

W. T. Beal is credited with making the first controlled cross in 1877 at Michigan State University (then the Agriculture College of Michigan). The genetic basis of hybridization, the exploitation of the phenomenon of hybrid vigor (heterosis) that makes the product of a cross between two inbred lines (repeated selfed) outperform its parents, was clearly elucidated by G. H. Shull and later E. M. East. Over the years, plant breeders have tried to find more efficient methods of breeding corn. In 1918, D. F. Jones proposed the double-cross hybrid method for commercial corn hybrid production (Figure 21–4). He was followed by many researchers—notably, F. D. Jones, H. K. Hayes, M. T. Jenkins, and G. F. Sprague—whose efforts significantly advanced corn breeding. In the 1950s and 1960s, the discovery of cytoplasmic male-sterility (cms), which eliminated the tedious process of detasseling plants in the field, further facilitated corn hybrid production. Double crosses were replaced by single-cross hybrids. In 1933, only about 1% of all U.S. corn producers used hybrid seed. By 1965, nearly 100% of all producers used hybrid seed. Single-cross hybrids are initiated by crossing inbred lines that have superior combining ability ("nick" well in a cross). Synthetic cultivars of corn are available. They are produced by randomly mating parents and increasing the product from such multiple crosses. Synthetic cultivars are developed especially for the agricultures of lesser-developed countries.

21.8: GROWTH STAGES

A summary of the growth staging in corn is presented in Table 21–1. The stages are based on the Leaf Collar Method, the system used by hail adjusters for hail damage assessment. Corn plants generally start to silk at the 14-leaf stage and tassel at the 16-leaf stage. Physiological maturity is marked by a black layer that forms under the outer layer of the kernel tip.

21.9: CULTURAL PRACTICES

21.9.1 TILLAGE

Corn producers prepare the land in various ways, by plowing, chiseling, disking, harrowing, or performing a combination of these tillage operations. Where soil compaction or hardpan occurs, deep tillage, deep chiseling, and in-row subsoiling may be helpful. The conventional tillage system entails stalk chopping, chisel plowing, and disk harrowing or use of field cultivation. The reduced till system includes stalk chopping and chisel plowing, whereas the no-till system may or may not involve stalk chopping.

Where erosion is a problem, the field may be chisel-plowed in the fall, leaving crop residue on the field in winter and starting seedbed preparation in late winter or very early spring. In the North and the Corn Belt, where the ground freezes in winter, the land can be plowed either in the fall or spring. Where the field carried alfalfa or pasture in the pre-

Table 21–1 Summary of the Growth Stages of Corn

Vegetative Stages			Reproductive Stages		
VE	=	emergence	R1	=	silking
V1	=	first leaf	R2	=	blister
V2	=	second leaf	R3	=	milk
V3	=	third leaf	R4	=	dough
Vn	=	nth leaf	R5	=	dent
VT	=	tasseling	R6	=	physiological maturity

vious season, plowing should occur in the fall in order to allow ample time for the material to decay. Fall plowing is not appropriate in the South, as it promotes water erosion.

Before seeding, the final seedbed preparation should be conducted to produce the tilth in the top 4 inches of soil, where conventional or reduced-till systems are adopted. The land can be double disked and harrowed to make it ready for seeding. Where drainage is poor, the land may be listed into a bed before seeding.

21.9.2 SEEDING

Seeding in conventional tillage may be accomplished by using a row-crop planter. A planter or grain drill may be used for seeding in a reduced-till or no-till environment. Listing provides a more economical method of weed control by tillage. Listed corn is better protected from spring frost. Listing is adapted to regions with limited rainfall and light soils.

Surface planting is suited to areas with abundant rainfall and heavy soil. Seeding is conducted at a depth of 1½–3 inches deep, deeper placement being used where soil is dry. Row spacing ranges from 36–44 inches apart, 40 inches being the most widely used. Spacing within rows may be 6 inches or wider. At 6 × 40 inches, the plant population per acre is 26,000; however, most hybrids produce well at 30-inch spacing. Narrower rows (15–22 inches) are known to be even more productive. Narrower rows are convenient if the spacings are matched with the equipment used to produce other crops (e.g., soybeans, sugar beets) that are used in the rotation.

21.9.3 DATE OF PLANTING

Corn should be planted into a warm soil. If soil conditions are not favorable for immediate germination, the seed may survive in the soil for about 3 weeks before emergence. Corn needs 100–150 growing degree days (heat units) to emerge. The early morning soil temperature of 50°F is warm enough for corn seeded at ½–2 inches to emerge. Seeding should not be conducted into wet soil, lest the soil become compacted.

Seeding corn begins about February 1 in southern Texas and proceeds northward at an average rate of 13 miles per day. Consequently, planting in the Corn Belt occurs about May 15 for early planting and about June for late planting. Planting dates should allow the corn to avoid specific adverse periods in the growing season. For example, planting in June in west Kansas allows the corn to tassel and silk after the extreme midsummer drought and heat period are over.

Growers should plant full-season hybrids first, then early-maturing and midseason varieties, so that the full-season varieties can take full advantage of the maximum heat units to be accumulated. Full-season varieties are more susceptible to yield reduction when planting is delayed, as compared with short- to midseason hybrids. It is advantageous for a producer to plant multiple hybrids of different maturities to reduce the damage from disease and environmental stress during the growing season. Furthermore, this practice spreads out the harvest time and the associated work load. A combination of hybrids to plant may be 25-50-25 (that is, 25% are early-maturing, 50% are midseason, and 25% are full-season). One study indicates that in central Ohio, for example, yields decline by approximately 1–1.5 bushels per day when planting is delayed beyond the first week of May. Early planting produces shorter plants with better standability. Delayed planting predisposes the plants to frost damage, diseases, and insects.

21.9.4 SEEDING RATES

Seeding rates for grain production range between 2,000 and 4,000 plants per acre. Corn for silage is seeded more closely than this rate. When growing corn on fertile soils that support about 160 bushels per acre under irrigated production, the producer should adopt a seeding rate that produces a final stand of 30,000 plants per acre. The plant population on fields that average 120 bushels per acre or less should be less (20,000–22,000 plants per acre).

When a high plant population is selected, it is important that the varieties be lodging-resistant (superior stalk quality for standability) and resistant to stalk diseases *(Anthracnose* and *Gibberella).* It is important for plants to be spaced and emerge evenly for optimal yields, regardless of plant population and planting date. Crowding promotes competition for growth factors and consequently barren plants and small ears. Similarly, gaps in the stands cause plants next to the gaps to compensate for the missing plants. However, they do so ineffectively, such that plants become crowded and hence produce poorly. A Purdue University study indicated that corn yield can be increased by 4–12 bushels per acre with optimal within-row spacing. Competition from large, early-emerging plants tends to decrease the yield of smaller, later-emerging plants. Uneven emergence is caused by several factors, including variability in the soil moisture in the seed depth zone, poor seed-to-soil contact, variable soil temperature in the seed zone, soil crusting, improper adjustment of planter closing wheels, and worn discopeners.

21.9.5 CROP ROTATION

Corn may be continuously cropped or in rotation with other crops. Research indicates that corn benefits from rotation in both no-till and plowed tillage systems. Common rotations are 2-year corn–small grain rotation, 3-year corn–small grain–clover rotation, 4-year corn-oats-wheat-clover, and 4-year corn–corn–small grain–clover rotations. Other crops used in rotations include sugar beets, potatoes, alfalfa, and meadow. The crops vary from one region to another. For example, corn-soybean is the most common cropping sequence in Ohio, with a 10% yield advantage over continuous cropping. In the South, a good rotation is cotton-corn, but these crops do not supplement each other well; hence, a legume should be included. Crop rotation to broadleaf crops (soybean, potatoes, dry bean) reduces the incidence of leaf diseases that overwinter or oversummer in corn debris.

21.9.6 FERTILIZATION

Corn is the most fertilized crop in U.S. production. It uses large amounts of nutrients, especially nitrogen. Nitrogen may be a preplant application or sidedress. Where rates exceed 50 lb/acre, this may interfere with phosphorus uptake. In this instance, banded application should be less than 50 lb/acre. High nitrogen increases the protein content of the kernel by increasing stalk susceptibility to green snap. In a rotation in which corn follows sugar beet or a fallow, high rates of phosphorus application are desirable. This is because the levels of mycorrhiza (fungi that aid phosphorus absorption by the roots) are reduced. Sandy soils may be low in potassium. Corn cropped on this soil type may benefit from potassium applied as a band application. In terms of micronutrients, some growing areas benefit from sulfur, zinc, and chlorine. Corn is moderately susceptible to salt damage.

21.9.7 IRRIGATION

Corn requires 18–22 inches of soil moisture during the growing season for optimal yield. Adequate moisture of at least 0.20 inch on average per day is required after emergence throughout the production season for optimal yield. However, the most critical periods in which higher daily moisture is needed are the early tassel, silking, and blister kernel stages (40–80 days after emergence), when the daily water use rate is about 0.30 inch per day. Corn is relatively drought-tolerant from emergence to the onset of tassels (about 40 days) and after blister kernel development. The effects of drought on corn yield are most serious at silk emergence and pollen shedding.

The highest water use demands occur in July and August, when the average water use may be 6–8 inches per month. When the temperature rises above 90°F, the water demand can be about 2.1 inches per week. The last irrigation of the growing season depends on the maturity of the corn kernels. Generally, irrigation is beneficial and should continue till about 75% of the plants have visible silks on the ears. It is important for irrigation to continue until the milk layer in the kernel has moved down to the tip of the kernel or the black layer has formed.

21.9.8   WEED MANAGEMENT

Weeds in a corn field may be managed by cultivation, crop rotation, or chemical methods. A weed management program is developed by taking into account the soil type, the kinds of weeds present, crop rotation, and the cost of implementation. Chemical applications may be preplant or postemergent. Herbicides approved for use in corn fields include Roundup® Ultra/RT, Lasso®, Dual®, atrazine, 2,4-D, and Basagram®. Atrazine, for example, may be used as a preemergence herbicide to control broadleaf weeds and germinating weedy grasses. 2,4-D is useful for controlling late-emerging weeds.

21.9.9   DISEASES

Corn diseases include rots and seedling blight, leaf diseases (e.g., eye spots, northern corn leaf blight, common rust, and stalk and ear diseases like common smut, head smut, ear rot). The characteristics of these diseases are summarized in Table 21–2.

Table 21–2   Common Diseases in Corn Production

1. Corn smut *(Ustilago maydis)*
   - Causes complete barrenness in many plants and reduces grain development
   - Galls form on aerial parts of the plant—ears, leaves, and stalks—the most destructive being those on the ears.
   - Galls begin as whitish structures that turn dark as black smut spores form inside.
   - Attack is more severe on vigorously growing plants.
   - Smut organisms live in the soil.
   - Control is by using smut-resistant cultivars.
2. Head smut *(Sphaceltheca reiliana)*
   - Fungus lives in the soil or on the seed.
   - Produces galls on the ear and tassel of the plant.
   - Control is by resting infected soil from corn production for at least 2 years and treating seed to reduce spread to new fields.
3. Rots: root (e.g., *Pythium arrhenomanes*), ear (e.g., *Gibberella zeae*), stalk (e.g., *Diplodia zeae*)
   - Cause a variety of symptoms, including reduction in crop stand, reduction in plant vigor, barrenness, chlorosis, general blighting of the plant, and rotting of ears.
   - Molding of corn grain in storage may be due to the growth of *Penicillium, Aspergillus,* and *Fusarium* species of rot-causing fungi.
   - *Diplodia* develops on infested kernels, infected seeds appearing dry and brown.
   - *Gibberella* causes both root rot and seedling blight.
   - Control by using sound cultural and management practices, such as using healthy seed, planting after soil is warm, and treating seed.
   - Using resistant hybrids and disinfecting seed-corn are effective practices.
4. Leaf diseases
   - Include bacterial diseases such as bacterial wilt *(Bacterium stewarti)* and *Helimithosporium* leaf spots (*H. maydis, H. turcicum, H. carbonum*).
   - Rusts (e.g., *Puccinia sorghi, P. polysora*).

21.9.10 INSECTS

The European corn borer is one of the major insect pests of corn. One of the major applications of biotechnology in food agriculture is the development of the *Bt*-corn hybrids. These are hybrids that have been genetically engineered to express an insecticidal protein produced by a bacterium, *Bacillus thuringiensis (Bt)*. Genetically modified crops are discussed in detail in Chapter 13. Other insect pests in corn production are corn earworms, chinch bugs, cutworms, and weevils. The characteristics and management of these and other pests are summarized in Table 21–3.

21.9.11 HARVESTING

Corn is physiologically mature when the black layer forms under the outer layer of the kernel tip. The moisture content of the kernel at this stage averages about 34%. The heat units, or accumulated degree days, are between 1,750 and 1,850 for varieties maturing in 70 days, 2,150–2,250 for 90-day maturing varieties, and 2,400–2,500 for 105-day maturing varieties.

Harvesting begins in August in the Gulf states and early October in the northern states (Figure 21–5). Early harvesting reduces harvest losses from such causes as lodging (from stalk rot and severe storms), weather delays, ear droppage, and grain shelled (or impact of ear snapping rolls of combine). Other losses associated with the combine operation are cylinder losses (improperly shelled ears) and separating losses (loose kernels, not shaken out of the ears and kernels) and can be reduced by proper combine operation and maintenance.

Corn can be picked when the grain moisture is 20% or lower and placed into crib storage. Hard husking from standing stalks can be done by manual labor at the average rate of 6 hours per acre per person. Husked corn is usually stored in a crib to dry down. If the moisture content is high (30% or more), husked corn is susceptible to rapid spoilage.

Sweet corn is harvested when the kernels are at the milk stage. At this stage, the silks become brown and dry (about 21 days after silking starts). The produce is sold as fresh corn for eating as corn on the cob or is processed for canning. The fresh corn harvested at this stage contains about 70% moisture, 5 to 6% sugar, and 10 to 11% starch. Harvested sweet corn deteriorates rapidly, losing its sweetness in about 2 days in hot summer and about 5 days in cooler weather.

Corn average and yield have remained relatively the same over the last five years (Figure 21–6). Similarly, farm prices of corn has stayed stable over the same period (Figure 21–7).

21.9.12 CORN FOR FEED

Stover

When corn is husked or snapped from standing stalks, the remaining stalks and leaves are called the corn **stover.** The stover contains about 30% of the total nutrients of the corn plant. Stover can be harvested for livestock feed or bedding material after it has been shredded.

Fodder

Corn may be grown and harvested prematurely for fodder or grazed green in the field. In semiarid regions where the weather conditions may prevent corn from producing grain, the premature crop may be harvested for livestock feed.

Silage

Corn for silage is most nutritious when the plants are cut when the ears are glazed and most leaves are on the plant. A silage harvester is used to harvest the plants. Corn harvested sooner (in the milk or silking stage) has high moisture content and produces poor-quality

Table 21–3   Common Insects of Corn in Production

1. European corn borer *(Pyrausta nubilalis)*
   - One of most serious pests of corn.
   - Larvae bore within stalks, tassels, cobs, and ears of corn and eat their way through these parts.
   - Larvae overwinter in stalks, stubble, and other remnant materials.
   - Larvae pupate from April to July, emerging as moths after 2–3 weeks.
   - Affected plants lodge easily.
   - Control by destroying plant debris from previous planting.
   - Control by using resistant cultivars (*Bt* cultivars), stiff-stalked, disease-resistant hybrids; also plant as soon as soil is warm to establish vigorous plants.
2. Southwestern corn borer *(Diatraea grandiosella)*
   - Damage is similar to European corn borer.
   - Affected plants lodge.
   - Control by rotation with grain sorghum or late planting.
3. Chinch bug *(Blissus leucopterus)*
   - Corn crop is most susceptible when planted next to barley or small grain and planted late.
   - Bugs overwinter in tall grasses and breed in small grain, migrating to corn in June or early July.
   - Chemicals or use of resistant cultivars are effective control measures.
4. Corn earworm *(Heliothis obsolete)*
   - Most destructive in southeastern production states.
   - Known by other names (tomato fruitworm, cotton bollworm) and affects many other crops.
   - The worm overwinters in pupal stage in the soil.
   - Larvae destroy developing kernels, feeding on leaves, tassels, and silks.
   - Prevent by growing hybrids with long, heavy husks to retard or prevent larvae from reaching the ear.
   - Planting early allows crop to escape the period of high population of insects.
   - Control by using resistant cultivars and chemicals.
5. Corn rootworm: southern corn rootworm *(Diabrotica undecimpunctata howardin)*, corn rootworm *(D. longicornis)*, and Colorado corn rootworm *(D. virgifera)*
   - Larvae bore into bases of seedlings and ruin growing point.
   - They feed on the roots of plants, causing lodging.
   - Control by chemicals or resistant hybrids.
6. Weevils: rice or black weevil *(Sitophilus oryza)* and Angoumois grain moth or fly weevil *(Sitotroga cerealella)*
   - These are storage pests and field pests.
   - Use hybrids with long, heavy husks for control.
7. Cutworms *(Agrotis orthogonia)*
   - Cut off the culm of young plants near the soil surface.
   - Use summer fallow to starve larvae, as well as chemicals.
8. Armyworms *(Cirphis unipuncta)*, fall armyworms *(Laphygma frugiperda)*
   - Pests devour leaves and sometimes growing point.
   - Control by insecticides.

FIGURE 21–5 A corn field (top). Harvesting corn (bottom). (Source: USDA)

FIGURE 21–6 Corn acreage and yield have remained relatively the same over the past five years. Drawn from USDA data.

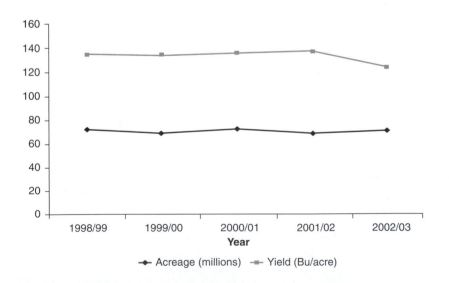

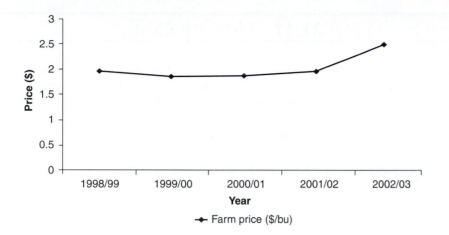

FIGURE 21–7 The farm price for corn has remained relatively the same over the past five years with a projected increase in 2002. Drawn from USDA data.

silage. The structure in which the materials will be stored determines the proper moisture content of the corn. In sealed, airtight silos, the appropriate moisture is 55 to 68%; in bag silos, 60 to 70%; in upright silos, 62 to 68%; in trench silos, 65 to 70%.

21.10: CORN GRADES AND GRADE REQUIREMENTS

The grades and grade requirements of corn as prescribed by the USDA Grain Inspection Service are summarized in Table 21–4. Grain that does not meet the standards set for any of the grades (i.e., 1 to 5) is classified as sample grade.

Table 21–4   Corn Grades and Grade Requirements

Grading Factors	GRADES OF U.S. NOS.				
	1	2	3	4	5
Minimum lb limits of					
Test weight (lb/bushel)	56.0	54.0	52.0	49.0	46.0
Maximum % limits of					
Damaged kernels					
Heat	0.1	0.2	0.5	1.0	3.0
Total	3.0	5.0	7.0	10.0	15.0
Broken corn and foreign material	2.0	3.0	4.0	5.0	7.0

U.S. sample grade =
  a. Corn that does not meet the requirements for U.S. Nos. 1, 2, 3, 4, or 5.
  b. Contains stones that have an aggregate weight in excess of 0.10% of sample weight, two or more pieces of glass, three or more crotalaria seeds (*Crotalaria* spp.), two or more castor beans (*Ricinus communis* L.), four or more particles of an unknown foreign susbtances) or a commonly recognized harmful or toxic substance(s), eight or more cockleburs (*Xanthium* spp.), or similar seeds singly or in combination, or animal filth in excess of 0.20% in 1,000 grams.
  c. Has a musty, sour, or commercially objectionable foreign odor.
  d. Is heating or of distinctly low quality.
Source USDA.

21.11: USES OF CORN

Corn is used for food, feed, and various industrial products. The major industrial products are high-fructose syrups, glucose/dextrose sugars, starch, and alcohol. To obtain these products, corn is either wet- or dry-milled. Dry-milling is simply grinding the grain. Wet-milling is a more complex industrial process, producing starch and starch-derived chemicals, as well as oil, bran, and protein products.

REFERENCES AND SUGGESTED READING

Agricultural Outlook. October 2001. Economic Research Service, United States Department of Agriculture, Washington DC.

Cobley, L. S., and W. M. Steele. 1976. *An introduction to the botany of tropical crops.* New York: Longman.

Galinat, W. C. 1988. The origin of corn. In G. F. Sprague, and J. W. Dudley, eds. *Corn and corn improvement.* 3d ed. Madison, WI: American Society of Agronomy, Crop Science Society of America, and Soil Science Society of America.

Martin, J. H., and W. H. Leonard. 1971. *Principles of field crop production.* New York: Macmillan.

Wych, R. D. 1988. Production of hybrid seed. In G. F. Sprague and J. W. Dudley, eds. *Corn and corn improvement.* 3d ed. Madison, WI: American Society of Agronomy, Crop Science Society of America, and Soil Science Society of America,

SELECTED INTERNET SITES FOR FURTHER REVIEW

http://www.aphis.usda.gov/bbep/bep/bp/corn.html

us corn production statistics

http://www.ag.ohio-state.edu/~ohioline/b472/corn.html

general corn production information

http://www./ext.nodak.edu/extpubs/plantsci/rowcrops/a1130-2.htm

general corn production information

OUTCOMES ASSESSMENT

PART A

Answer the following questions true or false.

1. T F   Corn is the single most important crop in United States agriculture.
2. T F   Corn is a warm season crop.
3. T F   Corn is a dioecious plant.

4. T F   The opaque-2 gene increases the oil content of the corn kernel.
5. T F   Most of the corn varieties grown in the United States are single-cross hybrids.
6. T F   Most of the corn grown in the United States is of the flint type.
7. T F   The *shrunken-2 (sh2)* gene makes corn supersweet.

PART B

Answer the following questions

1. Give three of the states of the Corn Belt of the United States.

   _____ _____ _____

2. Give the full scientific name of corn. _____

3. Give the seven types of corn according to endosperm and glume characteristics.

   _____ _____ _____

   _____ _____ _____

   _____

4. The immediate effect of the pollen plant on the characteristics of the endosperm, embryo, or scutellum is _____.

5. Give three of the top corn _____ _____
   _____ producing states in the United States.

6. Give the three top corn producing nations in the world. _____
   _____ _____

7. _____ is the mutant that is responsible for the floury endosperm of corn.

8. When corn is harvested by snapping from standing stalks, what remains of the plant in the field is called _____.

PART C

Write a brief essay on each of the following topics.

1. Discuss the relative importance of corn as a food crop versus other crops.

2. Briefly discuss the origin of corn.

3. Describe the general botanical features of the corn plant.

4. Discuss the time of seeding and seeding rate of corn.

5. Discuss the use crop rotation practices in corn production.

PART D

Discuss or explain the following topics in detail.

1. Briefly discuss the origin of corn.

2. Discuss the environmental conditions best suited to the production of corn.

3. Discuss the growth and development of corn.

4. Discuss the major insect pests in the production of corn.

5. Discuss the major diseases in corn production.

6. List and describe the seven types of corn.

7. Discuss the development of high lysine corn.

8. Discuss the corn varieties used in corn production.

9. What is the black layer?

10. What makes popcorn pop?

22
Sorghum

TAXONOMY

Kingdom	Plantae	Family	Poaceae
Subkingdom	Tracheobionta	Genus	*Sorghum* Moench—sorghum
Superdivision	Spermatophyta		
Division	Magnoliophyta	Species	*Sorghum bicolor* (L.) Moench—sorghum
Class	Lilliopsida		
Subclass	Commelinidae	Subspecies	*Sorghum bicolor* (L.) Moench ssp. *bicolor*—grain sorghum
Order	Cyperales		

KEY TERMS

Dhurrin	Feterita	Hegari
Prussic acid	Durra	Shallu
Kafir	Milo	

22.1: ECONOMIC IMPORTANCE

Sorghum was the world's number 4 most important grain crop in 1995, accounting for 4% of total cereal production and a total harvest of 53 million metric tons. Sorghum and millet are two of the major world food crops that originate from Africa. Even though

603

sorghum has become important in the agricultural production of developed countries, it is still primarily a developing country crop, with 90% of the world's acreage found in Africa and Asia.

In the United States, 470,525,000 bushels were produced in 2000 on 9,195,000 acres. The leading producer was Kansas, with 3,500,000 acres and production of 188,800,000 bushels, followed by Texas, with an average of 3,000,000 acres and production of 143,350,000 bushels (Figure 22–1). Other important producers were Nebraska, South Dakota, Colorado, Oklahoma, Missouri, Louisiana, New Mexico, and Arizona (Table 22–1). The varieties grown in the United States were primarily (78%) for grain. However, sorgho (sweet sorghum) is grown for forage, silage, and syrup (stalks crushed). The important sorghum states are Alabama, Mississippi, Georgia, Tennessee, and Iowa. Dual-purpose varieties (grain and forage) are produced in states such as Texas, Kansas, Nebraska, and Oklahoma.

FIGURE 22–1 General sorghum production regions in the United States. (Source: USDA)

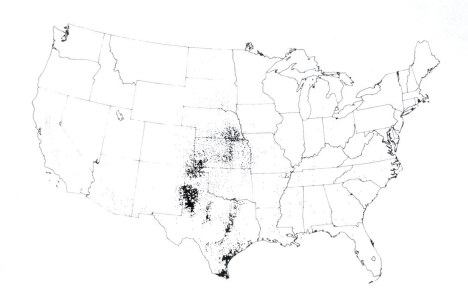

Table 22–1 Leading Sorghum-Producing States in the United States in 2000

State	Acreage Harvested (× 1,000 Acres)	Yield (Bushels)	Production (× 1,000 bushels)
Kansas	3,200	59.0	188,800
Texas	3,000	61.0	143,350
Nebraska	500	70.0	35,000
Missouri	270	92.0	24,840
Oklahoma	450	38.0	13,680
Colorado	210	32.0	6,720
South Dakota	120	49.0	5,880
Tennessee	22	75.0	1,650
New Mexico	65	25.0	1,625
Louisiana	215	83.0	765
U.S. total	7,726	60.9	470,526

Source: USDA.

On the world scene, sorghum is produced principally in Africa, Asia, the Americas, and Australia. Important producers in Africa include Nigeria, Sudan, Burkina Faso, Cameroon, Chad, Mali, and Rwanda. About 74% of the sorghum produced in Africa is used for food. Most developing countries use sorghum for food, unlike the United States, where sorghum is grown mainly for feed. Even though most of the acreage for sorghum occurs in developing countries, the United States leads the world in production, followed by India, Nigeria, China, Mexico, and Sudan.

22.2: ADAPTATION

Sorghum is adapted to a wide range of environmental conditions, but especially to hot, semiarid tropics. The most favorable temperature for sorghum is about 80°F, with a minimum of 60°F. It is adapted to regions with a summer rainfall of 17–25 inches. The plant remains dormant in the event of drought and resumes normal growth when moisture becomes available. Sorghum is tolerant of waterlogging and can grow in regions with high rainfall.

Sorghum grows in a wide variety of soils but is well suited to the heavy Vertisols that occur commonly in the tropics. It can produce well on marginal soils where other crops fail. It is more tolerant of salinity than is corn and can grow well on soils with a pH ranging between 5.0 and 8.5. Sorghum is a shortday plant. Its ability to adapt to hot, dry conditions stems from a number of morphological and physiological characteristics. The plant has more extensive secondary roots than corn and other cereals, and it has a small leaf area per plant. The leaves also have a waxy cuticle, which reduces water loss. Sorghum leaves and stalk wilt and dry more slowly than corn and are able to withstand drought longer.

22.3: ORIGIN AND HISTORY

Sorghum originated in northeastern Africa (Ethiopia, Sudan, East Africa), where both wild and cultivated species occur. It was domesticated in Ethiopia and parts of the Congo between 5000 and 7000 B.C., with secondary centers of origin in India, Sudan, and Nigeria. It moved into East Africa from Ethiopia around A.D. 200. It was distributed along trade and shipping routes throughout Africa and through the Middle East to India at least 3,000 years ago. Sorghum was taken to India from eastern Africa during the first millennium B.C. It arrived in China along the silk route. It was introduced into the Americas as guinea corn from West Africa through the slave trade at about the middle of the nineteenth century.

22.4: BOTANY

Sorghum (*Sorghum bicolor* [L.] Moench) is known by common names such as milo, kafir, and guinea corn. It belongs to the grass family, Poaceae (Gramineae), and the tribe Andropogonae, to which sugarcane belongs. The annual sorghums have $n = 10$ and include the grain sorghum, sorgho, broomcorn, and sudangrass. *Sorghum halepense* (johnsongrass) is a perennial sorghum with $2n = 20$.

The sorghum plant has culms that stand 2–15 feet tall, depending on the type and variety. It may produce two or more tillers. The stalk is solid. The center of the stem can be dry or juicy, insipid or sweet to taste. The dry-stalked variety has leaves with a white or yellow midrib, whereas the juicy-stalked variety has a dull, green midrib because of the presence of the juice instead of air spaces in the pithy tissues.

The number of leaves on the plant varies between 7 and 24, depending on the variety. The leaf sheaths are 30–35 cm long with a waxy surface. The leaf margin may be flat or wavy. The sorghum inflorescence is a panicle that may be loose or dense. It is usually erect but may curve to form a gooseneck. The panicle has a central rachis, with short or long primary, secondary, and mature tertiary branches, which bear groups of spikelets. The length and closeness of the panicle branches determine the shape of the panicle, which varies from densely packed conical to oval spreading and lax. Sorghum is predominantly self-pollinated.

A fully developed panicle may contain 2,000 grains, each one usually partly covered by glumes. The grain is rounded and bluntly pointed. Its diameter ranges from 4–8 mm, and of varying size, shape, and color according to the variety. Pigments occur in the pericarp, a testa, or both. Varieties with a pigmented pericarp are yellow or red. When the pericarp is white and a testa is present, the seed color may be buff or bluish white. When a colored pericarp and a testa are present, the seeds tend to be dark brown or reddish brown.

22.5: SORGHUM RACES

Five major races of sorghum are recognized—durra, kafir, guinea, bicolor, and caudatum. They differ in panicle morphology, grain size, and yield potential, among other characterisitics. Durra sorghums developed primarily in Ethiopia and the Horn of Africa, from which they spread to Nigeria and the savanna region of West Africa. Kafir types originated in eastern and southern Africa. Guinea sorghums are grown mainly in West and Central Africa, whereas bicolor types are the least important to African production, growing in East Africa. Caudatum varieties originated in Kenya or Ethiopia.

22.6: GRAIN SORGHUM GROUPS

Most of the grain sorghum in cultivation consists of hybrids, derived from kafir-milo crosses. The major groups of grain sorghum are kafir, hegari, milo, feterita, durra, shallu, kaoliang, broomcorn, and sudangrass: (Figure 22–2).

1. *Kafir.* These have thick, juicy stalks; relatively large, flat, dark green leaves; and awnless, cylindrical heads. The seed color may be white, pink, or red.
2. *Hegari.* These types have a more nearly oval head and more abundant leaves and sweeter juice than kafir; hence, they are more desirable for forage.
3. *Milo.* This group has a less juicy stalk; curly, light green leaves; smaller leaves and stalks. The head is short, compact, and oval, with large yellow or white seeds. The plant tillers more than kafir and is more drought-tolerant.
4. *Feterita.* This group has few leaves; relatively dry stalks; an oval, compact head; and very large, chalk-white seeds.

FIGURE 22–2   Panicles of grain sorghum groups: 1) Shallu, 2) Milo, 3) Kafir, 4) Feterita.

5. *Durra.* This group has dry stalks, flat seeds, and very pubescent glumes. The panicles are erect but may be compact or loose. The varieties are chiefly grown in North Africa, India, and the Near East.
6. *Shallu.* This group is characterized by tall, slender, dry stalks; a loose head; and pearly white seeds. The varieties are late-maturing.
7. *Kaoliang.* The varieties in this group have dry, stiff, slender stalks; an open, bushy panicle; and small brown or white seeds. They are grown exclusively in China, Korea, Japan, and southeastern Siberia.
8. *Broomcorn.* This group has a head with fibrous branches used for making brooms.
9. *Sudangrass.* This group has slender leaves and stalks, with loose heads and small, brown seeds.

22.7: PRUSSIC ACID POISONING

The young plants and leaves of sorghum, sudangrass, and johnsongrass contain a glycoside called **dhurrin.** This cynogenic glucoside breaks down to release a poisonous substance called **prussic acid** (or hydrocyanic acid). Processing these plants into silage reduces the poisonous substance. However, when they are eaten as a forage or fodder, cattle, sheep, and goats have suffered sorghum poisoning. The prussic acid content of plants decreases with maturity. Tillers are younger than the main stalk and hence have high prussic acid content. The upper leaves contain higher concentrations of prussic acid than lower leaves. Sudangrass generally has lower amounts of prussic acid than the other sorghums and hence are safer for animals to graze. Freezing tends to accelerate the release of the acid from the glucoside form. Consequently, frosted sorghum is more toxic than fresh leaves. Similarly, plants that have been subjected to drought and have stunted growth have high prussic acid content because they consist primarily of leaves.

22.8.1 LAND PREPARATION

A warm seedbed that provides good seed-to-soil contact is desired for sorghum seeding. Various tillage and planting systems are used in the production of sorghum. Conservation tillage (e.g., no-till, ridge till, zero-till, mulch till, reduced till) are required for highly erodible lands. The tillage option depends on the amount of residue on the soil. An 80 bushels/acre harvest will leave the land about 90% covered with crop residue. No-till planting is best suited to moderately and well-drained soils. Limited tillage involving one chiseling, one disking, and one field cultivation can reduce a 90% residue-covered soil to only 30% covered.

22.8.2 PLANTING PRACTICES

The row width and seeding rate depend on the kind or purpose of production—for grain, fodder, silage, or hay. The row width for most equipment is 30 inches. Row crops generally perform well at this spacing. Spacing in the row is about 6–8 inches for non-tillering varieties. When tillering varieties are used, the spacing in the row is increased to 12–15 inches. For forage or silage, spacing may be close (4–6 inches).

Where rainfall is limited (less than 20 inches), the plant population should be reduced (24,000 plants per acre). Under irrigation, the population could be increased to 100,000 plants per acre.

A key consideration in the timing of seeding is to avoid the flowering time coinciding with the period of the severest moisture stress in the growing season (hottest, driest, and time of summer). Planting in the southern United States may occur over a wide period, from late February to August. It is safe to plant sorghum no earlier than about 2 weeks after the usual time for seeding corn. Producers in Arizona and California plant in July. Medium to late planting produces better stands, taller stalks, larger heads, and a shorter growing period than early planting. Late planting may predispose the crop to damaging fall freeze.

22.8.3 CROP ROTATION

Sorghum is believed to intensively mine the soil of nutrients. It persists until killed by frost, depleting the topsoil of moisture and nutrients. In the Great Plains, sorghum produces higher yields in a rotation in which it follows a fallow, winter small grains, and cowpea or cotton than in continuous culture. The high sugar content of sorghum stalks (sometimes as high as 50%) is implicated in reduced yield in the continuous culture of sorghum. The sugar promotes microbial growth by supplying energy. The high microbial population competes for soil nitrogen and thereby slows the growth of sorghum plants in subsequent seasons.

22.8.4 FERTILIZATION

Sorghum has an extensive, fibrous root system that enables it to exploit the soil in the root zone for water and nutrients. Because sorghum is adapted to dry, marginal soil, it does not respond substantially to fertilizer under rainfed production. However, irrigated sorghum responds well to fertilization. The plant needs most nitrogen supplementation during the period of rapid growth and development (after the five-leaf stage). By the boot stage, 65 to 70% of the nitrogen has been taken into the plant. Applications of 80–160 lb/acre have been found beneficial to sorghum on irrigated fields. When nitrogen is applied as pre-plant, it is best to do so in late fall or spring, except on sandy soils. Nitrogen applied as sidedress is best applied soon after the five-leaf stage. When soil tests indicate phosphorus to be very low, an application of 30–50 lb/acre of phosphorus may be desirable. Similarly, an application of 60–80 lb/acre of potassium is beneficial.

22.8.5   IRRIGATION

Even though sorghum is a drought-tolerant plant, limited irrigated production is quite common. In Kansas, it is among the top five irrigated crops (others being corn, wheat, soybean, and alfalfa). About 18–25 inches of water are required to produce a normal crop in most places. The highest water use occurs in the boot head stage and peaks at about 0.3 inch/day. Water needs are more frequent and greater on sandy soils. Water needs for grain sorghum production are higher than for forage. Under fully irrigated production, corn is generally preferred to sorghum because of increased yield potential.

It is generally recommended that soil water be maintained at or greater than 50% available. However, for sorghum, the soil water can be depleted to an average of 30 to 40% available water before grain yields are adversely impacted.

22.8.6   WEED MANAGEMENT

The field should be weed-free at planting time. Weeds can be controlled by various practices, including crop rotation, cultivation, and application of herbicides. Crop rotation reduces the weed population by varying the timing and types of tillage and herbicide used. Herbicides may be used in no-till production to control emerged weeds early in crop establishment and at planting. Cultivation can be between herbicide applications for controlling later-emerging weeds.

The herbicides approved for weed control in sorghum include those applied at planting (e.g., Dual®, Frontier®, and Partner®) and those applied postemergence (e.g., atrazine, Shotgun®, Marksman®, Peak®, Permit®, dicamba®, and 2,4-D). Postemergence application should be done early enough for optimal weed control. Where sorghum follows the previous year's wheat crop, producers may use Roundup® and 2,4-D or dicamba®, followed by an atrazine application, to control weeds in wheat stubble in late summer.

22.8.7   DISEASES

Integrated pest management may be used by growers to keep diseases under control. This includes planting resistant hybrids, rotating crops, removing infected plant debris, purchasing high-quality seeds (disease-free, high percentage germination and viability), preparing the seedbed properly, and applying properly and safely pesticides as needed. Sorghum is attacked by smuts, leaf spots, and root and stalk rots. A summary of these diseases is presented in Table 22–2.

22.8.8   INSECTS

Sorghum is attacked by a variety of insects, including the chinch bug, sorghum midge, corn leaf aphid, greenbug, and corn earworm. The major insect pests are described in Table 22–3.

22.8.9   HARVESTING

Grain sorghum is a perennial plant. Consequently, the plant remains green and alive, even after the grain is matured, until killed by tillage or freezing temperatures. The grain dries slowly. At hard dough stage, the grain contains about 18 to 20% moisture. For effective combining, the grain moisture content should be about 13% or lower. Waiting for the grain to dry in the field delays harvesting and increases the risk of damage by weather factors and birds. Furthermore, delayed harvesting also delays the rotation of sorghum

Table 22-2 Common Diseases of Sorghum in Production

1. Kernel smut: covered-kernel *(Sphacelotheca sorghi),* loose-kernel *(S. cruenta)*
   - Affects kernel yield without significantly impacting forage yield
   - All sorghos, kafirs, and broomcorns are susceptible.
   - Infected ovules become filled with dark-colored spores instead of seed.
   - Control with seed treatment or resistant varieties.
2. Head smut *(Sphaceletheca reiliana)*
   - Entire head is replaced by gall of smut mass.
   - Spores occur in the soil or are occasionally carried by seed.
   - To control, burn the diseased plants or use resistant hybrids.
3. Root and stalk rots (root—*Periconia circinati;* charcoal rot of stalks—*Macrophomina phaseoli)*
   - Caused by soil-borne fungi
   - Stalk rots attack the base of the stalk and larger roots, rotting the pith and causing plants to lodge.
   - Drought-stressed plants are susceptible.
   - To control, use resistant cultivars.
4. Leaf spots (bacterial: e.g., *Pseudomonas andropogoni, Xanthomonas holcicola;* fungal: e.g., *Collectotrichum graminicolum, Puccinia purpurea)*
   - Numerous fungi and bacteria cause leaf diseases.
   - Control by using resistant cultivars, chemicals, or crop rotation.

Table 22-3 Common Insects in Sorghum Production

1. Chinch bug *(Blissus leucopterus)*
   - The bug moves from sorghum to other, nearby small grains after the sorghum crop has been harvested.
   - Most serious injuries occur during seasons of low rainfall.
   - Adults hibernate in plant debris or bunches of grass.
   - Control by using resistant cultivars, using chemicals, or planting early in the season.
2. Sorghum midge *(Centarinia sorghicola)*
   - Larvae cause injury to developing seed.
   - Most of the head appears blasted.
   - Flies breed on johnsongrass in spring and infest sorghum plants as soon as they head.
   - The midge overwinters in the larval stage on sorghum heads, in johnsongrass, or in plant debris.
   - Control by using chemicals.
3. Corn leaf aphid *(Aphis maidis)*
   - Honey-dew secretion from aphids promotes the growth of molds.
   - Aphids winter in the southern states and migrate to the north.
   - Control by chemicals.
4. Greenbug *(Toxoptera graminum)*
   - Causes reddening of small seedlings and sometimes death.
   - Excessive honey-dew deposit may occur.
   - Control with chemicals.
5. Corn earworm *(Heliothus obsoleta)*
   - Cause significant damage to compact-headed grain sorghum.
   - Overwinters in pupal stage in the soil.
   - Control with chemicals.

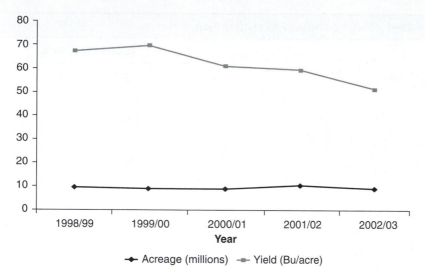

FIGURE 22–3 Sorghum acreage has remained relatively the same over the past five years, whereas the yield has declined. Drawn from USDA data.

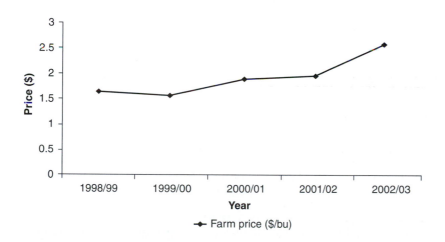

FIGURE 22–4 The farm price for sorghum has risen over the past five years. Drawn from USDA data.

with a winter crop (e.g., wheat). Desiccants (e.g., Diquat®, 28% nitrogen urea-ammonium nitrate) are applied as a pre-harvest treatment of grain sorghum. Roundup Ultra® may be used as a defoliate when the grain is for feed.

Sorghum acreage in the United States has remained relatively unchanged over the past five years, whereas the yield has declined (Figure 22–3). However, sorghum price at the farm has risen over the same period (Figure 22–4).

22.9: SORGHUM GRADES AND GRADE REQUIREMENTS

The grades and grade requirements of sorghum as prescribed by the USDA Grain Inspection Service are summarized in Table 22–4. Grain that does not meet the standards set for any of the grades (i.e., 1 to 4) is classified as sample grade.

Table 22–4 Sorghum Grades and Grade Requirements

Grading Factors	Grades U.S. Nos.			
	1	2	3	4
Minimum pound limits of				
Test weight (lb/bushel)	57.0	55.0	53.0	51.0
Maximum percent limits of				
Damaged kernels				
Heat (part of total)	0.2	0.5	1.0	3.0
Total	2.0	5.0	10.0	15.0
Broken kernels and foreign material				
Foreign material (part or total)	1.5	2.5	3.5	4.5
Total	4.0	7.0	10.0	13.0
Maximum count limits of				
Animal filth	9	9	9	9
Castor beans	1	1	1	1
Crotalaria seeds	2	2	2	2
Glass	1	1	1	1
Stones	7	7	7	7
Unknown foreign substance	3	3	3	3
Cockleburs	7	7	7	7

U.S. sample grade is sorghum that
  a. Does not meet the requirements for U.S. Nos. 1, 2, 3, or 4.
  b. Has a musty, sour, or commercially objectionable foreign odor (except smut or garlic odor).
  c. Is badly weathered, heating, or distinctly low quality.
Source: USDA.

REFERENCES AND SUGGESTED READING

Bennett, W. E., A. B. Maunder, and B. B. Tinker. 1990. *Modern grain sorghum production.* Blackwell Publishing. Ames, IA.

Kansas State Experimental Station. *Grain Sorghum Production handbook.* KSU Agricultural Experimental Station and Cooperative Extension Service. Manhattan, KS.

SELECTED INTERNET SITES FOR FURTHER REVIEW

http://www.oznet.ksu.edu/library/crpsl2/c687.pdf

Grain sorghum production handbook

http://www.psu.missouri.edu/CropSys/GrainU.S.Sorghum/

Grain sorghum production guide

http://sorghum.tamu.edu/genU.S.production/prodU.S.guides.html

General aspects of grain sorghum production

http://www.mgo.umn.edu/crops/Sorghum.htm

Links to various sorghum production sites

OUTCOMES ASSESSMENT

PART A

Answer the following questions true or false.

1. T F   Sorghum originated in Asia.
2. T F   Young leaves of sorghum contain a cynogenic glucoside called dhurrin.
3. T F   Sorghum is predominantly cross-pollinated.
4. T F   Kafir is a kind of sorghum.
5. T F   Grain sorghum is a perennial plant.
6. T F   Sorghum belongs to the family Poaceae.
7. T F   Sorghum plant has a taproot system.

PART B

Answer the following questions.

1. Give three of the top sorghum producing states in the United States.
   _____   _____   _____

2. Give three of the top sorghum producing nations in the world.
   _____   _____   _____

3. Give the scientific name of sorghum. _____

4. Give the five major races of sorghum. _____
   _____   _____   _____
   _____

5. _____ is the poisonous substance produced when dhurrin gluco-side breaks down in sorghum.

PART C

Write a brief essay on each of the following topics.

1. Discuss the relative importance of sorghum as a food crop versus other crops.
2. Briefly discuss the origin of sorghum.
3. Describe the general botanical features of the sorghum plant.
4. Discuss the time of seeding and seeding rate of sorghum.
5. Discuss the use of crop rotation in sorghum production.

PART D

Discuss or explain the following topics in detail.

1. Discuss the problem of prussic acid poisoning in sorghum production.

2. Discuss the races of sorghum.

3. Discuss the adaptation of sorghum in production.

23

Barley

TAXONOMY

Kingdom	Plantae	Subclass	Commelinidae
Subkingdom	Tracheobionta	Order	Cyperales
Superdivision	Spermatophyta	Family	Poaceae
Division	Magnoliophyta	Genus	*Hordeum* L.
Class	Lilliopsida	Species	*Hordeum vulgare* L.

KEY TERMS

two-row barley	feed barley
six-row barley	malt barley

23.1: ECONOMIC IMPORTANCE

Barley ranks as the world's fourth most important cereal crop after wheat, corn, and rice. It is grown widely all over the world in both tropical and temperate climates. In 2000, over 160 million metric tons of barley were produced on 80 million hectares of farmland in the world, the leading producing nations being Russia (40%), Canada, the United States, and China. Other major producers include France, the United Kingdom, and Turkey. In the third-world countries, barley is grown as the poor man's crop on marginal soils in dry and cool environments. Farmers in the high mountains of Tibet, Nepal, Ethiopia, and the Andes grow the crop. It is grown in North Africa, the Middle East, Afghanistan, Pakistan, Eritrea, and Yemen as a rain-fed crop. Developing countries produce about 28 million tons (18% of the world's production) on about 18.5 million hectares (25%) of the harvested area.

In the United States, barley is grown in nearly every state (Figure 23–1). The major producers are North Dakota (22.8%), Montana (14.9%), California (12.2%), Minnesota (9.8%), Indiana (8.7%), and South Dakota (8.0%) (Table 23–1). The total cropped land in 2002 was about 4 million hectares, yielding about 434 million bushels and averaging about 55 bushels per acre.

23.2: ORIGIN AND HISTORY

Barley is considered by some to be the oldest cultivated grain. Its origin is debatable; Asia and Ethiopia are suggested by some experts as probable centers of origin. Cultivated barley is believed to have derived from a **two-row barley** type, *Hordeum spontaneum,* which grows wild in Southeast Asia and northern Africa. Both wild and domesticated barleys are diploid ($2n = 14$). Barley was the main bread plant of the He-

FIGURE 23–1 General barley production regions in the United States. (Source: USDA)

Table 23–1 Top Barley Production States in the United States 1997–2001

| State | Thousands of Bushels | | | | |
	2001	2000	1999	1998	1997
North Dakota	79,750	97,350	59,520	106,150	101,250
Idaho	50,250	55,480	53,820	59,280	59,250
Montana	29,520	38,000	57,500	57,600	60,950
Washington	21,000	34,300	28,910	33,800	35,520
Colorado	8,560	12,075	9,030	9,430	9,612
Minnesota	7,975	15,360	8,460	22,825	23,460
California	5,830	5,780	6,400	7,500	8,550
Wyoming	6,970	7,885	7,310	7,140	8,400
Pennsylvania	4,200	5,325	4,970	5,025	4,556

Source: USDA.

brews, Greeks, Romans, and many European communities during the sixteenth century. The English and Dutch settlers were known to have made additional introductions during the following three decades. The English brought two-row types, whereas the Dutch introduced the **six-row barley types.** In 1900, more introductions were made from Sweden and Germany into the United States. The Spanish missionaries introduced the Coast barley types, a group of barleys found in northern Africa and Spain, into north Mexico, Arizona, and California, where they remain the dominant types.

23.3: ADAPTATION

Barley is a temperate crop, preferring cool, semiarid climates. It tolerates more heat under semiarid conditions than under humid conditions. It is well adapted to high altitudes with cold, short growing seasons. It succumbs to a temperature of 12°F or lower. Even though resistant to cold, winter barley is less winter hardy than winter wheat, triticale, or rye. High temperatures coupled with high humidity favor rust, mildew, and scab.

Barley performs best on well-drained, fertile loams but also grows well on light clays. It produces poorly on poorly drained soils, as well as on sandy soils. Drought conditions promote premature ripening. However, barley is more productive than wheat, oats, and rye under dryland agriculture. Hence, it is often said to be the most dependable cereal in dryland farming. Barley tolerates alkalinity but grows poorly when the pH is less than 6.0. It has the highest salt tolerance of all cereal crops and is sometimes used for reclaiming saline soils. Six-row barley types are more salt-tolerant than two-row types. Barley cultivated in the United States may be divided into four groups, as shown in Table 23–2.

Table 23–2 Cultivated Variety Groups of Barley

---

1. Manchuria-OAC-Aderbrucker group
   - Six-rowed, awned, spring-sown, medium-sized kernels
   - Probably originated in Manchuria
   - Tall plants with open and nodding heads
   - Used widely for malting
   - Grown primarily in the upper Mississippi Valley
2. Coast group
   - Six-rowed, awned, large kernels
   - Short to medium height, erect to semierect
   - Not prone to shattering
   - Originated in North Africa
   - Grown in California and Arizona
3. Tennessee winter group
   - Originated in the Balkan-Caucasus region or Korea
   - Six-rowed, awned, spikes tend to nod
   - Fall-seeded
   - Grown in the southeastern United States
4. Two-rowed group
   - Originated in Europe and Turkey
   - Grown mainly in the Pacific and Intermountain States and some northern Great Plains areas
   - Predominantly spring-type varieties
   - For malting or feed

FIGURE 23–2 Barley plants in a field. (Source: USDA)

23.4: BOTANY

Barley is an erect annual. It tillers more than oats and is the most susceptible among the cereals to lodging. Barley can attain a height of about 90 cm. It has a strong, fibrous root system, some xerophytic types being able to penetrate the soil to a depth of over 190 cm. Barley can be distinguished from other cereals by the presence of pronounced clasping auricles on the leaf (Figure 23–2). The inflorescence is a spike with a zigzag rachis. Each spike contains 3 spikelets (i.e., 3 spikelets per node), 1 floret per spikelet, and 10–30 nodes per spikelet. The florets are self-pollinating and are composed of a lemma, palea, and caryopsis when fertile. In 2-row barley, all 3 florets are fertile. Each spikelet is sub-tended by 2 narrow glumes, which are normally narrow and nearly hairlike.

About 10 to 15% of the barley kernel is hull (the lemma and palea adhere tightly to the caryopsis and do not fall off upon threshing). The test weight is 48 lb/bushel for hulled barley and 60 lb/bushel for naked (threshed) barley. The grain is 8–12 mm long, 3–4 mm wide, and 2–3 mm thick. The barley grain may be white, black, red, purple, or blue, the red, purple, and blue being due to anthocyanin.

23.5: CULTIVATED SPECIES OF BARLEY

Barley may be categorized into three species based on the fertility of the lateral spikelets—*Hordeum vulgare, H. distichum,* and *H. irregulare.*

1. *H. vulgare.* This species is six-rowed, in which all the flowers are fertile. It has a spike notched on opposite sides, with spikelets on each notch. There are two kinds of six-row barleys: (1) ordinary, which has lateral kernels that are slightly smaller than the central one, and (2) intermediate, which has lateral kernels that are smaller than the central one.
2. *H. distichum.* This species has only central fertile florets (i.e., two-row barleys). There are two types: (1) the common two-row (lateral florets consist of lemma,

palea, rachilla, and reduced sexual parts) and (2) the deficiens two-row (rarely has palea and is sterile or without sexual parts.)

3. *H. irregulare.* This species originates in Ethiopia and is sometimes called Abyssinian intermediate. The central florets are reduced, whereas the remainder may be sterile, fertile, or sexless.

Most of the cultivated barleys used in production are six- or two-row types.

23.6: CULTURAL PRACTICES

The cultural practices used in barley production are similar to those used in wheat production.

23.6.1 TILLAGE SYSTEM

Conventional tillage is the most common system for barley production. Many producers (about 30%) in the northern Plains use conservation tillage, whereas about 10% of the producers in the southern plains use this practice. Conservation tillage is used more frequently in the northern Plains probably because the soils in the Plains are more fragile. Over 50% of the producers in South Dakota use conservation tillage. Producers generally make an average of four field passes during tilling and planting operations. An average of 12 to 22% residue remains on the soil surface after these operations.

23.6.2 SEEDING DATES

Planting dates for barley range from October to January. Just like wheat, there are spring and winter varieties. Spring varieties are best planted from April to May. In the northern Great Plains (Minnesota, North Dakota, South Dakota, Colorado) barley is planted after April 1, whereas growers in the Pacific Northwest often plant before April 1. As previously indicated, barley is less winter-hardy than wheat and other cereals. Consequently, barley is not suited to areas that are prone to winter injury. It should be seeded earlier than wheat to establish good roots before winter dormancy. Winter barley is planted mainly in the southern states. It is planted from September 1 in Pennsylvania through early October in the southern states and California region. Barley is often planted earlier than normal when it is grown for pasture and later for feed barley (to avoid attack from barley yellow dwarf virus).

23.6.3 DEPTH AND RATE OF SEEDING

Being less winter-hardy, barley should be planted 1–1.5 cm deep into a firm seedbed. Shallow seeding may be done when the soil moisture is high. The recommended rate of seeding is 60–90 lb/acre and sometimes up to 100 lb/acre. Higher densities are suited to irrigated production. Row spacing is about 15–20 cm apart and is seeded with hoe- or furrow-type drills in irrigated production. Crop densities are higher where rainfall is heavy or irrigation is adopted. Some farmers use homegrown seed. Most (about 70%) of the producers in South Dakota, for example, use homegrown seed, whereas about 10% of the producers in Washington State and Wyoming use homegrown seed.

23.6.4 CROP ROTATIONS

Barley is most commonly grown in rotation with crops such as corn, wheat, or a fallow. About 30% of the growers in the Northwest grow barley in rotation with wheat. Similarly, producers in the northern Plains also use wheat in their rotations. However, corn is popular in the Northeast in barley rotations. In the Corn Belt, barley often follows oats, wheat, or soybean instead of corn, to reduce the incidence of scab diseases that are residual on corn. Growers in the West, where irrigation is common, often use barley as a companion crop in alfalfa production. In the South, producers use barley in place of wheat in a rotation. In eastern Washington and Oregon, barley may alternate with fallow or field peas. Barley is often grown in rotation with wheat or wheat fallow in Montana, North Dakota, Oregon, and Washington.

23.6.5 DOUBLE-CROPPING AND FALLOW

About 75% of barley acreage in the Northeast is double-cropped with corn. This practice is very common in places such as Pennsylvania, where there is about 50% adoption among producers. However, growers in Oregon adopt fallow cropping as a practice in barley production. Double-cropping is less important or negligible in this area. Fallow cropping is adopted at an average rate of less than 5% in all the states in the United States.

23.6.6 VARIETIES

Barley is grouped into two varieties, based on their use as either *feed barley* or *malt barley*. Most of the production (over 90%) in the Northeast and Southwest is for feed, whereas over 75% of the barley acreage in the Northwest involves feed varieties. Producers in the Northeast and Southwest grow more of the feed varieties because these regions are feed-deficit. The highest acreage occurs in Wyoming and North Dakota. Malting barley sells for a higher price than feed barley. Production in Pennsylvania, Utah, Washington, Nebraska, and Oregon are entirely or mostly for feed. On the other hand, more than 50% of all the barley produced in the northern Plains is for malt.

Most of the farmers in the northern Plains and Northwest grow barley mainly for cash (i.e., cash grain), whereas most of the producers in the Northeast and Southwest specialize in barley for livestock feed.

23.6.7 FERTILIZATION

Producers in the Northeast tend to use the largest amount of fertilizer in the production of barley, partly because of the adoption of double-cropping and large amounts of straw production. The nutrients used mostly in this region are phosphorus, potassium, and manures with less nitrogen. Barley is more susceptible to lodging, hence, nitrogen fertilization should be undertaken with great care. Growers in Idaho and Wyoming apply very high rates of fertilizer, partly because of heavy irrigation. In these states, about 90 lb/acre are applied, compared with about 30–50 lb/acre in other states. Spring-applied nitrogen is most important in production systems in which the barley was pastured the previous fall. Phosphorus is needed for good early root development and growth for better winter survival.

23.6.8 IRRIGATION

Barley is moderately drought-tolerant and suited to semiarid regions. Most barley is produced on dryland in the United States. Over 90% of the production in the northern Plains and Northeast regions is irrigated. About 50% of the producers in the Southwest grow barley under irrigation. In Wyoming and Utah, about 90% of the crop is irrigated. On the other hand, about 90% of the production in states such as North Dakota, Pennsylvania, South Dakota, and Washington is entirely or nearly devoted to dryland production. About 30% of the barley acreage in the Northwest is irrigated. When barley is grown as a cover crop, producers usually irrigate the crop.

23.6.9 DISEASES

Barley is attacked by a variety of diseases that can drastically reduce crop yield. Seed and seedling diseases include head blight, black point, and scab. Foliar diseases of barley include powdery mildew, net blotch, leaf blotch, and stripe blight. Viral attacks of barley include barley yellow dwarf virus, and the important head diseases include loose and covered smuts. The characteristics and control of these diseases are presented in Table 23–3.

Table 23–3   Common Diseases of Barley in Production

1. Covered smut *(Ustilago hordei)*
   - Causes barley heads to form hard, dark lumps of smut instead of kernels
   - The diseased head often is borne on shorter stems and appears later than the normal head.
   - Found in threshed grain as black, irregular, hard masses
   - Control by using resistant cultivars or chemicals.
2. Brown and black loose smuts *(Ustilago nuda, U. nigra)*
   - Afflicted plants bear heads in which there is a loose, powdery mass of smut in place of normal spikelets.
   - Control with hot-water treatment of infected seeds, especially foundation seed lots.
3. Barley stripe *(Helminthosporium gramineum)*
   - The disease causes long white or yellow stripes on the leaves, which turn brown.
   - Plants become stunted, and the grain is discolored and shrunken.
   - Control with resistant cultivars.
4. Spot blotch *(Helminthosporium sativum)*
   - Causes seedling blight, which results in stand reduction
   - Causes black point or blight on the kernels, which reduces the market quality of barley
   - Any part of the plant can be infected by the disease.
   - Control by crop rotation or field sanitation, such as burying crop residues.
   - Use of resistant cultivars and chemicals also is effective.
5. Scab or fusarium blight *(Gibberella saubinetti)*
   - Favorable temperatures for the disease occur in the southern production regions.
   - It is a wet-season disease.
   - Grain yield is reduced and kernels blighted.
   - Diseased kernels ripen prematurely and turn pinkish to dark brown and are shrunken.
   - The Scab causal organism overwinters in corn stalks and small-grain stubble.
   - Control with crop rotation, plowing under of plant debris, and seed treatments.
   - Resistant cultivars are available.
6. Powdery mildew *(Erysiphe graminis hordei)*
   - Growth of mildew occurs on leaves and stems.
   - Resistant cultivars are available.
7. Viral diseases
   - Include stripe mosaic that is seed-borne and soil-borne virus
   - Symptoms include yellow or light-green stripes or complete yellowing of leaves, culminating in stunting of plants.
   - Control by sowing disease-free seed on clean soil.
   - Yellow dwarf is transmitted by several species of aphids.
   - Control by using resistant cultivars.

23.6.10 INSECTS

The important insect pests of barley include wire worms and aphids. The characteristics of some of these pests are summarized in Table 23–4.

23.6.11 HARVESTING

Barley may be direct-combined or swathed prior to combining when the grain is in the hard dough stage. The grain is physiologically mature when the kernel moisture reaches about 40% (but is best harvested at 35%). However, delays in harvesting predispose barley to shattering in susceptible varieties. Growers in the Southwest benefit from planting shatter-

---

Table 23–4 Common Insect Pests of Barley in Production

1. Aphids
   - Most important insect pest in regions such as Idaho
   - Six aphid species are of economic importance: Russian wheat aphid *(Diuraphis noxia)*, English grain aphid *(Sitobion avenae)*, and greenbug *(Schizaphis graminum)* are most commonly associated with significant yield loss. The rose grass aphid *(Metopolophium dirhodum)*, corn leaf aphid *(Rhopalosiphum maidis)*, and bird cherry-oat aphid *(Rhopalosiphum padi)* can spread barley yellow dwarf virus; however, these species normally do not require control, unless populations develop during the first- or second-leaf stage.
   - Greenbugs damage spring barley though the transmission of barley yellow dwarf virus, particularly in the high mountain valleys of eastern Idaho, and by feeding on stems beneath the emerging head while still in the boot stage.
   - Control with chemicals.
2. Cereal leaf beetles *(Oulema melanopus)*
   - Both larvae and adults cause economic damage.
   - Adults overwinter; larval populations peak in mid- to late June.
   - One larva per flag leaf in boot stage is an economic threshold for the pest.
3. Barley thrips *(Limothrips dentricornis)*
   - Both adults and larvae cause economic loss.
   - Loss is more severe in drought conditions.
4. Wireworms
   - They are present in early production season.
   - Pest incidence is hard to predict.

FIGURE 23–3 The acreage of barley has remained relatively the same over the past five years, whereas yields appear to have declined slightly. Drawn from USDA data.

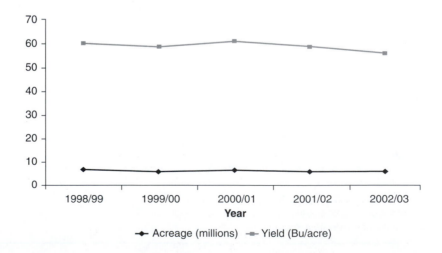

resistant varieties, because relative humidity in this region is subject to rapid drop after the crop matures. Yields of grain average between 55 and 75 bushels per acre, with yield in the Northeast being higher (because of high fertilization) than those in the Southwest. Yields are also higher on irrigated fields (e.g., in regions such as Wyoming and Utah).

The acreage of barley has remain relatively constant over the past five years. However, barley yield has shown some decline over the period (Figure 23–3). Prices at the farm gate has increased sharply over the last five years (Figure 23–4).

23.6.12   GRADES AND GRADE REQUIREMENTS

The grades and grade requirements of barley as prescribed by the USDA Grain Inspection Service are summarized in Table 23–5. Grain that does not meet the standards set for any of the grades (i.e., 1 to 5) is classified as sample grade.

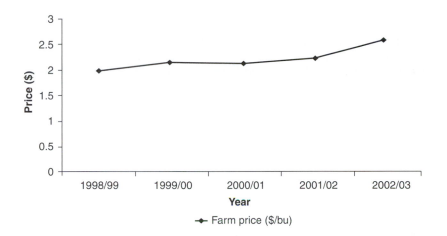

FIGURE 23–4 The farm price of barley increased slightly over the past five years. Drawn from USDA data.

Table 23–5   Barley Grades and Grade Requirements

Grading Factors	Grades U.S. Nos.				
	1	2	3	4	5
Minimum pound limits of					
Test weight (lb/bushel)	47.0	45.0	43.0	40.0	36.0
Sound barley (percent)	97.0	94.0	90.0	85.0	75.0
Maximum percent limits of					
Damaged kernels (percent)	2.0	4.0	6.0	8.0	10.0
Heat-damaged kernels (percent)					
Foreign material (percent)	1.0	2.0	3.0	4.0	5.0
Broken kernels (percent)	4.0	8.0	12.0	18.0	28.0
Thin barley (percent)	10.0	15.0	25.0	35.0	75.0

U.S. sample grade =
  a. Corn that does not meet the requirements for U.S. Nos. 1, 2, 3, 4, or 5.
  b. Contains stones that have an aggregate weight in excess of 0.10% of sample weight, two or more pieces of glass, three or more crotalaria seeds (*Crotalaria* spp.), two or more castor beans (*Ricinus communis* L.), four or more particles of an unknown foreign substance(s) or a commonly recognized harmful or toxic substance(s), eight or more cockleburs (*Xanthium* spp.), or similar seeds singly or in combination, ten or more rodent pellets, bird droppings, or equivalent quantity of other animal filth per 1 1/8 to 1 1/4 quarts of barley.
  c. Has a musty, sour, or commercially objectionable foreign odor.
  d. Is heating or of distinctly low quality.
Source: USDA.

23.7: USES

Barley is grown mainly for feed, malting, or brewing. Barley may be harvested for hay. It has the feed value of 95% of corn.

SELECTED INTERNET SITES FOR FURTHER REVIEW

http://www.ambainc.org/pub/prod/2001/prodU.S.081001.htm

Barley production statistics

http://www.agls.uidaho.edu/cerealsci/newU.S.pageU.S.4.htm

General information on barley production

http://cipm.ncsu.edu/cropprofiles/docs/IDbarley.html

General information on barley production

http://www.extension.umn.edu/distribution/cropsystems/DC2548.html

Barley growth and development

http://www.fas.usda.gov/wap/circular/2002/02-11/Wap%2011-02.pdf

World agricultural production statistics

REFERENCES AND SUGGESTED READING

Rasmussen, Donald C., 1985. Barley. *Soil Science Society of America*: Madison WI.

Weaver, John Carrier. 1950. *American Barley production: A study in agricultural geography*. Burgess Pub. Co. MN.

OUTCOMES ASSESSMENT

PART A

Answer the following questions true or false.

1. T F Most of the barley in the United States is grown for feed.
2. T F Barley is more winter-hardy than wheat.
3. T F North Dakota leads the nation in barley production.
4. T F Cultivated barley varieties are tetraploids.
5. T F *Sorghum vulgare* comprises six-row barley with all fertile flowers.

PART B

Answer the following questions.

1. Give the full scientific name of barley. _____

2. Based on use, barley is classified as either _____ or
   _____ .

3. The color of barley grain is caused by the pigment _____ .

4. Barley plants can be distinguished from other cereals by the presence of
   _____ .

PART C

Write a brief essay on each of the following topics.

1. Discuss the relative importance of barley as a food crop versus other crops.

2. Briefly discuss the origin of barley.

3. Describe the general botanical features of the barley plant.

4. Discuss the time of seeding and seeding rate of barley.

5. Discuss the use of crop rotation practices in barley production.

PART D

Discuss or explain the following topics in detail.

1. Discuss the environmental conditions best suited to the production of barley.

2. Discuss the growth and development of barley.

3. Discuss the major insect pests in the production of barley.

4. Discuss the major diseases in barley production.

5. Discuss the cultivated species of barley.

6. Discuss the use of barley in crop rotations.

7. Distinguish between the production of malt and feed barley in the United States.

24

Soybean

TAXONOMY

Kingdom	Plantae	Subclass	Rosidae
Subkingdom	Tracheobionta	Order	Fabales
Superdivision	Spermatophyta	Family	Fabaceae
Division	Magnoliophyta	Genus	*Glycine* Willd.
Class	Magnoliopsida	Species	*Glycine max* (L.) Merr.

KEY TERMS

Maturity classes
Inoculant

Symbiotic nitrogen
   fixation

Rhizobium

24.1: ECONOMIC IMPORTANCE

Soybean (*Glycine max* L. Merrill) is the major world oil seed. It is also a major source of meal used for livestock feed. It consists of about 35 to 40% protein and less than 20% oil. The United States is the world's leading producer of soybean, accounting for about 50% of the world's total production and about 40% of the cultivated acreage. The average yield of U.S. soybean was only about 11 bushels/acre in 1924, but it had increased to 25.4 bushels/acre by 1966. In 1994, the production reached a new high of 41.9 bushels/acre. In that year, producers in Iowa recorded a yield of 50.5 bushels/acre. Similarly, the acreage of soybean in 1924 was 1.8 million acres, 18.9 million acres in 1954, and 63.4 million acres in 1996. It rose to 72.7 million acres in 2000.

Soybean production in the United States occurs primarily in the North Central states (Figure 24–1), which overlaps the Corn Belt states. The major producing states are Iowa, Illinois, Minnesota, Indiana, Ohio, Missouri, and Nebraska, together accounting for 72% of the U.S. total production in 2000 (Table 24–1). Iowa and Illinois produce more than 30% of the U.S. crop. Soybean is important in the southern and southeastern states of Arkansas, Mississippi, North Carolina, Kentucky, Tennessee, Louisiana, and Alabama, which together produce about 10% of the total U.S. crop. Other minor producers are South Dakota, Kansas, Missouri, Wisconsin, and North Dakota.

On the world scene, the United States has dominated production since the 1950s, growing more than 75% of the world crop by the 1970s. By 2000, the United States, although still the world's leader, had a share of 45%, Argentina and Brazil taking up about 15% and 21%,

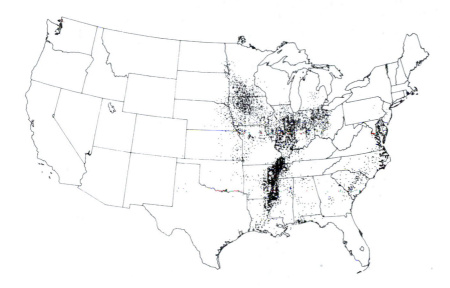

FIGURE 24–1 General soybean production regions of the United States. (Source: USDA)

Table 24–1 Leading Soybean-Producing States in the United States in 2000

State	Harvested (×1,000 acres)	Yield (Bushels)	Production (×1,000 Bushels)
Iowa	10,680	43.5	464,580
Illinois	10,450	44.0	459,800
Minnesota	7,150	41.0	293,150
Indiana	5,480	46.0	252,080
Ohio	4,440	42.0	186,480
Missouri	5,000	35.0	175,000
Nebraska	4,575	38.0	173,850
South Dakota	4,370	35.0	152,950
Michigan	2,030	36.0	73,080
U.S. total	72,408	38.1	2,757,810

Source: USDA.

FIGURE 24–2 General soybean production regions of the world. (Source: USDA)

respectively. China ranks fourth in the world, with about 12% of the world's total production. Other producers in the world include Japan, Indonesia, and Russia (Figure 24–2).

24.2: HISTORY AND ORIGIN

The soybean is considered to be among the oldest cultivated crops. The first record of the crop is contained in a 2838 B.C. Chinese book in which Emperor Cheng-Nung described the plant. Soybean was a "Wu Ku," one of the sacred five grains (the others being rice, wheat, barley, and millet) considered essential for the existence of the Chinese civilization. Cultivated soybean is believed to have derived from a wild progenitor, *Glycine ussuriensis,* which occurs in eastern Asia (Korea, Taiwan, Japan, the Yangtze Valley of Central China, the northeastern provinces of China, and the adjacent areas of Russia). The plant was first domesticated in the eastern half of northern China in the eleventh century B.C. It was introduced into Korea from this region and then into Japan between 200 B.C. and A.D. 300.

Soybean was grown in Europe in the seventeenth century. Its first introduction into the United States is traced to Samuel Bowen, an employee of East India Company, a seaman, who brought it to Savannah, Georgia, from China via England. Between 1804 and 1890, numerous soybean introductions were made into the United States from China, India, Manchuria, Korea, Taiwan, and Japan. In 1852, J. J. Jackson is reported to have first planted soybean as an ornamental plant in Davenport, Iowa. Most of the production of soybean in the United States prior to the 1920s occurred in the South, mostly for hay, and then spread to the Corn Belt after about 1924.

24.3: ADAPTATION

Soybean is a subtropical plant. However, it is grown over a wide range of ecological zones, ranging from the tropics to 52°N. Its climatic requirements are similar to those of corn. The minimum temperature for growth is about 50°F, and it is less susceptible to frost injury than is corn. Soybean can tolerate a short period of drought after germination. However, a combination of high temperature and low precipitation leads to low yields of grain and oil, as well as poor quality of oil. Soybean is a shortday plant. When northern varieties are grown in the southern latitudes, they mature sooner, producing less vegetative growth.

In terms of edaphic factors, soybean grows well on soils suited to corn production. Even though the crop will grow well on a wide variety of soil types, fertile, medium-textured, well-drained loams are best suited to the crop, as applies to corn. A desirable soil should have 2 to 6% organic matter. Compared with corn, soybean is better adapted to less fertile soils largely because of its ability to benefit from symbiotic nitrogen fixation. When grown in soils of pH higher than 7.5, soybean may exhibit iron chlorosis and other nutrient-deficiency problems.

24.4: BOTANY

Soybean (*Glycine max* L. Merrill) is an annual summer legume. Cultivated soybean is usually erect, with a well-defined main stem and branches and numerous leaves. Both determinate and indeterminate varieties are used in production. The leaves and stems are usually pubescent. The flowers are either purple or white and are borne in axillary racemes on peduncles at the nodes. The plant produces a large number of flowers, but only about two-thirds to three-fourths of them produce pods. The pods are also pubescent (Figure 24–2). They range in color from light straw to black, containing one to four seeds (occasionally five). The seeds are usually unicolored and may be straw-yellow, greenish-yellow, green, brown, or black. Bicolored seeds exist, such as yellow with a saddle of black or brown. The hilum is also colored in various patterns—yellow, buff, brown, and black. Soybean is primarily self-pollinated. In the proper soil environment, soybean is infected by the bacterium *Rhizobium,* resulting in roundish nodules on the roots, in which the nitrogen-fixing bacteria live.

24.5: VARIETIES

Soybean may be produced for forage, the varieties for this purpose generally being small-seeded, finer-stemmed, and very leafy. There are also varieties for edible, dry or green-shelled beans. These varieties usually have straw-yellow or olive-yellow seed and a yellow, brown, or black hilum. Soybean grown for grain is grouped into 13 *maturity classes,* ranging from 000 to X (Figure 24–3). The 000 group consists of the earliest-maturing varieties, whereas the X group consists of the latest-maturing varieties. Groups 000–IV are considered indeterminate, whereas groups V–X are determinate varieties. Further, early-maturing varieties (000–IV) are adapted to the northern production regions, whereas those with a high maturity class designation are adapted to southern production regions. The 000 varieties are adapted to the short summer growing seasons of the Northwest United States and

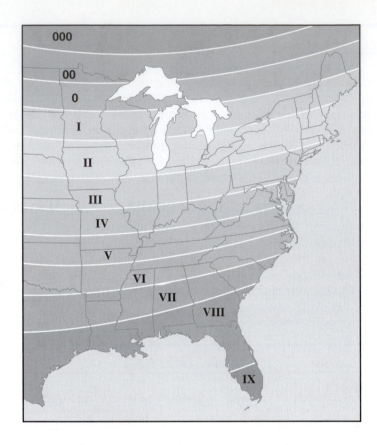

FIGURE 24–3 Soybean has 13 variety adaptation zones. Source: USDA

Canada. Varieties in groups II and III are best adapted to the Midwest growing area. Maturity groups VIII and higher are grown in the southern or Coastal Plain counties.

24.6: CULTURAL PRACTICES

24.6.1 TILLAGE

The seedbed should be firm and uniform for optimal stand establishment. Soybean seeds are smaller than corn grains and require more moisture imbibition than cereals for germination. Consequently, the seedbed should assure good seed-soil contact for effective moisture absorption for rapid germination and emergence. Quick emergence reduces the opportunities for disease infection. Soybean has epigeal seedling emergence and hence an intolerance for a compacted and crusty soil surface. Tilling while the soil is wet produces an undesirable, cloddy seedbed.

Shallow tilling in spring may be conducted to kill weeds in fall-planted fields. However, spring tillage may cause compaction of the soil below 5 inches depth, thereby restricting root growth and access to moisture and nutrients. No-till and reduced-till soybean production is on the increase. Special planters or drills may be needed to handle the surface residue under these tillage systems.

24.6.2 SEED INOCULATION

Soybean is capable of nodulating for *symbiotic nitrogen fixation,* provided the appropriate bacterium is available. It is infected by the bacterium *Rhizobium japonicum.* Native

Rhizobium is present in moist soil. However, for effective nodulation, a commercial *inoculant* is applied as a seed treatment just before planting. Soils that have carried soybean within the past 3 years have adequate bacteria for infection. However, for commercial production in a field that has not carried soybean for 5 years, there is a need for artificial inoculation.

Commercial inoculants are formulated in various ways. **Peat-based inoculants** (i.e., peat is the carrier material) are applied as a slurry to seeds before planting. **Clay-based inoculants** are applied dry to the seed, since they can become sticky and plug planters. **Granular formulations** are applied directly to the soil in a separate hopper. Some inoculants are liquid formulations. Applying directly to the soil is useful under conditions where the native *Rhizobium* population is low.

The bacteria in the inoculant can be killed by fungicidal seed treatment. Soybean inoculum is tolerant to thiram but susceptible to most of the other fungicides. Furthermore, the inclusion of molybdenum in the inoculum enhances the nitrogen fixation of the bacteria. Where the soil is acidic, liming to raise the pH to above 6.5 is necessary to promote bacterial activity.

Although inoculation can replace nitrogen application when it is effective and the plants produce sufficient nodules, poor nodulation will not support a good crop yield. In this instance, additional application of nitrogen (e.g., 60–80 lb of urea or ammonium nitrate) can increase yield by 8–12 bushels/acre.

24.6.3   SEEDING DATE

Planting date is perhaps the single most critical factor in soybean yield. It is critical to select the variety most suited to the maturity zone and even more critical to plant on time. Yield loss from delayed planting is estimated at 1/4 bushel/acre to about 1 bushel/acre per day (depending on the row width, date of planting, and plant type). Planting should be done as early as the frost date permits on fields where weeds are not a serious problem. Such timely planting allows the crop to take advantage of the full growing season for maximum yield. However, one should be careful not to plant sooner than 5 days before the average last killing frost. Planting early in cool, wet soil may produce low germination, high disease incidence, and poor crop stand. Soybean in the United States is planted in May–June. The soils are usually too cold to plant before May in the northern states. The planting dates are similar to those for corn, except that growers may plant soybeans a little later. Growers in Florida plant in June. If the purpose of the crop is pasture production, planting may be delayed further to July 1 in the northern states and August in the southern states.

24.6.4   ROW SPACING

Research indicates that narrow row spacing outyields wide-spaced soybean by an average of 3–10 bushels/acre. It is estimated that yield gains of 1/3 bushel/acre for each inch of row width reduction from 30 inches are achievable. Furthermore, thin-line varieties lose about 1/3 bushel/acre of yield potential for every inch of row width increase above 15 inches, whereas bushy varieties experience yield decline by 1/4 bushel/acre. Consequently, it is recommended that growers use bushy varieties for wide row spacing. Both plant types perform equally well under narrow spacing. The advantages of narrow row spacing of 10 inches or less include reduced erosion, increased harvesting efficiency, and early crop canopy, which helps control weeds. The convenience of using existing small-grain equipment for some of the planting and harvesting operations is most desirable. However, solid seeding works only if the weed infestation is low and the soil is fertile, with adequate moisture, especially during the critical pod filling stage.

24.6.5 SEEDING RATE

Soybeans do not seem to be responsive to changes in plant population. An optimal plant population is about 150,000 plants/acre, irrespective of row width. Low population (60,000 plants/acre) promotes reduced plant stature and increased branching, whereas excessive population (250,000 plants/acre) promotes increased plant height, weak stalks, and a higher predisposition to lodging. When planting occurs early (e.g., before May 15), a plant population of 80,000–140,000 is adequate. However, delaying planting (to early June) requires the plant population to be increased to 100,000–160,000 for desirable yields.

When producers adopt low seeding rates or narrow row spacing (less than 15 inches), they have to ensure that the plant spacing within rows is uniform to avoid yield loss.

24.6.6 PLANTING DEPTH

Soybean should be planted at a depth of 1–1½ inches. Shallow planting is critical for quick emergence, especially in soils that are prone to crusting. The hypocotyl of the soybean is delicate and cannot break through soil surface crusts, leading to seedling damage and mortality. This in turn leads to poor crop establishment. However, shallow planting predisposes seeds to herbicide damage.

24.6.7 CROP ROTATION

Soybean is grown in rotation with a variety of crops. In the Corn Belt, a rotation may be corn–soybean–small grains (e.g., wheat)–legumes. A rotation in the southern states may involve cotton, corn, or rice. Soybean can follow grass crops (e.g., corn, wheat, barley) in rotations, but not legumes (e.g., alfalfa, dry beans), canola, or sunflower in regions where white mold disease *(Sclerotinia sclerotiorum)* is known to be a problem. White mold uses soybean as a host, allowing the pathogen to carry over to other susceptible crops. White mold incidence can be reduced by adopting wider row spacing (30 inches or wider) to allow increased ventilation of the canopy formed by plants.

24.6.8 FERTILIZATION

Inoculation alone is capable of producing a yield of 70–80 bushels/acre if it is effective. Consequently, effective inoculation with a commercial soybean inoculant can replace commercial nitrogen fertilizer application. The bacteria work best when the soil pH is between 6.2 and 7.0. Soybean removes about 0.8 lb of phosporus/acre/bushel of harvest and 1.4 lb of potassium/acre/bushel of harvest. Moderate amounts of these nutrient elements are beneficial to soybeans.

24.6.9 IRRIGATION

Many soybean production areas do not receive adequate rainfall to sustain a yield of 40 bushels per acre. Soybean requires 20 inches of moisture during the critical V4–R7 growth stages to produce 40 bushels or more per acre of grain. Timely irrigation during the R5–R6 stages can increase yield by about 5 bushels per acre. Some general guidelines help the producer in deciding whether to irrigate or not. Soil moisture should not be allowed to drop below the 50 to 60% plant available water level during the germination and reproductive stages. During the early vegetative growth and maturation stages, soil moisture may be allowed to approach the 40% level.

24.6.10 WEED MANAGEMENT

Proper weed management in the first 2–4 weeks is critical for profitable soybean production. Planting at the right time, using quality seed, and proper spacing and plant population all contribute to the quick establishment of a good crop stand that suppresses weeds early in production. Seeding into a cold soil causes slow seed germination and seedling emergence, making soybeans poor competitors with weeds. Seeding into warm soil has the opposite effect, of quick and vigorous early growth, which helps soybeans compete well with weeds.

To reduce weeds in the early part of the production season, the final seedbed preparation should be done immediately before seeding to kill weeds and give soybeans a headstart on weeds. If cultivation is used to control weeds, it is best to conduct it in the warm part of the day when the soybeans are slightly wilted. This strategy will reduce physical damage (breakage) to the soybeans while weeds die quickly. Furthermore, the cultivation should be shallow (1–2 inches deep) to avoid excessive root pruning (reduces yield) and ridging of the rows, which causes excessive stubble height (increases harvest loss).

Pre- and postemergence herbicides are effective for controlling weeds in soybean production. However, soybean is susceptible to injury from herbicides such as 2,4-D, MCPA,® and Dicamba.® Furthermore, some varieties are intolerant of residual Atrazine® and Simazine® applied to the preceding crop in the preceding cropping season.

24.6.11 DISEASES

Soybean is affected by a variety of stem, root, and seed rots. These include *Phytophthora* root rot, *Pythium* and *Rhizoctonia* root rot, *Sclerotinia* stem rot, brown stem rot, and phomopsis seed rot. Other diseases are bacterial postule, bacterial blight, and soybean cyst nematode. The characteristics of some of these diseases and their management are summarized in Table 24–2. The economic importance of these diseases varies from one production region to another. For example, *Phytophthora* root rot is the most serious soybean disease in Ohio, whereas mosaic is common in Illinois. Pod and stem blight can be serious in the Corn Belt, whereas *Fusarium* blight occurs in the sandy soils in the southern production regions.

24.6.12 INSECTS

Insect pest losses are not significant in U.S. production. Common insects include grasshoppers, leaf hoppers, and blister beetles. Some of the characteristics of the major insects are described in Table 24–3.

24.6.13 HARVESTING

Mature soybean should be harvested on time to avoid field losses. The crop is ready to harvest when the seeds are at the hard dough stage (Figure 24–4). The moisture content of the seed should be 12 to 14%. Drier seed (less than 12% moisture) increases the incidence of seed coat cracking and splitting and shattering of the seed. The crop may be harvested at a high seed moisture content (17 to 18%), provided postharvest drying is available. Soybean is best harvested by straight combining. However, the operation is associated with substantial field losses from shattering caused by swathing soybean. Various improvements are available to minimize this loss—for example, the use of floating headers, pickup reels, and love bars. Harvesting at a time of day when relative humidity is high coupled with appropriate adjustment to the combine (e.g., reel speed to ground speed, cutter bar level, hold-down clips, wear plates) will reduce gathering losses.

Losing just four beans or one to 2 pods/square foot is equivalent to a loss of one bushel per acre. Under conditions in which green weed growth threatens to delay harvesting, the producer may use desiccants (e.g., paraquat, sodium chlorate) prior to harvesting. Mechanical damage to seed is perhaps the greatest cause of poor germination

Table 24–2 Common Diseases of Soybean in Crop Production

1. Bacterial blight *(Pseudomonas glycinea)*
   - Bacteria are seed-borne.
   - Appears as small, angular leaf spots
   - Affected leaves eventually dry and drop.
   - Bacteria overwinter in plant remains.
   - Control with crop rotation and plowing under of plant remains.
   - Use resistant cultivars.
2. Bacterial pustule *(Xanthomonas phaseoli* var. *sojense)*
   - Primarily a leaf disease
   - Leaves have angular leaf spots that become large, irregular, and brown.
   - Bacteria overwinter in plant debris.
   - Control with crop rotation and the use of resistant cultivars.
3. Wildfire *(Pseudomonas tabaci)*
   - Serious disease in southern production regions
   - Bacteria cause black or brown spots surrounded by yellow halos on leaves.
   - Control by crop rotation and the use of resistant cultivars.
4. Brown stem rot *(Cephalosporium gregatum)*
   - Fungal agents are soil-borne.
   - They cause decays of the interior stem, causing lodging and withering of leaves.
   - Control by crop rotation.
5. Pod and stem blight *(Diaporthe phaseolorum* var *sojae)*
   - Stems and pods are heavily spotted with black, spore-filled sacs.
   - If the stem is girdled, the plant will die.
   - Fungus may penetrate the seed and prevent future germination.
   - The fungus overwinters in diseased stems and seeds.
   - Control with crop rotation and sanitation.
6. Brown spot *(Septoria glycines)*
   - Attacks leaves, stems, and pods, causing premature defoliation
   - Disease appears as small, reddish-brown spots.
   - The fungus overwinters in crop residues.
   - Control by crop rotation and sanitation.
7. Mosaic
   - Plants are stunted, with malformed leaves.
8. Nematodes *(Meloidogyne* sp.*)*, cyst nematode *(Heterodera glycines)*
   - Found in fields in many regions of the South
   - Control with crop rotation and the use of resistant cultivars.

and low seedling vigor during field establishment of soybean. Soybean may be harvested for hay, the best time for this being when the seeds are at about 50% development. The stems become woody and lose more leaves if harvesting is delayed past this stage.

The acreage and yield of soybeans have remained relatively constant over the past five years (Figure 24–5). Soybean farm prices have been on the decline over the past five years (Figure 24–6).

24.6.14 GRADES AND GRADE REQUIREMENTS

The grades and grade requirements of soybean as prescribed by the USDA Grain Inspection Service are summarized in Table 24–4. Grain that does not meet the standards set for any of the grades (i.e., 1 to 4) is classified as sample grade.

Table 24-3 Common Insect Pests of Soybean in Crop Production

Soybean is relatively free from insect pests of economic importance. Insects that can
cause economic loss under favorable conditions include the following:
- Grasshoppers
- Blister beetles
- Leaf hoppers
- Green clover worms
- Armyworms
- Cutworms

Other insect pests are various beetles (Mexican bean, Japanese, bean-leaf, flea).
Insect pests can be controlled with insecticides.

FIGURE 24-4 Soybean plants in the field, (left) Mature soybean plants, (middle) Open soybean pod
showing seeds (right). Source: USDA.

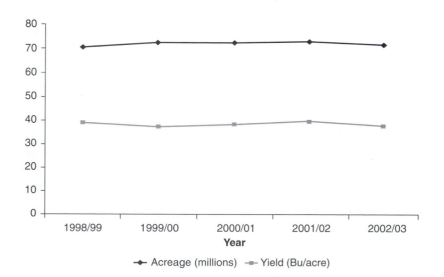

FIGURE 24-5 The acreage
and yield of soybeans have
remained relatively the
same over the past five
years. Drawn from USDA
data.

FIGURE 24–6 Except for a project rise in farm price in 2002, farm price received over the past five years has been on the decline. Drawn from USDA data.

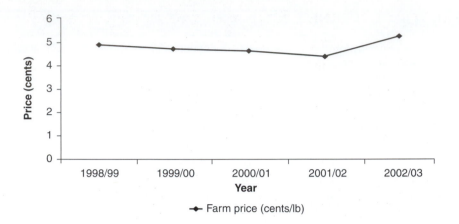

Table 24–4 Soybean Grades and Grade Requirements

Grade	Minimum Test Weight (lb/bushel)	Maximum limits of				
		Heat Damaged (percent)	Damaged Total (percent)	Foreign (percent)	Splits (percent)	Soybeans of (percent)
U.S. No. 1	56	0.2	2.0	1.0	10.0	1.0
U.S. No. 2	54	0.5	3.0	2.0	20.0	2.0
U.S. No. 3	52	1.0	5.0	3.0	30.0	5.0
U.S. No. 4	49	3.0	8.0	5.0	40.0	10.0

U.S. sample grade soybeans that do not meet the requirements of U.S. Nos. 1–4; or have a musty, sour, or commercially objectionable foreign odor (except garlic odor); or are heating or otherwise of distinctly low quality.

24.7: USE

Soybean is a multipurpose crop that is grown for its high seed oil and protein content. It is the world's leading source of vegetable oil. It has a good iodine number that ranges between 118 to 141, making it highly desirable for use in the manufacture of paint and dry oil products. Soybean cake and meal are the most important protein supplements in livestock feed ration in the world. Soybean flour is also widely used as a vegetable protein source for supplementing wheat flour used for bakery products. The crop is used widely in numerous popular oriental products, including curds, sauces, and breakfast foods.

REFERENCES AND SUGGESTED READING

Hoeff, E. G., E. D. Nafzinger, R. R. Johnson, and S. R. Aldrich. 2000. *Modern corn and soybean production*. MCSP publication. Savoy: IL.

Kansas State Experimental Station. Soybean production handbook. KSU Agricultural Experimental Station and Cooperative Extension Service. Manhattan, KS.

Selected Internet Sites for Further Review

http://www.ext.nodak.edu/extpubs/plantsci/rowcrops/a250w.htm

Soybean production in North Dakota

http://www.ag.ohio-state.edu/~ohioline/b472/soy.html

Soybean proction in Ohio

http://www.agron.iastate.edu/soybean/history/html

soybean production in Iowa

Outcomes Assessment

Part A

Answer the following questions true or false.

1. T F Soybean variety of maturity class V matures earlier than a variety of maturity class II.
2. T F The United States is the world's leading producer of soybean.
3. T F Soybean stem may be determinate or indeterminate in form.
4. T F Soybean is capable of symbiotic nitrogen fixation.
5. T F Soybean has hypogeal emergence.

Part B

Answer the following questions.

1. The wild progenitor of cultivated soybean is believed to be
   _____ .

2. Give three of the top soybean producing states in the United States.
   _____ _____ _____

3. Give three of the top soybean producing nations of the world.
   _____ _____ _____

4. Give the full scientific name of cultivated soybean. _____

5. Soybean maturity class VIII or higher are grown in what part of the United States? _____

6. Soybean infected by the bacterium _____ produces _____ on the roots in which _____ takes place.

7. Soybean consists of about _____% protein and _____% oil.

PART C

Write a brief essay on each of the following topics.

1. Discuss the relative importance of soybean as a food crop versus other crops.
2. Briefly discuss the origin of soybean.
3. Describe the general botanical features of the soybean plant.
4. Discuss the time of seeding and seeding rate of soybean.
5. Discuss the use of crop rotation practices in soybean production.

PART D

Discuss or explain the following topics in detail.

1. Discuss the environmental conditions best suited to the production of soybean.
2. Discuss the growth and development of barley.
3. Discuss the major insect pests in the production of barley.
4. Discuss the major diseases in barley production.
5. Discuss the importance of biological nitrogen fixation in soybean production.
6. Discuss the maturity classes of soybean varieties and their role in crop production.

25
Peanut

TAXONOMY

Kingdom	Plantae		Subclass	Rosidae
Subkingdom	Tracheobionta		Order	Fabales
Superdivision	Spermatophyta		Family	Fabaceae
Division	Magnoliophyta		Genus	*Arachis* L.
Class	Magnoliopsida		Species	*Arachis hypogaea* L.

KEY TERMS

Bunch peanut	Runner	Spanish
Runner peanut	Virginia	Hollow heart
Peg	Valencia	

25.1: ECONOMIC IMPORTANCE

Peanuts are an important legume crop in the warm climates of the world. The United States produces about 10% of the world's peanut crop, or about 3% of the total world peanut. This disproportionate share of the world's production is attributable to the high average yield per acre in the United States (2,800–3,000 lb/acre), compared with the world's average yield (800–1,000 lb/acre). Nine states in the United States account for 99% of the U.S. peanut crop: Georgia, Texas, Alabama, North Carolina, Florida, Oklahoma, Virginia, South Carolina, and New Mexico. Georgia alone produces 39% of total U.S. production.

FIGURE 25–1 General peanut production regions of the United States. (Source: USDA)

In 2001, the U.S. harvested acreage was 1,411,900 acres, a total production of 4,276,704,000 lb and an average yield of 3,029 lb/acre. There are three main peanut regions in the United States—Georgia–Florida–Alabama (Southeast), Texas–Oklahoma–New Mexico (Southwest), and Virginia–South Carolina–North Carolina. The Southeast region accounts for about 55% of all U.S. production (Figure 25–1).

On the world scene, peanuts are produced in Asia, Africa, Australia, and the Americas. India and China together account for more than 50% of the world's total production. Other substantial peanut-producing nations include Senegal, Sudan, Brazil, Argentina, South Africa, Malawi, and Nigeria (Figure 25–2).

25.2: ORIGIN AND HISTORY

The peanut is native to the Western Hemisphere. It probably originated in South America, the center of origin most likely being Brazil, where about 15 wild species are found. The Spanish explorers are credited with its spread throughout the New World. They introduced it to Europe, from which traders spread it to Asia and Africa. Peanuts reached North America via the slave trade. Commercial production of peanuts in the United States began in about 1876. The demand for the crop increased after the Civil War, transforming it from a regional (southern) food to a national food. Production began in the Cotton Belt after 1900. The expansion of the peanut industry was driven by advances in technology that resulted in the development of equipment and machinery for planting, harvesting, and processing the crop.

25.3: ADAPTATION

Peanuts are produced in both tropical and subtropical regions of the world, in regions where the rainfall is moderate, sunshine is abundant, and temperatures are high. Depending on the

FIGURE 25–2   General peanut production regions of the world. (Source: USDA)

variety, 120–160 frost-free days are required for peanut production. Optimal yields are obtained when rainfall during the growing season is 42–54 inches per annum. Seeds germinate when they imbibe moisture to about 35% of their weight. Peanut production in eastern New Mexico occurs between 86° and 91°F. Peanuts can be grown on a variety of soils. However, sandy loam and soils that are well drained and rich in calcium are ideal. When the soil is moist, rubbing between the index finger and thumb should not ribbon out but fall apart easily according to the ribbon test. The soil should be light-colored. Peanuts produced in this type of soil have bright, shiny, clean pods. Heavily textured soils tend to stick to the pods. Red or dark soils stain the pods and cause them to become darker after roasting. Loose soil favors the penetration of the peg and proper pod formation, and it facilitates harvesting. Peanuts grow best at neutral pH or a pH between 6.5 and 7.2.

25.4: BOTANY

The peanut *(Arachis hypogaea),* also called groundnut, earthnut, and other names, is technically a pea, not a nut. It is an unusual plant in the sense that it flowers above ground but fruits below ground. The cultivated plant is an annual, with a central, upright stem that may stand up to about 18 inches tall. It bears numerous branches, which vary from prostrate to nearly erect. It has pinnately compound leaves. The varieties in cultivation may be grouped into two, based on the arrangement of the nuts at the base of the stem. In **bunch peanuts,** the nuts are closely clustered about the base, whereas in **runner peanuts,** the nuts are scattered along prostrate branches that radiate from the base of the

FIGURE 25–3 Botanical types of peanuts: bunch (left) runner (right). (Source: USDA)

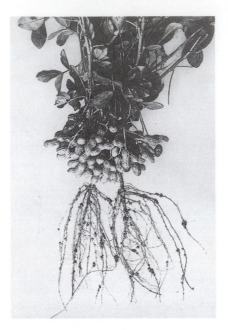
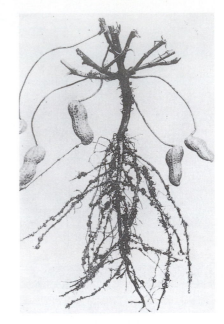

FIGURE 25–4 Peanut plants showing pods (left). Peanut seeds showing differences in sizes, shapes, and testa color (right). (Source: USDA)

plant to the top (Figure 25–3). Peanuts have a strong taproot system. Most roots are nodulated for biological nitrogen fixation.

The flowers arise in the leaf axils above the ground. They are self-pollinated. On fertilization, the ovary begins to enlarge, while the section behind it, called the **peg,** or **gynophore,** elongates to push the ovary into the soil for fruit development. The fruit is an indehiscent pod, which contains one to six (but usually one to three) seeds. The pods form only underground. Consequently, it is critical that the pegs reach the soil. The seed has a thin, papery testa that varies in color—brick red, russet, light tan, purple, white, black, or multi-colored Figure 25–4. Some varieties exhibit seed dormancy. Well-developed nuts have a shelling percentage of 70 to 80%.

25.5: VARIETIES

There are four basic market types of peanuts: Runner, Virginia, Spanish, and Valencia.

1. *Runner.* Runners have become the predominant peanut type, following the introduction of the variety the Florunner, which was responsible for the dramatic yield increase of the crop in the United States. They have uniform size and are grown mainly in Georgia, Alabama, Florida, Texas, and Oklahoma. About 54% of the crop is used for making peanut butter. Runners mature between 130 and 150 days, depending on the variety. The seeds are medium-sized (900–1,000 seeds/lb).

2. *Virginia (A. hypogaea hypogaea hypogaea).* Virginia varieties have dark green foliage and large pods. They have the largest seeds of all the types (about 500 seeds/lb) and have a russet testa. The pods usually have two seeds (occasionally three or four). They are grown mainly in Virginia and North Carolina. They mature in about 135–140 days and may have runner or bunch types. Large seeds are sold as snack peanuts.

3. *Spanish (A. hypogaea fastigiata vulgaris).* The Spanish peanuts are bunch types, with erect, light green foliage. The pods rarely contain more than two seeds, which are short with a tan testa. The seeds are small (1,000–1,400 seeds/lb). They have a higher oil content than other types. They are grown mainly in Oklahoma and Texas and are used mainly for making peanut candies, snack nuts, and peanut butter. They mature earlier than the runner types (about 140 days).

4. *Valencia (A. hypogaea fastigiata fastigiata).* The Valencia types typically bear many pods with three or four seeds and a bright red testa. They are erect and sparsely branching with dark green foliage. They are very sweet peanuts and are usually roasted and sold as in-the-shell or boiled. Valencias are grown mainly in New Mexico.

25.6: CULTURAL PRACTICES

25.6.1 TILLAGE

The goal of the tillage of soil for peanut production is to provide a soil with the proper tilth for the penetration of the pegs and for good pod growth, as well as to bury plants and debris to reduce disease incidence. The incorporation of residue should be done about 3 months before seeding to allow the thorough decomposition of the debris and to reduce the fungal population involved in the decomposition process. Some growers in regions where the soil may become too wet plant on raised beds. However, bedding up is not necessary, provided the soil is deep and well drained.

25.6.2 SEEDING DATE

Many peanut growers in the United States plant between April 10 and May 10. Earlier planting at about March 15 is recommended for growers in Georgia. In Texas, peanuts are planted at about the same time that cotton is planted or a little later, ranging from March 1 in the southern areas to May 15 in northern Texas. The time of planting should coincide with the occurrence of warm, moist soils for rapid germination and stand establishment. Planting into cool, wet soils, or planting too early or too deeply, may result in slow seed germination and seedling emergence and a high chance of seedling diseases, resulting in poor crop stand. The soil should maintain a temperature of 60°F for at least 5 days before it is safe to seed peanuts.

25.6.3 SEEDING RATE

Peanuts are seeded at a depth of 1 1/2–2 inches. Valencia varieties are seeded at 75–100 lb/acre. Lower seeding rates are adopted by growers using single-row beds, although the rate is higher where double-row beds are used. Plant population varies between 60,000 and 80,000 plants/acre. Row spacing is variable, ranging from 35 lb/acre in 36 inches for Spanish types up to 96 lb/acre or higher for large-seeded varieties in 24-inch rows. The average seeding rate in the United States is 67 lb/acre (unshelled kernels) and about 45 lb/acre of shelled kernels. Spacing in rows ranges from 3 to 8 inches in 30-inch rows, and 3 inches apart in 24-inch rows for small peanut types. Desirable stands for Spanish varieties are five or six plants per row foot; for Runners, desirable stands are four plants per row foot.

25.6.4 SEED INOCULATION

Just like soybeans, peanuts are capable of biological nitrogen fixation, provided the appropriate bacteria exist in adequate amounts in the soil. Native peanut bacteria live in substantial amounts in soils that have previously been sown to peanuts or other legumes, such as velvet beans and cowpeas. The application of artificial inoculants is especially beneficial when the crop is grown on fields that have not been previously sown to peanuts. The extent of nodulation and biological nitrogen fixation activity varies among species. There are different formulations of inoculants—peat-based granular inoculants or sterile peat-based inoculants. There are dry and liquid forms. The granular inoculants are placed in the seed furrow at planting while the sterile formulation is placed directly on the seed as a seed treatment. Yield increases of 30–40 percent have been attributed to artificial inoculant.

25.6.5 CROP ROTATION

Continuous cropping of peanuts on the same land is not recommended. Rotations are highly recommended, but attention should be paid to the crop that immediately precedes peanuts. For example, seedling diseases increase when peanuts follow peanuts or cotton. Similarly, leaf spot and web blotch diseases are more pronounced when peanuts follow peanuts. *Verticillium* and *Fusarium* wilts may increase if peanuts follow potatoes or other vegetables. Therefore, it is recommended that peanuts occupy the same land no more than once in 3–4 years. Grass-type crops (corn, grain sorghum, small grains) may immediately precede peanuts. The peanut crop benefits from the fertilization of the grass crop in the previous year. The taproot system of peanuts can access nutrients leached to lower depths. The incidence of southern blight, *Phytium,* and soil nematodes is lessened by including grass species in a rotation.

Some specific rotations include peanuts following cotton, tobacco, or truck crops (in Georgia); peanuts and corn and crimson clover; and peanuts with early potatoes followed by cowpeas (in Virginia).

25.6.6 FERTILIZATION

Peanuts are only slightly to moderately fertilized in production. They are capable of biological nitrogen fixation when adequately nodulated. Furthermore, when grown in rotation—and following highly fertilized crops, such as corn or legumes, which enrich the soil—peanuts are able to take advantage of the residual fertility to produce a good crop. Macronutrients—nitrogen (N), phosphorus (P), and potassium (K)—need to be applied only in small, balanced amounts (10–40 lb/acre) for high yields to be realized. On the other hand, high or excessive amounts of macronutrients are detrimental to yield and quality. High amounts of N promote vegetative growth at the expense of reproductive growth. Excessive K in the fruiting zone results in the calcium deficiency that is responsible for "pops" (unfilled shells). High P in the soil and low organic matter cause zinc deficiency.

Micronutrient deficiency causes yield and quality problems in peanuts. In addition to calcium effects, boron deficiency impairs normal seed development and causes an irregularly shaped, blackened cavity on the inner surface of the peanut kernel, called **hollow heart.** Copper deficiency is linked to shriveled kernels, whereas zinc deficiency causes stunting of plants. Micronutrient fertilization may be achieved through fertigation, the application of organic matter from the barnyard, or the incorporation of plant materials. Calcium requirements are usually met through the application of aglime or gypsum. Liming is necessary if soil pH is below 6.0. Nutrient-deficiency problems vary in occurrence and severity from one region to another. Peanut types differ in sensitivity to calcium in the fruiting zone. Virginia types need more calcium than Spanish and Runner types.

25.6.7   IRRIGATION

Peanuts use about 20–30 inches of water during the growing season. The peak daily water use varies from 0.20–0.30 inch/day. Water needs are critical during vegetative growth, flowering, and pod development. This level of need increases substantially on a very hot, dry day (to about 0.40 inch). Some provision for irrigation, at least as a standby, should be made in case of inadequate rainfall. Excessive moisture in the late season can promote diseases and sometimes precocious germination (germination of the kernels while in the pod).

25.6.8   WEED MANAGEMENT

Weed control in peanut production is by both chemicals and cultivation, the former more so than the latter. Some growers use only chemicals, but most growers in the Southwest use some cultivation for weed control. If the field is kept weed-free in the first 4–6 weeks, peanuts are often competitive with weeds in the later part of the season. Cultivation must be used very carefully. Flat sweeps or rolling cultivators may be used, provided the depth and lateral control of the cultivator are set properly. Soil should not be thrown onto the lower leaves and branches, which promotes southern blight and impedes flower, peg, and pod development.

Herbicides approved for use in peanut fields include pre-plant herbicides (e.g., Balan®, Prowl®, Treflan®, Vernam®), preemergence herbicides (e.g., Dual®, Pursuit®), and postemergence herbicides (e.g., Basagram®, Blazen®, Storm®, Tough®).

25.6.9   DISEASES

Peanuts are prone to injury from a wide variety of diseases, which can be controlled by chemicals and cultural practices. These diseases include those that destroy seedlings, such as seed rot and preemergence damping off, and postemergence death (e.g., *Rhizoctonia solani* and *Pythium myriotyl*). These soil-borne diseases are favored by planting into cool, wet soils. Damaged seed, broken seed, and seed planted too deeply into the soil also promote seed and seedling diseases. These diseases are controlled by measures such as seed treatment with fungicides prior to planting, crop rotation, and raised bed planting.

Foliar diseases include early leafspot and late leafspot, causing defoliation. Southern blight (by *Sclerotium rolfsii*) and *Sclerotinia* blight *(Sclerotinia minor)* are fungal diseases that affect the stem and branches. Other diseases include *Verticillium* wilt (by *Verticillium dahliae*), pod rot (by *Rhizoctonia solani*), and limb rot (by *Rhizoctonia solani*). These diseases (and others) and their effects are summarized in Table 25–1.

Table 25–1 Common Diseases of Peanuts in Crop Production

1. Blackhull *(Thielaviopsis basicola)*
   - Causes dark-colored areas on the peanut hull
   - Discoloration may extend into kernels in severe cases.
   - Disease is favored by low temperatures late in the season, heavy soil texture, soil alkalinity, and peanuts following cotton or peanuts in a rotation.
   - Control with Benlate.
2. *Verticillium* and *Fusarium* wilts
   - Fungi are soil-borne.
   - *Verticillium* occurs on potatoes; *Fusarium* occurs in vegetable crops.
   - Affected plants are stunted, with yellow leaves that wither and drop early.
   - Avoid planting peanuts after cotton or vegetable plants, including potatoes, in a rotation.
3. Leaf spot (*Cercospora* sp.)
   - Favored by high humidity
   - Brown to black, circular to irregular spots occur on leaflets, petioles, and stems, often surrounded by a yellow halo.
   - Control by applying fungicides before leaf spots are established and, better still, before the spores settle on the leaves.
4. Southern blight *(Sclerotium rolfsii)*
   - Causes thin, sinuous, white fungal filaments or hyphae on the plant and nearby soil
   - Fungi overwinter as sclerotia.
   - Control with applications of high rates of gypsum and sulfur.
   - Fungicides are also used in the control of the disease.
   - Observance of sanitation is also helpful.
5. Pod rots: *Pythium myriotylum, Rhizoctonia solani, Sclerotium rolfsii*
   - *Pythium* causes wet pod rot.
   - *Rhizoctonia* causes rots that penetrate the pods to the kernels inside, causing ulcers and pod discoloration.
   - *Sclerotium* usually affects a small area in the field.
   - Pod rots can be controlled by using chemicals.
6. Nematodes
   - Cause gall formations on peanut roots, pegs, and pods
   - Affected plants become stunted and lighter in color.
   - Control with crop rotations with small grains.
   - Use nematicides.

25.6.10 INSECTS

The major insect pests of peanuts include lesser cornstalk borers, granulate cutworms, foliage feeding insects, thrips, and spider mites. The management of some of these insect pests are summarized in Table 25–2.

25.6.11 HARVESTING

Peanuts are ready to be harvested when 65 to 70% is mature. Delaying the digging of peanuts by about 2 weeks may double profits. Early harvesting results in shriveled

Table 25-2 Common Insect Pests of Peanuts in Crop Production

Insects pests are relatively less important in peanut production than other crops. Control
is by use of chemicals.

Leafhoppers
Cutworms
Tobacco thrips
Velvetbean caterpillar
Corn earworms
Armyworms
White-fringed beetles

kernels. At proper maturity, the kernels display the distinct texture and color of the variety, the inside of the shell beginning to color and show dark veins. A digger-inverter is an implement that digs the plant and turns it over, so that pods are exposed to the sun. Another activity in the operation is windrowing. Modern equipment can perform all operations (digging, shaking, inverting, and windrowing) in one pass. Some growers do not recommend shaking ("fluffing") unless absolutely necessary. The goal of shaking is to remove the dirt. It is easier in sandy soils and occurs with less loss of pods with small-podded than with large-podded varieties. The peanut is windrowed for 3–4 days until it dries down to between 18 and 24% moisture content. At a moisture content above 24%, more aggressive threshing action is needed to separate pods from vines. This causes physical damage, which may go undetected until time of curing.

The curing of freshly harvested peanuts should start as soon as possible. The key to quality produce is slow drying at low temperature. The harvest produced is hauled in vented trucks or trailers to prevent the heating that promotes mold growth. Commercial curing is conducted at an air temperature at or below 95°F, relative humidity not exceeding 50%, and a drying rate of 1/2–1 1/2% moisture removal/hour. Curing can be done in the field in bulk by using a solar trailer. Field curing is susceptible to damage due to inclement weather.

25.6.12 GRADES AND GRADE REQUIREMENTS

The grades and grade requirements of peanuts as prescribed by the USDA Grain Inspection Service are summarized in Table 25–3.

25.7: USES

Peanuts are either harvested or used for livestock feed. The harvested peanuts may be shelled, crushed, and processed into oil, paste, or butter or consumed as salted nuts. Peanuts make nutritious hay for livestock feed.

Table 25-3 Peanut Grades

a. For cleaned Virginia-type peanuts in the shell

Grade and Minimum Screen Size	% Fall-Thru Prescribed Screen	% Cracked or Broken Shells, Pops, Paper, and Foreign Material	% Damaged Kernels	Count/lb
Jumbo (37/64 × 3″)	5.00	10.00	3.50	176
Fancy™ (37/64 × 3″)	5.00	11.00	4.50	225

b. For shelled Runner-type peanuts

Grade and Minimum Screen Size	% Fall-Thru Prescribed Screen	% Fall-Thru 16/64 × 3/4″ Slot Screen	% Other Types	% Splits	% Damage	% Damage and Minor Defects	% Foreign Material	% Moisture
Jumbo 21/64 × 3/4″	5.00	3.00	1.00	3.00	1.00	2.00	0.10	9.00
Medium 18/64 × 3/4″	5.00	3.00	1.00	3.00	1.00	2.00	0.10	9.00
Select 16/64 × 3/4″	———	3.00	1.00	3.00	1.00	2.00	0.10	9.00
No. 1	———	3.00	1.00	3.00	1.50	2.00	0.10	———
Mill run 16/64 × 3/4″	———	3.00	1.00	3.00	1.00	2.00	0.10	9.00
Splits 17/64	———	———	2.00	———	2.00	2.00	0.2	———

c. For shelled Virginia-type peanuts

Grade and Minimum Screen Size	% Fall-Thru Prescribed Screen	% Other Types	% Sound Splits or Broken Kernels	% Damage	% Damage and Minor Defects	% Foreign Material	% Moisture	Count per lb Maximum
Extra large 20/64 × 1″	3.00	0.75	3.00	1.00	1.75	0.10	———	512
Medium 18/64 × 3/4″	3.00	1.00	3.00	1.25	2.00	0.10	———	640
No. 1 15/64 × 1″	3.00	1.00	3.00	1.25	2.00	0.10	———	864
Virginia splits 20/64	3.00	2.00	Not less than 90.00	———	2.00	0.20	———	———
No. 2 17/64	6.00	2.00	As graded	———	2.50	0.20	———	———

d. For shelled Spanish-type peanuts

Grade and Minimum Screen size	% Fall-Thru Prescribed Screen	% Other Types	% Splits	% Damage	% Damage and Minor Defects	% Foreign Material	% Moisture
No. 1 15/64 × 3/4″	2.00	1.00	3.00	0.75	2.00	0.10	———
Splits 16/64″	———	2.00	———	2.00	2.00	0.20	———
No. 2 16/64″	6.00	2.00	———	———	2.50	0.20	———

REFERENCES AND SUGGESTED READING

Martin, J. H., and W. H. Leonard. 1971. *Principles of field crop production.* New York: Macmillan.

SELECTED INTERNET SITES FOR FURTHER REVIEW

http://www.cahe.nmsu.edu/pubs/U.S.h/h-648.html

Peanut production guide

http://lubbock.tamu.edu/WebCD/docs/other/b1514.pdf

Texas peanut production guide.

http://corn.agronomy.wisc.edu/AlternativeCrops/Peanut.htm

General guide to peanut production.

http://virtual.clemson.edu/groups/peanuts/greenhtml.htm

Peanut production information.

http://www4.ncsu.edu:8030/~bshew/extension/Production%20Guide%202003.pdf

Peanut production guide

OUTCOMES ASSESSMENT

PART A

Answer the following questions true or false.

1. T F   Valencia peanuts tends to bear pods with three to four seeds.
2. T F   Peanuts are best produced on clay soils.
3. T F   Runner peanut types are the predominant peanut types in United States production.
4. T F   "Pops" or unfilled shell are caused by calcium deficiency.
5. T F   Peanut is windrowed in production.
6. T F   Peanuts are capable of biological nitrogen fixation.
7. T F   Virginia peanuts have the smallest seeds among the peanut market types.

PART B

Answer the following questions.

1. Give the full scientific name of peanut. _____

2. Based on the arrangement of the nuts on the base of the stem, peanuts may be classified as either _____ or _____ type.

3. Give the four basic market types of peanuts. _____
   _____ _____ _____

4. Give three of the top peanut producing states in the United States.

   _____ _____ _____

5. Give three of the top peanut producing nations in the world.

   _____ _____ _____

6. After fertilization, the _____ elongates and enters the soil to form the _____.

7. _____ is a symptom caused by the deficiency of boron.

PART C

Write a brief essay on each of the following topics.

1. Discuss the relative importance of peanut as a food crop versus other crops.

2. Briefly discuss the origin of peanut.

3. Describe the general botanical features of the peanut plant.

4. Discuss the time of seeding and seeding rate of peanut.

5. Discuss the use crop of rotation practices in peanut production.

PART D

Discuss or explain the following topics in detail.

1. Discuss the environmental conditions best suited to the production of peanut.

2. Discuss the growth and development of peanut.

3. Discuss the major insect pests in the production of peanut.

4. Discuss the major diseases in peanut production.

5. Discuss the importance of biological nitrogen fixation in peanut production.

6. Describe the basic market types of peanuts.

7. Distinguish between runner and bunch peanut types.

8. Describe hollow heart abnormality and how it occurs in peanuts.

26
Cotton

TAXONOMY

Kingdom	Plantae	Subclass	Dilleniidae
Subkingdom	Tracheobionta	Order	Malvales
Superdivision	Spermatophyta	Family	Malvaceae
Division	Magnoliophyta	Genus	*Gossypium* L.
Class	Magnoliopsida	Species	*Gossypium hirsutum* L

KEY TERMS

Square	Fuzz	Ginning
Xenia metaxenia	Old world cotton	
Lint	New world cotton	

26.1: ECONOMIC IMPORTANCE

Cotton is the most important natural fiber in the world for textile manufacture, accounting for about 50% of all fibers used in the industry. It is more important than the various synthetic fibers, even though its use is gradually decreasing. It is grown all over the world in about 80 countries. The United States is the second largest producer of cotton after China, producing 16.52 million bales on 14.6 million acres in 1999. Most of the cotton is produced in the Cotton Belt (Figure 26–1). Most of the production currently occurs

Table 26–1 Leading Cotton-Producing Countries in the World

Country	Yield (Metric Tons)	
	1996	**1999**
China (mainland)	4.2	4.0
United States	4.1	3.6
India	3.0	2.7
Pakistan	1.6	1.6
Uzbekistan	1.1	1.2
Turkey	0.8	0.8
Others	4.8	5.2
Total	19.6	19.1

Source: USDA.

west of the Mississippi River, thanks to the spread of the devastating boll weevil attack in the eastern states. The leading production states in 2000 were Texas, Mississippi, Arkansas, and California, in order of decreasing importance. Other important cotton-producing states are Alabama, Arizona, Tennessee, Georgia, Louisiana, South Carolina, North Carolina, Missouri, Oklahoma, and Minnesota.

On the world scene, 88.7 million bales of cotton were harvested from 32.2 million hectares in 2000. The leading producing countries in the world are China (Mainland), the United States, India, Pakistan, Uzbekistan, and Turkey (Table 26–1). China and the United States produce nearly 50% of the total world production. Other major producers are Mexico, Brazil, Egypt, Greece, and Colombia (Figure 26–2).

26.2: ORIGIN AND HISTORY

The exact origin of cotton is not known. However, two general centers of origin appear to have been identified—Indochina and tropical Africa in the Old World (Mohenjo Daro, the Indus Valley of Asia, in 2500 B.C.) and South and Central America in the New World

FIGURE 26–2 General cotton production regions of the world. (Source: USDA)

(Huaca Prieta, Peru, in 8000 B.C.). Cotton was cultivated in India and Pakistan, and in Mexico and Peru, about 5,000 years ago. Cotton was grown in the Mediterranean countries in the fourteenth century and shipped from there to mills in the Netherlands in western Europe for spinning and weaving. Wool manufacturers resisted its introduction into Europe until the law banning the manufacturing of cotton was repealed in 1736. Similarly, the English resisted the introduction of cotton mills into the United States until Samuel Slater, who previously had worked in a cotton mill in England built one in 1790. Three types of cotton are grown in the United States—Sea Island, American-Egyptian, and Upland. These types probably originated in America, the first two believed to have come from South America. Upland cotton may have descended from Mexican cotton or from crosses of Mexican and South American species.

26.3: ADAPTATION

World production of cotton occurs between latitudes 45°N and 30°S, where the average temperature in summer is at least 77°F. Cotton requires a frost-free production period (175–225 days), moisture, and abundant sunshine for good plant growth, development, and ripening. To produce cotton under rainfed conditions, the annual rainfall should be at least 16 inches. The crop is usually irrigated in drier regions, such as California, Arizona, and Nevada. Similarly, cotton in Egypt and Sudan is an irrigated crop.

The ideal conditions for producing cotton are mild spring weather with light, frequent rain showers; a warm, moderately moist summer; and a dry, cool, prolonged fall season. Erratic moisture during production causes uneven development of the fiber, leading to uneven fiber strength. Rainfall during boll development is undesirable, delaying boll maturity and harvesting operations. The optimal temperature for germination and early growth is about 93°F. Germination and growth are hindered below 60°F or above 102°F. Average summer temperatures of 81° to 83°F have produced the highest yields in the United States.

Soils that are suited for cotton production should be moderately fertile and have a pH of 5.2–8.0 or higher. The Mississippi Delta and the irrigated valley of the Southwest have some of the best lands for cotton production in the United States. Sandy loams are preferred for cotton production. Clay soils tend to promote delayed maturity, excessive vegetative growth, and increased susceptibility to boll weevil damage. It is important that the growing conditions allow the crop to mature early enough to avoid the adverse impact of the environmental stresses that may occur late in the season.

26.4: BOTANY

Cotton (*Gossypium* spp.) belongs to the family Malvaceae, the mallow family. There are about 40 species in this genus, but only four species are grown for their economic importance as fiber plants. It is considered an annual plant, but it grows as a perennial in tropical areas where the average temperature for the coldest months stays above 65°F. The plant has a central stem, which attains a height of 2–5 feet. It has a deep taproot system. The leaves are arranged spirally around the stem. They are petioled and lobed (3–7 lobes). The stem and leaves are pubescent.

The flowers have five separate petals, with the stamens fused into a column surrounding the style. Three large, leaflike bracts occur at the base of the flower. The ovary develops into a capsule or boll (the fruit). The fruit bud (young fruit) is called a **square**. When dry, the capsule splits open along the four or five lines. Bolls average about 1 1/2–2 inches long (Figure 26–3). Only about 45% of the bolls produced are retained and develop to maturity. The plant is predominantly self-pollinated. When pollinated by foreign pollen, the phenomenon of **xenia metaxenia** causes a reduction in fiber length.

An open, mature cotton boll reveals the economic product, a fluffy mass of fibers surrounding the seeds. Each fiber is a single-cell hair, which grows from the epidermis of the seed coat. The long hair is called **lint**; the short hair, **fuzz**. Most wild species of cot-

FIGURE 26–3 Cotton plants showing unopened bolls (left). Opened cotton bolls showing fluffy lint (right). (Source: USDA)

ton do not have lint. The fibers attain full length within 18 days of initiation. Thereafter, there are about 25 days or more of thickening. Under unfavorable growing conditions, the resulting fiber is thin-walled and weak, and it produces poor, knotty yarns. A pound of lint is estimated to contain at least 100 million fibers. Fiber length ranges between 1/4 inch and more than 2 inches, whereas thickness ranges between 0.0006 and 0.0008 inch. All things being equal, long fibers produce smoother and stronger yarn than short fibers.

Genetically, the cultivated cotton varieties can be divided into two major groups:

1. **Old world cotton** ($2n = 26$) has the AA genome. It is comprised of (a) *Gossypium herbaceum,* which has five races, that originate in Africa and Asia, and (b) *G. arboreum,* which has six races of tree cotton and is found in India.
2. **New world cotton** ($2n = 52$) has the genome AADD (13 pairs of each of large and small chromosomes). The dominant species are (a) *Gossypium barbadense* (Sea Island and Egyptian cotton) and (b) *G. hirsutum* (Upland cotton).

26.5: VARIETIES

The varieties of cotton in cultivation derived from three species:

1. *G. hirsutum.* About 87% of all cultivated cotton is derived from this species. It is grown in America, Africa, Asia, and Australia. About 99% of all U.S. cotton is this species, which is also classified as Upland cotton. The varieties in this species produce fiber of variable length and fineness. The plant can attain a height of 6.5 ft.
2. *G. barbadense.* This species accounts for 8% of the world's cotton production and is grown in America, Africa, and Asia. The plant can reach a height of 8 ft and has yellow flowers and small bolls. It is also classified as Egyptian cotton and has long, fine, strong fibers used for manufacturing sewing threads. It is also called Pima cotton. New Pima cultivars are identified by number (e.g., Pima S-1 and Pima S-6).
3. *G. aboreum.* This species constitutes about 5% of total production and is grown mainly in East Africa and Southeast Asia. The plant can attain a height of 6.5 ft and has red flowers. The varieties from this species are also classified as Asiatic cotton. *G. herbaceum* is also classified as Asiatic cotton. The fibers produced are short (less than 1 inch) and poor-quality. This type of cotton fiber is used for manufacturing surgical supplies.

Another classification of cotton based on fibers, from longest to shortest, is Sea Island > Egyptian > American Upland long staple > American Upland short staple > and Asiatic.

26.6: CULTURAL PRACTICES

26.6.1 TILLAGE

Cotton may be planted on the flat, on ridges, or in beds. The plant has deep roots; hence, the land should be free from impervious layers near the top. Deep tillage or subsoiling has been found to be helpful for increasing yield on soils with hardpan problems. Where the land carried a previous crop, the residue should be incorporated into the soil. Growers in the humid Cotton Belt usually plant on ridges, whereas those in the western part of the belt often plant on level land or adopt furrow planting. Where beds are used, low ones are desirable for weed control and moisture conservation.

26.6.2 PLANTING DATE

Cotton should be seeded such that a good stand will be obtained and the crop will be ready for harvesting in favorable weather. The soil temperature should be warm (at least 60°F). The top 4 inches should stay continuously warm for at least 3 days, because a drop in temperature (41°F or below) can cause injury to imbibed seed, whereas temperatures below 50°F can cause chilling injury to emerging seedlings. Most of the cotton crop is planted in March, April, or May. Rains may come in late August and early September, causing boll rot when they coincide with boll opening. Irrigated cotton should be planted after May 1 to eliminate the danger of inadequate moisture associated with early planting. Further, the possibility of diseases with August rain is also reduced. This late planting will allow harvesting to occur in November and September.

Spacing and Depth of Planting

Cotton is sensitive to deep planting. A planting depth of 3/4 inch to 1 1/4 inches is preferred for "dusting in" in dry soil or cool weather planting. Under hot weather, while moisture is adequate, seeding depth may be deep (up to 2 inches). A planter with open center press wheels and low press loading may be used to reduce the crusting of the soil that impedes seedling emergence. Linted or delinted seeds may be used to establish a cotton crop. Delinting of the seeds may be mechanized (flame) or accomplished by use of chemicals (acid). Delinted seeds have better germination. About 7–10 lb/acre are needed for planting in semiarid regions (16–19 lb/acre are needed under irrigated production).

Population and Seeding Rate

Spacing in rows may be 3–8 inches in 40-inch rows. A final stand of two or three plants per foot of row is optimum. Rows of GM varieties are often wider (two plants per foot is economical).

26.6.3 CROP ROTATION

The types of crops involved in rotations with cotton depend on the production region and include soybean, corn, sorghum, and alfalfa. A continuous cotton production system is the norm in semiarid production regions, because the crop does not fit in well in sequence with winter wheat or sorghum, the other important crops in this ecosystem. It is common to grow winter legumes in cotton rotations to benefit from the biological nitrogen fixation provided by the legumes that reduces the nitrogen needs in production of cotton. Yield increase in excess of 250 lb/acre have been recorded in rotations in which cotton has followed a legume. Where cotton is grown under irrigation, the crops may follow alfalfa, sorghum, or an oil-seed crop, depending on the area. Crop rotation is beneficial where cotton root rot is a problem.

26.6.4 FERTILIZATION

Cotton responds to moderate amounts of balanced fertilization. The soils in the southeastern part of the Cotton Belt are generally deficient in available nitrogen (N), phosphorus (P), and potassium (K); hence, fertilization is a critical part of the cultural practice for cotton production. Generally, research has shown that an application of 18–20 lb/acre of nitrogen, N, P_2O_5, and K_2O increases cotton yield about 40 lb/acre. In conservation tillage systems, a starter application of 15–20 lb N plus 20–40 lb P_2O_5 is beneficial. Nitrogen application rates of 60–90 lb/acre are common, the lower rates occurring on heavier soils and the higher rates on sandy soils. Where legumes are included in a rotation, the rates are reduced by 20–30 lb/acre. Nitrogen rates can be applied all at once in clay soils or as a split application (50% preplant, 50% before first bloom as a sidedress). About 60 lb/acre of N is needed to produce one bale of cotton per acre. Excessive N application leads to excessive vegetative growth and delayed fruiting, harvesting problems, and reduced yield. About 60% of the mineral nutrients are taken up by cotton plants between the developmental stages of squares (boll buds) formation and boll formation.

The rich alluvial soils of the Mississippi Basin may need only moderate amounts of N (80–120 lb/acre). Similarly, such rates of N are needed in the western half of the Cotton Belt where production is under irrigation. However, under rainfed production in drier parts of Oklahoma and Texas, fertilizers are not usually part of cotton production practices.

26.6.5 IRRIGATION

Cotton production in California and Arizona is completely irrigated. Irrigated production on substantial scale occurs in New Mexico, Texas, Oklahoma, Arizona, Louisiana, and other states. The most effective methods of irrigation are furrow and basin irrigation. The goal in cotton irrigation is to maintain the soil at 50% field capacity or higher. About 24–42 inches of water is desired during the growing season for a good harvest. The maximum daily usage of water by cotton ranges between 0.25 and 0.40 inch at peak needs in midsummer.

26.6.6 GROWTH AND DEVELOPMENT

Cotton matures in about 150 days after planting. Cotton emerges 7–10 days after planting, blooming in early July and showing first open boll in mid- to late August.

It is important for cotton to set an early fruit load and to retain the load to suppress excessive vegetative growth. Growers should manage growth factors (nutrients, moisture) and use cultural practices (pest control, plant population) that promote proper plant development for balanced vegetative and reproductive growth. Some growers use growth regulators to reduce vegetative growth. Once such growth regulator is mepiquat chloride (or Pix®, Mepex®, Topit®, Mepichlor®), which has multiple effects, such as shortening internode length and reducing leaf area. Cotton growers do not universally apply this growth regulator.

26.6.7 WEED MANAGEMENT

Scouting is an important part of weed management in cotton production. Cotton grows slowly and is not competitive with weeds during the early stages of crop establishment. Early-season weeds reduce yields. Late-season weeds do not reduce yield but can interfere with the harvesting operation and reduce the quality of the lint through contamination. Consequently, weed control should be effective all season long.

Various herbicides are approved for use in cotton production —Prowl®, Poast®, Plus®, Cotoran®, Lorax®, and Fusilade 2000®. However, cotton is extremely susceptible to certain popular herbicides, such as 2,4-D and other auxin-type herbicides. It may be prudent to let growers nearby know about the cotton operation to avoid injury from herbicide drift. Furthermore, cotton is injured by residual effects stemming from application of herbicides such as atrazine, Glean®, Amber®, Canopy®, Pursuit®, Finesse®, Scepter®, and Tordon®.

Growers should avoid growing cotton in fields that have long histories of cocklebur and velvet leaf, weeds that adversely affect cotton harvest but are difficult to control with the herbicides approved for cotton production. Weeds in cotton fields are generally controlled by tillage and the use of herbicides. The most cost-effective methods are preplant and between-row cultivation.

Table 26-2   Common Diseases of Cotton in Crop Production

1. Root rot *(Phymatotrichum omnivorum)*
   - Disease is most destructive on calcareous soil of the Southwest and the black way soils of Texas.
   - Leaves and roots are yellow or bronze and eventually wither and die.
   - The fungus attacks dicot trees, shrubs, and herbs, but not the plants in the grass family.
   - The fungus can survive in the soil in plant remains or as dormant sclerotia.
   - Control by crop rotation with grass crops and deep tillage.
   - Crop debris should be plowed under.
2. Fusarium wilt *(Fusarium oxysporium* f. *vasinfectum)*
   - It is problematic in light, sandy soils from the Gulf Coast plain to New Mexico.
   - The disease causes stunting of the plants and yellowing of the leaves.
   - A cross section of the stem shows brown or black vascular tissue.
   - Control is achieved by using resistant cultivars and fertilization with large amounts of potassium.
3. Verticilium wilt *(Verticilium alboatrum)*
   - The fungus lives in the soil and infects the roots of the cotton plant.
   - Affected plants wilt, become mottled, and shed their leaves.
   - Use resistant cultivars for control.
4. Bacterial blight *(Xanthomonas malvacearum)*
   - This is seed-borne.
   - It is also called angular leaf spot or black arm.
   - Bacteria overwinter in cotton plant residue.
   - Control by crop rotation and observance of sanitation.
5. Cotton-boll rots *(Aspergillus, Fusarium, Nigrospora, Rhizopus)*
   - Some cause fiber deterioration in damp cotton.
   - Control by topping and defoliation of the plant to keep plants drier and reduce boll rotting.
6. Root-knot nematode *(Meloidogyne incognita)*, reniform nematode *(Rotylenchulus reniformis)*, meadow nematode *(Pratylenchus* sp.)
   - It can devastate a cotton crop.
   - Control with fumigation of soil and crop rotation.
   - Some resistant cultivars are available.

26.6.8   DISEASES

Cotton is susceptible to many diseases, some of which may not become widespread and economically important until the field has been sown to the crop for several years. The key diseases are fungal in origin and include root rot (by *Phymatotrichum omnivorum*), fusarium wilt *(Fusarium oxysporium),* and verticillium *(Verticillium alboatrum).* These and other diseases are described further in Table 26–2. Diseases that damage young seedlings (seedling blight and damping off) are important in cotton production.

26.6.9   INSECT PESTS

The most destructive insect pest in the United States in cotton fields is the cotton boll weevil *(Anthonomus grandis).* It is devastating to crops in production areas where rainfall exceeds 25 inches per annum. Other important insects are the pink bollworm *(Pectinophora gossypiella),* cotton leafworm *(Alabama orgillaceae),* and bollworm *(Heliothis armigera).* Some of these insect pests are described further in Table 26–3.

Table 26–3 Common Insect Pests of Cotton in Crop Production

1. Boll weevil *(Anthonomus grandis)*
   - This is perhaps the most destructive insect pest of cotton.
   - It is most destructive in areas where rainfall is more than 25 inches.
   - The adult female punctures the squares (bolls) and lays eggs inside them, destroying the lint.
   - Control by chemicals.
2. Pink bollworm *(Pectinophora gossypiella);* bollworm *(Heliothis armigera)*
   - The bollworm is also called corn earworm; it attacks corn, sorghum, and other crops.
   - It destroys the contents of the boll.
   - Control by chemicals and the use of resistant cultivars.
3. Webworm *(Loxostege similaris)*
   - It destroys leaves and causes boll development to be retarded.
   - Control with chemicals.
4. Cotton aphid *(Aphis gossypii)*
   - It causes the curling and shedding of leaves.
   - It secretes honey dew, which may contaminate the lint of opened bolls.
   - Control with chemicals.

26.6.10   HARVESTING

Harvesting Aids (Defoliation, Desiccation, and Topping)

The mechanical harvesting of cotton is a one-pass operation, which requires all bolls to be mature and dry at the time of harvesting (Figure 26–4) Harvesting aids are needed to hasten the maturation and drying of cotton, especially in production areas with a short growing season. Chemicals are used to remove or dry the leaves to facilitate mechanical harvesting. The treatment reduces fiber staining from chlorophyll and other plant pigments and reduces the plant materials that tend to clog harvesting machines. To be effective, these chemicals should be applied such that the leaves fall off rather than burn off. The weather at the time of application should be warm (90° to 100°F), sunny, and calm. The soil should be dry and low in nitrogen. The plants should not be experiencing moisture stress. The leaves should be fairly active, and the plants should not have secondary growth.

The grower should be aware that the chemical treatment may cause later-maturing bolls to be sacrificed. Consequently, time of application of the treatment is critical to reducing harvest loss. The crop is ready for chemical application for defoliation when the topmost first position boll that is harvestable is only four nodes above the cracked boll in the highest first position. A simpler index is when 75% of the bolls are open. Defoliation causes an abscission layer to form at the point of attachment between the leaf and the branch. This layer precedes leaf drop. Defoliants are necessary to remove the leaves, so that the desiccants can reach the bolls. Desiccants are used to prepare plants for harvesting by stripper-type harvesters. Defoliation facilitates hand picking. The necessity of defoliation varies annually and is determined by numerous factors including growing season, location, cotton market value, and the costs of production inputs.

Calcium cyanamide may be applied as a defoliant at the rate of 20–40 lb/acre in a 57% calcium cyanamide mixture. It is effective in humid regions where dew occurs for 2–4 hours after application. Another defoliant is a mixture of sodium chlorate with sodium metaborate that is applied as a dust or liquid. Some growers use magnesium chlorate-hexahydrate as a defoliant. For desiccation, some growers use pentachlorophenol. Defoliation reduces the canopy of the plant and thereby reduces the lodging of plants. However,

(a)

(b)

(c)

FIGURE 26–4 Harvesting cotton: (a) mechanical harvester picking cotton, (b) lint being loaded into a compactor (c) giant bale of cotton ready to be moved to the processing plant. (Source: USDA)

excessive vegetative growth makes plants top-heavy and predisposes them to lodging. The top 6–12 inches of plants may be mechanically removed with high clearance cutters. Topping also facilitates defoliation and desiccation.

Ethephon boll ripening agent (marketed as Prep®, Ethephon 6®, Pluck,® Super-Boll®) is used to speed boll opening. Other ethephon-based products available are Cotton Quik and Finish. They have some effect as defoliants but are best mixed with proper defoliants (e.g., DEF/Florex®, Dropp®, or Harvade®).

Cotton acreage and yield increased only slightly over the past five years (Figure 26–5). On the other hand cotton prices at the farm gate have declined over the past five years (Figure 26–6).

Cotton Ginning

After harvesting, the cotton is carried to a gin, where the fibers are removed from the seeds, a process called **ginning**. Other foreign materials (soil, hulls, plant debris) are also removed. Cotton is ginned when dry. Dry lint has a moisture content of about 8% or less. There are different kinds of gins (e.g., saw gin, roller gin). After ginning, the lint is pressed into bales weighing 480–500 lb and shipped to textile mills. The seeds are shipped to oil mills for processing.

Grading

Cotton grade is determined by evaluating the lint color, the ginning preparation, the maturity, and the presence of leaf and other foreign matter. Color is determined by using a

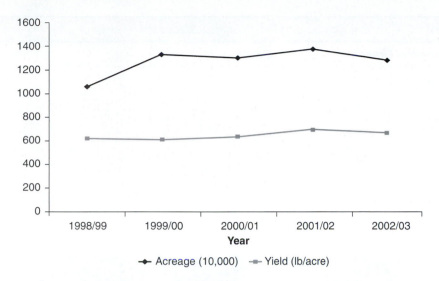

FIGURE 26–5 Cotton acreage and yield appear to have increased slightly over the past five years. Drawn from USDA data.

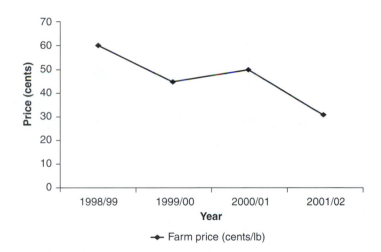

FIGURE 26–6 Farm price for cotton has declined over the past five years. Drawn from USDA data.

Nickerson-Hunter colorimeter. This instrument assigns values Rd (light or dark) and +b (yellowness). The recognized color classes are extra white, white-stained, and yellow-stained. There are 32 quality grades assigned within each of the color classes. American Upland white cottons are classified according to the commercial grades in Table 26–4. Similar grading is done for light spotted, spotted, and other classes. Pima cottons are graded 1–9.

Cotton Quality Specifications

The most important fiber quality is fiber length (Figure 26–7). It is measured by an instrument called a fibrograph. The staple classification may be short, medium, long, or

Table 26–4 Cotton Grades and Grade Characteristics for American Upland

Numerical Grade	Grade
No. 1	Middling fair
No. 2	Strict good middling
No. 3	Good middling
No. 4	Strict middling
No. 5	Middling
No. 6	Strict low middling
No. 7	Low middling
No. 8	Strict good ordinary
No. 9	Good ordinary

FIGURE 26–7 Cotton fibers vary in length (from left to right) Sea Island, Egyptian, American Upland long staple, American Upland short staple, and Asiatic (Source: USDA)

extra long (Table 26–5). Cotton fibers are also evaluated for length uniformity. Uniformity is measured by an HVI (high value instrument). Uniformity index interpretation is summarized in Table 26–5. Fiber strength is measured by using an HVI instrument and is given a value. The results are given in terms of grams/tex. Fiber strength is essential for making stronger yarns and using higher processing speeds. Fiber fineness and maturity are evaluated by using a micronaire instrument. The values range between 2.6 and 7.5 for the varieties. Finer fibers give stronger yarns but are susceptible to more neppiness (from immature fibers or entanglement of fibers) of the yarn because of the lower maturity. The micronaire values go up with maturity but decline with fiber thickness. Maturity impacts the yarn strength and evenness, as well as response to dyeing.

Table 26–5 Quality Specification of Cotton

1. Fiber Length

Staple Classification	Length (mm)	Length (inches)	Spinning count
Short	Less than 24	15/16–1	Coarse below 20
Medium	24–28	1.1/32–1.3/32	Medium count 20s–34s
Long	28–34	1.3/32–1.3/8	Fine count 34s–60s
Extra long	34–40	1.3/8–1.9/16	Superfine count 80s–140s

2. Fiber Strength—Measured in HVI (High Value Instrument)

G/tex	Classification
Below 23	Weak
24–25	Medium
26–28	Average
29–30	Strong
Above 30	Very strong

3. Length Uniformity

Uniformity Index	Classification
Below 77	Very low
77–79	Low
80–82	Average
83–85	High

4. Essential Cotton Quality Characteristics

Quality Factor	Description
Staple length	Long fiber gives stronger yarn and can be spun into finer counts of yarn.
Strength	Stronger fibers give stronger yarns.
Fiber fineness	Finer fibers produce finer count of yarn and stronger yarns.
Fiber maturity	Mature fibers give better yarns and better dye absorbency.
Uniformity ratio	High ratio means yarn is more even.
Elongation	High value helps reduce end-breakage in spinning.
Non-lint content	Low percentage of trash desired.
Sugar content	High sugar content creates stickiness of fiber and problems in spinning.
Moisture content	It should be 8.5 percent.
Feel	A smooth feel is desired for smooth yarn and high-quality fabric.
Class	High class is desired.
Grey value	High reading of calorimeter means cotton will reflect light better and produce yarn with good appearance.
Yellowness	Low value is desired for high-quality yarn.
Neppiness	Immaturity causes high neppiness and entangled fibers in the ginning process.

References and Suggested Reading

Smith, C. W., and J. J. Cothren (eds.). 1999. *Cotton: Origin, history, technology and production.* Wiley and Sons, New York: NY.

Gillham, F. E. M., J. M. Bell, T. Arin, G. A. Matthews, C. LeRumeur, and A. B. Hearn. 1995. *Cotton production prospects for the next decade.* The World bank, Washington: DC.

Selected Internet Sites for Further Review

http://www.griffin.peachnet.edu/caes/cotton/2000/contents.html

Cotton production guide

http://hubcap.clemson.edu/~blpprt/constill.html

Cotton production with conservation practices

http://msucares.com/crops/cotton/

General cotton production information with links

http://n-fl-bugs.ifas.ufl.edu/more/north_florida_bugs_page_3.htm

Cotton production guides

http://www.ces.uga.edu/pubs/PDF/RB428.pdf

Cotton production in Georgia

Outcomes Assessment

Part A

Answer the following questions true or false.

1. T F Cotton is the most important fiber in the world.
2. T F Old world cotton has 2n = 52.
3. T F Old world cotton has a genome AA.
4. T F Sea Island cotton fiber is longer in length than American Long Staple cotton fiber.
5. T F A fibrograph measures cotton fiber strength.
6. T F Most of the cultivated cotton varieties are *Gossypium hirsutum.*
7. T F The United States is the number one cotton producer in the world.

Part B

Answer the following questions.

1. Give the full scientific name of cotton. _____

2. The fruit bud or young fruit of cotton is called a _____.

3. Xenia metaxenia in cotton causes _____.

4. There are two major groups of cotton, _____ and _____.

5. Give two states of the Cotton Belt of the United States. _____ _____

6. _____ is an instrument for measuring fiber fineness and maturity.

7. The long cotton fiber is called _____, whereas the short fiber is called _____.

8. Give three of the major cotton producing nations in the world. _____ _____ _____

9. Give four of the essential cotton fiber quality characteristics. _____ _____ _____ _____

PART C

Write a brief essay on each of the following topics.

1. Discuss the importance of cotton as a fiber crop.

2. Briefly discuss the origin of cotton.

3. Describe the general botanical features of the cotton plant.

4. Discuss the time of seeding and seeding rate of cotton.

5. Discuss the use of crop of rotation practices in cotton production.

PART D

Discuss or explain the following topics in detail.

1. Discuss the environmental conditions best suited to the production of cotton.

2. Discuss the growth and development of cotton.

3. Discuss the major insect pests in the production of cotton.

4. Discuss the major diseases in cotton production.

5. Distinguish between old world and new world cotton varieties.

6. Discuss the use of harvesting aids in cotton production.

7. Discuss the grading of cotton fiber.

8. Discuss the relative importance of the three varieties of cotton in commercial production.

27
Potato

TAXONOMY

Kingdom	Plantae		Subclass	Asteridae
Subkingdom	Tracheobionta		Order	Solanales
Superdivision	Spermatophyta		Family	Solanaceae
Division	Magnoliophyta		Genus	*Solanum* L.
Class	Magnoliopsida		Species	*Solanum tuberosum* L

KEY TERMS

Solanine   Potato balls   Irish potato

27.1: ECONOMIC IMPORTANCE

Potato is among the top five crops that feed the world, the others being wheat, corn, sorghum, and rice. In 1988, potatoes ranked as the fourth most important food crop in the United States. The total harvest in 1988 was 1.388 million acres and a total yield of 475.8 million cwt. About 35% of the production was processed into frozen products (primarily fries). Per capita consumption of potatoes in the United States in 1999 was 144.7 lb. The top five potato-producing states in 1996 were Idaho, Washington, Oregon, Wisconsin, and North Dakota; Idaho led all production, with about 450 million cwt, followed by Washington (Figure 27–1).

FIGURE 27–2 General potato production regions of the world. (Source: USDA)

On the world scene, 293 million tons of potatoes were produced on 18 million hectares worldwide in 1998. With the breakup of the former Soviet Union, China became the world's leading producer of potatoes. Developing countries accounted for 36% of total production in 1998. In Asia, the key producers are China, India, Indonesia, and Nepal; in Africa, Egypt, South Africa, Algeria, and Morocco account for 80% of the region's total production (Figure 27–2). Latin American countries that produce a substantial amount of potatoes include Ecuador, Peru, Brazil, and Mexico. In Europe, the leading countries are Russia, Poland, Germany, and France.

27.2: ORIGIN AND HISTORY

Potato originates in the Andes Mountains of Peru and Bolivia, where the plant has been cultivated for over 2,400 years. The Aymara Indians developed numerous varieties on the Titicaca Plateau, some 10,000 feet above sea level. The first written account of potato was made in 1553 by Spanish conquistador Pedro Cieza de Leon in his journal "Chronicle of Peru." The Spanish introduced potato to Europe between 1565 and 1580. It was taken from England to Bermuda and later to Virginia in 1621. It was also introduced into Germany in the 1620s, where it had become a part of the Prussian diet by the time of the Seven Year War (1756–1763). Antoine Parmentier, a prisoner of war in Prussia, introduced the crop to France after the war. Potato was also later brought to North America when Irish immigrants (hence the name "Irish potato") brought it to Londonderry, New Hampshire, where large-scale production occurred in 1719.

By the 1840s, the devastation by the fungus *Phytophthora infestans* (causes late blight) and heavy rains had brought untold hunger and starvation to Ireland. This sparked a mass immigration of about 2 million Irish, mostly to North America. In America, Luther Burbank was first to undertake improvement of the plant, subsequently releasing the "Burbank" potato to the West Coast states in the late 1800s. A disease-resistant mutation of the Burbank potato was discovered in Colorado. It had reticulated skin and became known as the Burbank Russet and is grown on most farms in Idaho.

27.3: ADAPTATION

In the United States, potato is grown in every state, although commercial production is highly localized. Potato is a cool season crop. The optimal temperature for shoot growth and development is 72°F. In the early stages, soil temperature of about 75°F is ideal. However, in later stages, a cooler temperature of about 64°F is desired for good tuberization. Tuber formation and development are slowed when the soil temperature rises above 68°F, and they cease at 84°F. Above this temperature, the effect of respiration exceeds the rate of accumulation of assimilates from photosynthesis. July temperature should not exceed 70°F. Consequently, production in the warmer regions of the South occurs in early spring, or in fall or winter when cool temperatures prevail.

Potato shoot is sensitive to hard frost in fall, winter, or early spring. Tubers will freeze at about 29°F and lose quality on thawing. Tuberization is best under a short photoperiod, cool temperature, and low nitrogen, whereas vegetative shoot production is favored by long days, high temperature, low light intensity, and high amounts of nitrogen. However, tuberization can occur at 54°F at night. Tuber yield decreases by 4% for each 1°F above optimal temperature. The best potato production occurs in regions with daily growing season temperatures averaging between 60° and 65°F. High soil temperatures result in knobby and malformed tubers. Potato flowers and sets seed best when long days and cool temperatures prevail. Consequently, potatoes set seed when grown in the northern states, but not the southern states.

Soil moisture is critical to proper tuber formation. Moisture deficit results in poor tuber set and enlargement. Similarly, erratic soil moisture content causes various abnormalities, including the cracking of tubers and the formation of knobs on the eyes resulting from secondary growth. On the other hand, excessive moisture produces low tuber yield, excessive vine growth, low dry matter content, and disease problems. When a dry spell is followed by abundant moisture during tuberization, tuber malformation occurs, resulting in pear-shaped tubers. The best potato production occurs in the arid regions of the northern United States.

Because potato is a tuberous plant, the soil physical characteristics are critical for successful production. Well-drained, loamy soils are best for the production of potatoes. Fertile clay soil may be used, but sticky soils interfere with harvesting and processing for marketing. A soil pH of 5.0–5.5 is ideal. Above a pH of 5.0, the incidence of scab disease increases. Soils in the Great Plains region tend to be alkaline.

Light intensity is important in potato production. The exposure of tubers to light in the field or storage results in sunscaled tubers and the formation of **solanine**, a colorless, toxic alkaloid, as well as the appearance of chlorophyll (the tubers develop green patches, reducing market value). Most solanine is removed when tubers are peeled.

27.4: BOTANY

Potato *(Solanum tuberosum)* is an annual plant, with short (1–2 feet), erect, and branched stems. Its compound leaves can be 1–2 feet long, with a terminal leaflet. The flowers are borne in compound, terminal cymes with long peduncles. The flower color may be white, rose, lilac, or purple. The plant bears fruits (berries) called **potato balls**. The underground commercial part is a modified stem (or tuber) that is borne at the end of a stolon (Figure 27–3). The "eyes" on the tuber are actually rudimentary leaf scars formed by lateral branches. Each eye contains at least three buds protected by scales. When potatoes sprout, the sprouts are lateral branches with several buds. A section across a tuber reveals a pithy central core, with branches leading to each of the eyes.

FIGURE 27–3 Potato is swollen stem. (Source: USDA)

The pith is surrounded by parenchymatous tissue, which accumulates starch. A vascular ring toward the outer part of the tuber contains the cambium and cortex. The cortex contains the pigment (may be pink, red, or purple) found in potato varieties with colored skins. The skin (periderm) is corky and has lenticels (openings for gaseous exchange). Under the proper conditions, the surface of a cut potato develops a corky layer (suberization) to protect the exposed surface from decay.

27.5: VARIETIES

Four varieties accounted for about 75% of the potato acreage in the United States in the 1960s—Russet Burbank, Katahdin, Kennebeck, and Red Pontiac. These old varieties represent the most major shapes of potato—the long, cylindrical, russet skin of the Russet Burbank, the red, short, rounded shape of Triumph, and the oblong tubers of Green Mountain. The ideal potato tuber shape appears to be short, wide, and flat with shallow eyes. Katahdin is commonly grown in the regions where late potato is grown. Green Mountain is also a late variety that is grown in the northeastern states, while Red Pontiac is grown in the North Central States and the North East. The Russet Burbank, commonly grown in southern Idaho, Washington, Oregon, and Montana, is what is sometimes called the Idaho baking potato. The tuber shapes are variable—round, oval, elliptical, long, oblong, and blunt (Figure 27–4). In cross section, the tubers are either flat or round. Skin color is also variable—white, cream, buff, red, or yellow. They may be smooth, flaky, netted, or russeted in texture. The flesh of the tuber may be white, cream, buff, purple, or yellow. Light flesh colors are common in the United States. The Yukon Gold is an old popular yellow-fleshed potato cultivar in the United States.

Many new varieties have been developed for various markets—baking, frying, cooking, canning, creaming, dehydrating, and chipping. These varieties have characteristics that make them suitable for their specific uses. The round, smooth-skinned white eastern varieties are used for chipping (potato chips) and cooking (boiling), whereas the mutated western types are used for baking and frozen products (mainly French fries). The russet potato varieties have higher dry matter content than the eastern types. Dry matter is measured by the specific gravity of the tuber. High dry matter (1.085 specific gravity or higher) is desired for baking and processing. For frying, boiling, or mashing, the tuber should have a specific gravity of 1.080 or higher and at least 19.8% solids plus 14% starch.

27.6: CULTURAL PRACTICES

27.6.1 LAND PREPARATION

The economic part of potato is the tuber, which is borne underground. Consequently, the soil environment should be given high consideration in cultural practices. The soil preparation should promote good aeration, drainage, and tuber formation without deformation. Land preparation should not cause compaction; hence, excessive tillage should be avoided. The soil should be plowed below any compacted layer within the normal root zone of the potato plant (most potato roots occur within the top 12 inches of the soil). Plowing is done in the fall to incorporate plant residue. Plowing in the spring should be done early enough to allow incorporated plants to decompose before planting (about 4 weeks). In the South, some growers bed up the land with listers to improve drainage.

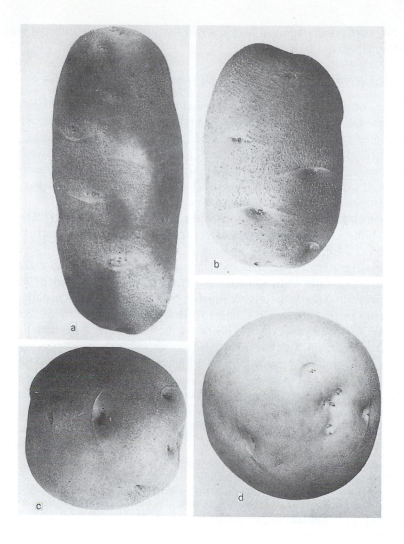

FIGURE **27–4** Common potato tuber shapes: (a) Russet Burbank (b) Green Mountain (c) Triumph and (d) Katahdin. (Source: USDA)

27.6.2   SEED POTATOES AND THEIR PREPARATION

Planting certified seed reduces the incidence of disease transmitted through the seed. Small tubers (1 1/2–2 inches diameter), called grade 1B potatoes, are used for potato seed. This grade of potatoes is produced by potato seed growers who plant at high plant populations to deliberately decrease the size of tubers to be used mainly for seeding the crop. Alternatively to using whole small tubers, growers often divide large tubers into small ones weighing between 1 1/4 and 2 ounces. Each piece should have one to three eyes.

Preparing cut pieces ahead of schedule requires suberization before planting. Suberization takes about 10 days at about 60°F and high humidity of about 85% in sacks. Suberized potato can be held safely for up to 30 days before planting. However, producers usually plant freshly cut pieces, dusting the cut surfaces with lime or gypsum to keep the pieces from sticking together.

In some regions, seed potatoes are treated with disinfectants to protect against soil-borne diseases, such as soft rots, common scab, rhizoctonia, and canker. These seed treatments include fungicides, formaldehyde, and mercuric chloride. Some researchers have shown that such a treatment is beneficial when the seeds are stored more than 3 days before planting. Using clean seed requires no preplant seed treatment. Furthermore, such practice does not protect against disease when the soil is heavily infected with scab or rhizoctonia pathogens.

27.6.3 SPACING AND PLANT POPULATIONS

Several types of planters are available for planting potatoes in the field. Some of these planters place the seed and apply fertilizer and systemic insecticides in one operation. The picker-planter jabs the seed with a pointed spike and may spread disease from an infected piece of potato to a disease-free piece. The cup-type planters are suitable for placing whole seeds. The most uniform seed placement is achieved with the assisted feed planter, which requires one or two workers to assist in distributing the seed in planter compartments, so that one piece is dropped each time.

Seeds are planted at a depth of 4–5 inches, deeper in sandy or dry soil than in heavy or wet soil. Similarly, seeds are planted deeper in late planting than early planting and when the soil is warm rather than cold.

The spacing adopted depends on various factors, including the variety planted, the size of the tubers desired, and whether irrigation will be provided. Varieties that produce few tubers per hill (e.g., Kennebec and Katahdin) are usually planted at closer spacing in rows (about 6 inches) for proper tuber size for the market. Row spacing may be 34–42 inches wide. When 34-inch rows are used, and 1 1/2 oz seed are planted at 8 inches apart in rows, 36 bushels (2,200 lb) of seed per acre are needed. On the whole, closer spacing (6–8 inches) promotes higher tuber yield and higher tuber quality. For the same spacing, using larger seed pieces gives higher yield per acre. In one study, the yield in a 36 × 8 inch planted field produced 20.4 cwt/acre with 1 1/2 oz seed size and 27.2 cwt/acres with 2 oz seed size. Planting at closer spacing with large seeds reduces the incidence of oversized tubers, hollow heart, and growth-cracked tubers.

27.6.4 TIME OF PLANTING

It takes about 3 1/2–4 months for an early potato crop to be ready for harvesting and 3–3 1/2 months for the late or fall potato crop to be ready for the market. Potato will sprout when the temperature is 40°F. Planting in the late winter and spring is usually undertaken when the air temperature is 55° to 58°F. The goal is to plant a late crop such that the tubers will be fully mature at or before the date of the average first killing frost. In the northern production regions, this temperature is attained about 10–14 days before the average date of the last killing frost. In the southern latitudes, potato can be planted earlier with little danger of frost damage. Winter potatoes are planted in October or November in southern California and Florida. Spring planting begins about January 20 to February 1 in southern Georgia. Some regions produce two crops per year, sometimes on the same field. This occurs in areas such as Maryland and Kentucky.

27.6.5 CROP ROTATION

Various rotations are adopted, based on the production region. Generally, potato performs best when it follows a green manure crop or a legume (e.g., clover, alfalfa, peas). A rotation in the Corn Belt may be corn–potato–small grains, whereas in the northeastern states it is often potato–small grains–clover. Small grains follow potato because a potato crop leaves the field in a condition in which no plowing is needed for planting small grains. In the southern region, early spring potato follows an early fall–sown green manure crop, such as crimson clover, vetch, and winter peas.

27.6.6 FERTILIZATION

Potato responds to both organic and inorganic fertilizers. Excessive amounts of nitrogen may decrease tuber yield, quality, and market grade. High nitrogen decreases the mealiness of potato and increases the skinning and bruising of the tubers. Manures (usually with phosphorus added at the rate of 40–50 lb/acre) are known to reduce scab injury and boost crop yields. Nitrogen application rates range between 30 and 100 lb/acre and are much higher when the crop is irrigated. A split application may be useful, applying 15–50 lb/acre at planting and sidedressing or topdressing with 30–40 lb/acre at blossom stage or when tubers begin to form.

Phosphorus application rates range between 30 and 80 lb/acre when soils are fertile. Where soil tests reveal low phosphorus, applications of 180 lb/acre and even higher may be profitable. Potassium is needed for heavy starch production. When soil tests indicate potassium levels of 0–99 lb/acre, an application of 300 lb/acre is profitable. Generally, applications of 50–150 lb/acre are part of fertilizer management in potato production.

27.6.7 IRRIGATION

Depending on the production area, 12–24 inches of water will be needed to produce a potato crop. Supplemental moisture application should be applied judiciously for a high yield of quality potato. Secondary growth and growth cracks occur when potato is irrigated or rain falls after a period of moisture stress. The soil should be kept uniformly moist until tubers have attained full size. Consequently, the soil should not be allowed to dry below 65°F field capacity, a condition that is difficult to maintain on sandy soils. To reduce the incidence of leaf diseases, supplementary moisture is best applied by furrow irrigation. The water needs of potato are highest after blooming, when the tubers are growing most rapidly.

27.6.8 WEED MANAGEMENT

Straw mulch may be used in potato production to control weeds and lower soil temperature in hot weather. Various herbicides have been approved for application in potato production. These may be applied as preplant, post-plant, preemergence, or postemergence herbicides.

27.6.9 DISEASES

Fungal diseases of potato include early blight (caused by *Alternaria*) and *Fusarium* and *Verticilium* wilts. Common scab is caused by a soil-borne fungus *(Streptomyces scabies)*, which causes corky pits on the stems of potato tubers. The blackleg bacterium causes blackening of the stems and yellowing and curling of the leaves. Viral diseases include mosaics and leaf roll. These diseases and others are described in Table 27–1.

27.6.10 INSECTS

The Colorado potato beetle is one of the most widespread insect pests of potato. The adults overwinter in the soil and emerge just about when potatoes are emerging. Other insect pests include fleabeetles, leafhoppers, and cutworms. The effects and management of these and other pests are summarized in Table 27–2. Planting potato in a field that previously carried sod should be avoided, since such soils increase the incidence of wireworm and grubworm attacks.

Table 27-1   Common Diseases of Potato in Crop Production

1. Common scab *(Streptomyces scabies)*
   - Fungus inhabits the soil and is carried on tubers.
   - Affected tubers exhibit round, corky pits on the skin.
   - Disease is promoted by soil alkalinity.
   - Control by raising soil acidity by applying 300–600 lb of sulfur per acre.
   - Resistant cultivars are available.
2. Rhizoctonia canker (black scurf) *(Rhizoctonia solani)*
   - Fungus is soil-inhabiting.
   - Crop stand is reduced and consequently yield loss occurs.
   - Disease occurs as irregular, small, black crusts on the skin.
   - Planting in the cool part of the year favors the disease.
   - Treat infected seed potato with chemicals (e.g., hot formaldehyde).
   - Planting on the ridge warms the soil quickly and reduces the population of the fungus.
3. Ring rot *(Corynebacterium sepedonica)*
   - It is caused by seed-borne bacteria.
   - Afflicted plants show wilting of the tips of the leaves and branches, followed later by leaf rolling, mottling, and color changing of leaves to pale yellow. Eventually, the plants die.
   - The region immediately below the skin begins to decay, causing a ring-rot appearance.
   - Do not plant infected seed potatoes.
   - Use resistant cultivars.
4. Late blight *(Phytophthora infestans)*
   - This was the first cause of potato famine in Ireland in 1845–1846.
   - Blight occurs first on the margins of lower leaves, spreading inward and eventually killing the leaves.
   - Affected leaf margins show irregular, water-soaked spots.
   - Dead leaves exude an odor.
   - Tubers are attacked, producing slightly sunken brownish or purplish spots.
   - The interior of the tuber is granular, brick-red blotches.
   - Control by using resistant cultivars or chemicals.
5. Early blight *(Alternaria solani)*
   - Disease is serious usually several weeks before harvest.
   - Affected leaves show small, scattered, dark, circular spots, followed by yellowing.
   - Spots enlarge, causing affected tissue to die.
   - Control by chemicals.
6. Mild mosaic (potato viruses A + X)
   - Affected part is mottled and yellowish or light-colored.
   - Yield is drastically reduced.
   - Use resistant cultivars.
7. Rugose mosaic (potato viruses Y + X)
   - Diseased leaves are crinkled, with blackening of the lower sides and death of the veins.
   - Plants are dwarfed with mottled leaves.
   - Use resistant cultivars.

Table 27–2 Common Insect Pests of Potato in Crop Production

1. Colorado potato beetle (*Leptinotarsa decemlineata*)
   - It is very widespread in the United States.
   - The larvae and beetles feed on plant parts.
   - Control by using insecticides.
2. Potato fleabeetle (*Epitrix cucumeris*)
   - The adult beetle feeds on foliage, whereas the larvae attack tubers.
   - Control by using insecticides.
3. Potato and tomato psylid (*Paratrioza cockerelli*)
   - Nymphs feed on the underside of the leaves.
   - Leaves curl and become yellow.
   - Control with chemicals.
4. Potato tuber worm (*Gnorimoschema operculella*)
   - Small, whitish worms mine leaves, stems, and tubers in storage and the soil.
   - Control by using chemicals.
5. Potato aphids
   - Various kinds of aphids occur in potato fields, transmitting various viral and other diseases.
   - Control by using chemicals.
6. Potato leafhopper (*Empoasca fabae*)
   - It causes a diseaselike condition called hopper burn.
   - Infected leaves curl and become yellow, dry, and brittle.
   - Control by using chemicals.
7. Nematodes: root-knot (*Meloidogyne* sp); golden nematode (*Heterodera rostochiensis*)
   - These attack potato tubers.
   - Control with chemicals as soil treatments, liquid, or fumigants.

27.6.11 HARVESTING

Potatoes are edible at any stage of tuber maturity. Growers decide when to harvest to meet specific market types. Often, early potatoes are harvested before optimal maturity to take advantage of the market demand and high prices. These premature tubers usually have watery flesh, bruise easily, and sustain early damage in handling. Processors may require potato to be harvested based on tuber chemistry (e.g., reducing sugar content) and chipping quality.

Prior to harvesting, the potato vines are killed by vine beaters or chemicals (e.g., sodium meta arsenite). Killing the vines hastens maturity and checks the development of oversized tubers and the incidence of hollow heart. This treatment also leads to good skin quality and facilitates harvesting. Harvesting usually occurs within 10–15 days after vine killing. Mechanical diggers are used to mechanically harvest commercial potatoes (Figure 27–5). They are washed with jets of water to remove all soil, air dried, sorted, and graded. The tubers are packaged (usually in burlap sacks) for the market.

27.6.12 STORAGE

If potatoes will not be marketed immediately, some storage is needed to keep tuber quality. First, the tubers are held at about 60°F and about 85% relative humidity for about 14 days for any bruises sustained during harvesting and handling to heal. The temperature is then lowered to 36–38°F to keep them from sprouting. Potatoes can remain dormant for up to about 3–5 months if storage room temperature is maintained at 40°F. Storing at

FIGURE 27–5 Harvesting potato (Source: USDA)

low temperatures can be done in bulk containers without danger of heat build-up. However, sacks should be used if temperatures are milder. The tubers respire in storage, being living things, and hence the warehouse environment needs to be ventilated to prevent heat and carbon dioxide buildup that promotes tuber deterioration. Furthermore, the relative humidity should be maintained at 85–90% to prevent tuber shrinkage.

When storage temperatures fall below 50°F, the starch accumulated in the tuber begin to be converted into sugar, and the tubers become sweet and soggy rather than the desired mealy texture for good cooking quality. Sugar-rich potatoes produce dark-colored French fries, as a result of the caramelization of the sugar. At 36°F, stored tuber can experience 25–35% starch-sugar conversion. At 60–70°F, most of the initial sugar in the tuber is used up in respiration, making tubers mealier. Small-scale storage of potato (e.g., home storage) can be enhanced by including apples in the potato sack to retard sprouting.

27.6.13   GRADES AND GRADE REQUIREMENTS

The grades and grade requirements of potato are summarized in Table 27–3.

Table 27–3 Potato Grades and Grade Requirements

Grade	Description
U.S. Extra No. 1	Has typical characteristics of variety, firm, clean, well mature, well shaped; free from freezing and diseases (blackheart, late blight, bacterial wilt, soft rot) and wet breakdown. Should also be free from injury caused by sprouts, internal defects, and other causes. Size shall not be less than 2¼ inches in diameter or 5 ounces in weight, and shall not vary more than 1¼ inches in diameter or more than 6 ounces in weight.
U.S. No. 1	Same as for extra No. 1, but size shall not be less than 1⅞ inches in diameter (unless otherwise specified in connection with grade).
U.S. commercial	Same as for No. 1 but free from serious damage caused by dirt or other foreign matter, Russet scab, and *Rhizoctonia*.
U.S. No. 2	Meet requirements of variety but not seriously misshapen, free from freezing or injury and diseases (blackheart, late blight, etc). Size should not be less than 1½ inches in diameter (unless otherwise specified in connection with the grade).
Unclassified	Not assigned any grade in accordance with the above description.

27.7: USES

Some of the potato crop is harvested as seed for planting. The remainder of the crop is used for food, feed, or industrial products. Dehydrated potatoes are used for food or whiskey. Potatoes are also used for starch or alcohol. Potato used for food is processed for frying, freezing, dehydrating, or canning.

REFERENCES AND SUGGESTED READING

Martin, J. H., and W. H. Leonard. 1971. *Principles of field crop production.* New York: Macmillan.

SELECTED INTERNET SITES FOR FURTHER REVIEW

http://www.potatohelp.com/potato101/statistics.asp

Potato production statistics

http://www.cipotato.org/market/potatofacts/growprod.htm

General potato production information

http://www.cals.nscu.edu/sustainable/peet/profiles/botpotato.html

Potato production in North Carolina

http://ipcm.wisc.edu/piap/potato.htm

Potato production information

http://edis.ifas.ufl.edu/BODY_CV131

Potato production information

http://www.cals.ncsu.edu/sustainable/peet/profiles/c15potat.html

Potato production guide

http://www.uga.edu/vegetable/potato.html

Overview of potato and its production

OUTCOMES ASSESSMENT

PART A

Answer the following questions true or false.

1. T F Irish potato is a root crop.
2. T F Potato originated in the Andes mountains of Peru and Bolivia.
3. T F Low dry matter of tuber is desired for baking and processing qualities of the tuber.
4. T F Potato is a cool-season crop.
5. T F Nitrogen increases the mealiness of potato tuber.

Answer the following questions.

1. Exposure of potato tubers to light in the field or storage causes the formation of _____, a colorless and toxic alkaloid.

2. The potato plant bears fruits called _____.

3. The scientific name of potato is _____.

4. The American potato producer who discovered a disease-resistant mutant potato in the 1800s was _____.

5. _____ is the potato disease that caused the migration of Irish people to America.

PART C

Write a brief essay on each of the following topics.

1. Discuss the importance of potato as a food crop.

2. Briefly discuss the origin of potato.

3. Describe the general botanical features of the potato plant.

4. Discuss the time of seeding and seeding rate of potato.

5. Discuss the crop rotation practices in potato production.

PART D

Discuss or explain the following topics in detail.

1. Discuss the environmental conditions best suited to the production of potato.

2. Discuss the growth and development of potato.

3. Discuss the major insect pests in the production of potato.

4. Discuss the major diseases in potato production.

6. Discuss the use of harvesting aids in potato production.

7. Discuss the grading of potato.

28

Alfalfa

TAXONOMY

Kingdom	Plantae	Subclass	Rosidae
Subkingdom	Tracheobionta	Order	Fabales
Superdivision	Spermatophyta	Family	Fabaceae
Division	Magnoliophyta	Genus	*Medicago* L.
Class	Magnoliopsida	Species	*Medicago sativa* L.

KEY TERMS

Common alfalfa   Turkistan alfalfa   Variegated alfalfa
Flemish alfalfa   Chilean alfalfa

28.1: ECONOMIC IMPORTANCE

Alfalfa is the oldest crop grown solely for forage. In 2001, it was grown as a pure stand or mixtures with other forage crops on 23.8 million acres, yielding a total of 80.2 million tons and an average of 3.38 tons/acre. Most of the U.S. acreage is in the North Central states, where the crop is produced largely in mixtures with forage grasses. The leading alfalfa-producing states are Wisconsin, California, Minnesota, Iowa, and Nebraska (Figure 28–1). Alfalfa seed is an important produce. It is produced largely in the western states, with California leading production with over 100,000 acres. Other important producers are Washington and Idaho.

FIGURE 28–1 General alfalfa production regions of the United States. (Source: USDA)

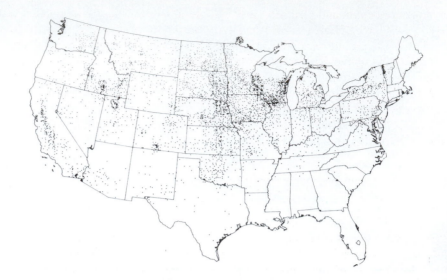

Alfalfa yields average about 150–200 lb/acre. However, some producers average about 500–1,100 lb/acre. Alfalfa acreage and production declined by about 10% between 1986 and 1997, in spite of an increase in price of over 30%. This decline is attributed partly to production constraints. The crop does not have a well-established market, and harvesting is difficult to schedule. Further, quality control for alfalfa is limited. The crop suffers yield and quality losses during drying in the field.

On the world scene, Argentina produces large acreages of alfalfa. The crop is produced to varying extent in Europe.

28.2: ORIGIN AND HISTORY

The name *alfalfa* is derived from the Arabic, meaning "best fodder." It is known as lucerne in Europe and is native to western Asia and eastern Mediterranean countries. The crop was first cultivated in Iran, from which it was introduced into Arabia and the Mediterranean countries and subsequently to the New World. Alfalfa is used for medicinal purposes in the ancient Chinese tradition, by the Ayuvedic physicians of India, and by North American Indians for a variety of ailments, including jaundice, blood clotting, digestive disorders, arthritis, and many more. Alfalfa sprouts feature predominantly in the modern health food industry.

In the United States, alfalfa was first grown in Georgia in 1736. It was introduced into California from Pennsylvania in 1841, and it came from Chile in about 1850.

28.3: ADAPTATION

Alfalfa is widely adapted but grows best where the climate is dry and irrigation is available. Although it can withstand prolonged drought, alfalfa is not productive under such conditions. It is tolerant of extreme heat and cold. Yellow-flowered alfalfa is known to have survived at −84°F, whereas common alfalfa varieties have survived about 120°F temperatures in Arabia and Death Valley, California. It grows in the mountains at altitudes of 300–900 feet. High altitude (mountain) alfalfa is more potent than cultivated alfalfa varieties. The cold-resistant variegated varieties are suitable for production in latitude 40°N and above. The non-hardy varieties are adapted to the southern latitudes where temperatures are more than 18°F. The plants depend on stored food for maintenance and regrowth. Winter damage and disease deplete the organic root reserves, resulting in subsequent stand reduction.

Alfalfa grows best in deep loams that are well drained and slightly acidic to slightly alkaline (pH 6.5–7.5). It is intolerant of high alkaline conditions.

28.4: BOTANY

Alfalfa is an herbaceous perennial legume that grows from a semiwoody base or crown. Under ideal conditions, it can live for 20 years or more. The most commonly cultivated species of alfalfa is *Medicago sativa L.* of the family Poaceae (Leguminoseae). It has purple flowers. Yellow-flowered alfalfa *(M. falcate)* is considered by some to be a subspecies of *M. sativa* (i.e., *M. sativa* spp. *falcate*). This species is not produced commercially in the United States.

The mature plant attains a height of 2–4 feet, depending on the variety. The crown produces few to numerous (5–20) thin, succulent, leafy, multi-branched stems. The leaves are pinnately trifoliate. Each stem terminates in a raceme or cluster of 10–100 flowers. The color of the corolla is variable—from purple through shades of blue, green, yellow, and cream to white. Flowers also occur in axillary racemes (Figure 28–2).

The flowers of commercially produced alfalfa varieties are slightly to completely self-incompatible. Cross-pollinated flowers set more pods and more seeds per pod than selfed flowers. The sexual part of the flower must be tripped (usually by insects) in order for seed set to occur in most cases. Tripping can be spontaneous or by the wind, rain, or sun. The most effective trippers are wild bees—leaf-cutter bees (*Meqachile* spp.) and ground bees (*Paranomia* spp). The stigmatic surface is ruptured by the tripping action to allow the penetration of the pollen tubes.

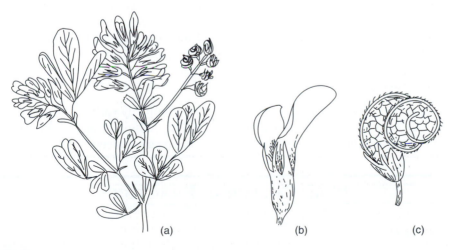

(a)   (b)   (c)

FIGURE 28–2   Alfalfa: (a) leaves, (b) flower, (c) seed.

On fertilization, the ovules in the ovary develop to produce a tightly curled pod. The number of curls varies from 1–5, according to the number of ovules that develop into mature seeds. A pod may contain 1–8 or even more kidney-shaped seeds. The seeds are normally light green in color. Alfalfa has a deep taproot system. The seed of alfalfa can be impermeable. Hard or impermeable seed coat occurs at the rate of 0 to 70% or more in varieties. The percentage of hard seed increases with maturity.

28.5: ALFALFA VARIETIES

All U.S. alfalfa varieties derive from three major alfalfa types—common, variegated, and yellow-flowered alfalfas.

1. Common alfalfa *(M. sativa* ssp. *sativa)*. These alfalfa types are usually distinguished by the name of the country (or state) of adaptation. In the United States, for example, are Montana Common, Arizona Common, and Kansas Common. On the international scene are Chilean (Spanish), Turkistan, and Flemish alfalfa:
   a. Chilean (Spanish) alfalfa. These were imported from Chile. The first alfalfa in Kansas was derived from this group. The strains derived from Chilean alfalfa vary in adaptation to cold and fall- and spring-growth habits.
   b. Turkistan alfalfa. These types include those grown in southern Russia, Iran, Afghanistan, and Turkey. The varieties vary in winter hardiness from moderately hardy to hardy. They are generally susceptible to leaf and stem diseases but resilient to some insects and crown and root diseases.
   c. Flemish alfalfa. These alfalfas were developed in northern France. They are known to be vigorous and stemmy in growth habit. They are early-maturing and recover quickly after cutting. The varieties are moderately resistant to some foliar diseases.
2. Variegated alfalfas *(M. sativa* ssp. *x varia)*. The variegated alfalfa varieties derive from natural crosses between the common and yellow-flowered types. The predominant flower color is purple, but brown, greenish-yellow, green, yellow, and nearly white flowers occur. They are more cold-resistant than common alfalfas and are largely responsible for the successful production of alfalfa in some of the northern states of the United States.
3. Yellow-flowered alfalfas *(M. sativa* ssp. *falcate)*. These types are cold-resistant. They are not grown commercially in the United States, except for uses as breeding material to develop hardy varieties. They are variable in growth habit and provide genetic resources for breeding for resistance to certain diseases.

Other types of alfalfa in production include the non-hardy varieties that are adapted to the Deep South and the creeping alfalfas, which can multiply from rhizomes. Hybrid alfalfa is clonally propagated after the right cross has been obtained.

28.6: CULTURAL PRACTICES

28.6.1 TILLAGE

A good seed bed for alfalfa should allow good soil-seed contact for water imbibition, especially since the crop has hard seed problems. Some growers practice no-till alfalfa culture.

28.6.2 SEEDING DATES

Alfalfa may be planted in spring or late summer. In Kansas, planting occurs in April to mid-May. It may be earliest in southern and eastern parts of the state when seeded with spring oats as a nurse crop. For late summer planting, it should occur by mid-August to allow seedlings to become well established before the winter dormancy.

28.6.3 SEEDING METHOD

Some growers adopt cross drilling, whereby 50% of the seed is drilled in one direction and the other 50% at right angles to the direction of the initial seeding. Broadcast seeding is also done but not desirable. For better success with this method, a soil packer may be used to run over the field after broadcasting. Some growers drill alfalfa or broadcast into winter wheat in early spring. The wheat crop should not be tall or of thick stand for this to be successful.

28.6.4 SEEDING RATES

Seeding rates depend on the soil type and moisture conditions. Growers producing alfalfa for hay under rainfed conditions seed at the rate of 8–12 lb/acre. On irrigated sandy soils, the seeding rate is higher (15–20 lb/acre). The plant population after the first season is about 8–10 plants per square foot. After several years, this may decline to 3–5 plants per square foot. Alfalfa compensates for thinner stands by producing more stems. Consequently, yields decline gradually. As the alfalfa stands thin to 2–3 plants or fewer, weeds invade the crop, resulting in a substantial reduction in yield and quality. When an alfalfa field reaches a stand of less than 50 stems per square foot, it should be replanted.

28.6.5 SEEDING DEPTH

Being small-seeded, alfalfa should be planted no deeper than 0.5–0.75 inch, less in heavier soils.

28.6.6 ALFALFA SEED PRODUCTION

Alfalfa seed is produced in drier regions, such as the western United States, where California leads the nation in production. Row spacing varies between 20 and 40 inches. The seeding rate is low (1–3 lb/acre), with plant space about 12 inches in rows. Plant population should not exceed 1,000 plants. The moisture supply should promote slow vegetative growth until the plant comes into blooms. Thereafter, irrigation should be modest to prolong the blooming period without promoting excessive vegetative growth. Sprinkler irrigation interferes with pollination and may damage the flowers; hence, it should not be used during the critical period. Alfalfa for seed should come into bloom in July for high yields. The second cutting is best suited for seed production. The first cutting is often light in bloom because of the cooler winter and short photoperiod, coupled with the low activity of pollinating insects. The three or four cuttings are usually too late for good seed set and maturation. The alfalfa flower must be tripped to set seed; hence, bees are required in seed production.

Alfalfa is cut for seed when about two-thirds to three-fourths of the pods are brown or black. The crop may then be direct-combined from windrows. Where the crop matures uniformly, it can be combined directly. Chemical desiccation may be used 3–4 days before combining. Seed yields average about 200 lb/acre, but some products average 500–1,100 lb/acre.

28.6.7 CROP ROTATION

Alfalfa is grown in rotations with small grains as companion crops. After 2–3 years of hay or seed production, the alfalfa crop is incorporated into the soil. This practice enriches the soil with nitrates that benefit the next crop, provided soil moisture is adequate. Crop burning may result if cropping after incorporating alfalfa occurs under dry conditions. Alfalfa has deep roots and exploits the soil extensively of nutrients and water.

28.6.8 FERTILIZATION

After alfalfa is established in the field, there is no opportunity for liming if the need arises. It is important to apply the needed lime (and sometimes phosphorus) before planting. Alfalfa is capable of biological nitrogen fixation when well nodulated. The seeds may be treated with *Rhizobium* inoculant before seeding. A pre-plant application of 15–20 lb of nitrogen per acre will provide the needed nitrogen while the nodules mature and become active. When necessary, phosphorus and potassium may be applied as top dressing.

28.6.9 IRRIGATION

Alfalfa is a drought-tolerant perennial with deep roots (8–12 inches) for exploring a large volume of soil for water. It has a seasonal water use of 40 inches. Alfalfa requires about 3–4 acre-inches of water to produce 1 ton of yield. This increases to 6–7 acre-inch per ton in summer. Regrowth is slowed if plants are stressed at the time of cutting. Consequently, it is desirable to water prior to harvesting or immediately thereafter. However, this practice also stimulates weed growth. Alfalfa is irrigated by a variety of methods, including corrugation, bedded-furrow, graded borders, and sprinkler irrigation.

28.6.10 WEED MANAGEMENT

Good alfalfa stands are able to suppress weeds. Weed infestation reduces forage quality and yields. Weeds should be controlled prior to seeding alfalfa by tillage or by chemicals. The grower who adopts reduced-tillage has to select from approved herbicides for weed control. These are Roundup® and Gramoxone Extra®. These are effective against perennial weeds. Summer annual weeds (foxtails, lambsquarters, crabgrass, pigweeds) grow very fast and can reduce alfalfa stands. Winter weeds (cheatgrass, pennycress) do not seriously affect fall-seeded alfalfa. Herbicides that can be used include Eptam®, Balan®, Poast®, Pursuit®, and Select®. Alfalfa that has been sprayed with herbicides such as Pursuit® should not be grazed or harvested within 30 days after treatment.

With years, established alfalfa stands become infested with weeds as thinning and open spaces occur. Dormant-season tillage can be conducted to control weeds but must be done carefully to avoid damaging the crowns. Herbicides that may be applied during the active growing season include Butyrac®, Poast Plus®, and Select®. Gramoxone Extra® and other non-selective herbicides should be applied during the dormant-season. Other dormant season herbicides included Karmex®, Lexone®, Sencor®, Sinbar®, and Velpar®. Alfalfa is sensitive to residual herbicides from previous crops (e.g., Glean®, Amber®, atrazine, Finesse®, Peak®).

28.6.11 DISEASES

Alfalfa is affected by a variety of diseases that impact the yield and quality of the harvested product. Seedling disease causes incomplete stand, reduced early canopy coverage, and reduced competitive ability against weeds. Wilt attacks cause thinning of the

Table 28–1 Common Diseases of Alfalfa in Crop Production

1. Bacterial wilt
   - This is the most important economic pest of alfalfa in the United States.
   - It is more common in moist conditions and less common in hot, dry regions.
   - Afflicted plants are stunted; crop stand is reduced.
   - Plants have small and yellowish leaves.
   - Afflicted plants usually die in the second year after infection.
   - Mowers and irrigation water spread disease.
   - Control by using resistant cultivars.
2. Leaf diseases
   - *Leaf spot* is the disease most common in humid regions in cool weather.
       Symptoms include small, circular, dark brown spots on the leaves.
       Leaf drop usually follows infection.
       Harvest the crop as soon as infection becomes severe.
       Resistant cultivars may be used to control the disease.
   - *Leaf blotch* is characterized by long, yellow blotches sprinkled with minute brown dots.
3. Crown wart
   - It is common in California.
   - It is characterized by galls on the crown at the base of the stem.
   - Use resistant cultivars for control.
4. Blackstem
   - It is most severe in the Southeast and West production regions.
   - Large, irregular brownish or blackish lesions appear on the leaves and petioles.
   - Use resistant cultivars for control.

stand, leading to reduced yields, whereas stem and leaf diseases cause leaf loss and reduced forage harvest. Alfalfa is also attacked by root and crown rots, which weaken the plants. Weakened plants have reduced persistence in the field, resulting in yield decline. Soil-borne diseases may be reduced by crop rotation, good soil drainage, and the use of resistant varieties. Major diseases and their characteristics are summarized in Table 28–1.

28.6.12 INSECTS

Alfalfa is affected by a number of insect pests that reduce leaf area or thin out the stand, leading to reduced yield. Important insects include the spotted alfalfa aphid, army cutworm, alfalfa weevil, blister beetle, and alfalfa caterpillar. The characteristics of these insects and others are summarized in Table 28–2.

28.6.13 HARVESTING

Alfalfa for forage is harvested once or twice during the growing season, even though two or three cuttings are common in the northern states, whereas three to five cuttings are common in the central and south central states. Under irrigation, alfalfa can be harvested as frequently as seven or eight times as practiced by producers in California. The timing of cutting is critical to the yield and quality of the crop. The fiber content increases with age, whereas protein decreases as the crop matures. Consequently, the best stage to harvest alfalfa for high yield and high protein is between one-tenth bloom and one-third to one-half bloom stages, when yield is about 3.35 and 3.19 tons, respectively. Most (70%) of the protein occurs in the leaves. Improper handling of the plants can lead to excessive leaf drop and, consequently, reduced quality and yield. To reduce this loss, the cut material should be windrowed and promptly baled or stored before the leaves dry to a stage

Table 28–2   Common Insect Pests of Alfalfa in Crop Production

1. Spotted alfalfa aphid
   - It is common in southern and central production states.
   - Eggs survive in winter in the Central States.
   - Control by using resistant cultivars or insecticides.
2. Alfalfa weevil
   - It lives in many western states.
   - Both larvae and adults are pests on the plants.
   - The adult weevil overwinters in plant debris.
   - Control by cutting the first crop promptly to starve the larvae or use insecticides.
3. Potato leafhopper
   - It is a problem in the eastern states.
   - Control by harvesting the first crop near the full bloom, so that the eggs attached to the plants of the first crop are prevented from developing further on the cured hay.
   - Insecticides may also be used in the early stages.
4. Alfalfa hopper
   - The problem occurs south of latitude 36°.
   - Control by sanitation—cleaning debris and weeds—to keep the insect population low.

at which such shattering can occur. Alfalfa may also be harvested, dehydrated, and ground into meal. The meal may be pelleted.

REFERENCES AND SUGGESTED READING

Martin, J. H., and W. H. Leonard. 1971. *Principles of field crop production.* New York: Macmillan.

OUTCOMES ASSESSMENT

PART A

Answer the following questions true or false.

1. T F   Alfalfa is capable of biological nitrogen fixation.
2. T F   The percentage of hard seed in alfalfa increases with maturity of the crop.
3. T F   Alfalfa is a herbaceous annual plant.
4. T F   Flemish alfalfas were developed in northern France.

PART B

Answer the following questions.

1. Give three of the top alfalfa producing states in the United States.

   _____ _____ _____

2. Give three of the top alfalfa producing nations in the world.

   _____ _____ _____

3. Give the three major alfalfa types in the world. _____

_____   _____

4. Alfalfa is Arabic for _____.

5. _____ is the most important economic pest of alfalfa in the
United States.

PART C

Write a brief essay on each of the following topics.

1. Discuss the importance of alfalfa as livestock feed.

2. Briefly discuss the origin of alfalfa.

3. Describe the general botanical features of the alfalfa plant.

4. Discuss the time of seeding and seeding rate of alfalfa.

5. Discuss use of alfalfa in crop rotation practices.

PART D

Discuss or explain the following topics in detail.

1. Discuss the environmental conditions best suited to the production of alfalfa.

2. Discuss the growth and development of alfalfa.

3. Discuss the major insect pests in the production of alfalfa.

4. Discuss the major diseases in alfalfa production.

5. Discuss the use of harvesting aids in alfalfa production.

APPENDIX *A*

A Comparison of Primitive and Modern Crop Production Activities

Activity	Primitive	Modern
Source of seed	Gather from wild; save and replant seed from harvest; low-yielding seed; poor quality	Gather from wild (but by scientists for breeding); improved seed from breeding; seed companies produce high-yielding-quality seed; fresh seed available each season
Land clearing	Manual operations; partial vegetation removal; slash and burn	Land often cleared of primary vegetation; stumps removed; operations often mechanized
Land preparation	Limited tillage; hand tools π	Terrain may be modified for various purposes (drainage, irrigation, erosion control); mechanized tillage—to varying degrees (no-till to conventional tillage)
Cropping system	Polycultures used widely	Monocultures used widely
Farm size	Usually small; subsistent production	Usually large; commercial production
Planting date	Crops grown in season; timing based on experience only	Production in season; more precise timing of operations; information available to avoid adverse weather; off-season production of especially horticultural crops under controlled environment
Pest management	Weeds removed with hand tools; diseases and insect pests managed by using adapted primitive varieties; no pesticides used	Agrochemicals used to manage pests; pest-resistant cultivars used; biological pest management
Provision of supplemental nutrients and water	Production mainly rainfed; inorganic fertilizers not used; organic matter used to limited extent	Rainfed; irrigation used widely in production; inorganic fertilizers widely used; green manures, compost, and other organic manures used
Harvesting	Manual; hand tools	Often mechanized
Processing	Sun drying; fireside drying	Artificial dryers used
Storage	Primitive barns/structures	Storehouse; sometimes climate-controlled storage

APPENDIX *B*

USDA Conservation Programs

Environmental Quality Incentives Program (EQIP)

Established in 1996. Purpose is to encourage farmers and ranchers to adopt practices that reduce environmental and resource problems (e.g., nutrient management, tillage management, and grazing management).

Wildlife Habitat Incentives Program (WHIP)

Established in 1996. Purpose is to provide cost-sharing assistance to landowners for developing upland wildlife, wetland wildlife, threatened and endangered species, fish, and other types of wildlife.

Conservation Farm Option (CFO)

Established in 1996 for producers of wheat, feed grains, cotton, and rice. Purpose is for qualified operators to have the option to receive one consolidated annual USDA conservation payment in lieu of separate payments from CRP, WRP, and EQIP.

Flood Risk Reduction Program

Established in 1996. Purpose is to offer risk reduction contracts to producers with frequently flooded contract acreage under the Agricultural Market Transition Act.

Conservation of Private Grazing Land Initiative

Established in 1996. Purpose is for the NRCS to coordinate technical, educational, and related assistance programs for owners and managers of non-federal grazing lands, including rangeland, pastureland, grazed forestland, and hay land.

Conservation Technical Assistance (CTA)

Established in 1936. Purpose is to provide technical assistance to farmers for planning and implementing soil and water conservation and water quality practices.

Water Bank Program (WBP)

Established in 1970. Purpose is to preserve, restore, and improve high-priority wetlands.

Emergency Conservation Program (ECP)

Established in 1978. Purpose is to provide financial assistance to farmers in rehabilitating cropland damaged by natural disasters and for conserving water during severe drought.

Emergency Watershed Protection Program

Established in 1950. Purpose is to provide technical and financial assistance to local institutions for removal of storm and flood debris from stream channels and for restoration of stream channels and levees to reduce threats to life and property.

Extension Education

Cooperative State Research, Extension, and Education Service (CSREES). Purpose is to provide information and recommendations on soil conservation and water quality practices to landowners and farm operators in cooperation with the State Extension Services and state local offices of USDA agencies and Conservation Districts.

Conservation Loans and Farm Debt Cancellation Easements

Purpose is to provide loans through the FSA to farmers for soil and water conservation, pollution abatement, and the building or improvement of water systems.

Forestry Incentives Program (FIP)

Established in 1975. Purpose is to provide cost-sharing up to 65% for tree planting and timber stand improvement for private forestlands of no more than 1,000 acres.

Forest Stewardship Program (FSP)

Established in 1990. Purpose is to provide grants to state forestry agencies for expanding tree planting and improvement and for providing technical assistance to owners of non-industrial private forestlands in developing and implementing forest stewardship plans to enhance multi-resource needs.

Pesticide Record Keeping

Established in 1990. Purpose is to require private applicators of restricted-use pesticides to maintain records accessible to state and federal agencies regarding produce applied, amount, and date and location of application.

Resource Conservation and Development Program (RC&D)

Established in 1962. Purpose is to provide technical assistance and limited financial assistance for planning and installation of approved projects.

Small Watershed Program

Established in 1954. Purpose is to assist state agencies and local units of government in flood prevention, watershed protection, and water management.

Data and Research Activities

Purpose is for the Agricultural Research Service (ARS) to conduct research on new and alternative crops and agricultural technology to reduce agriculture's adverse impacts on soil and water resources.

APPENDIX C

Agricultural Productivity

Productivity is a measure of the relationship between outputs and inputs in production.

CONCEPTS OF PRODUCTIVITY

U.S. farmers, like their counterparts everywhere, use inputs and natural resources to make products or generate outputs. The efficiency with which this conversion occurs depends on the managerial ability of the producer and the level of technology employed in a production operation. Economists normally express the relationship as *total factor productivity (TFP),* a measurement that considers the total output and inputs. TFP is calculated as a ratio of the index of outputs to the index of inputs. Economists are also interested in measuring the efficiency of production by noting the trends in TFP. An increasing trend indicates that more outputs are generated from a given input level.

In production theory, the concept of production function describes the amount of outputs that can be generated from a bundle or set of inputs, using a certain production technology. The TFP can change with time because of several factors, major ones including differences in efficiency, scale of production, and change in technology. Sometimes, maximum output is not realized for a given set of inputs. The method of production in this case is said to be inefficient.

TFP varies as the scale of production changes, due to the effect of economies of scale. As production inputs are transformed into outputs (i.e., technology) changes, TFP also changes over time. U.S. and world agriculture today is more high-tech than it was in the early 1900s.

TRENDS IN AGRICULTURAL PRODUCTIVITY, INPUTS, AND OUTPUTS

A vibrant economy depends on increased productivity. In terms of U.S. productivity growth, the agricultural sector of the economy has been most successful, posting one of the highest productivity growths of all industries. During the period from 1948 to 1994, agricultural productivity increased at an annual rate of 1.94%.

Output growth in agriculture was completely accounted for by productivity growth. However, productivity growth was less important as a source of output growth in non-farm business and manufacturing, accounting for only about 33% and 40%, respectively.

PRODUCTIVITY TRENDS

Productivity growth was generally slow between 1948 and 1994. The conditions under which such trends occurred were varied:

1. *Early 1950s.* The major conditions that faced the U.S. agricultural sector in this period included the exit of farm labor. However, U.S. agricultural producers began to move toward mechanization and the use of intermediate inputs (e.g., pesticides, fertilizers, and seed).
2. *1960s.* During the 1960s, growth in agricultural productivity was about 1.45%. Labor continued to be a problem. The use of intermediate inputs did not increase significantly.
3. *1970s.* The annual productivity growth increased to over 2.2% per annum. This was due largely to an increase in demand for U.S. agricultural exports. Producers increased their use of secondary inputs to boost production outputs. The oil embargo did not decrease agricultural energy consumption.
4. *1980s.* In the 1980s, U.S. agriculture experienced a period of surpluses that prompted the government to intervene with measures aimed at preventing the negative effects of a glut. Programs included payment-in-kind program and idling of agricultural lands. Production growth declined to 1.68%. This was also the period of financial restructuring of the agricultural industry. Both capital and the use of intermediate inputs declined.
5. *Early 1990s.* In spite of very low input levels, output growth rose to the highest level it had ever been.

OUTPUT AND INPUT TRENDS

Output trends shadowed the productivity trend, the most dramatic increase occurring in the early 1990s. During 1948–1994, the annual growth in output was about 1.88%. Whereas the weather was relatively mild and stable in the 1960s, the 1970s to 1990s were characterized by periods of adverse weather (drought, floods, early frost).

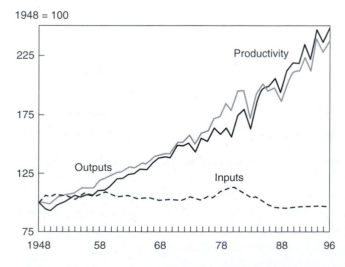

Input use in agriculture was stable between 1948 and 1994. Intermediate inputs only increased by about 1.42% per year, while energy increased by less than 0.9%. The greatest increase in input was realized with pesticides, the increase being about 5% per year. Labor input declined between 1948 and 1994.

FACTORS AFFECTING GAIN IN AGRICULTURAL PRODUCTIVITY

Gains in agricultural productivity are affected by internal and external factors.

RESEARCH AND DEVELOPMENT

The primary purposes of agricultural research and development are to increase the yield potential or productivity of plants and animals through breeding efforts and to improve the production environment through improved agronomic and animal husbandry practices. Agricultural research and development is also expected to protect the environment through the development of technologies that reduce soil erosion and non-point pollution. Both private and public sectors undertake agricultural research and development. Funding for private research and development currently exceeds public funding. Research and development benefit both farmers (through lower production costs and high profits) and consumers (through lower food prices and higher quality).

The benefit of public research to the taxpayer has been estimated at 20 to 60% of investment. The country stays competitive in the global market while improving the standards of living of the citizenry. Low-income earners are able to have access to high-quality food at affordable prices, due to lower food prices caused in part by agricultural research.

EXTENSION

Extension is an activity of transferring research and development information to producers. Such information includes the availability of new cultivars, new agronomic practices, government programs, and others. Much of this is funded by the public sector, even though the private sector has significant involvement. The rate of return to extension is difficult to estimate because of inadequate data.

EDUCATION

It is through education that the scientists who develop new technologies are trained. It was estimated that gains in education accounted for 8.6% of the increase in output during the period of 1948 to 1994. It is also known that farmers with education are more likely to adopt improved practices.

INFRASTRUCTURE

The impact of infrastructure on agricultural productivity is sketchy. Since agricultural production depends significantly on transportation, a good network of good roads positively impacts agricultural productivity. Since agricultural products are largely perishable, it is important that products are moved to consumers soon after harvesting.

GOVERNMENT PROGRAMS

The government impacts agricultural productivity through the allocation of resources and outputs. Data on impact, however, are limited. Nonetheless, a probable area of impact is tax policies, where the government may encourage private sector involvement in research and development by giving tax incentives. Further, the private sector will be more inclined to undertake research and development if there is intellectual property rights protection against unauthorized use of discoveries and inventions.

IMPORTANCE OF PRODUCTIVITY GROWTH

Growth in productivity has several distinct benefits to the economy:

1. Products are produced at higher efficiency.
2. Higher productivity in one sector of the economy frees up funds for use by other sectors.
3. Real prices of goods and services are lowered.
4. Real agricultural output prices decline through improving U.S. competitiveness in the global market.

In spite of the outstanding level of agricultural productivity (performance relative to other sectors in the U.S. economy), it accounts for only 1.5% of the GDP of business. Its impact on the productivity of the overall economy is thus small.

APPENDIX **D**

Federal Government Conservation and Environmental Programs

The activities of farming can cause the soil to lose its qualities (physical, chemical, and biological) that are essential for crop production and productivity. Soil health is thus of paramount importance to the producer. Since 1930, the USDA has administered a large number of programs intended to protect the earth's natural resources and the general environment from destruction from the adverse consequences of agricultural activities. The interventive activities are classified as *Conservation and Environmental Programs.* These programs are designed to assist producers (ranchers, farmers) and non-operators (landowners) in natural resource (soil, water, air, etc.) conservation and improvement for a healthy environment and sustainable production.

POLICY APPROACHES TO PROGRAM ADMINISTRATION

Each of the USDA's conservation and environmental programs is implemented by following one or more of six basic policy approaches:

1. *Technical assistance and education.* The USDA conducts workshops and educational or informational activities for producers. The USDA places its expertise at the disposal of producers, to assist in the decision-making process for high productivity.
2. *Financial assistance.* This may be in the form of monetary incentives for producers who embrace conservation and environmental programs. In some cases, the cost involved in participating in conservation and environmental programs is shared between the producer and the USDA.
3. *Public works projects.* These are activities undertaken by the USDA for the benefit of the general public (not only producers), such as watershed protection and flood plains prevention.
4. *Rental and easement.* In this approach, the USDA pays producers, for example, for idling land or placing lands in conservation programs.
5. *Compliance provision.* Under this approach, the producer has to implement certain approved conservation and environmental programs on the property in order to continue to qualify for certain USDA program benefits.

696

6. *Conservation data research.* The goal of this approach is to gather pertinent information through research to create a database for use in designing and implementing conservation and environmental programs and for effective and efficient program delivery.

RESOURCE CONSERVATION AND RELATED PROGRAMS AFFECTING AGRICULTURE

The U.S. federal government spent about $6.7 billion in 1996 on various resource conservation efforts. Of this amount, $3.2 billion (47%) consisted of USDA conservation and environmental activities. In addition to the USDA, there are three other federal entities that administer resource conservation and related programs that affect agriculture. These are the U.S. Environmental Protection Agency (EPA), the Army Corps of Engineers (ACE), and the U.S. Department of the Interior (USDI). The programs administered by the USDI and the ACE are focused on the areas of water resources conservation and management (e.g., irrigation, flood control, wetlands). EPA efforts are concentrated in the area of surface water quality, drinking water protection, groundwater protection, and the use of pesticides. USDA programs assist farmers in implementing management practices that reduce soil erosion and chemical applications. A list of specific resource conservation and related programs administered by the federal government is presented in Table 1.

Table 1 Resource Conservation and Related Programs Affecting Agriculture and the Responsible Agencies

1. **U.S. Department of Agriculture (USDA) programs**
   a. Conservation Reserve Program (CRP)
   b. Wetlands programs
   c. Other conservation programs
2. **U.S. Environmental Protection Agency (EPA) programs**
   a. Water quality programs
   b. Drinking water programs
   c. Pesticide programs
3. **Army Corps of Engineers programs**
   a. Dredge and Fill Permit Program (wetlands)
   b. Flood control programs
4. **U.S. Department of the Interior (USDI) programs**
   a. Range improvement
   b. Water development and management
   c. Water resources investigation
   d. Wetlands conservation
   e. Endangered species conservation
   f. Natural resources research

Source: USDA, ERS.

CONSERVATION RESERVE PROGRAM (CRP)

This USDA program was established by Congress in Title XII of the Food Security Act of 1985. The primary objective of the *Conservation Reserve Program (CRP)* is to reduce soil erosion on highly erodible lands (HEL). In achieving this purpose, other secondary objectives associated with reducing soil erosion are reducing sedimentation, improving water quality, fostering wildlife habitat, and ultimately improving producer incomes through a high and sustainable capacity for agricultural production.

Participation in the program is voluntary but long-term. The farm operator or owner retires cropland that is environmentally sensitive (HEL) for a period of 10 to 15 years. In exchange for this action, the federal government provides the participant with an annual per-acre rent plus half the cost of installing a vegetation cover on the land. There are signup dates for participation in the program. After the contract period, the operator may reenroll. Sometimes, participants are released from their contracts prematurely or reduce their acreage in the program. In 1996, about 33 million acres of land were under CRP contract. The Corn Belt had the most participation.

CONSERVATION COMPLIANCE

The Food Security Act of 1985 introduced two programs: *Conservation Compliance* and the *Sodbuster Program* (this was later eliminated). Instead of land idling, these two programs permit the operator to cultivate HEL, provided a USDA-approved soil conservation system (or combination of systems) was implemented. Failure to comply with regulations will result in ineligibility for federal assistance (e.g., commodity credit corporation price support, disaster payments, federal crop insurance, FHA loans). Examples of conservation systems are presented in Table 2. Conservation systems consist of one or more conservation practices. Systems vary in effectiveness in controlling soil erosion. Climate, topography, soil, predominant crops, and pre-existing production practices influence the choice of a particular conservation system. Factors such as the frequency of precipitation, strength of wind, plant material or soil, and exposure of land to erosive force affect the effectiveness of a system.

These programs that reduce soil erosion help to maintain soil quality and productivity. They reduce pollution of water bodies due to runoff and leaching of chemicals, as well as sedimentation, which increase the cost of maintenance of irrigation facilities and waterways.

WETLAND PROGRAM

According to the ACE, a *wetland* is an area that is inundated or saturated with water by surface or groundwater at a frequency and duration sufficient enough to support a prevalence of vegetation typically adapted for life in saturated soil conditions.

In 1992, there were approximately 124 million acres of land classified as wetlands belonging to non-federal owners in the 48 contiguous states.

Values of Wetland

Wetlands have a variety of uses to society, including the following: they store flood water, trap nutrients and sediments from surface erosion, recharge groundwater, provide habitat for fish and wildlife, buffer shorelines from wave damage, produce timber, pro-

Table 2 Most Frequently Used Conservation Management Systems and Technical Practices on Cultivated Highly Erodible Lands (HEL) Subject to Conservation Compliance

Item	Percent of Cultivated HEL
Management Systems	
Conservation cropping sequence/crop residue use	30.2
Conservation cropping sequence/conservation tillage	10.0
Conservation cropping sequence only	6.9
Crop residue use only	4.4
Conservation cropping sequence/conservation tillage/grassed waterways	2.2
Conservation cropping sequence/conservation tillage/contour farming/grassed waterways/terrace	2.2
Conservation cropping sequence/contour farming/crop residue use/terrace	2.1
Conservation cropping sequence/crop residue use/wind stripcropping	1.9
Conservation cropping sequence/contour farming/crop residue use/grassed waterways/terrace	1.8
Conservation cropping sequence/conservation tillage/crop residue use	1.8
Technical Practices	
Conservation cropping sequence	83.1
Crop residue use	53.1
Conservation tillage	31.3
Contour farming	19.8
Terrace	14.1
Grassed waterway	11.9
Field border	4.9
Wind stripcropping	3.9
Cover and green manure	3.5
Surface roughing	3.3
Grasses and legumes in rotation	2.7
Stripcropping/contour	1.9
Critical area planning	1.7
Pasture and hay land management	1.2

Source: USDA, ERS, 1995.
Note: Percentages will sum up to more than 100% because many conservation systems include multiple technical practices. CRP is excluded in the estimates.

vide grazing for livestock, provide outdoor recreation, and support educational and scientific activities.

Wetland Conversion

Wetlands are lost through conversion to other uses, including agricultural. Similarly, converted lands can be reconverted to wetlands. In the 1980s, such a conversion occurred quite frequently. For every acre of wetland converted to other uses, 2 acres were converted back to wetlands. Wetland losses have occurred mainly on the East Coast and especially in the Florida region, where a significant amount of wetland has been converted for use in crop production.

Factors Affecting Wetland Quality

Considering the importance of wetlands to society, it is critical that they be preserved. The four major factors that affect wetland quality are soil erosion, irrigation, forest cover, and urbanization. Soil erosion has been decreased through the implementation of programs such as CRP change production practices through the implementation of conservation tillage and conservation compliance.

Tree canopy protects watersheds from runoff and erosion and moderates water temperature through shading. It protects sensitive aquatic species from destruction. Harvesting of forest products reduces the tree cover effect and increases soil erosion.

Urbanization has played a significant role in wetland conversion. With such developments come increased surface runoff from construction activities (roads, houses) and industrial pollution of the general environment through discharge of toxic chemicals.

Factors Affecting Wetland Dynamics

Certain factors stem the conversion of wetlands to other uses. These include the following economic factors and federal protection and restoration programs:

1. Decreased profitability of conversion for agricultural use
2. Government programs such as Clean Water Act Section 404 program
3. Increased public awareness and interest in protection of the environment
4. Implementation of wetland restoration programs
5. Swampbuster provisions passed

FEDERAL AGRICULTURAL IMPROVEMENT REFORM ACT

In 1996, the U.S. Congress passed the *Federal Agricultural Improvement Reform Act (FAIR)*, which ended government restrictions on planting of certain crops (corn, cotton, rice, and wheat) and reduced federal subsidies to farmers. The impact of this act was the liberty granted farmers to plant crops to meet market demands. Unfortunately, in 1998, grain production dropped to a 10-year low, due to natural disasters (droughts, excessive heat, floods), causing an undesirable drop in exports. To keep exports flowing, the government intervened with a $5.9 billion emergency farm aid package in 1998. This included tax cuts for farmers, grain give-aways, and export credits. Additional government subsidies were necessary because the protection against bad harvests that was supposed to be provided under the U.S. crop insurance program was not effective when prices fell. Only about 65% of U.S. cropland was protected by the crop insurance program.

The effect of FAIR was also realized in shifts in farm production. There was a record harvest for soybean (2.76 billion bushels) in 1998, the second largest crop for corn, and the third largest for rice.

APPENDIX E

What Is Farm Management?

Farm management is the decision-making and problem-solving process by which a producer allocates limited resources to a number of production alternatives to organize and operate the production enterprise for profitability or to attain desired objectives.

CATEGORIES OF MANAGEMENT

Management is critical to the success of a production enterprise. Farms in the United States are becoming larger and more concentrated. Management is a dynamic process. The factors that affect production are constantly changing. These factors may be categorized as follows:

1. Those that are within total control of the operator
2. Those that can be manipulated by the operator
3. Those that are outside the control of the producer

The effectiveness of a farm manager and the difference in profitability between farms depend largely on the manager's ability to respond correctly and in a timely fashion to change and risk.

BASIC FARM MANAGEMENT PROBLEMS

All farm management problems may be classified under one of three basic types: what to produce, how much to produce, and how to produce it:

1. *What crop to produce.* The producer needs to select a crop or crops to produce in the enterprise. This choice is guided by several factors, including adaptation to the region, profitability (ready market, good prices), knowledge of the crop, cost of production, producer's preference or objective, and producer's experience (certain crops are not recommended for first-time farmers).
2. *How much to produce.* After choosing the crop(s) to produce, the producer should determine how much to produce. The amount to produce will determine the resources needed. The key resource in field production is land. If adequate land is not available, the producer may rent or buy additional land. The types of inputs

and input levels will determine productivity. The producer sets yield goals and purchases the appropriate inputs. These include seed (cultivar), genetically modified (GM) plants, and conventional cultivars, fertilizer rates, and use of irrigation, pesticides, machinery, and labor.

3. *How to produce.* The method of crop production is the third basic management decision. The producer should select combinations of production inputs that will minimize production costs for a given productivity goal. Crops may be produced with high or low inputs, mechanized, or produced through human labor. The producer should decide how weeds should be managed, mechanically or by use of chemicals.

FUNCTIONS OF A MANAGER

The crop producer as a manager performs three basic functions: *planning, implementation,* and *control:*

1. *Planning.* Planning is the key to the success of any business undertaking and thus is the most urgent role of a manager. A plan for the enterprise entails goals and objectives of the enterprise, resources needed, and strategies for identifying and solving problems. The operation of an enterprise is guided by an effective plan.

2. *Implementation.* Once developed, a plan should be implemented correctly. The resources needed should be acquired in a timely fashion and activities commenced according to schedule.

3. *Control.* The producer should monitor the implementation of the plan by keeping accurate records. These records should be analyzed and used to improve the implementation of the plan. Crop production is a high-risk undertaking. It is greatly influenced by the environment, hence the need to continuously monitor the implementation of a production plan.

ECONOMIC PROBLEMS IN PRODUCTION

The producer faces economic problems in a decision-making process. Problem solutions are generally short-term. The problems themselves and solutions change with time. An economic problem is characterized by three basic characteristics: production goals and objectives, limited resources, and alternative uses.

PRODUCTION GOALS AND OBJECTIVES

A producer should, as a matter of first priority, establish specific goals and objectives for the production enterprise. Goals provide guidelines for decision making in crop production and are the yardstick by which the effectiveness of management decisions are measured. They are also used in the planning of future long-term goals. Goals change as the producer grows older or realizes changes in the financial condition of the production enterprise. Common production goals are profit maximization, attainment of a certain business level, a stable income over time, and debt reduction. Profit maximization is a primary goal for most enterprises but, for crop producers, survival of the business is of paramount importance.

LIMITED RESOURCES

The amount of land, labor, and capital available affects production goals. These primary resources are exhaustible. The manager of a production enterprise should be aware that these resources change over time. Management decisions should include an identification of current production limits and the acquisition of additional resources.

ALTERNATIVE USES

To maximize profit from the total production enterprise, the manager may allocate the limited resources to production of alternative products. Some cropping land may be allocated to the production of hay or the grazing of livestock. New technology may change how resources are used.

STEPS IN THE DECISION-MAKING PROCESS

The crop producer should be skilled at allocating limited resources:

1. *Identify and define the problem.* The basic problems that face all crop producers are what to produce, how much to produce, and how to produce. In this context, individual producers may face specific and unique additional problems. For example, after identifying and producing a crop, the producer may realize that the yield per acre is below the county average. This provides an opportunity to scrutinize how the crop was produced. It may be necessary to adjust plant populations, time of planting, rate of fertilization, and other production activities. The key to success is early identification and clear definition of the problem. This will enable a timely solution to be found.

2. *Collect data and information.* After the producer has identified and clearly defined a problem, the next step is to collect data and information related to the problem. Some data may be generated anew by the producer (e.g., soil tests). Other pieces of information may be obtained from the producer's previous records on production decisions. Other sources of pertinent information are the county extension offices and agricultural publications from research stations and agricultural institutions. The Internet is a major source of information.

3. *Identify and analyze alternative solutions.* The data and information assembled are used in deciding how the problem should be solved most cost-effectively. Solutions adopted should be environmentally responsible. In searching for best solutions among alternatives, the producer's previous experiences and good judgment are valuable assets.

4. *Make a decision.* Choosing from among alternative solutions may not always be straightforward. Sometimes, a clearly superior alternative is available. The best alternative may not be cost-effective. Alternative solutions differ in risk level and practicality. The producer may find it best, in certain cases, to refrain from changing the current production strategies. He or she may also not have the resources or be willing to adopt the best alternative solution.

5. *Implement the decision.* A good choice of a solution to a problem should be implemented correctly and in a timely manner to be effective. First, the resources needed should be ascertained.

6. *Evaluate the decision.* It is important to evaluate the effectiveness of a decision in solving the problem. Adequate data and information should be collected to enable

the producer to determine the impact on the business goals as a result of the implementation of a solution to a problem. The evaluation will help the producer make adjustments for increased productivity.

TYPES OF DECISIONS

The crop producer as a decision maker makes two basic types of decisions: *organizational* and *operational.*

ORGANIZATIONAL DECISIONS

This level of decision making involves decisions that are long-lasting in effect. The producer decides what to grow, how much to grow, and how to grow it. These decisions are not revisited frequently. The producer decides on the resources needed to accomplish the production goals. This includes land to rent or purchase, capital to borrow, and cultivars to grow.

OPERATIONAL DECISIONS

Operational decisions are made more frequently and relate to the short-term implementation of organizational decisions. Some of these decisions are made on a daily, weekly, monthly, or seasonal basis. The producer may modify some decisions during the course of crop production in a particular growing season. For example, supplemental irrigation rates may be modified according to the weather conditions. Tillage, planting, and harvesting dates are also subject to local weather conditions.

FACTORS AFFECTING DECISION MAKING

The crop producer, as a decision maker, operates within an environment that is impacted by natural, physical, and biological factors. Some of these factors are fixed and unchangeable, while others can be manipulated to varying extents.

APPENDIX *F*

Standard Ways of Measuring Plant Growth

In order for scientists to manipulate crop productivity, it is important that they understand how the component units of the crop stand, the individual plants, accumulate biomass as well as how they perform in a group. Plant growth is affected by a host of environmental factors. Some of these factors can be manipulated in production, but not until scientists know how productivity will be affected by varying the amount of the factor. There are different techniques used by scientists to measure plant growth.

The *classical growth analysis* is used for studying the growth of single plants. The common measurements of this parameter are *relative growth rate, unit leaf rate, leaf area ratio, leaf area index,* and *crop growth rate.*

RELATIVE GROWTH RATE

Absolute growth rate (AGR) is calculated as the total increase in plant weight over a given period of time as follows:

$$AGR = (W_2 - W_1)/(t_2 - t_1)$$

where W_1 and W_2 are plant weights taken at two corresponding times, t_1 and t_2, respectively.

Relative growth rate (RGR) is the rate at which dry matter accumulates in a plant relative to the total dry weight at a given point in time of the plant's growth cycle. RGR in practice is measured as a mean RGR between two points in time as follows:

$$RGR = \frac{\ln W_2 - \ln W_1}{(t_2 - t_1)}$$

where $\ln W_1$ and $\ln W_2$ are the natural logs of plant weights at the first and second harvests, respectively, and t_1 and t_2 the corresponding times of the harvests. RGR follows an exponential pattern and declines as the size and absolute growth rate increases. Therefore, in order for a comparison of the RGR of different genotypes to be meaningful, calculations should be made on plants of similar size. Further, the genotypes should be evaluated under the same environmental conditions, such as temperature, irradiance, moisture, and photoperiod. In spite of the fact that this is the most widely calculated measure of growth in classical growth analysis, there is no evidence as yet of its significant impact on breeding superior highly productive cultivars.

Unit Leaf Rate

Another plant growth measurement is *unit leaf rate (ULR)*, or *net assimilation rate (NAR)*, which measures the increase in plant biomass (total organic matter) per unit of assimilatory leaf area. Since leaf area changes with time, it is calculated as a mean value between two points in the lifecycle of the plant as

$$ULR = \frac{(W_2 - W_1)(\ln A_2 - \ln A_1)}{(A_2 - A_1)(t_2 - t_1)}$$

where W_1 and W_2 = plant weights at the times t_1 and t_2. $\ln A_1$ and $\ln A_2$ are the natural logs of the corresponding leaf areas at harvests 1 and 2. The units depend on the amount of material involved and the period between the two measurement times (e.g., g cm^{-2} day^{-1} or kg m^{-2} week^{-1}).

Leaf Area Ratio

The *leaf area ratio (LAR)* is a growth measurement that is calculated as a ratio between leaf area and the total biomass accumulated in the plant at a given point in time, as follows:

$$LAR = \frac{A}{W}$$

where A = leaf area and W = plant weight. Like RGR and ULR, LAR between two points in time in the life cycle of the plant can be calculated as

$$\text{Rate increase in LAR} = \frac{(A_2 - A_1)(\ln W_2 - \ln W_1)}{(W_2 - W_1)(\ln A_2 - \ln A_1)}$$

where W_1 and W_2 are plant weights, and A_1 and A_2 are their corresponding leaf areas, respectively. The three single plant growth measures are related as follows:

$$RGR = ULR \times LAR$$

Leaf Area Index

RGR, ULR, and LAR are based on single plants. Growth analysis may be conducted on plants growing in a stand. The methods are modifications of those used for single plant growth analysis. The leafiness of a crop stand may be estimated at a point in crop development. *Leaf area index (LAI)* is a ratio of the total leaf area per unit area, as follows:

$$LAI = \frac{A}{P}$$

where A = leaf area and P = plot size.

In practice, a sample of plants is selected for leaf area measurement. The results are extrapolated to the entire plot by finding the population of plants in the given area (plant density).

CROP GROWTH RATE

Growth analysis using plants in groups approaches what the crop producer encounters in the field. An important indicator of crop productive capacity is the *crop growth rate (CGR)*. CGR measures the increase in crop biomass per unit ground area per unit time. This measure takes into account both photosynthetic gains and respiratory losses, as well as other crop architectural attributes that impact photosynthesis, such as crop height, and leaf shape and inclination.

CGR is calculated as follows:

$$CGR = \frac{W_2 - W_1}{(t_2 - t_1)}$$

where W_2 and W_1 are plant weights taken at two corresponding times, t_1 and t_2, respectively. Just like RGR, CGR may be meaningfully compared among genotypes only if the canopies of the genotypes are fully intercepting radiation, among other factors. LAI has significant effect on CGR prior to closing of the crop canopy. CGR, ULR, and LAI are related as follows:

$$CGR = ULR \times LAI$$

C_4 species, such as corn and sugarcane, have high CGR.

APPENDIX G

Selected Weeds, Insect Pests, and Diseases

SELECTED WEEDS

The following weeds represent some of the most important, persistent, and difficult-to-control species.

NUTSEDGE

The two species of nutsedge that are noxious are purple nutsedge *(Cyperus rotundus)* and yellow nutsedge *(Cyperus esculentum)*. These are considered by some as perhaps the world's worst weeds. Once purple nutsedge is present, it is nearly impossible to irradicate. Nutsedge is a perennial with prominent rhizomes and tubers (hence sometimes called nutgrass). It thrives in poorly drained areas. It is found all over the United States, especially the Southeast. Deep tillage brings the tubers above ground and discourages formation of new ones. Solarization and drainage are helpful control measures.

CRABGRASS

Important crabgrass species that are weeds include large crabgrass [*Digitaria sanguinalis* (L.) Scop] and smooth crabgrass [*Digitaria ischaemum* (Screb)]. These are both summer annual grasses that reproduce by seed. Large crabgrass is widespread in the United States. Smooth crabgrass is not common in the Southwest. These species can be controlled by preventing seed dispersal and rooting. Crabgrass may be spread by mechanical cultivation.

ANNUAL BUNCHGRASS, BARNYARDGRASS, AND JUNGLE RICE

An annual bunchgrasses, barnyardgrass *(Echinochloa crusgalli)* and jungle rice *(Echinochloa colomum)* are weeds in rice fields, pastures, and moist areas. They look much like rice in early growth and thus are difficult to control mechanically in early stages. To control, the plant should not be allowed to seed.

JOHNSONGRASS

Johnsongrass *(Sorghum halepense)* is a perennial grass that reproduces by both seed and rhizomes. It can attain a height of about 6.5-2 ft to 10 ft 3 meters. It is known to have allelochemicals that are released upon decomposition, thus inhibiting other plants from growing. Tillage is helpful in the control of this weed.

BINDWEED

Field bindweed *(Convolvulus arvensis* L.) and hedge bindweed *(C. sepium)* are perennial broad-leaf weeds. They reproduce by seed and rhizome and have twining habits. Field bindweed is sometimes called morning glory. Its seeds have exceptional longevity. Hedge bindweed is less drought-tolerant than field bindweed and is less widely distributed in the United States, occurring mainly in the eastern half.

COMMON LAMBSQUARTER

Lambsquarter *(Chenopodium album)* is an annual summer herb. It is most prevalent in the eastern half of the United States, occurring in cultivated fields and wastelands.

COMMON PURSLANE

Purslane *(Portulaca oleracea)* is an annual and succulent weed. It prefers hot and dry regions and sandy soil.

RAGWEED

Common ragweed *(Ambrosia artemisiifolia)* and giant ragweed *(A. trifida)* (also called horseweed or buffaloweed) are annual weeds of cultivated fields and waste areas. They produce copious amounts of pollen that contribute to the incidence of hayfever.

CANADA THISTLE

A perennial weed, Canada thistle *(Cirsium arvense)* is deep-rooted and reproduces by both seed and rhizomes. It infests hay crops, pastures, and cultivated fields.

FOXTAIL

Green foxtail *(Setaria viridis)* and yellow foxtail *(S. glauca)* are annual grassy weeds found in cultivated fields, pastures, hay fields, and wastelands.

PIGWEED

Pigweed *(Amaranthus retroflexus)* is an annual weed with a shallow taproot system. It reproduces by seed.

OTHER WEEDS

Other important weeds include common cocklebur *(Xanthium pensylvanicum)*, velvetleaf *(Abutilon theophrasti)*, wild buckwheat *(Polygonum convolvulus)*, horsenettle *(Solanum carolinese)*, quackgrass *(Agropyron repens)*, and cheatgrass, or downy brome *(Bromus tectorum)*.

SELECTED INSECT PESTS OF CROPS

Crop production is plagued by insect pests from all categories of insects.

GRASSHOPPERS

Grasshoppers damage all crops by feeding on leaves and other succulent parts. They destroy corn silk, leading to decreased effective pollination and consequently incomplete filling of ears. Important species are the redlegged grasshopper *(Melanoplus femurrubrum)* and differential grasshopper *(M. differentialis)*. They vary in color from greenish to brown. The adult stage is most destructive. The nymphs are hatched from eggs laid in plant debris or soil in late summer. Spraying soon after eggs hatch, to kill the nymphs, effectively controls grasshoppers.

BEAN LEAF BEETLE

The bean leaf beetle feeds on beans, soybean, and corn. Both larvae and adults are destructive. In corn, the larvae destroy roots and silk. The adult beetle is yellowish with dark grey or black markings. It lays eggs at the base of young plants. The adults stay alive through winter. The beetle is controlled by spraying to kill larvae in the ground or spraying leaves to kill adults.

ARMYWORM

The armyworm *(Pseudoletia unipuncta)* feeds on corn, soybeans, cotton, forages, and small grains. It is capable of having up to six generations a year in certain regions in the South. In small grains, the larvae cut the culms. Adults lay their eggs in weeds; hence, weed control is part of armyworm control strategy.

ALFALFA WEEVIL

The alfalfa weevil *(Hypera postica)* mainly attacks the first cutting of the plant. The larvae feed on the plant, boring holes in leaves. The adults are brown with a darker brown streak on the back.

APHIDS

Aphids attack all crops. They typically produce many generations per year, overwintering as either adults or eggs. Adults are greenish to blue-green in color. They are sucking insects that may be vectors for especially viral diseases. Important species include the pea aphid *(Acrythosiphon pisum)* and the spotted alfalfa aphid *(Therioaphis maculata)*. They are preyed upon by lady beetles and are adversely affected by high temperature (above 70°F).

EUROPEAN CORN BORER

The European corn borer *(Pyrausta nubilalis)* attacks the stem of corn, boring holes in it and making it prone to lodging. The larvae are pale-colored with brownish to black heads. They pupate inside the stalks and overwinter in them. The European corn borer is controlled by planting resistant cultivars (stiff-stalked). The *Bt* corn is genetically engineered to resist stalk borers.

CORN EARWORM

The corn earworm *(Heliothis zea)* attacks cotton plants as well as corn. It bores into the ear or the cotton boll. In corn, the tassels are damaged as the larva bores into the ear. The color is variable, including brown, greenish, and yellowish with stripes. The use of chemicals prior to the larvae boring into the ears is effective in controlling this pest. In cotton, rows of corn may be planted as trap plants to divert the larvae from the economic crop.

HESSIAN FLY

The Hessian fly *(Phytophaga destructor)* lays its eggs on young leaves of wheat and sometimes barley and rye. The adults are killed by frost. Wheat producers plant winter wheat on the day after a frost in autumn (called the "fly-free-day"). Resistant cultivars are also available.

OTHER INSECTS

Other important insects of crop plants are wireworms *(Melanotus* spp.) that attack roots of most crops. The angoumois grain moth *(Sitotroga cerealella)* is an important pest of the storehouse. Seed corn maggots *(Hylemya platura)* attack the seed of soybean, corn, and beans.

SELECTED FUNGAL DISEASES

Important fungal diseases of crop plants include the following.

PYTHIUM SEED ROT AND DAMPING OFF

Several species of *Pythium* are involved in pre- and post-emergence death of seeds and seedlings. The common species are *P. ultimum* and *P. debaryanum.* Infected seeds rot while young seedlings are killed (called damping off). This disease is often manifested as incomplete stand or gaps in the nursery plots or trays. If infected seedlings are transplanted, they die later in the field. Conditions that favor the disease include prolonged wet soil, improper soil temperature (i.e., too low a temperature for seeds and seedlings requiring high temperature for optimum growth and vice versa), excessive soil nitrogen, and lack of crop rotation.

    The pathogen occurs in moist cultivated soils and overwinters in the soil or plant residue. Cold, wet soils favor the development of disease. Soils used for crop rotation should be well drained and aerated. Planting of seed and seedlings should be done when soil temperature is appropriate. Seed dressing prior to planting is helpful, especially where field sterilization is impractical.

PHYTOPHTHORA ROTS

Many species of *Phytophthora* cause rots of leaves, fruits, tubers, and other plant parts. The most common species is *P. infestans,* the causal agent of late blight of potatoes and tomatoes. Infected plants show decay of the affected plant part. When roots are affected, the leaves of plants may be sparse and yellow and die back. The fungus overwinters as spores in plant remains. The disease may be controlled by planting resistant cultivars, using well-drained and aerated soils or systemic fungicides applied as seed treatment, sprays, or transplant dips. Examples of *Phytophthora* rots are pink rot of potato, stem rot of soybean, and black pod of cocoa.

LATE BLIGHT OF POTATO

Late blight afflicts potato, tomato, and other species in the family Solanaceae. Under most favorable conditions, late blight can devastate an entire field of crops in about a week or two. Symptoms start as water-soaked spots that enlarge rapidly. It is caused by *P. infestans.* Once established, late blight is difficult to control, unless the weather turns hot and dry (about 35°C or higher). The pathogen overwinters as mycelium in infected potato tubers. The use of resistant cultivars, observance of sanitation, and proper use of chemicals can control the disease. If chemicals are used, timing is of the essence. They should be applied at least 10 days before late blight is usually observed in the area. This should be repeated at intervals of about 5 to 10 days until foliage dies naturally.

DOWNY MILDEWS

Downy mildews affect mainly foliage but can spread to other young and succulent parts of the plant. The world's first fungicide, *Bordeaux mixture,* was discovered in 1885 as a result of a devastating downy mildew epidemic that destroyed the grape and wine industry in France and other parts of Europe. Similarly, the blue mold of tobacco devastated production from Florida to Canada in 1979. Downy mildew in cereals and grasses (e.g., rice, wheat, and corn) is caused by *Sclerophthora macrospora,* while *Peronosclospora sorghi* and *P. maydis* cause the disease in sorghum and corn, respectively. The pathogen overwinters mainly in dead leaves. It can be controlled most effectively by spraying fungicides.

POWDERY MILDEWS

Powdery mildews affect all kinds of plants, including weeds, grasses, fruit trees, forest trees, and others. They are the most common, visible, and readily recognizable plant disease. The symptoms are white spots or patches of white to grayish, powdery growth on young plant tissue. Sometimes, the entire plant is covered with the powdery growth, mostly on the upper side of leaves. All together, powdery mildews cause more economic loss than any single disease of crop plants. Their devastation comes about through parasitic feeding on a host that reduces plant growth and productivity. Chemical control is difficult or impractical in cereals (especially wheat and barley), making them among the most devastated of crop plants when infected. The pathogen involved in powdery mildew of cereals and grasses is *Erysiphe graminis. E. polygoni* affects legumes (soybean, beans).

ALTERNARIA *DISEASES*

Alternaria diseases are very common and include early blight of potato and tomato and leaf spot of bean, as well as many other horticultural diseases. They also cause damping off of seedlings and rots of tubers and fruits. *Alternaria* spots are generally dark brown to black. When underground parts are infected (e.g., potato tubers), sunken spots develop in lesions. Pathogenic species overwinter in infected plant debris as mycelium or spores on seeds. The spores produced are easily dislodged into the air, making them one of the fungal causes of hay fever allergies. The diseases can be controlled by using resistant cultivars, disease-free seeds, sanitation (burning of infected debris), and chemicals.

ERGOT

Ergot affects small grains and grasses, especially rye. It is caused by *Clivceps purpurea*. The kernels in the head are replaced by a large, purplish-black fungal mass. If eaten, infected grain can be poisonous to livestock and humans. Spores infect the open flowers and the sticky stage (honey dew) in the development of the kernel. Sclerotia remain in the soil from the previous year. This obligate parasite is controlled by using ergot-free seed, crop rotation, and removal of grass host.

SOUTHERN CORN BLIGHT

Infected plants show elliptical lesions with reddish-brown borders on the leaves. Lesions also later appear on the stalk and ears. The fungus *Helminthosporium maydis* causes the disease. The pathogen overwinters in infected corn tissue and spreads through spores. It is favored by warm and moist conditions. It is controlled by cytoplasmic genes. Race T of the pathogen devastates corn hybrids produced by the use of cytoplasmic male sterility (CMS), involving the Texas (T) cytoplasm. Hybrids with the normal (N) cytoplasm are resistant to the toxin produced by race T that kills susceptible T cytoplasm tissue ahead of fungal spread. Hybrids with N cytoplasm are susceptible to direct damage by the fungus. Early planting, sanitation, and the use of fungicides help to control the disease.

STALK ROTS

The pathogens involved in stalk rots include the fungi *Diplodia zea, Gibberella zea,* and *Nigrospora oryzae.* The inside of stalks that are infected is decayed, sometimes with pink discoloration. This causes stems to lodge. The problem is intensified by cultural factors such as inadequate potassium, excessive nitrogen, frost or hail, and high plant population, among others. Control may be effected through the use of resistant cultivars, crop rotation, sanitation, and good spacing.

EAR AND KERNEL ROTS

Fungal rots caused by pathogens such as *Gibberella zea* and *Fusarium moniliforme* affect corn ear and grain. Kernel rots appear as pink discolorations. Toxins produced by these pathogens may be toxic to livestock such as swine.

RUSTS

Rusts are among the most destructive plant diseases. They can devastate grain crops, especially wheat, oats, rye, and barley. Infection appears as numerous rusty, reddish-brown, orange, or yellow spots on especially the leaves and stems. There are about 4,000 species of rust fungi. One of the best known is *Puccinia* (e.g., *P. graminis,* which infects grain by causing stem rust). The causal organisms are very specialized, specific forms attacking specific species.

STEM RUST

Different forms of *P. graminis* attack different grains. *P. graminis tritici* attacks wheat, *P. graminis secalis* attacks barley, and *P. graminis avenae* attacks oats.

LEAF RUST

Leaf rust is similarly caused by a variety of species of *Puccinia. P. coronata* attacks oats, *P. recondita* attacks wheat and rye, and *P. hordei* attacks barley. Infected leaves have orange-yellow spores that occur mainly on the leaves. Other characteristics are similar to stem rust.

SMUTS

Many smuts attack the cereal grains themselves, filling the kernels with a mass of black spores. Certain kinds attack other plant parts as well. Affected plants, in addition to damaged kernels, may also be stunted in growth.

CORN SMUT

This disease is common in warm and dry areas. Galls form on the ears, tassels, and other above-ground parts of the plant. It is caused by *Ustilago maydis.* The fungus can remain viable for several years in the soil or plant debris. Upon invasion of the tissue, it induces hyperplasia, resulting in galls. Sanitation helps to control the disease.

LOOSE SMUT

Loose smut affects small grains (oat, barley, rye, wheat) and is more serious a disease problem in humid and subhumid production regions. Seeds of infected plants are destroyed as they become filled with black spores. *U. avenae* causes the disease in oats, while *U. tritici, U. nuda,* and *U. nigra* cause the disease in wheat, rye, and barley, respectively. The pathogen overwinters in seed. Spores of infested heads are spread to other plants by the wind.

COVERED SMUT

Also called *bunt* or *stinking smut,* covered smut afflicts small grains. Smutted kernels are bluish-green and exude a foul odor. The pathogens responsible for the disease in various crops are *U. hordei* (barley), *U. kolleri* (oats), and *Tilletia foetida* (wheat and rye). Infected plants are stunted and may tiller excessively.

SELECTED DISEASES CAUSED BY
MYCOPLASMA-LIKE ORGANISMS

Diseases caused by mycoplasma-like organisms in crop plants include the following.

ASTER YELLOWS

Aster yellows affects a wide range of plants, including tomato, common bean, potato, carrot, and aster. Plants affected become yellow or chlorotic and dwarfed. Carrot is particularly susceptible to the disease, which affects its roots as well. Symptoms of infection include the formation of adventitious shoot, creating an appearance like witches' broom disease. Mycoplasma-like organisms are very sensitive to antibiotics. Drenching the soil with tetracycline or foliar application is not an effective control measure.

LETHAL YELLOWING OF COCONUT PALM

This is a very devastating disease that kills palm trees in about 3 to 6 months after initial infection. It destroyed numerous trees in Florida in the 1970s. First sign of infection is premature drop of coconuts. Then, the lower leaves turn yellow and older ones die. Eventually, the top of the palm tree falls away, leaving the trunk.

CORN STUNT DISEASE

Corn stunt afflicts plants in the southern United States. Infected plants show faint yellowish streaks in young leaves and then turn reddish or purplish on the upper leaves. Symptoms appear 4 to 6 weeks after initial infection. Afflicted plants produce multiple small ears, but with few or no seeds. The tassels of infected plants are usually sterile. The disease is transmitted by leafhoppers *(Dalbulus elimatus)*.

SELECTED BACTERIAL DISEASES

Bacterial diseases of economic importance include the following.

BACTERIAL BLIGHTS OF BEAN

Three different bacteria cause blights in beans with symptoms that are difficult to distinguish. *Xanthomonas campestris* pv *phaseoli* causes common blight, while *Pseudomonas syringae* pv *phaseolicola* causes halo blight. The third disease, bacterial brown spot, is caused by *P. syringae* pv *syringae*. Symptoms start as small, water-soaked spots appearing on stems, pods, seeds, and leaves. Small, yellowish halo develops around the spot in halo blight. At later stages, legions become dark brown and die. The dead tissue may be blown out to leave holes. Sanitation, plowing under of crop residues, crop rotation, and use of resistant seed are recommended control measures.

BACTERIAL PUSTULES

Bacterial pustules appear on raised, yellowish-greenish spots surrounded by a lesion and yellow halo. It is caused by *Xanthomonas phaseoli* pv *sojensis*. The disease is seed-borne, and the bacteria overwinter in diseased leaves. Disease development is faster under warm, wet weather conditions. The recommended control is the same for bacterial blights.

ANGULAR LEAF SPOT

Angular leaf spot (ALS) is so-called because the first small, circular spots formed later enlarge and assume an angular shape. Tissue death occurs in later stages of the disease, causing holes and tears to occur. *Pseudomonas syringae* pv *lachrymans* causes ALS in cucumber, while *Xanthomonas campestris* pv *malvacearum* causes the disease in cotton. In cotton, the affected stems develop long, black lesions called "black arm." The bacteria overwinter in seed or plant debris.

RING ROT OF POTATO

This devastating disease is caused by *Clavibacter (Corynebacterium) michiganense* subsp. *Sepedonicum*. Infested tubers show a ring of yellow vascular discoloration in the early stages when cross-sectioned. Advanced stages of the disease show rots. Plants do not show any foliar symptoms until they are fully grown. The bacteria do not overwinter in soil but in infected tubers and tuber materials left on crates, farm machinery, and knives used to cut them. Not all tubers from one plant are infected.

COMMON SCAB

The causal agent of potato common scab is *Streptomyces scabes*. The disease is most prevalent in soils that are sandy and slightly alkaline to neutral. Tubers develop brownish, raised spots and lesions that later become corky. The blemishes are superficial on the tubers and thus the economic loss caused is due to value rather than yield. When peeling infested tubers, more skin needs to be removed, decreasing the usable part. Increasing soil acidity to a pH of about 5.3 with sulfur is an effective control of the disease.

BACTERIAL WILT

The bacterium *Corynebacterium insidiosum* causes wilting in alfalfa in advanced stages. Infected fields show reduced and dwarfed plants with small leaves that are yellowish and curled. Infection usually occurs in spring and early summer. Soils that are poorly drained are more prone to the bacteria. The disease infests plant roots and crown.

BACTERIAL SOFT ROT

The disease afflicts plants that have fleshy tissue (e.g., vegetables, potatoes, carrots, onions). Soft rots occur both in the field and in storage. Symptoms start off as small, water-soaked lesions that spread rapidly, sometimes converting the whole tuber into a mushy, watery mass. Bacterial soft rot is caused by *Erwinia carotovora* pv *carotovora* and other pathovars. The bacteria infect through wounds. Sanitation and proper cultural practices are key to control of the disease. Products to be stored should be dry and the storage temperature low (4°C).

OTHER BACTERIAL DISEASES

Other bacterial diseases of crop plants of economic importance include crown gall *(Agrobacterium tumifaciens)*, bacterial leaf spots, and blights of cereals.

SELECTED VIRAL DISEASES

Some of the viral diseases important to field crop production include the following.

TOBACCO MOSAIC VIRUS (TMV)

This is one of the many tobacco viruses that cause 20 to 50% yield reduction. It has worldwide distribution, affecting over 150 genera of plants. It affects plants by damaging leaves, flowers, and fruits. Afflicted plants show some chlorosis, leaf curling, mottling, dwarfing, and necrosis, among other symptoms. Most important, afflicted plants show a thicker, dark green area elevated over chorotic areas. TMV is very thermostable, being able to survive thermal inactivation at 93°C. It overwinters in infected plant stalks and leaves. Control is by use of resistant cultivars and sanitation.

POTATO VIRUS Y (PVY)

The potato virus Y also affects pepper, tobacco, and tomato. It is worldwide in distribution. This potyvirus (has threadlike particles) is transmitted through infected seed tubers and aphids. Affected plants are mottled. It is controlled by using certified virus-free seed tubers.

BEAN COMMON MOSAIC AND BEAN YELLOW MOSAIC

These two diseases occur wherever beans are grown, the bean common mosaic being more widespread than the yellow mosaic. The two diseases frequently occur on the same plant. Affected plants are stunted, with mottled and malformed leaves and pods.

Yellow mosaic causes more yellowing in the mottling of leaves. Depending on the stage in the life cycle of the plant when infection occurs, yield losses may be about 35% to even total loss. Bean common mosaic virus overwinters in perennial hosts such as clovers and gladiolus. To control the disease, the grower may use virus-free seed or resistant cultivars.

BARLEY YELLOW DWARF VIRUS (BYDV)

This virus can also infect oats, wheat, rye, and other grasses, the economic effect being most severe in oats. There are numerous strains of barley yellow dwarf virus. Affected plants are stunted, have reduced tillering, suppressed heading, sterility, and unfilled kernels. An early symptom is the appearance of yellowish, reddish, or purplish coloration along the leaf margins, tips, or lamina. The symptoms later spread to other parts of the leaf. Most serious infection occurs when plants are in seedling stage.

MAIZE DWARF MOSAIC

Maize dwarf mosaic affects corn, sorghum, and certain grasses. It may cause up to 40% yield reduction. Corn plants affected show stripped mottling, mosaic, or narrow streaks on young leaves. The ears are partially filled.

CUCUMBER MOSAIC

Cucumber mosaic virus has a wide host range and attacks a variety of species including vegetables, ornamentals, and other plants. Affected plants show mottling, distortion, and discoloration of leaves, fruits, and flowers. Field infection usually occurs when plants are about 6 weeks old. Fruits produced on infected plants have pale green or white areas mottled with dark green and appear rough and malformed. The virus overwinters in many perennial weeds and crop plants. Eliminating weed hosts, such as milkweed, ragweed, and nightshade, helps in the control of the disease.

OTHER VIRAL DISEASES

Other important field crop viruses include tobacco streak, tobacco ring spot, tomato ring spot, tomato spotted wilt, potato leaf roll, potato virus X, soybean mosaic, peanut mottle, and alfalfa curly top of sugar beets.

APPENDIX H

Percent Composition of Products of Crops Discussed in Part II of This Book

Crop/Product	Moisture %	Ash Protein %	Crude Extract %	Ether Fiber %	Crude Free %	Nitrogen-Extract %	Ca %	P %
Barley	9.6	2.9	12.8	2.3	5.5	66.9	0.07	0.32
Corn (kernel)	12.9	1.3	9.3	4.3	1.9	70.3	0.01	0.28
Cottonseed (meal)	7.4	5.2	36.6	5.6	15.3	29.9	0.28	1.30
Peanut (kernel)	5.5	2.3	30.2	47.6	2.8	11.6	0.06	0.38
Rice (rough)	9.7	5.4	7.3	2.0	8.6	67.0	0.10	0.10
Sorghum	12.8	2.1	9.1	3.6	2.6	69.8		
Soybean	8.0	4.8	38.9	18.0	4.8	25.5	0.22	0.67
Wheat	10.6	1.8	12.0	2.0	2.0	71.6	0.05	0.38
Potato	72.0	1.0	2.1	0.1	0.6	17.3	0.01	0.06
Alfalfa (silage)	68.9	2.7	5.7	1.0	8.8	12.9		

Source: USDA.

APPENDIX *I*

Selected Field Crop Production Ecosystems

Ecosystems of the world may be placed into two broad categories: *terrestrial* and *aquatic* ecosystems. An ecosystem may be natural or managed. Field crop ecosystems, or agroecosystems, are human constructs that are characterized by purposefulness. Global distribution of types is primarily under climatic control. Some systems have been developed in many ecologically diverse regions, while others are concentrated in certain areas. Further, some systems are well developed and stable, while others are evolving in response to ecological as well as socioeconomic factors. Selected examples of agroecosystems are described very briefly in this appendix. They show contrasts in the level of management and technology infusion into the agroecosystem.

COTTON CROPPING SYSTEMS

The cotton ecosystem is distributed over diverse climatic regions with different economic experiences. On the basis of temperature, there are either temperate or tropical systems. There are also differences due to rainfall, some regions being semiarid and some humid. There are differences in cultural intensity, some being high inputs and some low inputs. The cotton ecosystem occurs in the United States (especially the Mississippi Delta region), Africa (in Uganda, Sudan, Malawi, and Egypt), northern New South Wales, Australia, China, Pakistan, India, and Russia.

The soils vary in texture and other characteristics, some being heavy clay and some lighter loams. The systems in the United States are highly intensive cultural systems. They are highly mechanized and production is generally large-scale. In some African systems, partnership with the government occurs. Production sites are less intensive and have smaller acreages than in the United States, and they are less mechanized, some being subsistence-level.

CEREAL SYSTEMS OF THE NORTH AMERICAN CENTRAL GREAT PLAINS

The North American Great Plains is defined as the area east of the Rocky Mountains in Colorado and Wyoming and the western 80% of Kansas and Nebraska. This region has total cropland of over 40 million hectares, of which about 30 million hectares are cropped to cereal grains. About 99% of the cereal crop in this region is winter wheat.

This region receives 300 to 800 mm of precipitation per annum. The rainfall is erratic, and thus the region is described as semiarid in climate. About 97% of the cultivated soils are classified as Paleustalfs (old development and dry). They are predominantly calcareous. They are also sandy and low in organic matter. The area is prone to drought. The most common way of water management is by means of fallow.

The main cropping systems adopted by farmers include the following:

1. Continuous cropping
2. Fall-seeded crop–fallow
3. Fall-seeded crop–spring-seeded crop–fallow

Continuously cropped species are winter wheat, winter barley, spring-seeded barley, corn, oats, sorghum, and soybean. Winter rotations are conducted in a 3-year cycle, with winter wheat followed by grain sorghum, corn, or soybean. Most of the wheat is hard red winter wheat. While it is grown mainly for grain in this region, some of it is grazed.

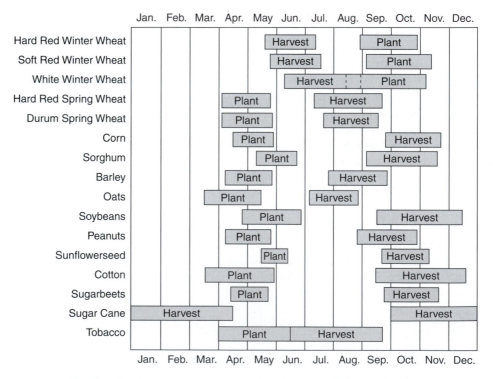

Cropping schedules of selected crops in the United States. (Source: USDA)

Blade plow is commonly used for tillage in the Central Great Plains. This leaves the stubble erect on the soil and consequently results in higher water storage by reducing evaporation. This practice became known as *stubble mulching*. Chisel plow or sweep plow is also used. Winter wheat is commonly rotated with field bean.

Wheat is planted at about August 15 and harvested between June and August the following year. The major forms of fertilizer used in the area are urea, anhydrous ammonia, and ammonium nitrate.

MIXED ROOT-CROP SYSTEMS OF WET SUB-SAHARAN AFRICA

The sub-Saharan region includes West Africa, East Africa, and Central Africa. The climate ranges from dry to subhumid. Rainfall is also variable, the drier Savannah region having less rainfall than the hot, humid, and wet tropical rain forest. Rainfall duration may be between 5 and 8 months, followed by a distinct dry season. The major soils in this region are poor in nutrients, ranging from non-acid and kaolinitic Alfisols of the Savannah to acid kaolinitic Ultisols and Oxisols of the humid forest. The soils of the forest are severely leached.

The major staple crops of this region are cassava *(Manihot esculenta)* and yams *(Dioscorea* spp.). Minor crops include cocoyam *(Xanthosoma saggitifolium* and *Colocasia esculenta)* and sweet potato *(Ipomea batatas)*. Root crops are cultivated in mixed systems in rotation with natural bush fallow. The land is cropped for about 1 to 5 years and fallowed for about 3 to 15 years. Land preparation and general crop production activities are conducted by means of simple hand-held tools such as machete and hoe. The bush is cleared, leaving the tree stumps. The cleared material is then burned (i.e., slash-and-burn agriculture). Cassava sticks are prepared and planted by loosening a spot in the ground and sticking in the cassava cutting. Sometimes, a low mound or ridge is prepared for planting. Yams are planted in mounds of varying sizes. These crops are further intercropped with short-season crops such as pepper, bean, corn, peanut, and melon by planting in the spaces between mounds or usually on the mound. This pattern of cropping creates a complex mixture of plants. Crop production in this region is low-input, relying on the fallow period for the soil to become rejuvenated. The scale of crop production is primarily subsistence.

APPENDIX *J*

Temperature: Converting Between Celsius and Fahrenheit Scales

°C	Known Temperature (°C or °F)	°F	°C	Known Temperature (°C or °F)	°F
−73.33	**−100**	**−148.0**	5.00	41	105.8
−70.56	−95	−139.0	5.56	42	107.6
−67.78	−90	−130.0	6.11	43	109.4
−65.00	−85	−131.0	6.67	44	111.2
−62.22	−80	−112.0	7.22	45	113.0
−59.45	−75	−103.0	7.78	46	114.8
−56.67	−70	−94.0	8.33	47	116.6
−53.39	−65	−85.0	8.89	48	118.4
−51.11	−60	−76.0	9.44	49	120.2
−48.34	−55	−67.0	**10.0**	**50**	**122.0**
−45.56	**−50**	**−58.0**	10.6	51	123.8
−42.78	−45	−49.0	11.1	52	125.6
−40.0	−40	−40.0	11.7	53	127.4
−37.23	−35	−31.0	12.2	54	129.2
−34.44	−30	−22.0	12.8	55	131.0
−31.67	−25	−13.0	13.3	56	132.8
−28.89	−20	−4.0	13.9	57	134.6
−26.12	−15	5.0	14.4	58	136.4
−23.33	−10	14.0	15.0	59	138.2
−20.56	−5	23.0	**15.6**	**60**	**140.0**
−17.8	**0**	**32.0**	16.1	61	141.8
−17.2	**1**	**33.8**	16.7	62	143.6
−16.7	2	35.6	17.2	63	145.4
−16.1	3	37.4	17.8	64	147.2
−15.6	4	39.2	18.3	65	149.0
−15.0	5	41.0	18.9	66	150.8
−14.4	6	42.8	19.4	67	152.6
−13.9	7	44.6	20.0	68	154.4
−13.3	8	46.4	20.6	69	156.2
−12.8	9	48.2	**21.1**	**70**	**158.0**
−12.2	**10**	**50.0**	21.7	71	159.8
−11.7	11	51.8	22.2	72	161.6

°C	Known Temperature (°C or °F)	°F	°C	Known Temperature (°C or °F)	°F
−11.1	12	53.6	22.8	73	163.4
−10.6	13	55.4	22.3	74	165.2
−10.0	14	57.2	23.9	75	167.0
−9.44	15	59.0	24.4	76	168.8
−8.89	16	60.8	25.0	77	170.6
−8.33	17	62.6	25.6	78	172.4
−7.78	18	64.4	26.1	79	174.2
−7.22	19	66.2	**26.7**	**80**	**176.0**
−6.67	**20**	**68.0**	27.2	81	177.8
−6.11	21	69.8	27.8	82	179.6
−5.56	22	71.6	28.3	83	181.4
−5.00	23	73.4	28.9	84	183.2
−4.44	24	75.2	29.4	85	185.0
−3.89	25	77.0	30.0	86	186.8
−3.33	26	78.8	30.6	87	188.6
−2.78	27	80.6	31.1	88	190.4
−2.22	28	82.4	31.7	89	192.2
−1.67	29	84.2	**32.2**	**90**	**194.0**
−1.11	**30**	**86.0**	32.8	91	195.8
−0.56	31	87.8	33.3	92	197.6
0	32	89.6	33.9	93	199.4
0.56	33	91.4	34.4	94	201.2
1.11	34	93.2	35.0	95	203.0
1.67	35	95.0	35.6	96	204.8
2.22	36	96.8	36.1	97	206.6
2.78	37	98.6	36.7	98	208.4
3.33	38	100.4	37.2	99	210.2
3.89	39	102.2	**37.8**	**100**	**212.0**
4.44	**40**	**105.2**			

To make conversions not included in the table use:

(a) $°F = 9/5°C + 32$, (b) $°C = 5/9(°F − 32)$

APPENDIX *K*

Metric Conversion Chart

Known	Multiplier	Desired
LENGTH		
inches	2.54	centimeters
feet	30	centimeters
feet	0.303	meters
yards	0.91	meters
miles	1.6	kilometers
AREA		
sq. inches	6.5	sq. centimeters
sq. feet	0.09	sq. meters
sq. yards	0.8	sq. meters
sq. miles	2.6	sq. kilometers
acres	0.4	hectares
MASS (WEIGHT)		
ounces	28	grams
pounds	0.45	kilograms
short ton	0.9	metric ton
VOLUME		
teaspoons	5	milliliters
tablespoons	15	milliliters
fluid ounces	30	milliliters
cups	0.24	liters
pints	0.47	liters
quarts	0.95	liters
gallons	3.8	liters
cubic feet	0.03	cubic meters
cubic yards	0.76	cubic meters
PRESSURE		
$lb/in.^2$	0.069	bars
atmospheres	1.013	bars
atmospheres	1.033	kg/cm^2
$lb/in.^2$	0.07	kg/cm^2
RATES		
lb/acre	1.12	kg/hectare
tons/acre	2.24	metric tons/hectare

APPENDIX *L*

English Units Conversion Chart

Known	Multiplier	Desired
LENGTH		
millimeters	0.04	inches
centimeters	0.4	inches
meters	3.3	feet
kilometers	0.62	miles
AREA		
sq. centimeters	0.16	sq. inches
sq. meters	1.2	sq. yards
sq. kilometers	0.4	sq. miles
hectares	2.47	acres
MASS (WEIGHT)		
grams	0.035	ounces
kilograms	2.2	pounds
metric tons	1.1	short tons
VOLUME		
milliliters	0.03	fluid ounces
liters	2.1	pints
liters	1.06	quarts
liters	0.26	gallons
cubic meters	35	cubic feet
cubic meters	1.3	cubic yards
PRESSURE		
bars	14.5	lb/in.^{2}
bars	0.987	atmospheres
kg/cm^2	0.968	atmospheres
kg/cm^2	14.22	lb/in.^{2}
RATES		
kg/hectare	0.892	lb/acre
metric tons/hectare	0.445	tons/acre

Glossary

A

Aerosol. Pesticide containing low concentration of active ingredient and propelled by an inert, pressurized gas.

Agricultural productivity. A measure of the relationship between inputs and outputs in agricultural production.

Alternative farming. The system of farming in which emphasis is placed on natural processes such as nutrient cycling, nitrogen fixation, and pest-predator relationships in agricultural production.

Anaerobic respiration. Respiration that occurs under an oxygen-deficient environment.

Annual weeds. Weeds that complete their lifecycle in one growing season or year (summer annuals, winter annuals).

B

Basin irrigation. A method of flood irrigation in which water pools over the land because of slightly raised boundaries.

Bast fiber. Fiber obtained from the bark of certain herbaceous plants (also called *soft fiber*).

Biennial weeds. Weeds that complete their lifecycle in two growing seasons.

Bin burn. Heat damage caused to grain in a storage bin.

Biofuels. Fuels from organic materials other than fossils.

Biological pest management (control). Pest management in which one organism is used to prey on another to reduce the latter's population and destructive effect.

Biological yield. Total biomass produced by a plant.

Biopharming. A biotechnology strategy for using animals to produce pharmaceuticals (e.g., in their milk).

Breeder seed. A small quantity of seed representing the first product of a breeding program that is maintained by the breeder.

Broadcasting. A method of application of materials (e.g., seed, fertilizer, pesticide) in a random fashion.

Broadleaf. A plant with broad lamina.

Bulk density. The mass (weight) of dry soil per unit bulk volume.

C

C_3 pathway. The photosynthetic pathway in which the first stable product is a three-carbon sugar (glyceraldehy-3-phosphate).

C_4 pathway. The photosynthetic pathway in which the first stable product is a four-carbon sugar (oxaloacetate).

Calvin cycle. Also called the C_3 pathway, it is the pathway by which carbon dioxide is fixed in which the first stable product is a three-carbon compound.

Capillary water. Water held in capillary pores (micropores) after drainage of larger pores (macropores) has occurred under gravity.

Carbon dioxide fertilization. The deliberate enrichment of an enclosed area with carbon dioxide.

Carbon dioxide fixation. The photosynthetic process by which carbon dioxide is reduced to carbohydrates.

Central dogma of molecular biology. Genetic information transfer is from DNA to protein.

Certified seed. Seed in the final stage of the certification process that farmers can purchase.

Chemical dormancy. Seed dormancy caused by an accumulation of chemical inhibitors in the seed.

Climate. The pattern of meteorological conditions over a broad area developed over many years.

Cohesion-tension theory of water movement. Water moves in conducting vessels in plants due to strong cohesive bonding among water molecules that allows water to withstand tension and retain continuity in a column.

Commercial farm. A farm with gross annual sales of more than $50,000.

Complementary DNA (cDNA). DNA synthesized in vitro by using an mRNA as template.

Complete metamorphosis. An insect lifecycle in which all stages are dramatically different.

Curing. The treatment by which harvested material acquires certain qualities through exposure to certain environmental conditions (e.g., left in the field under sunlight for a period, called sun-curing).

D

Deep tillage. A tillage operation that turns the soil at depths of about 3 to 4 feet.

Disease triangle. The concept that disease occurs when three factors (pathogen, susceptible host, and favorable environment) are present.

DNA (deoxyribonucleic acid). The hereditary material of most organisms.

Dockage. A measure of the presence of foreign matter in a grain sample.

E

Economic injury level. An estimate of the pest population density at which the value of the crop yield loss prevented is equal to the cost of implementing a treatment.

Economic yield. The part of a plant that is the reason for the production operation and that is harvested for sale or use.

Ecosystem. All biotic and abiotic factors in a defined area.

Emulsified concentrates. A pesticide formulation in which an active ingredient and an emulsifier are dissolved in petroleum solvents and prepared in high concentration.

Epidermis. The protective outer covering of the individual or part of it.

Escape. The lack of disease development in the presence of a pathogen due to lack of appropriate environment.

Evapotranspiration. The total moisture loss from transpiration (from leaves) and evaporation (from soil and other surfaces).

F

Farm. According to the Office of Management and Budget and the Bureau of Census, any place from which $1,000 or more of agricultural products were sold, or normally would have been sold, during the census year.

Farmyard manure. Organic material comprised predominantly of animal waste products from an animal farm.

Fermentation. The storage or preservation of organic material under anaerobic conditions (e.g., silage) resulting in the production of an alcoholic flavor.

Field capacity. The maximum amount of water a soil can hold after wetting and free drainage.

Flocculation. The aggregation, or clumping together, of tiny soil particles, especially fine clay.

Forward (futures) contract. An agreement to deliver a specified amount of grain or turn over ownership of grain in commercial storage at a predetermined price to a specified location.

Furrow irrigation. A surface irrigation method in which water flows between close corrugations or rows.

G

Gene-for-gene concept. The concept that, for each resistance gene in the host, there is a corresponding virulence gene in the pathogen.

Grade S. Designations assigned to reflect the quality of a commodity.

H

Harvest maturity. The stage at which a crop is harvested to meet the quality desired for its intended use.

Heritability. The degree of phenotypic expression of a trait that is under genetic control.

Horizontal resistance. Resistance of a host to pathogens conditioned by numerous genes (also called minor gene resistance, non-specific resistance, multi-genic resistance).

Host range. The variety of plants a pathogen can grow on.

Hydraulic conductivity. The readiness with which a liquid (e.g., water) flows through a solid (e.g., soil) in response to a given potential gradient.

Hygroscopic water. Water (in dry soil) held so tightly around soil particles that it is not available to plants.

Hypersensitive reaction. A response of a host to pathogen invasion in which tissue at the site of invasion dies (necrosis).

I

Igneous rock. Rock formed from the cooling and solidification of magma and that has not been changed appreciably since its formation.

Incomplete metamorphosis. The insect lifecycle in which certain stages are adult forms of previous stages and hence look similar.

K

Kaolinite. An aluminosilicate mineral of the 1:1 crystal lattice group.

L

Legislative pest management (control). The method of pest management in which laws are used to restrict the movement of infected organisms.

Lethal dose (LD_{50}). The milligrams of a toxicant per kilogram of body weight of an organism that are capable of killing 50% of the organism under test conditions.

Liming. The application of agricultural lime (calcium oxide) to the soil to correct soil acidity (raise soil pH).

Loess. Material transported and deposited by wind and consisting primarily of silt-sized particles.

M

Marketing contracts. Agreements by which a producer offers commodities for sale.

Matric potential. The portion of the total soil water potential caused by the attractive forces between water and soil solids as represented through adsorption and capillarity.

Mechanical pest management (control). The method of pest management in which mechanical devices or physical agents are used for destroying pests.

Mendelian inheritance. The inheritance of a gene according to the laws of Mendel, in which alleles segregate and assort independently.

Metamorphic rock. A rock that has been greatly altered from its previous condition through the combined effect of heat and pressure.

Microirrigation. Also called drip or trickle irrigation. The irrigation technique of delivering water in very small amounts in drips or sprays.

Micropropagation. The technique of multiplying plant materials in vitro in tissue culture.

Middleman. An intermediary acting as a broker between the producer and the consumer who may resell to consumers as purchased or value-added.

Minimum tillage. The tillage operation involving minimal soil manipulation for crop production.

Monoculture. The cultivation of one crop on the same piece of land year after year.

Monoecious plant. A plant in which the female and male organs occur on the same plant but on different parts of the plant.

Morill Act of 1862. The act that established land grant institutions in the United States.

N

No-till (also zero tillage). The procedure whereby a crop is planted directly into a seedbed not tilled since the harvest of the previous crop.

Nutrient cycling. The sequence of chemical and biological changes undergone by plant nutrients as they move from the atmosphere into water, soil, and living organisms and, upon death, are decomposed and mineralized back to inorganic components again.

O

Organochlorines. Pesticides containing chlorinated hydrocarbons as active ingredients.

Orographic effect. The effect of geographic surface features on climate.

Overwinter. The strategy by which pests survive the winter.

Ozone layer. A layer of O_3 in the upper atmosphere that protects life on earth from harmful ultraviolet rays in sunlight.

P

Parallel venation. The pattern of veins in the leaf in which minor veins run parallel to the central vein (characteristic of grass leaves).

Parent material. The unconsolidated and more or less chemically weathered mineral or organic matter from which soils are formed.

Partitioning. The allocation of assimilates to various parts of the plant.

Ped. A unit of soil structure such as an aggregate, a crumb, or a granule.

Percolation. The downward movement of water through soil.

Perennial weeds. Weeds with perpetual lifecycles that reappear without reseeding.

Physiological maturity. The plant maturity stage at which no new dry matter is added to the plant.

Phytoalexins. Chemicals exuded by certain plants upon injury to ward off pest attack.

Plant quarantine. The use of specific laws to control the movement of infected organisms.

Plant taxonomy. The science of classifying plants.

Polyculture. The practice of cultivating more than one crop on the same piece of land.

Primary tillage. A tillage operation in which the soil is inverted to bury vegetation and debris to a depth of about 6 to 14 inches, leaving a rough finish that is unsuited for seeding.

Q

Quantitative trait. A trait that is under the control of many genes.

R

Registered seed. An increase of the foundation seed.

Reticulate venation. A weblike pattern of arrangement of the veins in a leaf (characteristic of dicots).

S

Saturated flow. The movement of water in saturated (both macropores and micropores filled with water) soil.

Secondary tillage. A tillage operation that generally follows primary tillage and is conducted to a shallower depth of 2 to 6 inches and produces a finer tilth for seeding.

Sedimentary rock. Rock formed from materials deposited from suspension or precipitated from solution and usually being more or less consolidated.

Seed dormancy. The inability of viable seed to germinate in the presence of favorable environmental conditions.

Seed scarification. The technique of scratching seed with a hard seed coat problem to facilitate water imbibition for germination.

Seed viability. The capacity of a seed to germinate.

Seeding rate. The number of seeds planted per unit area.

Sigmoid growth curve. The characteristic s-shaped pattern associated with growth of organisms.

Silage (ensilage). A forage crop preserved in succulent condition by the process of fermentation.

Smectite. A group of silicate clays having a 2:1 lattice structure (e.g., montmorillonite).

Soil aggregate. Many particles held in a single mass or cluster such as a clod, block, or prism.

Soil permeability. The ease with which gases, liquids, and plant roots penetrate or pass through a bulk mass of soil or layer of soil.

Soil porosity. The volume percentage of the total soil bulk not occupied by solid particles.

Soil separates. The individual particle groups of mineral soil particles—sand, silt, and clay.

Soil solarization. The sterilization of soil in the field by using heat from the sun.

Soil taxonomy. The science of classification of soils.

Soil test. A method of ascertaining the nutrient status of the soil.

Soil texture. The relative proportions of the various soil separates in a soil.

Soil-plant-atmosphere continuum. The pathway of water movement from soil through plants into the atmosphere.

Sprinkler irrigation. A technique of irrigation in which water is moved under pressure through pipes and distributed through nozzles as a spray.

Subirrigation. An irrigation technique in which water is delivered to plant roots through pipes laid underground.

Subsoiling. A land preparation technique for breaking up impervious layers in the soil (e.g., pans).

Surface irrigation. A technique of irrigation in which water is spread over the soil surface.

Surge flow. A surface irrigation technique in which water efficiency is improved by delivering water intermittently to the soil surface.

Synthetic seed. In vitro produced seed comprised of somatic embryos encapsulated in a biodegradable protective coating.

T

Terracing. Land preparation geared toward making land suitable for tillage and preventing accelerated erosion through reduction in slope.

Tillage. The mechanical manipulation of soil for any purpose (e.g., for crop production).

Tillage system. The nature and sequence of tillage operations used in preparing a seedbed for planting.

Tilth. The physical conditions of the soil as they relate to its ease of tillage, fitness for a seedbed, and impedance to seedling emergence and root penetration.

Toxicity. The capacity of a chemical compound to kill living organisms.

Transplanting. The establishment of a crop stand by using seedlings.

U

Unsaturated flow. The movement of water through soil that is unsaturated with water.

V

Vegetable fiber. Fiber derived from parts of a plant.

W

Water-holding capacity. The ability of soil to hold water against drainage.

Weather. The meteorological conditions prevailing in a locality that is subject to short-term variation.

Weathering. All physical and chemical changes produced in rocks at or near the earth's surface by atmospheric factors.

Index

Subsoiling, 451
Sulfur, 250
Surface irrigation, 293
Sustainable agriculture, 368, 369
Sustainable model, 368
Sweep plow, 454
Symbiotic nitrogen fixation, 224
Synergism, 259
Synthetic seeds, 136
Systemic action, 340

T

Tame pasture, 432
Taproot system, 81
Taxon, 52
Taxonomic groups of organisms, 52–57
TCA cycle, 112
Temperature, 188
Temperature inversions, 180
Temperature stress, 182
Terracing, 446
Terrain modification, 446
Test weight, 532
Tetrazolium test, 470, 471
Theory of domestication by crowding, 6
3-way cross hybrid, 158
Tillage, 446, 193
Tillage implements, 452–463
Tillage systems, 444–452
    choosing a system, 457, 458
    conservation tillage, 449–452
    conservation tillage equipment, 457
    conventional tillage, 447–449
    deep tillage, 456, 457
    subsoiling, 456
    tillage implements, 452
Tilth, 446
Tissue, 71–73
Tissue culture, 168
Tissue test, 258, 259
Topography, 204
Topsoil, 208
Tower silos, 514
Trachaeophytes, 53
Translocation, 118
Transgenic crop production, 408–42
    acreage, 409
    crops, 409

traits, 410, 411
    herbicide tolerance, 411–41
    insect-resistance, 415–418
    fruit ripening, 420, 421
    benefits, 421
Transgenic plants, 166, 167, 408
Transpiration, 116
Trench silos, 514
Triglycerides, 12, 13
Troposphere, 176
True resistance, 319

U

U.S. agriculture:
    classification of farms, 24, 25, 27
    computer use, 17
    debt-to-asset ratio, 39
    energy use, 458–461
    farm labor, 38
    farm machinery, 461–463
    farm ownership, 33–35
    farm size, 25
    farm-related accidents, 462
    farming sections, 35, 36
    farms operated by individuals, 32
    irrigation, 290–304
    major agricultural states, 38
    number of farms, 25, 27
    off-farm revenue sources, 39–41
    organic farming, 394–405
    profile of farm operators, 40–45
    structure, 24–34
Unsaturated oil, 12, 13
Upland cotton, 13
USDA-defined farm typology groups, 27
USDA plant hardiness zones, 189
USDA soil texture triangle, 212

V

Vacuoles, 70
Value added, 525
Variable rate technology (VTR) strategy, 289
Variety, 54
Vascular bundles, 78
Vascular plants, 53
Vavilov, 18
Vegetable fibers, 13
Vegetable oils, 12

Venation, 63
Vernalization, 138, 135
Vertebrates, 326
Vertical resistance, 320, 319
Viral diseases, 313
Virulence, 321
Visual examination, 253

W

Warm season plant, 189, 190
Water. *See* Plant and soil water
Water distribution, 186
Water erosion, 229
Water holding capacity, 281
Water-use efficiency (WUE), 284, 285
Weather. *See also* Climate and weather, 176
Weathering, 206, 207
Weeds, 327–331
Wetland program, 412–14
Wheat production, 551–569
Wilmsen, E. N., 4
Windbreaks, 383–386
Wind erosion, 231, 233
Windrowing, 500, 501
WUE, 284, 285

X

Xerophytes, 283
Xylem, 72, 73

Y

Yellow foxtail, 423
Yield, 100, 101, 126, 125
Yield compensation, 131
Yield components, 130–131
Yield per se, 125
Yield potential, 131
Yield risk, 533
Yield stability, 161, 162

Z

Zero tillage, 449
Zinc, 252
Zinc deficiency, 252
Zone of accumulation, 209
Zone of elluviation, 209
Zone of illuvation, 209
Zone of leaching, 209